RISK AND UNCERTAINTY ASSESSMENT FOR
NATURAL HAZARDS

Recent natural disasters remind us of our society's increasing vulnerability to the consequences of population growth and urbanisation, economic and technical interdependence,
and environmental change. Assessment of risk and uncertainty is crucial for natural hazard
risk management, both in the evaluation of strategies to increase resilience, and in facilitating risk communication and successful mitigation.

Written by some of the world's leading experts on natural hazard science, this book
provides a state-of-the-art overview of risk and uncertainty assessment in natural hazards.
Using clearly defined terminology, it presents the core statistical concepts in a unified
treatment applicable across all types of natural hazards, and addresses the full range of
sources of uncertainty: from sparse and noisy measurements to imperfect scientific and
societal knowledge and limited computing power. The role of expert judgement and the
practice of uncertainty elicitation are discussed in detail. The core of the book provides
detailed coverage of individual hazards: earthquakes, volcanoes, landslides and avalanches,
tsunamis, weather events such as flooding, droughts and storms (including the consequences
of climate change) and wildfires, with additional chapters on risks to technological facilities,
and on ecotoxicological risk assessment. Concluding chapters address the wider context of
risk management, studying societal perceptions of natural hazard risk and human responses.

This is an invaluable compendium for academic researchers and professionals working in
the fields of natural hazards science, risk assessment and management, and environmental
science, and will be of interest to anyone involved in natural hazards policy.

JONATHAN ROUGIER is a Reader in Statistics at the University of Bristol. He specialises in
uncertainty assessment for complex systems, notably environmental systems such as climate and natural hazards. He has made several important contributions in the statistical field
of computer experiments, including general approaches for representing model limitations,
informal and formal approaches to model calibration and multivariate emulation for expensive models such as climate models. Dr Rougier's recent and current collaborations include
climate prediction and palaeo-climate reconstruction, ice-sheet modelling and sea-level rise
and inference for dynamical systems such as glacial cycles, avalanches and hydrocarbon
reservoirs.

STEVE SPARKS is a Professorial Research Fellow at the University of Bristol. He is a
volcanologist with interests in hazard and risk assessment, and his research includes the
physics of volcanic eruptions and fluid dynamics of hazardous volcanic flows. He is the
world's most highly cited scientist in volcanology and a former President of the International

Association of Volcanology and Chemistry of the Earth's Interior. Professor Sparks has been involved in hazard and risk assessment with advice for governments for volcanic emergencies, including during the eruption of the Soufrière Hills Volcano, Montserrat, and the emergencies related to volcanic ash from Iceland in 2010. He was on the planning committee of the International Research into Disaster Reduction of the International Council for Science (ICSU), and is currently joint leader of the Global Volcano Model project.

LISA HILL is a Research Manager at the University of Bristol and also works as an independent researcher. She has worked with researchers to explore the interface between environmental science and social science for many years, initially at the UK Research Councils, and later at the University of Bristol. Dr Hill's research interests are in human geography, archaeology and the environment, using non-representational theory to explore relations between people and the material world, particularly in the context of post-industrial and post-disaster landscapes.

RISK AND UNCERTAINTY ASSESSMENT
FOR NATURAL HAZARDS

JONATHAN ROUGIER

Department of Mathematics,
University of Bristol, UK

STEVE SPARKS

School of Earth Sciences,
University of Bristol, UK

LISA J. HILL

University of Bristol, UK

CAMBRIDGE
UNIVERSITY PRESS

CAMBRIDGE
UNIVERSITY PRESS

University Printing House, Cambridge CB2 8BS, United Kingdom

One Liberty Plaza, 20th Floor, New York, NY 10006, USA

477 Williamstown Road, Port Melbourne, VIC 3207, Australia

4843/24, 2nd Floor, Ansari Road, Daryaganj, Delhi - 110002, India

79 Anson Road, #06-04/06, Singapore 079906

Cambridge University Press is part of the University of Cambridge.

It furthers the University's mission by disseminating knowledge in the pursuit of education, learning and research at the highest international levels of excellence.

www.cambridge.org
Information on this title: www.cambridge.org/9781108446679

© Cambridge University Press 2013

First published 2013
First paperback edition 2017

A catalogue record for this publication is available from the British Library

Library of Congress Cataloging in Publication data
Risk and uncertainty assessment for natural hazards / Jonathan Rougier,
Lisa J. Hill, Steve Sparks, [editors].
p. cm.
Includes index.
ISBN 978-1-107-00619-5
1. Natural disasters. 2. Risk assessment. I. Rougier, Jonathan, 1966–
II. Hill, Lisa J. III. Sparks, R. S. J. (Robert Stephen John), 1949–
GB5014.R57 2013
363.34′2–dc23
2012023677

ISBN 978-1-107-00619-5 Hardback
ISBN 978-1-108-44667-9 Paperback

Contents

Contributors

W. P. Aspinall
Cabot Institute, BRISK and Department of Earth Sciences, Bristol University, Bristol BS8 1RJ, UK

P. Bates
School of Geographical Sciences, University of Bristol, Bristol BS8 1SS, UK

K. J. Beven
Lancaster Environment Centre, Lancaster University, Lancaster LA1 4YQ, UK *and* Department of Earth Sciences, Uppsala University, Uppsala, Sweden *and* ECHO, EPFL, Lausanne, Switzerland

P. G. Challenor
National Oceanography Centre, European Way, Southampton SO14 3ZH, UK

N. A. Chapman
MCM Consulting, Täfernstrasse 11, CH 5405 Baden-Datwil, Switzerland

R. M. Cooke
Chauncey Starr Chair for Risk Analysis, Resources for the Future, 1616 P St. NW, Washington, DC 20036-1400, USA

S. E. Cornell
Stockholm Resilience Centre, Stockholm University

H. S. Crosweller
Bristol Environmental Risk Research Centre (BRISK), Department of Earth Sciences, University of Bristol, Bristol BS8 1RJ, UK

T. L. Edwards
School of Geographical Sciences, University of Bristol, Bristol BS8 1SS, UK

J. Freer
School of Geographical Sciences, University of Bristol, Bristol BS8 1SS, UK

J. Hall
Environmental Change Institute, Oxford University Centre for the Environment, Oxford, UK

A. Hart
The Food and Environment Research Agency, Sand Hutton, York YO41 1LZ, UK

G. L. Hickey
Department of Mathematical Sciences, Durham University, Science Laboratories, South Road, Durham DH1 3LE, UK

B. E. Hill
US Nuclear Regulatory Commission, NMSS/HLWRS, EBB 2-02, Washington, DC 20555-0001, USA

L. J. Hill
University of Bristol, Bristol BS8 1TH, UK

T. K. Hincks
Bristol Environmental Risk Research Centre (BRISK), Department of Earth Sciences, Bristol University, Bristol BS8 1RJ, UK

E. A. Holcombe
School of Geographical Sciences, University of Bristol, Bristol BS8 1SS, UK

M. S. Jackson
School of Geographical Sciences, University of Bristol, Bristol BS8 1SS, UK

M. Kern
Federal Department of Home Affairs FDHA, Bern, Switzerland

D. J. Kerridge
British Geological Survey, Murchison House, Edinburgh EL9 3LA, UK

T. J. Lynham
Canadian Forest Service, Great Lakes Forestry Centre, 1219 Queen St. East, Sault Ste. Marie, P6A 2E5 Ontario, Canada

B. D. Malamud
Department of Geography, King's College London, London WC2R 2L, UK

J. Neal
School of Geographical Sciences, University of Bristol, Bristol BS8 1SS, UK

J. Pooley
Cabot Institute, Bristol Environmental Risk Research Centre and Department of Earth Sciences, University of Bristol, Bristol BS8 1RJ, UK

J. C. Rougier
Department of Mathematics, University of Bristol, Bristol BS8 1TW, UK

G. Schumann
School of Geographical Sciences, University of Bristol, Bristol BS8 1SS, UK

R. S. J. Sparks
Cabot Institute, Bristol Environmental Risk Research Centre and Department of Earth Sciences, University of Bristol, Bristol BS8 1RJ, UK

C. A. Taylor
Cabot Institute, Earthquake Engineering Research Centre and Department of Civil Engineering, University of Bristol, Bristol BS8 1TR, UK

J. Wilmshurst
School of Experimental Psychology, University of Bristol, 12A Priory Road, Bristol BS8 1TU, UK

M. J. Wooster
Department of Geography, King's College London, London WC2R 2L, UK

The Editors would also like to thank **Alex Clarke** for her help in preparing the index for this volume.

Preface

This collection originated in a scoping study commissioned in 2009 by the UK Natural Environment Research Council (NERC) on uncertainty and risk in natural hazards. This study brought together natural hazards experts and specialists in uncertainty assessment, perception and communication, in compiling a report with sections that covered each of the major hazards, cross-cutting themes and related but non-hazard risks. It found that there was a substantial opportunity for greater integration in natural hazards risk assessment, both *horizontally*, across hazards which shared common features, and *vertically* within a hazard, from the hazard event itself, to risk assessment and decision support. More recently, the study members have updated their contributions, to provide more detail than was possible at the time, and to take account of more recent progress. This volume is the result.

Jonathan Rougier
Steve Sparks
Lisa Hill

Cabot Institute
University of Bristol, UK

1

Risk assessment and uncertainty in natural hazards

L. J. HILL, R. S. J. SPARKS AND J. C. ROUGIER

1.1 Introduction

This edited volume concerns the topical and challenging field of uncertainty and risk assessment in natural hazards. In particular, we argue for the transparent quantification of risk and uncertainty so that informed choices can be made, both to reduce the risks associated with natural hazards, and to evaluate different mitigation strategies. A defensible framework for decision-making under uncertainty becomes a vital tool in what has been termed the era of 'post-normal science', wherein 'facts are uncertain, values in dispute, stakes high and decisions urgent' (Funtowicz and Ravetz, 1991; also see Hulme, 2009). Natural hazards, like the environmental systems of which they form a part, are rich and complex, and full of interactions and nonlinearities. Our understanding of their nature and our ability to predict their behaviour are limited. Nevertheless, due to their impact on the things that we value, the effects of natural hazards should be managed, which is to say that choices must be made, despite our limited understanding. As such, it is crucial when scientists contribute to decision-making or the formation of policy that their uncertainties are transparently assessed, honestly reported and effectively communicated, and available for scrutiny by all interested parties.

In this book we explore the current state-of-the-art in risk assessment and uncertainty for the major natural hazards. As we acknowledge, some uncertainty assessment methods require a level of scholarship and technique that can be hard for hazards experts to acquire on top of the demands of their own discipline. Most risk assessments and uncertainty analyses in natural hazards have been conducted by hazard experts, who, while acknowledging the full range of uncertainties, have tended to focus on the more tractable sources of uncertainty, and to accumulate all other sources of uncertainty into a lumped margin-for-error term. We would not claim that all sources of natural hazards uncertainty can be treated within a formal statistical framework, but we do claim that some very large uncertainties currently accumulating in the margin for error can be treated explicitly using modern statistical methods, and that the resulting uncertainty assessment will be more transparent, defensible and credible.

In this opening chapter, we consider the role of the natural hazards scientist during periods of quiescence, imminent threat, the hazard event itself and recovery, all of which present considerable challenges for the assessment and communication of uncertainty and risk. These topics are explored to varying degrees in the chapters that follow. First, let us examine the scale of the problem.

Risk and Uncertainty Assessment for Natural Hazards, ed. Jonathan Rougier, Steve Sparks and Lisa Hill. Published by Cambridge University Press. © Cambridge University Press 2013.

1.2 Vulnerability to natural hazards

We live in times of increasing vulnerability to natural hazards. Data on loss estimates for natural disasters (Smolka, 2006) reveal a trend of increasing catastrophe losses since 1950. Likewise, a World Bank assessment of disaster impacts in the periods 1960 to 2007 indicates an increase in absolute losses but in approximate proportion to increases in global GDP (Okuyama and Sahin, 2009). The reasons for these trends are many, including the concentration of (increasing) populations and critical infrastructure in urban areas, the development of exposed coastal regions and flood plains, the high vulnerability of complex modern societies and technologies and changes in the natural environment itself, including the possible impacts of climate change. Climate change increases both our vulnerability to natural hazards (e.g. through sea-level rise) and also our uncertainty about future natural hazard frequency (Mitchell *et al.*, 2006; Cutter and Finch, 2008; Jennings, 2011). Increases in the reporting of disasters coupled with increases in the uptake of insurance are other factors that might make these trends appear more pronounced.

Consider the development of coastal regions. On 29 August 2005, Hurricane Katrina struck the Gold Coast, with devastating consequences for the city of New Orleans. Although it was only the third most intense hurricane to hit the United States in recorded history, as measured by central pressure, it was almost certainly the country's costliest natural disaster in financial terms, with total damage estimates of US$75 billion (Knabb *et al.*, 2005), and the loss of around 1300 lives.

Yet these losses were small in comparison with the catastrophic effects of the 2011 Eastern Japan Great Earthquake and Tsunami. On 11 March 2011, a magnitude 9.0 earthquake occurred in the international waters of the western Pacific. Damage associated with the earthquake itself was limited, but a massive and destructive tsunami hit the eastern coast of Honshu within minutes of the quake, causing heavy casualties, enormous property losses and a severe nuclear crisis with regional and global long-term impact. As of 13 April 2011 there were 13 392 people dead nationwide and a further 15 133 missing (Japan National Police Agency, 2011; Norio *et al.*, 2011). An early evaluation by analysts estimated that the disaster caused direct economic losses of around US$171–183 billion, while the cost for recovery might reach US$122 billion (Norio *et al.*, 2011; Pagano, 2011). Japan is among the most exposed countries to natural hazards and the influences of climate change, being prone to earthquakes, tsunamis and typhoons, as well as sea-level rise; but the 11 March disasters also raised questions about the country's exposure to 'cascading' threats. The concept of cascading threats refers to the 'snowball effect' of crises that in their cumulative impact can cause major, and often unforeseen, disasters. For example, a primary hazard can cause a series of subsequent hazards, such as the radioactive pollution released by the damaged Fukushima Dai-ichi nuclear power plant. This event had repercussions around the world and raised many fundamental issues regarding the adequacy and transparency of technological risk assessments, especially for extreme natural hazards.

The vulnerability of critical infrastructure as a result of such cascade effects was felt to a lesser extent in the UK in 2007, revealing the interdependence and vulnerability of the nation's infrastructure. Exceptional rainfall in the summer of 2007 caused extensive flooding in parts of England, especially in South and East Yorkshire, Worcestershire, Gloucestershire and Oxfordshire. Following a sustained period of wet weather starting in early May, extreme storms in late June and mid-July resulted in unprecedented flooding of properties and infrastructure. There were 13 fatalities, and thousands were evacuated from their homes. Public water and power utilities were disrupted, with the threat of regional power blackouts. The resulting disruption, economic loss and social distress turned the summer 2007 floods into a national concern. Broad-scale estimates made shortly after the floods put the total losses at about £4 billion, of which insurable losses were reported to be about £3 billion (Chatterton *et al.*, 2010).

Yet arguably the most significant recent disruption to daily life in the UK and Europe was caused in April 2010, by the volcanic ash cloud arising from the eruption of the Icelandic volcano Eyjafjallajökull. Although the eruption was of low intensity and moderate magnitude, it produced a widespread cloud of fine ash, which was blown by north-westerly winds over Central Europe, Great Britain and Scandinavia. As a threat to aviation, the fine and quickly moving ash led aviation authorities to declare no-fly zones over European airspace. During the week of 14–21 April, 25 European countries were affected. The direct loss to airlines is estimated to exceed €1.3 billion, with more than four million passengers affected and more than 100 000 cancelled flights (Oxford Economics, 2010). This crisis reveals the extent to which social demands for free mobility, the movement of foodstuffs and other goods have grown in recent decades, and thus the extent to which our vulnerability to natural hazards, like volcanic ash eruptions, has increased as well. The probability of major disruption as a result of volcanic eruptions is likely to increase in the near future because of the seemingly inexorable increase in air traffic. Ours is a highly interconnected, globalised world, increasingly vulnerable, both socially and economically, to the effects of natural hazard events.

While the risk of economic losses associated with natural disasters in high-income countries has significantly increased (UN-GAR, 2011), the effects of urbanisation and population growth also increase vulnerability and the probability of unprecedented disasters in the developing world. Although the absolute economic losses associated with such events may be smaller, the relative effects on GDP and development are much greater in low-income countries (Okuyama and Sahin, 2009), and the tragic loss of life is often on a massive scale. There are also less quantifiable but equally significant effects on those caught up in disasters in terms of lost livelihoods, trauma and political instability.

As an illustration, the Sumatra-Andaman earthquake, 26 December 2004, was one of the largest ever recorded at magnitude 9.2. The associated tsunami caused an estimated 280 000 deaths in countries bordering the Indian Ocean, but the majority occurred in close proximity to the megathrust rupture, in northern Indonesia (Sieh, 2006). Earthquakes of this magnitude are rare events and are difficult to predict (see Chapter 8 of this volume). As an extreme event, the Asian tsunami was described as a wake-up call for the world (Huppert and Sparks, 2006). Yet global population growth and the expansion of megacities in the developing world continue to increase human exposure to such events (Bilham, 1995, 1998, 2004). Half the world's megacities of more than ten million people are located in earthquake-prone regions, and it is only a matter of time before one of them suffers an extreme catastrophe (Jackson, 2006). Poor building materials, regulations and planning, together with corruption (Ambraseys and Bilham, 2011), will exacerbate the impact.

In this respect, developed nations like Japan usually have higher levels of adaptive capacity to hazards than developing countries. Fatalities would have been much higher if the 2011 Japan earthquake and tsunami had occurred in the Philippines or Indonesia, for example. As Bilham (2010) notes, the Haiti earthquake of 12 January 2010 was more than twice as lethal as any previous magnitude 7.0 event; the reason for the disaster was clear: 'brittle steel, coarse non-angular aggregate, weak cement mixed with dirty or salty sand, and the widespread termination of steel reinforcement rods at the joints between columns and floors of buildings where earthquake stresses are highest' – the outcome of decades of unsupervised construction, coupled with inadequate preparedness and response. Indeed, corruption is evidently a major factor in loss of life from earthquakes (Ambraseys and Bilham, 2011), and there are important lessons for scientists, risk managers and the international development community.

1.3 Natural hazards science

If we are to minimise loss of life, economic losses and disruption from natural hazards in the future, there is an imperative for scientists to provide informed assessments of risk, enabling risk managers to reduce social impacts significantly, to conserve economic assets and to save lives. However, to be truly effective in this role, environmental scientists must explicitly recognise the presence and implications of uncertainty in risk assessment.

One of the key emergent issues in natural hazard risk assessment is the challenge of how to account for uncertainty. Uncertainty is ubiquitous in natural hazards, arising both from the inherent unpredictability of the hazard events themselves, and from the complex way in which these events interact with their environment, and with people. Uncertainty in natural hazards is very far removed from the textbook case of independent and identically distributed random variables, large sample sizes and impacts driven mainly by means rather than higher moments. In natural hazards, processes vary in complicated but often highly structured ways in space and time (e.g. the clustering of storms), measurements are typically sparse and commonly biased, especially for large-magnitude events, and losses are typically highly nonlinear functions of hazard magnitude, which means that higher moment properties such as variance, skewness and kurtosis are crucial in assessing risk.

At the same time, we have observed that there is often a lack of clarity in modelling approaches, which can lead to confusion or even exaggeration of hazard and risk by, for example, the incorporation of the same uncertainty in two or more different ways. For example, a forecast of a hazard footprint using a computer model might include inputs that are precautionary and err on the side of high hazard. If this approach is then promulgated across all inputs, it is possible to end up with values that individually are plausible but that collectively are implausible. Similar outcomes might arise where 'factors of safety' are built into models. This problem can only be addressed, in our view, by very careful analysis of how uncertainties and factors of safety are built into models and assessments. For many of the more extreme natural hazards, where data or scientific understanding that informs models are limited, the assessment of uncertainty is likely to require very careful assessment of tails of statistical distributions.

The experience of researching this book and the lack of transparency in many hazard models, and consequently derivative risk assessments, indicate to us that it is much better *not* to apply the precautionary principle or include factors of safety at the modelling stage. Rather, the hazard model needs to be developed with a full analysis of uncertainty. The systematic assessment of uncertainty can then be used to inform what factors of safety might be adopted or to apply the precautionary principle. It also appears unlikely that a deterministic model will be able to include such an analysis in a satisfactory way, strengthening the view that probabilistic modelling of hazard risk is both appropriate and necessary. These are recognised as challenging problems in statistics and earth systems science, and progress on them requires the close collaboration of natural hazards experts and professional statisticians.

1.3.1 Role of the natural hazard scientist

In exploring the role of the natural hazard scientist, it is useful to consider four stages in the natural hazards event: (1) quiescence; (2) imminent threat; (3) the event itself; and (4) the recovery stage back to 'life as normal'. The relative timescales vary greatly between different natural hazards. Each stage poses different challenges for natural hazards scientists, but the assessment and communication of uncertainty and risk is central to all.

1.3.1.1 Quiescence

Prior to an event there is typically a long interlude during which hazards scientists can contribute to the increasing resilience of society, through informing regulations, actions and planning. Two approaches are common. In the first, individual scenarios are analysed in detail; these 'what if' scenarios concern events judged to be possible and to have a high impact, such as a large earthquake at a nearby fault, or a high-tide storm surge. Often such assessments are driven by concern over the vulnerability of a specific installation, such as a nuclear reactor, a railway line or dwellings, and may be written into regulatory requirements. The assessment may lead to changes in policy, regulation, mitigation steps or plans for emergency response, such as evacuations. The primary contribution of natural hazards science is to map the potential footprint of such an event in space and time, and then to quantify the impact of that footprint in terms of simple measures of loss, such as structural damage or mortality.

The second approach generalises the first to consider a range of possible hazard events with their probabilities of occurrence. These probabilities, representing the inherent uncertainty of the hazard, are combined with the footprint for each event to derive hazard maps.

Commonly, such maps show the probability of some summary measure of the hazard footprint exceeding some specified threshold at different locations in a region. Where loss is quantifiable, the total loss in the region can be represented as an exceedance probability (EP) curve; the area under this curve is one simple way to define the *risk* of the hazard (i.e. the mathematical expectation of loss). Very often, the total loss in the region is the sum of the losses in each location, and in this case the hazard can also be represented in the form of a risk map, showing the probability of loss exceeding some specified threshold. More details are given in Chapter 2 of this volume.

In both of these approaches, scientific modelling combines observations, physical principles and expert judgements. By their very nature, though, natural hazards make such modelling extremely challenging. For example, rare events cannot be repeatedly observed, and so it is hard to assess their probabilities as a function of location and timing, magnitude and intensity. Expert judgement is often required to assess the extent to which probabilities can be extrapolated from smaller magnitude events and from events happening at different locations and different times (e.g. the 'ergodic assumption' that underpins many seismological aspects in earthquake engineering).

1.3.1.2 Imminent threat

In most natural hazards, with the exception of earthquakes in most circumstances, there is a period when the threat becomes imminent: the dormant volcano goes into a period of unrest; a large hurricane or typhoon is developing offshore; intense rainfall is forecast or has started to create a flood threat; recent weather favours forest fires, avalanches or landslides. Such precursors have the effect of raising the probability that an event will happen in the near future. Often these precursors will be diverse, and the main role of the natural hazard scientist at this point is to gather evidence and information from a wide range of sources, and to combine it effectively for the purposes of risk management. This may take the form of an established framework, like an early warning system based on in-place instruments, or a probabilistic network or a decision tree, but it may also involve *in situ* scientists making their best determination on the basis of all of the evidence available, much of which will be qualitative or poorly quantified.

At this stage, the risk manager may prepare to implement emergency plans, such as putting the emergency services on alert, cancelling leave, clearing arterial roads and carrying out evacuations. Effective communication of uncertainty is crucial, both between the scientists and the risk manager, and between the risk manager and the general population, given that some disruption to 'life as normal' is inevitable. As the situation may be rapidly developing, this communication needs to be selective and focused, for example using visualisations. These can include updated hazard and risk maps, but may also use less formal methods because maps are not always well-understood.

Commonly there is an added problem of false alarms: the hazard event may not materialise at all, or may be significantly weaker or stronger than forecast. In the public mind these outcomes may be interpreted as scientific 'failures', and this can undermine the credibility and effectiveness of future responses. Communicating the uncertainty of the imminent threat is absolutely vital, but also extremely challenging.

1.3.1.3 The event itself

Once the event has started, the *in situ* scientific team has a key role in interpreting the evidence for the risk manager. Most natural phenomena have complex time histories: floods may rise and fall, the level of volcanic activity can suddenly increase or move into apparent quiescence after intense eruptions, aftershocks can occur after a major earthquake, hurricanes can rapidly intensify or change course. Quite commonly, the primary event is associated with secondary hazards, such as landslides and floods following a hurricane, or landslides, tsunamis and fires following a major earthquake. The quality of information at this stage varies widely. For floods in metered catchments, the information is usually of sufficient quality to allow real-time numerical modelling of the event and its consequences (data assimilation); this is also helped by the long lead-time of many flood events (though not flash floods), which allows real-time systems to be activated. A similar situation applies for long-duration hazards, like wildfires. But in most rapid-onset and short-duration events, real-time information is of uneven quality, and therefore requires expert analysis and communication. The real challenge here is to quantify the uncertainty, in situations where numerical calculations have to be rapid and adaptive.

From a research standpoint, documenting the event itself is very important. Such research does not necessarily help in the unfolding crisis, but is invaluable for improving understanding of the natural hazard, and of the process of natural hazard risk management. A common theme in assessing natural hazards is that lack of good event data hinders the building and testing of physical and statistical models. This case history documentation needs to go beyond observations of the event to include the inferences that followed and decisions that were made as information arrived, in order to support a forensic reconstruction at a later stage.

1.3.1.4 The recovery stage

The recovery stage starts after an event has started to wane or has finished. There may be an intermediate period where it is unclear whether the event has really finished. Will there be more aftershocks? Will the volcano erupt again? Will there be more rain and further flooding? How long will it take for the flood water to subside? What about secondary hazards, like fire after an earthquake, or other contingent events, like the spread of diseases after flooding? These issues are all uncertain and the *in situ* scientific team must assess the probabilities as best they can, based upon the available evidence.

Once the event is clearly over, the initial recovery period offers another opportunity for scientists to document the impact of the event and to improve understanding, by compiling a database of structural damage following an earthquake, or by mapping flood extents, or the runout region for an avalanche or landslide. Later, scientists will attempt to reconstruct what happened, allowing for a better understanding of the event and its consequences, and also for improved calibration of the scientific models used to assess the hazard footprint and the loss. The importance of this type of forensic post-event analysis cannot be overstated given the complexity and rarity of large hazards events, and it is crucial in revealing previously unaccounted-for and possibly unknown phenomena. In principle there should be research into lessons learned, ideally as part of an objective investigation where actors can identify what went right and what went wrong. However, post-event analysis can be inhibited by concerns about 'who is to blame', preventing actors in the emergency from being completely candid. Eventually the recovery stage will turn, usually imperceptibly, into the first stage of 'life as normal' and the lessons learned can be used to improve resilience in the future.

1.3.2 Accounting for model limitations

Models play a central role in natural hazards science. Statistical models are used to describe the inherent uncertainty of the hazard. Physical theories are used to inform those statistical models, and to map out the hazard footprint; they are also used for some aspects of quantifying loss, such as assessing structural damage. More general qualitative models are used to describe public perceptions of uncertainty and risk, and to represent the way in which people will respond to evidence of an imminent or occurring natural hazard. Here we will focus, for simplicity, on the physical modelling of the hazard footprint (the subsequent effect of a hazard event in space and time), but the same comments apply equally to other modelling. Examples of footprint modelling based on physical principles include: hydraulic models for flooding; weather models for hydrometeorological hazards; plume models for volcanic ash deposition; fluids models for volcanic pyroclastic flows and lahars, as well as tsunamis; granular flow models for snow avalanches and landslides; and elastic wave models for earthquakes.

In all of these cases, the complexity of the underlying system is only partially captured by the model, and further simplifications may be imposed for tractability or due to computational limitations. Many hazards involve movement of waves or fluids (often in multiple phases) through the atmosphere, hydrosphere and lithosphere, involving highly nonlinear interacting processes operating at many different scales. Additionally, the environments are often characterised by complex topographies and micro-scale variations that are simply not knowable. Therefore even the most advanced hazards models have shortcomings in terms of structural simplifications and truncations of series expansions, with empirically determined parameterisations of the 'missing' physics. Likewise, the prescribed boundary conditions are invariably highly artificial. It is hard to think of any natural hazard process where the physics is adequately understood or the boundary conditions are well observed. In fact, the challenge of modelling the system is so great that physical models are often replaced wholesale by explicitly phenomenological models. These are designed to reflect observed regularities directly, rather than have them emerge as a consequence of underlying principles. Thus earthquake footprints are often imputed using simple empirical distance/

magnitude relationships, flooding footprints using transfer functions from precipitation to river flow, and avalanche footprints using an empirical model of runout angle against mean slope angle.

The challenge of model limitations is ubiquitous in natural hazards, and more generally in environmental science, which deals almost exclusively with complex systems. One response is to invest in model improvement. Typically this involves introducing more processes, or implementing a higher resolution solver; i.e. 'business as usual' for the modellers. This does not quantify uncertainty, of course, and it is not even clear that it reduces uncertainty given that including extra processes also introduces more uncertain model parameters. Experience suggests that doing more science and modelling often increases overall uncertainty, although it is helpful in these situations to distinguish between the level of uncertainty and our ability to quantify it. More complex models may involve more uncertain components, but if the resulting model is more realistic, uncertainty about these components may be easier to specify.

A complementary and underutilised response is to attempt to quantify the epistemic uncertainty arising from model limitations for existing models. This uncertainty has three components: parametric uncertainty, arising from incomplete knowledge of the correct settings of the model's parameters; input uncertainty, arising from incomplete knowledge of the true value of the initial state and forcing; and structural uncertainty, which is the failure of the model to represent the system, even if the correct parameters and inputs are known. Together, these three components represent a complete probabilistic description of the informativeness of the model for the underlying system, but in practice all are extremely challenging to specify and their specification will invariably involve a degree of subjectivity, which many natural scientists feel uncomfortable with. Consequently, they are often not specified, or specified rather naively.

For example, parametric uncertainty is often represented by independent and marginally uniform distributions on each parameter, given specified end-points. This seldom concurs with well-informed judgements, in which, say, central values of each parameter are likely to be more probable than extreme ones. One explanation is that outside of statistics, the uniform distribution is often viewed, quite wrongly, as 'less subjective' than other choices. A more sophisticated justification is that the choice of distribution does not matter as, under certain conditions, a large amount of observational data will dominate and the result obtained will thus be robust to the choice, so one might as well choose a uniform distribution. Input uncertainty is often ignored by replacing the uncertain boundary values with a 'best guess' based upon observations; for example, using mean climate instead of weather. Structural uncertainty is often ignored, or rolled into natural variability (in chaotic models) or measurement error. A recent development has been to address structural uncertainty through multiple models (with different parameter spaces), notably in climate and seismic hazard assessment.

Quantifying the epistemic uncertainty arising from model limitations typically involves making many model evaluations, and also replications for stochastic models. Thus it is a direct competitor for additional resources (e.g. computing resources) with model

L. J. Hill et al.

improvement, which uses the extra resources for more processes or increased resolution. Some areas of environmental science, such as climate science, have a strong culture of allocating additional resources to model improvement, and the quantification of epistemic uncertainty suffers as a consequence. Natural hazards models tend to be much smaller than climate models (excepting weather models for some hydrometeorological hazards like storms), and there is less expectation that modest model improvements will lead to an obviously more realistic model output. Therefore there is a better prospect for the use of experimental methods to help quantify epistemic uncertainty in existing models.

We believe that a more careful treatment of model limitations should be a research priority in natural hazards, and also more widely in environmental science. Naive treatments of parametric and input uncertainty, and neglect of structural uncertainty, compromise the tuning of model parameters to observations and lead us to understate predictive uncertainty. Overconfidence in a particular model may lead to misleading forecasts and assessments with potentially disastrous consequences for decision-making (which is often sensitive to the length of the right-hand tail of the loss distribution). They also limit the effectiveness of model criticism, which needs to be based on a joint understanding of the model and the system. Our current inability to demonstrate that environmental models are useful tools for risk management, particularly in high-profile areas like climate science, is devaluing the scientific contribution and provides an easy target for special interest groups. Within environmental science, there is a growing perception that modelling failures in specific areas are symptomatic of a general inability to provide quantitative predictions for system behaviour. There is no real basis for this drastic conclusion, but it points to an urgent need to think more deeply about the limitations of models, how these might be represented quantitatively and how model-based findings are communicated.

1.3.3 Gaps in current practice

What currently limits the impact of natural hazards science on natural hazards risk management? We see three gaps in current practice: (1) between the hazard process and the hazard loss distribution; (2) between the actions, uncertainties and losses and the choice of action; and (3) between the intention to act and the successful completion of the action. Informally, we might refer to these as the 'science gap', the 'action gap' and the 'completion gap'. These gaps can be collectively referred to in the 'last mile' concept (Shah, 2006), which suggests that knowledge is not being implemented effectively and that there is a wide gap between what is known and is practised. Indeed, the studies of particular hazards that led to this book, and are documented in the chapters that follow, indicate that practice commonly lags well behind state-of-the art methods and knowledge. In addition it is now widely appreciated that successful risk management involves much more than excellent science and its application. Responses to natural hazard threats and events are firmly within the human world, where many factors relating to collective and individual human behaviour influence the scale of an emergency and the extent to which it might become a disaster.

The science gap exists between the study of the hazard, which is what hazard scientists do, and the distribution of the loss, which is what risk managers need. To close this gap requires: first, a careful assessment of hazard uncertainty, including not just the intrinsic variability of the hazard itself, but also the epistemic uncertainty that reflects limitations in our knowledge of the hazard process in space and time. Second, it requires that hazard events be linked through to loss assessments. The loss itself will depend on which stakeholder the risk manager represents. For some combinations of hazard and stakeholder, this linkage from event to loss is fairly straightforward. For example, for earthquakes there is a reasonable link between peak ground acceleration and property damage. If the stakeholder is an insurance company, then the science gap is small. But if the stakeholder is the mayor, whose primary concern is human casualties, then the science gap is much larger, because there are many factors to account for, such as the varying disposition of the population of the city through the day.

Once the science gap has been addressed, the next stage is for the risk manager to choose between actions on the basis of the information in the risk assessments. For this gap we assume that, whatever action is chosen, it will be seen through to completion. This is an artificial assumption which we make in order to stress that choosing the action is already very challenging, without considering the issue of completion. The challenge arises from the presence of many recognised but unquantified sources of uncertainty that were not included in the formal assessment of risk. These are often represented as scenarios, inevitably ambiguous, and often involving qualitative representations of vulnerability that defy our desire to attach probabilities. Furthermore, the risk values will often have to balance incommensurable values – for example, losses measured in lives, and costs in dollars.

Choice between actions and scenarios is one part of the large and diverse field of statistical decision theory (as covered in a text such as Smith, 2010). But that is not to say that a single action pops out of the analysis. Decision theory provides a framework to examine how choices are sensitive to judgements; judgements, for example, concerning the quantifiable uncertainties used to compute the risks, the inclusion of additional scenarios or the attachment of tentative probabilities to scenarios. We interpret a 'robust' action as an action which is favoured across a range of reasonable judgements regarding the scenarios. This is the statistical notion of robustness. It is not at all the same as 'low regret', because it uses the science intensively: natural hazards science, statistics and decision theory. Therefore science can help to close the action gap through sensitivity analysis.

The third gap concerns the implementation and completion of the chosen action. As already explained, it is instructive to separate this from the choice of action itself, although a more sophisticated analysis would allow the probability of completion to depend on the action and the scenario, so that the actions would in fact only be intentions. The challenge here is persuading all of the stakeholders to commit to the action, even though some will perceive themselves to be disadvantaged, and some will be materially disadvantaged. Here there is certainly a role for expertise in risk perception and communication. But the developing literature on 'wicked' problems (see Conklin, 2005) suggests that solutions in the traditional sense may be unattainable. The best that can be achieved is a sense of shared ownership, in which the stakeholders genuinely feel that they have a role, and alternative actions are discussed and contrasted without a hidden agenda. Our view is that transparency in the risk assessment, and in the choice between actions, is crucial to promoting this shared sense of ownership, and that the application of well-established methods, such as the statistical treatment of uncertainty, is the best way to promote this transparency.

1.4 Outline of the chapters

This book brings together the current state-of-the-art in risk assessment and uncertainty. In Chapter 2, Rougier describes a framework for representing uncertainty in terms of probabilities. Drawing on the perspective of the 'risk manager', an individual responsible for making decisions regarding hazard intervention and mitigation, and answerable to an auditor, his focus is the inherent uncertainty of natural hazards, what is termed their 'aleatory' uncertainty. Although natural hazards are fundamentally diverse, Rougier argues that they share a number of commonalities, and as such that they merit a standard terminology.

In Chapter 3, Rougier and Beven consider the more complex issue of 'epistemic' uncertainty, which arises from limitations within the analysis itself in terms of input, parametric and structural uncertainty. Again, the emphasis is on the role of the risk manager, and the chapter proceeds by developing a general framework for thinking about epistemic uncertainty. As the authors suggest, in a standard risk assessment for natural hazards the EP curve captures aleatory uncertainty, and everything else is external. Of course, external uncertainty is not forgotten but is accounted for using a lumped adjustment, such as a margin for error. The authors suggest that it is feasible and beneficial to move some aspects of epistemic uncertainty into the EP curve, i.e. to transfer them from external to internal uncertainty. In this way, the uncertainty accounted for by the margin for error diminishes, and the difficulties associated with specifying its size become less important in the overall risk assessment.

In Chapter 4, Aspinall and Cooke outline a complementary approach to the use of statistical methods in the form of structured elicitations and pooling of expert judgements. Commonly in the assessment of natural hazards both quantitative data and our understanding of some of the key processes will be inadequate. Understanding the potential magnitude and timing of natural hazards for the purposes of taking mitigating actions inevitably requires scientists and decision-makers to make simplifications and assumptions in data analysis and applications of process models. As the authors explain, structured expert judgement elicitation coupled with a formalised mathematical procedure for pooling judgements from a group of experts provides a rational basis for characterising relevant scientific uncertainties and incorporating these into probabilistic hazard and risk assessments. The aim of the elicitation process is to provide reasoned quantification of uncertainty, rather than to remove it from the decision-making process.

In Chapters 5 and 6, Edwards and Challenor explore the challenges of risk assessment for hydrometeorological hazards. Chapter 5 summarises the present-day risks and uncertainties associated with droughts, heat waves, extreme precipitation, wind storms and ocean waves. As the authors note, most hydrometeorological hazards are true extremes, in the sense that they are not distinct events but a consequence of entering the high or low end of local climatic variation. As such, these hazards do not have triggers or inception events, although particular atmospheric or oceanic states may increase the likelihood of occurrence. Nonlinear feedbacks in the climate system make the prediction of hydrometeorological events extremely challenging, and this is further complicated by changes in climate forcing (Chapter 6), which reduces the value of historical information, and introduces a substantial new source of uncertainty. The net effect of feedbacks and the potential for tipping points are difficult to simulate, which means that long-term projections are subject to significant model uncertainties.

In Chapter 7, Freer *et al.* explain how climate change may also lead to an intensification of the hydrological cycle and an increase in flooding events in many parts of the world. The science of risk and uncertainty analysis has seen more activity in hydrology than in many other fields, and yet the emergence of new concepts, techniques and debates has served to highlight the difficulties of providing effective guidance, and agreement on best practice. As with other natural hazards, hydrological risks are characterised by extreme events, where observed data sources are often very limited and are rarely able to characterise the overall behaviour of events in detail. In addition to an overview of flood risk quantification, flood hazard and risk mapping, flood alert and flood warning, the use of new technologies to improve flood prediction, and economic and social impacts and insurance assessments, the authors provide examples of current predictive capability and future research challenges.

In Chapter 8, Aspinall highlights the fundamental role that uncertainties play in probabilistic seismic hazard assessment, and makes evident their wide seismological variety and differing quantitative extents in all areas of the problem. For low-probability high-consequence situations especially, such assessments are constrained by limitations in data and understanding, such as the short span of the historical record for extreme earthquakes, the existence of unknown, hidden active faults, and shortfalls in our generic understanding of complex earthquake processes. Thanks mainly to the acute need of the nuclear industry worldwide for guidance on earthquake hazards, and consequent investment in detailed studies, this is perhaps the most developed of all natural hazard probabilistic assessment methodologies. There is clear potential for the many methodological insights gained in this domain to be transferred to risk assessment approaches for other natural hazards, in terms of tenets and concepts for the treatment and representation of uncertainties, principles and subtleties in data analysis, and fallacies and pitfalls in judgement and interpretation.

In Chapter 9, Hincks *et al.* assess the current state-of-the art in landslide hazard and risk assessment. As the authors suggest, recent literature on landslide hazard assessment has demonstrated significant advances in the numerical modelling of preparatory and triggering processes at the scale of individual slopes. At the same time, sophisticated physics-based approaches have been developed in order to model the emplacement dynamics of landslides at this localised scale. In common with other hazards examined in this volume, the limiting component in deterministic landslide prediction is the acquisition of adequate data to parameterise the model. Recent probabilistic approaches aim to account for the uncertainties associated with the lack of data and to incorporate the natural variability of slope parameters. However, the science of landslide prediction remains highly empirical due to the formidable challenges of understanding complex multiphase earth materials in motion. At the same time, models lack statistically robust calibrations, as well as formal assessments of their validity in terms of structural errors.

In Chapter 10, Hincks *et al.* explain how effective risk management for tsunamis requires cooperation on a global scale, and rapid communication of data and hazard information. Most major tsunamis typically result from large-magnitude, shallow-focus earthquakes, which produce a large vertical or sub-vertical movement on the fault plane, but they can also be produced by very large landslides and volcanic eruptions. There are no explicit, globally adopted standards for tsunami modelling and forecasting tools, although increasing cooperation and data sharing between tsunami warning centres is resulting in a more unified approach to the provision of alerts. Because inundation on land is difficult to predict (particularly in real-time), pre-event tsunami hazard information for other parts of the globe typically consists of travel time maps and probability estimates for wave height at the coast. As the trigger is often an earthquake, it is not yet possible to predict precisely when and where a tsunami will occur. Existing warning systems are rule-based, and do not account for uncertainties (or missing data) in seismic source information. However, as the authors note, probabilistic forecasts using Bayesian Networks and logic trees are starting to be developed.

In Chapter 11, Sparks *et al.* suggest that current improvements in volcano forecasting are largely driven by an improved understanding of the physics of volcanic processes and advances in data-led conceptual models. In general, however, precise prediction is not achievable. Signals precursory to volcanic eruptions take place on a wide range of time-scales, from many years to a few tens of minutes. As a result, precise prediction about the onset of an eruption is rarely possible. The main basis for forecasting the nature of a future eruption is the historical or geological record of past eruptions. By mapping young volcanic deposits, it is possible to generate a hazard zonation map, in which the area around a volcano is divided into zones of decreasing hazard, enabling the relevant authorities to identify communities at high risk and to help in the management of volcanic crises. As the authors stress, the position of hazard zone boundaries is implicitly probabilistic, but it is only recently that more rigorous approaches to locating such boundaries have been developed.

In Chapter 12, Hincks *et al.* consider wildfire hazards, which have become prominent in recent years, largely due to increasing vulnerability, but the early effects of climate change may also be a factor. Wildfire is a complex process: from combustion in which fuel chemistry and fluid dynamics dominate, to convection, where factors such as weather conditions, landscape, topography, fuel properties and human intervention control behaviour and extent of the burn. All of these factors are highly variable and pose difficult challenges for risk forecasting. Operational wildfire models that predict fire spread and behaviour are largely empirically based, and are thus constrained by the data used to construct the model. A primary source of epistemic uncertainty is measurements made on the fires themselves, which can have significant implications for wildfire frequency–size statistics. For process-based wildfire models, uncertainties arise in accuracy and resolution, both temporal and spatial. Uncertainties are also associated with model parameterisation and methods of data assimilation. It is only in the last 5–10 years that probabilistic models to account for parameter variability have become more widespread.

In Chapter 13, Sparks *et al.* consider the interplay between natural hazards and technological facilities. We live in a globalised and highly interdependent world, with sophisticated technological facilities and networks that support all facets of life. Yet these complexities and interdependencies make us less resilient to natural hazards, as technological facilities and critical infrastructure are vulnerable to complex, cascading event sequences. Responding to such emergent phenomena is a major challenge. Too many hazard and risk assessments are based on narrow, deterministic approaches, which commonly fail to anticipate extremes and to take full account of the complexity of the systems. The authors argue that the only way forward is through robust systematic hazard and risk assessment that is probabilistic in character.

In Chapter 14, Hickey and Hart provide a useful parallel in risk assessment and uncertainty analysis from the field of ecotoxicology. When assessing the safety of a chemical substance it is necessary to determine the risk and possible consequences to the environment. Such an assessment comprises many components, including human health, persistent, bio-accumulative and toxic assessment and ecological risk. There are a number of international regulatory and scientific technical guidance documents pertaining to ecotoxicological risk assessments, and as such this is a fairly advanced field. Importantly, the authors advocate a tiered approach to risk assessment, such that increasing the level of assessment leads to a refinement of uncertainty handling and risk characterisation.

In Chapter 15, Cornell and Jackson argue that many of the most pressing challenges for uncertainty analysis in the context of natural hazards relate to the human dimensions of risk. It is widely accepted within contemporary social science that modern societies negotiate and conceive risk as a function of specific industrial, technological and scientific capacities. As such, risk can be understood as a function of changing social values, and social research is of central importance to the framing and understanding of contemporary risk and uncertainty. However, mainstream risk research has primarily focused on the environmental and technical dimensions of risk, as the subject matter of this book attests. The kind of cross-disciplinary engagement and channelling of academic research into practice advocated by the authors represents a radical new challenge, including the need to integrate a greater range of expert evidence into the suite of risk calculation and response strategies.

In Chapter 16, Crosweller and Wilmshurst consider the difficult subject of human decision-making. As they note, human responses to a given risk scenario will vary greatly. Humans do not behave in rational ways, and often display what might be deemed 'unexpected behaviours'. It is these behaviours that risk managers try to change, by providing more information about the nature of the risk, for example, yet often this does not have the desired effect. As the authors note, the link between risk perception and human behaviour is not straightforward, and apparent resistance by local communities to being moved out of an at-risk area can be attributed to a range of possible psychological factors, cultural norms and beliefs.

1.5 Outlook

This overview sets the scene for the book, which explores these matters in much more detail and in application to specific hazards. A central message of the book is that much can be done, especially to address the gaps and to travel the last mile through systematic uncertainty and risk assessment. We hope that the book will promote dialogue regarding the value that follows from transparent and defensible assessment of uncertainty, even if not all sources of uncertainty can be accounted for. Below, we briefly summarise our own view about the outlook for uncertainty and risk assessment in natural hazards.

First, there are underlying principles that can be applied across all hazards, and, with appropriate local modifications, in individual hazard areas. One way forward is to establish standardised definitions and methodologies with a formal statistical basis, to remove needless ambiguity, and to promote the sharing of good practice across hazard areas. Certain types of hazard map and hazard product should be easily recognised, no matter whether they concern volcanoes, earthquakes, landslides, or floods. A threshold exceedance hazard map is one such example: such a map needs a clearly defined time interval and a clearly defined threshold, with contours to indicate probabilities of threshold exceedance over the interval. A major challenge is how to capture a richer description of uncertainty in hazard maps and products, but in principle this too should be amenable to some standardisation. We recognise, of course, that different hazards have very different time and space domains, and there will also be unique aspects of how hazard and risk assessment is conducted and communicated. But we think there is a need for much greater standardisation than exists at present.

Second, the probabilistic treatment of aleatory uncertainty can be extended to cover some sources of epistemic uncertainty, notably in statistical modelling of aleatory processes, and in accounting for model limitations. Broader statistical models can be used to widen the catalogue of calibration data for assessing both aleatory and epistemic uncertainty, and to allow for non-stationarity. The mature and very successful field of experimental design provides tools that could be much more widely used, both in the field and the laboratory, and in statistical inference (in the design and analysis of computer experiments). Expert elicitation methods provide a formal structure in which to embed hazard and risk assessment, and allow for consideration of aspects of the assessment where there are deficiencies of data or process models. We anticipate that expert elicitation will become much more prominent as its efficacy for a holistic approach becomes increasingly clear.

Third, more use could be made of a formal decision framework as a starting point for choosing between actions. The quantification of hazard and risk, wherever possible, seems the key starting point. Of course, significant simplifications are required, but a rigorous, systematic approach in which well-defined risks are clearly quantified has the advantages of reproducibility and transparency. Of course, these simplifications have to be well understood if the quantification is to be used intelligently to inform decision-making, and some risks associated with natural hazard events are not easy to define to the point where they can be quantified. This is particularly the case in assessing how human behaviour affects risk. But the presence of these risks does not detract from the benefit of applying a scientific approach to hazard and risk assessment wherever possible. As we argue above, in Section 1.3.3, such an approach can help to narrow all three 'gaps' in current practice.

References

Ambraseys, N. and Bilham, R. (2011) Corruption kills. *Nature* **469**: 53–155.

Bilham, R. (1995) Global fatalities from earthquakes in the past 2000 years: prognosis for the next 30. In *Reduction and Predictability of Natural Disasters*, ed. J. Rundle, F. Klein and D. Turcotte, Reading, MA: Addison Wesley, pp. 19–31.

Bilham, R. (1998) Earthquakes and urban development. *Nature* **336**: 625–626.

Bilham, R. (2004) Urban earthquake fatalities: a safer world, or worse to come? *Seismological Research Letters* **75**: 706–712.

Bilham, R. (2010) Lessons from the Haiti earthquake. *Nature* **463**: 878–879.

Chatterton, J., Viviattene, C., Morris, J., *et al.* (2010) *The Costs of the Summer 2007 Floods in England*. Project Report, Environment Agency, Bristol.

Conklin, J. (2005) *Dialogue Mapping: Building Shared Understanding of Wicked Problems*, New York: John Wiley and Sons.

Cutter, S. L. and Finch, C. (2008) Temporal and spatial changes in social vulnerability to natural hazards. *Proceedings of the National Academy of Sciences* **105** (7): 2301–2306.

Funtowicz, S. O. and Ravetz, J. R. (1991) A new scientific methodology for global environmental issues. In *Ecological Economics: The Science and Management of Sustainability*, ed. Robert Costanza, New York, NY: Columbia University Press, pp. 137–152.

Hulme, M. (2009) *Why We Disagree About Climate Change: Understanding Controversy, Inaction and Opportunity*, Cambridge: Cambridge University Press.

Huppert, H. E. and Sparks, R. S. J. (2006) Extreme natural hazards. *Philosophical Transactions of the Royal Society* **364**: 1875–1888.

Jackson, J. (2006) Fatal attraction: living with earthquakes, the growth of villages into megacities, and earthquake vulnerability in the developing world. *Philosophical Transactions of the Royal Society* **364**: 1911–1925.

Japan National Police Agency (2011) *Damage and Police Responses to the Northeast Pacific Earthquake*. 13 April. http://www.npa.go.jp/archive/keibi/biki/index.htm (accessed 25 July 2012).

Jennings, S. (2011) Time's bitter flood: trends in the number of reported natural disasters. Oxfam Research Report.

Knabb, R. D., Rhome, J. R. and Brown, D. P. (2005) Tropical Cyclone Report Hurricane Katrina 23–30 August 2005. National Hurricane Center, Miami, FL. http://www.nhc.noaa.gov/pdf/TCR-AL122005_Katrina.pdf (accessed 25 July 2012).

Mitchell, J. F. B., Lowe, J. A., Wood, R. A., *et al.* (2006) Extreme events due to human-induced climate change. *Philosophical Transactions of the Royal Society* **364**: 2117–2133.

Norio, O., Ye, T., Kajitani, Y., Shi, P., *et al.* (2011) The 2011 Eastern Japan great earthquake disaster: overview and comments. *International Journal of Disaster Risk Science* **2** (1): 34–42.

Okuyama, Y. and Sahin, S. (2009) Impact estimation for disasters: a global aggregate for 1960 to 2007. World Bank Policy Working Paper 4963, pp. 1–40.

Oxford Economics (2010) The economic impacts of air travel restrictions due to volcanic ash. Report for Airbus.

Pagano, M. (2011) Japan looks for market stability after quake. *The Independent*, 13 March. http://www.independent.co.uk/news/business/news/japan-looks-for-market-stability-after-quake-2240323.html (accessed 25 July 2012).

Shah, H. C. (2006) The last mile. *Philosophical Transactions of the Royal Society* **364**: 2183–2189.

Sieh, K. (2006) Sumatran megathrust earthquakes: from science to saving lives. *Philosophical Transactions of the Royal Society* **364**: 1947–1963.

Smith, J. Q. (2010) *Bayesian Decision Analysis: Principles and Practice*, Cambridge: Cambridge University Press.

Smolka, A. (2006) Natural disasters and the challenge of extreme events: risk management from an insurance perspective. *Philosophical Transactions of the Royal Society* **364**: 2147–2165.

UN-GAR (2011) *Global Assessment Report on Disaster Risk Reduction*, Geneva: United Nations.

2

Quantifying hazard losses

J. C. ROUGIER

2.1 Introduction

This chapter considers the natural hazard risk manager as an agent for choosing between interventions that affect the impact of future hazards. As such, it is possible to proceed broadly quantitatively, and to set aside the many less-quantifiable concerns that arise during a period of real-time hazard management. Thus the aim is to describe a framework in which it is possible to address questions such as 'Should we build a firebreak at this location, or leave things as they are?'. This chapter only considers one source of uncertainty, which is the inherent uncertainty of the hazard itself, often termed *aleatory* uncertainty. Chapter 3 of this volume considers the extent to which a quantitative analysis can be extended to include more general *epistemic* uncertainties.

The objective is a precise definition of the common notions of natural hazard risk and uncertainty assessment, within the framework of probability. To this end, Section 2.2 provides a brief justification for the use of probability as the calculus of uncertainty that is appropriate for natural hazards. Section 2.3 then defines quantitative risk in a limited but precise way. Section 2.4, the heart of the chapter, considers the three stages in which one passes from the hazard itself to an evaluation of risk, noting that at each stage there are opportunities for the risk manager's intervention. Section 2.5 considers the implications of using simulation as the primary computational tool for assessing risk, the uncertainties engendered by limits in computing resources, and ways to quantify these uncertainties. Section 2.6 describes different types of hazard map. Section 2.7 concludes with a summary.

Note that I have not provided references for natural hazard examples; plenty can be found in the rest of this volume. Most of the references are to statistics books.

2.2 Why probability?

Hazard losses are uncertain, and therefore a calculus that accommodates uncertainties is necessary to quantify them. The role of quantification within a formally valid framework was discussed Chapter 1 of this volume. This chapter and Chapter 3 advocate the use of probabilities and the probability calculus, and this is in fact the dominant method for quantifying uncertainty, not just in natural hazards, but in almost every endeavour with substantial uncertainty. However, probability is not the only calculus for uncertainty. A standard reference such as Halpern (2003) covers alternative approaches such as lower and

Risk and Uncertainty Assessment for Natural Hazards, ed. Jonathan Rougier, Steve Sparks and Lisa Hill. Published by Cambridge University Press. © Cambridge University Press 2013.

upper probabilities, Dempster–Shafer belief functions, possibility measures, ranking functions and relative likelihood. Therefore this section provides a brief justification for the use of the probability calculus in natural hazards risk assessment.

The first point to note is that no uncertainty calculus can do a complete job of handling uncertainty. Each calculus represents a normative description of how, on the basis of certain axioms and principles, one can make inferences about uncertain events, and adjust those inferences in the light of additional information. The need for such a framework is attested by the fact that people are demonstrably not good at handling uncertainty in even quite simple situations. Gigerenzer (2003) provides an enjoyable and informative introduction, while Lindley (1985) provides an introduction to the use of probability in reasoning and decision-making.

To illustrate, consider the axioms and principles of the probability calculus. The three axioms concern probabilities defined on events (or propositions) that are either true or false. They are:

(1) All probabilities are non-negative.
(2) The probability of the certain event is 1.
(3) If events A and B cannot both be true, then $\Pr(A \text{ or } B) = \Pr(A) + \Pr(B)$.

These axioms, and their generalisations, are discussed in most textbooks in probability and statistics (see, e.g. Grimmett and Stirzaker, 2001: ch. 1; DeGroot and Schervish, 2002: ch. 1).

There are really two principles in probability. The first is a principle for extending probability assessments from a limited set of events to a more complete set that is appropriate for the information that might be collected and the decisions that must be made. In fact, formal treatments using the probability calculus usually sidestep this issue by starting from the notion that probabilities have already been assigned to all events in a *field*. A field is a collection of sets of events, including the null event, which is closed under complements and unions. The assertion is that this field is appropriate in the sense given above; i.e. it contains all the events of interest. But there are more general approaches to extending probability assessments, discussed, for example, by Paris (1994: ch. 6).

The second principle of the probability calculus is that inferences are adjusted by conditioning, so that if event B were found to be true, then the adjusted probability of A would be $\Pr(A \text{ and } B)/\Pr(B)$. The axioms and the principle of conditionalisation have several different justifications, including: probabilities as relative frequencies, as subjective degrees of belief, as logical consequences of more primitive axioms for reasoning. These are outlined in Hacking (2001), with a much more forensic but technical assessment in Walley (1991), especially chapters 1 and 5.

Halpern (2003: p. 24) lists the three most serious problems with probability as: (1) probability is not good at representing ignorance; (2) an agent may not be willing to assign probabilities to all events; and (3) probability calculations can be expensive. In the context of natural hazards, I think (2) is the most concerning, and (3) can be an issue, although it is becoming less so as statistical computational techniques continue to improve, along with computer power. The first item, as Halpern himself notes, can be handled (in principle and at least partially in practice as well) within the probability calculus by extending the set of events. This is effectively the approach explored in Chapter 3 of this volume.

All uncertainty calculuses have axioms and principles. In each case, the agent will find herself asking, of the axioms and principles: Do I believe that? Should I believe that? These questions do not have simple answers, which should not be surprising in the light of the proliferation of competing approaches to uncertainty representation. Consequently, an agent does not sign up for a particular uncertainty calculus with the conviction that it is exactly what is required for her situation. It is very important for her to understand the limitations of her chosen calculus, so as not to overstate the result, and to have methods for addressing the limitations informally.

In natural hazards things become more complicated, because the agent is the risk manager, who must satisfy an auditor. What seems to be a compelling calculus to the risk manager, possibly after detailed study and reflection, may seem much less so to the auditor, and to the stakeholders he represents. Overall, this line of reasoning suggests favouring calculuses with simple axioms and easily stated limitations.

On this basis, the probability calculus is highly favoured: its axioms are the simplest, and its main limitation is that it is necessary to specify probabilities for all events in an appropriate field. In practice this limitation can be addressed informally using sensitivity analysis, in which alternative choices for hard-to-specify probability distributions are tried out. The tractability of probability calculations makes sensitivity analysis a feasible strategy for all but the largest problems. And the universality of probabilistic reasoning makes communication relatively straightforward. That is not to say, though, that these issues are clear-cut. It is important for all parties to appreciate that not all aspects of uncertainty can be expressed as probabilities, and that a probabilistic analysis may create the false impression that all uncertainty has been accounted for.

2.3 A quantitative definition of risk

Risk is a multivalent and multivariate concept, yet it is often treated as though it is a well-defined scalar – for example, in statements comparing the 'riskiness' of different hazards. This indicates the nature of quantified risk as a summary measure whose role is to represent the gross features of a hazard, and to operate at the first stage of a 'triage' of hazards and actions.[1] In their review of risk terminology, the UK Central Science Laboratory (CSL) recommends that the definition of risk 'should include both probability and the degree of effect, including its severity, but in a way that keeps them distinct and gives rise to a single dimension' (Hardy *et al.*, 2007: p. 70). The presumption in this recommendation is that uncertainty will be quantified in terms of probabilities, as discussed in the previous section.

---

[1] 'Triage' in the sense of sorting by priority and expediency, as in the emergency room of a hospital.

I will adopt 'loss' as the cover-all term for the quantifiable aspects of harm and damage that follow from a natural hazard event.

To illustrate, supervolcanoes are more risky than asteroid strikes because the loss in the two cases is the same order of magnitude (i.e. catastrophic), but the probability of a supervolcano is higher than an asteroid strike on human timescales (Mason *et al.*, 2004).

If risk is a summary term then it should be derived from a more detailed analysis. The second stage of a triage is then to refer back to this detailed analysis for those hazards for which such an assessment is required. To my mind, this constrains the operational definition of risk to be the mathematical expectation of loss, and the detailed analysis for which it is the summary is the loss probability distribution function. The usual feature of this distribution for natural hazards would be a long right-hand tail (positive skewness), indicating that small losses are common, but occasionally very large losses occur. If X denotes the unknown loss that a hazard induces over a specified time interval, then its distribution function is denoted

$$F_X(x) = \Pr(X \leq x),$$

where 'Pr' denotes probability, and 'x' denotes the abscissa. This distribution function is usually visualised in terms of the loss exceedance probability (EP) curve, which is the graph of $1 - F_X(x)$ on the vertical axis against x on the horizontal axis, as shown in Figure 2.1 (in statistics this would be called the 'survivor function'). This graph does not necessarily start at $(0, 1)$; in fact, it starts at $(0, 1-p)$ where p is the probability that there are no losses during the time interval. The EP curve must attain zero at some finite value for x because there is always a limit to how much can be lost; therefore, having bounded support, X must have finite moments, and so its mathematical expectation and variance certainly exist.

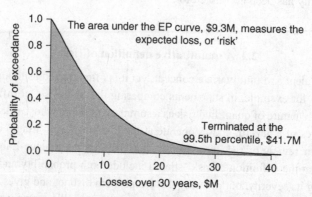

Figure 2.1 Example of an exceedance probability (EP) curve. EP curves are computed for losses from *hazard outcomes* over some specified period, such as 30 years; see Section 2.4.1 for the distinction between events and outcomes.

Risk may then be defined as the mathematical expectation of loss, i.e. the sum of the product of each possible loss amount and its probability. This is also referred to as the 'mean loss'. This definition has the features outlined in the CSL recommendation above: it is a scalar quantity measured in the same units as the loss, and incorporates both the loss estimates and their probabilities. The same-sized risk could indicate a hazard that will probably induce a medium-sized loss, or a hazard that will probably induce a small loss but occasionally induce a very large loss. These two cases cannot be distinguished in a scalar summary, but they can be distinguished in the underlying probability distributions.

Within statistics this definition of risk as expected loss has a long provenance in the field of decision theory (see, e.g. Rice, 1995: ch. 15). In catastrophe modelling for insurance, the EP curve and risk focus on financial loss for a one-year time interval. The risk is termed the 'average annual loss' (AAL), and represents the 'fair price' for an insurance premium.

There is a very strong connection between this definition of risk and the EP curve: *risk is the area under the EP curve*. The fact that the expected loss is equal to the area under the EP curve is a purely mathematical result (see, e.g. Grimmett and Stirzaker, 2001: 93), but it is, from our point of view, a very useful one as well. It indicates that the gross comparison of different hazards, or of different actions for the same hazard, can be done by plotting their EP curves on the same graph and comparing the area underneath them. Then, in cases where this gross comparison is not sufficient, a more detailed assessment of the EP curve can be made, comparing, for example, the probabilities of very large losses.

2.3.1 Summary

The loss due to a hazard for a specified time interval is an uncertain quantity that is represented by a probability distribution; this is usually visualised in terms of an EP curve. Risk is defined to be the expected (or mean) loss; it is a mathematical result that this is equal to the area under the EP curve.

2.4 Structural modelling of hazard outcomes

This section follows the flow of information and judgement from the hazard to the loss. It proceeds in three stages: representing the aleatory uncertainty of hazard outcomes; representing the 'footprint' of a hazard outcome in space and time; and representing the loss of this hazard outcome as a loss operator applied to the footprint. Probabilities assigned to hazard outcomes 'cascade' through the stages to induce a probability distribution for loss.

It is worth pausing to examine the need for such a tripartite representation. After all, it would be much easier simply to construct a catalogue of 'similar' hazard events, and then use the collection of losses from those events as a proxy for the loss distribution. Why bother with all the extra modelling?

Two of the reasons are driven by the need for the scientific analysis to inform the risk manager's decisions: non-stationarity in the hazard domain, and the intention to intervene. The natural hazard risk manager has to operate on time intervals of 30 years or more, over which many aspects of the hazard loss can be expected to change substantially; this is the problem of *non-stationarity*. For example, increasing populations result in increasing city size, increasing population density in cities, changes in the quality of buildings and infrastructure, changes in land-surface characteristics in catchments. In this situation historical losses are not a reliable guide to future losses over the whole of the time interval, and additional judgements are needed to perform the extrapolation.

Second, the risk manager is concerned explicitly with evaluating actions, which represent *interventions* to change the loss that follows a hazard event. For example: constructing defences, changing building regulations or relocating people living in hazardous areas. For this purpose, it is important to model the hazard in a causal rather than a statistical way. The key role of the tripartite representation is that it provides a framework within which interventions can be modelled and compared. Each possible intervention generates its own EP curve and its own risk. Then the risk manager has the task of choosing between interventions according to their EP curves, taking account of the costs and benefits of intervening and the probability of completing the intervention successfully.

A third reason is that the catalogue of observed hazard events is likely to be incomplete. This is a concern shared by insurers, who otherwise operate on short time intervals and are not concerned with interventions. There is a large probability that the next event will be unlike the catalogued ones, and so physical insights are required to extrapolate from the catalogue in order to build up a picture of what might happen. In some ways this is also an issue of non-stationarity. The catalogue can be extended by widening the criteria under which events can be included. In earthquake catalogues, for example, the events observed at a specified location may be augmented with events observed at other similar locations. Effectively this makes a judgement of spatial and temporal stationarity; hence non-stationarity is one reason why catalogues might be very incomplete.

It is questionable to what extent any framework can encompass all features of natural hazards. The framework presented here seems sufficiently general, even if some 'shoe-horning' is necessary for particular hazards. It is sufficient to illustrate the main concepts, and there is also value in considering how well a particular hazard does or does not fit. It also corresponds closely with the modular approach adopted by catastrophe-modelling companies.

2.4.1 The hazard process

The specification of the hazard process comprises three stages:

(1) specification of the hazard domain;
(2) enumeration of the hazard events; and
(3) assignment of probabilities to hazard outcomes.

In Step 1, the hazard domain is simply the spatial region and the time interval over which the hazard is being considered. This has to be specified a priori in order that the probabilities in Step 3 are appropriately scaled, particularly over the time interval. It would be common for the time interval to start 'now'. As already noted, the risk manager is likely to be considering time intervals of many years: 30, for example, for a hazard such as an earthquake, longer for nuclear power installations, and longer again for nuclear waste repositories.

In Step 2, each hazard event is described in terms of a tuple (a tuple is an ordered collection of values). This tuple will always include the inception time of the event, so it is written (t, ω), where ω is a tuple that describes the hazard event that starts at time t. For an earthquake, ω might comprise an epicentre, a focal depth and a magnitude; or, perhaps, a marked point process representing a local sequence of shocks with specified epicentres, focal depths and magnitudes. For a rain storm, ω might comprise a time-series of precipitations, where each component in the time-series might itself be a spatial map. Likewise with a hurricane, although in this case each component of the time-series might be a spatial map of wind velocities. The set of all possible values of ω is denoted as Ω.

In Step 3 it is important to distinguish between a *hazard event* and a *hazard outcome*. The hazard outcome is the collection of hazard events that occur in the hazard domain. Therefore $\{(t, \omega), (t', \omega')\}$ would be a hazard outcome comprising two events: at time t, ω happened, and at time t', ω' happened. The collection of all possible hazard outcomes is very large, and, from the risk manager's point of view, each outcome must be assigned a probability. In practice this assignment of probabilities can be 'tamed' as follows, recollecting that ω describes all features of the hazard event bar its inception time (i.e. including location and magnitude).

2.4.1.1 *Simplifying choices*

(1) Different ω correspond to probabilistically independent processes; and
(2) Each process is a homogeneous Poisson process with specified rate λ_ω.

Under these simplifying choices, the probability of any hazard outcome can be computed explicitly, once the rates have been specified. The simplifying choices imply that the time between hazard events is exponentially distributed with a rate equal to $\lambda = \sum_\omega \lambda_\omega$, and that the probability that the next event is ω is equal to λ_ω / λ (see, e.g. Davison, 2003 examples 2.35 and 2.36).

2.4.1.2 *Interventions*

The risk manager has few opportunities to intervene in the hazard process, which in this case would take the form of changing the probabilities on the hazard outcomes. Examples would be controlled burning for forest fires and controlled avalanches. In situations where the trigger is sometimes man-made, the probabilities can be reduced by information campaigns, regulations and physical exclusion. Again, this applies mainly to forest fires and avalanches; also perhaps to terrorism. One can also consider more speculative possibilities; for example, geoengineering solutions such as cloud-seeding to reduce the probability of large rain storms over land.

2.4.2 *The footprint function*

The footprint function represents the 'imprint' of a hazard outcome at all locations and times in the hazard domain. The purpose of the distinction between the hazard outcome and its footprint is to separate the hazard outcome, which is treated as *a priori* uncertain, from its effect, which is determined *a posteriori*, largely by physical considerations. For example, for inland flooding the hazard outcome would describe a sequence of storm events over a catchment. The footprint function would turn that sequence of storm events into a sequence of time-evolving maps of water flows and water levels in the catchment.

Often the footprint function will be expressed in terms of an initial condition and a hazard event. The initial condition describes the state of the hazard's spatial domain just prior to the hazard event. In the case of inland flooding, this might include the saturation of the catchment, the level of the reservoirs and rivers and possibly the settings of adjustable flood defences such as sluice gates. The footprint function of a hazard *outcome* is then the concatenation of all of the hazard event footprint functions, where the initial condition for the first event depends on its inception time, and the initial conditions for the second and subsequent events depend on their inception times and the hazard events that have gone before. This is more sophisticated than current practice, which tends to be event-focused, but it shows that there is no difficulty, in principle, to generalising to hazard outcomes that comprise multiple events; nor, indeed, to outcomes that involve multiple hazards.

Most natural hazards modelling currently uses deterministic footprint functions. For flows (volcanoes, landslides, avalanches, tsunamis and coastal flooding) these are often based on the shallow-water equations, single- or multiphase, or, for homogeneous flows, sliding-block models. Earthquakes use the equations of wave propagation through elastic media. For inland flooding, there is a range of different footprint functions, from topographically explicit models, through compartmental models, to empirical models. Sometimes physical insights are used to derive parametric relationships between a hazard event and particular features of the hazard footprint, and the parameters are then statistically fitted to a catalogue of similar events. Woo (1999) contains many examples of these types of relationships, based on dimensional or scaling arguments (often implying power laws).

In this chapter we treat the footprint function (and the loss operators, below) as known. In practice they are imperfectly known, which is an important source of epistemic uncertainty, discussed in Chapter 3 of this volume.

2.4.2.1 *Interventions*

The footprint function provides an important route through which the risk manager can model the effect of interventions. These would be interventions that interfered physically with the impact of the hazard on the region. For example, building a firebreak, a snow dam or a levée. Typically, these interventions change the topography and other land-surface features local to the hazard event, and thus change the evolution of the event's impact in space and time; for example, diverting a flow or retarding it to allow more time for evacuation.

2.4.3 The loss operator

The footprint of the hazard is neutral with respect to losses. Loss is a subjective quantity, and different risk managers with different constituencies will have different notions of loss, although these will typically centre on loss of life or limb, loss of ecosystem services and financial loss due to damage to property. The purpose of the functional separation between the footprint function and the loss operator is to allow for this distinction between what is generic (the hazard footprint) and what is subjective (the risk manager's loss). Any particular loss operator transforms a hazard outcome's footprint into a scalar quantity. Typically, this will take the form of the addition of the losses for each hazard event, where it is likely that the losses from later events will depend on the earlier ones. For simplicity I will treat the loss operator as deterministic, but the generalisation to an uncertain loss operator is straightforward, and discussed in Chapter 3 of this volume.

The nature of the loss operator depends on the hazard and the risk manager. Consider, as an illustration, that the hazard is an earthquake and the risk manager represents an insurance company. The buildings in the insurance portfolio are each identified by location, type and value. The type indicates the amount of ground acceleration the building can withstand, and a simple rule might be that accelerations above this amount destroy the building, while accelerations at or below this amount leave the building intact. To compute the loss of an event, the footprint function is summarised in terms of a spatial map of peak ground acceleration, and this would be converted into a loss (in millions of dollars) by determining from this map which of the insured properties were destroyed. (Naturally, this is a rather simplified account of what actually happens.)

In the same situation, if the risk manager's loss operator concerns loss of life, then the same peak ground acceleration map from the footprint function might be combined with maps of building type and population density. A peak ground acceleration map is an example of an *extremal hazard map*, discussed in Section 2.6.

2.4.3.1 Interventions

Interventions in the loss operator concern changes in vulnerability. For example, the risk manager for the insurance company can change the loss distribution by modifying the premiums for different types of property or by refusing to insure certain types of building. The risk manager concerned with loss of life could change building regulations or re-zone the city to move people away from areas with very high peak ground accelerations. Another example of an intervention to affect the loss operator is the installation of an early warning system (e.g. for tsunamis) or the implementation of a phased alert scheme (e.g. for wildfires). This might be accompanied by a public education programme.

2.4.4 Summary

The chain from hazard to the distribution of loss has three components. First, the hazard process itself, which comprises the enumeration of different hazard events, and then the

Table 2.1 *Summary of the three components of the hazard risk assessment, with key concepts and opportunities for the risk manager's intervention*

	Concepts	Opportunities for intervention
Hazard process	Hazard's spatial-temporal domain, events and outcomes, probabilities, simplifying choices	Reducing probability of human triggers: wildfires, avalanches
Footprint function	Objective, typically implementing physical equations or statistical regularities. Shows the imprint of the hazard outcome in space and time, often in terms of individual hazard events	Change the topography of hazard's spatial domain: levées, dams, firebreaks
Loss operators	Subjective, depending on constituency and risk manager. Typically measuring loss of life, loss of ecosystem services or financial damage	Reduce vulnerability: regulations, re-zoning and relocation, harden critical infrastructure

assignment of probabilities to hazard outcomes. The distinction between hazard events and hazard outcomes is crucial (see Section 2.4.1). Second, the footprint function, which represents a hazard outcome in terms of its impact on the hazard domain. Third, the loss operator, which varies between risk managers, and maps the hazard footprint into a scalar measure of loss. The risk manager has the opportunity to intervene in all three components: by changing the probabilities of the hazard outcomes, by changing the local topography and therefore the hazard footprint, or by changing the vulnerability, and therefore the loss operator. This is summarised in Table 2.1.

2.5 Estimating the EP curve

The EP curve (see Section 2.3) represents the distribution function of the risk manager's loss. There is a standard formula for computing the EP curve, which is simply to sum the probabilities for all outcomes which give rise to a loss greater than x, for each value of x. In practice, this calculation would usually be done by simulation. This has important implications for how the EP curve and quantities derived from it are reported, particularly high percentiles.

2.5.1 Uncertainty and variability

There is a very important distinction in statistics between *uncertainty* and *variability* (see, e.g. Cox, 2006, especially ch. 5), and a failure to understand this distinction lies at the heart of many suspect attempts to quantify uncertainty using variability. Uncertainty about the loss induced by a hazard is represented in the form of a probability distribution function, the function denoted F_X in Section 2.3. From a probabilistic point of view, this probability distribution function is a complete description of uncertainty. In practice, we are aware that when we construct such a function we are obliged to introduce various simplifications. In this chapter we have considered the aleatory uncertainty of the hazard outcomes as the *only* source of uncertainty. This is very great simplification! In Chapter 3 of this volume, we extend our consideration to incorporate epistemic sources of uncertainty, such as incomplete information about the hazard outcome probabilities, the footprint function or the loss operator(s). The effect of this extension is to make a better assessment of the probability distribution function of loss. Or, as discussed in Chapter 3, to produce several different probability distributions, each one specified conditionally on certain simplifications.

But representing our uncertainties probabilistically is not the only challenge. The computational framework, in which uncertain hazard outcomes cascade through the footprint function and the loss operator(s), can always be simulated, but often this simulation will be expensive. With limited resources, we will have an incomplete knowledge of the probability distribution of loss based upon the simulations we have – say, n simulations in total. To put this another way, were we to have done the n simulations again with a different random seed, the sample of losses would have been different, and the summary statistics, such as the expected loss, would have different values. This is the problem of *variability*. Expensive simulations mean that our knowledge of the loss distribution is limited, and that our numerical descriptions of it are only approximate.

In statistics, variability is summarised by *confidence intervals*. It is very important to appreciate that a confidence interval is *not* a representation of uncertainty. The distribution function F_X is a representation of uncertainty. A confidence interval is a representation of variability of our estimate of some feature of F_X (like its mean), which reflects the fact that n, the number of simulations we can afford to do in order to learn about this feature, is not infinite. To state that the risk (i.e. mathematical expectation) of a hazard is $20.9 million is a statement about uncertainty. To state that the 95% confidence interval for the risk of a hazard is [$17.1 million, $25.3 million] is a statement about both uncertainty and variability: the location of this interval is an assessment of uncertainty, while its width is an assessment of variability. The latter will go down as n increases.

I will focus here on the variability that arises from not being able to perform a large number of natural hazards simulations. But in some situations, variability can also arise from the limited information about the hazard itself. For example, a flood engineer might use the historical distribution of annual maximum river heights at some specified location to estimate the probabilities in the hazard process. If the record does not go back very far, or if only the recent past is thought to be relevant because of changes in the environment, then estimates of the probabilities based on treating the historical measurements as probabilistically independent realisations from the hazard process will be imprecise. I prefer to treat this as a manifestation of epistemic uncertainty about the hazard process, and this is discussed in Chapter 3 of this volume.

This section is about assessing confidence intervals for properties of the loss distribution. My contention is that variability ought to be assessed because natural hazards loss simulations are expensive, and some important assessments of uncertainty, such as high percentiles of the loss distribution, will tend to be highly variable.

2.5.2 Monte Carlo simulation and EP curve variability

By simulating hazard outcomes we can estimate the EP curve; technically, simulation is an implementation of Monte Carlo integration (see, e.g. Evans and Swartz, 2000). Better estimates, i.e. ones that tend to be closer to the true value, can be achieved at the expense of a larger number of simulations, or a more careful experimental design (e.g. applying variance reduction techniques such as control variables, or antithetic variables). For complex applications, where resources constrain the number of simulations, measures of variability are required to indicate the closeness of the estimated EP curve to the true EP curve, and likewise for other quantities related to the EP curve, like the mean, or the 95th or 99.5th percentiles of the loss distribution.

A Monte Carlo simulation is constructed for a specific hazard process, footprint function and loss operator: uncertainty about these is the subject of Chapter 3 in this volume. It has the following steps:

(1) simulate a hazard outcome;
(2) evaluate the footprint function for the outcome;
(3) evaluate the loss operator for the footprint.

Repeating this many times for probabilistically independent simulations of the hazard outcome will build up a histogram of losses, which is an estimator of the probability distribution of losses. If the footprint function or loss operator are stochastic, then they too can be simulated rather than simply evaluated.

Step 1 can be complicated, bearing in mind the distinction between a hazard event and a hazard outcome: the latter being the concatenation of many events over the hazard domain. But accepting the simplifying choices described in Section 2.4.1 makes this step straightforward. The sequence of events that make up a hazard outcome may also affect the footprint function and the loss operator. Simplifying choices in the same vein would be to treat the interval between hazard events as sufficiently long for the hazard domain to 'reset', so that the footprint and loss of later events does not depend on earlier ones. This seems to be a common choice in catastrophe modelling.

The variability of a Monte Carlo estimator can be quantified in terms of confidence intervals or confidence bands. A 95% confidence band for the EP curve comprises a lower and upper curve with the property that the true EP curve will lie entirely within these curves at least 95% of the time. Confidence bands for the EP curve can be computed using Dvoretzky–Kiefer–Wolfowitz Inequality, as described in Wasserman (2004, ch. 7). For a $1 - \alpha$ confidence band the upper and lower curves are $\pm$ the square root of $\ln(2/\alpha)/(2n)$

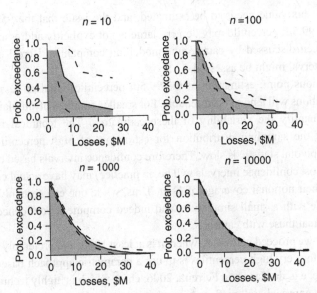

Figure 2.2 Effect of different numbers of simulations on the estimated EP curve, showing a 95% confidence band. The simulations in the four panels are nested so that, for example, the case with. $n = 100$ simulations extends the previous ten simulations with 90 new ones.

vertically about the empirical EP curve, where n is the number of simulations. For example, with $n = 1000$, a 95% confidence band is ± 0.043. Figure 2.2 shows estimated EP curves and 95% confidence bands for different numbers of simulations.

2.5.3 Estimating the risk

The risk is defined to be the expected loss, as previously discussed. This is easily estimated from a random sample, and a normal confidence interval will likely suffice, or a Student t-interval if the number of simulations is small. For the four simulations given in Figure 2.2, the estimated risk and a 95% CI is: $n = 10$, \$9.8 million (5.2, 14.4); $n = 100$ \$9.2 million (7.6, 10.8); $n = 1000$, \$9.7 million (9.2, 10.2); $n = 10\,000$, \$9.5 million (9.4, 9.7). With 10 000 simulations, the proportional uncertainty is only a few per cent.

2.5.4 Estimating high percentiles

The Solvency II Directive[2] on solvency capital requirement is described in terms of the 99.5th percentile of the loss distribution for a one-year time interval (Paragraph 64, page 7). In the section 'Statistical quality standards' (Article 121, page 58) there is no guidance about

---

[2] http://eur-lex.europa.eu/LexUriServ/LexUriServ.do?uri=OJ:L:2009:335:0001:0155:EN:PDF

how accurately this value needs to be estimated, and the issue that the 95% confidence interval for the 99.5th percentile may be very large is not explicitly addressed, although it has certainly been discussed by catastrophe-modelling companies. Here I outline briefly how such an interval might be assessed.

First, an obvious point: estimators of the 99.5th percentile based on less than at least $n = 1000$ simulations will tend to be unreliable. For smaller numbers of simulations it will be very hard to quantify the variability of these estimators. The technical reason is that convergence of the sampling distribution for estimators of high percentiles to a form which admits a pivotal statistic is slow. Therefore confidence intervals based on asymptotic properties (as most confidence intervals will be, in practice) may have actual coverage quite different from their notional coverage of $1 - \alpha$. Thus, while one can certainly estimate the 99.5th percentile with a small simulation, and indeed compute a confidence interval, an auditor should treat these with caution.

So how might we proceed in the case where n is at least 1000, and preferably much larger? One possibility for very high percentiles is to use a parametric approach based on extreme value theory (see, e.g. de Haan and Ferreira, 2006: ch. 4, which is highly technical). Often a more transparent approach will suffice. A simple estimator of Q, the desired percentile, is to interpolate the ordered values of the simulated losses. A 95% confidence interval for Q based on this estimator can be assessed using the bootstrap. By a loose analogy with the exponential distribution, a nonparametric bootstrap on the logarithm of Q should be effective (see Davison and Hinkley, 1997: §§ 2.7 (notably example 2.19) and 5.2). But one must bear in mind that the bootstrap is a very general technique with lots of opportunities for both good and bad judgements. It would be wise to treat the bootstrap 95% confidence interval of the 99.5th percentile as merely indicative, unless the number of simulations is huge (tens of thousands).

For the illustration, the estimated 99.5th percentile and approximate 95% confidence intervals are: $n = 1000$, \$43.8 million (36.9, 53.8); $n = 5000$, \$42.3 million (40.5, 45.2); $n = 10\,000$, \$41.7 million (40.3, 43.4). Here, even 10 000 simulations gives a confidence interval with a width of several million dollars, which is a proportional uncertainty of about 10%.

2.6 A digression on terminology

In natural hazards, one often finds statements such as 'ω is an event with a return period of k years', where k might be 100 or 200, or 1000. Or, similarly, 'ω is a one in k year event': Personally, I do not find such terminology helpful, because it is an incomplete description of the event, and, in order to be useful, must apply within a strongly parameterised hazard process. I would rather see these issues made explicit.

The explicit statement would be 'Considering the hazard process over the next t years, the marginal distribution of event ω is a homogeneous Poisson process with arrival rate $1/k$.'. This makes the time interval over which the description holds explicit. It also translates the absence of any further information into an explicit statement of homogeneity. If such homogeneity were not appropriate, then more information should have been given in the first place. The fact that such a statement is not qualified by any reference to other events suggests that the conditional distribution of ω is judged to be unaffected by the other events. Hence such statements embody completely the simplifying choices of Section 2.4.1.

Perhaps we can take it as conventional that all statements about return periods embody these simplifying choices. But what is still missing in the original statement is an explicit time domain: for how long into the future does event ω have an arrival rate of $1/k$? There is no convention about such an interval, and so all such statements need to be qualified. But then we are in the realm of 'ω is an event with a return period of k years, over the next t years'. Many people will find two references to different numbers of years confusing. But considering that this is a more precise statement than before, one wonders how much confusion there is currently. Do people, for example householders, perhaps think that a return period of 1000 years is good for the next 1000 years? Or only for the next year? Or for the next few years? This is a source of needless ambiguity that can be totally eliminated with a little more precision.

The same language crops up in another context as well. The '1 in k year loss' is synonymous with the $(1 - 1/k)$th quantile of the one-year loss distribution. So why not say this? Rather than the '1 in 200 year loss', state '99.5th percentile of the one-year loss distribution'.

Personally, I think this evasive language reflects a reluctance to use explicit probabilistic constructions. Such constructions would show clearly how subjective these assessments are, depending as they do on very strong judgements that would be hard to test with the limited data available. The simplifying choices for the hazard process are very strong; and one does not get to the 99.5th percentile of the loss distribution without also making some strong judgements about distributional shapes. But concealing this subjectivity is the wrong response if our intention is to inform the risk manager. Rather, we should get it out in the open, where it can be discussed and refined. In some cases we will be forced into strong judgements for reasons of tractability. In this case, too, it is necessary to acknowledge these judgements explicitly, so that we can decide how much of a margin for error to include in the final assessment. This is discussed in more detail in Chapter 3 of this volume.

2.7 Creating hazard maps

There is no general definition of a hazard map. Although the term tends to have recognised meaning within individual natural hazard fields, methods for producing such maps vary widely. One important distinction is whether the map shows features of the hazard footprint or of the hazard loss. For losses it makes sense to add losses over events (although the size of the loss from later events may depend on the earlier ones). This is not true for aspects of the hazard footprint, even though these may relate strongly with losses per event.

For example, for an earthquake event, peak ground acceleration at a given location is often taken to be strongly related to damage to buildings at that location. But the addition of peak ground accelerations at a given location across the events in a hazard outcome does not largely determine the loss from a hazard outcome at that location: many small shocks do not incur the same loss as one large one, even if later losses are not affected by earlier ones.

Therefore I favour reserving the term *risk map* for showing losses, in situations where the loss over the hazard domain is the sum of the losses over each location. In this case, a risk map would show the expected loss at each location, and the total risk of the hazard would be the sum of the risks in each location. (Technically, this follows from the linear property of expectations: if risk were not defined as an expectation, there would be no reason for risk maps to have this appealing property.) In other words, *risk maps summarise the losses from hazard outcomes*.

What about maps that summarise probabilistic aspects of the hazard footprint, for which *(probabilistic) hazard map* seems appropriate? Even within a single hazard, such as volcanoes, there are a number of different ways of visualising the uncertain spatial footprint of the hazard. This variety seems strange, because in fact it is very clear what features a probabilistic hazard map ought to have. These are summarised in Table 2.2.

The first stage of constructing a hazard map is to reduce the hazard footprint to a spatial map. This means summarising over the time index, if one is present. A standard summary would be to take the maximum of some component over time, and for this reason I refer to the resulting maps as *extremal hazard maps:* note that these are constructed for a specific outcome (or event). Whatever summary is taken, the general principle for informing the risk manager is to use the value that relates best to loss at the location, and this will typically be a maximum value. For flooding, for example, the summary might be the maximum depth of inundation over the hazard outcome, at each location. An illustration for an earthquake is shown in Panel A of Figure 2.3.

The second stage is then to attach probabilistic information from the hazard process to each extremal hazard map. If the extremal hazard maps show hazard outcomes then attaching probabilities is straightforward, since we have assigned probabilities directly to hazard outcomes over the hazard domain. Likewise, if the simplifying choices have been used, then any particular concatenation of events into a hazard outcome can be assigned an explicit probability based on the rates of each outcome. Other situations, in which more

Table 2.2 *Features to be found in a probabilistic hazard map*

(1) Clearly stated time interval over which the probabilities apply.
(2) Clearly stated extremal operator, which summarises hazard outcomes over the time index.
(3) Clearly stated threshold for which probabilities of exceedance will be computed in each pixel.
(4) Contours or shading indicating regions of similar probability.

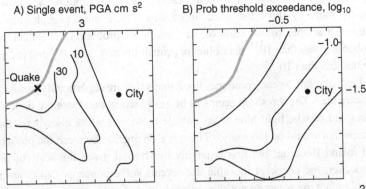

A) Single event, PGA cm s^2

B) Prob threshold exceedance, log$_{10}$

Mag. 5.1 event, focal depth 9.3 km   Thirty-year period, PGA threshold 5cm s^2

Figure 2.3 Two different hazard maps, an illustration for earthquakes (with made-up numbers). Map A shows a summary measure for a single earthquake event (peak ground acceleration, or PGA), which is typically a deterministic calculation that depends on the event itself (location, magnitude, depth) and on the physical properties of the domain. This is an example of an *extremal hazard map*. Map B is a *probabilistic hazard map*, showing the probability of exceedance of a specified threshold over a specified interval (see Table 2.2). This map is derived from many extremal hazard maps, after the incorporation of a probabilistic description of the hazard outcomes.

complicated hazard processes have been used, or where the footprint of a later event depends on the earlier ones, will probably have to be handled through simulation.

The third stage is to display the resulting set of probability-weighted extremal hazard maps as a single map. Each pixel has its own distribution for the summary value, but this is too much information to show on one map (although a visualisation tool would give the user the option to click on a pixel and see a distribution). Therefore further information reduction is required. The most useful approach, in terms of staying close to losses, seems to be to construct a map showing the probability of exceedance of some specified threshold, at each location; this is what I term a *probabilistic hazard map*. It will often be possible to identify a threshold at which losses become serious, and then this map can be interpreted loosely as 'probability of serious loss from the hazard over the next *t* years', where *t* is specified. Such a map identifies *hazard zones*, but it would be incorrect, in my treatment, to call them risk zones (because there is no explicit representation of loss). An illustration of a probabilistic hazard map for earthquakes is shown in Panel B of Figure 2.3.

2.7.1 UK Environment Agency flood maps

What would one expect to find in a probabilistic flood hazard map? I would expect the map to reflect a specified time interval that was relevant to risk managers and households, something between 1–5 years (or else produce maps for different time intervals), and to show the probability of inundation at each location; i.e. a threshold set at zero inches, or maybe slightly more to allow for some imprecision. Then these probabilities could be displayed using a simple colour scale of, say, white for probability less than 10^{-3}, light blue for probability less than 10^{-2}, dark blue for probability less than 10^{-1} and red or pink for probability not less than 10^{-1}.

The UK Environment Agency produces flood maps that are superficially similar to this, at least in appearance.[3] The threshold seems to be zero inches, and there are dark blue areas around rivers and coasts, light blue areas outside these and white elsewhere (there is no pink – perhaps that would be too scary). There is a distinction between the probabilities of coastal and inland flooding, but this is largely immaterial given the scale of the colour scheme. However, the checklist in Table 2.2 reveals some concerns. No time interval is specified, and in fact the zones do not show probabilities, but instead the extent of 1 in 100-year (probability = 0.01 in one year, dark blue) and 1 in 1000 year (probability = 0.001 in one year, light blue) events. This focus on events rather than outcomes makes it impossible to infer probabilities for hazard outcomes for a specific time interval except under the very restrictive condition *that these are the only two flooding events that can occur*. In this case, the interpretation in Section 2.6 would allow us to convert the Environment Agency flood map into a probabilistic hazard map for any specified time interval. But, of course, this is a totally indefensible condition.

The origin and development of the Environment Agency flood maps is complicated (see, e.g. Porter, 2009). They were never meant to be probabilistic hazard maps, and this focus on two particular events was driven in part by the requirements of the Planning Policy Statement 25.[4] Unfortunately, from the point of view of risk management, town planners, actuaries, businesses and householders cannot inspect the map and infer a probability for flood inundation over a specified period. But a probabilistic hazard map would have been much more expensive to compute, because a much wider range of events would have had to have been assimilated.

2.8 Summary

Ambiguity and imprecision are unavoidable when considering complex systems such as natural hazards and their impacts. That is not to say, though, that one cannot be systematic in developing a formal treatment that would serve the needs of the risk manager. Such a treatment removes needless ambiguity by the use of a controlled vocabulary. Natural hazards, for all their diversity, show enough common features to warrant a common controlled vocabulary. This chapter is an attempt to specify such a vocabulary within a probabilistic framework, defining: hazard domain, hazard event, hazard outcome, footprint function, loss operator, EP curve, risk, extremal hazard map, probabilistic hazard map and risk map. The entry points where judgements are required are the probabilities of the hazard

---

[3] http://www.environment-agency.gov.uk/homeandleisure/37837.aspx
[4] http://www.communities.gov.uk/documents/planningandbuilding/pdf/planningpolicystatement25.pdf

outcomes, the footprint function and the loss operator(s). All other aspects of the hazards analysis presented here then follow automatically, and their production is effectively an issue of statistical technique and computation.

Two issues are worth highlighting. First, there is a crucial distinction between hazard events and hazard outcomes, which encompasses both the distinction between one event and many, and between a single hazard type and different hazard types. This follows from the specification of a hazard domain in which several different hazard events spanning more than one hazard type might occur. This is central to the role of the risk manager, who is concerned with losses *per se*, but less so to the insurer, who is able through a contract to limit losses to certain hazard types. This distinction infiltrates other aspects of the analysis; for example, the EP curve is a description of outcomes, not events; likewise risk maps and hazard maps as defined here. The probabilistic link between events and outcomes is complicated, but can be tamed using the two simplifying choices described in Section 2.4.1. Similarly, footprint functions and loss operators are defined on outcomes not events, but there are obvious simplifications that allow definitions on events to be extended to outcomes.

Second, the use of simulation to construct EP curves and their related quantities should be seen as an exercise in statistical estimation. Hence EP curves should be presented with confidence bands to assess variability, and risk and quantiles should be presented with confidence intervals. The high percentiles required by regulators, e.g. 99.5th percentile for Solvency II, present serious problems for estimation, and the confidence intervals of such percentiles may be both wide and unreliable.

It is also important to reiterate the point made in Section 2.2. The probability calculus seems to be a good choice for quantifying uncertainty and risk in natural hazards. But it is by no means perfect, and nor is it the only choice. Sensitivity analysis with respect to hard-to-specify probabilities is crucial when assessing the robustness of the analysis. The importance of sensitivity analysis favours a simpler framework that can easily be replicated across different choices for the probabilities (and also for other aspects such as structural and parametric choices in the footprint function – see Chapter 3 of this volume), over a more complicated framework which is too expensive to be run more than a handful of times. From the point of view of the risk manager, and bearing in mind how much uncertainty is involved in natural hazards, I would favour simpler frameworks that function as tools, rather than a more complicated framework that we are forced to accept as 'truth'. This issue pervades environmental science; Salt (2008) is a good reality check for modellers who may be in too deep.

Finally, with the tools to hand, the risk manager is ready to start making difficult decisions. Her role is not simply to report the risk of doing nothing, but to manage the risk by evaluating and choosing between different interventions. Each intervention, in changing the hazard outcome probabilities, the footprint function or the loss operator, changes the EP curve. Ultimately, therefore, the risk manager will be faced with one diagram containing an EP curve for each intervention. As each intervention also has a financial and social cost, and also a probability of successful completion, the risk manager is not able to proceed on the basis of the EP curves alone, but must perform a very demanding synthesis of all of these aspects of the problem. It is not clear that anything other than very general guidelines can be given for such a challenging problem. But what is clear is that the risk manager will be well-served by a set of EP curves and probabilistic hazard maps that are transparently and defensibly derived. The purpose of this chapter has been to make this derivation as transparent and defensible as possible.

Acknowledgements

I would like to thank Thea Hincks and Keith Beven, whose detailed and perceptive comments on previous drafts of this chapter lead to many improvements.

References

Cox, D. R. (2006) *Principles of Statistical Inference*, Cambridge: Cambridge University Press.

Davison, A. C. (2003) *Statistical Models*, Cambridge: Cambridge University Press.

Davison, A. C. and Hinkley, D. V. (1997) *Bootstrap Methods and their Application*, Cambridge: Cambridge University Press.

DeGroot, M. H. and Schervish, M. J. (2002) *Probability and Statistics*, 3rd edn, Reading, MA: Addison-Wesley.

de Haan, L. and Ferreira, A. (2006) *Extreme Value Theory: An Introduction*, New York, NY: Springer.

Evans, M. and Swartz, T. (2000) *Approximating Integrals via Monte Carlo and Deterministic Methods*, Oxford: Oxford University Press.

Gigerenzer, G. (2003) *Reckoning with Risk: Learning to Live with Uncertainty*, London: Penguin Books Ltd.

Grimmett, G. R. and Stirzaker, D. R. (2001) *Probability and Random Processes*, 3rd edn, Oxford: Oxford University Press.

Hacking, I. (2001) *An Introduction to Probability and Inductive Logic*, Cambridge: Cambridge University Press.

Halpern, J. Y. (2003) *Reasoning about Uncertainty*, Cambridge, MA: MIT Press.

Hardy, A. R., Roelofs, W., Hart, A., *et al.* (2007) Comparative review of risk terminology. Technical Report, The Central Science Laboratory, doc. S12.4547739 (November).

Lindley, D. V. (1985) *Making Decisions*, 2nd edn, London: John Wiley and Sons.

Mason, B. G., Pyle, D. M. and Oppenheimer, C. (2004) The size and frequency of the largest explosive eruptions on Earth. *Bulletin of Volcanology* **66**: 735–748.

Paris, J. B. (1994) *The Uncertain Reasoner's Companion: A Mathematical Perspective*, Cambridge: Cambridge University Press.

Porter, J. (2009) *Lost in* translation: managing the 'risk' of flooding–the story of consistency and flood risk mapping. Unpublished conference paper. http://www.kcl.ac.uk/sspp/departments/geography/research/hrg/researchsymposium/1aporter.pdf (accessed 25 July 2012).

Rice, J. A. (1995) *Mathematical Statistics and Data Analysis*, 2nd edn, Belmont, CA: Duxbury Press.

Salt, J. D. (2008) The seven habits of highly defective simulation projects. *Journal of Simulation* **2**: 155–161.

Walley, P. (1991) *Statistical Reasoning with Imprecise Probabilities*, London: Chapman & Hall.

Wasserman, L. (2004) *All of Statistics: A Concise Course in Statistical Inference*, New York, NY: Springer.

Woo, G. (1999) *The Mathematics of Natural Catastrophes*, River Edge, NJ: Imperial College Press.

3

Model and data limitations: the sources and implications of epistemic uncertainty

J. C. ROUGIER AND K. J. BEVEN

3.1 Introduction

Chapter 2 focused entirely on aleatory uncertainty. This is the uncertainty that arises out of the randomness of the hazard itself, and also, possibly, out of the responses to the hazard outcome. That chapter chased this uncertainty through the footprint function and a loss operator to arrive at an exceedance probability (EP) curve. Such a structured approach (e.g. as opposed to a purely statistical approach) was motivated by the need to evaluate different interventions for choosing between different actions; and by the possibility of non-stationarity in the boundary conditions on policy-relevant timescales measured in decades.

Different risk managers will have different loss operators, and hence different EP curves. Likewise, the same risk manager will have different EP curves for different actions. A very simple summary statistic of an EP curve is the area underneath it, which corresponds to the expected loss ('expectation' taken in the mathematical sense), which is defined to be the *risk*. This presupposes a well-defined EP curve, i.e. a probability function F such that, if X is the loss, then

$$F_X(x) = \Pr(X \leq x),$$

adopting the standard convention that capitals represent uncertain quantities and lower case represent ordinates or specified values. Given F_X, the EP curve is defined as $EP(x) = 1 - F_X(x)$.

This chapter considers the conditions under which the EP curve is itself uncertain, and examines the consequences. This is the *epistemic uncertainty* that arises in both model representations and the data used to drive and evaluate models. The source of epistemic uncertainty is limitations in the analysis – it could, in principle, be reduced with additional time and resources. Thus epistemic uncertainty arises partly out of the need for the risk manager to take actions on the basis of what she currently knows, rather than being able to wait. For simplicity, the presentation in this chapter is in terms of computing an EP curve for a specific action; i.e. the analysis will need to be replicated across actions if the task is to choose between actions. Section 3.2 presents a general framework for thinking about epistemic uncertainty, and examines its main limitations. Sections 3.3, 3.4 and 3.5 consider particular sources of epistemic uncertainty, and how they might be incorporated into the EP

Risk and Uncertainty Assessment for Natural Hazards, ed. Jonathan Rougier, Steve Sparks and Lisa Hill. Published by Cambridge University Press. © Cambridge University Press 2013.

curve, and into the decision analysis. Section 3.6 concludes with a summary. The interested reader will also want to consult Spiegelhalter and Riesch (2011), a wide-ranging review of current practice in uncertainty assessment, with recommendations for good practice.

As in Chapter 2, the emphasis here is on the needs of the risk manager, whose role involves managing the decision process and recommending actions. She is answerable to several different stakeholders, each of whom is entitled (and advised) to hire an auditor. The risk manager must be able to demonstrate to the auditor(s) that her analysis is valid, and that it is suitable. Validity can be ensured by operating within a probabilistic framework, and this also promotes transparency, making it easier to demonstrate suitability. The same comments on 'Why probability (as opposed to some other formal method for uncertainty quantification)?' made in Chapter 2 apply here as well. Statistics is a framework for assessing the impact of judgements on choices. It provides a set of rules, demonstrably valid, by which the mechanics of turning judgements into choices is to a large extent automated. This allows us to focus on the judgements, and, with efficient mechanics, to subject them to stress testing in the form of sensitivity analysis, which will reveal the extent to which different defensible judgements would have yielded the same choice of action.

It is understood, however, that probability, although powerful, also makes very strong demands that are hard to meet in complex situations such as natural hazards, and which entail many simplifications. This is especially true of epistemic uncertainty. Therefore this chapter suggests a somewhat pragmatic approach. It is crucial that the various sources of epistemic uncertainty are clearly identified. But the degree to which each one can be fully incorporated into the EP curve will vary. Some are straightforward (uncertainty about the hazard process, for example), some moderately challenging (uncertainty about the footprint function and loss operators), and some extremely demanding (uncertainty about external factors, such as greenhouse gas emissions in the current century).

One useful way of picturing the treatment of epistemic uncertainty advocated here is in terms of 'internal' and 'external' uncertainty, as discussed by Goldstein (2011: § 3.1). He states

Internal uncertainties are those which arise directly from the problem description. Many analyses in practice are carried out purely on the basis of assessing all of the internal uncertainties, as these are an unavoidable component of any treatment of the problem. External uncertainties are all of the additional uncertainties which arise when we consider whether the treatment of the internal uncertainties indeed provides us with a satisfactory uncertainty description.... Most of the conceptual challenges ... arise in the appropriate treatment of the external uncertainties.

In the standard risk assessment for natural hazards, the EP curve captures aleatory uncertainty, and everything else is external. Of course, external uncertainty is not forgotten. Instead, it is accounted for using a lumped adjustment, such as a margin for error. In reinsurance, for example, it appears as a loading which increases the premium above that of the expected loss. In flood defence design it appears as an 'freeboard' allowance. In other areas of engineering it appears as an acceptable factor of safety. The message of this chapter is that it is feasible and beneficial to move some aspects of epistemic uncertainty into the EP curve, i.e. to transfer them from external to internal uncertainty. In this way, the uncertainty accounted for by the margin for error diminishes, and the difficulties associated with specifying its size become less important in the overall risk assessment, and in the choice between different actions.

3.2 Epistemic uncertainty in the EP curve

3.2.1 Recapitulation

We briefly recapitulate the main points from Chapter 2. Loss, denoted X, is expressed as a random quantity dependent on the hazard outcome, denoted ω. Probabilities are attached to each outcome, and this uncertainty 'cascades' down to the loss. This calculation requires the enumeration of all possible hazard outcomes, and their probabilities, and for this reason is often not very practical. Instead, a random simulation approach can be used to approximate the probability distribution F_X. In this chapter, random simulation is termed 'sampling' according to standard statistical usage, and one such random sample is termed a 'realisation'. This is because the word 'simulator' has a different technical meaning, as explained in Section 3.4.1 below.

In the random sampling approach, n realisations are independently sampled from the hazard process, and for each realisation a loss is sampled. The resulting set of sampled losses approximates the loss distribution, and summaries of this distribution such as the expectation (the *risk*) or the 99.5th percentile can be estimated, including confidence intervals to quantify variability. This was discussed in Section 2.5 of this volume. In this chapter we proceed as though n can be set large enough that variability can be ignored, purely to simplify the presentation.

3.2.2 Relaxing the suppositions

Now consider this calculation in more detail. While providing a general framework for estimating the EP curve, it involves a series of suppositions. The simplest way to conceptualise epistemic uncertainty is that it allows the risk manager to relax the stringent condition that the suppositions have to be true. As already discussed, she will relax this condition anyhow in her use of the EP curve for decision analysis. For example, by including a lumped margin for error which will be larger the more unlikely the suppositions underlying the EP curve are judged to be. But by thinking probabilistically the risk manager can relax some of the suppositions within a formal quantitative framework, which enables her to guarantee formal validity, and also enhances transparency.

In principle, epistemic uncertainty can be fully incorporated into the EP curve, as demonstrated by the following thought experiment. Note that this thought experiment is designed to point out the challenges for quantifying epistemic uncertainty: it is *not* a template for how to proceed. First, group all of the suppositions underlying a particular EP curve under the label 'θ'. In this chapter θ will be used wherever there is a source of epistemic uncertainty. Any particular distribution function for loss should be written not as $F_X(x)$ but as $F_X(x \mid \theta)$, that is, the distribution function for loss conditional on θ. We refer to $1 - F_X(x \mid \theta)$ as a *conditional exceedance probability* curve, or CEP.

Now for the thought experiment. Imagine that it is possible to enumerate all the different sets of suppositions under which a CEP curve might be constructed, denoted by the set Θ. Attach a probability to each θ in Θ, where this probability represents the risk manager's judgement that this θ is the appropriate θ ('right θ' is too strong here, even in a thought experiment), and where the probabilities sum to 1 over Θ. Then the probability calculus asserts that a collection of CEP curves should be combined into an EP curve by a vertically weighted average, where the weights are given by the probabilities over the θ values in Θ.

One difficulty with this thought experiment is that it is generally not possible to enumerate the set Θ. By the very nature of epistemic uncertainty, the enumeration of the possible θs might be incomplete (an example is the small number of assumed scenarios for future emissions that underlie the IPCC and UKCP09 projects of climate change in this century). We may appreciate quite clearly in assessing CEP curves that there are other possibilities (as in the case of emissions scenarios) but accept that, for practical limitations of computer resources or availability of measurement techniques, not all possible θs can be incorporated in the analysis.

A second difficulty is that, even were we able to enumerate a representative set Θ, it is extremely challenging to assign a probability to each member, such that the probabilities sum to 1. This is even more challenging if the set Θ is clearly incomplete, for then there will be some probability left over with nowhere to go.

But we reiterate the point made at the beginning of this chapter: the objective is to transfer some of the uncertainty in the risk from 'external' to 'internal'. Not being able to enumerate all possible members of Θ, or not being able to assign probabilities, should not prevent the risk manager from benefiting from those situations where she can identify some of the members of Θ, and/or provide approximate probabilities for some of the members. Therefore we describe the benefits that follow from being able to enumerate Θ, and from being able to sample from Θ. The risk manager must decide whether these benefits are worth the extra effort, in terms of providing a more clear justification for her choice of action. And then she must justify her decision to the auditor.

3.2.2.1 *Benefits of enumeration*

Suppose that it is possible to enumerate Θ, or, more likely, to identify a representative collection of values for θ which roughly span the possible outcomes for epistemic uncertainty. In this case quite a lot can be achieved without having to assign probabilities to the individual values, if it is possible to compute the CEP curve corresponding to $F_X(x \mid \theta)$ for each θ. In particular, it is possible to compute lower and upper envelopes for the EP curve, and thus lower and upper bounds on the risk. It is a basic property of probabilities that the unconditional probability of an event always lies within the range of the conditional probabilities. Expressed in terms of EP curves,

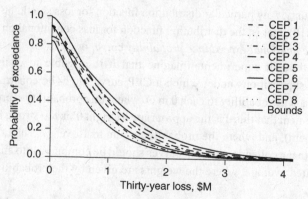

Figure 3.1 The lower and upper envelope of the CEP curves provides a lower and upper bound on the EP curve, irrespective of the probability attached to each CEP curve. These in turn provide lower and upper bounds on the risk.

The EP curve always lies within the lower and upper envelope created by the CEP curves.

This is illustrated in Figure 3.1.

This result allows the EP curve to be bounded without having to make any reference to the probabilities of each of the possible outcomes for epistemic uncertainty. Therefore it may be possible for the risk manager to choose between actions entirely in terms of the bounds. If it is clear that the lower bound of the EP curve for action A is above the upper bound of the EP curve for action B then the probabilities that are attached to the epistemic uncertainties are immaterial as far as the risk ordering is concerned: the risk of A will always be greater than that of B. Technically to be sure of this all possible values of θ must be enumerated, but a high level of confidence will follow from a reasonable attempt to span the range of possible values for θ.

It is also sometimes possible to rule out certain choices as being *inadmissible*. This will be discussed further below, in Section 3.5. But, briefly, if the CEP for action A is above that of the CEP curve for action B for each θ, then action A would never be preferred to B, and A can be removed from further consideration.

3.2.2.2 Benefits of sampling

Suppose now that it is possible to sample values of θ from the distribution $\Pr(\theta)$. In this case it is also possible to proceed by enumeration, as above. The sampled values are amalgamated into a representative set, and then a CEP curve is computed for each θ in the representative set. Thus lower and upper bounds on the EP curve are available. Alternatively, it is possible to combine the CEP curves into a single EP curve, by vertical averaging, using probabilities approximated by the relative frequencies in the sample.

A more standard and less laborious calculation would simply be to attach the sampling of θ to the front end of the sampling procedure for the loss X. First, sample θ, then sample the hazard outcome ω conditional on θ, then sample the loss X conditional on θ and ω.

Repeating this process many times generates a distribution for losses that incorporates epistemic uncertainty about θ, and aleatory uncertainty about ω. Hidden inside this calculation is a collection of CEP curves, but they are never explicitly constructed. Sometimes this is the natural way to proceed, if there is a ready mechanism for generating probabilistic samples for θ. In the next section, for example, sampling θ and then ω given θ is a very natural extension to just sampling ω given θ.

Sampling the epistemic uncertainty as well as the aleatory uncertainty is the incremental update to a procedure that was already sampling aleatory uncertainty. From this point of view, it is a cheap upgrade, and unlikely to induce complications such as coding errors. Uncovering the underlying CEP curves, though, provides additional information which can be used to assess robustness and sensitivity.

3.2.3 Epistemic uncertainty and risk

The discussion so far has been in terms of the relationship between the CEP curves and the EP curve. This has obvious implications for the relationship between the conditional risk and the (unconditional) risk. The risk of the EP curve could be lower or higher than the conditional risk of a specified CEP curve. So generalising a CEP curve to account for epistemic uncertainty does not necessarily increase the risk. The risk manager could discover, for example, that other possible choices for θ implied much lower losses.

As well as a risk, each EP or CEP curve also has a variance, which is a summary of the uncertainty of the loss. The variance of a specified CEP curve is a conditional variance. It is a standard probabilistic result that the overall variance is equal to the variance of the conditional risk, plus the mean of the conditional variance (see, e.g. Grimmett and Stirzaker, 2001: ch. 3, § 7). A single CEP curve only contributes to the second term, and therefore generalising a CEP curve to account for epistemic uncertainty will tend to increase uncertainty about the loss, exactly as might be expected. As has already been stressed, though, this is not 'new' uncertainty. This is uncertainty that has been 'internalised', i.e. moved from a margin for error into the EP curve for a more transparent analysis. Consequently the margin for error ought to be reduced, to compensate. To pose the very difficult question 'How much should the margin for error be reduced?' is simply to emphasise the importance of putting as much epistemic uncertainty as possible into the EP curve.

3.3 Uncertainty about the hazard process

The hazard process is characterised by a set of outcomes ω in Ω, and a probability for each outcome. Typically, these probabilities are derived from a statistical model fitted to a catalogue of previous events. Given the scarcity of large hazard events, and their lack of homogeneity, such statistical models will be somewhat crude (there may, for example, be several competing distributional forms), and the estimated parameters ill-constrained. The need to treat parametric uncertainty carefully, as possibly constituting a major source of epistemic uncertainty about the EP curve, suggests operating within a Bayesian statistical framework for estimating the parameters, and simulating the hazard process.

By way of illustration, consider earthquakes. First, suppose that it is possible to specify a stochastic process for earthquakes in a specified region. The nature of this process is not important; what is important is that the stochastic process is represented as a member of a well-specified ensemble of models indexed by a parameter θ. This θ is uncertain, but the catalogue of earthquakes in the region is informative about it. A common approach is to derive a point-estimate for θ by finding that value which best explains the observations; maximum likelihood methods for parameter estimation are discussed by Davison (2003: ch. 4, 7); stochastic processes, including for earthquakes, are also covered (Davison, 2003: ch. 6). Then it is common to use the point estimate for θ as though it were the true θ. In other words, to sample the hazard process using its estimated parameter values. This does not allow for the epistemic uncertainty associated with our inability to fully constrain θ using the catalogue. A two-pronged approach can be taken to rectify this. The first is to perform the sampling within a Bayesian statistical approach, explicitly allowing for uncertainty about θ, and the second is to extend the process of learning about θ to other regions, also following a Bayesian approach.

In the Bayesian approach θ is treated as epistemically uncertain, and given its own initial distribution, the 'prior' distribution, which incorporates general judgements about θ formed 'prior' to consulting the catalogue. Learning about θ takes the form of conditioning θ on the catalogue. Sampling a realisation for future earthquakes is a two-stage process:

(1) Sample one realisation of θ from the distribution of θ conditional on the catalogue;
(2) Sample one realisation of the earthquake process for this realisation of θ.

By allowing θ to remain uncertain, Bayesian simulation of the earthquake process incorporates both aleatory and epistemic uncertainty. If the EP curve was already being computed by sampling then this is a 'free upgrade'. Note that if the catalogue is very detailed and we can treat the underlying process as stationary, then there will be very little epistemic uncertainty about θ after conditioning, in which case the Bayesian approach reduces to the simple simulation method. Bayesian estimation and sampling is discussed by Davison (2003: ch. 11).

One limitation of this approach is that it only uses the catalogue from one site or region to learn about the θ for that region. It is tempting to augment the catalogue with events from other regions, because increasing the size of the catalogue is likely to reduce epistemic uncertainty about θ. But, at the same time, it does not seem defensible to insist that θ is the same across all regions, especially where the observational data might be of varying quality and subject to epistemic errors. Bayesian hierarchical modelling provides a statistical solution. Each region is allowed to have its own value of θ, say θ_r for region r. The θ_r are assumed to be similar, in the sense that they are treated as being drawn independently from the same underlying distribution. By allowing this underlying distribution itself to be uncertain, information from the different regions can be pooled. This allows us to learn about the θ_r of a region for which we have observations, which will be strongly affected by the catalogue from that region, but also affected by the catalogues from the other regions. It also allows us to learn about the θ_r of a region for which we do not have observations.

This same framework extends to considering the role of historical catalogues in describing the future hazard process when boundary conditions may have changed. More sophisticated approaches replace the simple exchangeability across regions described here with a more detailed treatment that can include regional covariates, and explicit representations of space and time. Again, as in the special case of just a single region, the output from this type of modelling is a simulation of the hazard process, which accounts for both aleatory and epistemic uncertainty. Therefore it can be plugged into the front of a calculation that estimates the EP curve by sampling.

Bayesian hierarchical models are discussed in many textbooks. See, e.g. Davison (2003: § 11.4) for an introduction, or Gelman *et al.* (2004) for a more detailed treatment, including computation and diagnostics. Banerjee *et al.* (2004) cover hierarchical modelling of spatial data. Sahu and Mardia (2005) review spatial-temporal modelling.

For simplicity, the above explanation was for an uncertain parameter within a specified statistical model. In a more general treatment, both the model and the parameter (which is then expressed conditionally on the model) are allowed to be uncertain. This is an important generalisation, given the typical absence of really large events in a typical catalogue, and the convexity of the loss operator (itself very uncertain for large hazard events). So, for example, switching from a Frechet to a Gumbell distribution for extreme values in flood assessment may well make a large difference to the risk, even though both curves fit equally well over the catalogue. This case could be handled by embedding both alternatives within the generalised extreme value (GEV) family of distributions, but only if parametric uncertainty is explicitly included.

Other situations are more complicated. Draper (1995) provides an example in which two different statistical models fitted on the same data (the Challenger O-ring data) give non-overlapping predictions. More recently, the ongoing debate about climate reconstruction for the last millennium has been largely about the effect of using different statistical models on the same database of historical temperature proxies; see, e.g. McShane and Wyner (2011), and the discussion that follows. Issues of model choice and model criticism are a core part of statistical theory and practice.

3.3.1 Uncertainty about the loss operator

Exactly the same comments also apply to the loss operator; if this is represented as a stochastic relationship between the hazard outcome and the loss that follows: (1) simulate within a Bayesian framework to allow for epistemic uncertainty in the relationship between hazard and loss; (2) use hierarchical models to extend the calibration inference to other datasets which, while not identical, are sufficiently similar to be informative; (3) allow for alternative statistical models.

3.4 Uncertainty about the footprint function

The footprint function maps a hazard outcome into a trajectory in space and time, from which loss can be computed. The footprint function will often be influenced by the risk manager's choice of action. For example, the decision to increase the height of seawalls will restrict the footprint of large storm surges.

3.4.1 Three sources of epistemic uncertainty

In the simplest case, which is also the most common, the footprint function is treated as a deterministic function of the hazard outcome, usually expressed in terms of a particular hazard event. Thus, the hazard event might be a storm, described in terms of a time-series of precipitation over a catchment, and the footprint function might depend on a hydrological model describing how that storm affects river-flow in the catchment and consequent inundation of the flood plain, predicted using a hydrodynamic model based on the shallow-water equations. Or the hazard event might be a volcanic eruption, described in terms of location, orientation and magnitude, and the footprint function might be a dynamical model describing the evolution of pyroclastic, lahar or lava flows, e.g. using multiphase versions of the shallow-water equations. We refer to such functions as (deterministic) *simulators*. This is to avoid another instantiation of the word 'model', which is already heavily overloaded.

Simulators are limited in their ability to represent the impact of the hazard. In general, the 'gap' between the simulator output and the system behaviour can be thought of as the accumulation of three sources of epistemic uncertainty. First, simulators usually require the input of initial and boundary conditions in addition to the description of the hazard event, and these inputs may be imperfectly known, giving rise to *input uncertainty*. For example: a compartmental hydrological simulator requires an initial specification of the saturation of each of the boxes; a fire simulator requires the dryness of the vegetation. In many cases, initial condition uncertainty can be somewhat finessed by running the simulator through a spin-up period prior to the hazard event, during which the effect of uncertainty about the condition at the start of the spin-up period is reduced through the assimilation of observational data. This is the basis of weather forecasting (see, e.g. Kalnay, 2003). Boundary condition uncertainties, such as the pattern of rainfall over a catchment in an extreme flood event, are more difficult to finesse and, for a given measurement system, might vary significantly in their characteristics from event to event. As in weather forecasting, data assimilation can sometimes help compensate for such epistemic uncertainties for real-time forecasting purposes (e.g. Romanowicz *et al.*, 2008), but it can often be difficult to construct realisations of epistemic boundary condition uncertainties in simulation.

The second source of epistemic uncertainty is *parametric uncertainty*. Within the simulator there are typically parameters that do not have well-defined analogues in the system, or whose analogues are not measurable. This is an inevitable feature of simulators, which, through their abstraction, introduce ambiguity into the relationship between the simulator state vector and the system. So, for example, in a compartmental hydrological simulator the rates connecting the different compartments will be uncertain, and usually catchment-dependent. Parameters are often tuned or calibrated in an exercise designed to reduce the misfit between simulator output and system behaviour, for events in which the inputs were approximately known (e.g. previous storms on the same catchment). Here we distinguish between 'tuning', which describes a more-or-less ad hoc procedure, and 'calibration', which is explicitly statistical and aims to produce some assessment of parametric uncertainty.

The third source of epistemic uncertainty is *structural uncertainty*. This accounts, crudely, for all of the limitations of the simulator that cannot be eliminated by calibrating the parameters of the simulator. Formally, and this is another unrealistic thought experiment, we suppose that the inputs are known exactly, and that the 'best' values of the parameters have been identified in some way. The structural uncertainty is the uncertainty about the system that remains after running the simulator with the correct inputs and best parameter values. In the simplest situation, structural uncertainty is quantified without reference to the precise value of the inputs and parameters, as a lumped value. Often it is given in very crude terms. An avalanche simulator, for example, might predict steady-state velocity as a function of slope and snow density; it might be considered to have a structural uncertainty of ± 1 m s, where this is stated without reference to any particular slope or density. Note that this kind of assessment can be improved with more resources, but initially a crude estimate will often suffice in terms of identifying where the major sources of uncertainty lie. Structural uncertainty is discussed further in Section 3.4.3.

3.4.1.2 The loss operator again

In situations where the loss operator is a deterministic function of the hazard outcome, then exactly the same concerns apply.

3.4.2 Parametric uncertainty

It is a cannon of uncertainty assessment that one must start, if possible, with quantities that are operationally defined. The 'best' value of the parameters of a simulator, however, are not operationally defined and, as a consequence, nor is the discrepancy between the system and the simulator output at its best parameterisation. Of course science is beset by such difficulties, and we do the best we can. In the case of uncertainty about the simulator, *it is surely far better to include a crude assessment of parametric and structural uncertainties than simply to ignore them*. To ignore them is effectively to set these uncertainties to zero, and relegate them to the external uncertainty and the margin for error. On this basis, we proceed to outline a simple assessment, advocating that such an assessment is much better than nothing.

The first step in addressing parametric uncertainty is to identify hard-to-quantify parameters, and to assess plausible ranges for each of them. Notionally, the parameter ranges represent the possible values that the 'best' parameterisation might take. These ranges could be assessed on the original scale, or on a log-scale for parameters that are strictly positive; in the latter case they might often be assessed proportionately, for example ±10% around the standard value. Experience in a wide range of environmental modelling applications suggests that the effective value of a parameter, in terms of generating acceptable simulator outputs, can be well away from the system equivalent. Sometimes this is for well-understood reasons, as with viscosity in ocean simulators, where it is common to distinguish molecular viscosity, which is a property of water, from eddy viscosity, which is determined partly by the solver resolution. Generally, however, a good strategy is to be prepared to revise the parameter ranges after a pilot study of simulator runs have been compared to system measurements. Proceeding sequentially is absolutely crucial, even though it might be cheaper and more convenient to submit the entire experiment as one batch.

Once ranges have been assigned (and perhaps revised), the shape of the parametric uncertainty distribution can be specified to construct simple probability representations. This is a more subtle issue than it might first appear. It is natural (although by no means necessary) to treat uncertainty about the parameters as probabilistically independent, in the absence of strong information to the contrary. However, the joint distribution of large numbers of probabilistically independent random variables has some counter-intuitive properties. It is also natural to use a peaked distribution rather than a uniform one to reflect the common judgement that the best value of each parameter is more likely to be found in the centre of its range than on the edge. However, if there are more than a handful of parameters, these two judgements made together will cause the joint probability over all the parameters to be highly concentrated at the centre of the parameter space.

For example, if two parameters are each given independent symmetric triangular distributions on the unit interval, the inscribed sphere (i.e. in two dimensions, the circle with radius 0.5) occupies 79% of the volume, but contains 97% of the probability. For five parameters, this becomes 16% of the volume and 68% of the probability; for ten parameters, 0.2% of the volume and 14% of the probability. On the other hand, if independent uniform distributions are used, then for ten parameters 0.2% of the volume will contain 0.2% of the probability; hence with ten parameters switching from uniform to triangular causes a substantial increase in the probability near the centre of the parameter space, and corresponding substantial decrease in the probability in the corners and along the edges. The moral of this story is to think carefully about using marginal distributions that have sharp peaks if there are lots of parameters. It might be better, in this situation, to focus initially on just a few of the key uncertain parameters and treat the others as fixed; or to use a much less-peaked marginal distribution.

Once a probability distribution for the parameters has been quantified, parametric uncertainty in the simulator can be incorporated into the EP curve by sampling: sample the hazard outcome ω, sample the simulator parameters θ, run the simulator with arguments ω and θ, compute the loss; and then repeat. If a statistical model of input uncertainty is postulated, then input uncertainty can also be introduced by sampling. This simple Monte Carlo approach can be substantially improved by statistical techniques for variance reduction, which will be particularly important when the simulator is slow to evaluate. For many natural hazards simulators, however, for which the evaluation time tends to be measured in seconds rather than hours (compare to weeks for a large climate simulator), the balance currently favours simple Monte Carlo simulation when taking into account the possibility of coding and computation errors. This will shift, though, as simulators become more complex (e.g. 3D multiphase flow simulations for volcanic pyroclastic flows, taking hours), which seems to be inevitable. Variance reduction methods to improve on simple Monte Carlo simulation are discussed by Davison (2003: ch. 3, § 3).

The choices for parametric uncertainty, and structural uncertainty, discussed next, can both be assessed and possibly revised when there are system measurements that correspond, either directly or indirectly, to simulator outputs. This is discussed in Section 3.4.4.

3.4.3 *Structural uncertainty*

It is usually not hard to construct a list of the main sources of structural uncertainty in a simulator; indeed, the chapters on individual hazards in this volume do exactly that. But it is very difficult to quantify these, either jointly or individually (as the chapters illustrate). Mostly, structural uncertainty is ignored at the sampling stage (to be relegated to the margin for error) or, in the case of simulator calibration, to be discussed in Section 3.4.4, treated as small relative to measurement error.

Where structural uncertainty is incorporated, the method that is favoured in statistics is to use a lumped additive 'discrepancy' (possibly additive on a log-scale for strictly positive values). An example for a simulator with spatial outputs would be

$$\text{system}(s) = \text{simulator}(s; \theta^*) + \text{discrepancy}(s),$$

where s is spatial location, and θ^* is the 'best' value of the simulator parameters, itself uncertain (a different illustration is given below). The discrepancy is treated as probabilistically independent of θ^*. This framework is discussed by Rougier (2007) in the context of climate modelling. It is undoubtedly restrictive, and one generalisation is discussed by Goldstein and Rougier (2009), designed to preserve the tractability of the additive discrepancy in the context of several different simulators of the same underlying system. Here we will just consider the simple case, as not ignoring the discrepancy would already be a large improvement for most environmental science simulations.

The key point to appreciate when thinking about the discrepancy is that when a simulator is wrong, it is typically systematically wrong. This is because 'wrongness' corresponds to missing or misrepresented processes, and the absence of these processes does not cause random errors that vary independently across the space and time dimensions of the simulator output, but instead causes errors that vary systematically. So if the simulator value is too low at location s then it is likely to be too low at s' for values of s' in the neighbourhood of s. This indicates that the discrepancy does not behave like a regression residual or a measurement error, in which values at s and s' are uncorrelated. Instead, the correlation between the discrepancy at s and at s' depends on the locations of these two points in space and/or time.

The separation between s and s' at which the discrepancy is effectively uncorrelated is termed the 'decorrelation distance'. In general the decorrelation distance will depend on s and on the direction away from s. But simplifying choices can be made. The simplest of all is that the discrepancy is stationary and isotropic, i.e. its properties can be described by just two quantities: a standard deviation and a correlation length. For natural hazards, stationarity and isotropy are likely to conflict with judgements about simulator limitations, and with known features of the spatial domain. That is not to say, however, that these simple statistical models are not useful. They may still be better than not accounting for the discrepancy at all.

Little can be said about how to specify the discrepancy in general: this depends far too much on the particular application. However, an illustration might be helpful. Suppose that the purpose of the simulator was to compute the steady-state velocity profile of an avalanche (Rougier and Kern, 2010), denoted $v(z)$ where $z \geq 0$ is the slope-normal height. We might judge that the dominant source of structural uncertainty was an additive term that shifted the whole profile relative to the simulator output, with a secondary source of uncertainty that tilted the profile relative to the simulator output. Hence the discrepancy might be represented as

$$\text{discrepancy}(z) = u_1 + u_2(z - z_m),$$

where both v_1 and v_2 are uncorrelated random quantities with zero means, and the sizes of the two standard deviations σ_1 and σ_2 control the degree of uncertainty concerning shifting and tilting; z_m is the central value in the range of z, around which the tilt pivots. So setting $\sigma_1 = 0$ insists that there is no shifting in the discrepancy, and setting $\sigma_2 = 0$ insists that there is no tilting. The value of σ_2 will have units of $1 / \text{sec}$; in fact, this represents uncertainty about what avalanche scientists refer to as the 'shearing rate'. The best way to choose σ_1 and σ_2 is to try some different values and sample v_1 and v_2 using, say, two normal distributions. This discrepancy is illustrated in Figure 3.2.

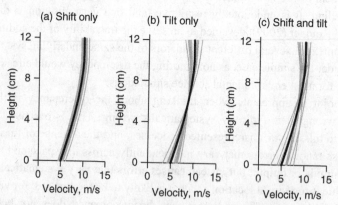

Figure 3.2 Simple representations of structural uncertainty for a simulator of avalanche steady-state velocity profile. The thick black line shows the simulator output, and the thinner lines show a sample of ten possible realisations of the actual system value, incorporating structural uncertainty based on $\sigma_1 = 1\,\text{m s}$ and $\sigma_2 = 20\,\text{m s}$ (see text of Section 3.4.3).

The point of this illustration is to show that specifying the discrepancy does not have to be any more complicated than judgements about simulator limitations allow. One could go further, for example by including higher-order terms in $(z - z_m)$, but only if judgements stretched that far. It is easy to become overwhelmed with the complexity of specifying a stochastic process for the discrepancy, and not see the wood for the trees. The discrepancy is there to account for those judgements about simulator limitations that are amenable to simple representations. To return to the theme of this chapter, the discrepancy offers an opportunity to internalise some aspects of epistemic uncertainty, by moving them out of the margin for error and into the EP curve. Just because an expert does not feel able to quantify all of his judgements about a simulator's structural uncertainty does not mean that those judgements that *can* be quantified need to be discarded. Any discrepancy representation that is more realistic than treating the simulator as though it were perfect is an improvement.

Once a representation for the discrepancy has been specified, then it can be incorporated into the EP curve by simulation, effectively by adding a random realisation of the discrepancy to the outcome of running the simulator. Although 'noising up' a deterministic simulator in this way might seem counter-intuitive, it must be remembered that the purpose of the discrepancy is to avoid treating the simulator as though it were perfect when it is not. Adding some noise to the simulator output is a simple way to achieve this.

3.4.4 Evaluating the simulator when measurements are available

In many cases there will be system measurements available, and these can be used to check the various judgements that have been made, and, if everything looks okay, to refine judgements about the simulator parameters and the structural uncertainty. However, it is very important in this process that systematic errors in the measurement process do not get fed into the simulator evaluation and calibration.

For example, Beven *et al.* (2011) have shown how basic rainfall-runoff observations might be inconsistent in certain events in the available records. It is quite possible, for example, that in a flood the volume of discharge estimated from observed water levels might be significantly greater than the observed volume of rainfall estimated from raingauge or radar-rainfall data. This can be because of errors in the rating curve used to convert water levels to discharges (particularly in extreme events that require extrapolation from measurements at lower flows) or the conversion of radar reflectivity to patterns of rainfall intensities or the interpolation from a small number of raingauges to the rainfall volume over a catchment. The nature of these errors might vary from event to event in systematic ways (see, e.g. Westerberg *et al.*, 2011). They might also have carry-over effects on subsequent events depending on the relaxation time of the catchment system. If a rainfall-runoff model is used as a simulator in estimating the footprint of such an event, it is clear that it will necessarily provide a biased estimate of the discharge observations for such events (at least if that model is consistent in maintaining a water balance). This is one of the reasons why there has been an intense debate about model calibration techniques in the hydrological modelling literature in recent years; between those approaches, treating the residuals as if they had no systematic component, and those incorporating non-stationarity effects that can arise from a common source of uncertainty such as the rating curve (e.g. Mantovan and Todini, 2006; Beven *et al.*, 2008). Similar issues will arise in many areas of natural hazard simulation.

3.4.4.1 Model criticism

If the structural uncertainty is represented as an additive discrepancy, as outlined above, then the focus is on the residuals at different values for the parameters

$$\text{residual}(s_i; \theta) = \text{measurement}(s_i) - \text{simulator}(s_i; \theta),$$

where s_i ranges over the locations where there are measurements, and we are ignoring the time-dimension, for simplicity (although in fact time often presents more challenges than space). More complicated situations are also possible, in which there are multiple experiments with different inputs, or different control variables in the simulator. In the snow rheology analysis of Rougier and Kern (2010), for example, there are ten experiments, with varying environmental conditions summarised primarily by snow density.

Suppose that each measurement has an additive error e_i, of the form

$$\text{measurement}(s_i) = \text{system}(s_i) + e_i,$$

where it is usual to treat the measurement errors as uncorrelated in space. The residual at the best parameter θ^* comprises

$$\text{residuals}(s_i; \theta^*) = \text{discrepancy}(s_i) + e_i.$$

This relation can be used to examine the various modelling choices that have been made, about the simulator, the simulator parameters and the discrepancy. This is *model criticism*, where it is important to appreciate that 'model' here represents all judgements, not just the physical ones that go into the simulator, but also the statistical ones that link the simulator and the system.

The main thing that complicates model criticism is that the discrepancy is unlikely to be uncorrelated across the measurements, or, to put it another way, the decorrelation length of the discrepancy is likely to be longer than the spatial separation between measurements.

There are various solutions to this problem, but the easiest one is to thin or clump the measurements. Thinning means deleting measurements, and ideally they would be deleted selectively, starting with the least reliable ones, to the point where the spacing between measurements was at least as wide as the decorrelation length of the discrepancy. *Clumping* means amalgamating individual measurements into a smaller number of pseudo-measurements, and ideally this would be done over regions that are at least as wide as the decorrelation length of the discrepancy; the simulator outputs will need to be amalgamated to match. There is no requirement that the clumping regions be the same size: in time they might reflect the dominant relaxation time of the system (which might include multiple events); in spatial domains they might also reflect discontinuities due to features such as a change in surficial or bedrock geology.

Both thinning and clumping discard information, but sometimes one does not need much information to identify bad modelling choices. The advantage of clumping is that it does not rely too strongly on distributional choices, which would be necessary for a more sophisticated analysis. This is because the arithmetic mean of the measurement errors within a clump will tend to a normal distribution (see, for example, the version of the central limit theorem given by Grimmett and Stirzaker (2001: 194). This supports the use of summed squared residuals as a scoring rule for choosing good values for θ^*, providing that the measurement errors can be treated as probabilistically independent.

For the rest of this section we suppose that the measurement errors can be treated as probabilistically independent (our caution at the start of this subsection notwithstanding) and that the measurements have been clumped to the point where the clumped discrepancies can be treated as probabilistically independent. The variance of the clumped residual for each region is the sum of the variance of the clumped discrepancy and the variance of the mean of the measurement errors. We denote this residual variance as σ_i^2 for pseudo-measurement i, and the standard deviation as σ_i. This is a value that we must specify from our judgements about the discrepancy and our knowledge of the measurement. In a nutshell, it quantifies how much deviation we expect to see between the simulator output at its best parameterisation, and the values of the measurements we have made (but see also the approach based on limits of acceptability in Section 3.4.5). Clearly, this is something we ought to have a judgement about if our intention is to use the simulator to make statements about the system. What we have done here is try to find an accessible way to represent and quantify these judgements.

Now consider running the simulator over a range of possible values for θ, termed an *ensemble* of simulator runs. Each member of this ensemble can be scored in terms of the sum of its squared scaled residuals, where the scaled residual at location s_i equals residual $(s_i; \theta) / \sigma_i$. The larger the sum of the squared scaled residuals, the worse the fit. One member of the ensemble, say θ^+, will have the least-bad fit. Now if θ^+ was actually θ^*, we would expect most of the scaled residuals for θ^+ to lie between ± 3 according to the 'three sigma' rule (see Pukelsheim, 1994). But of course θ^+ will not be θ^*. But the simulator output at θ^+ will be close to that at θ^* if (1) the ensemble is quite large, (2) the number of parameters is quite small and (3) the simulator output is quite flat around θ^*. In this case, the scaled residuals for θ^+ should lie between ± 3, or a little bit wider. Much larger residuals are diagnostic evidence of suspect modelling choices, such as ignoring the possibility of systematic errors in the measurements. Likewise, if all the residuals are very small, then this can be interpreted as possibly an over-large discrepancy variance.

The key feature that makes this diagnostic process work for a large variety of different systems, different natural hazards in our case, is the clumping. To explore this a bit further, one interpretation of clumping is that it is focusing on those aspects of the simulator that are judged reliable. Simulators based on solving differential equations are often not very reliable at spatial or temporal resolutions that are nearly as high as the solver resolution, but become more reliable at lower resolutions. So the expert may have good reasons for thinking that the simulator output for location s is a bad representation of the system at location s, but may still think that the simulator output averaged over a region centred on s is a good representation of the system averaged over the same region. There is no need to restrict this process to averages. The expert can decide which aspects of the simulator output are reliable, and focus on those. These may be averages, or they may be other features: the timing of a particular event, such as the arrival of a tsunami wave, or the scale of an event, such as peak ground acceleration in an earthquake. There is no obligation to treat all of the simulator outputs as equally reliable, and this can be incorporated into the diagnostic process. It should also be reflected in the way that the simulator is used to predict system behaviour, if possible; i.e. to drive the loss operator from the reliable aspects of the simulator output.

3.4.4.2 Calibration

Let us assume that the model criticism has been satisfactory. Calibration is using the (pseudo-) measurements to learn about θ^* and possibly about the structure of the discrepancy as well. Again, it is not necessary to restrict attention to pseudo-measurements for which the residuals can be treated as uncorrelated, but it is simple and less dependent on parametric choices.

We distinguished in Section 3.4.1 between tuning and calibration. Tuning is aimed at identifying a good choice for θ^* in an *ad hoc* way, prior to plugging in this choice without any further consideration of parametric uncertainty. Really, though, the notion of plugging in values for the parameters is untenable when modelling complex systems, such as those found in natural hazards, as has been stressed in hydrology (see, e.g. Beven, 2006, 2009; Beven *et al.*, 2011; Beven and Westerberg, 2011). With complex systems, model limitations (and, very often, input data limitations) severely compromise our ability to learn about θ^*. Therefore assessing (i.e. not ignoring) parametric uncertainty is crucial, and this favours a statistical approach in which judgements are made explicit, and standard and well-understood methods are applied.

Calibration is an example of an *inverse problem*, for which there are massive literatures on both non-statistical and statistical approaches (see, e.g. Tarantola, 2005). The simplest statistical approach is just to rule out bad choices of θ in the ensemble of runs according to the residuals, and then to treat the choices that remain as equally good candidates for θ^*. This ruling out can be according to the residuals, with different types of rules being used in different situations. Where there appear to be lots of good candidates then ruling out all choices for which one or more of the scaled residuals exceeds, say, 3.5 in absolute size might be a reasonable approach. In other cases, a more cautious approach would require more large residuals before a candidate was ruled out. There are no hard-and-fast rules when epistemic uncertainty plays a large part in the assessment: the distinction between ruled-out and not-ruled-out has to smell right according to the judgements of the expert. Looking at scaled residuals is a good starting point, but ought to be backed-up with a qualitative assessment, such as 'parameters in this region get the spatial gradient wrong'. This combination of quantitative and qualitative assessment should be expected because we do not expect to be able to quantify our epistemic uncertainty with any degree of precision.

Craig *et al.* (1997) developed the idea of 'history matching' by ruling out bad choices for θ^*, and considered the different ways of scoring the vector of residuals. Vernon *et al.* (2010) provide a very detailed case study in which bad choices for θ^* in a large simulator are eliminated through several phases of an experiment, where each phase 'zooms in' to perform additional simulator evaluations in the not-ruled-out region of the parameter space. Both of these papers are adapted to large applications, using an *emulator* in place of the simulator, to account for the very long run-times of their simulators (in Vernon *et al.* (2010), the simulator encompasses the entire universe). Where the run-time of the simulator is measured in seconds then an emulator is not required.

This section has described offline calibration, but there are advantages to combining calibration and prediction, in which calibration measurements from previous events are used directly in the prediction of future events, which allows us to take account of systematic effects in the discrepancy that may span both the calibration and the prediction. The key references here are Kennedy and O'Hagan (2001) for the fully Bayesian approach, and Craig *et al.* (2001) for the Bayes linear approach, which is more suitable for large simulators. Goldstein and Rougier (2006) extend the Bayes linear approach to explicit calibration. Rougier (2007) outlines the general fully probabilistic framework for calibrated prediction in the context of climate.

3.4.5 Less probabilistic methods

The diagnostic and calibration approaches outlined above have supposed that (1) structural uncertainty and measurement error can be treated additively, and (2) clumping can be used to reduce the effect of correlations within structural uncertainty. The focus on additive residuals after clumping is a natural consequence of these two suppositions. But in some applications the epistemic uncertainties associated with inputs, simulator limitations and system measurements can be impossible to separate out, and can result in residuals that are effectively impossible to decorrelate (Beven, 2002, 2005, 2006; Beven *et al.*, 2010). For example, latent (i.e. untreated) error in the simulator inputs creates a very complicated pattern in the discrepancies, in which the decorrelation length will be related to the dynamics of the system, and where the decorrelation length may be longer than the event duration. This makes it very hard to define formal statistical models that link the simulator output, the system behaviour and the measurements, and has led to alternative approaches to uncertainty estimation.

Beven and co-workers, for example, advocate a less statistical approach, informed by the particular challenges of hydrological modelling (see, e.g. Beven, 2006, 2009: ch. 2, 4). Hydrological simulators for flood risk assessment are forced by noisy and imperfectly observed inputs, namely the precipitation falling on the catchment. The epistemic uncertainty in such inputs reflects the differences between the actual precipitation and that observed by raingauges and radar-rainfall, which will vary from event to event in complex ways. Similarly, discharge measurements used to calibrate simulators can be subject to epistemic uncertainties as rating curves vary over time (e.g. Westerberg *et al.*, 2011), or need to be extrapolated to flood peaks involving overbank flows. Furthermore, both measured and predicted discharges can change rapidly, e.g. through several phases of a storm, which means that 'vertical' residuals (observations less simulator output at each time-point) can become very large simply through small errors of phase in the simulator.

As an alternative, the generalised likelihood uncertainty estimation (GLUE) approach (Beven and Binley, 1992) replaces the standard misfit penalty of the sum of squared scaled residuals with a more heuristic penalty that can be carefully tuned to those aspects of the simulator output that are thought to be reliable indicators of system behaviour. This does not rule out the use of penalties that are derived from formal statistical models; but it does not oblige the expert to start with such a statistical model. In recent applications, parameter evaluation has been based on specifying prior limits of acceptability ranges for simulator outputs and, within these ranges, defining weighting schemes across the runs in the ensemble that will tend to favour good fits over bad ones (e.g. Blazkova and Beven, 2009; Liu *et al.*, 2009).

Although the GLUE approach has been criticised within hydrology for being insufficiently statistical, in fact it preceded by over a decade a recent development in statistics along exactly these lines, generally termed 'approximate Bayesian computation', or ABC (see, e.g. Beaumont *et al.*, 2002; Toni *et al.*, 2009). Like the GLUE approach, ABC uses summary statistics to create ad hoc scoring rules that function as likelihood scores. However, it should be noted that it is an active area of research in statistics to understand how the replacement of formal with informal likelihoods affects the inference, notably asymptotic properties such as the consistency of the estimator of the best values of the simulator parameters. Similar approximate methods are used in the generalised method of moments approach, also termed the 'indirect method' (see, e.g. Jiang and Turnbull, 2004).

3.5 Uncertainty about external features

'External features' encompass all those things that do not occur inside the calculation of the EP curve, but which nonetheless affect the environment within which the EP curve is computed. These tend to be large and unwieldy. For example: responsiveness of the affected population to an early warning system; effectiveness of encouraged or forcible relocation; changes in population demographics; rate of uptake of new building regulations; future greenhouse gas emissions; changes in government policy. These are sometimes referred to as 'Knightian uncertainties', following Knight's (1921) distinction between what he termed 'risk', which could be quantified (effectively as chance), and 'uncertainty', which could not. It is surely a common experience for all of us that some uncertainties leave us totally nonplussed.

The characteristic of an uncertain external feature is that we can enumerate its main possible outcomes, and possibly compute a CEP curve under each possible outcome, but we cannot attach a probability to each outcome. Greenhouse gas emissions scenarios for the twenty-first century provide a clear example. Climate simulators are run under different emissions scenarios, and these can be used to generate realisations of weather, for example as done in the UK Climate Impacts Programme[1] (UKCIP; see also Chapter 6). These realisations of weather (the hazard process) can be used to drive a hydrological model of a catchment (the footprint function), which in turn can be used to evaluate losses, and to compute a CEP curve. So the emission scenario lives at the very start of the process, and different emissions scenarios give rise to different CEP curves. But, intriguingly, the attachment of probabilities to the emissions scenarios was explicitly discouraged (see Schneider, 2002 for a discussion).

Section 3.2 demonstrated that progress can be made if external features can be enumerated; in particular, EP curves and risks can be bounded. Continuing in this vein, decision support can often continue, despite not being able to specify a probability for each outcome. Each possible action generates a set of CEP curves, one for each possible combination of external factors. So each pair of (action, external feature) has its own risk, measured as the area under its CEP curve. These risks can be collected together into a matrix where the rows represent actions and the columns represent external features. This is the classic tableau of decision analysis (see, e.g. Smith, 2010: § 1.2).

First, it is possible that some actions can immediately be ruled out as being *inadmissible*. An inadmissible action is one in which the risk of some other action is no higher for every possible combination of external features, and lower in at least one combination.

Second, optimal decisions often have the property of being relatively robust to changes in probabilities (Smith, 2010: § 1.2), and so a sensitivity analysis over a range of different probability specifications for the external features may indicate an outcome that is consistently selected over a sufficiently wide range of different probability specifications.

Finally, there are non-probabilistic methods for selecting an optimal action. One such method is *minimax loss*. Minimax loss suggests acting on the assumption that nature is out to get you. The minimax action is the action which minimises the maximum risk for a given action. In other words, each action is scored by its largest risk across the external features, and the action with the smallest score is chosen. Another non-probabilistic method is *minimax regret* (see, e.g. Halpern, 2003: § 5.4.2). The point about these types of rules, however, is that they generate actions that can be hard to justify in the presence of even small amounts of probabilistic information. As Halpern (2003: 167–168) notes, 'The disadvantage of [minimax loss] is that it prefers an act that is guaranteed to produce a mediocre outcome to one that is virtually certain to produce an excellent outcome but has a very small chance of producing a bad outcome'.

Overall, when accounting for external features in choosing between actions, it might be better to assign probabilities, even very roughly, and then to examine the sensitivity of the

[1] http://www.ukcip.org.uk/

chosen action to perturbations in the probabilities. The decision to adopt a non-probabilistic principle such as minimax loss is itself a highly subjective and contentious one, and such an approach is less adapted to performing a sensitivity analysis. A more formal alternative to a sensitivity analysis with probabilities is to use a less prescriptive uncertainty calculus. An example might be the Dempster–Shafer approach, in which belief functions are used to provide a more general uncertainty assessment, which preserves notions of ignorance that are hard to incorporate into a fully probabilistic assessment. Note, however, that the properties of such approaches are less transparent than those of probability, and any type of calculation can rapidly become extremely technical. It is emphatically *not* the case that a more general uncertainty calculus leads to a more understandable or easier calculation. Readers should consult Halpern (2003: ch. 2, 3) for compelling evidence of this.

3.6 Summary

Epistemic uncertainty is the uncertainty that follows from constraints on our understanding, our information or our resources (e.g. computing, or time for reflection). It is contrasted with aleatory uncertainty, which is the inherent uncertainty of the hazard itself. This chapter has characterised epistemic uncertainty in terms of a multiplicity of EP curves, where each EP curve can be thought of as conditioned on the value of one or more uncertain quantities, and is denoted here as a CEP curve ('C' for 'conditional'). While inspecting such a collection of curves is extremely revealing, it is necessary for communication and for decision-making to reduce them, where possible, to a single summary EP curve.

Three main sources of epistemic uncertainty have been examined, and it is efficient and sometimes necessary to treat them in three different ways.

First, on the understanding that the EP curve will be computed by sampling the hazard outcome from the hazard process, epistemic uncertainty about the hazard process itself can be handled by performing that sampling within a Bayesian framework, which allows the parameters of the hazard process to be uncertain. These parameters can be conditioned on a catalogue of historical and laboratory measurements. The range of suitable measurements can be extended in some situations by using a Bayesian hierarchical model in which different classes of measurement (e.g. measurements from different locations) are treated as exchangeable (see Section 3.3). This extension will tend to reduce uncertainty about the hazard process. The Bayesian approach can also be extended to incorporate alternative hazard processes, through 'model averaging', or by enumeration and bounding. The same comments apply to stochastic loss operators.

Second, epistemic uncertainty in the footprint function (referred to above as the 'simulator') can be removed by additional sampling, over different realisations of the simulator parameters, over different realisations of the simulator's structural uncertainty, and also input uncertainty if this is thought to make a large contribution. Structural uncertainty will tend to be systematic rather than uncorrelated, which involves the specification of decorrelation lengths (Section 3.4.3). Where there are system measurements available, these can be used to check the model specification ('model' here incorporating both the simulator and the statistical model), and to learn about the parameters. A simple and largely non-parametric approach is to 'clump' the measurements together within regions (typically spatial or temporal) about as wide as the decorrelation length of the structural uncertainty. This justifies the use of the simulator residuals as a diagnostic of fit and as a penalty function for ruling out poor choices of the simulator parameters. Other approaches, such as GLUE, use a more heuristic penalty to provide a more flexible description of those aspects of the footprint function that are considered reliable indicators of system behaviour. The same comments apply to deterministic loss operators.

Third, epistemic uncertainty also applies to external features of the problem: these tend to be enumerable but unwieldy, so that attaching probabilities is too subjective to be generally defensible. In this case the set of EP curves, one for each possible combination of external features, cannot be collapsed into one summary EP curve.

Clearly, these three sources of epistemic uncertainty can be combined, and actions can also be introduced. The result will be one EP curve for each (action, external feature) pair, with the first and second sources of epistemic uncertainty having been averaged out. In the presence of external features, it may still be possible to select actions with small risks using methods from statistical decision theory.

Finally, we stress that accounting for epistemic uncertainty is challenging, particularly in defining appropriate scores for model criticism and calibration. But the risk manager must act, taking the best decision with her available resources, and justifying that decision to the auditor. We reiterate the point made at the start of this chapter. Failure to incorporate epistemic uncertainty explicitly into decision-support tools such as the EP curve leads to it being represented implicitly, as a lumped margin for error, or, especially in policy, as a cautionary attitude which tends to favour inertia.

We hope we have shown that some aspects of epistemic uncertainty can be represented explicitly, and that this confers substantial advantages to the risk manager. Within the framework that we have outlined, it is possible to make reasonable judgements about many of the quantities required, such as ranges for parameters or the accuracy of the simulator. These judgements can be further tuned through sensitivity analysis and often, if there are calibration data available, the process can be put onto a more formal footing. This outline will seem somewhat subjective, involving judgements that can be hard to trace back to anything more than accumulated experience. But it should be recognised that, for example, to ignore a simulator's limitations and treat it as perfect is also a judgement: the value of zero is just as subjective as any other value, but much harder to defend. After all, zero is definitely wrong, whereas, say, ± 1 m s is at least worth discussing.

Acknowledgement

We would like to thank Thea Hincks for her detailed and perceptive comments on an earlier draft of this chapter.

References

Banerjee, S., Carlin, B. P. and Gelfand, A. E. (2004) *Hierarchical Modeling and Analysis for Spatial Data*, Boca Raton, FL: Chapman & Hall/CRC.

Beaumont, M. A., Zhang, W. and Balding, D. J. (2002) Approximate Bayesian Computation in population genetics. *Genetics* **162**: 2025–2035.

Beven, K. J. (2002) Towards a coherent philosophy for environmental modelling. *Proceedings of the Royal Society London A* **458**: 2465–2484.

Beven, K. J. (2005) On the concept of model structural error. *Water Science and Technology* **52** (6): 165–175.

Beven, K. J. (2006) A manifesto for the equifinality thesis. *Hydrological Processes* **16**: 189–206.

Beven, K. J. (2009) *Environmental Modelling: An Uncertain Future?*, London: Routledge.

Beven, K. J. and Binley, A. M. (1992) The future of distributed models: model calibration and uncertainty prediction. *Hydrological Processes* **6**: 279–298.

Beven, K. J. and Westerberg, I. (2011) On red herrings and real herrings: disinformation and information in hydrological inference. *Hydrological Processes* **25**: 1676–1680.

Beven, K. J., Smith, P. J. and Freer, J. (2008) So just why would a modeller choose to be incoherent?. *Journal of Hydrology* **354**: 15–32.

Beven, K. J., Leedal, D. T. and Alcock, R. (2010) Uncertainty and good practice in hydrological prediction. *Vatten* **66**: 159–163.

Beven, K. J., Smith, P. J. and Wood, A. (2011) On the colour and spin of epistemic error (and what we might do about it). *Hydrology and Earth System Sciences* **15**: 3123–3133.

Blazkova, S. and Beven, K. J. (2009) A limits of acceptability approach to model evaluation and uncertainty estimation in flood frequency estimation by continuous simulation: Skalka catchment, Czech Republic. *Water Resources Research* **45**: W00B16.

Craig, P. S., Goldstein, M., Seheult, A. H., *et al.* (1997) Pressure matching for hydrocarbon reservoirs: a case study in the use of Bayes linear strategies for large computer experiments. In *Case Studies in Bayesian Statistics* vol. 3. New York, NY: Springer-Verlag, pp. 36–93.

Craig, P. S., Goldstein, M., Rougier, J. C., *et al.* (2001) Bayesian forecasting for Complex systems using computer simulators. *Journal of the American Statistical Association* **96**: 717–729.

Davison, A. C. (2003) *Statistical Models*, Cambridge: Cambridge University Press.

Draper, D. (1995) Assessment and propagation of model uncertainty. *Journal of the Royal Statistical Society, Series B* **57**: 45–97.

Gelman, A., Carlin, J. B., Stern, H. S., *et al.* (2004) *Bayesian Data Analysis*, 2nd edn, Boca Raton, FL: Chapman & Hall/CRC.

Goldstein, M. (2011) External Bayesian analysis for computer simulators. In *Bayesian Statistics 9*, ed. J. M. Bernardo, M. J. Bayarri, J. O. Berger, A. P. Dawid, D. Heckerman, A. F. M. Smith and M. West, Oxford University Press.

Goldstein, M. and Rougier, J. C. (2006) Bayes linear calibrated prediction for complex systems. *Journal of the American Statistical Association* **101**: 1132–1143.

Goldstein, M. and Rougier, J. C. (2009) Reified Bayesian modelling and inference for physical systems. *Journal of Statistical Planning and Inference* **139**: 1221–1239.

Grimmett, G. R. and Stirzaker, D. R. (2001) *Probability and Random Processes*, 3rd edn, Oxford: Oxford University Press.

Halpern, J. Y. (2003) *Reasoning about Uncertainty*, Cambridge, MA: MIT Press.

Jiang, W. and Turnbull, B. (2004) The indirect method: inference based on intermediate statistics – a synthesis and examples. *Statistical Science* 19: 239–263.

Kalnay, E. (2003) *Atmospheric Modeling, Data Assimilation and Predictability*, Cambridge: Cambridge University Press.

Kennedy, M. C. and O'Hagan, A. (2001) Bayesian calibration of computer models. *Journal of the Royal Statistical Society, Series B* 63: 425–464.

Knight, F. H. (1921) *Risk, Uncertainty, and Profit*, Boston, MA: Hart, Schaffner & Marx; Houghton Mifflin Company.

Liu, Y., Freer, J. E., Beven, K. J., *et al.* (2009) Towards a limits of acceptability approach to the calibration of hydrological models: extending observation error. *Journal of Hydrology* 367: 93–103.

Mantovan, P. and Todini, E. (2006) Hydrological forecasting uncertainty assessment: incoherence of the GLUE methodology. *Journal of Hydrology* 330: 368–381.

McShane, B. B. and Wyner, A. J. (2011) A statistical analysis of multiple temperature proxies: are reconstructions of surface temperatures over the last 1000 years reliable?. *Annals of Applied Statistics*, forthcoming.

Pukelsheim, F. (1994) The three sigma rule. *The American Statistician* 48: 88–91.

Romanowicz, R. J., Young, P. C., Beven, K. J., *et al.* (2008) A data based mechanistic approach to nonlinear flood routing and adaptive flood level forecasting. *Advances in Water Resources* 31: 1048–1056.

Rougier, J. C. (2007) Probabilistic inference for future climate using an ensemble of climate model evaluations. *Climatic Change* 81: 247–264.

Rougier, J. C. and Kern, M. (2010) Predicting snow velocity in large chute flows under different environmental conditions. *Applied Statistics* 59 (5): 737–760.

Sahu, S. and Mardia, K. V. (2005) Recent trends in modeling spatio-temporal data. In *Proceedings of the Special Meeting on Statistics and Environment*, Società Italiana di Statistica, Università Di Messina, 21–23 September, pp. 69–83. http://www.south-ampton.ac.uk/~sks/revsis.pdf

Schneider, S. H. (2002) Can we estimate the likelihood of climatic changes at 2100?, *Climatic Change* 52: 441–451.

Smith, J. Q. (2010) *Bayesian Decision Analysis; Principles and Practice*, Cambridge: Cambridge University Press.

Spiegelhalter, D. J. and Riesch, H. (2011) Don't know, can't know: embracing deeper uncertainties when analysing risks. *Philosophical Transactions of the Royal Society, Series A* 369: 1–21.

Tarantola, A. (2005) *Inverse Problem Theory and Methods for Model Parameter Estimation*, Philadelphia, PA: SIAM.

Toni, T., Welch, D., Strelkowa, N., *et al.* (2009) Approximate Bayesian computation scheme for parameter inference and model selection in dynamical systems. *Journal of the Royal Society Interface* 6: 187–202.

Vernon, I., Goldstein, M. and Bower, R. G. (2010) Galaxy formation: a Bayesian uncertainty analysis. *Bayesian Analysis* 5 (4): 619–670.

Westerberg, I., Guerrero, J.-L., Seibert, J., *et al.* (2011) Stage-discharge uncertainty derived with a non-stationary rating curve in the Choluteca River, Honduras. *Hydological Processes* 25: 603–613.

4

Quantifying scientific uncertainty from expert judgement elicitation

W. P. ASPINALL AND R. M. COOKE

Now the geologists Thompson, Johnson, Jones and Ferguson state that
our own layer has been ten thousand years forming. The geologists
Herkimer, Hildebrand, Boggs and Walker all claim that our layer has
been four hundred thousand years forming. Other geologists, just as
reliable, maintain that our layer has been from one to two million years
forming. Thus we have a concise and satisfactory idea of how long our
layer has been growing and accumulating.

Mark Twain, A Brace of Brief Lectures on Science, *1871*

4.1 Introduction

Most scientists would like to see scientific advice used more in government decision-making
and in all areas of public policy where science is salient, and many would welcome the
opportunity to sit on expert review panels or scientific advisory committees. When it comes
to taking such decisions in many areas of hazard and risk assessment, the traditional
committee approach still holds sway. Often in a committee setting, however, the role of
scientific uncertainty is not an item on the agenda, and seldom a prominent component of the
discussion. But misunderstanding its importance or misstating its extent will contribute to
poor decisions.

The slow, deliberative committee process, seeking a wide range of opinions with majority
voting on outcomes, offers some parallels with the scientific process itself, but only in as
much as a show of hands can equate to strength of argument. But as a means of gathering
expert opinion it is inadequate under many conditions, such as an urgent civil emergency
arising from an incipient natural disaster such as a hurricane or volcanic eruption – situations
demanding prompt scientific advice. Usually, such advice would be the prerogative of a
chief scientist, with all the contingent stresses and personalisation issues involved, including
the pressure to be extremely cautious. When people's lives are at risk, the responsibilities are
massive.

But, as with climate change modelling or forecasting severe weather events, in such cases
there is always inherent randomness in the complex dynamic processes involved – usually

Risk and Uncertainty Assessment for Natural Hazards, ed. Jonathan Rougier, Steve Sparks and Lisa Hill. Published by
Cambridge University Press. © Cambridge University Press 2013.

denoted as 'aleatory uncertainty' (see Chapter 2). Even the most sophisticated scientific models do not capture fully the range and extents of these stochastic variations – only Nature herself can do that! Understanding the potential scale and timing of natural hazards for the purposes of taking actions to protect the public inevitably requires simplifications and assumptions which, in turn, necessitate expert judgements, some of them intellectually brave.

How can this be done in a balanced, neutral but, in some sense, rationally optimised way? A formal procedure for eliciting and pooling judgements from a group of experts can help to quantify scientific uncertainty for high-stakes decision-making. There are several methods for doing this, each with their own flaws and biases (Kahneman *et al.*, 1982; Tversky and Kahneman, 2005; Kynn, 2008; Kahneman, 2011). In the 1960s, for example, methods such as Delphi surveying aimed to extract a consensus from a group of experts for informing decisions.

Contributing to the difficulties in obtaining expert views in the presence of scientific uncertainty is the vexed issue of the distinction – if such exists – between 'objective' frequency- or statistics-based probability and 'subjective' degree-of-belief probability (see, e.g. Vick, 2002) – each having its own philosophical underpinnings, protocols and research literature. For most extreme, and difficult, hazard assessment problems, the former is rarely tenable – the data simply do not exist. This being so, recourse to belief-based expert judgements is the only viable option, and the means of eliciting such views becomes the real challenge.

In such circumstances, it is the decision-makers who are responsible – and answerable – for taking decisions under uncertainty but, not infrequently, they try to offload the onus onto scientists. As a failure of leadership, this sometimes is done covertly. Former EPA administrator Christine Todd Whitman didn't see it that way and roundly admitted:[1] 'A big part of my frustrations was that scientists would give me a range. And I would ask, "please just tell me at which point you are safe, and we can do that".'

However, it is neither rational nor appropriate to expect total consensus among experts when they are asked to make judgements on ill-constrained complex problems. In fact, the scientific method is the only legitimate way to seek to get scientists to agree. If the scientific method is unable to resolve an active issue, then the scientists *should* disagree, and any committee or other 'consensual' process that attempts to square this circle will promote confusion between consensus and certainty.

Thus, a valid goal of structured elicitation is to quantify uncertainty, not remove it from the decision process. For safety-critical decisions, individual experts are as likely to be overcautious as overconfident in their opinions. Ultimately, however, decision-makers want the best advice, not the most cautious and not the overconfident either, and this is where uncertainty muddies the water. A formal procedure for pooling the judgements of a small group of experts can give each the opportunity to express their 'true' opinion. In high-stakes

---

[1] Quote is from *Environmental Science and Technology Online*, 20 April 2005.

situations, the scientists involved can also defend their advice as coming from a collective decision process, rather than a single personal view.

A fundamentally distinct approach for doing this has been formulated by Cooke (1991), which differs from earlier elicitation procedures in that it does not try to impose total consensus or absolute agreement on a group. Instead, it is a structured procedure for obtaining uncertainty judgements from experts, measuring their individual judgement capabilities with a performance-based metric, and then applying mathematical scoring rules to combine these individual judgements into a 'rational consensus' that can inform the deliberations of policy-makers.

This particular approach features prominently in the current chapter, and crops up most frequently in Sections 4.2 and 4.3, on expert judgement and expert elicitations in practice, respectively. The discussion then moves on to consider issues in communicating expert uncertainty, with a focus on policy setting (Section 4.4), followed by a brief examination of possible, or desirable, future directions in research on expert elicitation, theoretical and methodological (Section 4.5).

4.2 Expert judgement elicitation

While expert opinion is almost invariably sought when scientific or technical uncertainty impacts on an important decision process, such uncertainty is ubiquitous with scientific knowledge – if it were not, any decision related to the scientific issue at hand would be obvious. Thus, there is the inescapable corollary that the experts themselves cannot be absolutely certain, and thus it is almost impossible that they will ever be in total agreement with one another. This is especially true where such uncertainty is substantial, or where the consequences of the decision are particularly serious or onerous.

In circumstances where scientific uncertainty impinges on the resolution of an issue, soliciting expert advice is not new. However, the forms which this can take continue to evolve. The main landmarks in this progression are grouped below according to whether the operable methodology is structured or unstructured. Very roughly, 'structured' means that the questions which experts answer have clear operational meaning, and the process of arriving at a defined position is traceable and open to review.

Operational meaning is of paramount importance. Questions of substantial cognitive depth often involve ambiguities ('what is meant by 'severe weather'?). In order to focus on the underlying uncertainties about nature, opacity or ambiguity in the meaning of words must be resolved. Left to their own devices, different experts will do this in different ways, resulting in unnecessarily noisy answers. Ambiguity can be reduced to an acceptable level by casting the questions in terms of thought experiments, or 'thought measurements' that could in principle, albeit not in practice, be performed. Translating the original questions into questions with clear operational meaning is often a difficult task. On the other hand, some questions have little or no cognitive depth ('In what year will man land on Mars'?).

Some examples of unstructured and structured methods are:

Unstructured expert judgement:
- expert oracles: courtroom experts
- expert pundits: talking-head experts
- expert surveys: the wisdom of crowds
- 'Blue ribbon'[2] panels: greybeard experts

Structured expert judgement
- Delphi
- nominal group techniques
- equal weighting
- self-weighting
- peer weighting
- performance weighting
- technical integrator based
- citation based
- likelihood based
- Classical Model.

A number of good books approach expert judgement with a wide compass and discuss various of these approaches: Kahneman *et al.* (1982); Morgan and Henrion (1990); Cooke (1991); Woo (1999, 2011); O'Hagan *et al.* (2006). The influential thinking of Tversky and Kahneman (2005) about judgements under uncertainty is reprised and extended by Kahneman (2011), with more contemporary psychological and neuroscience insights.

Cooke and Goossens (2008) compiled an archive of case histories, termed the TU Delft database, comprising expert judgement studies in which experts were asked questions from their field for which true values were known post hoc. This has provided a resource for testing various expert combination schemes. We offer a brief sketch of the expert judgement approaches, and illustrate the strengths and weakness of the most relevant of these with case studies and anecdotal experience. The unstructured approaches are treated summarily.

The above synoptic summary may surprise those readers who expected to see listed something denoting a Bayesian approach. Bayesian approaches fall under likelihood-based methods and are briefly referenced in that context.

4.2.1 Unstructured elicitations

Expert oracles are experts who have a hotline to the truth, denied to the rest of us. One such expert is as good as any other, provided (s)he is really an expert. The archetype is the ballistics expert giving courtroom testimony on which bullet came from which gun. There is no second opinion, no dissenting view and no uncertainty with oracular experts.

---

[2] 'Blue Ribbon' is a term invented in the United States; if it had been coined in the UK one presumes it would have been Blue Riband. Reluctantly, we feel obliged to stick with contemporary American usage.

Expert pundits are found on TV and radio news shows and in newspapers where experts are conjured up (or volunteer themselves) to shed light on whatever is newsworthy. The word derives from the ancient Sanskrit word *pandita*, meaning learned or wise. Without structured constraint, oracular status gets lost in the cacophony of acrimonious interruption, and the term pundit has acquired an unfavourable aroma. Nonetheless, these 'experts' are widely perceived as knowing more than the rest of us.

Expert surveys may appear, superficially, to be structured, but usually have some promotional aspect, as in '99% of all doctors prescribe Damnital for pain relief'. Unlike other opinion surveys, however, credibility is not always enhanced by including ever more experts. What would we say to '99% of financial experts say a financial crisis is not imminent'? The crowd may often be wise, but the herd sometimes stampedes over a cliff.

Blue ribbon panels of greybeards draw credibility from their scientific status and, purportedly, from the disinterest associated with their age and station – a position where knowledge marries honesty. On questions impacting the stature and well-being of a field to which such a person has devoted his/her career, is it naive to think that the marriage of knowledge and honesty is really made in heaven? Convening committees of the 'great and good' are commonplace in government, whereas blue ribbon panels in science are infrequent and, by science's usual pecuniary standards, expensive undertakings. In early 2010, President Obama convened the Blue Ribbon Commission on America's Nuclear Future,[3] and Chief White House Energy Advisor Carol Browner said 'It is time to move forward with a new strategy based on the best science and the advice of a broad range of experts.' US Energy Secretary Steven Chu said the commission will have a free hand to examine a 'full range of scientific and technical options' on waste storage, reprocessing and disposal – with one exception: the once scientifically favoured Yucca Mountain underground repository. Thus, all too often, independent scientific thought becomes totally marginalised in highly politicised matters.

4.2.2 Early structured elicitations and advances

To better understand where things stand now, a brief reiteration of the evolution of structured expert judgement methodologies from primeval forms to present praxis may be helpful; some thematic variations are mentioned along the way, for completeness.

The *Delphi method* was pioneered in the 1950s at the RAND Corporation (Helmer, 1966, 1968) and is the method that most laypeople have heard of. It was originally crafted for technology assessment ('In what year will commercial rocket transport become available?'), but quickly branched into many fields. Consensus is achieved by repeatedly feeding assessments back to a set of expert respondents and asking outliers to bolster their eccentric views with arguments. Delphi studies are frequently conducted by mail, and questions typically do not involve a great deal of cognitive depth. Sackman's withering *Delphi Critique* (1975)

---

[3] The BRC Final Report was released on 26 January 2012 (available at http://www.brc.gov on 27 July 2012); a draft with tentative conclusions was discussed in a Congressional joint hearing in October 2011.

noted that, without reporting the drop-out rate, the suspicion cannot be allayed that 'consensus' results from simply pestering the outliers away. Delbecq *et al.* (1975), and Gough (1975) in separate studies found that the Delphi method performed worst of all techniques analysed. Fischer (1975) and Seaver (1977) found no significant differences among any methods. The latter study also found that group consensus tended to produce more extreme probability estimates. The scoring variables used in these studies are not always directly comparable, and there is no direct quantification of uncertainty. Woudenberg (1991) reviewed the literature concerning quantitative applications of the Delphi method and found no evidence to support the view that Delphi is more accurate than other judgement methods or that consensus in a Delphi study is achieved by dissemination of information to all participants. Data then available suggested that any consensus achieved is mainly due to group conformity pressure, mediated by feedback to participants of statistics characterising group response rather than information *per se*, corroborating Sackman (1975). For more background on the Delphi method, see Cooke (1991).

Nominal group techniques (Delbecq *et al.*, 1975) involve bringing the experts together to debate the issues under the guidance of a skilled facilitator. Ideally, differences are fully aired during the first day, some bonding occurs after dinner and on the morrow a group consensus emerges. These were among the techniques tested by Fischer (1975) and Seaver (1977) with inconclusive results. The fact that experts must come together for a day or more constitutes a serious limitation of a practical nature. High-value experts have high-value agendas, and scheduling such sessions can easily cost six months of calendar time, and some will inevitably cancel at the last minute. While nominal group techniques were developed to obtain consensus summaries in specific circumstances, such methods were found to have significant limitations (Brockhoff, 1975; Seaver, 1978).

Equal weighting is the simplest of the mathematical aggregation methods. In a panel of N experts, each expert's probability, or probability distribution, is assigned a weight $1/N$, and the weighted probabilities or distributions are added to produce a 'decision-maker' (DM). Equal weighting of forecasts (as opposed to probabilities) emerges in Bayesian updating models under certain conditions (Cooke, 1991), with Clemen and Winkler (1987) putting forward arguments in favour of the approach. Its primary appeal is its simplicity. While other weighting schemes require justification, rightly or wrongly, equal weighting is usually accepted uncritically and without demur. Evidence (Cooke and Goossens, 2008) shows that the equal weight combination is usually statistically acceptable – usually, but not always. In seven out of the 45 studies examined by them, the hypothesis that the equal weight combination was statistically accurate would be rejected at the 5% level. Another feature of equal weighting is that the combined distributions can become quite diffuse as the number of experts increases. Figure 4.1 shows the 90% central confidence bands and median values assigned by 30 BA pilots for one particular target item: the wide spread on the credible interval for the equal weight (EQUAL) DM is by no means atypical (also shown is the corresponding Classical model performance weight (PERF) DM – see below). For instance, we see that the equal weight DM 90% confidence spread is much broader than that of any of the individual experts. The performance-based DM, which in this analysis distributes most

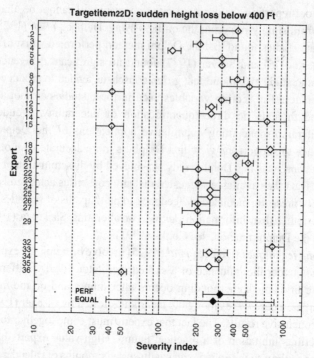

Figure 4.1 Range graph plot showing the median values and 90% confidence bands provided by 36 airline pilots asked to assign a relative severity index to sudden height loss below 400 ft altitude, a potential flight operations safety factor. At the bottom is the combined equal weights solution (EQUAL), with the corresponding classical model performance-based weights solution above (PERF). The equal weights credible interval is much wider than that of any individual expert. The performance-based DM distribution takes most support from the three top-ranked pilots, producing a much narrower uncertainty spread than equal weights. (Three other pilots, 10, 15 and 36, provided responses that deviated significantly from the majority, which would prompt a facilitator to explore their reasoning; six pilots provided null responses for this item.)

weight over the three top-ranked pilots, shows a distribution which is more 'humanoid'. (We note that three other pilots in this case provide responses that depart significantly from the majority, a diagnostic trigger for the facilitator to review, with them, their reasoning.)

Hora (2004) showed that equal weighting of statistically accurate experts always results in a statistically *inaccurate* combination. A procedure for uncertainty quantification based on equal weighting was developed for US nuclear risk applications (Hora and Iman, 1989). In the field of seismic risk, the SSHAC Guidelines (SSHAC, 1997) expound an elicitation approach that has the goal of assigning equal weights to experts (Hanks et al., 2009). The notion is that, as long as the evaluators have equal access to information and have fully participated in the workshops, 'equal weights are expected' (Hanks et al., 2009: 14). This said, within the higher-level SSHAC procedures the technical facilitator integrator (TFI) has the option to use different weights for different evaluator teams.

Self-weighting was originally used in some Delphi studies, in which experts were asked to rank their own expertise on a scale of 1 to 7 – the units were not given. Brockhoff (1975) found that women consistently ranked themselves lower than men, and that these rankings failed to predict performance. After Kahneman *et al.* (1982) highlighted the phenomenon of overconfidence, the inclination to use expert self-weights lost force. An early study of Cooke *et al.* (1988) found a strong negative correlation between informativeness and statistical accuracy. Roughly, that translates to: the tighter the 90% confidence bands, the more likely that the realisation falls outside these bands. The correlation is not strict, however, and there are often experts who are both statistically accurate and informative.

Peer weighting is an option that can be tried if the experts in a panel all know each other or each other's work. It was proposed independently by De Groot (1974) and by Lehrer and Wagner (1981). Real applications are sparse, but, in a study for dam erosion risk modelling (Brown and Aspinall, 2004), a comparison was made between performance-based expert weightings for an expert panel with mutual peer weighting by colleagues in the group. There were some major differences in ranking between the two: for example, some experts scored significantly less well on the performance-based measure than their colleagues might have anticipated, while others did much better (see Figure 4.2). If

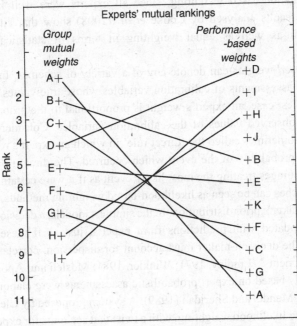

Figure 4.2 Dam erosion risk panel elicitation: comparison between experts' ordering by group-assigned mutual weights (nine initial experts: A–I) and the ranking of all 11 experts in the eventual elicited group (A–K) from performance-based scoring using Cooke's (1981) Classical Model (right-hand column). Note several apparent reversals of reputational authority are evident for the nine (see text).

anecdotal evidence can be relied upon, this sort of bias can be expected in peer weighting with almost any group of specialists, whatever the discipline: typically, some experts are well-regarded but tend to be strongly opinionated, while others – perhaps more reflective, and considered indecisive or diffident by their peers – are, in fact, better estimators of uncertainty. In the dam erosion study the quantification of model parameter uncertainties was the main objective, so it was decided appropriate by the problem owner that the better-calibrated experts should be rewarded with greater weight, even though this appeared to conflict with the peer appraisals.

Performance weighting denotes any system in which weights are assigned to experts' views according to their performance. These come in a number of variants, with two principal flavours dominating current praxis (see Section 4.2.3); other, less often used, variations include citation-based and likelihood-based weights.

Citation-based weights, which are based on the relative number of times that experts cite each other's work. Many complications arise in implementing this deceptively simple idea. Should self-citations be counted? How do we deal with joint authorship? Do we discount the citations of older experts, as these tend naturally to grow with age? The uncertainty analyses conducted in the Joint US Nuclear Regulatory Commission–European Union uncertainty analysis of accident consequence codes afforded an opportunity to test citation-based weighting, since all experts were well known and well published. The results analysed by Cooke *et al.* (2008) show that citation weighting performed about as well as equal weighting in terms of statistical accuracy and informativeness.

Likelihood-based weights can denote any of a variety of schemes. In any case, such methods require assessments of calibration variables whose true values are known post hoc. In the simplest case, an expert's weight is proportional to the probability which he assigned to the observed values of the calibration variables. Considered as a scoring variable, this is sometimes called the 'direct rule' in which an expert's reward is proportional to his/her probability of the event which occurred. The direct rule is notoriously improper: it encourages treating the most likely event as if it were certain (Cooke, 1991). Bayesian approaches can be seen as likelihood-based weighting methods, with admixture of the decision-maker's prior distribution. In the simplest models, a decision-maker's prior distribution is updated with predictions from experts, just as if these were physical measurements. The decision-maker must account for dispersion, correlation and bias in his 'expert-instruments' (Lindley, 1971; Winkler, 1981; Mosleh and Apostolakis, 1982). Bayesian methods based on expert probabilistic assessments were elaborated by Morris (1974, 1977) and Mendel and Sheridan (1989). A system proposed by Clemen and Reilly (1999) is similar to likelihood weighting, with a twist that accounts for expert dependence. A check, using studies from the TU Delft database, showed its performance was rather indifferent (Kallen and Cooke, 2002). In more recent work, Cooke *et al.* (2008) were able to test a version of likelihood weights using cases from the TU Delft database, and also found very uneven performance.

4.2.3 *Current mainstream approaches to structured elicitation*

In current practice, there are two major but different *performance weighting* schemes that are often utilised for structured elicitation, in which weights are assigned to experts' views. In one, weights are ascribed according to the appraisal of performance by a technical facilitator or facilitators, and, in the other, performance weights are determined by applying a scoring rule measure to an empirical test of judgement skill.

4.2.3.1 *The SSHAC elicitation procedure*

The pre-eminent approach to expert elicitation using a technical facilitator approach is found in the methodology espoused by the US Department of Energy (DoE) for estimating site-specific probabilistic seismic hazard curves (see also Chapter 8). In order to accommodate uncertainties in earthquake occurrence rates and in ground motion attenuation processes that influence effects at a site, the DoE favours a method[4] for expert elicitation that is based mainly on work for the US NRC, articulated by the Senior Seismic Hazard Analysis Committee (SSHAC, 1997; Budnitz *et al.*, 1998).

The SSHAC committee report outlines a formalised expert elicitation procedure with four levels of intensity of application. The most extreme, SSHAC Level 4, utilises the concept of the 'Technical Facilitator Integrator' to integrate the information from groups of subject-matter specialists. The procedure also involves the use of participatory reviews, requiring additional specialist experts to follow progress in the different topics that contribute to the hazard assessment from the beginning of the study, rather than waiting until the end of the whole project to provide comments. It is held that this approach creates the proper conditions to provide for a defensible decision to be made, based on a commonality of information and data among all the experts involved.

The SSHAC thinking is intended to promote a methodological culture that aspires to direct incorporation of epistemic uncertainty into a 'total' risk assessment process, by 'finding the center, body, and range of the informed technical community' (Hanks *et al.*, 2009). In pursuit of this, a TFI seeks to represent the whole community of experts by elicitation of the views of a sample group. The TFI may seek to correct the capriciousness of chance in the expert group selection: 'to represent the overall community, if we wish to treat the outlier's opinion as equally credible to the other panelists, we might properly assign a weight (in a panel of 5 experts) of 1/100 to his or her position, not 1/5'.[5] Thus differential weighting is based on the integrator's assessment of the 'representativeness' of an expert's views, and in that sense relates to the expert's 'performance' or, more accurately, the TFI's perception of it.

The SSHAC approach positively embraces the concept of equal weights, endeavouring to make a virtue of democratic principles in this way of thinking. However, this takes no account of the precept that not all experts are equally adept at making judgements, especially where this concerns uncertainty. It might be argued that a method that assesses the relative proficiency of experts in this regard will furnish a more robust approach for decision support.

---

[4] US DoE Standard: Natural phenomena hazards assessment criteria; DOE-STD-1023–95, April 2002.
[5] NUREG/CR-6372, at p. 36.

4.2.3.2 The Classical Model

One of the main goals of a formal elicitation of expert judgement is to remove as much subjectivity from decision-making as possible. In the context of scientific advice, misunderstanding can arise about objectivity and subjectivity in relation to expert judgement for two reasons. One is judgements are based on an individual expert's personal degree of scientific belief on an issue and, because this is 'personal', it is sometimes construed as being 'subjective', in some ill-defined sense. The second is that such judgements are often referred to as 'expert opinion' – while *anyone* can have an opinion, meaningful scientific judgement requires specialist knowledge, practised reasoning skills and real experience. An expert is someone who not only knows a great deal about a discipline, but also understands how that discipline is organised, its rules, semantics, culture and methodologies.

Thus, genuine expertise entails the key attribute of objectivity, and eliciting expert judgement calls for a structured, rational process that captures this objectivity in a neutral and impartial manner. In many elicitations, such as those undertaken in connection with the Montserrat volcanic activity (Section 4.3.5), emphasis is placed on ascertaining scientists' judgements about the extent of uncertainties on parameters, variables or probabilities, for the purpose of including these in probabilistic hazard and risk assessments.[6] For scientists assessing natural hazards, emergent diseases, climate change or any number of other challenges that may turn out to be major threats to public safety, concern ought to be focused on the systematic rational interpretation of all available evidence and associated uncertainties. Some of this evidence and many of the relevant uncertainties can only be characterised on the basis of expert judgement.

However, it is important to recognise that experts are not necessarily equally adept and proficient at judging scientific uncertainty so, for decision support, a constructive method is needed which can differentiate individual judgements. One way in which this can be done is to use a procedure which theoretically maximises objectivity in the face of uncertainty, utilising a method that can weigh experts' judgements on the basis of empirically validated performance metrics, measured within the specialism domain.

In what follows, a detailed description is given of the Cooke Classical Model (Cooke, 1991), the only structured group elicitation approach that has this empirical control, an important credential of the scientific method. In addition, the scheme offers reproducible and auditable tracking of how a particular scientific issue has been characterised quantitatively by a group of experts for input to a decision. In increasingly litigious times, this ought to be seen as beneficial by participating experts when they are called upon to give science-based advice in support of crucial, possibly life-or-death, decisions. It is in such circumstances that the concept and principles of the Classical Model come to the fore.

---

[6] Elsewhere, the apparently simpler, and hence more appealing, deterministic approach is sometimes followed. However, the extents of these multiple, and multiplied, uncertainties are almost impossible to accommodate rationally into decision-making under the deterministic prescription – sometimes the consequences of this shortcoming can be disastrous: for instance, deterministic underestimation of the upper bound magnitude for great earthquakes off Tohoku, northeast Japan, was a key factor in the tragic surprise caused by the earthquake and tsunami of 11 March 2011.

The Classical Model method relies on the use of proper scoring rules for weighting and combining expert judgements through statistical accuracy and information scores, measured on calibration variables (see Cooke, 1991), and operationalises principles for rational consensus via a performance-based linear pooling or weighted averaging model. The weights are derived from experts' calibration and information scores, as measured on seed item calibration variables. Calibration variables serve a threefold purpose:

(1) to quantify experts' performance as subjective probability assessors;
(2) to enable performance-optimised combinations of expert distributions; and
(3) to evaluate and hopefully validate the combination of expert judgements.

The name 'Classical Model' derives from an analogy between calibration measurement and classical statistical hypothesis testing. An expert's *calibration* score is the p-value (i.e. significance level) at which one would falsely reject the hypothesis that the expert's probability statements were correct. If an expert gives 90% confidence bands for several items which are later observed (or known otherwise), and if a substantial proportion of the observed values fall outside those confidence bands, then the calibration score is derived from the probability that such an outcome could have arisen purely by chance. The designation 'classical' denotes a contrast with various Bayesian models, which require prior assessments of expert and variables by the decision-maker.

The Classical Model performance-based weights also rely on a second quantitative measure of performance, *information*, which measures the degree to which a distribution is concentrated.

In practice, both of these measures can be implemented for either discrete or quantile elicitation formats. In the discrete format, experts are presented with uncertain events and perform their elicitation by assigning each event to one of several pre-defined probability bins, typically 10%, 20%, ... 90%. In the quantile format, experts are challenged with an uncertain quantity taking a value in a continuous range, and they give pre-defined quantiles, or percentiles, for their corresponding subjective uncertainty distribution, typically 5%, 50% and 95%. In application, this format is found to have distinct advantages over the discrete format, not least in the effort demanded of experts to express their uncertainty judgements.

For *calibration* in the case of three quantiles, each expert divides the quantity range into four inter-quantile intervals within which they can judge the relative probabilities of success, namely: $p_1 = 0.05$: the quantity is less than or equal to the 5% value; $p_2 = 0.45$: greater than the 5% value and less than or equal to the 50% value, and so on.

If N such quantities are assessed, each expert may be regarded as a statistical hypothesis, namely that each realisation of the N falls in one of the four inter-quantile intervals with probability vector corresponding to those defined intervals, from which can be formed a sample distribution of the expert's inter-quantile intervals. If the realisations are indeed drawn independently from a distribution with quantiles as stated by the expert, then the likelihood ratio statistic is asymptotically distributed as a chi-square variable with 3° of freedom (see Cooke, 1991; Cooke and Goossens, 2008).

For such a distribution, it is straightforward to derive the familiar chi-square test statistic for goodness of fit, and this test scores the particular expert as the statistical likelihood of the hypothesis:

H_e: the inter-quantile interval containing the true value for each variable is drawn independently from the expert's defined probability vector.

Using the likelihood ratio statistic, the analyst computes a test p-value, the latter being the probability that a deviation at least as great as that observed could occur on N realisations if H_e were true; this p-value forms the basis for determining the expert's statistical accuracy or calibration score. Calibration scores are absolute and can be compared across studies. However, before doing so, it is appropriate to equalise the power of the different hypothesis tests by adjusting the effective number of realisations in each when these differ from one study to another.

Although the calibration score uses the language of simple hypothesis testing, it must be emphasised that expert-hypotheses are not being rejected; rather, this language is used to measure the degree to which the data support the hypothesis that the expert's probabilities are accurate. Low calibration scores, near zero, mean that it is unlikely that the expert's probabilities are correct.

Under the Classical Model, the second scoring variable is *information* or *informativeness*. Loosely, the information in a distribution is the degree to which the distribution is concentrated. Information cannot be measured absolutely, but only with respect to a background measure. Being concentrated or 'spread out' is measured relative to some other distribution – commonly, the uniform and log-uniform background measures are used (other possible background measures are discussed by Yunusov *et al.*, 1999).

Measuring information requires associating a density with each quantile assessment of each expert. To do this, the unique density is found that complies with the experts' quantiles and is minimally informative with respect to the background measure. For a uniform background measure, the density is constant between the assessed quantiles, and is such that the total mass between quantiles i and $i+1$ agrees with the corresponding relative probability p_i. The background measure is not elicited from experts but must be the same for all experts; it is chosen by the analyst.

The *information score* for an expert is calculated from the average relative information of the expert's joint distribution given the selected background, under the assumption that the variables involved are independent. As with calibration, the assumption of independence reflects a desirable property for the decision-maker process, and is not an elicited feature of the expert's joint distribution. Furthermore, the information score does not depend on the realisations: an expert can give himself a high information score by choosing his quantiles very close together but, evidently, his or her information score also depends on the ranges and assessments of the other experts in the particular group. Hence, information scores cannot be compared across studies.

Of course, other measures of concentratedness could be contemplated. The above information score is chosen because it is tail-insensitive, scale invariant and 'slow', in the sense that large changes in the expert assessments produce only modest changes in the information score. This contrasts with the likelihood function in the calibration score, which is a very fast function. This causes the product of calibration and information to be dominated by the calibration score.

To illustrate the ways in which the calibration and information traits of individual experts can vary in practice, a hypothetical quartet is presented schematically in Figure 4.3. In that example, four mythical experts respond to the same ten seed item calibration questions. While seed items are questions for which the true values are known to the facilitator, or will become known, experts are not expected to know these values precisely but, on the basis of their expertise, they should be able to define quantised credible ranges for each, within which the true value falls with some assignable probability.

Figure 4.3 illustrates some typical expert judgement traits which, in practice, can affect individual expert's uncertainty assessments to one degree or another; ideally, such propensities should be controlled for when eliciting opinions for decision support. However, it is very difficult to determine a priori from other criteria, such as publication record or reputation, which expert has what tendency in this sense – such insight only emerges from an empirical judgement calibration analysis.

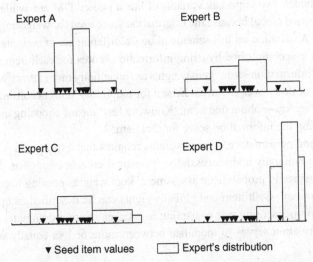

Figure 4.3 Schematic representation of four experts' uncertainty distributions in relation to a set of ten seed items used for calibration scoring with the Classical Model. Each expert defines an uncertainty range for each seed item by nominating three quantiles which should enclose the realisation value (small extensions are added in the tails to create four location probability bins). Over the ten items: Expert A is statistically coherent – his distributional spreads capture and take support from the realisation values – and he is also informative and hence scores well; Expert B is equally accurate, statistically, but is less informative than A, and would be ranked with a lower weight; Expert C is superficially similar to B, but his judgements are systematically shifted relative to realisation support and would be penalised for this bias; Expert D is very confident in his judgements yet extremely poorly calibrated, statistically – he should be surprised by the realisation values – and his weighting would be very low, even zero, as a consequence (see text for a more detailed description of scoring concepts).

Under the Classical Model concept, the combination of assessments by several experts is called a *decision-maker* (DM), and is an example of linear pooling.[7] The Classical Model is essentially a method for deriving weights in a linear pool. 'Good expertise' corresponds to good calibration (high statistical likelihood, high p-value) and high information. Weights are wanted which reward good expertise and which pass these virtues on, through the pooling DM.

The reward aspect of weights is very important. An expert's influence on the DM should not appear haphazard, and experts should be discouraged from gaming the system by tilting their assessments in an attempt to achieve a desired outcome. These goals are achieved by imposing a strictly proper scoring rule constraint on the weighting scheme. Broadly speaking, this means that an expert gains his or her maximal expected weight by, and only by, stating assessments in conformity with his or her true beliefs. Some might be tempted to attempt to game the process by deliberately expanding their credible ranges beyond genuine personal assessments, as underestimating uncertainty is a recognised trait. But being lured into this – even marginally – engenders the risk of being penalised via the information scoring: an expert cannot be sure a priori that he or she has a confirmed tendency to systematically understate uncertainties.

Within this model, two important variants of linear pooled DM are available: the *global weight DM* is termed global because the information score used for weighting is based on all assessed items. A variation on this scheme allows a different set of weights to be used for each item. This is accomplished by using information scores for each item rather than the overall average information score. Item weights are potentially more attractive as they allow an expert to up- or down-weight him or herself for individual items according to how much he or she feels they know about that item. 'Knowing less' means choosing quantiles further apart and lowering the information score for that item.

Of course, good performance of item weights requires that experts can perform this up/down weighting rationally and successfully. Anecdotal evidence suggests that, as experts gain more experience in probabilistic assessment, item weights pooling improves over the global weights option. Both item and global weights can be described as optimal weights under a strictly proper scoring rule constraint: in each, calibration dominates over information, while information serves to modulate between more or less equally well calibrated experts.

Since any combination of expert distributions yields assessments for the calibration variables themselves, any DM combination can be evaluated on the calibration variables. Specifically, the calibration and the information of any proposed DM can be computed with justifiable expectation that the best pooled DM would perform better than the result of

---

[7] Cooke (1991: ch. 8, 9) discusses asymptotically strictly proper scoring rules in formal mathematical terms, and also expounds on the properties of such rules for combining probabilities by linear pooling (ch. 11). Suffice it here to say that much work has been done on different ways of merging quantitative judgments and that there are strong reasons for forming (weighted) combinations of experts' density functions (or cumulative density functions) by linear pooling – rather than averaging variable values with the same probabilities (i.e. quantile functions). Counterfactual examples are easily constructed which demonstrate that alternatives, other than linear pooling, often result in combination averages that assign mass to forbidden probability intervals or otherwise produce irrational or impossible marginalisation or zero preservation exceptions (see Cooke, 1991, ch. 11).

simple averaging (i.e. an *equal weight DM*), and that the proposed DM is not significantly worse than the best expert in the panel – although this sometimes happens. Figure 4.1 shows a typical example of a performance-based weighted target item solution from an elicitation of 30 senior civil airline pilots concerning flight operation safety factors compared with the corresponding *equal weights* solution (see Section 4.3.4).

Thus, the Classical Model process is designed to enable rational consensus: participants pre-commit to a process bearing the essential hallmarks of the scientific method, before knowing the outcome. These hallmarks are:

Scrutability/accountability: all data, including experts' names and assessments, and all processing tools are available for peer-review and results must be open and reproducible by competent reviewers.

Empirical control: quantitative expert assessments are subjected to empirical quality controls.

Neutrality: the method for combining/evaluating expert opinion should encourage experts to state their true opinions, and must not bias results.

Fairness: experts' competencies are not pre-judged, prior to processing the results of their assessments.

The Classical Model approach is a bit more complicated than others, with the main precept that the weights reward good performance in probabilistic assessment. This is *not* the same as commanding a wealth of knowledge, or giving accurate predictions. Rather, a good probabilistic assessor is one who is able to quantify his or her uncertainty in a way which is informative and statistically accurate – considered as a statistical hypothesis, an expert would be credible.

The Classical Model system has been widely applied and documented (see Cooke and Goossens, 2008, and references therein). Independent espousals have come from practitioners across various fields (Woo, 1999, 2011; Aspinall, 2006, 2010; Klügel, 2007; French, 2008, 2011; Lin and Bier, 2008).

4.3 Expert elicitations in practice

There is a steadily burgeoning catalogue of case histories of expert-informed decisions in practice: in seismic hazard assessment, volcanic hazards assessment, emerging infectious disease risk modelling, flight safety and dam safety, finance, biosecurity and nuclear safety, all of which have utilised either structured elicitations (to different degrees of formalisation) or unstructured elicitations. A very informative generic discussion on how to conduct an expert elicitation workshop, targeted at facilitators mainly, is provided by the Australian Centre of Excellence for Risk Analysis (ACERA, 2010), a report which has a valuable literature review companion (ACERA, 2007). These two reports represent an excellent start point for novice facilitators and domain experts, alike.

Here we recount a few insights from examples of the formalised structured type, starting with two prominent SSHAC case histories, before adding to the narrative experiences with the Classical Model.

4.3.1 SSHAC and the PEGASOS Project

The SSHAC approach (see Section 4.2.3) was applied in the PEGASOS Project, the first European trial application of the procedure at its most complex level (level 4), with the goal of developing site-specific seismic hazard estimates for the sites of Swiss nuclear power plants (Abrahamson et al., 2004; Zuidema, 2006; Coppersmith et al., 2009). The power plant owner sponsored the study in response to a request from the Swiss regulator, following the latter holding discussions with US consultants and with US NRC officers. The PEGASOS Project was subdivided into four subprojects:

(1) SP1 – Seismic source characterisation (four groups of experts, each group consisting of three experts).
(2) SP2 – Ground motion characteristics (five experts).
(3) SP3 – Site-response characteristics (four experts).
(4) SP4 – Hazard quantification.

Therefore, the study followed the convolution approach, separating source, ground motion and site-response characteristics, and commissioned two TFIs to manage the elicitations, one for SP1 and another for SP2 and SP3. The study was based on the use of comprehensive logic trees to reflect the epistemic uncertainties related to source, ground motion attenuation and site characteristics relevant for the evaluation of seismic hazard models for the Swiss sites. Equal weights were used to combine the different expert opinions.

Seismologically, an interesting feature of the project was the attempt to constrain maximum ground motion levels by estimating an upper limit for this variable (see also Chapter 8). The final review of the project results by the sponsor (see Klügel, 2005a, 2007), as well as by the Swiss nuclear safety authority, identified the need for a further development of the probabilistic seismic hazard analysis, and a 'refinement' project is underway.

One of Klügel's (2005a, 2009) main criticisms (of many) of the original project was that the SSHAC expert opinion elicitation concept is philosophically suspect and logically flawed, and relies on inadequate elicitation and aggregation methods that are based on political consensus principles or on census principles, rather than on principles of rational consensus (Klügel, 2005d). His initial intervention (Klügel, 2005a) engendered a major debate in the literature (Budnitz et al., 2005; Klügel, 2005b–f, 2006, 2009; Krinitzsky, 2005; Lomnitz, 2005; Musson et al., 2005; Wang, 2005). Judging by the ensuing furore, not to mention the huge related costs and the sheer difficulty of coping with a logic tree that had at least 10^{15} retained branches to represent all the models and alternative parameterisations that the evaluator panels could envisage, it seems the SSHAC approach has not been an unqualified success in application in the PEGASOS Project. Nor, as discussed next, has the approach proved convincing in volcanological hazard assessments undertaken for the Yucca Mountain Project (YMP).

4.3.2 SSHAC and the Yucca Mountain Project

An important case history in terms of the use of expert judgement is provided by the proposed high-level radioactive waste repository at Yucca Mountain, Nevada, where studies of the geologic basis for volcanic hazard and risk assessment had been underway for nearly three decades. Probabilistic hazard estimates for the proposed repository depend on the recurrence rate and spatial distribution of past episodes of volcanism in the region, and are of particular concern because three episodes of small-volume, alkalic basaltic volcanism have occurred within 50 km of Yucca Mountain during the Quaternary.

The use of expert elicitations within the YMP was extended to cover more specific issues, such as the Near-Field/Altered Zone Coupled Effects Expert Elicitation Project (NFEE), and similar issues requiring expert judgement inputs. In addressing such technically complex problems, expert judgement had to be used because the data alone did not provide a sufficient basis for interpreting the processes or for the outputs needed for subsequent analyses. For example, experiments such as the 'single heater test' did not provide a direct estimate of the spatial distribution of fracture permeability changes: site-specific experimental data had to be interpreted and combined with other data and analyses before they could be used in performance assessments.

In such cases, experts have to integrate and evaluate data in order to arrive at conclusions that are meaningful for technical assessments of particular geophysical and geochemical effects, including quantitative and qualitative expressions of the uncertainties. In principle, the process which is followed should be the same, however abundant or scarce the data, the only difference being the level of uncertainty involved. In this sense, expert judgement is not a substitute for data; rather, it is the process by which data are evaluated and interpreted. Where data are scarce and uncertainties high, the uncertainties expressed by each expert and the range of judgements across multiple experts should reflect that high degree of uncertainty.

The YMP studies closely followed the procedural guidance first set out in SSHAC (1995), both in spirit (e.g. a belief in the importance of facilitated expert interactions) and, in many cases, in details of implementation (e.g. suggestions for conducting elicitation interviews). For example, the NFEE process was designed – in accordance with SSHAC guidance – to result in assessments that represent the 'range of technical interpretations that the larger informed technical community would have if they were to conduct the study'. However, in as much as SSHAC professes to be 'non-prescriptive' in specifying a single way of implementing the process, it would be wrong to suggest that all the YMP expert elicitations conformed precisely with the SSHAC procedures. In some cases, the processes were specifically tailored to be appropriate for relatively modest studies involving fewer experts.

The various guidance documents which sought to constrain work associated with the high-level waste programme signified that the goal was not to establish a rigid set of rules for eliciting expert judgement but, rather, to draw from experience – both successes and failures – criteria for identifying when expert judgement should be used, and to outline approaches for motivating, eliciting and documenting expert judgements. In this, there was recognition that alternative approaches to the formal or informal use of expert judgement are available (e.g. Meyer and Booker, 1991, republished as Meyer and Booker, 2001).

It is outside the scope of the present chapter to go into all the detailed procedures that were used for the YMP studies, which are fully described in the literature (e.g. Kotra *et al.*, 1996).

4.3.3 *The Classical Model: a foundational example from meteorology*

In October 1980, an experiment involving subjective probability forecasts was initiated at a weather station of the Royal Dutch Meteorological Institute. The primary objective of the experiment was to investigate the ability of the forecasters to formulate reliable and skilful forecasts of the probability that certain critical threshold values of important meteorological variables would (or would not) be exceeded. The trial involved prognostications of wind speed, visibility and rainfall, and went on for more than six years; the experts' judgements on future values of these variables were scored and combined with the *Classical Model* (Cooke, 1987, 1991).

The most important conclusions to be drawn from this investigation were that probabilistic expert assessments can provide a meaningful source of information, and that the quality of expert assessments drops off as the events being assessed extend further and further into the future. Also, the chosen model for combining expert judgement leads to a predictive performance that is markedly better than those of the individual experts themselves. Thus, it was found that there is every reason to consult more than one expert and, then, combine their judgements. Curiously, it was also found that the judgements of an individual expert with a good information foundation can be improved when they are combined with the assessments of an expert with an inferior information source. While this result seems counterintuitive, it emerges overwhelmingly from the study, regardless of which framework was used for combining judgements.

Cooke (1991) emphasises the importance of defining appropriate scoring variables to measure performance: proper scoring rules[8] for individual assessments are really tools for eliciting probabilities. Generalisations of such scoring variables are applied in the Classical Model whose 'synthetic decision-maker' was seen to outperform results of simple arithmetic or geometric averaging. This result is significant in light of the work of Clemen and Winkler (1986), for instance, according to which simple models, such as the arithmetic and geometric averages, were considered to be more reliable on large datasets, including weather forecasting, than other more complicated combination models. From the very comprehensive Dutch meteorological study, however, it is apparent that it is possible to do better than simple averaging or equal weights.

---

[8] In decision theory a scoring rule is a measure of the performance of a person who makes repeated decisions under uncertainty. A rule is said to be 'proper' if it is optimised for well-calibrated probability assessments and 'strictly proper' if it is uniquely maximised at this point.

4.3.4 The Classical Model and airline pilots

Another extensive application of the Classical Model was undertaken with a large corps of senior airline pilots, in connection with the elicitation of opinions on civil aviation safety.[9] In this study, over 40 captains were calibrated using the scoring rules of the Classical Model. The basis of the calibration exercise was a suite of ten seed questions related to the topic of interest for which precise quantified realisations were available. The pilots were tested for their ability to estimate swiftly these quantities and their uncertainties on the basis of their experience, reasoning and beliefs. While there were wide scatters of opinions on questions (sometimes surprisingly so for professionals uniformly trained and working under identical controlled conditions), the application of the scoring rules to generate the weighted combination 'decision-maker' invariably produced central estimates that were within a few per cent of true values. This outcome demonstrated, to the satisfaction of the airline's flight operations management, that the technique provides decision guidance that is sufficiently accurate for most practical purposes.

A similar conclusion, regarding the utility of the classical model methodology in respect of provision of scientific advice during a volcanic eruption crisis (see Section 4.3.5), was reached by Aspinall and Cooke (1998).

4.3.5 Montserrat volcano and the Classical Model

Expert elicitation has been an important component in the provision of scientific advice during the volcanic eruption crisis (Aspinall, 2006) in Montserrat, West Indies, where a volcano has been erupting, occasionally violently, on a small, populated island for more than 15 years. Initially, when the volcano first became restless in 1995, about ten volcanologists became involved in advising the local government on what possible hazards might materialise, so the authorities could determine warning levels, travel restrictions and evacuations. As activity increased over subsequent years, when lethal hot block-and-ash flows spewed from the volcano at speeds of 150 km hr, destroying everything in their path, more and more scientists became involved in the challenge of assessing the volcano's possible behaviour (Aspinall *et al.*, 2002).

This was the first time a formalised elicitation procedure was used in a live volcanic crisis. It allowed the disparate group of specialists to make rapid joint assessments of hazard levels as conditions at the volcano changed, sometimes dramatically, helping forecast conditions a few days or weeks in the future. In most cases, these turned out to reasonably match what happened (Aspinall and Cooke, 1998).

The relative speed with which such elicitations can be conducted, compared with traditional consensus processes, is one of its great advantages in such situations. Another is that it provides a route for experts uneasy about participating, or inexperienced with offering policy advice, to get involved in decision-making processes under the guidance of a neutral

---

[9] Unpublished report to British Airways by Aspinall & Associates, 1997 (WP Aspinall, pers. comm.).

facilitator, without feeling obliged to strive for complete agreement. Because no absolute consensus *sensu stricto* is imposed on the group by the Classical Model, an individual expert is not compelled to adopt the outcome as his or her personal belief. In practice, however, it is rare to meet substantial dissent; a more typical reaction is: 'Those aren't exactly the probability values I would have chosen but, for our present purposes, I can't object to any of them' (Aspinall and Cooke, 1998).

The Montserrat example evinces a spread of differential expert weights, typical of Classical Model results: using a set of seed items concerning facts about volcanic hazards, expert scores for 18 volcanologists are shown in Figure 4.4, together with a simple ordinal index of their experience of such crises. Characteristically (and comfortingly for grey-beards), veteran experts by and large achieve the highest scores, as might be hoped – over the years, they have been imbued with a functioning and reliable feel for uncertainty. But, there are clear exceptions to the principle – in Figure 4.4, experts 17 and 11 punch above

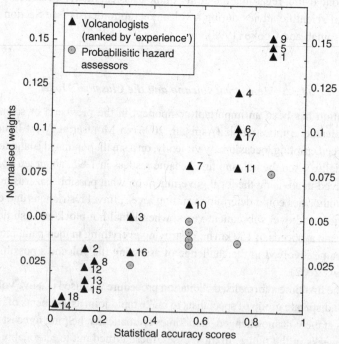

Figure 4.4 Profile of expert scores for 18 volcanologists (triangles) involved in the early stages of the Montserrat volcano crisis (see text). The horizontal axis shows individual statistical accuracy (calibration) scores obtained for a set of seed items on volcanic hazards, while the vertical axis (normalised weights) plots the resulting overall weight ascribed to each expert when individual statistical accuracy is factored by informativeness score; the weights are normalised to sum to unity across the whole group. A rudimentary ordinal ranking of individual volcanologist's experience is given (1–18). Circular markers show the corresponding scores achieved by a group of seven earth scientists, knowledgeable about probabilistic hazard assessment but not specialists in volcanology.

their expected weights, given their relative inexperience, while senior experts 2 and 3 fall down the rankings, being penalised for failing adequately to assess uncertainty ranges or for being statistically inaccurate – or both!

In a comparison exercise, the same elicitation seed items were presented to seven earth scientists who specialised in probabilistic seismic hazard assessment, but not volcanology *per se*. All seven of these non-specialists fared better than a few low-weight volcanologists (see Figure 4.4), but even the best is out-scored by the eight top-weighted volcanologists; the interpretation here is that the hazard analysts' deficit of subject-matter knowledge was offset by their understanding of how uncertainties play into probabilistic reasoning – but not to the extent that this made any of them into genuine experts on uncertainties relating to volcanic hazards.

This last case exemplifies a common denominator in any elicitation of expert judgement: there is no way of knowing for certain, a priori, who will achieve a worthwhile performance-based score until experts are objectively calibrated on some empirical basis, and whose opinions should be marked down for want of sound reasoning about uncertainty (for corroboration of this point, see Flandoli *et al.*, 2011). Establishing what can, and should, be said collectively about scientific uncertainties can, in turn, have important potential impact on how such uncertainties are communicated beyond the confines of an expert group.

4.4 Communicating expert uncertainty

Communicating scientific uncertainty to policy-makers is an area where there is perhaps a significant obstacle to wider adoption of structured expert judgements in decision-support. Some feel that better information from an expert perspective – more complete in terms of quantified uncertainty, but more complex overall – may confound decision-makers rather than promote better decisions, so that improvements in capturing uncertainty in risk assessments have to be matched with improvements in communication.

For instance, Gigerenzer *et al.* (2005), discussing weather forecasts, point out that the quantitative statement 'there is a 30% chance of rain' is assumed to be unambiguous and to convey more information than does a qualitative statement like 'It might rain tomorrow'. Because the forecast is expressed as a single-event probability, however, it does not specify the class of events it refers to, and a bald numerical probability like this can be interpreted in multiple, mutually contradictory ways. To improve risk communication, Gigerenzer *et al.* suggest experts need to operationalise forecasts[10] by specifying, quite precisely, the reference class they are working with – that is, the class of events to which a single-event probability refers.

Other studies suggest most decision-makers welcome more information, not less. If technical information is presented in context, alongside training in uncertainty concepts, it can help decision-makers appraise the confidence they should have in their decisions. As discussed in Chapter 8 of this volume, given the depth and variety of uncertainties involved

[10] In agreement with the discussion in Section 4.2.3, *supra*.

in a probabilistic seismic hazard assessment, experts' views are an unavoidable component of the assessment process. Where a decision has to be made on the basis of very uncertain evidence – which is inevitably the case for low-probability scenarios – the only practical recourse is to make use of relevant expert judgement. Recognising this to be the case, the use of expert judgement is, nowadays, an accepted feature of such studies (see, for example, IAEA, 2010) – although how this is mobilised is not laid down, and remains a matter of debate, and some contentiousness.

This said, it is also important to distinguish expert judgement from engineering or other technical judgement, as the differences between them can be sufficient to affect system safety in a tangible way (Skipp and Woo, 1993).

4.4.1 Expert judgement and engineering judgement

Expert judgement, properly elicited, is not arbitrary but must satisfy various fundamental principles. In particular, the judgement must be coherent with the axioms of probability theory which provide systematic rules for the synthesis of subjective opinions.

By contrast, engineering judgement, in standard civil engineering practice for instance, is a less tangible and less well-defined concept, used in a more holistic fashion (Skipp and Woo, 1993). Thus, engineering judgement may be captured in colloquial statements of the form 'To do it that way will get you into difficulties', or, 'If you do it this way you will be alright.'

The recursive use of engineering judgement in estimating appropriate factors of safety defies quantification, except in the binary sense that inadequacy is ultimately exposed by failure. The main characteristics of engineering judgements are:

- not constrained by formal logic;
- cannot be easily calibrated;
- may incorporate conservatism;
- tend to be heuristic and holistic.

Differences of opinion over safety issues will always be present, but terminological confusions should be avoidable. Where an 'engineering judgement' is unwittingly offered in place of an 'expert judgement', or vice versa, problems can be expected. As probabilistic or Bayesian methods become more widely accepted as methods of weighing evidence, there will be an increasing need to clarify the nature of the judgement being offered (as note by Gigerenzer et al., 2005).

In the case of expert judgement in hazard assessments, more is required than some general awareness of the topic: the judgements need to be made by experts with extensive experience of doing hazard assessments in the relevant region. In principle, a rigorous mathematical check on the consistency of expert judgements with the fundamental laws of probability is afforded by comparative hazard assessments for a range of different sites across a region.

In practice, it often seems that the full ramifications of the use of expert judgement are not properly appreciated. Two salient comments can be made in this regard:

(1) the proper use of expert judgement means far more than simply adopting *ipso facto* a particular point of view which happens to have been expressed by an 'expert'; and
(2) even where formal procedures are used for making some decisions, other, similarly significant, decisions are very often afforded no special attention at all.

These points are both of considerable importance.

In the civil nuclear power industry, the way in which expert judgement is employed to come to decisions on all issues involving uncertainty is a matter which has long been a topic of considerable concern to the Nuclear Safety Directorate of HSE in the UK, and to their American counterparts, the United States Nuclear Regulatory Commission (US NRC).

Nowadays, regulatory authorities expect not only the process by which expert judgement is elicited and different opinions resolved to be satisfactory, they also require appropriate attention to be paid to the reporting of that process (in terms of an auditable 'trail') because legitimate divergence of opinion and lack of scientific consensus have repercussions on regulatory decision-making (Kelly *et al.*, 1991).

Perhaps more than anywhere else, it is in regulatory areas that the search for formal methods of eliciting expert judgement and aggregating individual opinions has been motivated, with a need to find ways of handling divergent opinions within safety cases and for underpinning risk-informed policy setting.

4.4.2 Scientific uncertainty and policy setting

While it should be self-evident that uncertainty is an unavoidable ingredient in every instance of science-informed decision-making, this is not always recognised adequately, and can involve challenging complications for scientists and policy-makers alike. For instance, on one side there may be very knowledgeable scientists who are reluctant to offer opinions or expert judgements when the topic entails direct and important societal decisions which they may consider marginal to their expertise or inimical to their independence.

With scientists, too, there can be a reluctance that is best described as aversion to 'epistemic risk' – the fear of being proved 'wrong' (Parascandola, 2010). Empirical scientists, in particular, traditionally have taken a conservative approach towards epistemic risk by minimising the extent to which they push inferences beyond observations. Twentieth-century scientists and philosophers developed methods for managing epistemic risk within the scientific endeavour, by quantifying degrees of uncertainty and evidential support. In the biomedical sciences, for example, the principal strategy has been the avoidance of Type I error (i.e. rejecting the null hypothesis when it is true, or finding a false positive in other test situations), using the p-value and some statistical significance level as a criterion (usually $p < 0.05$).

Nearly all scientists (with some unscrupulous exceptions) wish to avoid a situation where fallacies or errors turn out to be present in their work. So, for some, there can be a reluctance to state something which might be crucial for a decision-taker because they fear it may be overturned eventually by new data or evidence. For tricky problems, academic scientists sometimes have the luxury of suspending judgement to mitigate epistemic risk but, once taken outside science into, say, wider society or a policy forum, this aversion can lead to bias, or even advice paralysis, when the issue concerns extremely low-probability events with significant consequences. In such circumstances, data are inevitably sparse and frequently the decision to be made is stark and dichotomous – evacuate the population or not – and scientific conservatism may be a mistaken strategy. The weakness of evidence linking special volcano-seismic events called tornillos to explosive eruptions at Galeras volcano, Colombia, prevented their significance being appreciated in 1993, with fatal results for scientists in the crater; the difficulty was one of weighing the possible implications of novel and meagre evidence, with large uncertainties, in a non-scientific context (Aspinall *et al.*, 2003).

There is an almost endless list of life-and-death situations where such science-related concerns apply: approving a clinical trial for a new drug for human use; directing research in biomedicine, where ethical and controversial issues arise and deciding costly but only partially effective measures for the mitigation of natural hazards. It is, therefore, imperative in all such cases that scientific uncertainty is rationally evaluated, and not exaggerated or dismissed to minimise epistemic risk, nor left untreated for decisions to fall prey to the precautionary principle.

The counterpart to epistemic risk in the domain of the decision-maker is 'volitional risk'. Put crudely, this risk amounts to weighing the likelihood that a decision will appear foolish or stupid post hoc – an aversion surely endemic in politicians. In the face of this risk, a decision-taker will be strongly influenced to avoid putting himself or herself in a position where their judgement can be questioned. To do so, some decision-takers may choose to discount the implications of scientific uncertainty in a risk assessment or, as noted earlier, find a way to offload responsibility onto the hapless scientist. Others may reject uncertainty considerations because of fears they will undermine confidence in policy decisions or, where scientific doubt clearly exists, open up related regulations to legal challenge.

While volitional risk has been aired in the context of the recent banking crisis (where usual restraining effects were seemingly offset by 'moral hazard' – the prospect of being not allowed to fail), considerations of epistemic and volitional risk have received little attention for their role in, and influence on, the provision of scientific advice; this is a topic that might merit research.

Formalised expert elicitation, when it utilises a performance-based pooling algorithm, can help bridge these various positions and different perspectives, and can strengthen confidence in, and the credibility of, scientific advice in a policy arena. As just noted, politicians in particular are averse to uncertainty or ambiguity, so when expert elicitations indicate margins of uncertainty on scientific questions that are wider than hoped for, that information can be difficult to digest. For example, expert assessments of real-world problems with strong political ramifications, such as SARS exposure risks to healthcare workers or vCJD infectivity from blood products (e.g. Tyshenko *et al.*, 2011), often manifest more uncertainty about scientific knowledge than individual experts might express or problem owners might infer. This said, such exercises are found beneficial by problem owners; in the case of assessing risks due to vCJD infection via blood products, the organisers of a follow-on policy workshop indicated strong approbation of the method.[11]

In addition to the valuable attributes listed above, a structured elicitation habitually identifies certain key issues that can get overlooked with other decision-support procedures.

Another important trait of the performance-based Classical Model procedure is that the estimated overall uncertainty on a particular scientific judgement from a group of experts is often smaller than that derived from equal weights pooling (see, e.g. Figure 4.1), thereby increasing credibility with policy-makers. But providing a candid, unbiased presentation of the true spread of expert uncertainty remains essential for the decision process. Such uncertainty cannot be reduced to zero, even if individual experts in oracular mode (Section 4.2.1) may be inclined to play it down for self or professional aggrandisement.

If an elicitation has been geared purposely to quantifying scientific uncertainty as reliably as possible – given the nature of the problem and the current state of understanding – then communicating this to a problem owner as an authentic expression of the true extent of uncertainty can sometimes lead to acceptance issues: the scientific message, for his or her purposes, may be less clear-cut than hoped for!

4.5 Future directions

While different forms of structured expert elicitations have been tried in certain difficult regulatory decision domains – where scientific or engineering uncertainty is a fundamental element – such elicitations are also starting to take root in other areas, such as climate-change impact on fisheries (Rothlisberger *et al.*, 2009), and invasive species threats (Neslo *et al.*, 2009). Recently, the US EPA formed a task force (EPA, 2009) to recommend ways of using expert elicitations in regulatory issues whenever data are sparse, such as deciding dose criteria for air quality standards; some pioneering expert judgement work has been applied in this context (Tuomisto *et al.*, 2008).

This expanding world of structured elicitation of experts is bringing to light a number of topics meriting direct research on the methodologies concerned.

---

[11] 'The primary benefit being that the [elicitation] exercise itself communicated several previously undetermined values with ranges for infectious disease uncertainty gaps. Elicitation results targeted iatrogenic transmission through surgical instruments reuse, genotype effects, incubation times, infectivity of blood and a ranking of various iatrogenic routes. The usefulness of the structured expert elicitation that uses a mathematical formalism was seen as a highly useful process that could be applied to other public health risk issues. During the open discussion the benefits identified by the workshop participants included its: 1) utility to help prioritize issues within and between different government departments; 2) application for emerging near-term threats such as pandemic influenza, swine flu and avian flu; 3) utility for future, longer-term foresight exercises; 4) value as a process in increasing overall transparency; 5) ability to reduce bias in decision making; 6) use as a tool to provide proxy values for risk assessors until evidence-based data become available; 7) capacity as a recorded exercise to help support policies decisions; 8) application to improve risk assessments and risk modelling for which data may be missing, inadequate or questionable; and 9) usefulness in delivering values in a shorter time for issues with high uncertainty.' Accessed January 2010 on the PRIONET website, http://www.prionetcanada.ca/

4.5.1 Research on expert judgement modelling

The following are just a few sample topics relating to expert elicitations that merit further research.

4.5.1.1 Scoring rules

Interest in single-variable scoring rules for combination (as opposed to the Classical Model, which is a grouped-variable scoring rule) is reviving after some of Hora's earlier publications. For instance, in 2004, Hora showed that an equal weight combination of well-calibrated experts is not well calibrated. Lin and Cheng (2008) have started using the Brier score, a forecast skill metric from meteorology. Wisse (2008) developed a linear Bayes moments method approach, advocating expectations as a basis, rather than probabilities; partly this was to reduce the computational burden that decision models involving continuous probability distributions carry, although this latter method has, thus far, failed to gain traction in practice.

4.5.1.2 Seed item choice

Further research would also be helpful into what properties make a set of seed questions suitable for calibration. Does an expert's performance on the seed questions faithfully predict their performance when it comes to the real problem under discussion? This is difficult to study with probabilistic predictions, such as in the volcano eruption problem, but appropriate seed questions need to enhance the credibility of the final result.

4.5.1.3 Empirical expert judgement studies

There have been a few, albeit limited, efforts to use the TU Delft database as a research resource, e.g. in the RESS special issue (Cooke, 2008). However, Lin and Cheng (2009) have been following this up. There is much more still to do here, especially in studying the properties of remove-one-at-a-time cross validation. Also needed are more published 'stories from the trenches' about how to configure an expert elicitation problem, how to structure the expert group – analyst interaction, how to provide feedback to the experts and problem owners, and finally, how to communicate uncertainty to the various stakeholders.

4.5.1.4 Fitting models with expert judgement

In nuclear safety, various authorities base radiation exposure guidelines on biological transport models, foreseen with numerous transport coefficients. However, the following problem presented itself: formal uncertainty analysis required a joint distribution over these transfer coefficients, but the experts involved were unable or unwilling to quantify their uncertainty over these parameters, claiming these were too far removed from their empirical knowledge. The solution was to quantify uncertainty just on certain observable quantities, which could be predicted by the models, and then pull this uncertainty back onto the model parameters in a process known as probabilistic inversion. The potential application of such techniques is huge, and there is much room for advancing the procedure where expert judgements are involved. Kraan and Bedford (2005), and Kurowicka and Cooke (2006) treat probabilistic inversion in detail.

4.5.1.5 Dependence modelling/dependence elicitation

Another issue that emerged in the nuclear safety work is dependence elicitation. Prior to this, dependence in elicited uncertainties was largely ignored, as if all important dependencies were captured in functional relationships. Of course, this is not remotely true, as best illustrated with a few examples from the Joint USNCR, ERU Studies:

- the uncertainties in effectiveness of supportive treatment for high radiation exposure in people over 40 and in people under 40;
- the amount of radioactivity after one month in the muscle of beef and dairy cattle;
- the transport of radionuclides through different soil types.

The joint study had to break new ground in dependence modelling and dependence elicitation, and these subjects are treated extensively in the documentation (Goossens and Harper, 1998). The format for eliciting dependence was to ask about joint exceedence probabilities: 'Suppose the effectiveness of supportive treatment in people over 40 was observed to be above the median value, what is your probability that also the effectiveness of supportive treatment in people under 40 would be above its median value?' Experts became quickly familiar with this format. Dependent bivariate distributions were found by taking the minimally informative copula which reproduced these exceedence probabilities, and linking these together in a Markov tree structure. These techniques can and should be developed further.

4.5.1.6 Stakeholder preference modelling

A sharp distinction can be made between uncertainty quantification and preference modelling. The latter has become an active topic of late. Expert discrete choice data can be elicited in the form of rankings or pair-wise comparisons, and probabilistic inversion used to fit some model, typically a multicriteria decision model (MCDM). These models represent the value of an alternative as a weighted sum of scores on various criteria. With MCDM, a distribution over criteria weights is obtained that optimally recovers distributions of stakeholder rankings. The discrete choice data are split into a training set and a validation set, thereby enabling out-of-sample validation. There are many interesting research issues here, including how best to configure the discrete choice elicitation format, and how best to perform validation.

4.5.2 Research on elicitation methodology

There are several open research questions on expert elicitation methodological issues that will need to be answered if this fledgling field is to build its body of practice and become recognised more widely as a robust methodology. French (2011) provides a valuable and timely review.

4.5.2.1 Expert panel selection

For example, the selection and number of experts to compose a group for elicitation needs careful thought. The problem owner may want the selection to be as impartial and independent as possible to prevent criticism of bias, or they may restrict the selection to company employees to ensure confidentiality and relevance.

To-date, experience with expert panels suggests that 8–15 experts is a reasonable, and viable, number for tackling a given problem, but this has not been rigorously tested. With more than 20 people involved, a situation of diminishing returns seems to ensue; a limited-size group, selected meticulously by the problem owner, can capture the wider collective view without compromising informativeness or calibration accuracy, and thus helps constrain investment of people's time and commitment in an elicitation exercise.

4.5.2.2 New protocols for distance elicitation

To-date, significant departures from established elicitation protocols have not been a big issue. This said, the existing procedures can be labour intensive and group elicitations, in particular, can be time-consuming and, in the case of a global discipline such as volcanology, often difficult to convene for reasonable cost. Eventually, opportunities to take advantage of new media technologies and to find ways and means of using them to conduct tele-elicitations – without loss of formalism and structure – will need to be explored. Then, a good understanding of, and perhaps training in, the principles and concepts will be even more essential – for facilitators and experts, alike.

4.5.2.3 Target item question framing

Sometimes an elicitation reveals two camps of opinion within a group, making it difficult to derive a rational consensus value. In such cases, the dichotomy usually arises either from an ambiguity in question framing, or because the two groups have different experience references. Detecting that this condition may exist within a group elicitation is a powerful diagnostic property of the structured elicitation process, and the difficulty can usually be resolved by facilitated discussion or by reframing the target question. It is very common to encounter the need for such clarification among a group of experts, however close-knit they are in terms of specific scientific interests (e.g. seismologists) or professional training (e.g. airline pilots).

4.5.2.4 Other methodology issues

Additional existing concerns that deserve research include:

(1) Could training experts improve their performance?
(2) What are the pros and cons of different structures for conducting an elicitation?

(3) Can more qualitative aspects of human behaviour (e.g. social vulnerability information) be factored into assessments of risk using expert judgement?

(4) Can expert elicitation be studied usefully in terms of group dynamics and psychology?

Lastly, are there alternatives to quantitative elicitation based on the use of a few defining quantiles – e.g. ranking, ordering or probabilistic inversion? And under what circumstances would such alternative formulations or frameworks be regarded as more appropriate than the quantitative distributional approach?

4.5.3 Teaching and expert learning

It is tempting to think that an expert might be able to improve his or her expert judgement performance in the Classical Model sense through participating in elicitations. Conventionally, expert learning is not a proper goal for an expert judgement elicitation; rather, the problem owner should want experts who are proficient in judging uncertainties, independent of the elicitation process itself. Nor, in principle, should they be learning the fundamentals of the subject-matter from the elicitation. However, in discussing and framing target questions in the preamble to elicitation, experts may develop a better sense of the relevant uncertainties from a facilitated exchange of new information, status reviews and data summaries.

If, after responding to seed questions, an expert were to learn that all, or many of the actual realisations fell outside his 90% confidence intervals, he might conclude that his intervals had been too narrow, and try to broaden them in subsequent assessments. Weak anecdotal evidence from repeated uncertainty assessments – as in the volcano case above – suggests that, with a few notable exceptions, self-directed training in this regard is difficult: expert leopards do not find it easy to change their uncertainty spots!

Perhaps this is right and proper, when experts are enjoined to express their true beliefs, but as applications of the Classical Model method proliferate, expert learning in this context should become a topic meriting detailed research. New findings could inform approaches to teaching and training.

Some recent pioneering efforts have been made towards teaching the principles of expert elicitation techniques at undergraduate and post-graduate level (Stone *et al.*, 2011). This limited experience suggests that introducing the concepts to young students, letting them actively apply the elicitation procedures and use performance analysis algorithms, such as the Classical Model, are a powerful way of schooling them in how demanding it can be to evaluate scientific uncertainty convincingly, reliably and defensibly.

It is clear that concepts like calibration, entropy and mutual information – and their roles in quantifying scientific uncertainty in a decision support context – are widely unfamiliar to the majority of scientists, young and old alike, so training and teaching initiatives are needed.

4.6 Summing up

Notwithstanding all the diversity of topics touched on above, and the many excursions that the discussion has taken, what can be said with confidence is that combining a structured elicitation of expert opinion with a formalised scoring process is an affirmative way of catalysing scientific discussion of uncertainty, and provides a rational approach for characterising and quantifying hazard and risk assessments in a robust way. Structured elicitation procedures can help identify, pose and respond to many natural hazards questions where scientific uncertainty is pervasive: our judgement is that the range and compass of the approach is almost unlimited.

Experience indicates also that the structured elicitation process that feeds into formalised judgement pooling with the Classical Model is a highly effective way of framing scientific debates, and that some mutual learning takes place among a group with consequent improved generic understanding of uncertainty. These are central and valuable features of this particular expert elicitation procedure, helping improve scientific advice for decision-making.

As mentioned earlier in Section 4.2.2, one impediment to the wider take-up of such procedures can be the cost of an elicitation in terms of human time and resources, and funding for participation, travel and meetings. The time of high-level specialists, especially those internationally recognised, is always at a premium. Thus cost and time frequently prove practical limitations, with problem owners often reluctant to furnish sufficient funding to do the job properly and comprehensively. Moreover, with many important science-based decision issues it can be a challenge both to constrain a problem owner's expectations to manageable proportions, and to rein in the experts' almost inevitable desires to tackle a wide scope and large number of target questions. Extensive experience indicates that a well-designed exercise with a practised facilitator can tackle 10–30 scientifically tricky target items over a two-day period; attempting more may entail 'elicitation fatigue'. This said, the approach has been used successfully in more than 50 studies and is rapidly gaining academic credibility, professional acceptance and even political standing.

Acknowledgements

Our thanks go to R. S. J. Sparks in particular, and to J. C. Rougier and J. Freer (University of Bristol) and G. Woo (Risk Management Solutions Ltd.) for discussions, suggestions, improvements and generous encouragement. As with any errors, the views expressed are those of the authors, and should not be construed as representing the opinions or positions of any group or organisation with whom they may be or have been associated. WPA was supported in part by a European Research Council grant to Prof. R. S. J. Sparks FRS: VOLDIES Dynamics of Volcanoes and their Impact on the Environment and Society.

References

Abrahamson, N. A., Coppersmith, K. J., Koller, M., *et al.* (2004) Probabilistic seismic hazard analysis for Swiss nuclear power plant sites, PEGASOS Project, Vols 1–6, NAGRA.

ACERA (2007) Eliciting expert judgments: literature review. http://www.acera.unimelb. edu.au/materials/endorsed/0611.pdf (accessed 28 December 2011).

ACERA (2010) Elicitation tool: process manual. http://www.acera.unimelb.edu.au/materi-als/software.html (accessed 28 December 2011).

Aspinall, W. P. (2006) Structured elicitation of expert judgment for probabilistic hazard and risk assessment in volcanic eruptions. In *Statistics in Volcanology*, ed. H. M. Mader, S. G. Coles, C. B. Connor and L. J. Connor, London: The Geological Society for IAVCEI, pp. 15–30.

Aspinall, W. (2010) Opinion: a route to more tractable expert advice. *Nature* **463**: 294–295.

Aspinall, W. and Cooke, R. M. (1998) Expert judgment and the Montserrat Volcano eruption. In *Proceedings, 4th International Conference on Probabilistic Safety Assessment and Management PSAM4*, ed. Ali Mosleh and Robert A. Bari, vol. 3, New York, NY: Springer, pp. 2113–2118.

Aspinall, W. P., Loughlin, S. C., Michael, A. D., *et al.* (2002) The Montserrat Volcano Observatory; its evolution, organization, role and activities. In *The Eruption of Soufrière Hills Volcano, Montserrat from 1995 to 1999*, ed. T. H. Druitt and B. P. Kokelaar, London: Geological Society of London, pp. 71–91.

Aspinall, W. P., Woo, G., Voight, B., *et al.* (2003) Evidence-based volcanology; application to eruption crises. *Journal of Volcanology and Geothermal Research: Putting Volcano Seismology in a Physical Context; In Memory of Bruno Martinelli* **128**: 273–285.

Brockhoff, K. (1975) The performance of forecasting groups in computer dialogue and face to face discussions. In *The Delphi Method, Techniques and Applications*, ed. H. A. Linstone and M. Turoff, Reading, MA: Addison Wesley, pp. 291–321.

Brown, A. J. and Aspinall, W. P. (2004) Use of expert elicitation to quantify the internal erosion processes in dams. In *Proceedings of the British Dam Society Conference: Long-term Benefits and performance of Dams*, Canterbury: Thomas Telford, pp. 282–297.

Budnitz, R., Apostolakis, G., Boore, D. M., *et al.* (1998) Use of technical expert panels: applications to probabilistic seismic hazard analysis. *Risk Analysis* **18**: 463–469.

Budnitz, R. J., C. A. Cornell and Morris, P. A. (2005) Comment on J.U. Klugel's 'Problems in the application of the SSHAC probability method for assessing earthquake hazards at Swiss nuclear power plants,' in Engineering Geology, vol. 78: 285–307. *Engineering Geology* **82**: 76–78.

Clemen, R. T. and Reilly, T. (1999) Correlations and copulas for decision and risk analysis. *Management Science* **45**: 208–224.

Clemen, R. T. and Winkler, R. L. (1986) Combining economic forecasts. *Journal of Business & Economic Statistics* **4** (1): 39–46.

Clemen, R. T. and Winkler, R. L. (1987) Calibrating and combining precipitation probability forecasts. In *Probability & Bayesian Statistics*, ed. Viertl, R.., New York, NY: Plenum Press, pp. 97–110.

Clemen, R. T. and Winkler, R. L. (1999) Combining probability distributions from experts in risk analysis. *Risk Analysis* **19**: 187–203.

Cooke, R. (1987) A theory of weights for combining expert opinions. Delft University of Technology, Department of Mathematics and Informatics, Report 87–25.

Cooke, R. M. (1991) *Experts in Uncertainty:Opinion and Subjective Probability in Science.* Oxford: Oxford University Press.

Cooke, R. M. (2008) Special issue on expert judgment. *Reliability Engineering & System Safety* **93**: 655–656.

Cooke, R. M. and Goossens, L. L. H. J. (2008) TU Delft expert judgment data base. *Reliability Engineering & System Safety* **93**: 657–674.

Cooke, R. M., Mendel, M. and Thijs, W. (1988) Calibration and information in expert resolution. *Automatica* **24**: 87–94.

Cooke, R. M., ElSaadany, S. and Huang, X. (2008) On the performance of social network and likelihood-based expert weighting schemes. *Reliability Engineering & System Safety* **93**: 745–756.

Coppersmith, K. J., Youngs, R. R. and Sprecher, C. (2009) Methodology and main results of seismic source characterization for the PEGASOS Project, Switzerland. *Swiss Journal of Geosciences* **102**: 91–105.

De Groot, M. (1974) Reaching a consensus. *Journal of the American Statistical Association* **69**: 118–121.

Delbecq, A. L., Van de Ven, A. and Gusstafson, D. (1975) *Group Techniques for Program Planning.* Glenview, IL: Scott, Foresman.

EPA (2009) Expert elicitation white paper: external review draft, January. http://www.epa. gov/spc/expertelicitation/index.htm (accessed 6 November 2009).

Fischer, G. (1975) An experimental study of four procedures for aggregating subjective probability assessments. Technical Report 75–7, Decisions and Designs Incl., McLean, Virginia.

Flandoli, F., Giorgi, E., Aspinall, W. P., *et al.* (2011) Comparison of a new expert elicitation model with the Classical Model, equal weights and single experts, using a cross-validation technique. *Reliability Engineering & System Safety* **96**: 1292–1310.

French, S. (2008) Comments by Prof. French. *Reliability Engineering & System Safety Expert Judgment* **93**: 766–768.

French, S. (2011) Aggregating expert judgment. *Revista de la Real Academia de Ciencias Exactas, Fisicas y Naturales. Serie A. Matematicas (RACSAM)* **105**: 181–206.

Gigerenzer, G., Hertwig, R., van den Broek, E., *et al.* (2005) 'A 30% chance of rain tomorrow': how does the public understand probabilistic weather forecasts? *Risk Analysis* **25**: 623–629.

Goossens, L. H. J. and Harper, F. T. (1998) Joint EC/USNRC expert judgment driven radiological protection uncertainty analysis. *Journal of Radiological Protection* **18**: 249–264.

Gough, R. (1975) The effect of group format on aggregate subjective probability distributions. In *Utility, Probability and Human Decision Making*, ed. D. Wendt and C. Vlek, Dordrecht, Reidel.

Hanks, T. C., Abrahamson, N. A., Boore, D. M., *et al.* (2009) Implementation of the SSHAC Guidelines for Level 3 and 4 PSHAs: experience gained from actual applications. US Geological Survey Open-File Report 2009–1093. http://pubs.er.usgs.gov/publication/ofr20091093 (accessed 27 July 2012).

Helmer, O. (1966) *Social Technology*, New York, NY: Basic Books.

Helmer, O. (1968) Analysis of the future: the Delphi method, and The Delphi Method–an illustration. In *Technological Forecasting for Industry and Government*, ed. J. Bright, Englewood Cliffs, NJ: Prentice Hall.

Hora, S. C. (2004) Probability judgments for continuous quantities: linear combinations and calibration. *Management Science* **50**: 597–604.

Hora, S. C. and Iman, R. L. (1989) Expert opinion in risk analysis: the NUREG-1150 experience. *Nuclear Science and Engineering* **102**: 323–331.

IAEA (2010) *Seismic Hazards in Site Evaluation for Nuclear Installations.* Vienna: International Atomic Energy Agency.

Kahneman, D. (2011) *Thinking, Fast and Slow,* London: Allen Lane.

Kahneman, D., Slovic, P. and Tversky, A. (eds) (1982) *Judgment Under Uncertainty: Heuristics and Biases,* Cambridge: Cambridge University Press.

Kallen, M. J. and Cooke, R. M. (2002) Expert aggregation with dependence. In *Probabilistic Safety Assessment and Management,* ed. E. J. Bonano, A. L. Camp, M. J. Majors, R. A. Thompson, London: Elsevier, pp. 1287–1294.

Kelly G. B., Kenneally R. M. and Chokshi N. C. (1991) Consideration of seismic events in severe accidents. In *Proceedings of Probabilistic Safety Assessment and Management* **2**, pp. 871–876.

Klügel, J.-U. (2005a) Problems in the application of the SSHAC probability method for assessing earthquake hazards at Swiss nuclear power plants. *Engineering Geology* **78**: 285–307.

Klügel, J.-U. (2005b) Reply to the comment of Krinitzsky on J.U. Klugel's 'Problems in the application of the SSHAC probability method for assessing earthquake hazards at Swiss nuclear power plants', in Engineering Geology, vol. 78: 285–307. *Engineering Geology* **82**: 69–70.

Klügel, J.-U. (2005c) Reply to the comment on J.U. Klugel's 'Problems in the application of the SSHAC probability method for assessing earthquake hazards at Swiss nuclear power plants', in Engineering Geology, Vol. 78: 285–307, by Budnitz, by, J.U, Klugel. *Engineering Geology* **82**: 79–85.

Klügel, J.-U. (2005d) Reply to the comment on J.U. Klugel's: 'Problems in the application of the SSHAC probability method for assessing earthquake hazards at Swiss nuclear power plants,' Eng. Geol. Vol. 78: 285–307, by Musson *et al. Engineering Geology* **82**: 56–65.

Klügel, J.-U. (2005e) Reply to the comment on J. U. Klugel's: 'Problems in the application of the SSHAC probability method for assessing earthquake hazards at Swiss nuclear power plants,' in Engineering Geology, vol. 78: 285–307, by Lomnitz, by J.U. Klugel. *Engineering Geology* **82**: 74–75.

Klügel, J.-U. (2005f) Reply to the comment on J. U. Klugel's: Problems in the application of the SSHAC probability method for assessing earthquake hazards at Swiss nuclear power plants, in Engineering Geology, Vol. 78: 285–307, by Wang, by J.U, Klugel. *Engineering Geology* **82**: 89–90.

Klügel, J.-U. (2007) Error inflation in probabilistic seismic hazard analysis. *Engineering Geology* **90**: 186–192.

Klügel, J.-U. (2009) Probabilistic seismic hazard analysis for nuclear power plants: current practice from a European perspective. *Nuclear Engineering and Technology* **41**: 1–12.

Klügel, J. U., Mualchin, L. and Panza, G. F. (2006) A scenario-based procedure for seismic risk analysis. *Engineering Geology* **88**: 22.

Kotra, J. P., Lee, M. P., Eisenberg, N. A., *et al.* (1996) *Branch Technical Position on the Use of Expert Elicitation in the High-Level Radioactive Waste Program.* NUREG-1563. Washington, DC: US Nuclear Regulatory Commission.

Kraan, B. C. P. and Bedford, T. J. (2005) Probabilistic inversion of expert judgments in the quantification of model uncertainty. *Management Science* **51** (6): 995–1006.

Krinitzsky, E. L. (2005) Comment on J.U. Klugel's 'Problems in the application of the SSHAC probability method for assessing earthquake hazards at Swiss nuclear power plants', in Engineering Geology, vol. 78: 285–307. *Engineering Geology* **82**: 66–68.

Kurowicka, D. and Cooke, R. (2006) *Uncertainty Analysis with High Dimensional Dependence Modelling*. Chichester: John Wiley and Sons.

Kynn, M. (2008) The 'heuristics and biases' in expert elicitation. *Journal of the Royal Statistical Society: Series A* **171**: 239–264.

Lehrer, K. and Wagner, C. G. (1981) *Rational Consensus in Science and Society*. Dordrecht: Reidel.

Lin, S.-W. and Bier, V. M. (2008) A study of expert overconfidence. *Reliability Engineering & System Safety* **93**: 775–777.

Lin, S.-W. and Cheng, C.-H. (2008) Can Cooke's Model sift out better experts and produce well-calibrated aggregated probabilities? 2008 *IEEE International Conference on Industrial Engineering and Engineering Management*, Singapore: IEEE Singapore Section, pp. 425–429.

Lin, S.-W. and Cheng, C.-H. (2009) The reliability of aggregated probability judgments obtained through Cooke's classical model. *Journal of Modelling in Management* **4**: 149–161.

Lindley, D. V. (1971) *Making Decisions*, London: John Wiley and Sons.

Lindley, D. (1983) Reconciliation of probability distributions. *Operations Research* **31**: 866–880.

Lomnitz, C. (2005) Comment on J.U. Klugel's 'Problems in the application of the SSHAC probability method for assessing earthquake hazards at Swiss nuclear power plants', in Engineering Geology, vol. 78: 285–307. *Engineering Geology* **82**: 71–73.

Mendel, M. and Sheridan, T. (1989) Filtering information from human experts. *IEEE Transactions on Systems, Man and Cybernetics* **19** (1): 6–16.

Meyer, M. A. and Booker, J. M. (2001) *Eliciting and Analyzing Expert Judgment: A Practical Guide*. Philadelphia, PA: ASA-SIAM.

Morgan, M. G. and Henrion, M. (1990) *Uncertainty: A Guide to Dealing with Uncertainty in Quantitative Risk and Policy Analysis*. New York, NY: Cambridge University Press.

Morris, P. (1974) Decision analysis expert use. *Management Science* **20**: 1233–1241.

Morris, P. (1977) Combining expert judgments: a Bayesian approach. *Management Science* **23**: 679–693.

Mosleh, A. and Apostolakis, G. (1982) Models for the use of expert opinions. Workshop on low probability high consequence risk analysis, Society for Risk Analysis, Arlington, VA, June.

Musson, R. M. W., Toro, G. R., Coppersmith, K. J., *et al.* (2005) Evaluating hazard results for Switzerland and how not to do it: A discussion of 'Problems in the application of the SSHAC probability method for assessing earthquake hazards at Swiss nuclear power plants' by J-U Klugel. *Engineering Geology* **82**: 43–55.

Neslo, R., Micheli, F., Kappel, C. V., *et al.* (2009) Modeling stakeholder preferences with probabilistic inversion: application to prioritizing marine ecosystem vulnerabilities. In *Real Time and Deliberative Decision Making: Application to Risk Assessment for Non-chemical Stressors*, ed. I. Linkov, E. A. Ferguson and V. S. Magar, Amsterdam: Springer: pp. 265–284.

O'Hagan, A., Buck, C. E., Daneshkhah, A., *et al.* (2006) *Uncertain Judgments: Eliciting Experts' Probabilities*. Chichester: John Wiley and Sons.

Parascandola, M. (2010) Epistemic risk: empirical science and the fear of being wrong. *Law, Probability and Risk* **9**: 201–214.

Rothlisberger, J. D., Lodge, D. M., Cooke, R. M., *et al.* (2009) Frontiers in ecology and the environment. *The Ecological Society of America*. www.frontiersinecology.org (accessed 27 July 2012).

Sackman, H. (1975) *Delphi Critique, Expert Opinion, Forecasting and Group Processes*, Lexington, MA: Lexington Books,.

Seaver, D. (1977) How groups can assess uncertainty. *Proceedings of the International Conference on Cybernetics and Society*, New York: Institute of Electrical and Electronics Engineers, pp. 185–190.

Seaver, D. A. (1978) Assessing probabilities with multiple individuals: group interaction versus mathematical aggregation. Technical Report SSRI-78–3, Social Science Research Institute, University of Southern California, Los Angeles.

Skipp, B. O. and Woo, G. (1993). A question of judgement: expert or engineering? In *Risk and Reliability in Ground Engineering*, London: Thomas Telford, pp. 29–39.

SSHAC (1995) *Recommendations for Probabilistic Seismic Hazard Analysis: Guidance on Uncertainty and Use of Experts; prepared by Senior Seismic Hazard Analysis Committee (SSHAC)*, Livermore, CA: Lawrence Livermore National Laboratory.

SSHAC (1997) [Senior Seismic Hazard Analysis Committee, R. J. Budnitz, Chairman, G. Apostolakis, D. M. Boore, L. S. Cluff, K. J. Coppersmith, C. A. Cornell and P. A. Morris] *Recommendations for Probabilistic Seismic Hazard Analysis: Guidance on Uncertainty and Use of Experts*, NUREG/CR-6372, Washington, DC: US Nuclear Regulatory Commission.

Stone, J., Aspinall, W. P. and Sparks, R. S. J. (2011) Poster presented at Soufriere Hills Volcano: 15 Years on Conference, Montserrat, West Indies, April.

Tuomisto, J. T., Wilson, A., Evans, J. S., *et al.* (2008) Uncertainty in mortality response to airborne fine particulate matter: combining European air pollution experts. *Reliability Engineering & System Safety Expert Judgment* **93**: 732–744.

Tversky, A. and Kahneman, D. (2005) Judgment under uncertainty: heuristics and biases. In *Social Cognition Key Readings*, ed. D. L. Hamilton, New York, NY: Psychology Press, chapter 10.

Tyshenko, G. M., ElSaadany, S., Oraby, T., *et al.* (2011) Expert elicitation for the judgment of Prion disease risk uncertainties. *Journal of Toxicology and Environmental Health, Part A: Current Issues* **74**: 261–285.

Vick, S. G. (2002) *Degrees of Belief: Subjective Probability and Engineering Judgment*, Reston, VA: ASCE Press.

Wang, Z. (2005) Comment on J.U. Klugel's: Problems in the application of the SSHAC probability method for assessing earthquake hazards at Swiss nuclear power plants, in Engineering Geology, vol. 78: 285–307. *Engineering Geology* **82**: 86–88.

Winkler, R. L. (1981) Combining probability distributions from dependent information sources. *Management Science* **27**: 479–488.

Wisse, B., Bedford, T. and Quigley, J. (2008) Expert judgement combination using moment methods. *Reliability Engineering & System Safety Expert Judgement* **93**: 675–686.

Woo, G. (1999) *The Mathematics of Natural Catastrophes*, River Edge, NJ: Imperial College Press.

Woo, G. (2011) *Calculating Catastrophe*, River Edge, NJ: Imperial College Press.

Woudenberg, F. (1991) An evaluation of Delphi. *Technological Forecasting and Social Change* **40**: 131–150.

Yunusov, A. R., Cooke, R. M. and Krymsky, V. G. (1999) Rexcalibr-integrated system for processing expert judgment. In *Proceedings 9th Annual Conference: Risk Analysis: Facing the New Millenium*. Rotterdam, 10–13 October, ed. L. H. J. Goossens, Netherlands: Delft University, pp. 587–589.

Zuidema, P. (2006) PEGASOS: A PSHA for Swiss nuclear power plants–some comments from a managerial point of view. OECD-IAEA Specialist meeting on the Seismic Probabilistic Safety Assessment of Nuclear Facilities, Seogwipo, Jeju Island, Korea.

5

Risk and uncertainty in hydrometeorological hazards

T. L. EDWARDS AND P. G. CHALLENOR

5.1 Introduction

Extreme weather and ocean hazards are inescapable. Weather is the state of the atmosphere, to which every part of the earth's surface is exposed. In coastal regions and at sea there are equivalent oceanic hazards, such as extreme waves. Every region has a local normal range for atmospheric and oceanic, or 'hydrometeorological', conditions, around which societal decisions such as habitation and infrastructure are made. But inevitably there are fluctuations outside these bounds that result in hazard events. In general it is too difficult and costly to be fully prepared for the rare extremes.

This chapter summarises the present-day risks and uncertainties associated with droughts, heat waves, extreme precipitation, wind storms and extreme ocean waves. How these may change under future climate change is discussed in Chapter 6. Other meteorological hazards, such as extreme cold, are not included here for reasons of space. Flooding and storm surge hazards are described separately in Chapter 7. Section 5.2 outlines the characteristics of the hazards. Section 5.3 discusses methods of assessing risks for hazards and their impacts. Section 5.4 describes organisations and tools of risk management and communication. Section 5.5 summarises.

5.2 Hydrometeorological hazards

This section gives an overview of the hazards (definition, triggers and scale) and their associated losses (exposure and vulnerability; types and quantification of loss).

5.2.1 Definition

The United Nations International Strategy for Disaster Reduction (http://www.unisdr.org) gives a general definition for a hydrometeorological hazard as a 'process or phenomenon of atmospheric, hydrological or oceanographic nature that may cause loss of life, injury or other health impacts, property damage, loss of livelihoods and services, social and economic disruption, or environmental damage'. Most hydrometeorological hazards are true 'extremes' in the sense that they are not distinct events but a consequence of entering the high or low end of local climatic variation (e.g. IPCC, 2011). Exceptions to this are cyclonic storms (such as hurricanes), which are discrete atmospheric patterns of circulating wind.

Risk and Uncertainty Assessment for Natural Hazards, ed. Jonathan Rougier, Steve Sparks and Lisa Hill. Published by Cambridge University Press. © Cambridge University Press 2013.

Hydrometeorological hazards thus do not have triggers or inception events, though particular atmospheric or oceanic states make them more likely (Section 5.2.2). These 'extremal' characteristics lead naturally to probabilistic risk assessments, quantified in terms of the probability of being in a particular region of climate state space (Chapter 6). The term 'risk' is defined in Chapter 2.

The definition of a hydrometeorological hazard varies by location and by the type of risk assessment. Extremes must be defined relative to regional climatology (mean climate) rather than fixed thresholds, because local acclimatisation and infrastructure affect the seriousness of the impacts. And for risk assessments, definitions range from simple climate-based metrics such as daily maxima, which are used for quantifying the probability of occurrence, to more complex multivariate definitions, for describing impacts on health or infrastructure.

Extreme events are typically defined in terms of threshold exceedance, where thresholds are defined relative to local climatology (such as the 90th percentile of daily maximum temperatures at a given location for a particular time period) so as to quantify events of a fixed rarity. Alternatively, an absolute threshold may be used for all locations so as to quantify events of a fixed intensity. Duration of hazard is also important, because the impacts of extreme events are greater when extreme conditions prevail over extended time periods. The joint CCI/CLIVAR/JCOMM Expert Team on Climate Change Detection and Indices (ETCCDI) recommended 27 indices of extreme temperature and precipitation so as to standardise comparisons globally (Alexander *et al.*, 2006). These have been widely used, in particular for the most recent assessment report by the Intergovernmental Panel on Climate Change (IPCC, 2007a).

The path into the extreme also affects the hazard. A slow increase in dryness before a drought may result in a more severe event than a sudden transition, due to the cumulative (integrated) impacts, while a heat wave at the end of the summer typically results in lower mortality rates than one at the start due to acclimatisation during the season and the 'harvesting' of weak individuals (Baccini *et al.*, 2008). This context is not always considered in risk assessments, partly because sub-dividing extreme data into categories exacerbates any problems of low numbers of observations.

Once a hazard is defined, there are two types of risk assessment: physical model-based forecasting (Section 5.3,2), which is the prediction of specific hazard events in the near future, and statistical analysis (Section 5.3.3), which is the estimation of frequency over the longer term. For the latter, two very commonly used metrics are the 'return period', which is the inverse of the probability of an event with specified magnitude at a given location occurring during the next year, and equivalently the 'return level', the magnitude associated with a given return period. A more precise characterisation is given in Chapter 2. Return periods are defined for a particular hazard magnitude and location, but there are other relevant characteristics such as duration, spatial extent and (in the case of storms) path. The return period may depend on how these aspects are combined (Della-Marta *et al.*, 2009). A return period is a statement of probability at a given time, not a projection for the future; if the data are stationary and independent, the return period can be interpreted as the expected waiting time until the next event. Return periods are usually estimated from the observational record, but are also estimated from re-analysis datasets (observations assimilated into a physical weather model simulation) such as ERA-40 (Uppala *et al.*, 2005) or from stand-alone simulations (Chapter 6).

Each hydrometeorological hazard has a typical definition, or in some cases several, according to the event type and the intended use of the information. These definitions are outlined below.

5.2.1.1 Extreme precipitation

Extreme precipitation events are usually defined simply by their magnitude, as an exceedance of a climatological threshold such as the 99th percentile (e.g. Barnett *et al.*, 2006). The definition may include a measure of persistence such as a five-day total (Beniston *et al.*, 2007). Frequency and ratio of extreme precipitation relative to local climatology are also important due to risk of inundating flood control systems (Chapter 7). Four indices of precipitation are shown in Figure 5.1.

5.2.1.2 Heat waves

Heat waves are persistent periods of high temperature. The persistence is significant for human health risk, particularly for individuals that are not usually vulnerable (e.g. Joacim *et al.*, 2010). Heat wave definitions always vary by region, because humans acclimatise to their local temperature and humidity. The simplest are expressed in terms of the duration of the daily maximum temperature, T_{max}. One commonly used definition is T_{max} exceeding climatology by 5°C for at least five consecutive days, either any such period (Frich *et al.*, 2002) or the longest such period in the year (Tebaldi *et al.*, 2006). A more complex definition

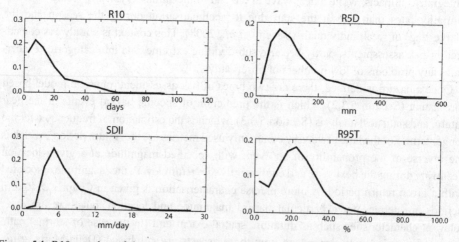

Figure 5.1 R10: number of days with precipitation $\geq$10 mm d^{-1}; SDII: simple daily intensity index: annual total / number of days with precipitation $\geq$1 mm d^{-1}; R5D: maximum five-day precipitation total; and R95T: fraction of annual total precipitation due to events exceeding the 1961–1990 95th percentile. Adapted from Frich *et al.* (2002) © Inter-Research.

using a critical threshold ($T_c = 25°C$ for Central Europe) is used by Kyselý (2009): a continuous period during which mean T_{max} exceeds T_c by 5°C in at least three days, the mean T_{max} for the whole period exceeds T_c by 5°C, and T_{max} does not fall below T_c. Temperature variability is much smaller in some regions than others, so fixed-excess definitions such as these are not appropriate for global studies: instead the excess should be set relative to local climatology (Alexander *et al.*, 2006).

Duration of high T_{max} is generally thought to be insufficient information for assessing human health risks. Persistently high night-time temperatures prevent core body temperatures from cooling, and high humidity levels generally reduce tolerance to heat; Gosling *et al.* (2009) discuss these issues in more detail. So warning systems increasingly use sophisticated metrics more suitable for human health. The Heat Index of the National Oceanic and Atmospheric Administration (NOAA) National Weather Service (NWS) combines temperature and humidity criteria (Robinson, 2001). This is a measure of 'how hot it feels': for example, an air temperature of 96°F (36°C) and relative humidity of 65% results in a Heat Index of 121°F (49°C). 'Synoptic' definitions of heat waves take into account the overall type of meteorological conditions and the previous impact of these conditions on mortality and health. One example used widely in the United States and in parts of Europe is the Spatial Synoptic Classification (SSC; Sheridan, 2002; Sheridan and Kalkstein, 2004). This uses several meteorological variables (temperature, dew point, cloud cover, pressure and wind) to classify weather into one of ten types and quantifies the persistence, relative timing within the summer season and particular meteorological characteristics of that type. Souch and Grimmond (2004) and Smoyer-Tomic *et al.* (2003) describe other heat wave and heat stress indices. Some studies define a heat wave according to the characteristics of a particular event, such as the 2003 European heat wave, to estimate the return period (Schär *et al.*, 2004) and changing risk of such events in the future (Beniston, 2004). However, the number of analogues found is sensitive to the definition of the event: whether in terms of absolute temperature or threshold exceedance, and whether individual or consecutive days (Beniston, 2004).

5.2.1.3 Droughts

Droughts are persistent dry periods that last from a few days to several years. There are various types (IPCC, 2007b; Burke and Brown, 2008), including meteorological drought (low precipitation), hydrological drought (low water reserves), agricultural drought (low soil moisture) and environmental drought (a combination of these). Droughts are quantified using one of various indices. Some are based solely on precipitation, such as the Standard Precipitation Index (SPI) and the China-Z Index (CZI; Wu *et al.*, 2001), while others take into account additional factors such as temperature, potential evaporation, runoff and soil moisture. A commonly used metric of 'cumulative departure of moisture supply' from the normal is the Palmer Drought Severity Index (PDSI: Palmer, 1965). This is based on precipitation, temperature and soil moisture, though has a number of different formulations (Robeson, 2008; Dai, 2011). In fact, more than 150 definitions of drought exist

(Nicolosi *et al.*, 2009). These have different strengths and weaknesses, and the most suitable choice depends on the problem, context and user (Kallis, 2008).

Drought severity is described using thresholds of an index such as the first (extreme), fifth (severe) or twentieth (moderate) percentiles of present-day climate. Different thresholds may be set for each region to account not only for local climatology but also environmental conditions that increase vulnerability. The PDSI is calibrated for the United States and has categories -1.0 to -1.9 for mild drought, -2 to -3 for moderate drought, -3 to -4 for severe drought, and beyond -4 for an extreme drought (Leary *et al.*, 2008).

The rate of onset and cessation of drought are also important. A slow transition to drought may allow more time for mitigation, but could also worsen the impacts due to the cumulative effects before and after the defined drought phase.

5.2.1.4 Wind storms

Wind storms include hurricanes, typhoons, tornados, thunderstorms and extratropical cyclones. Tropical wind storms are 'warm core' cyclones, generated by convection of warm air over warm sea surface waters, while mid-latitude storms are 'cold core' and form along weather fronts. Wind storms are usually defined in terms of wind speed, with an absolute or relative threshold. A hurricane is a tropical storm with wind speeds of 74 mph ($33 \, \mathrm{m \, s^{-1}}$, Category 1 on the Saffir–Simpson scale, or about 12 on the Beaufort scale). 'Storm force' winds in the UK have mean speed greater than 55 mph ($25 \, \mathrm{m \, s^{-1}}$, 10 on the Beaufort scale) and gusts greater than 85 mph ($38 \, \mathrm{m \, s^{-1}}$) affecting most of the country for at least six hours (UK Government Cabinet Office, 'UK Resilience': available from http://www.cabinetoffice.gov.uk). Della-Marta *et al.* (2009) compare five wind-based indices for storm magnitude and spatial extent, finding some are more suitable for quantifying strength relative to climatology and others for absolute strength.

Wind speed is not the only important metric for wind storms. The destruction caused by Hurricane Katrina in the United States in 2005 (Section 5.2.4.3) was due not to high wind speed but slow storm transit speed, heavy rainfall and a massive associated storm surge, as well as the vulnerability of the point at which it struck the coast. Storms are therefore often characterised with additional quantities such as sea-level pressure, vorticity, precipitation, lightning, storm surges or accumulated cyclone energy (ACE) (e.g. Beniston *et al.*, 2007).

Mid-latitude wind storms, or extratropical cyclones, in the Northern Hemisphere occur at a range of spatial scales. They are often studied at 'synoptic' (large) scales, 100–1000 km spatially and 6–72 hours in duration. Extreme synoptic storms have relatively moderate wind speeds compared with hurricanes or tornados, but cause great wind damage and economic loss due to the integrated effects of their large spatial extent. Smaller-scale structures may also develop within a large-scale storm. 'Sting-jets' (Browning, 2004) are characterised by very intense wind gusts and scales of order of tens of kilometres. 'Bombs' are rapidly deepening extratropical cyclones (Sanders and Gyakum, 1980) that develop in isolation from other vortices and are often characterised by spatial scales of around 100 km. Activities at these different spatial scales can interact and develop two-way feedbacks that may lead to explosive intensification and cyclogenesis (Rivière and Orlanski, 2007; Fink *et al.*, 2009).

5.2.1.5 *Extreme waves and currents*

Wind storms over the ocean generate large waves whose height depends on the wind speed, duration and fetch (distance over which the wind blows), and submarine earthquakes cause very large (tsunami) waves, but not all large waves are associated with storms and earthquakes. 'Freak waves' generated by nonlinearities in the wave field itself can be a significant hazard. Freak waves can also be generated when energy is focused by local currents and sea bed topography ('spatial focusing'). The Agulhas current along the east coast of South Africa, for example, meets swell from the Southern Ocean and creates 'holes in the sea' that cause severe damage to ships and loss of life. Freak waves are also created by nonlinear wave–wave interactions ('nonlinear focusing': Tucker and Pitt, 2001; Dysthe *et al.*, 2008). A further proposed mechanism is simple linear superposition of waves ('dispersive focusing': Kharif and Pelinovsky, 2003).

The wave field is regarded as a stationary random process over short periods of time (of order 30 minutes). This stochastic process is characterised by a frequency spectrum as a function of the direction of travel of the waves. Summary statistics are then derived from these spectra, such as 'significant wave height', which is four times the square root of the area under the spectrum (the total energy of the wave field). Extreme wave conditions are characterised in terms of the summary statistics: for example, freak waves are often defined as at least twice the significant wave height, which for linear waves is the expected height of the highest wave in a three-hour period (Dysthe *et al.*, 2008). An overview is given by Tucker and Pitt (2001). Although waves are the most important environmental force on offshore structures such as oil rigs, ocean currents can also be a serious hazard.

5.2.2 *Triggers*

Hydrometeorological hazards have no distinct trigger, but they can often be linked to certain states of the climate. Natural variability in the form of fluctuations and oscillations is present in the climate at all timescales, from hours to centuries and longer (Section 5.3.1), and the phases of this variability affect the likelihood of extreme events. If these are identified then particular climate states can be useful indicators of hazard risk in other regions or in the near future. One is the El Niño Southern Oscillation (ENSO), a measure of sea surface temperature anomalies in the tropical eastern Pacific, which affects the likelihood and location of extreme warm events, extreme precipitation and tropical cyclones (e.g. Rusticucci and Vargas, 2002; Camargo *et al.*, 2007). Another is the North Atlantic Oscillation (NAO), usually defined as the difference in the normalised surface air pressure between Iceland and the Azores or the Iberian peninsular. The NAO varies on decadal timescales and is a good proxy for the North Atlantic storm track. It has been linked with Atlantic cyclone intensity and with winter storms, precipitation and wave conditions in the eastern North Atlantic and Europe (e.g. Woolf *et al.*, 2002; Elsner, 2003; Mailier *et al.*, 2006; Vitolo *et al.*, 2009). Another example of a particular climate state is 'blocking', where an atmospheric pattern persists and prevents normal local weather from moving through the area, which results in very stable weather. If this happens to occur at a time of high temperature or precipitation, a period of extremes can last for days or weeks.

There are physical connections between the different hazards. The 'hot and dry' hazards (heat waves, drought) are linked to each other through meteorological processes such as evapotranspiration and atmospheric convection, while the 'wet and windy' hazards (precipitation, storms) are linked through low atmospheric pressure. The hydrological cycle and water conservation connect droughts with extreme precipitation (Held and Soden, 2006). Hydrometeorological hazards also trigger other types of hazard: drought and heat waves cause wildfires; persistent extreme precipitation causes flooding and landslides; and wind storms cause storm surges.

5.2.3 Scale

Different parts of the world experience hazards of different scales – i.e. severity and frequency – due to the dependence of local climatic variability on local atmospheric patterns (e.g. circulation belts, jet streams, monsoons), topography (elevation and proximity to mountains and coast) and land and water use (e.g. intensive agriculture, deforestation, water consumption and redirection).

The scale of 'wet and windy' hydrometeorological hazards is mostly determined by natural factors. Extreme precipitation events are most severe in the mid-latitudes, in monsoon regions and on the windward side of mountains, while storm events are most severe in the tropics and in open areas such as coasts. One human factor, aerosol pollution, affects precipitation by acting as cloud condensation nuclei, but the interactions are complex: heavy rain events may increase at the expense of light rain events, and the net suppression or enhancement depends on particle size (Levin and Cotton, 2008; Li *et al.*, 2011). In an average year, ten tropical storms develop in the Gulf of Mexico, Caribbean Sea or Atlantic Ocean, of which six develop into hurricanes (Townsend, 2006). Mid-latitude winter storms are the most frequent natural disaster for Central Europe, while sting-jets are also believed to be a major cause of wind damage in northern Europe (Fink *et al.*, 2009). In general the highest storm-related waves are in the North Atlantic in winter (Holliday *et al.*, 2006). Waves in the Southern Ocean are smaller but with little seasonal variability, so the Southern Ocean is always rough, whereas the North Atlantic and North Pacific are relatively calm during the summer months (e.g. Young, 1999; Woolf *et al.*, 2002).

The scale of 'hot and dry' hydrometeorological hazards is influenced by human and natural factors. Heat waves are most severe in areas of inland desert, semi-desert and with Mediterranean-type climates (i.e. tropical and subtropical regions), and are often associated with an unusual location of the jet stream. Urban areas experience higher temperatures than rural areas due to the heat island effect. Drought hazards are most severe in areas with low precipitation, strong winds, high temperature, vulnerable soil types (such as high sand content) and low water reserves. These regions are often subtropical, mid-continental or primarily supplied by glacial meltwater. Humans amplify droughts through deforestation, intensive farming and high population densities (IPCC, 2011), which leave soil exposed, increase runoff and deplete water reservoirs such as aquifers. An example of this is the Anglian region in eastern England, where low water availability is exacerbated by very high demand from agriculture. Areas that have suffered from severe drought include north-eastern and western Africa, northern and central China, western and mid-continental north America, the Amazon basin and Australia (e.g. Dai, 2011).

Regional variation in the magnitude of extremes may be illustrated with some examples. In the UK, the highest recorded daily rainfall during one of the most severe flooding events in living memory (November 2009) was 316 mm (UK Met Office website: http://www. metoffice.gov.uk). But the current world record for a 24-hour period is nearly six times greater, 1.825 metres, recorded in January 1966 by the island Réunion, east of Madagascar (World Meteorological Organization: http://wmo.asu.edu). The most significant UK wind storms in recent decades were in 1987, with a mean wind speed of 50 mph and gusts up to 115 mph, and in 1990, with mean wind speed of 69–75 mph and gusts up to 104 mph (UK Government Cabinet Office, 'UK Resilience': available from http://www.cabinetoffice.gov. uk). Hurricane Katrina in 2005 was classified as Category 5 (greater than 155 mph) but weakened to Category 3 (111–130 mph) before reaching land in Louisiana. The highest directly measured gust speed is 253 mph, during a tropical cyclone in April 1996 on Barrow Island off the north-west coast of Australia, and the highest recorded speed (inferred using radar) is 302 mph in May 1999 in Oklahoma (World Meteorological Organization: http:// wmo.asu.edu). The highest wave height measurements are around 30 m (Kharif *et al.*, 2009), though such high values are not necessarily classified as 'freak' according to the usual definition of twice the significant wave height (Section 5.2.1.5). During Hurricane Ivan over the Gulf of Mexico in 2005, for example, the maximum wave height was 26.3 m, while twice the highest significant wave height was 30.8 m (Forristall, 2005). During the UK heat wave of August 1990, temperatures of 32°C or more (a critical threshold: Section 5.4.1.2) were recorded in virtually all parts of England and some parts of Wales, 11.5°C hotter than the 1971–2000 average for August maximum temperatures (UK Met Office website: http://www.metoffice.gov.uk); the most persistent UK event was in 1976, with 32°C exceeded for 15 consecutive days (UK Government Cabinet Office, 2008). During the 2010 Northern Hemisphere heat wave, a temperature of 38.2°C was recorded in Moscow (Barriopedro *et al.*, 2011). The number of European extreme drought events during the twentieth century, with PDSI less than or equal to −4, is shown in Figure 5.2 (Lloyd-Hughes and Saunders, 2002).

5.2.4 Loss

This section gives an overview of the losses caused by hydrometeorological hazards, outlining the entities that are exposed and vulnerable, the types of loss that may occur and the quantification of these losses.

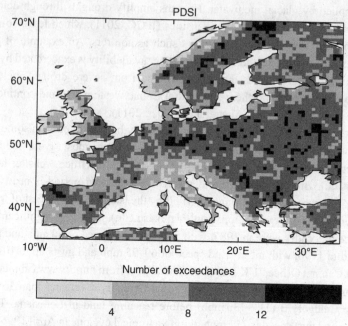

Figure 5.2 Number of extreme drought events (PDSI less than or equal to −4) by grid cell for Europe during the period 1901–1999. (Lloyd-Hughes and Saunders, 2002) © Wiley.

5.2.4.1 Exposure and vulnerability

Meteorological hazards have global reach: they affect everything under the sky. Hydrological hazards are almost as pervasive, because exposure comes from proximity to bodies of water (e.g. waves) and from limited water supply (droughts). Some types of hazard do occur mainly in one region, such as hurricanes in the tropical Atlantic or typhoons in the Pacific.

The entities exposed to hydrometeorological hazards are therefore wide-ranging: humans; agriculture (including crops, livestock and irrigation); infrastructure (including transport, power and housing); and the environment (including forests, surface soil, eco-systems and habitats) (e.g. Beniston *et al.*, 2007; UK Environment Agency, 2009; UK Government Cabinet Office, 2008). Vulnerability varies by location: poverty and inefficient infrastructure reduce the ability to prepare for hazards and to recover (IPCC, 2011); local architectural styles, construction practices, building regulations and enforcement affect the vulnerability of buildings (RMS, 2005); and vulnerability is greater if the climate is close to habitable limits, though prior conditioning to warm climate or recent high temperatures reduces vulnerability to heat waves.

Densely populated areas are particularly vulnerable, because not only are more people exposed but the effects are amplified: low water reserves in the case of drought, and mortality or injury due to damaged or collapsing buildings in the case of extreme precipitation and wind storms. Nearly 50% of the world's most populated areas are highly vulnerable to drought (USDA, 1994). However, rural areas are also vulnerable: for example, trees in densely wooded areas are felled by strong winds, causing damage and injury.

Social and cultural practices are also important, particularly for heat waves. The most vulnerable individuals are those living alone, ill, bedridden, housebound, elderly, over-weight, lacking nearby social contacts, lacking access to transportation or lacking sufficient domestic cooling (Klinenberg, 2002). Air pollution also exacerbates the health risk (Souch and Grimmond, 2004). The European heat wave in 2003 and the two Chicago heat waves during the 1990s mostly affected elderly people in urban areas (United Nations, 2006).

5.2.4.2 *Types of loss*

Humans are killed, injured or made ill by every type of hydrometeorological hazard: wind storms and waves cause physical injury; drought causes dehydration, related diseases and malnutrition due to famine; heat waves cause dehydration, heat stroke, oedema (swelling), rashes and cramps (reviewed by Kovats and Hajat, 2008); and extreme precipitation can lead to waterborne diseases (Curriero *et al.*, 2001) and, in combination with high temperatures, to malaria (Githeko and Ndegwa, 2001). But the impacts on humans are not limited to mortality and morbidity: sustained drought can cause mass migration, political unrest and violent conflict over natural resources (Gleick *et al.*, 2006), while heat waves decrease productivity and increase violent crime (Smoyer-Tomic *et al.*, 2003). Drought, heat waves and extreme precipitation reduce agricultural yields through high temperature, low water availability and soil erosion. Extended drought can lead to famine, amplifying problems of health and social stability (Gleick *et al.*, 2006).

Infrastructure, property and business are also harmed. Energy structures (including power stations, oil and gas platforms, refineries, pipelines and renewable energy devices) are damaged and power supplies interrupted or reduced by wind storms and extreme waves (physical damage), extreme precipitation (flooding), heat waves (overheating) or drought (overheating; reduced flow through hydroelectric dams). Transport and telecommunication links are damaged by wind storms, rain and (in the case of ships) extreme waves, and during heat waves road surfaces are susceptible to melting, vehicle engines to overheating and railway lines to buckling (De Bono *et al.*, 2004). Water supplies are affected by extreme precipitation (polluting), drought (decreasing sources) and heat waves (buckling pipes).

Hydrometeorological hazards may be natural, but can severely harm the natural environment. Drought and extreme precipitation cause habitat damage and soil erosion, and the latter can lead to desertification and dust storms. Wind storms also harm ecosystems (particularly forests, e.g. Bründl and Rickli, 2002). Wildfires are caused by drought and heat waves. Heat waves may affect glaciers, permafrost and rock stability (De Bono *et al.*, 2004). Water pollution from extreme precipitation has negative impacts on ecosystems as well as human health.

5.2.4.3 Quantification of loss

Quantification of loss is more straightforward for some types of harm and damage than others. The most often used metrics relate to human health and financial cost.

Mortality can be expressed in absolute numbers, though is more usefully expressed as 'excess death rate', the difference between reported and typical deaths for a given time period. But it can be difficult to quantify the effect of 'harvesting', the phenomenon by which the death of vulnerable individuals is hastened but overall mortality is not increased (e.g. Baccini *et al.*, 2008). Ill health can be quantified with hospital records, but these data are less available and less easily interpreted than mortality data (Sheridan and Kalkstein, 2004). Mortality rates in the United States are highest from extreme heat and hurricane hazards (Figure 5.3), though the latter are dominated by Hurricane Katrina, which caused over 1000 deaths. Negative impacts of heat extremes rose during the period 1995–2004, especially in Europe (United Nations, 2006), where the 2003 heat wave led to the loss of more than 70 000 lives (Robine *et al.*, 2008). Waves contributed to one of the most severe environmental disasters to occur in the UK, when flood defences on the east coast of England were breached in 1953 by a combination of high tides, storm surge and large waves: 307 people were killed in the area (UK Government Cabinet Office, 2008).

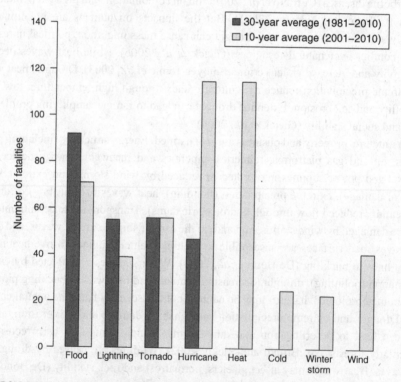

Figure 5.3 Mean annual weather fatalities for the United States, Puerto Rico, Guam and the Virgin Islands. Data from the National Weather Service.

Financial losses due to physical damage of property and infrastructure or interruption of services are typically quantified by insurance claims or economic effect. The former is not comprehensive because insurance coverage is rare in many parts of the world. Both insured and uninsured losses from natural hazards have risen during the twentieth century due to increases in exposure (population; habitation of susceptible areas such as coasts) and vulnerability (asset ownership; changes in social structure, infrastructure and environmental conditions) (Grossi and Kunreuther, 2005; Dlugolecki *et al.*, 2009). Climatic extremes caused more than 170 'billion-dollar events' (in insurance terms) globally during the second half of the twentieth century, compared with 71 due to earthquakes (Beniston *et al.*, 2007). The most important were tropical cyclones and mid-latitude winter storms (Fink *et al.*, 2009), followed by floods, droughts and heat waves. Hurricane Katrina is the most destructive natural disaster in the history of the United States (Townsend, 2006). It caused destruction or severe damage to around 300 000 homes; at least ten oil spills; an initial shutdown of more than 91% of oil and 83% of gas production in the Gulf of Mexico region (RMS, 2005); and estimated damages of US$96 billion (Townsend, 2006). In December 1999, the three storms Anatol, Lothar and Martin caused a total insured loss of more than €10 billion across Europe (Fink *et al.*, 2009) and in 2007 another, Kyrill, caused €4–7 billion of insured losses and the loss of electricity for at least two million homes (Fink *et al.*, 2009). Lawton (2001) reports an estimate that freak waves were responsible for 22 of the 60 losses of super-carriers (very large aircraft carriers) to sudden flooding between 1969 and 1994 (Kharif and Pelinovsky, 2003).

Harm to ecosystems and environmental services is harder to quantify, though its importance is increasingly recognised (e.g. the Millennium Ecosystem Assessment; Hassan *et al.*, 2005). Harm may be expressed in terms such as loss of biodiversity, habitat or natural resources. Mid-latitude storms have caused significant environmental losses: in 2007, storm Kyrill uprooted an estimated 62 million trees across Europe (Fink *et al.*, 2009).

5.3 Risk assessment

Risk assessment for hydrometeorological hazards has a very long history. Humans have attempted to forecast weather from patterns on the earth and in the stars since antiquity, recorded its extremes in historical documents and measured its variations since the invention of each meteorological instrument (e.g. thermometer, rain gauge, barometer). The longest instrumental record of temperatures is the Central England Temperature (CET), which began in 1659 (Parker *et al.*, 1992). The field of numerical weather prediction, proposing equations to describe the physical behaviour of the atmosphere, has existed since the turn of the twentieth century, and computers have been used to solve these equations since the 1950s (Lynch, 2008). Despite this long history, hydrometeorological hazards, and their impacts, are still challenging to predict (Section 5.3.1).

The word 'risk' is very often applied in this field, as in many others, only to the likelihood of a hazard occurring rather than in the full sense incorporating damage and loss; an exception to this is catastrophe modelling (Section 5.3.3). There are two types of risk assessment, based on physical modelling and statistical analysis. Physical models, also known as 'dynamical models' or 'simulators', are mathematical equations written in computer code to simulate the behaviour of the atmosphere, ocean and land. They are used to forecast specific hazard events – their magnitude, extent and timing – in order to issue warnings or take steps to mitigate the hazard (Section 5.3.2). Statistical analysis of hazard events (Section 5.3.3) assesses the long-term characteristics of a particular hazard type, expressed in terms of return period or dependence on other variables. Some risk assessments include both physical and statistical elements, in particular hurricane forecasts (Sections 5.3.2, 5.3.3.3).

5.3.1 Challenges in risk assessment

The short length of the observational records limits estimation of return periods for rare hydrometeorological events. For some variables the record length is 150 years or more, but in general observations or reliable re-analysis data are only available for a few decades. Hydrometeorological hazards also have their own particular challenges: the earth system's internal variability, and non-stationarity due to the influence of forcings. These are discussed below.

5.3.1.1 Internal variability

The earth's atmosphere and oceans are continuously changing due to the influence of natural and anthropogenic 'forcings' that produce a radiative imbalance at the top of the atmosphere (Section 5.3.1.2). But even if there were no imbalance, and the earth system were at equilibrium, there would still be fluctuations of unforced 'internal' variability. The peaks and troughs of these fluctuations give rise to extreme events. Internal variability derives from two sources: chaos and oscillations. These are challenging to simulate (Section 5.3.2) and must be accounted for in statistical analyses (Section 5.3.3).

Chaos arises from the sensitivity of the atmosphere's evolution to small changes in the 'initial conditions', a consequence of the earth system's complex and nonlinear nature. The atmosphere and ocean are turbulent, with nonlinear feedbacks involving exchanges of heat and water, chemical reactions, and so on, between every part of the air, sea, land, ice and life. Nonlinearity and complexity lead to chaos, but chaos does not imply randomness. A chaotic system can be deterministic: in such a system, perfect knowledge of the initial conditions and response would enable prediction over long timescales. But knowledge of the earth system cannot be perfect. Observations have finite coverage and are uncertain. Physical models have finite temporal and spatial resolution and are imperfect, because many processes are unknown or difficult to represent, and the observations available to evaluate them are limited. So chaos, combined with imperfect knowledge, results in 'limited predictability', where uncertainties in initial conditions are greatly amplified in predictions beyond a few days. The limit of predictability is thought to be about two weeks (Lorenz, 1965).

Limited predictability is equivalent, in practice, to a random element, so most forecasting models have stochastic components (Section 5.3.2).

Oscillations in the state of the atmosphere and ocean move around the earth at all timescales, from weeks (such the Madden–Julian Oscillation, MJO) to years (such as the ENSO and NAO; Section 5.2.2) and longer. These oscillations are difficult to simulate realistically in physical models because they must emerge spontaneously from the smaller-scale behaviour described by the equations; some physical models are more successful at this than others (IPCC, 2007a). Oscillations are also a challenge in statistical analysis of observations, because they are difficult to separate from long-term trends. But these longer-term changes do bring an important advantage: recent improvements in understanding and measuring their persistent influence on atmospheric conditions allow forecasts to be made of spatial and temporal averages (such as the number of hurricanes in the coming year) with much longer lead times than Lorenz's fortnight, albeit with larger uncertainties (Sections 5.3.2, 5.3.3.3).

5.3.1.2 Non-stationarity

The atmosphere and oceans are continuously 'pushed' towards warmer or cooler states by a radiative imbalance at the top of the atmosphere. This imbalance is the net result of 'forcings', both natural and anthropogenic in origin, which affect the amount of solar radiation that warms the planet or escapes to space. During the instrumental record (the past 150 years or so), the most important forcings are: greenhouse gases and tropospheric ozone that warm the planet (increasing due to industrial emissions and release from natural reservoirs such as forests and wetlands, and varying due to natural cycles); incoming solar radiation (varying in regular cycles and long-term trends); and sulphate aerosols that have a cooling effect (varying due to industrial emissions and volcanic eruptions). More details are given in Chapter 6. Over the twentieth century, the net forcing has resulted in an increase of about 0.7°C in global mean temperature (IPCC, 2007a), with larger changes over land and in regions with strong positive feedbacks such as the Arctic. The twentieth-century response is relatively small compared with internal variability such as ENSO, but reflects an increase in energy in the system that is thought to have shifted the statistical distribution of several aspects of weather (IPCC, 2011). The impacts of hazard events are also non-stationary, due to changing social and cultural factors: for example, mortality due to heat waves is decreasing in the United States due to increased use of air conditioning, improved healthcare and improved public awareness of impacts (Souch and Grimmond, 2004). As well as interannual non-stationarity, hydrometeorological hazards and their impacts vary with the seasonal cycle. The influence of these oscillations, like those of internal variability in the previous section, aids forecasting on seasonal timescales.

Non-stationarity should therefore be accounted for, where present, in statistical analysis of observational records. Furthermore, warming is expected to continue over the next few decades and longer, due to the slow response of the oceans to past forcing, the long lifetime of greenhouse gases, the cumulative effect of a (likely) slow rate of reducing greenhouse gas emissions, and a (likely) reduction in sulphate aerosols from pollution.

Hydrometeorological hazards are part of the climatic spectrum and likely to be affected by this warming: estimates of return periods, which inherently assume stationarity, have limited applicability for these hazards in the future. Estimating risk of hydrometeorological hazards under future climate change is discussed in Chapter 6.

5.3.1.3 Hazard- and impact-specific challenges

Some hydrometeorological hazards, or aspects of hazards, are more challenging to simulate or analyse than others. Temperature is the most straightforward to simulate because the dominant drivers of regional temperature patterns are latitudinal insolation patterns and the distribution of land around the globe, which are well known (IPCC, 2007a). Precipitation patterns are influenced by these and also by small-scale processes such as atmospheric instabilities, the flow of air over local topography and very small-scale processes such as formation of cloud condensation nuclei (IPCC, 2007a); these are difficult to simulate (Stephens *et al.*, 2010) and in general must be represented in a simplified form (Section 5.3.2). Precipitation also responds nonlinearly to forcings, which complicates statistical analysis. Forecasting hurricane tracks is improving over time, but forecasting their intensities and landfall is much less successful (Camargo *et al.*, 2007). Droughts and heat waves are very influenced by soil moisture, but a global observational network is currently at an early stage (Dorigo *et al.*, 2011), so this important driver must be inferred from precipitation deficits for hazard assessments (Hirschi *et al.*, 2011). However, the global extent of hydrometeorological hazards may be of some advantage: research techniques of developed nations can be transferred to regions with fewer resources, and larger datasets can aid statistical analysis.

Assessing impacts and loss is also more challenging for some vulnerable entities than others: human mortality and health impacts, for example, depend on complex interactions between many factors that are hard to predict, such as societal decisions (CCSP, 2008), and are quantified differently depending on the relative importance of individual behaviour (sociology and environmental psychology), financial security (economics) and national capability and decision-making (political science) (Alcamo *et al.*, 2008).

5.3.2 Physical modelling

This section describes physical modelling of hydrometeorological hazards and impacts and quantification of uncertainty in their predictions. Physically based, dynamical models – simulators that attempt to represent real-world processes – are essential tools in forecasting the specific evolution of a state through time. They can also, if run multiple times simultaneously (described below) or for a long time period (Chapter 6), be used to assess the probability distribution of a future hazard or impact. The advantage of using a physically based model to assess event probabilities, rather than a statistical analysis based on the historical record, is that it can account for non-stationarity, because the boundary conditions that drive its evolution (discussed below) can be varied in time and space. The disadvantage is that it is hugely expensive, infeasibly so for events that have return periods of several years. Therefore assigning probabilities to low-probability, high-magnitude hazards tends to be done using statistical analysis (Section 5.3.3). Statistical models are also used for forecasting if there are temporally persistent states (such as ENSO) that influence the hazard (such as hurricane intensity), by assessing the future probability distribution given the specific present conditions.

5.3.2.1 Types of physical model

Numerical weather prediction (NWP) models are complex, computationally expensive simulators of the atmosphere and ocean that are used to forecast day-to-day weather. A large number of variables are simulated, including temperature, pressure, precipitation and wind. Extreme weather hazards and modes of internal variability are not explicitly coded but are emergent properties of the individual processes. Physical modelling of large waves has much in common with atmosphere–ocean modelling: one example is described by Tolman (2009), and an overview can be found in Kharif et al. (2009). Forecasting of tropical cyclones and of weather-related impacts is discussed at the end of the section.

NWP models are initialised with 'today's weather' using current observations of the atmosphere and ocean, then run forward for a specific time. This produces a single prediction, or realisation, of weather, which can be converted to a binary forecast for a hazard event: for example, whether the forecast period includes a storm of a given magnitude at a given location. To generate many realisations and approximate an event probability, random elements are introduced and the simulator is run several times simultaneously (Section 5.3.2.2).

Forecast lead times range from a few hours or days (short-range), to a fortnight (medium-range), to 6–12 months (seasonal). The evolution of physical quantities, such as the transfer of heat, is calculated in discrete timesteps represented with a three-dimensional grid or spherical harmonics, with spatial domain and resolution chosen according to the lead time of the forecast, timescale of the physical processes and available computing resources. Short-range forecasts are often performed for a sub-continental region, at a horizontal resolution of around 5–25 km; vertical resolution ranging from tens of metres near the surface to several hundred metres in the stratosphere; and temporal resolution of a few minutes. Seasonal forecasts are made over a global domain with climate models (Chapter 6), with a reduced resolution (around 140 km horizontally in the mid-latitudes) to reduce computational expense. Physical processes that occur on smaller spatio-temporal scales than the simulator resolution, such as convective cloud formation, are represented in parameterised form (Chapter 6). Forecasts have improved since the addition of stochastic elements in the form of randomly varying parameters or processes (Section 5.3.2.2).

Forecasting models require both initial conditions to prescribe the initial state, derived from observations and previous forecasts, and 'boundary conditions' to control the system evolution, supplied by global model simulations (Chapter 6) and 'analysis' datasets (e.g. sea surface temperatures from combining satellite and *in situ* measurements). Because the atmosphere is chaotic (Section 5.3.1.1), forecast success is heavily dependent on the accuracy of model initialisation. NWP model simulations are therefore updated, or 'nudged', frequently by current observations in a process known as 'data assimilation' (Evensen, 2009), which takes into account the relative uncertainties of the observations and simulations. Data assimilation is an inverse method founded in Bayes theorem (Wikle and Berliner, 2007) in which NWP model predictions (samples from the prior distribution) are updated with observations to estimate the most likely atmospheric state (mode of the posterior distribution). The dimension of this state is extremely high so there are various simplifications, such as estimation of means or modes and variances rather than full probability distributions, and assumptions, such as linearity or Gaussian uncertainties, to make the problem tractable. 'Variational' methods of data assimilation (e.g. Rabier *et al.*, 2000) are gradient-based optimisation schemes, which minimise a cost function (the simplest being least squares) to find the mode of the posterior distribution; this requires an 'adjoint' model, constructed by differentiating the model equations by hand or with automatic software, which is very challenging for such complex models. 'Ensemble' methods (e.g. Evensen, 1994) use a set of about 100 simulations, each starting from slightly different initial states to approximate the mean and covariance of the prior distribution, but these are not yet used much in operational forecasting (e.g. Houtekamer *et al.*, 2005). Methods with few or no simplifying assumptions are currently only suitable for very low dimensional problems, though their efficiency is continuously being improved (van Leeuwen, 2009, 2010, 2011).

Seasonal forecasts for tropical cyclones are currently undergoing a sea change in methods and capabilities. Predictions were traditionally based on statistical analysis of past relationships with climate indicators such as Atlantic sea surface temperatures and ENSO (Section 5.3.3.3). Now a small number of forecasts use dynamical methods, applying criteria to climate model simulations for variables such as sea surface temperature, vertical shear of horizontal wind and low-level vorticity, to infer the presence of tropical storms (Vitart and Stockdale, 2001; Barnston *et al.*, 2003; Camargo and Barnston, 2009). A hybrid 'statistical-dynamical' method is described by Emanuel *et al.* (2006) in which synthetic storms are seeded randomly with tracks derived from statistical modelling of re-analysis data (Section 5.3.3) and intensities simulated by a high-resolution physical model. Dynamic modelling has only recently become comparably successful to statistical modelling at making seasonal predictions of tropical cyclone frequency (e.g. Vitart *et al.*, 2007). As computational power increases, model resolution can increase and this is likely to further improve prediction skill (Vitart *et al.*, 2007). Overviews of seasonal tropical cyclone prediction are given by Camargo *et al.* (2007) and Klotzbach *et al.* (2011).

Hazard impacts can be simulated if the important processes are understood. A well-developed hazard impact model can be chained, or 'coupled', to an NWP model to give a probabilistic assessment of loss, including real-time forecasts if the lead time is sufficiently long (days or longer), or the hazard onset slow (e.g. drought), or the impacts slow to emerge (e.g. cumulative impacts of extreme weather on vegetation: Senay and Verdin, 2003; Chapter 6). If the processes are not well understood, or the timescales short, statistical modelling of past events is used instead (Section 5.3.3.3): for example, impacts on human 'health and financial loss are too challenging to describe with physically based laws, so they are forecast with statistical models or expert synthesis of observations and physical understanding (e.g. Funk, 2011).

5.3.2.2 Uncertainty assessment in physical modelling

Uncertainty assessment is relatively well-developed in hydrometeorogical hazard risk analysis due to a long history of weather recording and forecasting. Forecast uncertainty is a function of the observational and NWP model uncertainties. The former arise from measurement uncertainty, incomplete coverage and uncertainty in the 'observation operators', which translate measured quantities into simulated (e.g. transformation, interpolation or integration). The latter are more difficult to estimate and arise from uncertain model parameters (representing sub-grid-scale processes) and model structure (due to missing or poorly understood processes and finite resolution). Forecast uncertainty is estimated by sampling from these uncertain quantities and running different variants of the forecast simultaneously to produce an ensemble prediction.

Ensembles are generated with several types of perturbation to explore this uncertainty. Weather forecasts are critically dependent on the accuracy of the initial state, so 'initial condition ensembles' have been used in numerical weather prediction since the 1990s. Initial state perturbations are either constructed randomly, or by sampling from the observational uncertainties, or to generate the fastest growing impact on the forecast (e.g. Persson and Grazzini, 2007). Parameter and structural uncertainty are estimated by perturbing the model configuration (e.g. Houtekamer *et al.*, 2009): for example, with varying resolution (Persson and Grazzini, 2007); stochastically varying parameters (Buizza *et al.*, 1999; Palmer, 2001; brief overview by McFarlane, 2011); or stochastic kinetic energy backscatter (Shutts, 2005; Berner *et al.*, 2009). These schemes explore both structural and parametric uncertainty simultaneously: for example, stochastic parameters allow the model to move into rarer states, such as blocking-type flow in Northern Hemisphere winters (Palmer *et al.*, 2005). Another approach to estimating structural uncertainty is with a multimodel ensemble (MME), a group of models from separate research centres (e.g. Palmer *et al.*, 2004; Hagedorn *et al.*, 2005; Weisheimer *et al.*, 2009). MMEs are mainly used for forecasts made over global domains (medium-range and longer lead times; see Chapter 6) because short-range forecasts are often made with regional models so there are fewer opportunities for comparison. Individual members of an MME are averaged together or considered as a group; they may be weighted equally or according to their degree of success in matching observations (discussed below). The mean of an MME is often found to outperform the individual members (e.g. Hagedorn *et al.*, 2005; Vitart, 2006), increasingly so with ensemble size (e.g. Ferro *et al.*, 2008), though it does not outperform the best member. Model structural uncertainty can also be estimated with non-ensemble approaches (Hollingsworth and Lönnberg, 1986; Parrish and Derber, 1992).

The frequency with which a given hazard appears in an ensemble forecast is treated as an estimate of its probability of occurrence: for example, a heat wave that occurs in 20% of the ensemble simulations is assigned a probability of 20%. If the NWP model is well-calibrated (below), then over the long term such an event will occur in 20% of such forecasts. Ensemble forecasts tend to be referred to as probabilistic, but they are more accurately described as frequency distributions due to their low sample size and incomplete representation of all uncertainties; further discussion can be found in Chapter 6.

Calibration of NWP forecasts is tuning of the ensemble perturbations so that forecast frequencies match the real world over the long term. Forecasts are repeatedly tested against observations to calculate skill scores, metrics that quantify the success of the model forecast relative to a reference such as local climatology or the weather of the previous time period (e.g. Brier, 1950; Talagrand *et al.*, 1997; Persson and Grazzini, 2007). Forecast verification is an extensive field with a long history: reference texts are Jolliffe and Stephenson (2003) and the relevant chapter in Wilks (2011), and a review of recent advances is given by Casati *et al.* (2008). Over the past decades, improvements in model structure, resolution, data assimilation methods and observational datasets have led to continuous improvement in skill scores: the US National Centers for Environmental Prediction (NCEP) 'S1' metric, for example, has been recorded since the mid-1950s and shows a doubling in forecast length for a given skill score over 1–2 decades (Kalnay, 2003). Ensemble forecasts are generally underdispersed: over the long term, they are more narrowly spread than the observations. This bias may be partially corrected with post-processing methods that re-weight the ensemble members according to their performance during a training period (e.g. Gneiting, 2005; Raftery *et al.*, 2005). Post-processing methods are currently being adapted to allow spatially varying calibration (Kleiber *et al.*, 2010) and spatio-temporal correlation structure (e.g. Berrocal *et al.*, 2007), but not yet multivariate correlation.

Forecasts of extreme weather events cannot be verified, of themselves, with the usual skill scores: not only does their rarity drastically decrease the number of tests and increase the uncertainty, but the skill scores tend towards trivial limits (such as zero) with increasing event rarity (Stephenson *et al.*, 2008). Various alternatives have been proposed, either by constructing new skill scores suitable for extremes (Casati *et al.*, 2008) or by adapting extreme value theory (Section 5.3.3.1) for verification purposes (Ferro, 2007), but in general calibration is performed with normal day-to-day weather. Some extreme weather events are much harder to predict than others (Section 5.3.1.3), which means, in effect, that NWP models are less well calibrated for these events, and that the assessed probabilities must be interpreted with caution.

5.3.3 Statistical analysis

The objective of statistical analysis is to assess the probability distribution of a future hazard or its impact. Statistical methods for assessing future hazard event probabilities are only valuable to the extent that a defensible relationship can be proposed between historical and future events. For hydrometeorological events, the biggest challenge is the non-stationarity of the earth system during the instrumental period. Return periods are interpreted under an assumption of stationarity, so if the drivers that determine hazard frequency and properties do change significantly then the predictions become less reliable. Some hazards are not significantly affected, but for most there are trade-offs if choosing subsets of the data that are approximately stationary, and difficulties if building non-stationarity into the statistical model.

Statistical analysis of hydrometeorological hazards takes many forms, according to the objectives of the analysis and the characteristics of the available data. The observational datasets analysed for extreme weather assessments are typically point data from weather stations. Other types of data include: satellite observations; re-analysis datasets such as ERA-40 (Uppala *et al.*, 2005), which are continuous gridded data generated by assimilating observations into NWP model simulations; or model simulations alone, often in relation to climate change (Chapter 6). Observational datasets of hazard impacts come from a wide range of sources: in particular, government and hospital records for mortality and morbidity, and insurance losses for financial impacts. Observational data may only be available as maxima of a given time period, usually one year, known as 'block maxima' (or equivalently minima). However, block maxima are, by definition, a small subset of the observations. Other data may contain useful information in the properties of, for example, the second- and third-largest events in each time period. If all the observational data are available, it can be preferable to analyse the *r*-largest maxima ($r > 1$), or the exceedances over a threshold ('peaks-over-threshold', POT), or the whole time-series. The relative advantages of these are discussed in the following sections.

Three common objectives for statistical analysis are: estimating return levels, with extreme value analysis; quantifying the trend or covariates of a hazard, with regression methods; and generating 'pseudo-hazard' (also known as 'synthetic hazard') datasets, with 'weather generation' and 'catastrophe modelling'. A brief introduction to the latter two is given here.

Weather generation is the stochastic simulation of a large number of synthetic long datasets of 'pseudo-weather' (e.g. precipitation), each statistically consistent with the original observations. The main purposes of this are to analyse their characteristics (e.g. frequency of rare extremes) and to use them as inputs for impact models (e.g. crop yields), in order to supplement observational records that are short and spatially or temporally incomplete or simulations that are short and have low spatio-temporal resolution.

Catastrophe modelling is stochastic simulation of a large number of synthetic long datasets of 'pseudo-events' (e.g. hurricanes), application of these to an inventory of vulnerable quantities (usually buildings) and estimation of the damage and associated losses (usually financial; sometimes also environmental and social). Catastrophe ('cat') models were first developed in the late 1980s by three US firms (AIR Worldwide, EQECAT and Risk Management Solutions) to improve insurance risk management. After several major hurricane and earthquake events worldwide, including Hurricane Hugo in 1989 and Hurricane Andrew in 1992, these firms made their models more sophisticated and their exposure databases more comprehensive (Cummins, 2007). Other companies subsequently developed in-house models, and US federal and state government supported development of two free open-source models: HAZUS-MH (Vickery *et al.*, 2000a, 2000b, 2006a, 2006b;

FEMA, 2007) and the Florida Public Hurricane Loss Model (FPHLM) (Powell *et al.*, 2005; Pinelli *et al.*, 2008). The details and assumptions of the free models are available in the peer-review literature and technical documents, because they are intended for risk mitigation, regulation and emergency preparation, but the majority of cat models are proprietary and intended for pricing insurance or reinsurance policies, so their details are not public.

Cat models are used to make probabilistic estimates of property losses: mostly residential, commercial and industrial buildings, but also forests and off-shore energy structures such as oil platforms and pipelines. These may be general assessments, such as expected losses for the following year, or specific forecasts for a recent hazard event (e.g. insured losses from Hurricane Katrina: RMS, 2005). Their comprehensive modelling of the entire hazard-to-loss causal chain – hazards, impacts, exposure, vulnerability, damage and loss – is unique in hydrometeorological hazard risk assessment. Cat models combine several modules: a hazard model (statistically based, but where possible physically motivated: Section 5.3.3.2); extensive databases that map exposure and vulnerability (such as property location, market value, reconstruction value, age, construction type and number of storeys); a damage model (structural damage as a function of hazard intensity per building type, derived from wind tunnel tests, post-event analysis and expert judgement); and a cost model (including cost of repairs/replacement, cost increases due to demand surge, business interruption, relocation and insurance policy details). Monte Carlo techniques are used to simulate tens of thousands of years of hazards, as in weather generation, which are propagated through the modules to estimate statistical properties of loss. Cat models are tailored to particular hazard types – the first were constructed for earthquakes, hurricanes and floods, but other natural hazards are now covered, including mid-latitude wind storms, winter storms, tornados and wildfires – and particular regions, such as the Gulf and Atlantic coasts of the United States. The region-specific information is not only the portfolio at risk but also local environmental conditions such as topography (for modelling precipitation), surface roughness (for wind dissipation) and location of trees (for tree debris). Cat model outputs include probability distributions of loss and exceedance probability (EP) curves for both total and insured losses. A reference text is Grossi and Kunreuther (2005).

The three objectives described above – return levels, covariates and 'pseudo-hazards' – are discussed here according to the characteristics of the data: stationary and independent (Section 5.3.3.1), correlated (Section 5.3.3.2) and non-stationary (Section 5.3.3.3). A reference text for extreme value modelling is Coles (2001). An overview of techniques for correlated or non-stationary hydrometeorological observations is given by Khaliq *et al.* (2006).

5.3.3.1 Stationary independent data

If observations can be treated as independent and stationary, estimation of the frequency and intensity of extremes is greatly simplified. How often this may be possible is discussed at the end of the section.

Under these assumptions, and if the record is sufficiently long, the extremes of continuous data may be characterised with the extreme value limit theorem, which is analogous to the

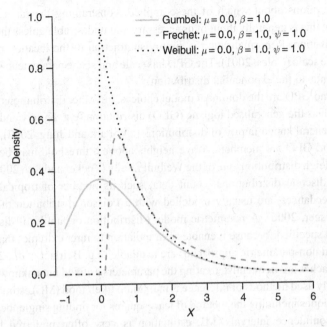

Figure 5.4 Examples of extreme value distributions, with values given for the location (μ), scale (β), and shape (ψ) parameters.

central limit theorem. This states that the renormalised (explained below) block maxima of random observations from the same distribution tend asymptotically, if they converge at all, to one of the generalised extreme value (GEV) family of distributions: Gumbel (exponential-tailed), Fréchet (heavy-tailed) or negative Weibull (bounded, light-tailed). Examples of these are shown in Figure 5.4. The *r*-largest maxima in a block can equivalently be described with a more general form of the GEV (see Coles, 2001). The extreme value limit theorem also states that the amounts by which a threshold is exceeded tend asymptotically to one of the generalised Pareto distribution (GPD) family of power law distributions. In the case of *r*-largest and POT, there is a bias–variance trade-off in which larger (typically longer) samples produce tighter predictions, but may introduce bias through more serious violations of modelling assumptions. The choice may be aided with a graphical method (mean residual life plot: Coles, 2001) or sensitivity studies (see below).

The strength of extreme value theory (EVT) is that it provides tools to estimate return levels without making a-priori assumptions about the probability distribution from which the observational data or their extremes are sampled. The limit theorem removes the need to model the full dataset, and the distribution of the extremes can be inferred empirically by estimating the characteristic parameters from the data. The GEV distribution has parameters for location, scale and shape, which control the mean, asymmetry and rate of tail decay of the distribution, respectively: a positive shape parameter corresponds to the Fréchet, zero to Gumbel and negative to the negative Weibull. Early extreme value analyses did make a-priori assumptions about which of these applied, constraining the range of the shape parameter, but this is generally thought unnecessary and undesirable unless there is a strong physical motivation. Renormalisation is simply subtraction of the location parameter and division by the scale (Coles, 2001). The GPD has scale and shape parameters: shape equal to zero corresponds to the exponential distribution.

The GEV and GPD are the dominant model choices, but other distributions are occasionally used, such as the generalised logistic (GLO) distribution (e.g. Kyselý and Picek, 2007) or the more general kappa family of distributions (e.g. Park and Jung, 2002), of which the GEV, GPD and GLO are members. Wave heights above a threshold may be fitted with a modified Rayleigh distribution, one of the Weibull family (Tucker and Pitt, 2001). EVT does not apply to discrete distributions: count data, such as number of tropical cyclones or threshold exceedances, are usually modelled with a Poisson distribution or similar (e.g. Jagger and Elsner, 2006). A parametric model (distribution or family of distributions) is almost always specified, because it enables extrapolation to rarer extremes than are present in the data, but non-parametric approaches are available (e.g. Beirlant et al., 2004).

Most approaches are frequentist, treating the parameters as fixed but unknown. One of the most commonly used methods of inference is maximum likelihood (ML) estimation, a broad range of techniques including the method of least squares for finding single 'best' parameter values with confidence intervals. ML estimation is very often preferred because it is intuitive, many software tools are available and it is flexible: for example, it is easy to include covariates (Section 5.3.3.3). For small samples, ML may give a poor estimate (Engeland et al., 2004) so moment-based approaches such as probability-weighted moments or L-moments (Storch and Zwiers, 2003) might be more appropriate, though moments do not always exist for extreme distributions. Model fitting may be repeated for sensitivity studies: for example, to assess the stability of parameter estimates to the choice of threshold or r value. Once the parameters are estimated, model 'goodness-of-fit' is assessed with graphical or other tests. If the model is thought to describe the data sufficiently well, it is used to estimate return levels, usually for more extreme quantiles than are present in the observations. Uncertainty in the return level estimates is assessed using likelihood or bootstrapping methods. Parameter and model uncertainty assessment are discussed in Section 5.3.3.4.

A judgement must be made about whether the data may be considered stationary. Records of extremes that vary seasonally may be treated as stationary by using the annual maxima, but long-term, interannual non-stationarity might seem hard to avoid, given the earth's climate is never at equilibrium. However, stationarity might be approximated by taking logarithms of the ratio of successive observations (Coles, 2001), or assumed if the observed quantities are not closely linked with global atmospheric temperature (e.g. Wimmer et al., 2006), or the record relatively short (e.g. Baccini et al., 2008), or the spatial domain small (e.g. Kyselý and Picek, 2007). The trade-off in using a spatial or temporal subset to approximate stationarity is that it leads to larger statistical parameter uncertainties and thus less reliable probability assessments, especially for large-magnitude hazards. Another important point to consider is whether changes in the measurement process – for example in spatial coverage, fraction of events recorded or systematic error – have created a false impression of non-stationarity (e.g. surface temperatures: Brohan *et al.*, 2006; rain gauges: McGregor and MacDougall, 2009; hurricanes: Vecchi and Knutson, 2011).

Non-stationarity is present to different degrees in different hazard observational records. The number of warm days (globally) and heat waves (in many regions) are thought to have increased since 1950, as have the number of heavy precipitation events and drought intensity and length in some regions (IPCC, 2011). Storm intensity in Europe does not seem to show a significant trend during the twentieth century (Della-Marta *et al.*, 2009). Hurricane frequency is thought not to have increased globally during the twentieth century, but there is disagreement about whether their intensity has increased with global mean temperature or is varying on natural cycles (Emanuel, 2005; Webster *et al.*, 2005; Pielke, 2005, 2006; Anthes *et al.*, 2006; Landsea *et al.*, 2006; Landsea, 2007; IPCC, 2011). Non-stationarity of the hurricane measurement process is a contributing factor in this debate.

Economic losses from extreme weather events have increased, largely due to increasing exposure and assets (IPCC, 2011). Losses due to hurricanes have been concluded to be stationary after normalising for inflation and increasing population, wealth and (in the case of the United States) settlement of coastal areas for US losses from 1900–2006 (Pielke *et al.*, 2008) and global losses from 1950–2005 (Miller *et al.*, 2008), though there may be a trend in the United States since 1970 due to increased hurricane activity in this region (Miller *et al.*, 2008; Schmidt *et al.*, 2009). It is difficult to account for all covariates, such as loss reporting and vulnerability, in these studies (Miller *et al.*, 2008).

A general note of caution on the assumption of stationarity: a failure to reject the null hypothesis that the historical process is stationary may simply reflect the low power of the test statistic for identifying a non-stationary process that may behave differently in the future. If there are underlying physical reasons for believing that the data are stationary, this may be more robust.

5.3.3.2 Correlated data

Correlations in data may occur in the form of multivariate, spatial or temporal dependencies. These are common in hydrometeorological hazards due to the teleconnections and feedbacks that link places and aspects of climate with each other, the large spatial extent of atmospheric and ocean currents and the temporal persistence of meteorological and environmental phenomena.

There are analogues of EVT for multivariate extremes (such as heat waves, droughts and storms), including component-wise block maxima and threshold excesses (Coles, 2001). However, this involves sequential ordering of each dimension: results depend on the ordering and there may not be an obvious choice for a given problem. An alternative is the copula approach, in which the joint distribution for continuous random variables is broken down into the individual, 'marginal' distributions and a copula function that models their dependence structure. A family of copulas is chosen (e.g. Gaussian) and their parameters estimated along with those of the marginal distributions. A general reference for copulas is given by Nelsen (2006), their application to risk management by McNeil *et al.*

(2005) and to natural hazards by Salvadori *et al.* (2007). Schölzel and Friederichs (2008) give a useful overview and call for copulas to be more widely used in weather and climate research.

Hydrometeorological observations are very often correlated in space and time because of spatially extensive or temporally persistent weather systems and modes of natural variability. In the case of continuous variables, this can lead to a very heavy-tailed distribution for which the GEV and GPD are no longer valid. Block maxima might be treated as independent, if the dependence in the full dataset is not too long-range, but threshold exceedances cannot due to the 'clustering' of events that exceed a given threshold. This may be dealt with by processing the data to approximate temporal independence or by adjusting the assumed frequency model to account for the dependence.

The simplest approach for approximating temporal independence is to apply a simple filter. Brown *et al.* (2008), for example, thin their temperature observations by using only the maximum threshold exceedance in each consecutive ten-day window. This fixed-window approach fails if there are some particularly persistent clusters in the data. More often used is a tailored pre-processing, known as 'declustering', which identifies individual clusters in the data in order to model their maxima. The most common of these is 'runs declustering', which uses interexceedance time (a minimum gap length between clusters; e.g. Della-Marta *et al.*, 2009). These methods necessitate a bias-variance trade-off between the number of observations and their independence, an obvious disadvantage for datasets of rare events. Their design may also be rather subjective. An alternative is decorrelation, or 'pre-whitening', in which dependence is modelled and removed. In general this is only appropriate if there are physical motivations for a particular correlation structure, and care must be taken to avoid removing an underlying trend (Storch, 1995; Khaliq *et al.*, 2006).

A simple approach to incorporating dependence in the model is to inflate the variance: for example, by replacing a simple Poisson model of count data with one that includes a dispersion parameter (e.g. negative binomial, or the generalised linear model family), thereby allowing for overdispersion (e.g. Vitolo *et al.*, 2009). Another commonly used method is to empirically estimate the 'extremal index' and use this to modify the parameters in the GPD (Coles, 2001). The extremal index is the reciprocal of the 'limiting mean cluster size': for example, an extremal index of 0.2 indicates a mean cluster size of five exceedances, where 'limiting' refers to clusters of exceedances of increasingly high thresholds. Extremal index is sensitive to cluster definition, but fortunately this does not always propagate to the estimated return levels (Coles, 2001). An alternative is to model the correlation explicitly and incorporate this in the statistical inference: for example, Baccini *et al.* (2008) use generalised estimating equations (GEE) for count data to allow the inclusion of correlation structure. Wavelet and kernel density estimation methods have also been proposed for dependent data (Khaliq *et al.*, 2006).

Weather generation has historically focused on precipitation, for which temporal persistence is important. In parametric weather generation, time dependence is described with an autoregressive or a clustering model (e.g. Jones *et al.*, 2009; Kyselý, 2009; Nicolosi *et al.*, 2009). Non-parametric methods, which sample from historical observations, must preserve temporal structure: for example, by taking multi-day sections of data. Models with spatial correlation have been developed (Wilks and Wilby, 1999; Mehrotra *et al.*, 2006), but in general weather generation is location-specific (e.g. Jones *et al.*, 2009).

In contrast, catastrophe modelling of hurricanes attempts to quantify their spatial extent, direction of movement, evolution through time and spatio-temporal propagation of their impacts. Spatio-temporal evolution is important for estimating damage: a low-category hurricane that lingers can be more destructive than a high-category one that moves swiftly though. This evolution can be difficult to infer from individual wind speed observations, if they are even available (anenometers often fail in high winds), but is straightforward from individual hurricane tracks. Cat modelling thus has a different philosophical grounding to EVT modelling: it attempts to capture the space-time behaviour of entire event systems (hurricane genesis, landfall and decay) and derive wind speeds from these, rather than characterising the wind speed point data directly.

Cat modelling of hurricanes, though physically motivated where possible, is fundamentally statistical analysis of historical data. Hall and Jewson (2007, 2008) describe various methods for stochastically generating synthetic storm data by calibrating with, or sampling from, historical observations of their location, intensity, direction and speed. Separate statistical models are used for each aspect of the storm: for example, a spatial Poisson process for genesis; a random walk for deviation from mean tracks; and, more recently, regression against time of year and sea surface temperature for evolution of intensity. The impact of the hurricane is then propagated with a physically motivated statistical model to generate region-wide wind fields (with wind magnitudes derived from hurricane direction, circular motion and speed of decay) and downscaled (Chapter 6) to derive winds at point locations; other quantities such as precipitation are also modelled. Correlation is incorporated by the sampling or modelling of entire hurricane systems through time and with model parameters that vary smoothly in space. Cat models typically assume individual hurricane events are temporally independent, although research methods are being developed to account for clustering. They also assume stationarity, but tend to be used for very short time horizons or else observe that the loss uncertainty dominates the hazard uncertainty.

5.3.3.3 Non-stationary data

Most datasets of hydrometeorological extremes are non-stationary. If a near-stationary subset cannot be selected, one solution is to build non-stationarity into the statistical model, but this raises the issue of how to ensure the model holds in the past and continues to hold in the future under changing forcings. Another solution is to regress the hazard or impacts onto covariates, but this also requires assumptions about the stability of these relationships into the future. Statistical modelling approaches to non-stationarity are discussed in this section; physical modelling approaches are discussed in Chapter 6.

Some kinds of non-stationarity are more difficult to deal with than others. The seasonal cycle is an important time-varying influence on many hazards, but a relatively predictable one. The effect of a mode of natural variability on a hazard can be quantified, and if it is long-range (in time) can be used as a predictor variable for forecasting. However, the long-term trends of climate change, and other driving factors such as land use and societal change, are only somewhat predictable in the near-term and much more uncertain after a few decades. For the hazards and impacts that are influenced by changing drivers, neither an assumption of stationarity nor a model of recent non-stationarity are likely to be valid for long.

There are a variety of strategies available to incorporate non-stationarity in a statistical model of historical observations. Typically these focus on changes in location (the mean of the data, or the location parameter of the distribution). They may also include the scale or other parameters, though it is more difficult to estimate the shape parameter when it is allowed to vary (Coles, 2001). Selection of the extremes may be affected: for example, in POT analysis the threshold of exceedance might be varied with time (e.g. Brown *et al.*, 2008; Della-Marta *et al.*, 2009). The simplest approach is to divide the data into subsets that can be assumed stationary and estimate the model parameters separately for each. The subsets may be determined by non-overlapping time windows (e.g. Coles *et al.*, 2003), moving time windows (e.g. Zhang *et al.*, 2001), an external factor such as 'high' or 'low' values of an index of climate variability (e.g. Jagger and Elsner, 2006) or abrupt jumps ('change points') identified with regression, non-parametric (e.g. Li *et al.*, 2005) or Bayesian methods (e.g. Elsner *et al.*, 2004; Zhao and Chu, 2010). An overview of change point detection is given by Kundzewicz and Robson (2004). Dividing the data into subsets is also used for assessing changes in return period and in generating pseudo-weather datasets for the present day and long-term future (Chapter 6). A more flexible approach to discrete subsets is to model the dependence of the parameters on time: as a function of the year, the seasonal cycle or a combination of these (e.g. Ribereau *et al.*, 2008; Kyselý, 2009; Menéndez *et al.*, 2009).

Non-stationarity may be diagnosed using non-parametric tests based on ranking or resampling of the data such as the Kendall's tau (e.g. Kunkel *et al.*, 1999; Zhang *et al.*, 2001; Chavez-Demoulin and Davison, 2005). This avoids the need to propose a model for the data, but does not allow extrapolation outside the data nor identify the shape of the trend. Kundzewicz and Robson (2004) outline some commonly used non-parametric tests of trends in hydrological data. A proposed semi-parametric approach is 'local likelihood', in which parameters are estimated separately at each time using the whole dataset and re-weighted so that nearby points in time assume more importance (e.g. Ramesh and Davison, 2002); this may be helpful in choosing a suitable parametric form (Khaliq *et al.*, 2006).

It may be desirable to remove non-stationarity to de-trend the data. Eastoe and Tawn (2009) point out that there can be problems in fitting data if the exceedance threshold is allowed to vary with time, so they propose a pre-processing approach where the trend is modelled and removed. Pre-processing, also known as 'pre-whitening' (as is decorrelation, Section 5.3.3.2), is discussed by Chatfield (2004).

A hazard, or parameters of a distribution such as the GEV, may be modelled as a function of other variables. These regression approaches typically aim to assess the influence of climatic factors on hazards or the influence of hazards on vulnerable entities. Covariates may be trends (such as global mean temperature) or oscillations (such as modes of natural variability), though over short timescales it may be difficult to distinguish between these.

The parameters of the regression model are estimated with ML or other methods (Sections 5.3.3.1, 5.3.3.4).

Regression of hazards on preceding climatic conditions can be used for near-term forecasting. This has been particularly important for tropical cyclones on seasonal timescales, for which physical modelling is challenging due to the small spatio-temporal scales of cyclogenesis and the initial condition uncertainties. Counts of hurricanes, for example, have been modelled using Poisson regression or the more flexible GLM with respect to sea surface temperatures (AIR Worldwide Corporation, 2006) or metrics such as the ENSO, NAO, SOI (Southern Oscillation Index), Atlantic Multidecadal Oscillation, and Sahel Rainfall Index (Elsner, 2003; Elsner and Bossak, 2004; Elsner and Jagger, 2006). Hurricane wind speeds have been modelled with respect to global mean temperature (Jagger and Elsner, 2006). Camargo *et al.* (2007) outline predictor variables for statistical seasonal hurricane forecasts from different meteorological agencies around the world. Forecasts of hurricane frequency are increasingly made with dynamical models or statistical-dynamical methods, but forecasts of their intensity and landfall are currently more successful from statistical methods. Regression onto modes of internal variability has also been used for mid-latitude storms (e.g. Mailier *et al.*, 2006; Vitolo *et al.*, 2009) and heat waves (e.g. Rusticucci and Vargas, 2002).

Regression-type approaches are important in impacts modelling, where the physical mechanisms are often poorly known or difficult to predict. It is used, for example, to quantify the relationship between extreme temperature or precipitation and human mortality (Githeko and Ndegwa, 2001; Sheridan and Kalkstein, 2004; Baccini *et al.*, 2008), and between hurricanes and economic losses (Saunders and Lea, 2005).

Regression onto modes of natural variability is a space-for-time substitution, deducing a relationship from recent spatial patterns and using it for the future. It can also be viewed as a device for handling spatial dependence. That is, if a hazard can be seen to depend on a large-scale feature such as the NAO, spatial structure can be modelled by quantifying the dependence on the NAO in each location, treating each location as conditionally independent, then inducing spatial dependence through variation in the NAO. For assessments of the far future, regression approaches do not avoid the problem of non-stationarity. They rely on an assumption of the stability of these relationships (e.g. teleconnections) even though they may be significantly altered by changing boundary conditions.

Non-stationary spatio-temporal modelling is very challenging, especially for extremes. This is a relatively new field in statistics, and some of the simpler approaches are rather ad hoc, so it is important to be cautious in interpreting the results. Ribereau *et al.* (2008: 1037), for example, warn that 'Extrapolating the trend beyond the range of observations is always a delicate and sometimes dangerous operation (since we assume that the trend will remain the same in the future), especially when dealing with extremes. Hence high return levels in a non-stationary context must be interpreted with extreme care.'

5.3.3.4 Uncertainty assessment in statistical analysis

Statistical analysis uncertainty can be divided into those from inputs (measurements), model parameters and model structure. Measurement uncertainties vary in space and time due to the changing coverage and quality of observation networks (e.g. surface temperatures: Brohan *et al.*, 2006; rain gauges: McGregor and MacDougall, 2009; and hurricanes: Vecchi and Knutson, 2011). Iman *et al.* (2005a, 2005b) describe the effects of perturbing the inputs of a hurricane catastrophe model.

Parameter confidence intervals are commonly estimated with likelihood-based methods (such as the 'delta' method, which relies on asymptotic and regularity assumptions, or 'profile likelihood', which is generally more accurate but requires more computation; Coles, 2001) or bootstrapping methods (either non-parametric, by resampling from the original data, or parametric, by resampling from the fitted distribution; Davison and Hinkley, 1997). Alternatively, parameter probability density functions can be estimated with Bayesian methods, which are becoming more popular in extremes analysis because they allow prior information to be incorporated, estimate the entire distribution (or summary characteristics such as moments or quantiles), do not require asymptotic and regularity assumptions and are flexible and transparent, This enables assimilation of disparate information (e.g. Jagger and Elsner, 2006; Baccini *et al.*, 2008), and makes it easier to 'keep track' of all sources of uncertainty (e.g. Coles *et al.*, 2003). Pang *et al.* (2001) give an introduction to two of the simplest computational methods in the context of extreme wind speed analysis. Generalised maximum likelihood (El Adlouni *et al.*, 2007) has been suggested as a compromise, reducing computational expense by focusing on the mode of the posterior distribution.

Quantification of uncertainties can be sensitive to the chosen subset of data and method: for example, confidence intervals may be smaller for *r*-largest maxima than block maxima (due to the larger dataset), but care must be taken that the model is still valid with the lower threshold; confidence interval estimates for some distributions may be very dependent on the method (e.g. short-tailed distribution for storms: Della-Marta *et al.*, 2009); non-parametric bootstrapping may underestimate uncertainties relative to parametric boot-strapping (Kyselý, 2008); and likelihood-based methods may give unphysical parameter estimates (Pang *et al.*, 2001).

Structural uncertainty in statistical analysis is associated with the choice of distribution or model and whether the necessary assumptions for that model are valid (e.g. independence, stationarity and regularity). Models are usually used to extrapolate to rarer extremes than are in the observations; GEV-based estimates of uncertainty for extrapolated values should be considered lower bounds (Coles, 2001). Robust testing of model reliability is essential. Goodness-of-fit, the success of a model in describing observations, is assessed using graphical methods such as quantile plots (Coles, 2001; Beirlant *et al.*, 2004) or statistical tests (Coles, 2001); the properties of these tests vary, which can affect the results (e.g. Lemeshko *et al.*, 2009, 2010).

In extreme value analysis, data selection affects model structural uncertainty: *r*-largest and peaks-over-threshold analyses use more of the observations than block maxima, and therefore provide more information with which to check the model, but if *r* is too large or the threshold is too low the model assumptions may no longer be valid. Long record length is required for the asymptotic limit, but a short record may be required for stationarity.

5.4 Risk management

Risk management comprises the steps taken to assess, mitigate and cope with risk in the periods between hazard events (monitoring and planning: Section 5.4.1) and during and after hazard events (communication, response, recovery and analysis: Section 5.4.2). Aspects of risk management that are relevant to climate change adaptation are discussed in Chapter 6.

5.4.1 Hazard monitoring and planning

Monitoring of hazards is the real-time forecasting and early warning systems of hazard events or their impacts (Section 5.4.1.1). Planning for hazards encompasses developing tools to enable rapid decision-making, such as event triggers or emergency procedures (Section 5.4.1.2), and making long-term decisions, such as land-use planning, regulation of the private sector or hedging financial risk (Section 5.4.1.3).

5.4.1.1 Monitoring and warning systems

The purpose of monitoring systems is primarily to inform decision-making before and during a hazard event. They provide information that can be translated into event probabilities, and are often assimilated into NWP models that generate warnings and trigger actions (such as communication or emergency response) when the forecast probability for a hazard of a given magnitude exceeds a threshold. Monitoring networks are well developed for most hydrometeorological hazards due to longstanding meteorological services around the world. Individual agencies such as the UK Met Office (UKMO) issue daily weather forecasts that include warnings of extreme precipitation, storms, heat waves and severe marine conditions. They also contribute to regional warning services such as the European Meteoalarm (http://www.meteoalarm.eu) and offer consulting services for users that require non-standard forecasts.

Monitoring systems are continuously updated with improved understanding of the underlying processes (e.g. for droughts, tropical cyclones) and links with impacts (e.g. for heat waves). Heat wave monitoring, for example, has traditionally been based on daily maximum temperatures, but since heat wave disasters such as the 1990s Chicago and 2003 European events a number of cities have introduced warning systems that take account of other conditions linked with excess mortality and ill health. The Heat Index of the NOAA NWS includes both temperature and humidity, and several warning systems use spatial synoptic classification (Sections 5.2.1.2, 5.4.1.2).

Lead times govern the effectiveness of early warning systems in minimising hazard impacts. Even if the inception of some hazards cannot be accurately simulated, the conditions under which they might form are known and this can be used to set a threat level. Tornado lead times are very short, up to 15–20 minutes, and warning systems are only operational in a few countries, but the atmospheric conditions in which a tornado is formed can be forecast several hours in advance, which enables people to move nearer to shelters

(United Nations, 2006). The number of tornado-related deaths significantly dropped in the United States during the last century, mainly as a result of the Doppler Radar Network (United Nations, 2006). Tropical cyclones such as hurricanes are monitored globally by the WMO Global Tropical Cyclone Warning System, with lead times ranging from 24 hours to several days (United Nations, 2006). A long lead time is necessary for large-scale evacuations to avoid widespread loss of life, but not sufficient: evacuation orders were given two days before Hurricane Katrina made landfall in Louisiana, but they were voluntary. This, along with other failings, led to avoidable deaths (Section 5.4.2.4).

For longer timescales, predictions of the overall severity of the coming tropical storm season are made with methods from across the spectrum of approaches: statistical, historically calibrated; the new generation of high-resolution dynamical models; and hybrid statistical-dynamical methods (Sections 5.3.2, 5.3.3; Camargo et al., 2007). Severity is measured in terms of, for example, frequency and ACE, and expressed as summaries (e.g. median, spread and long-term mean) or probabilistic forecasts with uncertainties (Section 5.4.2.1).

Drought is a multivariate and slow-developing hazard, so warning systems are more complex and less well-developed. Droughts do not require rapid response, but the continuation of an existing drought can be assessed with an NWP model in seasonal prediction mode. Droughts are typically monitored by national or regional centres in association with meteorological agencies, using observations of several quantities including precipitation, streamflow, groundwater levels and snow pack, together with historical observations and climate model simulations (United Nations, 2006). These are combined into multi-index (Section 5.2.1.3) drought monitors that include quantitative and graphical tools. The US National Drought Mitigation Center (NDMC), for example, together with the NOAA and the US Department of Agriculture, produces the US Drought Monitor based on the PDSI, SPI and other metrics. A Global Drought Monitor based on PDSI, SPI and the NDMC drought categories is produced by the University College London (UCL) Hazard Research Centre.

Impacts of hazards may also be monitored to aid decision-making and, in the longer term, policy-making. During summer months, for example, the UK Health Protection Agency monitors heat wave health impacts using the number of calls to the National Health Service advice phoneline and the number of consultations with family doctors ('Heatwave plan for England 2009', available from http://www.dh.gov.uk). In the United States, the NDMC monitors drought impacts with the Drought Impact Reporter, which shows the number of reported impact events by state and type (agriculture, water and energy, environment, fire and social). For regions vulnerable to famine, the Famine Early Warning Systems Network (FEWS NET) monitors a number of driving factors such as rainfall and vegetation to provide early warning for issues of food security.

Although monitoring and warning systems for hydrometeorological hazards are relatively well developed, a United Nations report commented that collaboration between the hydrometeorological community and the disaster risk management and response communities could be improved (United Nations, 2006).

5.4.1.2 Rapid decision-making

Quantitative thresholds have been defined to aid decision-making during an event for many of the hydrometeorological hazards. Heat waves, in particular, have clearly defined procedures. Some are based only on temperature, while others include factors important to health identified in statistical analysis of past observations. In the United States, for example, alert procedures are triggered when the NWS Heat Index (Section 5.2.1.2) is expected to exceed 105–110°F (41–43°C), depending on local climate, for at least two consecutive days. The UK 'Heatwave plan for England 2009' has four levels of decision-making and action, each a list of responsibilities for government departments and author-ities. The first is triggered by date (1 June), outlining general preparedness and urban planning measures; the second is triggered by a UKMO forecast of 60% probability of temperatures exceeding regional impact thresholds for maximum and minimum temper-atures on at least two consecutive days; the third when those thresholds are exceeded; and the fourth when the heat wave is severe and prolonged. Regional maximum and minimum temperature thresholds range from 28°C and 15°C for the north-east to 32°C and 18°C for London. This system draws on the World Health Organization's 'EuroHEAT' project (Michelozzi *et al.*, 2007; Baccini *et al.*, 2008).

Triggers based on synoptic heat wave classification (Section 5.2.1.2) and forecast impacts are used by the Toronto Heat Health Alert System. If an 'oppressive air mass' is forecast with a probability of excess mortality exceeding 65%, various actions take place, including: distribution of information through media, government and non-profit organisations; direct contacting of vulnerable individuals; opening of a Heat Information Line; and daytime opening of homeless shelters. If the probability of excess mortality is predicted to be greater than 90%, additional actions occur such as the opening of community 'Cooling Centres' and the extension of swimming pool opening hours.

Risk management for droughts also include quantitative thresholds. In the UK, the Environment Agency produces drought plans for water companies outlining a range of trigger levels based on drought status and calendar date, with corresponding actions to manage supplies and demand (UK Environment Agency, 2009).

Severe events trigger emergency procedures, which may include mandatory orders to protect public health or services. One example is the Assistance to Shelter Act introduced in 2009 by the Legislative Assembly of British Columbia, which states that a police officer may transport a person they consider to be at risk to an emergency shelter 'using reasonable force if necessary'. In the case of severe drought in the UK, interruptions to domestic water supply can be authorised by emergency drought order (EDO); these have been used three times in England and Wales since 1945, most recently in 1976 (UK Government Cabinet Office, 2008). For some catastrophic hazards such as hurricanes, local and national govern-ments have the authority to order mandatory evacuation, though in practice this is not, or cannot be, enforced (Fairchild *et al.*, 2006; Townsend, 2006). Mandatory orders are not only given by the state or police; in the UK, restrictions on non-essential water use, such as hosepipe and sprinkler bans, may be imposed during droughts by the private water companies without permission from the government or Environment Agency – these may be sufficient to avoid severe impacts (UK Government Cabinet Office, 2008).

Training exercises are used to inform and develop emergency plans. The US Federal Emergency Management Agency (FEMA) funded and participated in a five-day hurricane disaster simulation exercise in the New Orleans area called 'Hurricane Pam' about a year before the real Hurricane Katrina struck the same location. The hurricane category was the same and the simulated impacts very similar, including large-scale evacuation, breached levees and extensive destruction (US House of Representatives, 2006). Hurricane Pam was not a total success in preparing officials for Hurricane Katrina (Section 5.4.2.4), but the medical contingency plan it yielded, though unfinished, was judged 'invaluable' to the response effort (US House of Representatives, 2006).

5.4.1.3 Long-term planning

Long-term planning for risk management should ideally aim to use 'low-regrets' options (IPCC, 2011) that address a wide range of extreme events and other societal benefits such as improved livelihoods, sanitation, infrastructure and sustainability. Some examples are: for heat waves, improved social care networks for vulnerable groups; for droughts, improved irrigation efficiency and use of drought-resistant crop varieties; and for hurricanes, adoption and enforcement of building codes (IPCC, 2011).

Planning regulations exist to enforce good risk management in the private and public sectors and minimise losses for vulnerable entities and infrastructure. An important area of long-term planning is the setting of design codes. Buildings and other structures must be constructed to withstand extreme wind, and extreme waves and currents if offshore. Regional variations in building code levels and their enforcement can significantly affect vulnerability (RMS, 2005). Design codes are almost always expressed in terms of return values and thus inherently assume stationarity of risks. For safety reasons, the intended design life of a structure is invariably much shorter than the design value. The most commonly used design return values are 50 or 100 years. Ship certification agencies such as Lloyds and Norske Veritas demand that vessels can withstand 50-year return wave conditions, and limit their certification to those sea areas and seasons where this can be guaranteed.

Insurers manage risk using the predictions of catastrophe models (Section 5.3.3). A simple example given by RMS (2008) is summarised here. If a cat model (or the weighted or unweighted mean of several models) estimates the 99.6th percentile probability of exceedance at US$40 million insured losses for a given year, but the insurance company only has capital to cover US$30 million losses, it must reduce the portfolio at risk. Some options for this are: increasing premiums, refusing coverage or hedging the risk with reinsurance or catastrophe bonds (insurance-linked securities related to natural catastrophes). 'Cat' bonds are increasingly seen as more desirable than reinsurance because they provide access to a very large pool of capital and most investment portfolios have little correlation with natural catastrophes (RMS, 2008). More details on risk management with catastrophe models are provided by Grossi and Kunreuther (2005).

Insurance industry risks are regulated by external public bodies to help protect the public and public funding from profiteering or collapse. The Florida Commission, for example, certifies catastrophe models to ensure they give plausible predictions and that insurers have sufficient capital for the predicted exceedance probabilities. The 'Solvency II' regulation treaty is a recent development in the European insurance industry with global repercussions. European insurance and reinsurance companies must demonstrate to European Union regulators their solvency under the risk they retain, and catastrophic risk from hydro-meteorological hazards is a major contributor. 'Solvency II' is driven by the European Union, but companies with global portfolios must estimate their global risk. There are similar initiatives in the Japanese and American insurance and reinsurance markets. A further example of insurance industry regulation is described in Section 5.4.2.4.

5.4.2 Hazard communication, response and recovery

5.4.2.1 Communication

Risk communication for extreme weather hazards might have some advantage over other hazards because the public are exposed to daily weather forecasts, though familiarity could breed overconfidence in interpretation. There is evidence that the public infer their own probabilities from deterministic forecasts based on their past experiences, such as a decrease in forecast skill with increasing lead time, or a reduced skill for rainfall relative to temperature (Morss *et al.*, 2008; Joslyn and Savelli, 2010), but this blurs the distinctions between forecaster, decision-maker and user. Alternatively, a probabilistic forecast can be provided: since the 1960s, the NWS (known at the time as the Weather Bureau) has supplied probability of precipitation (PoP) forecasts. There is evidence to suggest that users make better decisions if forecast uncertainties are given (Roulston *et al.*, 2006; Roulston and Kaplan, 2009).

Extreme weather warnings summarise ensemble forecasts (Section 5.3.2.2) with varying degrees of information content (see Stephens *et al.*, 2012). The 'probability' or 'risk' of a hazard essentially corresponds to the fraction of the ensemble that predicts a given event, though additional processing and interpretation by human forecasters is usually involved. The fraction may be expressed in categories: the UKMO, for example, issues 'Severe Weather Warnings' using a traffic-light system with colours ranging from green for less than 20% ('Very low') risk to red for greater than or equal to 60% ('High') risk. A similar approach is a spatial map of probability contours, such as the NOAA Storm Prediction Center forecasts for wind storms, tornados and hail (Figure 5.5); in the wind storm maps, hatched areas correspond to 10% or greater probability of wind gusts 65 knots or greater within 25 miles of a point. The NOAA provides a 'conversion table' alongside these probability forecasts, with which users can convert the numbers to risk categories 'slight', 'medium' and 'high'.

For continuous variables (rather than discrete events such as storms), full ensemble forecasts can be displayed as 'spaghetti diagrams'. An example of this is provided by the NOAA NWS, though on the NCEP Environmental Modeling Center website rather than the main NWS forecast site. The ensemble members are shown as individual contours, which

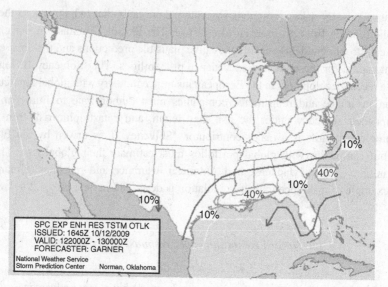

Figure 5.5 An example of the NOAA Storm Prediction Center Experimental Enhanced Resolution Thunderstorm Outlook: each forecast contains 10%, 40% and 70% contours for the probability of thunderstorms during the forecast period. From the National Weather Service.

diverge through time (Figure 5.6). Spaghetti diagrams are analogous to the 'smiley face' representations of risk often used in healthcare, because they show 'multiple possible futures' (Edwards *et al.*, 2002).

However, these forecasts of ensemble frequencies generally do not contain estimates of uncertainty or model skill. Tropical storm forecasts issued by the NOAA are one exception: their uncertainty is communicated in a variety of graphical representations, of which the most well-known is the National Hurricane Center (NHC) 'cone of uncertainty' for storm tracks. This includes: present storm location and type, forecast track line, a white-filled cone representing average forecast error out to three days, a black cone outline representing forecast error for days four and five, and coastlines under a watch/warning (Figure 5.7). The cone is constructed from a series of circles, each of which represents a forecast period (e.g. 36 hours), with the size of each circle set to enclose 67% of the previous five years' official forecast errors. The cone of uncertainty is intended for real-time decision-making by the public, government agencies and the private sector, and is widely used and largely success-ful. However, many users do misinterpret it. Some interpret the central track line as the only area of impact, while others interpret the cone as the area of impact rather than of storm track uncertainty (for more detail, see Broad *et al.*, 2007; Stephens *et al.*, 2012). This has serious implications because (by design) the cone only predicts the central path correctly about two-thirds of the time. Broad *et al.* (2007) show that 'more' is not always better than 'less' in the case of forecasting, and that uncertainty in one quantity (track location) may be misinter-preted as certainty in another (impact region). New visualisation options have been suggested (Figure 5.7), but the cone of uncertainty has not, as yet, substantively changed.

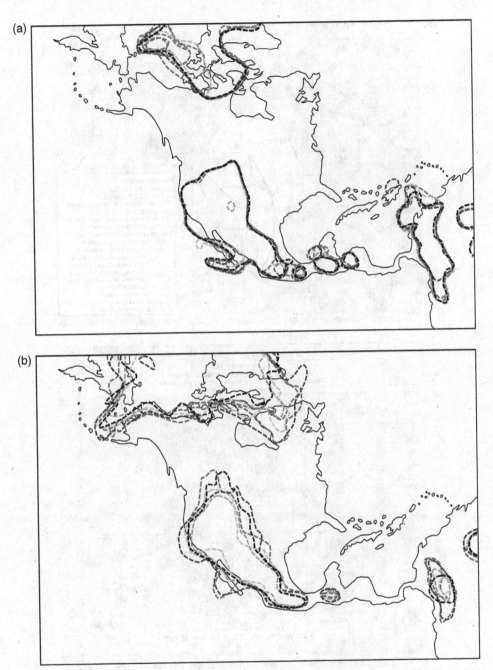

Figure 5.6 NCEP spaghetti diagrams of temperature at 850 hPa height, showing model predictions for contours −5°C and 20°C at 00 hours (a) and 192 hours (b). Adapted from the National Weather Service.

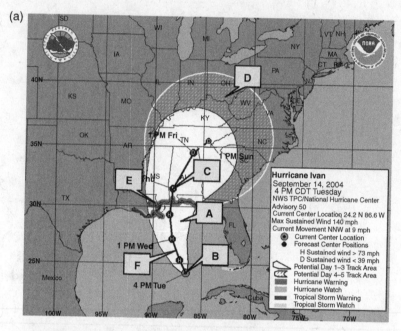

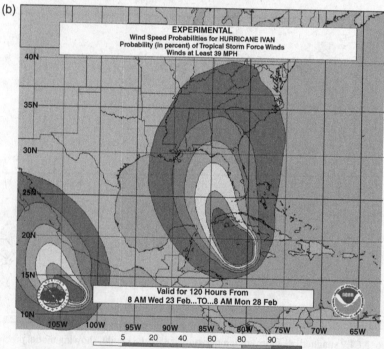

Figure 5.7 The forecast of Hurricane Ivan in 2004 displayed as: (a) The NHC cone of uncertainty (with annotations: see Broad *et al.*, 2007); and (b) an alternative experimental visualisation (Broad *et al.*, 2007) © American Meteorological Society.

Uncertainty estimates are more often provided for seasonal forecasts, reflecting the inherent challenges of predictability. For North Atlantic tropical storms, the UKMO includes uncertainty ranges for the number and ACE up to six months ahead. An example is:

Six tropical storms are predicted as the most likely number to occur in the North Atlantic during the July to November period, with a 70% chance that the number will be in the range three to nine. This represents below-normal activity relative to the 1990–2005 long-term average of 12.4. An ACE index of 60 is predicted as the most likely value, with a 70% chance that the index will be in the range 40 to 80 – which is below normal relative to the 1990–2005 average of 131.

Median, spread and long-term mean are also visualised as barcharts. The simultaneous use of different expressions (language, quantities and graphics) reflects lessons that have been learned about public understanding of risk: it is clear there is no 'one size fits all' to communication or visualisation of uncertainty (Broad *et al.*, 2007; Morss *et al.*, 2008; Joslyn and Nichols, 2009; Joslyn *et al.*, 2009; Spiegelhalter *et al.*, 2011; Stephens *et al.*, 2012).

During a hazard event, effective communication of the emergency response is also essential, to minimise negative impacts (such as lawlessness) that can arise when people are uncertain about their survival, rescue or prospects for evacuation (US House of Representatives, 2006).

5.4.2.2 Emergency response

Most extreme weather disaster events require responses on timescales of hours to days (droughts are an exception). Responses during a hazard event may be proactive (minimising impacts) or reactive (coping with impacts). The former reflect successful implementation of well-designed emergency procedures; the latter reflect poor implementation or poor planning, and may give rise to the need for humanitarian relief (IPCC, 2011). The training exercise Hurricane Pam, for example, was criticised for its 'emphasis on managing the aftermath of the catastrophe and not creating initiatives that would diminish the magnitude' (US House of Representatives, 2006).

Actions include evacuation, search and rescue, providing shelter and giving medical aid, many of which require complex operations including logistics (management of the supply, transport and provision of resources such as blankets and clean water) and effective public communication (Section 5.4.2.1). Hazard response may include aspects that are not essential to immediate survival such as care of mental health (e.g. farmers during drought: Hayes *et al.*, 2004). Police and military personnel may be required to enforce mandatory emergency orders or, in the case of temporary societal breakdown leading to theft or violence, the basic rule of law (e.g. Hurricane Katrina: US House of Representatives, 2006). Response efforts may be hampered as infrastructure can be damaged and services interrupted in unpredictable ways; this is particularly challenging if a disaster occurs on a large spatial scale. The emergency post-disaster phase of Hurricane Katrina was unusually long (emergency sheltering ended only after 14 weeks), 3.5 times longer than the comparable-scale disaster of the 1906 San Francisco earthquake (Kates *et al.*, 2006).

5.4.2.3 Post-impact recovery

Post-event recovery includes restoring infrastructure and services, repairing or replacing damaged buildings and clearing debris. If there are a large number of deaths, the recovery period may also include dealing with corpses. Environmental recovery may include reforestation or reintroduction of wildlife species. Recovery planning should aim to mitigate risk from future disasters (e.g. discouraging rebuilding in risk-prone areas) and if possible make other improvements, such as upgrading services or increasing sustainability (IPCC, 2011). Reddy (2000) examines aspects that make this more or less challenging, such as the level of involvement of stakeholders, in the context of the 1989 Hurricane Hugo. Most aspects of post-disaster recovery are similar to those of other natural hazards. One area in which they differ is that the spatial extent may be very large – an entire country or continent affected by the same hazard – which can hamper recovery of infrastructure and services.

5.4.2.4 Post-impact analysis and lessons learned

A catastrophic event provides a window of opportunity in which it is easier to make policy changes due to increased awareness and availability of resources (CCSP, 2008). Post-event evaluation of the forecasts and decision-making leads to improvements in risk assessment and management: Weisheimer *et al.* (2011), for example, find they can successfully 'hindcast' the 2003 European heat wave once several model parameterisation changes are made. New damage and loss datasets may be used to verify impacts model predictions to improve risk management; however, disaster-related datasets are generally lacking at the local level (IPCC, 2011).

As a result of the severe European heat wave in 2003, the UK introduced the Heat Health Watch System, and during the hot weather of July 2006 only 680 excess deaths were recorded (UK Government Cabinet Office, 2008). Lessons were similarly learned after the Chicago heat wave of 1995 (Klinenberg, 2002). The city had made disastrous mistakes: an emergency was not declared until hundreds of fatalities had occurred; the Fire Department refused paramedic requests for additional staff and ambulances; there was no system to monitor the hospital bypass situation; and the Police Department did not attend to elderly residents. Public relations apparatus was even used to deny there was a disaster and later to define the disaster as natural and unpreventable. When a second heat wave occurred four years later, the city acted quickly and effectively, issuing warnings and press releases to the media, opening cooling centres with free bus transportation, contacting elderly residents and sending police officers and city workers to visit the elderly that lived alone. These actions are thought to have drastically reduced the death toll in the second heat wave: 110 residents died, compared with an excess mortality of 739 four years earlier.

The US House of Representatives report on Hurricane Katrina described some successes in risk assessment and management, including the 'accuracy and timeliness of National Weather Service and National Hurricane Center forecasts' that 'prevented further loss of life'. But it also described a litany of failures and avoidable negative impacts. These included failures in:

(1) *long-term planning*: inadequate levees; inadequately trained and experienced staff; lack of a regional shelter database; an overwhelmed logistics system;

(2) *development of emergency procedures*: lessons not fully learned from the Hurricane Pam exercise; inadequate planning for massive damage to communications; lack of preparation to evacuate and provide medical care for vulnerable groups;

(3) *implementation of emergency procedures*: mandatory evacuation ordered only 19 hours before landfall, rather than at the time of early warning (56 hours); critical elements of the National Response Plan executed late, ineffectively or not at all; poor coordination between federal and state bodies; reactive, rather than proactive, deployment of medical personnel; subjective evacuation decisions for nursing homes that led to preventable deaths; collapse of local law enforcement and lack of effective public communication that led to civil unrest and further delay in relief (US House of Representatives, 2006).

Lessons of risk management may be learned long after the initial recovery period. Homeowners affected by Hurricane Katrina were typically covered by private insurers for wind damage and by the federal government's National Flood Insurance Program (NFIP), or not at all, for flood damage. This led to large uncertainties in forecasting insured losses for insurers, due to the difficulty in predicting damage assessments and disputes (RMS, 2005), and in the extent of coverage for policy-holders due to the cap on NFIP flood claims (Williams, 2008) or lack of flood insurance. Private insurers that issued NFIP flood insurance policies ('write-your-own', WYO) determined the apportioning damages between wind and flooding, which led to conflicts-of-interest, claim disputes and litigation (Williams, 2008). As a result of the flood-related claims of Hurricane Katrina, the NFIP's borrowing authority rose 14-fold to about US\$20.8 billion (US GAO, 2007). To ensure the NFIP did not pay too much in future storm events, the US GAO recommended increasing FEMA's access to the procedures and claims information from WYO insurers (Williams, 2008). At the time of writing, these provisions (Flood Insurance Reform Priorities Act of 2011; Flood Insurance Reform Act of 2011) are under consideration by the US Congress.

5.5 Summary

Risk assessment for hydrometeorological hazards builds on a long history of weather forecasting. The definition of these hazards is continuously evolving, from simple univariate metrics to complex multivariate, impacts-related indices, to improve their relevance and usefulness. Nonlinear feedbacks in the climate, and between climate and impacts, make their prediction extremely challenging, and there is a limit of predictability of about a fortnight for atmospheric phenomena. But physical and statistical modelling are strong, diverse research areas that complement each other: the former mainly for event forecasting and warning systems, the latter for longer-term risk management. Both make use of long-term oscillatory modes of natural variability correlated with extreme weather (such as the ENSO) to extend the predictability of risk to many months in advance. Impacts modelling is still catching up with hazard event modelling in terms of comprehensiveness and uncertainty quantification, which is perhaps inevitable given their later position in the causal chain.

Physically based models are generally well-resourced and very frequently tested against observations in operational numerical weather prediction, so their predictive skill is increasing decade-upon-decade: this is due to wider availability and better assimilation of observations, and improvements in models from the addition of physical processes and increases in resolution. Statistical modelling has advanced beyond the early extreme value analyses by incorporating model flexibility, spatial and temporal correlation and non-stationarity, though progress can still be made in these areas and caution must be used in interpreting estimates for the far future. Catastrophe modelling is still catching up with other types of modelling in terms of physical theory and non-stationarity, but leads others in terms of end-to-end assessment of the cascading risks from hazard to financial loss.

In risk management, hydrometeorological hazards are in the fortunate position of having extensive observational and forecasting networks, from longstanding national meteorological agencies and the regional collaborations between them. These give early warnings for extreme weather conditions with varying levels of information content and uncertainty quantification; lessons are still being learned about the most successful approaches. Quantitative decision-making thresholds are incorporated into emergency planning, and policy has long existed for risk mitigation in the form of building codes and other regulation. However, the implementation of risk management strategies has sometimes been lacking, notably during the catastrophic Hurricane Katrina in 2005. Insurers use catastrophe models to assess hurricane risk as a fraction of their capital, which is subject to regulation.

Hydrometeorological hazards have the disadvantage, then, of occurring everywhere under the sun. But longstanding interests in the weather, water availability and ocean conditions have driven the development of sophisticated computational and statistical techniques, and fairly comprehensive observational networks, so that a substantial amount of the risk may be managed.

Acknowledgements

Many thanks to Debbie Hemming and Christos Mitas for helpful reviews that improved the manuscript. Thanks to Michael Bründl, Joanne Camp, Andy Challinor, Ruth Doherty, Carina Fearnley, Bruce Malamud, Doug MacNeall, Mike Steel, Roger Street, Paul Valdes, Renato Vitolo and Nick Watkins for useful discussions and contributions. Thanks to the Natural Environment Research Council (NERC) and the Bristol Environmental Risk Research Centre (BRISK) for financial support.

References

AIR Worldwide Corporation (2006) Understanding climatological influences on hurricane activity. *The AIR Near-term Sensitivity Catalog.* pp. 1–16.

Alcamo, J., Acosta-Michlik, L., Carius, A., *et al.* (2008) A new approach to quantifying and comparing vulnerability to drought. *Regional Environmental Change* **8** (4): 137–149.

Alexander, L. V., Zhang, X., Peterson, T. C., *et al.* (2006) Global observed changes in daily climate extremes of temperature and precipitation. *Journal of Geophysical Research* **111** (D5): D05109.

Anthes, R. A., Corell, R. W., Holland, G., *et al.* (2006) Hurricanes and global warming: potential linkages and consequences. *Bulletin of the American Meteorological Society* **87** (5): 623–628.

Baccini, M., Biggeri, A., Accetta, G., *et al.* (2008) Heat effects on mortality in 15 European cities. *Epidemiology* **19** (5): 711–719.

Barnett, D., Brown, S., Murphy, J., *et al.* (2006) Quantifying uncertainty in changes in extreme event frequency in response to doubled CO2 using a large ensemble of GCM simulations. *Climate Dynamics* **26** (5): 489–511.

Barnston, A. G., Mason, S., Goddard, L., *et al.* (2003) Multimodel ensembling in seasonal climate forecasting at IRI. *Bulletin of the American Meteorological Society* **84** (12): 1783–1796.

Barriopedro, D., Fischer, E., Luterbacher, J., *et al.* (2011) The hot summer of 2010: redrawing the temperature record map of Europe. *Science* **332** (6026): 220–224.

Beirlant, J., Goegebeur, Y., Segers, J., *et al.* (2004) *Statistics of Extremes: Theory and Applications*, Hoboken, NJ: John Wiley and Sons.

Beniston, M. (2004) The 2003 heat wave in Europe: a shape of things to come? An analysis based on Swiss climatological data and model simulations. *Geophysical Research Letters* **31** (2): L02202.

Beniston, M., Stephenson, D., Christensen, O., *et al.* (2007) Future extreme events in European climate: an exploration of regional climate model projections. *Climatic Change* **81**: 71–95.

Berner, J., Leutbecher, M., Palmer, T., *et al.* (2009) A spectral stochastic kinetic energy backscatter scheme and its impact on flow-dependent predictability in the ECMWF ensemble prediction system. *Journal of the Atmospheric Sciences* **66** (3): 603–626.

Berrocal, V. J., Raftery, A. E. and Gneiting, T. (2007) Combining spatial statistical and ensemble information in probabilistic weather forecasts. *Monthly Weather Review* **135** (4): 1386–1402.

Brier, G. (1950) Verification of forecasts expressed in terms of probability. *Monthly Weather Review* **78** (1): 1–3.

Broad, K., Leiserowitz, A., Weinkle, J., *et al.* (2007) Misinterpretations of the 'Cone of Uncertainty' in Florida during the 2004 Hurricane season. *Bulletin of the American Meteorological Society* **88** (5): 651–667.

Brohan, P., Kennedy, J., Harris, S., *et al.* (2006) Uncertainty estimates in regional and global observed temperature changes: a new data set from 1850. *Journal of Geophysical Research-Atmospheres* **111** (D12): D12106.

Brown, S. J., Caesar, J. and Ferro, C. A. T. (2008) Global changes in extreme daily temperature since 1950. *Journal of Geophysical Research-Atmospheres* **113** (D5): D05115.

Browning, K. A. (2004) The sting at the end of the tail: damaging winds associated with extratropical cyclones. *Quarterly Journal of the Royal Meteorological Society* **130** (597): 375–399.

Bründl, M. and Rickli, C. (2002), The storm Lothar 1999 in Switzerland: an incident analysis. *Forest, Snow and Landscape Research*, **77**(1/2): 207–217.

Buizza, R., Miller, M. and Palmer, T. (1999) Stochastic representation of model uncertainties in the ECMWF Ensemble Prediction System. *Quarterly Journal of the Royal Meteorological Society* **125** (560): 2887–2908.

Burke, E. J. and Brown, S. J. (2008) Evaluating uncertainties in the projection of future drought. *Journal of Hydrometeorology* **9** (2): 292–299.

Camargo, S.J. and Barnston, A.G. (2009) Experimental dynamical seasonal forecasts of tropical cyclone activity at IRI. *Weather and Forecasting* **24** (2): 472–491.

Camargo, S.J., Barnston, A., Klotzbach, P., *et al.* (2007) Seasonal tropical cyclone forecasts. *WMO Bulletin* **56** (4): 297–309.

Casati, B., Wilson, L. and Stephenson, D. (2008) Forecast verification: current status and future directions. *Meteorological Applications* **15**: 3–18.

CCSP (2008) *Weather and Climate Extremes in a Changing Climate. Regions of Focus: North America, Hawaii, Caribbean, and U.S. Pacific Islands*. Washington, DC: Department of Commerce, NOAA's National Climatic Data Center.

Chatfield, C. (2004) *The Analysis of Time Series: An Introduction*, 6th edn, Boca Raton, FL: CRC Press.

Chavez-Demoulin, V. and Davison, A. (2005) Generalized additive modelling of sample extremes. *Journal of the Royal Statistical Society Series C-Applied Statistics* **54** (1): 207–222.

Coles, S. (2001) *An Introduction to Statistical Modeling of Extreme Values*, New York, NY: Springer.

Coles, S., Pericchi, L. and Sisson, S. (2003) A fully probabilistic approach to extreme rainfall modeling. *Journal of Hydrology* **273** (1–4): 35–50.

Cummins, J. (2007) Reinsurance for natural and man-made catastrophes in the United States: current state of the market and regulatory reforms. *Risk Management and Insurance Review* **10** (2): 179–220.

Curriero, F., Patz, J., Rose, J., *et al.* (2001) The association between extreme precipitation and waterborne disease outbreaks in the United States, 1948–1994. *American Journal of Public Health* **91** (8): 1194.

Dai, A. (2011) Drought under global warming: a review. *Wiley Interdisciplinary Reviews: Climate Change* **2**: 45–65.

Davison, A.C. and Hinkley, D.V. (1997) *Bootstrap Methods and their Application*, Cambridge: Cambridge University Press.

De Bono, A., Peduzzi, P. and Giuliani, G. (2004) Impacts of summer 2003 heat wave in Europe. *Early Warning on Emerging Environmental Threats* **2**: n.p.

Della-Marta, P.M., Mathis, H., Frei, C., *et al.* (2009) The return period of wind storms over Europe. *International Journal of Climatology* **29** (3): 437–459.

Dlugolecki, A., *et al.* (2009) Climate change and its implications for catastrophe modelling. In *Coping with Climate Change: Risks and Opportunites for Insurers*, London: Chartered Insurance Institute, pp. 1–23.

Dorigo, W.A., Wagner, W., Hohensinn, R., *et al.* (2011) The International Soil Moisture Network: a data hosting facility for global in situ soil moisture measurements. *Hydrology and Earth System Sciences Discussions* **8** (1): 1609–1663.

Dysthe, K., Krogstad, H.E. and Muller, P. (2008) Oceanic rogue waves. *Annual Review of Fluid Mechanics* **40**: 287–310.

Eastoe, E. and Tawn, J. (2009) Modelling non-stationary extremes with application to surface level ozone. *Journal of the Royal Statistical Society Series C-Applied Statistics* **58** (1): 25–45.

Edwards, A., Elwyn, G. and Mulley, A. (2002) Explaining risks: turning numerical data into meaningful pictures. *British Medical Journal* **324** (7341): 827–830.

El Adlouni, S., Ouarda, T., Zhang, X., *et al.* (2007) Generalized maximum likelihood estimators for the nonstationary generalized extreme value model. *Water Resources Research* **43** (3): W03410.

Elsner, J. (2003) Tracking hurricanes. *Bulletin of the American Meteorological Society* **84** (3): 353–356.

Elsner, J. and Bossak, B. H. (2004) Hurricane Landfall Probability and climate. In *Hurricanes and Typhoons: Past, Present and Future*, New York, NY: Columbia University Press, pp. 333–353.

Elsner, J. and Jagger, T. H. (2006) Prediction models for annual US hurricane counts. *Journal of Climate* **19** (12): 2935–2952.

Elsner, J., Niu, X. and Jagger, T. (2004) Detecting shifts in hurricane rates using a Markov chain Monte Carlo approach. *Journal of Climate* **17** (13): 2652–2666.

Emanuel, K. (2005) Increasing destructiveness of tropical cyclones over the past 30 years. *Nature* **436** (7051): 686–688.

Emanuel, K., Ravela, S., Vivant, E., *et al.* (2006) A statistical deterministic approach to hurricane risk assessment. *Bulletin of the American Meteorological Society* **87** (3): 299–314.

Engeland, K., Hisdal, H. and Frigessi, A. (2004) Practical extreme value modelling of hydrological floods and droughts: a case study. *Extremes* **7**: 5–30.

Evensen, G. (1994) Sequential data assimilation with a nonlinear quasi-geostrophic model using Monte Carlo methods to forecast error statistics. *Journal of Geophysical Research* **99** (C5): 10143–10162.

Evensen, G. (2009) *Data Assimilation: The Ensemble Kalman Filter*, New York, NY: Springer.

Fairchild, A., Colgrove, J. and Moser Jones, Marian (2006) The challenge of mandatory evacuation: providing for and deciding for. *Health Affairs* **25** (4): 958–967.

FEMA (2007) HAZUS-MH MR3 Technical Manual.

Ferro, C. (2007) A probability model for verifying deterministic forecasts of extreme events. *Weather and Forecasting* **22** (5): 1089–1100.

Ferro, C. A. T., Richardson, D. S. and Weigel, A. P. (2008) On the effect of ensemble size on the discrete and continuous ranked probability scores. *Meteorological Applications* **15** (1): 19–24.

Fink, A. H., Brücher, B., Ermert, V., *et al.* (2009) The European storm Kyrill in January 2007: synoptic evolution, meteorological impacts and some considerations with respect to climate change. *Natural Hazards and Earth System Science* **9**: 405–423.

Forristall (2005) Understanding rogue waves: Are new physics really necessary? Proceedings of the 14th 'Aha Huliko'a Winter Workshop 2005 on Rogue Waves, 25–28 January, Honolulu, USA, pp. 1–7.

Frich, P., Alexander, L., Della-Martin, P., *et al.* (2002) Observed coherent changes in climatic extremes during the second half of the twentieth century. *Climate Research* **19** (3): 193–212.

Funk, C. (2011) We thought trouble was coming. *Nature* **476** (7358): 7.

Githeko, A. and Ndegwa, W. (2001) Predicting malaria epidemics in the Kenyan highlands using climate data: a tool for decision makers. *Global Change & Human Health* **2** (1): 54–63.

Gleick, P. H., Cooley, H. and Katz, D. (2006) *The World's Water, 2006–2007: The Biennial Report on Freshwater Resources*, Washington, DC: Island Press.

Gneiting, T. (2005) Atmospheric science: weather forecasting with ensemble methods. *Science* **310** (5746): 248–249.

Gosling, S. N., Lowe, S., McGregor, G., *et al.* (2009) Associations between elevated atmospheric temperature and human mortality: a critical review of the literature. *Climatic Change* **92** (3–4): 299–341.

Grossi, P. and Kunreuther, H. (eds) (2005) *Catastrophe Modeling: A New Approach to Managing Risk*, New York, NY: Springer.

Hagedorn, R., Doblas-Reyes, F. J. and Palmer, T. N. (2005) The rationale behind the success of multi-model ensembles in seasonal forecasting: I – basic concept. *Tellus Series A – Dynamic Meteorology and Oceanography* **57** (3): 219–233.

Hall, T. M. and Jewson, S. (2007) Statistical modelling of North Atlantic tropical cyclone tracks. *Tellus Series A-Dynamic Meteorology and Oceanography* **59** (4): 486–498.

Hall, T. M. and Jewson, S. (2008) Comparison of local and basinwide methods for risk assessment of tropical cyclone landfall. *Journal of Applied Meteorology and Climatology* **47** (2): 361–367.

Hassan, R. M., Scholes, R. and Ash, N. (eds) (2005) *Ecosystems and Human Well-being: Current State and Trends – Findings of the Condition and Trends Working Group of the Millennium Ecosystem Assessment*, Washington,. DC: Island Press.

Hayes, M., Wilhelmi, O. and Knutson, C. (2004) Reducing drought risk: bridging theory and practice. *Natural Hazards Review* **5**: 106.

Held, I. M. and Soden, B. J. (2006) Robust responses of the hydrological cycle to global warming. *Journal of Climate* **19** (21): 5686–5699.

Hirschi, M., Seneviratne, S., Orlowsky, B., *et al.* (2011) Observational evidence for soil-moisture impact on hot extremes in southeastern Europe. *Nature Geoscience* **4** (1): 17–21.

Holliday, N. P., Yelland, M., Pascal, R., *et al.* (2006) Were extreme waves in the Rockall Trough the largest ever recorded? *Geophysical Research Letters* **33** (5): L05613.

Hollingsworth, A. and Lönnberg, P. (1986) The statistical structure of short-range forecast errors as determined from radiosonde data. Part 1: the wind field. *Tellus Series A – Dynamic Meteorology and Oceanography* **38** (2): 111–136.

Houtekamer, P., Mitchel, H., Pellerin, G., *et al.* (2005) Atmospheric data assimilation with an ensemble Kalman filter: results with real observations. *Monthly Weather Review* **133** (3): 604–620.

Houtekamer, P. L., Mitchell, H. L. and Deng, X. (2009) Model error representation in an operational ensemble Kalman filter. *Monthly Weather Review* **137** (7): 2126–2143.

Iman, R., Johnson, M. and Watson, C. (2005a) Uncertainty analysis for computer model projections of hurricane losses. *Risk Analysis* **25** (5): 1299–1312.

Iman, R., Johnson, M. E. and Watson, C. C. (2005b) Sensitivity analysis for computer model projections of hurricane losses. *Risk Analysis* **25** (5): 1277–1297.

IPCC (2007a) *Climate Change 2007: The Physical Science Basis. Contribution of Working Group I to the Fourth Assessment Report of the Intergovernmental Panel on Climate Change*, Cambridge: Cambridge University Press.

IPCC (2007b) *Climate Change 2007: Impacts, Adaptation and Vulnerability. Contribution of Working Group II to the Fourth Assessment Report of the Intergovernmental Panel on Climate Change*, Cambridge: Cambridge University Press.

IPCC (2011) *Special Report on Managing the Risks of Extreme Events and Disasters to Advance Climate Change Adaptation (SREX) Summary for Policymakers*, Geneva: IPCC.

Jagger, T. H. and Elsner, J. B. (2006) Climatology models for extreme hurricane winds near the United States. *Journal of Climate* **19** (13): 3220–3236.

Joacim, R., Kristie, E. and Bertil, F. (2010) Mortality related to temperature and persistent extreme temperatures: a study of cause-specific and age-stratified mortality. *Occupational and Environmental Medicine* **68** (7): 531–536.

Jolliffe, I. T. and Stephenson, D. B. (2003) *Forecast Verification: A Practitioner's Guide in Atmospheric Science*, Chichester: John Wiley and Sons.

Jones, P. D., *et al.* (2009) Projections of future daily climate for the UK from the Weather Generator. In *UK Climate Projections*, Exeter: Met Office Hadley Centre, pp. 1–48.

Joslyn, S. L. and Nichols, R. M. (2009) Probability or frequency? Expressing forecast uncertainty in public weather forecasts. *Meteorological Applications* **16** (3): 309–314.

Joslyn, S. and Savelli, S. (2010) Communicating forecast uncertainty: public perception of weather forecast uncertainty. *Meteorological Applications* **17** (2): 180–195.

Joslyn, S., Nadav-Greenberg, L. and Nichols, R. M. (2009) Probability of precipitation assessment and enhancement of end-user understanding. *Bulletin of the American Meteorological Society* **90** (2): 185–193.

Kallis, G. (2008) Droughts. *Annual Review of Environment And Resources* **33**: 85–118.

Kalnay, E. (2003) *Atmospheric Modeling, Data Assimilation, and Predictability*, Cambridge: Cambridge University Press.

Kates, R. W., Colten, C. E., Laska, S., *et al.* (2006) Reconstruction of New Orleans after Hurricane Katrina: a research perspective. *Proceedings of the National Academy of Sciences* **103** (40): 14653–14660.

Khaliq, M. N., Ouarda, T., Ondo, J., *et al.* (2006) Frequency analysis of a sequence of dependent and/or non-stationary hydro-meteorological observations: a review. *Journal of Hydrology* **329** (3–4): 534–552.

Kharif, C. and Pelinovsky, E. (2003) Physical mechanisms of the rogue wave phenomenon. *European Journal of Mechanics B – Fluids* **22** (6): 603–634.

Kharif, C., Pelinovsky, E. and Slunyaev, A. (2009) *Rogue Waves in the Ocean*, New York, NY: Springer.

Kleiber, W., Raftery, A., Baars, J., *et al.* (2010) Locally calibrated probabilistic temperature forecasting using geostatistical model averaging and local Bayesian model averaging. *Monthly Weather Review* **139** (8): 2630–2649.

Klinenberg, E. (2002) *Heat Wave: A Social Autopsy of Disaster in Chicago*, Chicago, IL: University of Chicago Press.

Klotzbach, P., Barnston, A. G., Bell, G., *et al.* (2011) Seasonal forecasting of tropical cyclones. In *Global Guide to Tropical Cyclone Forecasting*, ed. C. Guard, Geneva: World Meteorological Organization.

Kovats, R. S. and Hajat, S. (2008) Heat stress and public health: a critical review. *Annual Review of Public Health* **29** (1): 41–55.

Kundzewicz, Z. and Robson, A. (2004) Change detection in hydrological records: a review of the methodology. *Hydrological Sciences Journal* **49** (1): 7–19.

Kunkel, K., Andsager, K. and Easterling, D. (1999) Long-term trends in extreme precipitation events over the conterminous United States and Canada. *Journal of Climate* **12** (8): 2515–2527.

Kyselý, J. (2008) A cautionary note on the use of nonparametric bootstrap for estimating uncertainties in extreme-value models. *Journal of Applied Meteorology and Climatology* **47** (12): 3236–3251.

Kyselý, J. (2009) Recent severe heat waves in central Europe: how to view them in a long-term prospect? *International Journal of Climatology* **30** (1): 89–109.

Kyselý, J. and Picek, J. (2007) Regional growth curves and improved design value estimates of extreme precipitation events in the Czech Republic. *Climate Research* **33** (3): 243–255.

Landsea, C. (2007) Counting Atlantic tropical cyclones back to 1900. *Eos* **88** (18): 197–208.

Landsea, C., Harper, B., Hoarau, K., *et al.* (2006) Can we detect trends in extreme tropical cyclones? *Science* **313** (5786): 452–454.

Lawton, G. (2001) Monsters of the deep. *New Scientist* **170** (2297): 28–32. http://www. newscientist.com/article/mg17022974.900-monsters-of-the-deep.html.

Leary, N., Conde, C. and Kulkarni, J., *et al.* (ed.) (2008) *Climate Change and Vulnerability*, London: Earthscan.

Lemeshko, B., Lemeshko, S. and Postovalov, S. (2009) Comparative analysis of the power of goodness-of-fit tests for near competing hypotheses: I. The verification of simple hypotheses. *Journal of Applied and Industrial Mathematics* **3** (4): 462–475.

Lemeshko, B., Lemeshko, S. and Postovalov, S. (2010) Comparative analysis of the power of goodness-of-fit tests for near competing hypotheses: II. Verification of complex hypotheses. *Journal of Applied and Industrial Mathematics* **4** (1): 79–93.

Levin, Z. and Cotton, W. R. (eds) (2008) *Aerosol Pollution Impact on Precipitation: A Scientific Review*, New York, NY: Springer-Science Press.

Li, Y., Cai, W. and Campbell, E. (2005) Statistical modeling of extreme rainfall in southwest Western Australia. *Journal of Climate* **18** (6): 852–863.

Li, Z., Niu, F., Fan, J., *et al.* (2011) Long-term impacts of aerosols on the vertical development of clouds and precipitation. *Nature Geoscience* **4** (12): 888–894.

Lloyd-Hughes, B. and Saunders, M. (2002) A drought climatology for Europe. *International Journal of Climatology* **22** (13): 1571–1592.

Lorenz, E. N. (1965) A study of the predictability of a 28-variable atmospheric model. *Tellus* **17**: 321–333.

Lynch, P. (2008) The origins of computer weather prediction and climate modeling. *Journal of Computational Physics*, **227** (7): 3431–3444.

Mailier, P. J., Stephenson, D., Ferro, C., *et al.* (2006) Serial clustering of extratropical cyclones. *Monthly Weather Review* **134** (8): 2224–2240.

McFarlane, N. (2011) Parameterizations: representing key processes in climate models without resolving them. *Wiley Interdisciplinary Reviews: Climate Change* **2** (4): 482–497.

McGregor, P. and MacDougall, K. (2009) A review of the Scottish rain-gauge network. *Proceedings of the Institution of Civil Engineers – Water Management* **162** (2): 137–146.

McNeil, A. J., Frey, R. and Embrechts, P. (2005) *Quantitative Risk Management: Concepts, Techniques and Tools*, Princeton, NJ: Princeton University Press.

Mehrotra, R., Srikanthan, R. and Sharma, A. (2006) A comparison of three stochastic multi-site precipitation occurrence generators. *Journal of Hydrology* **331**: 280–292.

Menéndez, M., Mendez, F., Izahuirre, C., *et al.* (2009) The influence of seasonality on estimating return values of significant wave height. *Coastal Engineering* **56** (3): 211–219.

Michelozzi, P., Kirchmayer, U., Katsouyanni, K., *et al.* (2007) Assessment and prevention of acute health effects of weather conditions in Europe, the PHEWE project: background, objectives, design. *Environmental Health* **6**: 12.

Miller, S., Muir-Wood, R. and Boissonade, A. (2008) An exploration of trends in normalized weather-related catastrophe losses. In *Climate Extremes and Society*, ed. H. F. Diaz and R. J. Murnane, Cambridge: Cambridge Universiy Press, pp. 225–247.

Morss, R. E., Demuth, J. L. and Lazo, J. K. (2008) Communicating uncertainty in weather forecasts: a survey of the U.S. public. *Weather and Forecasting* **23** (5): 974–991.

Nelsen, R. (2006) *An Introduction to Copulas*, 2nd edn, New York, NY: Springer.

Nicolosi, V., Cancelliere, A. and Rossi, G. (2009) Reducing risk of shortages due to drought in water supply systems using genetic algorithms. *Irrigation and Drainage* **58** (2): 171–188.

Palmer, T. (2001) A nonlinear dynamical perspective on model error: a proposal for non-local stochastic-dynamic parametrization in weather and climate prediction models. *Quarterly Journal of the Royal Meteorological Society* **127** (572): 279–304.

Palmer, T., Alessandri, A., Andersen, U., *et al.* (2004) Development of a European multi-model ensemble system for seasonal-to-interannual prediction (DEMETER). *Bulletin of the American Meteorological Society* **85** (6): 853–872.

Palmer, T., Shutts, G., Hagedorn, F., *et al.* (2005) Representing model uncertainty in weather and climate prediction. *Annual Review of Earth and Planetary Sciences* **33**: 163–193.

Palmer, W.C. (1965) *Meteorological Drought*, Washington, DC: US Department of Commerce Weather Bureau.

Pang, W., Forster, J. and Troutt, M. (2001) Estimation of wind speed distribution using Markov chain Monte Carlo techniques. *Journal of Applied Meterology* **40**: 1476–1484.

Park, J. and Jung, H. (2002) Modelling Korean extreme rainfall using a Kappa distribution and maximum likelihood estimate. *Theoretical and Applied Climatology* **72** (1): 55–64.

Parker, D., Legg, T. and Folland, C. (1992) A new daily central England temperature series, 1772–1991. *International Journal of Climatology* **12** (4): 317–342.

Parrish, D. and Derber, J. (1992) The National-Meteorological-Center's spectral statistical-interpolation analysis system. *Monthly Weather Review* **120** (8): 1747–1763.

Persson, A. and Grazzini, F. (2007) *User Guide to ECMWF Forecast Products*, 4th edn, Reading: ECMWF.

Pielke, R. (2005) Are there trends in hurricane destruction? *Nature* **438** (7071): E11.

Pielke, R., Landsea, C., Mayfiled, M., *et al.* (2006) Reply to 'Hurricanes and global warming – Potential linkages and consequences'. *Bulletin of the American Meteorological Society* **87** (5): 628–631.

Pielke, R., Gratz, J., Landsea, C., *et al.* (2008) Normalized hurricane damage in the United States: 1900–2005. *Natural Hazards Review* **9**: 29.

Pinelli, J., Gurley, K. and Subramanian, C. (2008) Validation of a probabilistic model for hurricane insurance loss projections in Florida. *Reliability Engineering and System Safety* **93**: 1896–1905.

Powell, M., Soukup, G., Gulati, S., *et al.* (2005) State of Florida hurricane loss projection model: atmospheric science component. *Journal of Wind Engineering and Industrial Aerodynamics* **93** (8): 651–674.

Rabier, F., Järvinen, H.s and Klinker, E. (2000) The ECMWF operational implementation of four-dimensional variational assimilation: I. Experimental results with simplified physics. *Quarterly Journal of the Royal Meteorological Society* **126**: 1143–1170.

Raftery, A., Balabdaoui, F., Gneiting, T., *et al.* (2005) Using Bayesian model averaging to calibrate forecast ensembles. *Monthly Weather Review* **133** (5): 1155–1174.

Ramesh, N. and Davison, A. (2002) Local models for exploratory analysis of hydrological extremes. *Journal of Hydrology* **256**: 106–119.

Reddy, S. (2000) Factors influencing the incorporation of hazard mitigation during recovery from disaster. *Natural Hazards* **22** (2): 185–201.

Ribereau, P., Guillou, A. and Naveau, P. (2008) Estimating return levels from maxima of non-stationary random sequences using the Generalized PWM method. *Nonlinear Processes in Geophysics* **15** (6): 1033–1039.

Rivière, G. and Orlanski, I. (2007) Characteristics of the Atlantic storm-track eddy activity and its relation with the North Atlantic oscillation. *Journal of the Atmospheric Sciences* **64** (2): 241–266.

RMS (2005) *Hurricane Katrina: Lessons and Implications for Catastrophe Risk Management*, Newark, CA: RMS.

RMS (2008) *A Guide to Catastrophe Modeling*, Newark, CA: RMS.

Robeson, S. M. (2008) Applied climatology: drought. *Progress in Physical Geography* **32** (3): 303–309.

Robine, J., Cheung, S., Le Roy, S., *et al.* (2008) Death toll exceeded 70,000 in Europe during the summer of 2003. *Comptes Rendus Biologies* **331** (2): 171–178.

Robinson, P. (2001) On the definition of a heat wave. *Journal of Applied Meteorology* **40** (4): 762–775.

Roulston, M. S. and Kaplan, T. R. (2009) A laboratory-based study of understanding of uncertainty in 5-day site-specific temperature forecasts. *Meteorological Applications* **16** (2): 237–244.

Roulston, M., Bolton, G., Kleit, A., *et al.* (2006) A laboratory study of the benefits of including uncertainty information in weather forecasts. *Weather and Forecasting* **21** (1): 116–122.

Rusticucci, M. and Vargas, W. (2002) Cold and warm events over Argentina and their relationship with the ENSO phases: risk evaluation analysis. *International Journal of Climatology* **22** (4): 467–483.

Salvadori, G., Kottegoda, N., De Michele, C., *et al.* (2007) *Extremes in Nature: An Approach Using Copulas*, Berlin: Springer.

Sanders, F. and Gyakum, J. R. (1980) Synoptic-dynamic climatology of the 'bomb.' *Monthly Weather Review* **108** (10): 1589–1606.

Saunders, M. A. and Lea, A. S. (2005) Seasonal prediction of hurricane activity reaching the coast of the United States. *Nature* **434** (7036): 1005–1008.

Schär, C., Vidale, P., Lüthi, D., *et al.* (2004) The role of increasing temperature variability in European summer heatwaves. *Nature* **427** (6972): 332–336.

Schmidt, S., Kemfert, C. and Höppe, P. (2009) Tropical cyclone losses in the USA and the impact of climate change: a trend analysis based on data from a new approach to adjusting storm losses. *Environmental Impact Assessment Review* **29** (6): 359–369.

Schölzel, C. and Friederichs, P. (2008) Multivariate non-normally distributed random variables in climate research: introduction to the copula approach. *Nonlinear Processes in Geophysics* **15** (5): 761–772.

Senay, G. and Verdin, J. (2003) Characterization of yield reduction in Ethiopia using a GIS-based crop water balance model. *Canadian Journal of Remote Sensing* **29** (6): 687–692.

Sheridan, S. (2002) The redevelopment of a weather-type classification scheme for North America. *International Journal of Climatology* **22** (1): 51–68.

Sheridan, S. and Kalkstein, L. (2004) Progress in heat watch-warning system technology. *Bulletin of the American Meteorological Society* **85** (12): 1931–1941.

Shutts, G. (2005) A kinetic energy backscatter algorithm for use in ensemble prediction systems. *Quarterly Journal of the Royal Meteorological Society* **131** (612): 3079–3102.

Smoyer-Tomic, K., Kuhn, R. and Hudson, A. (2003) Heat wave hazards: an overview of heat wave impacts in Canada. *Natural Hazards* **28** (2–3): 463–485.

Souch, C. and Grimmond, C. (2004) Applied climatology: 'heat waves.' *Progress in Physical Geography* **28** (4): 599–606.

Spiegelhalter, D., Pearson, M. and Short, I. (2011) Visualizing uncertainty about the future. *Science* **333** (6048): 1393–1400.

Stephens, G. L., L'Ecuyer, T., Suzuki, K., *et al.* (2010) Dreary state of precipitation in global models. *Journal of Geophysical Research* **115** (D24): D24211.

Stephens, E. M., Edwards, T. L. and Demeritt, D. (2012) Communicating probabilistic information from climate model ensembles – lessons from numerical weather prediction. *Wiley Interdisciplinary Reviews: Climate Change* **3** (5): 409–426.

Stephenson, D., Casati, B., Ferro, C. A. T., *et al.* (2008) The extreme dependency score: a non-vanishing measure for forecasts of rare events. *Meteorological Applications* **15** (1): 41–50.

Storch, H. von (1995) Misuses of statistical analysis in climate research. In *Analysis of Climate Variability: Applications of Statistical Techniques*, ed., H. von Storch and A. Navarra, New York, NY: Springer Verlag, pp. 11–26.

Storch, H. von and Zwiers, F. W. (2003) *Statistical Analysis in Climate Research*, Cambridge: Cambridge University Press.

Talagrand, O., Vautard, R. and Strauss, B. (1997) Evaluation of probabilistic prediction systems. In *ECMWF Workshop on Predictability*, Reading: ECMWF.

Tebaldi, C., Hayhoe, K., Arblaster, J., *et al.* (2006) Going to the extremes. *Climatic Change* **79** (3–4): 185–211.

Tolman, H. (2009) User manual and system documentation of WAVEWATCH III TM version 3.14, NOAA/NWS/NCEP/MMAB Technical Note 276.

Townsend, F. F. (2006) *The Federal Response to Hurricane Katrina: Lessons Learned*, Washington DC: The White House.

Tucker, M. J. and Pitt, E. G. (eds) (2001) *Waves in Ocean Engineering*, New York, NY: Elsevier.

UK Environment Agency (2009) Drought plan for Anglian region.

UK Government Cabinet Office (2008) National risk register. http://www.cabinetoffice.gov.uk.

United Nations (2006) *Global Survey of Early Warning Systems: An Assessment of Capacities, Gaps and Opportunities Towards Building a Comprehensive Global Early Warning System For All Natural Hazards*, Geneva: United Nations.

Uppala, S. M., Kållberg, P., Simmons, A., *et al.* (2005) The ERA-40 re-analysis. *Quarterly Journal of the Royal Meteorological Society* **131** (612): 2961–3012.

USDA (1994) *Major World Crop Areas and Climatic Profiles*, Washington, DC: World Agricultural Outlook Board, U.S. Department of Agriculture.

US GAO (2007) GAO-07-169 National Flood Insurance Program: New Processes Aided Hurricane Katrina Claims Handling, but FEMA's Oversight Should Be Improved. pp. 1–68.

US House of Representatives (2006) A failure of initiative: final report of the Select Bipartisan Committee to Investigate the Preparation for and Response to Hurricane Katrina.

van Leeuwen, P. (2009) Particle filtering in geophysical systems. *American Meterological Society* **137**: 4089–4114.

van Leeuwen, P. (2010) Nonlinear data assimilation in geosciences: an extremely efficient particle filter. *Quarterly Journal of the Royal Meteorological Society* **136** (653): 1991–1999.

van Leeuwen, P. (2011) Efficient nonlinear data-assimilation in geophysical fluid dynamics. *Computers & Fluids* **46** (1): 52–58.

Vecchi, G. A. and Knutson, T. R. (2011) Estimating annual numbers of Atlantic hurricanes missing from the HURDAT database (1878–1965) Using Ship Track Density. *American Meteorological Society* **24** (4): 1736–1746.

Vickery, P., Skerlj, P., and Steckley, A., *et al.* (2000a) Hurricane wind field model for use in hurricane simulations. *Journal of Structural Engineering* **126** (10): 1203–1221.

Vickery, P., Skerlj, P. and Twisdale, L.s (2000b) Simulation of hurricane risk in the US using empirical track model. *Journal of Structural Engineering* **126** (10): 1222–1237.

Vickery, P., Lin, J., Skerlj, P., *et al.* (2006a) HAZUS-MH hurricane model methodology: I. Hurricane hazard, terrain, and wind load modeling. *Natural Hazards Review* **7**: 82–92.

Vickery, P., Skerlj, P., Lin, J., *et al.* (2006b) HAZUS-MH hurricane model methodology: II. Damage and loss estimation. *Natural Hazards* **7**: 94–103.

Vitart, Frédéric (2006) Seasonal forecasting of tropical storm frequency using a multi-model ensemble. *Quarterly Journal of the Royal Meteorological Society* **132** (615): 647–666.

Vitart, F. D. and Stockdale, T. (2001) Seasonal forecasting of tropical storms using coupled GCM integrations. *Monthly Weather Review* **129** (10): 2521–2537.

Vitart, F., Huddleston, M., Déqué, M., *et al.* (2007) Dynamically-based seasonal forecasts of Atlantic tropical storm activity issued in June by EUROSIP. *Geophysical Research Letters* **34** (16): L16815.

Vitolo, R., Stephenson, D., Cook, I., *et al.* (2009) Serial clustering of intense European storms. *Meteorologische Zeitschrift* **18** (4): 411–424.

Webster, P., Holland, G., Curry, J., *et al.* (2005) Changes in tropical cyclone number, duration, and intensity in a warming environment. *Science* **309** (5742): 1844.

Weisheimer, A., Doblas-Reyes, F., Palmer. T., *et al.* (2009) ENSEMBLES: a new multi-model ensemble for seasonal-to-annual predictions – skill and progress beyond DEMETER in forecasting tropical Pacific SSTs. *Geophysical Research Letters* **36**: L21711.

Weisheimer, A., Doblas-Reyes, F., Jung, T., *et al.* (2011) On the predictability of the extreme summer 2003 over Europe. *Geophysical Research Letters* **38** (5): L05704.

Wikle, C. and Berliner, L. (2007) A Bayesian tutorial for data assimilation. *Physica D: Nonlinear Phenomena* **230** (1–2): 1–16.

Wilks, D. S. (2011) *Statistical Methods in the Atmospheric Sciences: An Introduction*, 3rd edn, Oxford: Academic Press.

Wilks, D. S. and Wilby, R. L. (1999) The weather generation game: a review of stochastic weather models. *Progress in Physical Geography* **23** (3): 329–357.

Williams, O. (2008) *National Flood Insurance Program: Greater Transparency and Oversight of Wind and Flood Damage Determinations are Needed*, Washington, DC: US GAO.

Wimmer, W., Challenor, P. and Retzler, C. (2006) Extreme wave heights in the north Atlantic from altimeter data. *Renewable Energy* **31** (2): 241–248.

Woolf, D., Challenor, P. and Cotton, P. (2002) Variability and predictability of the North Atlantic wave climate. *Journal of Geophysical Research – Oceans* **107**: 3145.

Wu, H., Hayes, M., Weiss, A., *et al.* (2001) An evaluation of the Standardized Precipitation Index, the China-Z Index and the statistical Z-Score. *International Journal Of Climatology* **21** (6): 745–758.

Young, I. R. (1999) Seasonal variability of the global ocean wind and wave climate. *International Journal of Climatology* **19** (9): 931–950.

Zhang, X., Hogg, W. and Mekis, E. (2001) Spatial and temporal characteristics of heavy precipitation events over Canada. *Journal of Climate* **14** (9): 1923–1936.

Zhao, X. and Chu, P.-S. (2010) Bayesian changepoint analysis for extreme events (typhoons, heavy rainfall, and heat waves): an RJMCMC approach. *Journal of Climate* **23** (5): 1034–1046.

6

Hydrometeorological hazards under future climate change

T. L. EDWARDS

6.1 Introduction

The climate is always changing. Natural causes, such as volcano eruptions and variations in solar output, are continuously nudging the radiation balance of the atmosphere, causing it to warm or cool, and the oceans, land surface and cryosphere respond. Since the Industrial Revolution (IPCC, 2007a), perhaps much earlier (Ruddiman, 2003), humans have also been affecting climate by changing the surface of the land and the composition of the atmosphere. The net effect of these changes over the twentieth century has been a warming of the climate (IPCC, 2007a).

Climate change is a change in the statistical properties of weather, usually defined on timescales of three decades or more: not only in the mean ('what we expect', coined by science fiction writer Robert A. Heinlein in 1973), but the full distribution. Weather hazards lie in the extreme tails of this distribution. As mean global temperature increases, the distribution shifts towards some extreme weather hazards such as heat waves, and if it also changes shape (widens or skews) then the magnitude of these extremes could change more quickly than the mean. So far the picture appears to be mixed, with maximum temperatures changing more quickly than the mean since 1950 in some areas and more slowly in others (Brown et al., 2008). Further increases in global mean temperature are predicted to occur this century (based on inertia in the climate system, inertia in societal change and a range of plausible long-term changes), and several studies have predicted that the intensity of extreme high temperatures and precipitation are likely to increase more rapidly than the mean in many areas (e.g. Beniston et al., 2007; Kharin et al., 2007). Predictions for extreme weather events are summarised by the Intergovernmental Panel on Climate Change (IPCC) in the recent *Special Report on Managing the Risks of Extreme Events and Disasters* (IPCC, 2012).

Risk assessment and management for present-day hydrometeorological hazards are described in the previous chapter; for convenience some of the main points are summarised here. Extreme weather is often defined (Section 5.2.1) in terms of exceedances of a threshold, often multivariate (i.e. extremes of multiple variables), where the threshold is either fixed or relative to local climatology. Some alternative definitions are the highest value(s) of a fixed time period, such as annual maxima, or definitions based on synoptic classifications.

Risk and Uncertainty Assessment for Natural Hazards, ed. Jonathan Rougier, Steve Sparks and Lisa Hill. Published by Cambridge University Press. © Cambridge University Press 2013.

Extreme weather events have no distinct trigger but may be associated with persistent atmospheric states, such as blocking, or particular modes of natural variability, such as the El Niño Southern Oscillation (ENSO: Section 5.2.2). Forecasts are made with both physically based, computationally expensive computer models (Section 5.3.2) and statistical methods (Section 5.3.3), for lead times ranging from days to seasons. These are necessarily probabilistic, due to the chaotic nature of the atmosphere. For physically based models, ensemble forecasts of hazard frequencies are generated by perturbing initial conditions and other uncertain quantities. In statistical methods, probabilistic forecasts are generated by inferring a statistical relationship between the hazard and a persistent atmosphere–ocean state. Long-term return values are also estimated, typically extrapolating to rarer extremes and mostly under an assumption of stationarity. Both physically based model ensembles and statistical models are calibrated using historical and recent observations. Uncertainty assessment relies upon the quality and stationarity of the observational record, and our ability to quantify the limitations of the physical or statistical model.

Exposure to extreme weather is global, though exposure to particular hazard types depends on local factors such as topography, proximity to the coast and hydrological feedbacks; losses occur to all sectors, including human health, infrastructure and business and ecosystems (Section 5.2.4). Loss modelling is typically statistical, because the physical mechanisms are often not well understood. Catastrophe modelling (Section 5.3.3), both proprietary and open, incorporates the full causal chain from hazard to loss using various statistical modules, though most are limited in their ability to incorporate physical plausibility constraints. Risk management strategies for extreme weather encompass hazard-monitoring networks (weather forecasting and observational networks) that trigger real-time alerts and actions and long-term planning to reduce, for example, vulnerability of housing or insured assets (Section 5.4.1). Hazard communication and response are challenging for extreme weather events, because they are multivariate, probabilistic and depend on interactions at small spatial scales while affecting large regions; lessons are inevitably learned in post-event recovery and analysis (Section 5.4.2).

Extreme weather hazards have very serious consequences today, so assessment and management of the potential future changes in their risks is therefore of great importance to society. This chapter summarises the risks and uncertainties of droughts, heat waves, extreme precipitation and wind storms under future climate change. Freak ocean waves (Chapter 5) are not discussed. Attribution of individual extreme weather events to human causes is a rapidly developing research area (e.g. Stott *et al.*, 2004; IPCC, 2012; Min *et al.*, 2011; Pall *et al.*, 2011), and not covered here. Section 6.2 discusses methods of risk assessment for hazards and their impacts; Section 6.3 describes risk management and communication; and Section 6.4 summarises.

6.2 Risk assessment

This section describes the challenges and current methods of assessing risk for extreme weather hazards under future climate change. In climate science, the term 'risk' has often been used in the sense of 'probability of occurrence' of, for example, a particular hazard event. But true risk assessments, which also incorporate hazard impacts, are becoming more common; this is the sense in which the word 'risk' is used in this and the previous chapter (a more detailed definition is given in Chapter 2). Risk assessment for weather hazards under future climate change encompasses multiple research areas in a long causal chain. First, there is consideration of the range of possible future driving factors, both human and natural, that influence climate. Second, there is prediction of the future response of climate to those factors, including not only mean climate but the extremes. Third, there is assessment of the future impacts of these hazards, including changes in exposure and vulnerability. Uncertainties must be propagated along this causal chain.

The main differences between extreme weather in the present day (Chapter 5) and under future climate change (this chapter) are: the long timescale on which climate is defined; the resulting importance of changing boundary conditions; and the unpredictability of long-term changes in those boundary conditions. Boundary conditions are the external controls of the system evolution, such as atmospheric composition (e.g. greenhouse gas concentrations), land surface properties (e.g. agricultural use) and configuration of the Earth's orbit. The first two of these differences, long timescale and changing boundary conditions, rule out the use of the historical record for model calibration, which shifts the focus from statistical to physically based models so as to explore the effect of boundary conditions different to today, while the third adds a further 'dimension' of uncertainty over which risk assessments must be made. These challenges are described in more detail in the next section.

6.2.1 Challenges in predictability

Prediction is 'very difficult, especially if it's about the future', as the physicist Niels Bohr is said to have pointed out, and the earth system has many particularly challenging features. Risk assessment for hydrometeorological hazards under future climate change has all the difficulties described in Chapter 5 added to those of predicting the distant future. In fact, it might at first seem faintly ludicrous to assert that statements can be made about the earth system decades from now, when weather forecasting has a predictability timescale of about a fortnight and seasonal forecasting is limited to about a year (Chapter 5). But long-term climate change concerns statistical properties rather than chronological order: it is a forced (boundary condition) problem, a problem of the 'second kind' (Lorenz, 1975), rather than one of initial conditions. Chaos (Chapter 5) is therefore not one of the principal difficulties, at least in long-term prediction. Instead, the focus is on predicting, or alternatively defining a plausible range of, future boundary conditions.

Recently there have been increased efforts to make short-term ('decadal') forecasts of year-to-year changes (such as ENSO: Chapter 5) over the next few years. These are time-scales of great interest for decision-making and policy-making. But decadal predictions are extremely challenging, because the timescales are too short to be dominated by boundary conditions, and too long to be dominated by initial conditions (Hawkins and Sutton, 2009).

Like seasonal forecasts (Chapter 5), they rely on accurate initialisation to make use of persistent and oscillatory modes of the climate. This newly emerging research field is showing much promise (Smith *et al.*, 2007). An overview is given by Hawkins (2011).

Challenges in predicting future boundary conditions (forcings) are described in Section 6.2.1.1; in the earth system response to these forcings in Section 6.2.1.2; and some hazard- and impacts-specific challenges in Section 6.2.1.3.

6.2.1.1 Forcings

A climate forcing is something that acts to perturb the radiative equilibrium of the atmosphere (with units of watts per square metre): by altering either the amount of incoming solar radiation (insolation), or the fraction of radiation that is reflected, or the amount of longwave radiation emitted to space (IPCC, 2007a). These perturbations cause warming or cooling of the atmosphere and/or surface, which is then propagated throughout the earth system. The term 'forcing' is often used interchangeably with '(change in) boundary conditions' because the latter induces the former.

Natural forcings are always acting on the earth system. The amount and distribution of solar radiation reaching the earth varies continuously, in periodic cycles and longer-term changes. Volcanic eruptions occur randomly, emitting sulphur dioxide gas which combines with water in the atmosphere to make sulphate aerosol particles that reflect solar radiation and change the reflectivity (albedo) and lifetime of clouds. Some of these natural forcings are predictable – in particular, the effect of the earth's orbital cycles and the length (though not amplitude) of the 11-year sunspot cycle – but others are unpredictable, such as long-term solar variations and volcanic eruptions.

Anthropogenic (human-caused) forcings arise from large-scale industrial, agricultural and domestic changes to the earth's atmosphere and land surface. Some important forcings are: greenhouses gases (GHGs: including carbon dioxide, methane, nitrous oxide and ozone), which decrease the amount of longwave radiation emitted by the earth to space and have a warming effect at the surface; industrial sulphate aerosol particles, which act similarly to those from volcanoes; and replacement of forests with crops, which increases the reflectivity of the surface and increases the GHG concentrations in the atmosphere, notably if the deforestation involves burning.

Future changes in anthropogenic forcings depend on population, policy, economics and growth, technology and efficiency, mitigation (Section 6.3.1.1) and geoengineering (Section 6.3.1.2). Geoengineering is intentional alteration of the earth system, particularly to counter anthropogenic climate change, with measures that reduce positive (warming) forcings, such as GHG extraction from the atmosphere, or increase negative (cooling) forcings, such as addition of reflective aerosol particles to the atmosphere (Irvine and Ridgwell, 2009; Lenton and Vaughan, 2009). In the short-term, GHGs and industrial aerosol emissions are somewhat predictable, because significant changes in infrastructure, behaviour and population growth take years or decades, and some GHGs have mean lifetimes of several decades. In the longer term, strong mitigation or industrial growth could occur and this is virtually impossible to predict at the centennial scale.

Limited predictability of future anthropogenic forcings is addressed by making predictions for a set of 'possible futures' (emissions scenarios: Section 6.2.4.1). These predictions, which are conditional on a given scenario ('what would happen if'), rather than attempting to encompass every aspect of the future ('what will happen'), are better described as 'projections', though the distinction is not always made. Forcing scenarios usually attempt to encompass the range of plausible future human influence on climate, because the motivation for their use is assessment of risk management strategies. Unpredictable natural forcings are usually set at present-day levels (solar output) or not included (volcanic eruptions).

6.2.1.2 Response to forcings

The statistical properties of weather are challenging to simulate in the present day (Chapter 5), let alone the future. Not only are natural and anthropogenic forcings always changing, but some parts of the earth system take centuries (the deep oceans), millennia (the cryosphere) and longer to respond so the earth system is never at equilibrium: in other words, stationarity can never be assumed except on short temporal and spatial scales. This means the earth's future response to forcings cannot be deduced or extrapolated from the past. Even if forcings were to stay fixed at present-day levels, an assumption of stationarity would have limited applicability.

Feedbacks (Chapter 5) modulate the climate response to forcings. The distinction between forcing and feedback may depend on the timescale of the processes under consideration: for example, on short timescales land vegetation and ice sheets might be treated as (approximately) static boundary conditions, but on long timescales as dynamically responding components of the earth system. Cloud–climate feedbacks are the main source of uncertainty in the time dependent response of surface temperature in climate models (Soden and Held, 2006), particularly in the shortwave radiation feedback of low-level clouds (Bony and Dufresne, 2005; Crucifix, 2006; Webb *et al.*, 2006). Feedbacks between vegetation, climate and carbon dioxide (CO_2) affect surface water runoff, soil moisture and carbon storage, but the magnitude and sign of the response vary between models (Cox *et al.*, 2000; Betts *et al.*, 2007; Alo and Wang, 2008). Some known positive feedbacks, such as methane clathrate release (Maslin *et al.*, 2010), are not routinely, or ever, included in predictions, which may lead to underestimation of future change (Scheffer *et al.*, 2006; Torn and Harte, 2006).

Modelling of climate extremes and their impacts is either combined (known as 'coupled', 'interactive' or 'dynamic') or separate ('offline', 'prescribed'), so the distinction between them can be somewhat subjective. The arbitrariness of these boundaries is largely due to the presence of feedbacks: for example, the two-way interactions between climate and vegetation are now much more commonly included in climate models, though in other studies the impacts of climate and weather extremes on vegetation are modelled offline (i.e. the feedback loop is cut). Feedbacks between hazards and impacts can either amplify or diminish the latter. Dlugolecki (2009) gives examples for insured losses from flooding: these may be diminished (negative feedback) with risk management by reduced habitation of at-risk areas or improved building structures, or amplified (positive feedback) in large-scale disasters by long-term neglect of properties (after evacuation) or political pressure on insurers to pay for uninsured losses and property improvements. Human-related feedbacks, such as the inter-actions between climate and the global economy, are represented in integrated assessment models (e.g. Leimbach *et al.*, 2010), though the climate modules are usually much simpler than the state-of-the-art.

6.2.1.3 Hazard- and impact-specific challenges

All weather hazard events are challenging to predict because of their rarity (limiting the number of observations and the ability to understand the relevant processes) and their position in the extremes of chaotic internal variability (Chapter 5). Some types are partic-ularly hard to predict. For future changes, extreme temperature is thought the most straight-forward, because climate models simulate the major driving processes (Chapter 5; IPCC, 2012), though the regional effects of feedbacks and changes in atmospheric and ocean circulation on temperature extremes are still poorly understood (Clark *et al.*, 2010). Extreme precipitation and tropical cyclones are far more challenging, because the processes and weather systems occur on small spatial scales (Knutson *et al.*, 2010; IPCC, 2011): for example, while warm sea surface temperatures drive the genesis of tropical cyclones, other factors such as vorticity and wind shear are thought to be important in determining their frequency (Kim *et al.*, 2010).

Predictions for some hazards do not even agree on the sign of change in many regions of the world (IPCC, 2012), in particular for multivariate hazards (such as drought) and modes of variability (such as ENSO). One of the difficulties with multivariate hazards is in their definition (IPCC, 2012): Burke and Brown (2008), for example, find the magnitude and sign of predicted drought changes depend on the drought index (Chapter 5) used. Even if predictions of the main drivers of drought (temperature, precipitation, evaporation and runoff) agree, predictions of the resulting soil moisture may be wildly different (Dai, 2011).

6.2.2 Physical modelling

The previous chapter described how physical models or 'simulators' of the earth system are used to simulate weather. Ensemble forecasts, generated by perturbing initial conditions and other uncertain quantities, are used to approximate probability distributions for hazard events. Ensemble forecasts are repeatedly tested against observations to assess whether the ensemble is 'well-calibrated' (explores all important uncertainties). Statistical modelling of past observations is also used, for seasonal forecasting and for estimating return periods, but this is not possible for the longer term: changing boundary conditions place physical modelling at the heart of predicting changes in extremes.

For long-term risk assessment of extremes, then, physical models are run for future scenarios (to make predictions) and for the past (to critique the model). This section describes these models and the difficulties with using them to make multi-decadal simu-lations of climate: their complexity, computational expense and inevitable limitations in representing the earth system. Added to these is the multi-decadal timescale of climate, meaning predictions cannot be repeatedly tested against observations to ensure a model is adequate and an ensemble well-calibrated, and the unpredictable future boundary conditions (Sections 6.2, 6.2.1). There are several approaches to overcoming these difficulties, involving hierarchies of physical models (described in this section) supplemented with statistical modelling (described in the next), along with expert judgement and other strategies. However, many of these methods are still under active development, and a consensus on approaches is far from emerging.

The most advanced, general circulation models (GCMs), are similar to numerical weather prediction (NWP) models, but have lower spatial and temporal resolution and include different processes (more details are given in Chapter 5). A small number of models, including the UK Met Office Unified Model, are used for both climate prediction and NWP. Climate models are used for much longer simulations (decades to centuries) than NWP models, so they incorporate more long-timescale processes such as deep ocean circulation, atmospheric chemistry and the carbon cycle. Components with very long time-scale responses, in particular the cryosphere, can also be included, though the coupling is usually asynchronous to reduce computational expense: for example, one-year simulations from the atmosphere–ocean model are alternated with ten-year simulations from the ice-sheet model. The most comprehensive GCMs, particularly those that include chemical and biological as well as physical processes, are also called 'Earth System Models'. The simulation length and complexity of GCMs result in significant computational expense. Lower resolution helps to mitigate this, though simulation rate may still be as slow as 50 model years per month, even on a supercomputer (Easterbook and Johns, 2009). A reference text on climate modelling is McGuffie and Henderson-Sellers (2005); an overview of their current abilities is given by the US Climate Change Science Program (CCSP, 2008); and the current state-of-the-art, multiple climate model dataset (CMIP5: Section 6.2.4.3) is described by Taylor *et al.* (2011).

Recently there has been greater emphasis on making predictions with a continuum of resolutions and timescales, from three-day weather predictions to decadal and centennial climate predictions, in order to share their strengths: for example, the climate prediction configuration of the Unified Model benefits from daily testing of the operational weather forecasting configuration. This is termed 'seamless prediction' (Rodwell and Palmer, 2007; Palmer *et al.*, 2008).

The low resolution of GCMs, hundreds of kilometres horizontally (Chapter 5), places great reliance on the parameterisation of small-scale processes (Section 6.2.2.1). A spectrum of model complexity exists, in which processes are represented in parameterised form to a greater or lesser extent (Section 6.2.2.2). Simpler, faster models are used for longer simulations and large ensembles, and complex models for shorter simulations, small ensembles or reduced domains. Predictions of regions (country to continental scale) suitable for impacts studies are made with high-resolution, limited-area models similar to those of operational NWP (regional climate models: RCMs), driven by boundary conditions derived from climate models. This is known as physical downscaling (Section 6.2.2.3).

6.2.2.1 Parameterisation

Physical models can never include every process, describe all included processes perfectly or have infinite resolution, so the behaviour of imperfectly calculated processes is represented using parameters equivalent to real-world or abstracted quantities. These uncertain parameters are given fixed standard values determined from calibration with observations, expert knowledge and specific high-resolution physical modelling studies (e.g. Randall *et al.*, 2003). An important example is convective cloud formation, where the relevant processes occur on sub-grid spatial scales. The clouds must be represented with bulk formulae: for example, relating fractional cloud cover in a grid box to relative humidity. Other examples involve interactions at the earth's surface, such as energy and moisture fluxes between the boundary layer, land and sea ice. Important parameterisation schemes in climate models are reviewed by CCSP (2008) and McFarlane (2011). A reference text is Stensrud (2009).

Model calibration (tuning) aims to find parameter values that bring the model simulation in closest agreement with the target datasets for a wide range of physical quantities. This is usually an impossible aspiration because the most successful values of some parameters are incompatible with those of others, or because improvements in simulating one physical quantity come at the expense of others. Climate models are tuned to reproduce mean observations, not extremes, due to the limited number of observations. Another important requirement of tuning (not necessary for NWP models) is to keep the global energy balance at the top of the atmosphere stable over long periods of time, to avoid artificial temperature drift (CCSP, 2008). Parameter uncertainty in climate simulations is explored by 'detuning' the model. The resulting ensembles are used directly or re-weighted with comparisons to observations (Section 6.2.4.2).

An alternative to parameter calibration is the ASK method (Stott and Forest, 2007), rooted in detection and attribution analysis, in which a model prediction is re-scaled by the error compared with historical observations; the model response to each type of forcing is separated out and scaled separately. The ASK method uses an assumption of a linear relationship between errors in simulating past climate change and predicting future climate change, and is most reliable if the relative fractions of different forcings remains the same (Stott and Forest, 2007); this is not the case for mitigation or geoengineering scenarios (Section 6.3.1). The ASK approach has been applied to regional change (Stott and Forest, 2007), but attribution methods for extremes are in their infancy (e.g. IPCC, 2012; Min *et al.*, 2011; Pall *et al.*, 2011).

Most climate models are deterministic, with fixed parameter values. But physical consistency can fail for deterministic parameterisations: for example, violating conservation of energy at small spatial scales. It is increasingly recognised that stochastic components, introduced through randomly varying parameters or processes (Chapter 5) can improve climate model predictions (Palmer, 2001; Palmer *et al.*, 2005; Palmer and Williams, 2009; McFarlane, 2011).

6.2.2.2 Spectrum of model complexity

Physical climate models vary from complex, computationally expensive GCMs and RCMs, to models of intermediate complexity that have even lower resolution and a larger fraction of parameterised processes and prescribed components (earth system models of intermediate complexity: EMICs), to simple conceptual ('toy') models. Some of this variation is a reflection of continuing model development over time, with state-of-the-art model versions used alongside older, faster versions, and some is due to design, with simpler models used to improve understanding or for low computational expense. A model is selected and configured according to the scientific problem and available resources: long-timescale processes may be neglected if simulations are short; simpler, faster models enable many replications for statistical analysis. In general, simple and intermediate complexity models cannot adequately simulate regional climate change and extremes (Stott and Forest, 2007), but they may be tuned to GCM or RCM simulations to scale their results for other forcing scenarios (Section 6.2.4.1). Even state-of-the-art models do not simulate some extremes well (Section 6.2.1.3): precipitation is one example for which models typically underestimate the frequency and intensity of extreme events (IPCC, 2007a).

Physical models of the impacts of extreme weather (health, environment and financial: Chapter 5) also exist for some research areas. Environmental models of varying complexity exist for the effects of drought, extreme temperature and extreme precipitation on crops (e.g. Challinor *et al.*, 2004), forests (Allen *et al.*, 2010), terrestrial ecosystems (e.g. Knapp *et al.*, 2008) and interactive vegetation–climate feedbacks (e.g. JULES: Best *et al.*, 2011; Clark *et al.*, 2011). Vegetation models can be adapted to simulate the impacts of heat waves on cities (Masson, 2006), and bio-physical models simulate the impacts of heat waves on health with human energy balance equations (McGregor, 2011). Physical models of structures also exist: for example, a model of the impacts of extreme precipitation on urban drainage systems is described by Wright *et al.* (2006). The significant advantage of physically based impacts models is that they do not assume the relationship between hazard and impact is fixed in the future: for example, the effect of changing carbon dioxide concentrations on crops is incorporated in vegetation models. However, physically based models do not exist for all types of impacts and this is a rapidly progressing field.

6.2.2.3 Dynamical downscaling

'Downscaling' means increasing the spatial or temporal resolution of a dataset, usually the output of a GCM. The term 'spatial downscaling' is generally reserved for methods that incorporate the effects of local topography, coastlines and land use on the signal, rather than simple linear interpolation to a higher-resolution grid. Downscaling is crucial for hazard impact assessment, because the hazards and impacts occur on much smaller spatio-temporal scales than GCM resolutions (e.g. extreme temperature and precipitation: Diffenbaugh *et al.*, 2005; Gosling *et al.*, 2009; crops: Tsvetsinskaya *et al.*, 2003). There are two very different approaches, based on statistical (Sections 6.2.3.2, 6.2.3.3) and physical ('dynamical') modelling.

Dynamical downscaling is another name for regional climate modelling (proprietary versions have also been developed for catastrophe loss modelling: Section 6.2.3.2). RCMs are very computationally intensive, which restricts the length and number of simulations, but unlike statistical downscaling they generate physically and spatially consistent results for a large number of variables and allow changes in boundary conditions to be made. For predictions of future climate, an RCM is either run offline with boundary conditions (such as temperature, wind and pressure) at the domain edges (and, if the RCM is used to simulate only the atmosphere, the ocean surface) from a pre-existing GCM simulation, which only allows movement of weather systems from the GCM to the RCM ('one-way nesting'), or coupled to a GCM, allowing weather systems to move between the domains ('two-way nesting'). The former is predominant, being more straightforward technically and computationally, and useful for comparing RCMs, but the latter is more physically realistic (e.g. Lorenz and Jacob, 2005). Variable-resolution global models are also used (Fox-Rabinovitz et al., 2008). Downscaling is reviewed by the CCSP (2008) and Rummukainen (2010).

Tropical cyclones are not resolved by the current generation of RCMs. They are simulated with tailored high-resolution physical models or statistical models driven by climate model simulations (Chapter 5).

6.2.2.4 Extremes and anomaly methods in physical modelling

GCMs are a 'necessary evil' in estimating the effect of changing boundary conditions on extremes, because they incorporate causal physical mechanisms that relate global forcing (e.g. CO_2 emissions) to regional climate (e.g. the distribution of European daily temperatures). The response of climate to forcing is estimated either with two steady-state (equilibrium) simulations – for example, one using pre-industrial boundary conditions and the other using doubled atmospheric CO_2 concentrations – or with a time-evolving (transient) simulation of the historical period and next one or two centuries. Climate simulations from GCMs are downscaled with regional climate modelling (previous section), weather generation (Section 6.2.3.2) or statistical spatial downscaling (Section 6.2.3.3) to increase their temporal and spatial resolution. This is necessary for estimating extremes at a given location and for providing inputs to impacts models.

The frequency and intensity of extremes in downscaled simulations can be quantified empirically (e.g. counting threshold exceedances) or, particularly for rarer extremes, modelled statistically (e.g. fitting a GPD model to those threshold exceedances: Section 6.2.3.1). These methods either use an assumption of stationarity (for equilibrium simulations or short sections of transient simulations) or quantify the time-evolving change (e.g. by allowing the parameters of the statistical model to vary with time).

Climate simulators inevitably have systematic biases. When studying the effect of changing boundary conditions on extremes, some of these may be cancelled out, to first order, by focusing on the change ('anomaly') between the simulations of the present-day and future scenarios, rather than the absolute values of the latter. Effectively the future distribution of extremes is inferred from a combination of the present-day observational record, present-day simulation and future scenario simulation (Ho et al., 2012). There are two methods, which are exactly equivalent in a special case: if the distributions of extremes in the three datasets all have the same scale and shape.

The first is 'bias correction', in which the future scenario simulation is corrected by the model discrepancy, which is the difference between the present-day simulation and observations. The second is the 'change factor' method, in which observations are changed by the simulated climate anomaly, which is the difference between the present-day and future simulations. The simplest approach to these is to apply only the mean change (mean discrepancy or mean anomaly). This is appropriate if the distributions in the three simulations have the same scale (e.g. variance) and shape (e.g. skewness). But typically they do not, so a better approach is to apply the changes to several quantiles (for empirically estimated distributions) or to all the parameters (for parametric approaches such as GPD modelling or statistical downscaling). Unfortunately, even if this is done, bias correction and change factors can give very different results (Ho *et al.*, 2012). The appropriate choice, including the underlying assumptions, is not obvious and this is an important area of future research.

6.2.3 Statistical modelling

Statistical modelling has two supporting roles in estimating future extremes from physical climate models such as GCMs. The first is to characterise properties of the extremes in a given simulation: to estimate return periods (extreme value analysis, Section 6.2.3.1) or generate pseudo-event datasets for impacts studies (weather generation and cat modelling, Section 6.2.3.2). The second is to supplement and extend a set of simulations, by quantifying relationships between high- and low-resolution information (statistical spatial downscaling, Section 6.2.3.3) or model parameters and output (emulation, Section 6.2.3.4).

Many impacts models are statistically based, because the mechanisms are not known (Chapter 5). These implicitly assume the relationship between hazard and impact (e.g. mortality) is fixed in the future, because the changing effect of external factors such as economic growth and infrastructure is impossible to predict. Catastrophe ('cat') models of loss, developed for insurance and reinsurance, incorporate some physical-based constraints but are principally statistical.

6.2.3.1 Extreme value analysis

Methods of analysing extreme values in observational records are described in the previous chapter: for example, fitting a generalised extreme value (GEV) distribution to annual maxima, or a generalised Pareto distribution (GPD) to threshold exceedances. These methods estimate long-term return periods for extremes of a given magnitude, usually extrapolating to rarer extremes than are in the observations and using an assumption of stationarity.

These extreme value theory (EVT) methods are increasingly used for analysing climate simulations (Naveau *et al.*, 2005; Katz, 2010; references therein). EVT relies on an asymptotic limit (Chapter 5): in principle, GCMs could be used to generate very long equilibrium simulations, tens of thousands of years, to approximately satisfy this assumption, but in practice the models are too computationally expensive to allow simulations of more than 10–20 decades (the problem is worse for the observational record: Chapter 5).

Extremes are either assessed in two equilibrium simulations, two near-equilibrium parts of a simulation or continuously through a transient simulation. An example of the second is to estimate the change in GEV location parameter (Chapter 5) in simulations of the present day and the end of the twenty-first century. Bias correction or change factor methods (Section 6.2.2.4) are used to remove some systematic simulator errors, and hypothesis tests (e.g. goodness-of-fit tests: Chapter 5) are applied to determine the significance of any change. Examples of the third, analysing extremes in non-stationary data, are described in Chapter 5: for example, GEV or GPD parameters may be allowed to vary through the simulation in step-changes (e.g. Min *et al.*, 2009), with a linear trend (e.g. Cooley 2009; Bjarnadottir *et al.*, 2011) or covarying with global mean temperature or an index of natural variability (e.g. Hanel *et al.*, 2009; Hanel and Buisand 2011).

6.2.3.2 *Weather generation and catastrophe modelling*

Weather generation is statistical modelling of a dataset to create many plausible realisations of the same time period with the same statistical properties as the original (e.g. Qian *et al.*, 2004; Kyselý and Dubrovský, 2005; Semenov, 2008; Jones *et al.*, 2009; Kyselý, 2009; Nicolosi *et al.*, 2009). This is a type of statistical downscaling applied to GCM or RCM output to increase temporal resolution. In parametric weather generation, the parameters of a time-series model are calibrated with observations or climate simulations; non-parametric weather generators use bootstrapping methods (Chapter 5). Each location is usually treated independently (Chapter 5). The pseudo-hazard datasets are used for risk assessments or as inputs to impacts models; they may need further processing for this (e.g. thermal simulation of buildings: Eames *et al.*, 2011). Bias correction or change factor methods (Section 6.2.2.4) are used to make projections: in the latter, parameters of the weather generator are calibrated with observations then modified by projected climate anomalies from the GCM (e.g. Jones *et al.*, 2009). For changes in extreme weather it is important to apply not only mean climate anomalies but also aspects of variability such as variance, skewness and correlation. An overview of weather generation is given by Wilks and Wilby (1999), and a more recent discussion in the context of extreme precipitation is given by Furrer and Katz (2008).

Catastrophe models (or 'cat models') of hurricanes and other hazards are statistical methods for generating pseudo-event datasets that incorporate the whole chain from hazard to financial loss (Chapter 5). In the past, cat models have been based on sampling or calibration with historical observations, implicitly assuming stationarity, but recently RCMs (Schwierz *et al.*, 2010) and proprietary numerical weather models (the 'US and Canada Winterstorm' and 'European Windstorm' models developed by Risk Management Solutions) have been incorporated into loss modelling so as to make risk assessments under changing boundary conditions.

6.2.3.3 *Statistical spatial downscaling*

Statistical spatial downscaling, sometimes referred to as statistical bias correction, uses empirical relationships between information at high spatial resolution (meteorological station data, such as precipitation) and low spatial resolution (GCM or mean station data, such as temperature, humidity and sea-level pressure). These relationships are derived using statistical modelling techniques such as regression (e.g. Wilby *et al.*, 2002), artificial neural networks (e.g. Mendes and Marengo, 2009) and synoptic classification schemes. Statistical temporal downscaling is known as weather generation (Section 6.2.3.2). Techniques for downscaling mean climate may not be suitable for extremes: for example, extremal indices (Chapter 5) are often non-Gaussian, so the underlying assumptions of linear regression are not valid and a more flexible form such as a generalised linear model may be more appropriate. Benestad *et al.* (2008) describe strategies and example code for downscaling extremes and probability distributions.

These methods are easy to implement and, compared with dynamical downscaling (Section 6.2.2.3), do not require the very significant computational expense, can access finer scales, are applicable to variables that cannot be directly obtained from RCM outputs and can be comparably successful (IPCC, 2007a; CCSP, 2008). However, they cannot include feedbacks, there is no spatial coherence between sites, physical consistency between climate variables is not ensured and, most importantly, they include an implicit assumption that these relationships remain unchanged under different boundary conditions (IPCC, 2007a).

If meteorological data with good spatial coverage are available, an alternative is to use a change factor approach (Section 6.2.2.4), adding GCM climate anomalies to an observed climatology (e.g. Hewitson, 2003). Using a mean change factor does not allow for changes in variability and temporal correlation and makes the assumption, like pattern scaling (Section 6.2.4.1), of an unchanging spatial pattern; for non-Gaussian variables such as precipitation, care must be taken to avoid unphysical results (Wilby *et al.*, 2004).

6.2.3.4 Emulation

Statistical representation of complex models, known as 'emulation' (or 'meta-modelling'), is used to make predictions for areas of parameter space not explored by a physical model (O'Hagan, 2006). Emulators are usually constructed using regression, with coefficients calibrated from an ensemble of simulations with different parameter settings (e.g. Murphy *et al.*, 2004, 2007; Kennedy *et al.*, 2006; Rougier and Sexton, 2007; Rougier *et al.*, 2009; Sexton *et al.*, 2011), though neural networks have also been used (e.g. Knutti *et al.*, 2003; Piani *et al.*, 2005; Sanderson *et al.*, 2008). The primary motivation is estimation of simulator uncertainty for models that are too computationally expensive to allow many replications. Expert judgement may also be incorporated by requiring that the outputs satisfy physical principles such as conservation of mass and water. Emulation is conceptually simple, but requires careful choices to avoid problems such as over-fitting (Rougier *et al.*, 2009; Sexton *et al.*, 2011). Success of an emulator can be assessed with diagnostic tests, such as leaving one or more model simulations out of the emulator calibration and predicting their output (Rougier *et al.*, 2009; Sexton *et al.*, 2011).

6.2.4 Uncertainty assessment

The previous two sections describe the difficulties with simulating long-term changes in extreme weather. Computational limitations restrict the spatial and temporal resolution, so that their realism and relevance is compromised, and this can only partially be addressed by parameterisation and tuning. Our understanding is incomplete for the physics, chemistry and biology of some aspects of the earth system, such as the cryosphere or vegetation, that might well be important for 100-year time intervals under changing boundary conditions. So, although climate simulators can be run in ensembles to approximate probabilities of extreme weather events, these approximations are very crude, being based on a small number of runs, and compromised by the limitations of the code. On top of this, future boundary conditions are unknown, adding a further uncertainty to be sampled.

The physical and statistical modelling methods described can address some of these issues of model inadequacy: bias correction or change factors to cancel some systematic errors; a hierarchy of physical model complexities and statistical emulation for different research questions (high complexity for improving representation of extremes; low complexity and emulation for increasing simulation length or number); physical and statistical downscaling to increase resolution; and extreme value analysis to extrapolate to rarer extremes than can reasonably be simulated. This section further describes these methods, and other tools, on the topic of quantifying uncertainties, including: ensembles that sample uncertainties in initial and boundary conditions, model parameters and structure; pattern scaling and emulation to extend ensemble size; introduction of a 'discrepancy' term to account for structural uncertainty; and approaches to calibration of parameters and discrepancy using the historical record.

The future of the earth system cannot be known exactly, because future forcings are unpredictable, but it is of paramount importance to assess uncertainties of climate model projections conditional on these forcings to inform decision-making and policy-making on potential options for mitigation (Section 6.3.1.1) and adaptation (Section 6.3.2). Uncertainties for extreme weather hazard projections can be partitioned into two kinds, associated with physical model predictions and the statistical modelling of those predictions. The first kind arise from uncertain physical model inputs (Section 6.2.4.1), parameters (Section 6.2.4.2) and structure (Section 6.2.4.3), and are explored and quantified using ensembles of predictions using different inputs (boundary and initial conditions), parameter values and physical models (an overview is given by Foley, 2010). The second kind are from statistical assessment of simulations and from statistical prediction of physical model output such as emulation. Uncertainties in statistical model parameters and structure are assessed with the usual methods, including graphical checks, bootstrapping and hypothesis tests, described in Chapter 5.

6.2.4.1 Model input uncertainty

Boundary condition uncertainty is sampled with a suite of predictions for different future 'storylines' (Section 6.2.1.1), scenarios that describe the effect on anthropogenic emissions of plausible future societal and technological changes, including economic growth, energy sources and efficiency and land use. Forcing scenarios are not predictions, and do not have relative probabilities assigned, so climate predictions are made conditionally upon each individual scenario. For the IPCC Fourth Assessment Report (IPCC, 2007a), projections are made based on the Special Report on Emissions Scenarios (SRES: Nakićenović *et al.*, 2000), which included storylines that ranged from rapid growth and fossil-intensive energy sources ('A1B') to global sustainability approaches ('B1'); the storylines are detailed descriptions of GHGs and sulphur dioxide emissions, which are translated into atmospheric concentrations using biogeochemical modelling (IPCC, 2007a). For the Fifth Asssessment Report the focus is 'representative concentration pathways' (RCP: Vuuren *et al.*, 2011); unlike SRES, these include scenarios with explicit climate policy intervention (Section 6.3.1.1) and are simpler, being expressed in terms of atmospheric concentrations so they can be used directly in climate models. Forcings due to possible geoengineering strategies (Section 6.3.1.2), including changes to stratospheric albedo (e.g. Ricke *et al.*, 2010), land surface albedo (e.g. Ridgwell *et al.*, 2009; Oleson *et al.*, 2010) and carbon uptake (e.g. Aumont and Bopp, 2006), are considered in separate studies. Future natural forcings (solar, volcanic) are 'known unknowns'. An early suite of climate projections (Hansen *et al.*, 1988) incorporated volcanic eruptions, but these are not usually included and no attempt is made to assess the effect of changes to solar forcing from the present day.

Climate models are so computationally expensive that simpler models are often used to approximate their predictions to increase the size of the boundary condition ensemble. This is known as 'pattern scaling': a simple climate model (SCM) simulates global mean temperature as a function of atmospheric forcing (i.e. for a range of scenarios), and GCM results from one scenario are scaled by the global mean temperature from the SCM. Pattern scaling relies on the assumption that simulated patterns of climate change scale linearly with global temperature; this is thought to be less reliable for precipitation than for temperature (Mitchell, 2003; IPCC, 2007a; Wilby *et al.*, 2009). Harris *et al.* (2006) estimate pattern scaling error by comparing the result of a 'slab' (simplified ocean) GCM scaled by a simple model with 'truth' from the dynamic ocean version of the GCM.

Initial condition uncertainty – incomplete knowledge of the initial state of the system – is also sampled with ensembles of simulations (Chapter 5). It is most important in decadal climate predictions, in which internal variability dominates (e.g. Smith *et al.*, 2007). Long-term projections focus on trends rather than year-to-year changes, so initial condition ensembles for these tend to be small and averaged over (e.g. Stainforth *et al.*, 2005; Meehl *et al.*, 2007) or else not performed at all.

6.2.4.2 Model parameter uncertainty

Uncertainty in climate model parameterisation schemes can be explored using perturbed parameter ensembles (PPEs), more usually and misleadingly named perturbed *physics* ensembles, which are groups of 'detuned' model versions created by changing the values of the control parameters (Section 6.2.2.1). Perturbed parameter ensembles offer the opportunity to sample uncertainty systematically in a well-defined space. They are often supplemented with statistical methods such as emulation (Section 6.2.3.4) to improve this sampling. A further motivation is to test detuned model versions against past observations, to avoid the circularity induced by tuning and validating with the same time period (Section 6.2.4.3). The full, unweighted ensemble of simulations may be analysed, or if the parameters are sampled from probability distributions (e.g. uniform, Gaussian) with sufficient sample size they can be integrated out using Monte Carlo methods to assess the contribution of parametric uncertainty to the predictions. The latter method can include likelihood weighting of ensemble members in a Bayesian statistical framework (Chapter 5) by comparing simulations of the recent past with observations (e.g. Sexton *et al.*, 2011).

The two most extensively studied PPEs use variants of the Hadley Centre climate model, which is a version of the UK Met Office Unified Model. The first is an ensemble of 17 versions, supplemented by an ensemble of about 300 using a simplified ocean model, used to generate projections for the UK Climate Impacts Programme (UKCIP: Murphy *et al.*, 2009; Sexton *et al.*, 2011). The second is a suite of ensembles with different model complexities, domains and experimental aims, each with hundreds to tens of thousands of members, generated with distributed (publicly donated) computing for the project *ClimatePrediction.net* (Stainforth *et al.*, 2005 and many subsequent studies). PPEs have also been performed with other models than the Hadley Centre model (e.g. Yang and Arritt, 2002; Lynn *et al.*, 2008; Sterl *et al.*, 2008; Yokohata *et al.*, 2010; Fischer *et al.*, 2010; Klocke *et al.*, 2011; Sanderson, 2011). PPEs can be performed with RCMs, but their computational expense limits them to small ensemble size (Murphy *et al.*, 2009; Buontempo *et al.*, 2009), short simulation length (Yang and Arritt, 2002) or both (Lynn *et al.*, 2008). PPEs are also generated with simpler, lower-resolution models, such as EMICs, but these typically focus on global mean temperature rather than regional patterns or extremes. An overview of PPEs is given by Murphy *et al.* (2011).

The original motivation for PPEs was quantification of uncertainty in the global mean temperature response to CO_2 forcing (Andronova and Schlesinger, 2001; Murphy *et al.*, 2004; Stainforth *et al.*, 2005), but they have since been used in risk assessments of future extreme temperature and precipitation (Barnett *et al.*, 2006; Clark *et al.*, 2006; Sterl *et al.*, 2008; Fischer *et al.*, 2010; Fowler *et al.*, 2010), drought (Burke and Brown, 2008; Burke *et al.*, 2010; Hemming *et al.*, 2010) and storms (Kim *et al.*, 2010). The disadvantage of studying extremes in PPEs is that for a given computational resource have to be made with simulation length: for example, the UKCIP simulations using a simplified ocean model are only 20 years long (e.g. Barnett *et al.*, 2006).

Climate model PPEs have not yet been much used in assessments of future hazard impacts. PPE methods have been used to some extent as inputs to impacts models (e.g. heat wave mortality: Gosling *et al.*, 2011), but propagation of the cascading parametric uncertainty in both climate and impacts models – the simplest method being a factorial experiment, with every combination of parameter values – is still relatively rare (e.g. crops: Challinor *et al.*, 2005, 2009).

6.2.4.3 *Model structural uncertainty*

Structural uncertainty arises from imperfect implementation or knowledge about the relevant physical processes: it may be considered the leftover error, or 'discrepancy', at the model's best (tuned) parameter values (Rougier, 2007). It is represented with a covariance matrix that describes the variations and correlations of model discrepancy across locations, times and variables. There are various methods of estimating this covariance matrix in numerical weather prediction (Chapter 5), which rely on repeated tests of the model forecasts against reality. These methods cannot be directly translated to climate model predictions, because of the multi-decadal definition of climate. Structural uncertainty is therefore extremely challenging to quantify, but necessary for assessing the success of climate model predictions and for decision-making under future climate change. A related problem is the appropriate interpretation of a range of predictions from different climate models. In the past, model structural uncertainty was considered so difficult to quantify it was essentially ignored, and all models treated as equally skilful, but the literature has flourished since the IPCC Fourth Assessment Report (IPCC, 2007a). The main points are summarised here.

Expert elicitation has been proposed as a method of estimating structural uncertainty (Kennedy and O'Hagan, 2001; Goldstein and Rougier, 2004). The state space of climate models is extremely high-dimensional, so many simplifying assumptions about the correlation structure (spatial, temporal and multivariate) would be necessary in order to elicit a covariance matrix and the complexity of GCMs may be too great for this to be practicable (Sexton *et al.*, 2011). Instead, structural uncertainty is inferred by comparing model output with expectations, observations and output from other models.

Simulator discrepancy arises from two sources: imperfect implementation of theoretical knowledge (e.g. numerical approximations, coding errors), and imperfect theoretical knowledge of the system (missing or imperfectly represented physical processes). The former are inaccuracies, checked with 'verification', while the latter are inadequacies, assessed with 'validation' (e.g. Roy and Oberkampf, 2011). Verification of climate models is challenging due to their complexity, and true verification may be impossible (Oreskes *et al.*, 1994). Both formal and informal methods are used, including bit-wise comparisons with previous versions, isolated testing of numerical methods and frequent comparison with observations and other models, though recommendations have been made to improve and extend these (Pope and Davies, 2002; Easterbrook and Johns, 2009; Clune and Rood, 2011). The UK Unified Model is used for operational NWP so it is tested very frequently, albeit under different configuration and timescales than for climate prediction.

Validation of climate models is even more challenging: the difficulties of their complexity are compounded by the long timescale of climate and the relative sparsity of observations of the high-dimensional, multi-decadal state. The model is evaluated with observations, where available, or with other models. If the model does not agree with observations, after accounting for observational uncertainty, the remaining discrepancy can inform estimates of the model structural uncertainty; conversely, model evaluation (e.g. of members of a PPE) requires an estimate of the model structural uncertainty to be meaningful (e.g. Sexton and Murphy, 2011). If a model agrees well with observations, it is not guaranteed to be successful in prediction; reasons are described below. Validation can therefore only be a partial assessment of adequacy (Oreskes *et al.*, 1994; Knutti, 2008a). Methods and interpretation are vigorously discussed and debated in the literature (e.g. Räisänen, 2007; Knutti, 2008a; Reichler and Kim, 2008; Hargreaves, 2010; Fildes and Kourentzes, 2011). The main challenges and strategies are discussed here briefly.

Climate modelling as a research field is only four or five decades old, not much longer than the timescale on which climate is defined, so projections of future climate cannot be repeatedly tested against observations to ensure ensemble predictions are well-calibrated (as in NWP: Chapter 5). Some out-of-sample climate forecasts exist and these have appeared to be quite successful (e.g. Smith *et al.*, 2007; Hargreaves, 2010), though these are single rather than repeated tests. Climate model validation therefore relies on multiple comparisons of 'hindcasts' rather than forecasts. Their skill in reproducing historical observations is held to be an indicator of their likely success at predicting future climate. There are a number of difficulties with this approach.

First, the availability of data. The observational record is 100–150 years long for a very limited number of variables, a few decades for most variables, and limited in spatial coverage; satellite records are 1–2 decades long. This is partially addressed with 'space-for-time' substitution: repeated model testing is performed across different locations instead of, or as well as, different times. One example is the recent application to climate model simulations of the rank histogram, a skill metric used in NWP, by replacing multiple forecast tests in time with multiple hindcast tests in space (Annan and Hargreaves, 2010). However, such tests require careful consideration of the spatial correlation structure of the model discrepancy. Datasets of extremes are, of course, particularly limited; evaluation of climate model extremes focuses on their magnitude and trend (CCSP, 2008).

A second difficulty is circularity: climate models are both tuned and evaluated with observations in the historical period. So if a model reproduces the historical observations, the best that can be said is that it is empirically adequate or consistent (Knutti, 2008a, 2008b). There are various strategies to mitigate this circularity. One is to decrease reliance on parameterisation by adding physical processes and increasing resolution, but this brings relatively slow improvements. A second is to consider agreement with observations as a constraint on the model parameter space (Knutti, 2008b) or, more formally, to detune climate models and incorporate this uncertainty in future predictions (PPEs: Section 6.2.4.2). Circularity could also be avoided by validating with independent observations, but these are not easy to come by: datasets are relatively few (CCSP, 2008), and new ones slow to appear (due to the global coverage and long timescale required), and model developers are generally aware of datasets even if they are not formally used in tuning. One solution is to validate the model with independent information from palaeoclimate reconstructions. These reconstructions are based on 'proxies' such as the response of vegetation (inferred from, for example, fossilised pollen or tree rings) to climate; they are more uncertain than observations, but the signal is often large enough to compensate. These datasets have been used to test the relative success of different climate models (e.g. Braconnot *et al.*, 2007) and perturbed parameter ensemble members (e.g. Edwards *et al.*, 2007). Model parameters have also been tuned to palaeoclimate information (e.g. Gregoire *et al.*, 2010). However, palaeoclimate reconstructions are usually of annual or seasonal mean climate, not extremes, and are not always provided with uncertainty estimates, which are required for these methods.

A third difficulty is the assumption that a model that is successful at hindcasting will also be successful at forecasting. It is proper to assume that the laws of physics are constant, but successful parameter values may depend on the state of the system. Tuning can also conceal modelling inadequacies such as compensating errors or boundary conditions (Knutti, 2008b). This point can also be addressed with palaeoclimate information: the more out-of-sample tests in which a model is found to be 'empirically adequate', the more likely it is to be adequate for the future state. In fact, there are indications from comparisons with palaeoclimate records that GCMs may be too stable (Valdes, 2011).

Model set-up can be changed to quantify the sensitivity of the results: in particular, the effect of changing resolution or domain. One example is a 'Big Brother/Little Brother' experiment, in which a large-domain RCM simulation is filtered to GCM resolution and used to drive a small domain, to compare with the original RCM result (Rummukainen, 2010). Changes in climate model set-up can also be propagated to impacts models, to test their effect on estimated hazard impacts (e.g. Gosling *et al.*, 2011).

But the primary method of comparing model results and exploring model structural uncertainty is with a multi-model ensemble (MME), a collection of climate models produced by different research and meteorological institutes around the world. These are 'ensembles of opportunity', rather than ensembles of design such as those described above. However, simulations generated by these MMEs are systematically designed and analysed: standardised boundary conditions are used and the simulations made publicly available for the World Climate Research Programme Coupled Model Intercomparison Projects (CMIP3: Meehl *et al.*, 2007; CMIP5: Taylor *et al.*, 2011), so as to allow model intercomparisons (e.g. PRUDENCE: Christensen *et al.*, 2002) and multi-model predictions (IPCC, 2007a). MMEs have been used extensively to assess extreme weather hazards under future climate change, including: extreme temperature (e.g. Meehl and Tebaldi, 2004; Weisheimer and Palmer, 2005; Kharin *et al.*, 2007; Fischer and Schär, 2008; Kyselý, 2009), extreme precipitation (e.g. Palmer and Räisänen, 2002; Meehl *et al.*, 2005; Räisänen, 2005; Tebaldi *et al.*, 2006; Kharin *et al.*, 2007), drought (e.g. Wang, 2005; Burke and Brown, 2008) and hurricanes (Emanuel *et al.*, 2008; references therein).

MMEs can be used to estimate structural uncertainty, so long as some assertion is made about their 'exchangeability': that is, that each is judged equally plausible by experts (Sexton *et al.*, 2011). This approach has been used to estimate the structural uncertainty of the Hadley Centre climate model, with the motivation that models in the MME are likely to incorporate at least some of the processes that are missing in the Hadley Centre model: emulation is used to search parameter space for the closest match to each member of the MME (treating each model in turn as 'truth'), the remaining discrepancy calculated and the structural uncertainty covariance matrix estimated from these discrepancies across the MME (Sexton *et al.*, 2011; Sexton and Murphy, 2011). The authors acknowledge this may be an underestimate due to common model inadequacies (described below) and test the sensitivity of their results to the method.

MMEs are not systematic samples of a well-defined space, so it is not straightforward to integrate over their future predictions in the same way as PPEs (Section 6.2.4.2). Various methods have been proposed for interpreting their predictions for policy-making (Tebaldi and Knutti, 2007; Knutti, 2010; Knutti *et al.*, 2010). First, with expert judgement: an early estimate of the range of climate sensitivity to a doubling of atmospheric CO_2, 1.5–4.5°C (Charney, 1979), was based on the judgement of a small group of experts about the uncertainty represented by a small number of climate model simulations that spanned the range 2–3.5°C, and this estimate remained unchallenged until recently (IPCC, 2007a). Second, as a 'probabilistic' distribution, analogous to NWP probabilistic forecasts (e.g. Räisänen and Palmer, 2001; Palmer and Räisänen, 2002). Third, as an unweighted mean: this has been found to outperform individual ensemble members (e.g. Lambert and Boer, 2001; Palmer *et al.*, 2004). Fourth, as a weighted mean, based on comparisons with observations (Giorgi and Mearns, 2003; Tebaldi *et al.*, 2005). However, ensemble averaging tends to produce smooth fields biased towards the climate mean state, so it may not be suitable for extremes (Palmer and Räisänen, 2002).

There are very many metrics of skill, and no model in the global MME performs best in all, or even the majority, of these metrics (IPCC, 2007a). But the main difficulty in combining MME predictions is that they are not independent: model components are inevitably similar due to the use of the best available theoretical knowledge, and its representation in code, as well as the movement of model developers and code between institutes. This results in common biases between models (Collins *et al.*, 2011). The degree of independence is extremely difficult to quantify, though several recent attempts have been made (Jun *et al.*, 2008; Pirtle *et al.*, 2010; Collins *et al.*, 2011; Masson and Knutti, 2011; Pennell and Reichner, 2011). A weighted mean based on skill metrics may be worse than an unweighted mean, if inter-model correlations are not taken into account (Weigel *et al.*, 2010).

6.3 Risk Management

Risk management for climate change is an extremely broad topic. Aspects relevant to changing risks of extreme weather are the main focus here; managing the risks of extreme weather hazards in the present climate (disaster risk management) is discussed in Chapter 5. One of the biggest challenges for extreme weather hazards in the medium- to long-term future is the mismatch in timescales under consideration between decision-makers and scientists: from the seasonal to multi-annual time frame of political, insurance and business concerns to the multi-decadal time frame for which climate change can be estimated (Dlugolecki, 2009; IPCC, 2012).

Mitigation and geoengineering are strategies that attempt to reduce the severity of climate change by modifying the projected future forcing (Section 6.3.1). The relative effectiveness of different strategies is estimated with the methods described in this chapter: for example, comparing a 'business-as-usual' emissions scenario with a mitigation scenario that strongly reduces emissions, or comparing a 'constant solar insolation' scenario with one that reduces insolation to approximate the effects of possible future geoengineering. Adaptation strategies attempt to reduce the severity of the impacts of climate change by modifying the future loss (Section 6.3.2). The relative effectiveness of different strategies is estimated with the methods described in this chapter, comparing 'no adaptation' losses with 'adaptation' losses. Communication strategies attempt to inform decision-makers and the public of the projected risks (Section 6.3.3).

6.3.1 *Mitigation and geoengineering*

The risks of extreme high temperature hazards (heat waves and droughts) increase, of course, with global temperatures. For extreme precipitation and storm hazards, atmospheric and ocean temperatures are also important drivers but the relationship is more complex (Section 6.2.1.3), so increasing temperature may lead to increased risk in some regions and decreased risk in others (IPCC, 2012). Mitigation and geoengineering strategies address the risk of anthropogenic climate change, and therefore the risk of hazards driven by temperature increases. A detailed synthesis report is given by the IPCC (2007c).

6.3.1.1 *Mitigation*

Mitigation strategies aim to reduce anthropogenic climate change by reducing the positive (warming) anthropogenic forcings. Mitigation is an extensive topic, global in reach (because GHGs have global effects) and politically charged (for example, developed nations are largely responsible for past GHG emissions but developing nations are most vulnerable to impacts). A few of the issues are outlined here; a more detailed discussion is given by the IPCC (2007c).

Most strategies for mitigating anthropogenic climate change focus on the sources of anthropogenic GHG emissions, by reducing usage of GHG-emitting processes (e.g. increased insulation of buildings, reduced air travel or reduced livestock consumption) or reducing their GHG intensity (e.g. energy efficiency, nuclear or renewable energy and fuel, carbon capture and storage, or low-methane agriculture). Some strategies focus on the sinks of GHGs, reducing atmospheric GHG concentrations by increasing uptake to terrestrial (e.g. trees, soil litter) or ocean reservoirs. Inertia in the climate system (due to long-timescale processes) and in societal change mean that most mitigation strategies are limited in the effect they can have on short-term (annual to decadal) climate change.

Some mitigation strategies have been put in place at the global scale. An international treaty was adopted in 1992 (United Nations Framework Convention on Climate Change, UNFCCC: http://unfccc.int), with legally binding GHG emissions targets agreed in 1997 and annual meetings to assess progress (IPCC, 2007c). National strategies of varying strength have also been put in place; the UK is currently unique in introducing a long-term legally binding framework, the Climate Change Act of 2008 (details can be found at http://www.legislation.gov.uk).

One of the controversial aspects of mitigation is the cost of large-scale societal and economic change motivated by predictions that are (inevitably) uncertain; for example, regional extreme temperature changes are poorly constrained by global warming targets (Clark *et al.*, 2010). The UNFCCC explicitly applies the precautionary principle: 'where there are threats of serious or irreversible damage, lack of full scientific certainty should not be used as a reason for postponing such measures' (IPCC, 2007c: Article 3.3:). However, it is difficult to ascertain the most effective courses of action, and the best time frame for their implementation. A well-known attempt at quantifying the relative costs of action and inaction is the *Stern Review on the Economics of Climate Change*, an extensive report prepared by economists for the UK government, which estimated that if no action were taken the overall costs and risks of climate change would be equivalent to losing 5–20% or more of global GDP each year and that spending around 1% of global GDP each year to reduce GHG emissions would avoid the worst impacts (Stern, 2007). The assumptions and methodology of the *Stern Review* have been criticised (e.g. Weitzman, 2007): in particular, the use of a low discount rate (reduction in future costs of taking action).

6.3.1.2 Geoengineering

Geoengineering is deliberately changing the climate (Lenton and Vaughan, 2009; Irvine and Ridgwell, 2009). It has not yet been undertaken, except in small-scale studies (e.g. Boyd *et al.*, 2000). Some proposed strategies aim to reduce the positive (warming) forcings, while others increase the negative (cooling) forcings. These correspond to the two possible methods of changing the earth's energy balance: increasing the outgoing (longwave) radiation, or reducing the incoming solar (shortwave) radiation.

The former methods decrease the forcing due to GHGs by extracting them from the atmosphere with technological (e.g. artificial trees: Lackner, 2009) or biological (e.g. fertilisation of ocean phytoplankton: Aumont and Bopp, 2006; afforestation, charcoal and wood storage) methods. These can also be considered mitigation strategies.

The latter type, known as solar radiation management (SRM), increase planetary reflectivity (albedo). Currently the most discussed method is injection of particles into the air to increase the reflectivity of the atmosphere or clouds (sulphate aerosols: e.g. Crutzen, 2006; water droplets: e.g. Ricke *et al.*, 2010); other suggestions include increasing the reflectivity of urban structures (Oleson *et al.*, 2010) or crops (Ridgwell *et al.*, 2009), or positioning reflective structures between the earth and sun (Angel, 2006). One advantage of SRM over GHG reduction is that it could potentially cool the climate much more quickly. Different SRM methods have different strengths and weaknesses: for example, structures in space are thought to be prohibitively expensive, while the reversibility of particle injection methods (due to their short lifetime in the atmosphere) is an advantage if unforeseen negative impacts were to occur.

Risk management with SRM has several disadvantages relative to GHG reduction. It might allow anthropogenic GHG emissions to continue at their present rate, avoiding large-scale societal and economic changes, but would need to be maintained indefinitely: if it were stopped, rapid warming from the underlying GHG forcing would occur. SRM does not address carbon dioxide dissolution in the ocean, which may have significant negative impacts on calcifying marine organisms (Turley *et al.*, 2010). Furthermore, SRM does not exactly cancel GHG forcing, because the earth system responds differently to different types of forcing (IPCC, 2007a). While global mean temperature could be returned to pre-industrial levels, this is unlikely to be the case for regional precipitation, drought, and storms (e.g. Ricke *et al.*, 2010).

The global impacts of geoengineering strategies are explored with climate models. There are therefore large uncertainties in their effectiveness and the likelihood of negative impacts, particularly for methods that alter complex, nonlinear parts of the earth system such as the balance of ocean ecosystems or the hydrological cycle (IPCC, 2007c). A multi-model intercomparison project to evaluate the effects of stratospheric geoengineering with sulphate aerosols is currently underway (Kravitz *et al.*, 2011). There are also enormous challenges in implementation, both practical and legal: for example, potentially setting the climatic needs and the rights of societies in different parts of the world against each other. It is therefore currently considered a method of last resort, or at least a method of support of, rather than a replacement for, mitigation.

6.3.2 Adaptation

Adaptation is management of risk by improving the ability to live with climate change: increasing resilience to minimise the severity of the impacts. Some categories of adaptation (Wilby *et al.*, 2009) include resource management (e.g. of reservoirs), new infrastructure (e.g. flood defences) and behavioural change (e.g. diversification and genetic engineering of crops: CCSP, 2008; IPCC, 2007b). A detailed synthesis report is given by the IPCC (2007b).

Adapting to the impacts of extreme weather is a continuous process through human history (Chapter 5; IPCC, 2007b), but in the future these events may move outside the bounds of past experience in severity, frequency and location, requiring greater action and perhaps 'transformational' changes (IPCC, 2012). Measures to adapt to hazards in a changing climate are taken either as a consequence of past changes (e.g. increased use of artificial snow-making by the ski industry: IPCC, 2007b) or projections of future change. But the latter are uncertain, sometimes contradictory (Wilby and Dessai, 2010), and adaptation measures expensive, so strategies motivated by projections are generally 'no-' or 'low-regrets' options ('climate-influenced' decisions: Willows and Connell, 2003) that address multiple types of hazard, present and future risks and societal benefits such as water resource planning, coastal defence, social welfare, quality of life and livelihoods (IPCC, 2007b; Wilby and Dessai, 2010; IPCC, 2012). This helps to avoid competition for resources between adaptation, mitigation and disaster risk management (IPCC, 2012). Public pressure for adaptation measures might increase with the occurrence of extreme weather events (IPCC, 2007c), though Hulme *et al.* (2011) argue that attempts to attribute extreme weather events to anthropogenic causes (e.g. Stott *et al.*, 2004; Min *et al.*, 2011; Pall *et al.*, 2011) are likely to 'increase the political and ethical complexities of these decisions'. Decisions are inevitably made with multiple objectives, which may be conflicting: 'maladaptations' are those that constrain adaptation by increasing exposure or vulnerability (e.g. developing housing in vulnerable areas: Willows and Connell, 2003; Chapter 5).

Adaptation has the potential to reduce future impacts even if the frequency or severity of hazards increase: for example, increasing temperatures and amplification of the hydrological cycle may increase the frequency and severity of droughts, but improvements in managing water resources mean this does not necessarily lead to more frequent restrictions in supply. The most successful adaptation strategies are iterative, with continuing dialogue between scientists and risk managers to improve the transfer, interpretation and usefulness of risk assessment information. Unfortunately, the cascading chain of uncertainties mean the use of climate scenarios for adaptation is 'lagging [their use in impact assessment] by nearly a decade' (Wilby *et al.*, 2009); the authors review current uncertainties in climate risk within an adaptation context.

Many risk management frameworks for climate change have been proposed (e.g. IPCC, 1994; Willows and Connell, 2003; Johnson and Weaver, 2009; IPCC, 2012). The 'risk, uncertainty and decision-making framework' (Figure 6.1; Willows and Connell, 2003) of the UK Climate Impacts Programme (http://www.ukcip.org.uk) is intended for users with some knowledge of climate risks that want to fully understand these risks and obtain a good understanding of adaptation options. The framework is described in a comprehensive technical document with detailed guidance, methods and options for implementation (Willows and Connell, 2003). Additional frameworks and tools are provided, based on the same underlying principles but aimed at different users, including an 'Adaptation Wizard' for users new to adaptation. Detailed case studies are given: for example, identifying the potential increasing risks of extreme weather for a UK container port (e.g. power supply disruption and damage, crane and pilot stoppages, port closures). The UKCIP framework has been used extensively by a wide variety of stakeholders to assess impacts and risk. A qualitative risk analysis with this framework can be followed with a quantitative analysis using the most recent UK Climate Projections (UKCP09: Murphy *et al.*, 2009), which contain more information and exploration of uncertainty than the previous projections (UKCIP02: Hulme *et al.*, 2002) and are expressed in probabilistic terms.

6.3.3 Communication

There are a number of problems in communicating the risks of future climate change and extreme weather hazards. A broad overview is given by Hulme (2009). The public receive conflicting messages from the media about whether the climate will change significantly (in contrast with the broad consensus within the scientific community on the large-scale

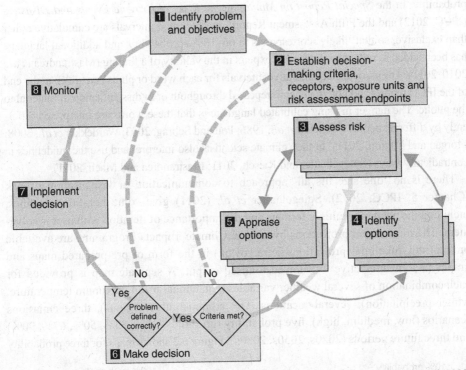

Figure 6.1 The UKCIP risk, uncertainty and decision-making framework (Willows and Connell, 2003 © UKCIP, 2011).

aspects: IPCC, 2007a) and whether individual extreme weather events are attributable to human causes (a reflection of this rapidly changing research area: e.g. Min *et al.*, 2011; Pall *et al.*, 2011). Communication of climate change risks is a wide-ranging and developing area. Some issues are outlined here.

Many difficulties arise from mental models and cognitive biases, where differences in framing, level of detail, personal and cultural experiences and narrative affect the interpretation of risk (Patt and Schrag, 2003; Marx *et al.*, 2007; Mastrandrea *et al.*, 2010; Morgan and Mellon, 2011; Sterman, 2011; Spiegelhalter *et al.*, 2011). Sterman (2011) describes some of the difficulties non-experts can have in understanding climate-related risks, including the long timescales, feedbacks and the complexity of the system, and makes a number of policy recommendations including a wider use of interactive simulations for learning; simple climate models with graphical user interfaces (e.g. Price *et al.*, 2009) could be used for this purpose.

Scientific literacy plays a part, but even well-informed users may be confused by the balance of consensus and uncertainty in the IPCC Assessment Reports (Socolow, 2011). These reports have guidelines that standardise and, over the years, refine the communication of uncertainty with calibrated language to reflect probabilistic predictions. In the Fourth Assessment Report (IPCC, 2007a) these included categories such as 'likely', corresponding to 60–90%

probability. In the *Special Report on Managing the Risks of Extreme Events and Disasters* (IPCC, 2012) and the Fifth Assessment Report, the calibrated intervals are cumulative rather than exclusive, so that 'likely' corresponds to 66–100% probability, and additional language has been added to reflect confidence of experts in the validity of a finding (Mastrandrea *et al.*, 2010, 2011). The associated quantitative intervals for each word or phrase are given at the end of the IPCC reports and not, in general, repeated throughout or in dissemination of material to the public. The danger in using calibrated language is that these words are interpreted differently by different people (Wallsten *et al.*, 1986; Patt and Schrag, 2003; Wardekker *et al.*, 2008; Morgan and Mellon, 2011). In fact, climate scientists also interpret and use the guidelines in contradictory ways (Spiegelhalter and Riesch, 2011; Mastrandrea and Mach, 2011).

There is no 'one size fits all' approach to communication or visualisation of risk (Chapter 5; IPCC, 2012). Spiegelhalter *et al.* (2011) give some helpful guidelines, including the need for multiple formats and the importance of iteration with user involvement. The most recent projections by the UK Climate Impacts Programme are available online (http://ukclimateprojections.defra.gov.uk) in the form of pre-prepared maps and graphs and an interactive tool for customised output. A separate map is provided for each combination of several weather variables (e.g. annual mean maximum temperature, winter precipitation), several locations (UK administrative regions), three emissions scenarios (low, medium, high), five probability thresholds (10%, 33%, 50%, 67%, 90%) and three future periods (2020s, 2050s, 2080s). Figure 6.2 shows maps for three probability

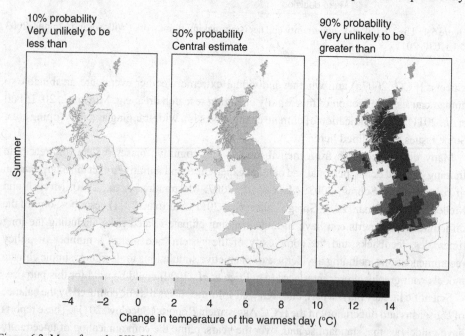

Figure 6.2 Maps from the UK Climate Impacts Programme showing 10%, 50% and 90% probability levels of changes to the temperature of the warmest day of the summer, by the 2080s, under the medium-emissions scenario (Murphy *et al.*, 2009 © UK Climate Projections, 2009).

thresholds for the change in temperature of the warmest day of the summer, by the 2080s, for the medium-emissions scenario. Explanatory phrases such as 'central estimate' and 'very unlikely to be greater than' are included to clarify the meaning of the probability levels. The separation of the different variables and scenarios makes large amounts of forecast information available, even without the customisation tool. These are used extensively by decision-makers such as local government agencies. However, there is little empirical evidence on the effectiveness of visualisation methods for climate change risks (Pidgeon and Fischhoff, 2011; Stephens *et al.*, 2012).

6.4 Summary

Risk assessment and risk management for extreme weather in a changing climate is essential for global society, but it has many challenges. The limited predictability of future anthropogenic forcings is addressed by making predictions for a set of plausible storylines and scenarios, but these might not cover the range of possible futures. The effects of feedbacks are difficult to simulate for some important parts of the earth system, such as clouds and the carbon cycle, which means long-term projections are subject to substantial uncertainties. Short-term decadal predictions are highly desirable for policy-making and planning; methods are being learned from numerical weather prediction, but forecasts are dominated by internal variability and this remains one of the most challenging research areas. Extremes of climate are even more difficult to simulate than the mean, because observations are rarer (for model calibration and validation) and state-of-the-art climate models too expensive to run multi-centennial simulations. The most difficult weather hazards to simulate are extreme precipitation and tropical cyclones, because they are driven by processes on small spatial scales, and drought, which is defined by multiple variables and indices.

Physical modelling is crucial to incorporate the effects of changing boundary conditions. A spectrum of physical climate models exists, varying from complex, high-resolution regional models to simpler and lower-resolution global models. There is a trade-off, given finite computational resources, between climate model complexity (to reduce biases) and the length and number of simulations (to simulate extremes and quantify uncertainty). Physical impacts models (e.g. the effect of extreme weather on health and the environment) exist, but most impacts models are statistical and implicitly assume the hazard–impact relationship does not change through time.

Quantification of uncertainty incorporates all the challenges described in Chapter 5 and more. In particular, the multi-decadal timescale on which climate is defined rules out traditional validation methods used in numerical weather prediction. Models are mainly evaluated with hindcasts rather than forecasts, and repeated tests are made across spatial locations to partially compensate for the lack of repeated tests in time. Ensembles explore uncertainty in climate forcing, initial conditions, model parameters, and model structure. The significant computational expense of climate models usually limits the size of these ensembles to dozens, but simpler models (e.g. pattern scaling) and statistical models (e.g. emulation) can supplement their predictions, and distributed computing has provided the resources to generate very large numbers of simulations. Not all ensembles are created equally: parametric and initial condition uncertainty (the former more important in centennial predictions, the latter in decadal) may be explored systematically and integrated out for a single model, but a suite of predictions for different scenarios or with different climate models must be treated individually or at least, in the case of multiple models, handled with care. It is essential, but very difficult, to estimate model structural uncertainty, because it represents the degree of realism of a climate model in an extremely high dimensional and correlated space; expert elicitation and multi-model ensembles are currently the main contenders. Model validation under uncertainty is further complicated by the low availability of observations. Reconstructions of past climates can help with this, though these require estimates of uncertainties to be useful. The causal chain under consideration is long – from emissions to concentrations, global climate, regional climate, extremes, impacts and, finally, risk management – and this is reflected in the relative uptake of methods to quantify and propagate uncertainty at each stage. It is rare for studies to propagate more than about three of these sources of uncertainty in any one assessment.

Potential risk management strategies are mitigation and geoengineering (reducing the human warming impacts on climate, or increasing the cooling impacts), adaptation (increasing resilience to extreme weather hazards today and in a changing climate) and communication (increasing awareness of the changing risks). Mitigation and geoengineering strategies are global in their effects, so they require international or global agreement to implement. The relative efficiency of mitigation, geoengineering and adaptation strategies is assessed with projections for different plausible future scenarios. However, the complex feedbacks and long causal chains of interactions between parts of the earth system and driving forces such as humans and the sun introduce many cascading uncertainties. This can lead to vague or contradictory risk assessments that are difficult to use in decision-making. To address this, risk management searches for 'no regrets' options that are flexible and/or yield additional benefits such as reduced costs (e.g. increased energy efficiency) or societal benefits (e.g. increased food security). Communication of risk is extremely challenging, given the long timescale, complex and statistical nature of climate and uncertainty in predictions. There have been improvements in the use of calibrated language, but little evaluation of the success of different visualisation methods.

The earth's climate is at once familiar and mysterious, and the weather hazards that occur in its extremes are among the most formidable aspects and the most challenging to predict. This chapter has outlined current methods and challenges of simulating the effect of changing boundary conditions on extreme weather, the statistical modelling that supports it and the types of risk management under consideration. The area cannot be done justice here, and is changing very rapidly, so the reader is encouraged to explore the literature further.

Acknowledgements

Many thanks to Simon Brown and Christos Mitas for helpful reviews that improved the manuscript. Thanks to Michael Bründl, Joanne Camp, Andy Challinor, Ruth Doherty, Carina Fearnley, Bruce Malamud, Doug MacNeall, Mike Steel, Roger Street, Paul Valdes, Renato Vitolo and Nick Watkins for useful discussions and contributions. Thanks to the Natural Environment Research Council (NERC) and the Bristol Environmental Risk Research Centre (BRISK) for financial support.

References

Allen, C., Macalady, A., Chenchouni, H., *et al.* (2010) A global overview of drought and heat-induced tree mortality reveals emerging climate change risks for forests. *Forest Ecology and Management* **259** (4): 660–684.

Alo, C. A. and Wang, Guiling (2008) Hydrological impact of the potential future vegetation response to climate changes projected by 8 GCMs. *Journal of Geophysical Research – Biogeosciences* **113** (G3): G03011.

Andronova, N. and Schlesinger, M. (2001) Objective estimation of the probability density function for climate sensitivity. *Journal of Geophysical Research* **106** (22): 605–622.

Angel, R. (2006) Feasibility of cooling the Earth with a cloud of small spacecraft near the inner Lagrange point (L1). *Proceedings of the National Academy of Sciences* **103** (46): 17184–17189.

Annan, J. D. and Hargreaves, J. C. (2010) Reliability of the CMIP3 ensemble. *Geophysical Research Letters* **37**: L02703.

Aumont, O. and Bopp, L. (2006) Globalizing results from ocean in situ iron fertilization studies. *Global Biogeochemical Cycles* **20** (2): GB2017.

Barnett, D., Brown, S., Murphy, J., *et al.* (2006) Quantifying uncertainty in changes in extreme event frequency in response to doubled CO_2 using a large ensemble of GCM simulations. *Climate Dynamics* **26** (5): 489–511.

Benestad, R. E., Hanssen-Bauer, I. and Chen, D. (2008) *Empirical-Statistical Downscaling*, Hackenssack, NJ: World Scientific Publishing Co. Pte. Ltd.

Beniston, M., Stephenson, D., Christensen, O., *et al.* (2007) Future extreme events in European climate: an exploration of regional climate model projections. *Climatic Change* **81**: 71–95.

Best, M. J., Pryor, M., Clark, D., *et al.* (2011) The Joint UK Land Environment Simulator (JULES) model description: Part 1 – Energy and water fluxes. *Geoscientific Model Development* **4** (3): 677–699.

Betts, R. A., Boucher, O., Collins, M., *et al.* (2007) Projected increase in continental runoff due to plant responses to increasing carbon dioxide. *Nature* **448** (7157): 1037–1041.

Bjarnadottir, S., Li, Y. and Stewart, M. (2011) A probabilistic-based framework for impact and adaptation assessment of climate change on hurricane damage risks and costs. *Structural Safety* **33** (3): 173–185.

Bony, S. and Dufresne, J. (2005) Marine boundary layer clouds at the heart of tropical cloud feedback uncertainties in climate models. *Geophysical Research Letters* **32** (20): L20806.

Boyd, P. W., Watson, A., Law, C., *et al.* (2000) A mesoscale phytoplankton bloom in the polar Southern Ocean stimulated by iron fertilization. *Nature* **407** (6805): 695–702.

Braconnot, P., Otto-Bliesner, O., Harrison, S., *et al.* (2007) Results of PMIP2 coupled simulations of the Mid-Holocene and Last Glacial Maximum: Part 1 – experiments and large-scale features. *Climate of the Past* **3** (2): 261–277.

Brown, S. J., Caesar, J. and Ferro, C. A. T. (2008) Global changes in extreme daily temperature since 1950. *Journal of Geophysical Research – Atmospheres* **113** (D5): D05115.

Buontempo, C., Lørup, J. and Antar, M. (2009) Assessing the impacts of climate change on the water resources in the Nile Basin using a regional climate model ensemble. *Earth and Environmental Science* **6**: 292017.

Burke, E. J. and Brown, S. J. (2008) Evaluating uncertainties in the projection of future drought. *Journal of Hydrometeorology* **9** (2): 292–299.

Burke, E. J., Perry, R. H. J. and Brown, S. J. (2010) An extreme value analysis of UK drought and projections of change in the future. *Journal of Hydrology* **388** (1–2): 131–143.

CCSP (2008) *Climate models: An Assessment of Strengths and Limitations, Department of Energy,* Washington, DC: Office of Biological and Environmental Research.

Challinor, A., Wheeler, T., Craufurd, P., *et al.* (2004) Design and optimisation of a large-area process-based model for annual crops. *Agricultural and Forest Meteorology* **124**: 99–120.

Challinor, A., Wheeler, T., Slingo, J., *et al.* (2005) Quantification of physical and biological uncertainty in the simulation of the yield of a tropical crop using present-day and doubled CO2 climates. *Philosophical Transactions of the Royal Society B – Biological Sciences* **360** (1463): 2085–2094.

Challinor, A. J., Wheeler, T., Hemming, D., *et al.* (2009) Ensemble yield simulations: crop and climate uncertainties, sensitivity to temperature and genotypic adaptation to climate change. *Climate Research* **38** (2): 117–127.

Christensen, J. H., Carter, T. R. and Giorgi F. (2002) PRUDENCE employs new methods to assess European climate change. *EOS* **83**: 13.

Clark, D. B., Mercado, L., Sitch, S., *et al.* (2011) The Joint UK Land Environment Simulator (JULES) model description: Part 2 – carbon fluxes and vegetation dynamics. *Geoscientific Model Development* **4** (3): 701–722.

Clark, R. T., Brown, S. J. and Murphy, J. M. (2006) Modeling northern hemisphere summer heat extreme changes and their uncertainties using a physics ensemble of climate sensitivity experiments. *Journal of Climate* **19** (17): 4418–4435.

Clark, R. T., Murphy, J. M. and Brown, S. J. (2010) Do global warming targets limit heatwave risk? *Geophysical Research Letters* **37**: L17703.

Clune, T. L. and Rood, R. B. (2011) Software testing and verification in climate model development. *IEEE Software* **28** (6): 49–55.

Collins, M., Booth, B., Bhaskaran, B., *et al.* (2011) Climate model errors, feedbacks and forcings: a comparison of perturbed physics and multi-model ensembles. *Climate Dynamics* **36**: 1737–1766.

Cooley, D. (2009) Extreme value analysis and the study of climate change. *Climatic Change* **97**: 77–83.

Cox, P., Betts, R., Jones, C., *et al.* (2000) Acceleration of global warming due to carbon-cycle feedbacks in a coupled climate model. *Nature* **408** (6809): 184–187.

Crucifix, M. (2006) Does the Last Glacial Maximum constrain climate sensitivity? *Geophysical Research Letters* **33** (18): L18701.

Crutzen, P. J. (2006) Albedo enhancement by stratospheric sulfur injections: a contribution to resolve a policy dilemma? *Climatic Change* **77** (3–4): 211–220.

Dai, A. (2011) Drought under global warming: a review. *Wiley Interdisciplinary Reviews: Climatic Change* **2**: 45–65.

Diffenbaugh, N., Pal, J., Trapp, R., *et al.* (2005) Fine-scale processes regulate the response of extreme events to global climate change. *Proceedings of the National Academy of Sciences of the United States of America* **102** (44): 15774–15778.

Dlugolecki A., *et al.* (2009) Climate change and its implications for catastrophe modelling. In *Coping with Climate Change: Risks and Opportunites for Insurers*, London: Chartered Insurance Institute, pp. 1–23.

Eames, M., Kershaw, T. and Coley, D. (2011) On the creation of future probabilistic design weather years from UKCP09. *Building Services Engineering Research & Technology* **32** (2): 127–142.

Easterbrook, S. M. and Johns, T. C. (2009) Engineering the software for understanding climate change. *Computing in Science & Engineering* **11** (6): 64–74.

Edwards, T. L., Crucifix, M. and Harrison, S. P. (2007) Using the past to constrain the future: how the palaeorecord can improve estimates of global warming. *Progress in Physical Geography* **31** (5): 481–500.

Emanuel, K., Sundararajan, R. and Williams, J. (2008) Hurricanes and global warming: results from downscaling IPCC AR4 simulations. *Bulletin of the American Meteorological Society* **89** (3): 347–367.

Fildes, R. and Kourentzes, N. (2011) Validation and forecasting accuracy in models of climate change. *International Journal of Forecasting* **27** (4): 968–995.

Fischer, E. M. and Schär, C. (2008) Future changes in daily summer temperature variability: driving processes and role for temperature extremes. *Climate Dynamics* **33** (7–8): 917–935.

Fischer, E. M., Lawrence, D. M. and Sanderson, B. M. (2010) Quantifying uncertainties in projections of extremes: a perturbed land surface parameter experiment. *Climate Dynamics* **37**: 1381–1398.

Foley, A. M. (2010) Uncertainty in regional climate modelling: a review. *Progress in Physical Geography* **34** (5): 647–670.

Fowler, H. J., Cooley, D., Sain, S., *et al.* (2010) Detecting change in UK extreme precipitation using results from the climateprediction.net BBC climate change experiment. *Extremes* **13** (2): 241–267.

Fox-Rabinovitz, M., Cote, J., Dugas, B., *et al.* (2008) Stretched-grid Model Intercomparison Project: decadal regional climate simulations with enhanced variable and uniform-resolution GCMs. *Meteorology and Atmospheric Physics* **100** (1–4): 159–178.

Furrer, E. M. and Katz, R. W. (2008) Improving the simulation of extreme precipitation events by stochastic weather generators. *Water Resources Research* **44** (12): W12439.

Giorgi, F. and Mearns, L. (2002) Calculation of average, uncertainty range, and reliability of regional climate changes from AOGCM simulations via the 'reliability ensemble averaging' (REA) method. *Journal of Climate* **15** (10): 1141–1158.

Goldstein, M. and Rougier, J. C. (2004) Probabilistic formulations for transferring inferences from mathematical models to physical systems. *SIAM Journal on Scientific Computing* **26** (2): 467–487.

Gosling, S. N., Lowe, J., McGregor, G., *et al.* (2009) Associations between elevated atmospheric temperature and human mortality: a critical review of the literature. *Climatic Change* **92** (3–4): 299–341.

Gosling, S. N., McGregor, G. R. and Lowe, J. A. (2011) The benefits of quantifying climate model uncertainty in climate change impacts assessment: an example with heat-related mortality change estimates. *Climatic Change* **112** (2): 1–15.

Gregoire, L., Valdes, P., Payne, A., *et al.* (2010) Optimal tuning of a GCM using modern and glacial constraints. *Climate Dynamics* **37** (3–4): 1–15.

Hanel, M. and Buishand, T. A. (2011) Analysis of precipitation extremes in an ensemble of transient regional climate model simulations for the Rhine basin. *Climate Dynamics* **36**: 1135–1153.

Hanel, M., Buishand, T. A. and Ferro, C. A. T. (2009) A nonstationary index flood model for precipitation extremes in transient regional climate model simulations. *Journal of Geophysical Research – Atmospheres* **114**: D15107.

Hansen, J., Fung, A., Lacis, D., *et al.* (1988) Global Climate changes as forecast by Goddard Institute for Space Studies 3-Dimensional model. *Journal of Geophysical Research – Atmospheres* **93**: 9341–9364.

Hargreaves, J. C. (2010) Skill and uncertainty in climate models. *Wiley Interdisciplinary Reviews: Climatic Change* **1** (4): 556–564.

Harris, G. R., Sexton, D., Booth, B., *et al.* (2006) Frequency distributions of transient regional climate change from perturbed physics ensembles of general circulation model simulations. *Climate Dynamics* **27** (4): 357–375.

Hawkins, E. (2011) Our evolving climate: communicating the effects of climate variability. *Weather* **66** (7): 175–179.

Hawkins, E. and Sutton, R. (2009) The Potential to narrow uncertainty in regional climate predictions. *Bulletin of the American Meteorological Society* **90** (8): 1095–1107.

Hemming, D., Buontempo, C., Burke, E., *et al.* (2010) How uncertain are climate model projections of water availability indicators across the Middle East? *Philosophical Transactions of the Royal Society A – Mathematical, Physical and Engineering Sciences* **368** (1931): 5117–5135.

Hewitson, B. (2003) Developing perturbations for climate change impact assessments. *Eos* **84** (35): 337–348.

Ho, C., Stephenson, D., Collins, M., *et al.* (2012) Calibration strategies: a source of additional uncertainty in climate change projections. *Bulletin of the American Meteorological Society* **93** (1): 21–26.

Hulme, M., O'Neill, S. J. and Dessai, S. (2011) Is weather event attribution necessary for adaptation funding? *Science* **334** (6057): 764–765.

Hulme, M. (2009) *Why We Disagree About Climatic Change*, Cambridge: Cambridge University Press.

Hulme, M., Jenkins, G. J., Lu, X. *et al.* (2002) Climate Change Scenarios for the United Kingdom: The UKCIP02 Scientific Report, Tyndall Centre for Climate Change Research, School of Environmental Sciences, University of East Anglia, Norwich, UK. 120pp.

IPCC (1994) *Technical Guidelines for Assessing Climate Change Impacts and Adaptations*. London: University College London and the Center for Global Environmental Research.

IPCC (2007a) *Climate Change 2007: The Physical Science Basis. Contribution of Working Group I to the Fourth Assessment Report of the Intergovernmental Panel on Climate Change*, Cambridge: Cambridge University Press.

IPCC (2007b) *Climate Change 2007: Impacts, Adaptation and Vulnerability. Contribution of Working Group II to the Fourth Assessment Report of the Intergovernmental Panel on Climate Change*, Cambridge: Cambridge University Press.

IPCC (2007c) *Climate Change 2007: Mitigation. Contribution of Working Group III to the Fourth Assessment Report of the Intergovernmental Panel on Climate Change*, Cambridge: Cambridge University Press.

IPCC (2012) Summary for Policymakers. In: *Managing the Risks of Extreme Events and Disasters to Advance Climate Change Adaptation* (C. B. Field, V. Barros, T. F. Stocker, *et al.* (eds.)). A Special Report of Working Groups I and II of the Intergovernmental Panel on Climate Change. Cambridge University Press, Cambridge, UK, and New York, NY, USA, pp. 3–21.

Irvine, P. and Ridgwell, A. (2009) 'Geoengineering': taking control of our planet's climate. *Science Progress* **92** (2): 139–162.

Johnson, T. E. and Weaver, C. P. (2009) A framework for assessing climate change impacts on water and watershed systems. *Environmental Management* **43** (1): 118–134.

Jones, P. D., Kilsby, C., Harpham, C., *et al.* (2009) Projections of future daily climate for the UK from the Weather Generator. In *UK Climate Projections*, Exeter: Met Office Hadley Centre, pp. 1–48.

Katz, R. W. (2010) Statistics of extremes in climate change. *Climatic Change* **100** (1): 71–76.

Kennedy, M. and O'Hagan, A. (2001) Bayesian calibration of computer models. *Journal of the Royal Statistical Society Series B – Statistical Methodology* **63**: 425–450.

Kennedy, M. C., Anderson, C., Conti, S., *et al.* (2006) Case studies in Gaussian process modelling of computer codes. *Reliability Engineering and System Safety* **91** (10–11): 1301–1309.

Kharin, V. V., Zwiers, F., Zhang, X., *et al.* (2007) Changes in temperature and precipitation extremes in the IPCC ensemble of global coupled model simulations. *Journal of Climate* **20** (8): 1419–1444.

Kim, J.-H., Brown, S. J. and McDonald, R. E. (2010) Future changes in tropical cyclone genesis in fully dynamic ocean- and mixed layer ocean-coupled climate models: a low-resolution model study. *Climate Dynamics* **37** (3–4): 737–758.

Klocke, D., Pincus, R. and Quaas, J. (2011) On constraining estimates of climate sensitivity with present-day observations through model weighting. *Journal of Climate*, **24**: 6092–6099.

Knapp, A. K., Beier, C., Briske, D., *et al.* (2008) Consequences of more extreme precipitation regimes for terrestrial ecosystems. *Bioscience* **58** (9): 811–821.

Knutson, T., McBride, J., Chan, J., *et al.* (2010) Tropical cyclones and climate change. *Nature Geoscience* **3**: 157–163.

Knutti, R. (2008a) Should we believe model predictions of future climate change? *Philosophical Transactions of the Royal Society A – Mathematical, Physical and Engineering Sciences* **366** (1885): 4647–4664.

Knutti, R. (2008b) Why are climate models reproducing the observed global surface warming so well? *Geophysical Research Letters* **35** (18): L18704.

Knutti, R. (2010) The end of model democracy? *Climatic Change* **102**: 395–404.

Knutti, R., Stocker, T., Joos, F., *et al.* (2003) Probabilistic climate change projections using neural networks. *Climate Dynamics* **21** (3–4): 257–272.

Knutti, R., Furrer, R., Tebaldi, C., *et al.* (2010) Challenges in combining projections from multiple climate models. *Journal of Climate* **23** (10): 2739–2758.

Kravitz, B., Robock, A., Bouchers, O., *et al.* (2011) The Geoengineering Model Intercomparison Project (GeoMIP). *Atmospheric Science Letters* **12** (2): 162–167.

Kyselý, J. (2009) Recent severe heat waves in central Europe: how to view them in a long-term prospect? *International Journal of Climatology* **30**: 89–109.

Kyselý, J. and Dubrovský, M. (2005) Simulation of extreme temperature events by a stochastic weather generator: effects of interdiurnal and interannual variability reproduction. *International Journal of Climatology* **25** (2): 251–269.

Lackner, K. S. (2009) Capture of carbon dioxide from ambient air. *The European Physical Journal Special Topics* **176** (1): 93–106.

Lambert, S. J. and Boer, G. J. (2001) CMIP1 evaluation and intercomparison of coupled climate models. *Climate Dynamics* **17** (2–3): 83–106.

Leimbach, M., Bauer, N., Baumstark, L., *et al.* (2010) Mitigation costs in a globalized world: climate policy analysis with REMIND-R. *Environmental Modeling & Assessment* **15** (3): 155–173.

Lenton, T. M. and Vaughan, N. E. (2009) The radiative forcing potential of different climate geoengineering options. *Atmospheric Chemistry and Physics* **9** (15): 5539–5561.

Lorenz, E. (1975) Climatic predictability. In *The Physical Basis of Climate and Climate Modeling*. Geneva: World Meteorological Organization, pp. 132–136.

Lorenz, P. and Jacob, D. (2005) Influence of regional scale information on the global circulation: a two-way nesting climate simulation. *Geophysical Research Letters* **32** (18): L18706.

Lynn, B. H., Healy, R. and Druyan, L. M. (2008) Quantifying the sensitivity of simulated climate change to model configuration. *Climatic Change* **92** (3–4): 275–298.

Marx, S. M., Weber, E., Orlove, B., *et al.* (2007) Communication and mental processes: experiential and analytic processing of uncertain climate information. *Global Environmental Change* **17** (1): 47–58.

Maslin, M., Owen, M., Betts, R., *et al.* (2010) Gas hydrates: past and future geohazard? *Philosophical Transactions of the Royal Society A – Mathematical, Physical and Engineering Sciences* **368** (1919): 2369–2393.

Masson, V. (2006) Urban surface modeling and the meso-scale impact of cities. *Theoretical and Applied Climatology* **84**: 35–45.

Mastrandrea, M. D. and Mach, K. J. (2011) Treatment of uncertainties in IPCC Assessment Reports: past approaches and considerations for the Fifth Assessment Report. *Climatic Change* **108** (4): 659–673.

Mastrandrea, M. D., Fields, C. B., Stocker, T. F., *et al.* (2010) Guidance note for lead authors of the IPCC Fifth Assessment Report on consistent treatment of uncertainties. Intergovernmental Panel on Climate Change (IPCC). http://www.ipcc-wg2.gov/meetings/CGCs/index.html#UR.

McFarlane, N. (2011) Parameterizations: representing key processes in climate models without resolving them. *Wiley Interdisciplinary Reviews: Climatic Change* **2** (4): 482–497.

McGregor, G. (2011) Human biometeorology. *Progress in Physical Geography* **36** (1): 93–109.

McGuffie, K. and Henderson-Sellers, A. (2005) *A Climate Modelling Primer*, 3rd edn, Chichester: John Wiley and Sons.

Meehl, G. A. and Tebaldi, C. (2004) More intense, more frequent, and longer lasting heat waves in the 21st century. *Science* **305**: 994–997.

Meehl, G. A., Arblaster, J. M. and Tebaldi, C. (2005) Understanding future patterns of increased precipitation intensity in climate model simulations. *Geophysical Research Letters* **32** (18): L18719.

Meehl, G. A., Cover, C., Delworth, T., *et al.* (2007) THE WCRP CMIP3 Multimodel dataset: a new era in climate change research. *Bulletin of the American Meteorological Society* **88** (9): 1383–1394.

Mendes, D. and Marengo, J. A. (2009) Temporal downscaling: a comparison between artificial neural network and autocorrelation techniques over the Amazon Basin in present and future climate change scenarios. *Theoretical and Applied Climatology* **100** (3–4): 1–9.

Min, S.-K., Zhang, X., Zwiers, F., *et al.* (2009) Signal detectability in extreme precipitation changes assessed from twentieth century climate simulations. *Climate Dynamics* **32** (1): 95–111.

Min, S., Zhang, X., Zwiers, F., *et al.* (2011) Human contribution to more-intense precipitation extremes. *Nature* **470** (7334): 378–381.

Mitchell, T. (2003) Pattern scaling: an examination of the accuracy of the technique for describing future climates. *Climatic Change* **60** (3): 217–242.

Morgan, M. G. and Mellon, C. (2011) Certainty, uncertainty, and climate change. *Climatic Change* **108** (4): 707–721.

Murphy, J. M., Sexton, D., Barnett, D., *et al.* (2004) Quantification of modelling uncertainties in a large ensemble of climate change simulations. *Nature* **430** (7001): 768–772.

Murphy, J. M., Booth, B., Collins, M., *et al.* (2007) A methodology for probabilistic predictions of regional climate change from perturbed physics ensembles. *Philosophical Transactions of the Royal Society A – Mathematical, Physical and Engineering Sciences* **365** (1857): 1993–2028.

Murphy, J. M., Sexton, D., Jenkins, G., *et al.* (2009) Climate change projections. In *UK Climate Projections.* Exeter: Met Office Hadley Centre, pp. 1–194.

Murphy, J., Clark, R., Collins, M. *et al.* (2011) Perturbed parameter ensembles as a tool for sampling model uncertainties and making climate projections. *Proceedings of ECMWF Workshop on Model Uncertainty,* 20–24 June 2011, 209–220. http://www.ecmwf.int/newsevents/meetings/workshops/2011/Model_uncertainty/presentations/Murphy.pdf.

Nakićenović, N., *et al.* (2000) *Special Report on Emissions Scenarios: A Special Report of Working Group III of the Intergovernmental Panel on Climate Change.* Cambridge: Cambridge University Press.

Naveau, P., Nogaj, M., Ammann, C., *et al.* (2005) Statistical methods for the analysis of climate extremes. *Comptes Rendus Geoscience* **337**: 1013–1022.

Nicolosi, V., Cancelliere, A. and Rossi, G. (2009) Reducing risk of shortages due to drought in water supply systems using genetic algorithms. *Irrigation and Drainage* **58** (2): 171–188.

O'Hagan, A. (2006) Bayesian analysis of computer code outputs: a tutorial. *Reliability Engineering and System Safety* **91**: 1290–1300.

Oleson, K. W., Bonan, G. B. and Feddema, J. (2010) Effects of white roofs on urban temperature in a global climate model. *Geophysical Research Letters* **37**: L03701.

Oreskes, N., Shrader-Frechette, K. and Belitz, K. (1994) Verification, validation, and confirmation of numerical models in the earth sciences. *Science* **263** (5147): 641–646.

Pall, P., Aina, T., Stone, D., *et al.* (2011) Anthropogenic greenhouse gas contribution to flood risk in England and Wales in autumn 2000. *Nature* **470** (7334): 382–385.

Palmer, T. (2001) A nonlinear dynamical perspective on model error: a proposal for non-local stochastic-dynamic parametrization in weather and climate prediction models. *Quarterly Journal of the Royal Meteorological Society* **127** (572): 279–304.

Palmer, T. and Räisänen, J. (2002) Quantifying the risk of extreme seasonal precipitation events in a changing climate. *Nature* **415** (6871): 512–514.

Palmer, T. and Williams, P. (eds) (2009) *Stochastic Physics and Climate Modelling,* Cambridge: Cambridge University Press.

Palmer, T., Alessandri, A., Andersen, U., *et al.* (2004) Development of a European multi-model ensemble system for seasonal-to-interannual prediction (DEMETER). *Bulletin of the American Meteorological Society* **85** (6): 853–872.

Palmer, T., Shutts, G., Hegedorn, R., *et al.* (2005) Representing model uncertainty in weather and climate prediction. *Annual Review of Earth and Planetary Sciences* **33**: 163–193.

Palmer, T., Doblas-Reyes, F., Weisheimer, A., *et al.* (2008) Toward seamless prediction: calibration of climate change projections using seasonal forecasts. *Bulletin of the American Meteorological Society* **89** (4): 459–470.

Patt, A. and Schrag, D. (2003) Using specific language to describe risk and probability. *Climatic Change* **61**: 17–30.

Pennell, C. and Reichler, T. (2011) On the effective number of climate models. *Journal of Climate* **24** (9): 2358–2367.

Piani, C., Frame, D., Stainforth, D., *et al.* (2005) Constraints on climate change from a multi-thousand member ensemble of simulations. *Geophysical Research Letters* **32** (23): L23825.

Pidgeon, N. and Fischhoff, B. (2011) The role of social and decision sciences in communicating uncertain climate risks. *Nature Climatic Change* **1** (1): 35–41.

Pirtle, Z., Meyer, R. and Hamilton, A. (2010) What does it mean when climate models agree? A case for assessing independence among general circulation models. *Environmental Science & Policy* **13** (5): 351–361.

Pope, V. and Davies, T. (2002) Testing and evaluating atmospheric climate models. *Computing in Science & Engineering* **4** (5): 64–69.

Price, A. R., Goswami, S., Johnson, M., *et al.* (2009) The Aladdin-2 launchpad: engaging the user community and broadening access to the GENIE model on local and remote computing resources. In UK e-Science All Hands Meeting 2009, Past Present and Future. Monday 7–9 December 2009, Oxford.

Qian, B., Gameda, S. and Hathoe, H. (2004) Performance of stochastic weather generators LARS-WG and AAFC-WG for reproducing daily extremes of diverse Canadian climates. *Climate Research* **26**: 175–191.

Räisänen, J. (2005) Impact of increasing CO2 on monthly-to-annual precipitation extremes: analysis of the CMIP2 experiments. *Climate Dynamics* **24**: 309–323.

Räisänen, J. (2007) How reliable are climate models? *Tellus Series A – Dynamic Meteorology and Oceanography* **59** (1): 2–29.

Räisänen, J. and Palmer, T. (2001) A probability and decision-model analysis of a multimodel ensemble of climate change simulations. *Journal of Climate* **14** (15): 3212–3226.

Randall, D., Krueger, S., Bretherton, C., *et al.* (2003) Confronting models with data: the GEWEX cloud systems study. *Bulletin of the American Meteorological Society* **84** (4): 455–469.

Reichler, T. and Kim, J. (2008) How well do coupled models simulate today's climate? *Bulletin of the American Meteorological Society* **89** (3): 303–311.

Ricke, K. L., Morgan, M. G. and Allen, M. R. (2010) Regional climate response to solar-radiation management. *Nature Geoscience* **3** (8): 537–541.

Ridgwell, A., Singarayer, J., Hetherington, A., *et al.* (2009) Tackling regional climate change by leaf albedo bio-geoengineering. *Current Biology* **19** (2): 146–150.

Rodwell, M. J. and Palmer, T. N. (2007) Using numerical weather prediction to assess climate models. *Quarterly Journal of the Royal Meteorological Society* **133** (622): 129–146.

Rougier, J. (2007) Probabilistic inference for future climate using an ensemble of climate model evaluations. *Climatic Change* **81** (3–4): 247–264.

Rougier, J. and Sexton, D. M. H. (2007) Inference in ensemble experiments. *Philosophical Transactions of the Royal Society A – Mathematical, Physical and Engineering Sciences* **365** (1857): 2133–2143.

Rougier, J., Sexton, D., Murphy, J., *et al.* (2009) Analyzing the climate sensitivity of the HadSM3 climate model using ensembles from different but related experiments. *Journal of Climate* **22** (13): 3540–3557.

Roy, C. and Oberkampf, W. (2011) A comprehensive framework for verification, validation, and uncertainty quantification in scientific computing. *Computer Methods in Applied Mechanics and Engineering* **200**: 2131–2144.

Ruddiman, W. (2003) The anthropogenic greenhouse era began thousands of years ago. *Climatic Change* **61** (3): 261–293.

Rummukainen, M. (2010) State-of-the-art with regional climate models. *Wiley Interdisciplinary Reviews: Climatic Change* **1** (1): 82–96.

Sanderson, B. M. (2011) A multimodel study of parametric uncertainty in predictions of climate response to rising greenhouse gas concentrations. *Journal of Climate* **24** (5): 1362–1377.

Sanderson, B. M., Knutti, R., Aina, T., *et al.* (2008) Constraints on model response to greenhouse gas forcing and the role of subgrid-scale processes. *Journal of Climate* **21** (11): 2384–2400.

Scheffer, M., Brovkin, V. and Cox, P. M. (2006) Positive feedback between global warming and atmospheric CO2 concentration inferred from past climate change. *Geophysical Research Letters* **33** (10): L10702.

Schwierz, C., Kollner-Heck, P., Zenklusen Mutter, E., *et al.* (2010) Modelling European winter wind storm losses in current and future climate. *Climatic Change* **101**: 485–514.

Semenov, M. (2008) Simulation of extreme weather events by a stochastic weather generator. *Climate Research* **35**: 203–212.

Sexton, D. and Murphy, J. (2011) Multivariate probabilistic projections using imperfect climate models. Part II: robustness of methodological choices and consequences for climate sensitivity. *Climate Dynamics* **38**: 2543–2558.

Sexton, D., Murphy, J., Collins, M., *et al.* (2011) Multivariate probabilistic projections using imperfect climate models. Part I: outline of methodology. *Climate Dynamics* **38**: 2513–2542.

Smith, D. M., Cusack, S., Colman, A., *et al.* (2007) Improved surface temperature prediction for the coming decade from a global climate model. *Science* **317** (5839): 796–799.

Socolow, R. H. (2011) High-consequence outcomes and internal disagreements: tell us more, please. *Climatic Change* **108** (4): 775–790.

Soden, B. J. and Held, I. M. (2006) An assessment of climate feedbacks in coupled ocean–atmosphere models. *Journal of Climate* **19** (14): 3354–3360.

Spiegelhalter, D. J. and Riesch, H. (2011) Don't know, can't know: embracing deeper uncertainties when analysing risks. *Philosophical Transactions of the Royal Society A – Mathematical, Physical And Engineering Sciences* **369**: 4730–4750.

Spiegelhalter, D., Pearson, M. and Short, I. (2011) Visualizing uncertainty about the future. *Science* **333** (6048): 1393–1400.

Stainforth, D. A., Aina, T., Christensen, C., *et al.* (2005) Uncertainty in predictions of the climate response to rising levels of greenhouse gases. *Nature* **433** (7024): 403–406.

Stensrud, D. J. (2009) *Parameterization Schemes: Keys to Understanding Numerical Weather Prediction Models*, Cambridge: Cambridge University Press.

Stephens, E. M., Edwards, T. L. and Demeritt, D. (2012) Communicating probabilistic information from climate model ensembles – lessons from numerical weather prediction. *Wiley Interdisciplinary Reviews: Climate Change* **3** (5): 409–426.

Sterl, A., Severijns, C., Dijkstra, H., *et al.* (2008) When can we expect extremely high surface temperatures. *Geophysical Research Letters* **35**: L14703.

Sterman, J. (2011) Communicating climate change risks in a skeptical world. *Climatic Change* **108**: 811–826.

Stern, N. H. (2007) *Stern Review on the Economics of Climate Change*, Cambridge: Cambridge University Press.

Stott, P. A. and Forest, C. E. (2007) Ensemble climate predictions using climate models and observational constraints. *Philosophical Transactions of the Royal Society A – Mathematical, Physical And Engineering Sciences* **365** (1857): 2029–2052.

Stott, P., Stone, D. A. and Allen, M. R. (2004) Human contribution to the European heatwave of 2003. *Nature* **432** (7017): 610–614.

Taylor, K. E., Stouffer, R. J. and Meehl, G. A. (2011) An overview of CMIP5 and the experiment design. *Bulletin of the American Meteorological Society* **93**: 485–498.

Tebaldi, C., Smith, R. L., Nychka, D., *et al*. (2005) Quantifying uncertainty in projections of regional climate change: a Bayesian approach to the analysis of multimodel ensembles. *Journal of Climate* **18** (10): 1524–1540.

Tebaldi, C., Hayhoe, K., Arblaster, J., *et al*. (2006) Going to the extremes. *Climatic Change* **79** (3–4): 185–211.

Torn, M. S. and Harte, J. (2006) Missing feedbacks, asymmetric uncertainties, and the underestimation of future warming. *Geophysical Research Letters* **33** (10): L10703.

Tsvetsinskaya, E., Mearns, L., Mavromatis, T., *et al*. (2003) The effect of spatial scale of climatic change scenarios on simulated maize, winter wheat, and rice production in the southeastern United States. *Climatic Change* **60**: 37–71.

Turley, C., Eby, M., Ridgwell, A., *et al*. (2010) The societal challenge of ocean acidification. *Marine Pollution Bulletin* **60** (6): 787–792.

Valdes, P. (2011) Built for stability. *Nature Geoscience* **4** (7): 414–416.

Vuuren, D. P. van, Edmonds, J., Kainuma, M., *et al*. (2011) The representative concentration pathways: an overview. *Climatic Change* **109**: 5–31.

Wang, G. L. (2005) Agricultural drought in a future climate: results from 15 global climate models participating in the IPCC 4th assessment. *Climate Dynamics* **25**: 739–753.

Wallsten, T. S., Zwick, R., Forsyth, B., *et al*. (1986) Measuring the vague meanings of probability terms. *Journal of Experimental Psychology – General* **115** (4): 348.

Wardekker, J. A., van der Sluijs, J. P., Kloprogge, P., *et al*. (2008) Uncertainty communication in environmental assessments: views from the Dutch science–policy interface. *Environmental Science & Policy* **11** (7): 627–641.

Webb, M. J., Senior, C., Sexton, D., *et al*. (2006) On the contribution of local feedback mechanisms to the range of climate sensitivity in two GCM ensembles. *Climate Dynamics* **27** (1): 17–38.

Weigel, A. P., Liniger, M. A., Appenzeller, C., *et al*. (2010) Risks of model weighting in multimodel climate projections. *Journal of Climate* **23** (15): 4175–4191.

Weisheimer, A. and Palmer, T. (2005) Changing frequency of occurrence of extreme seasonal temperatures under global warming. *Geophysical Research Letters* **32** (20): L20721.

Weitzman, M. L. (2007) A review of the Stern Review on the economics of climate change. *Journal of Economic Literature* **45** (3): 703–724.

Wilby, R. L. and Dessai, S. (2010) Robust adaptation to climate change. *Weather* **65** (7): 180–185.

Wilby, R., Dawson, C. and Barrow, E. (2002) SDSM: a decision support tool for the assessment of regional climate change impacts. *Environmental Modelling & Software* **17** (2): 147–159.

Wilby, R., Charles, S., Zorita, E., *et al*. (2004) Guidelines for use of climate scenarios developed from statistical downscaling methods. Available from the DDC of IPCC TGCIA.

Wilby, R. L., Troni, J., Biot, Y., *et al*. (2009) A review of climate risk information for adaptation and development planning. *International Journal of Climatology* **29** (9): 1193–1215.

Wilks, D. S. and Wilby, R. L. (1999) The weather generation game: a review of stochastic weather models. *Progress in Physical Geography* **23** (3): 329–357.

Willows, R. I. and Connell, R. K. (eds) (2003) *Climate Adaptation: Risk, Uncertainty and Decision-making*. Oxford: UKCIP.

Wright, G., Jack, L. and Swaffield, J. (2006) Investigation and numerical modelling of roof drainage systems under extreme events. *Building and Environment* **41** (2): 126–135.

Yang, Z. and Arritt, R. (2002) Tests of a perturbed physics ensemble approach for regional climate modeling. *Journal of Climate* **15** (20): 2881–2896.

Yokohata, T., Webb, M., Collins, M., *et al.* (2010) Structural similarities and differences in climate responses to CO_2 increase between two perturbed physics ensembles. *Journal of Climate* **23** (6): 1392–1410.

7

Flood risk and uncertainty

J. FREER, K. J. BEVEN, J. NEAL, G. SCHUMANN,
J. HALL AND P. BATES

7.1 Introduction

Extreme floods are among the most destructive forces of nature. Flooding accounts for a significant proportion of the total number of reported natural disasters occurring in the world (Figure 7.1a) and over the last 30 years this proportion has been increasing (Figure 7.1b). Reasons for this trend may not be clear; for each hazard there is a need to quantify whether this is an increase in the hazard itself, an increase in exposure to the hazard internationally or a change in the reporting of what constitutes a natural disaster. Internationally, the costs and scale of flooding are enormous but differ depending on the types of impact that are analysed and the databases used. Globally in 2007 it was estimated that annually 520 million people are affected by floods and that the death toll is approximately 25 000 people in any one year. Jonkman (2005) found for a study using data from 1974 to 2003 (from data maintained by the Centre for Research on the Epidemiology of Disasters in Brussels) that floods are the most significant natural disaster type in terms of the number of people affected – some 51% of the total of that period of approximately five billion people affected by natural disaster (droughts are second with 36%, and earthquakes third at 2%). However, in terms of overall estimated deaths flooding accounts for 10% of the approximately two million reported deaths associated with natural disasters over the 1974–2003 period (droughts 44% and earthquakes 27%). In monetary terms an assessment by Munich RE for the period 1980–2010 determined that at 2010 prices the losses totalled US$3000 billion from ~19 400 events with 2.275 million fatalities. Of these, hydrological catastrophes (flooding and mass movement, i.e. landslips and debris flow in this case) accounted for 24% of these monetary losses, from 35% of the total events, and 11% of the fatalities. Other categories of natural disasters included in these totals were geophysical, meteorological and climatological (NatCatSERVICE, 2011). In a different report, 'water-related' disasters were estimated to cost some US$50–60 billion to the world economy each year (UNESCO, 2007).

The impact of flooding is disproportionally felt throughout the world due to differences in the frequency of large flooding events and the vulnerability of the population and associated infrastructure (see Figure 7.2). A UNESCO report (UNESCO, 2007) concluded that about 96% of flood deaths in 1997–2007 were from developing countries. In the United States floods kill more people than any other severe weather event and cause damage estimated at

Risk and Uncertainty Assessment for Natural Hazards, ed. Jonathan Rougier, Steve Sparks and Lisa Hill. Published by Cambridge University Press. © Cambridge University Press 2013.

(a)

(b)

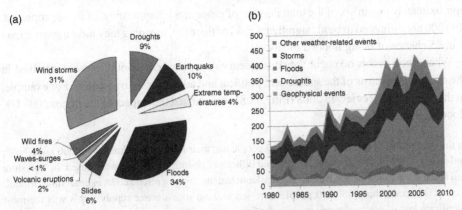

Figure 7.1 Global number of reported disasters by proportion (a) 1974–2003 averages; and (b) trends by year 1980–2010 (source: EM-DAT: The OFDA/CRED International Disaster Database – www.emdat.net – Université catholique de Louvain – Brussels – Belgium).

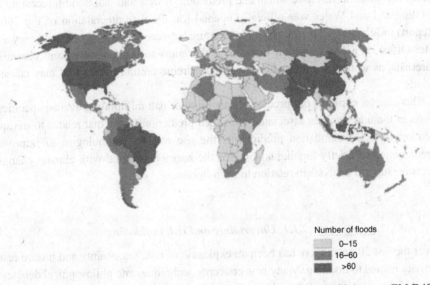

Figure 7.2 The number of occurrences of flood disasters by country: 1974–2003 (source: EM-DAT: The OFDA/CRED International Disaster Database – www.emdat.net – Université catholique de Louvain – Brussels – Belgium).

US$3.5 billion each year. Even in the UK, where deaths from flooding events are fortunately rare, a recent report by the Environment Agency in the UK (Environment Agency, 2009) determined that two million of the homes in England and Wales are built on a flood plain, near a river or where there is a risk of surface water flooding during heavy rainfall (this is approximately one in six of the total number of properties). Importantly, of those properties, 490 000 are estimated to be at 'significant risk' of flooding, meaning they have a greater than 1 in 75 chance of being flooded in any given year.

When flood events do occur the consequences are significant, both for those involved in the event and in terms of the subsequent political and administrative response. For example, in the UK the Pitt Review in 2008 (Pitt, 2008) summarised the impact of the major 2007 UK flooding event as follows:

In the exceptional events that took place, 13 people lost their lives, approximately 48,000 households and nearly 7,300 businesses were flooded and billions of pounds of damage was caused. In Yorkshire and Humberside, the Fire and Rescue Service launched the 'biggest rescue effort in peacetime Britain'. Across Gloucestershire, 350,000 people were left without mains water supply – this was the most significant loss of essential services since the Second World War.

The report concluded that major improvements were needed at local and national level and made 92 recommendations that spanned all aspects of flood mitigation, adaptation, parliamentary acts, infrastructure needs, planning, prediction and forecasting needs, emergency services and risk assessment and prediction. A new national flood forecasting centre for England and Wales was officially opened (on the recommendation of the 2008 Pitt Report) in July 2009. The centre, a joint venture between the Environment Agency and the Met Office, is a £10.7 million investment that aims to improve river and coastal flood forecasts, as well as warning about possible extreme rainfall events that may cause flash floods.

This chapter explores our predictive capability of natural flooding events; specifically it looks at the uncertainties associated with such predictions, how that relates to our quantification of flood inundation predictions, the use of new technologies to improve our predictions and briefly highlights some of the issues associated with climate change and decision support analysis in relation to such floods.

7.1.1 Uncertainty and risk in flooding

Over the last 20 years there has been an explosion of risk, uncertainty and hazard research activity related to flooding. Many new concepts, techniques and philosophical debates have detailed the difficulties of providing effective guidance and an agreement on best practice. Emerging conflicts are a consequence of our inability to construct appropriate models, to obtain high-quality data that are representative of flooding processes and to provide predictions and forecasts that are robust and have been evaluated effectively. As with other natural hazards, hydrological risks are characterised by extreme events, where observed data sources are often very limited and are rarely able to characterise the overall behaviour of events in detail. These issues are also generic to the prediction of extreme natural events, commonly over large spatial domains. In this chapter we focus on floods due to natural extreme events.

The overall concepts of addressing uncertainties in the modelling and flood modelling process has been widely highlighted (e.g. Bates and Townley, 1988; Romanowicz *et al.*, 1996; Apel *et al.*, 2004; Pappenberger and Beven, 2006; Blazkova and Beven, 2009b; Beven, 2011) and is the subject of considerable debate in the flood-risk community and more widely (see the debates and discussions in Andréassian *et al.*, 2007; Beven *et al.*, 2007, 2008a; Hall *et al.*, 2007; Mantovan and Todini, 2006; Mantovan *et al.*, 2007; Montanari, 2007; Sivakumar, 2008).

There are different types of uncertainty that occur in natural hazards. Merz and Thieken (2005) state 'Natural uncertainty has also been termed (basic) variability, aleatory uncertainty, objective uncertainty, inherent variability, (basic) randomness, and type-A uncertainty. Terms for epistemic uncertainty are subjective uncertainty, lack-of-knowledge or limited-knowledge uncertainty, ignorance, specification error, prediction error, and type-B uncertainty.' Aleatory uncertainties can be treated formally by statistical probabilistic methods, although it can be difficult to identify an appropriate statistical model. Epistemic uncertainties imply that the nature of the uncertainty may not be consistent in time or space, and may therefore be difficult to treat formally by probabilistic methods. While we may often choose to treat epistemic uncertainties *as if* they were aleatory in nature, this choice will be an approximation that may not always be appropriate.

To illustrate the difficulties in applying these concepts to real problems it should be noted that there is epistemic uncertainty about both estimates of flood peak discharges and the form of an appropriate probabilistic distribution for describing the aleatory natural variability of floods. Epistemic uncertainties are likely for the 'observed' flood peak discharges on which a frequency analysis is based, especially if these are based on extrapolations from measurements at much lower discharges. Both sources of uncertainty are traditionally treated by a probabilistic analysis conditional on a particular distribution (the generalised logistic or generalised Pareto distribution for annual maximum or peak of threshold series in the UK Flood Estimation Handbook; Institute of Hydrology, 1999). Statistical estimators for these distributions make assumptions about independence that may not be justified in hydrological time-series (due to longer-term catchment memory). When empirically analysing the asymptotic behaviour of flood probability distributions for French rivers, Bernardara *et al.* (2008) found the occurrence of heavy tail distributions, which is important since the characterisation of heavy tails considerably affects the frequency of extremes, they have many practical and societal consequences for flood risk management.

Further complications arise when model simulations of nonlinear responses are involved, such as rainfall-runoff models used in flood estimation and real-time flood forecasting. In this case it is known that the antecedent conditions in a catchment prior to an event are not easily observed and, if estimated by running a model, can be prone to large errors. In addition, the inputs for that event may be well or poorly estimated depending on the way in which the inputs are observed and the particular nature of the event. Therefore, epistemic errors arise associated with the antecedent conditions, the input data and the model structure being used. Some models are often evaluated by comparison with some observations in time (and sometimes space) that are not, themselves, error free. The resulting series of residuals will reflect the conflation of these different sources of error. Consequently these residuals will in general have a complex structure that will not be easily represented by a statistical error model. Assuming that they can could consequently lead to bias in inference and over-conditioning of parameter values (Beven, 2006; Beven *et al.*, 2008a). It also means that it is very difficult to disaggregate the different sources of uncertainty, and in particular to estimate uncertainty due to model structure. This also then implies that it may be difficult to differentiate between competing model structures and parameter sets, termed the *equifinality* problem by Beven (1993, 2006).

Some input–output couples for this type of model may be so much in error or incorrectly characterised that they are *disinformative* in the model identification process (Beven *et al.*, 2011a), whereas assuming a statistical error model implies that every residual is informative. Information content of the input–output data are consequently overestimated. This has been widely observed in the past, being one of the factors in the frequent significant reduction in performance in split-record evaluations of calibrated hydrological models. Assessing the real information content of data series in natural hazards models is a major challenge, but because of the epistemic nature of the sources of uncertainty will not be easily addressed within a wholly statistical framework (Beven, 2006).

Therefore there is a clear need to improve the quantification of flood hazard and flood risk, which in addition requires functions defining the vulnerability/consequences. This involves complex interactions of multiple processes and sparse data which is inherently uncertain. This will require the engagement of more established flood risk scientists and centres regarding the application of their technical expertise in flood defence, forecasting methods and the quantification of flood risk in less developed countries. The International Flood Initiative, a cooperation between UNESCO and the World Meteorological Organization (UNESCO, 2007), has highlighted these issues, suggesting a new, holistic approach is required, drawing on collaborative efforts between various communities (NGOs, research centres, national capabilities, etc.).

7.1.2 Future changes in flooding

In the future, huge changes in large-scale urbanisation with uncontrolled growth of megacities, population growth in flood plain areas and changes in land use mean that the general vulnerability of communities to flooding will continue to increase. Despite this, we must recognise floods are also an important driver of biodiversity, and nurture the sustainability of some ecosystems. Some human activities indeed depend upon flooding, such as for sustaining the fertility of soils over large flood plains. Future changes in flooding can be primarily attributed to land-use change factors, infrastructure changes and potential climate change impacts.

Floods exhibit extreme behaviour and are a combination of a set of complex and interacting processes. The evidence to prove clear linkages with land-use change is not easily quantified, although there is plenty of anecdotal evidence. A positive correlation has been found with flooding and significant de-forestation at the global scale. Bradshaw *et al.* (2007) determined shifts in flood frequency, although measures of flood severity (flood duration, people killed and displaced and total damage) showed some weaker, albeit detectable correlations to natural forest cover and loss.

Climate change might also lead to an intensification of the hydrological cycle and an increase in floods in many parts of the world (Huntington, 2006). In the UK, for example, a possible increase in winter precipitation may lead to an increase in flooding (Hulme and Dessai, 2008). A rigorous estimate of the possible increase in flood hazard and risk is therefore a crucial task for planning future adaptation strategies.

As a consequence of the huge impact of floods the science of risk and uncertainty analysis has seen more activity in hydrology than in many other science disciplines. In this chapter we survey some of the main types of flooding risk and uncertainty assessment.

7.1.3 Flood risk management decisions

Quantification and reduction of uncertainty is of inherent scientific interest. It is also of great applied interest to flood risk managers. The last couple of decades have seen risk becoming the dominant paradigm for the management of floods, as risk provides a rational basis for evaluating and deciding between management options (Hall *et al.*, 2003). Uncertainty is of relevance to decision-makers because it implies that circumstances cannot be excluded from consideration where their preference ordering between risk management options may change.

Increasingly, governments are requiring a careful consideration of uncertainty in major planning and investment decisions. For example, the USWRC (1983) (quoted by Al-Futaisi and Stedinger (1999)) state that: 'Planners shall identify areas of risk and uncertainty in their analysis and describe them clearly, so that decisions can be made with knowledge of the degree of reliability of the estimated benefits and costs and of the effectiveness of alternative plans.' The UK Department of the Environment, Food and Rural Affairs (Defra) provide guidance on flood and coastal defence which repeatedly calls for proper consideration of uncertainty in appraisal decisions. Guidance document FCDPAG1 (Defra, 2001) on 'good decision-making' states: 'Good decisions are most likely to result from considering all economic, environmental and technical issues for a full range of options, together with a proper consideration of risk and uncertainty.' As Pate-Cornell (1996) states, in the context of quantified risk analysis 'Decision makers may need and/or ask for a full display of the magnitudes and the sources of uncertainties before making an informed judgment.'

However, the practice of uncertainty analysis and use of the results of such analysis in decision-making is not widespread, for several reasons (Pappenberger and Beven, 2006). Uncertainty analysis takes time, so adds to the cost of risk analysis, options appraisal and design studies. The additional requirements for analysis and computation are rapidly being (more than) compensated for by the availability of enhanced computer processing power. However, computer processing power is only part of the solution, which also requires a step change in the approach to managing data and integrating the software for uncertainty calculations (Harvey *et al.*, 2008). The data necessary for quantified uncertainty analysis are not always available, so new data collection campaigns (perhaps including time-consuming expert elicitation exercises) may need to be commissioned. Project funders need to be convinced of the merits of uncertainty analysis before they invest in the time and data collection it requires.

Naturally, therefore, the effort invested in risk and uncertainty analysis needs to be proportionate to the importance of the decision(s) under consideration. Less significant decisions may merit no more than a qualitative review of the main sources of uncertainty and some sensitivity analysis, to appraise whether the chosen course of action is particularly vulnerable to uncertainties. More substantial decisions will merit the full panoply of uncertainty, sensitivity and decision robustness analysis. One such case is the Thames Estuary 2100 (TE2100) project, which appraised the options for management of flooding in the Thames Estuary during the twenty-first century, including the possibility of relocation of the Thames Barrier that currently protects London from surge tides. The sources of uncertainty in TE2100 were summarised by Hall and Solomatine (2008). In view of the severe uncertainties surrounding sea-level rise, future economic growth in the estuary, and the cost of flood dikes, Hall and Harvey (2009) applied info-gap robustness analysis (Ben-Haim, 2001) in order to identify options that were robust to these critically important sources of uncertainty.

7.2 Types of flooding

7.2.1 Inland flooding

In very general terms the flooding process in most cases involves intense precipitation in catchment networks that lead to excess stream discharge or urban storm drainage. The capacity of established rivers and streams to cope, defined as the bankfull discharge, or the capacity of a storm drainage network, is exceeded. As a consequence excess water moves into flood plains or into neighbouring areas that are normally unaffected by the river or stream system. There are a large number of physical controls on flooding, including catchment area, topography, precipitation rates, infiltration rates, pre-existing surface and groundwater conditions and characteristics of the river system itself. These factors in turn are affected by climate, season, ground conditions, vegetation, land use, geology, soil structure and existence (or not) of flood protection infrastructure, to name a few of the more important. The controlling processes may vary in both space and time, making floods very complicated phenomena to understand, capture in models and to forecast. From the perspective of risk there are also complex interactions of flood waters with the natural and human environment. We note that some inland floods can be caused by other processes (e.g. snow melting, volcanic eruptions on glaciers, glacier bursts, dam failure, landslides into dams), but intense precipitation in convective or frontal storm systems is by far the most important.

7.2.2 Coastal and estuarine flooding

Coastal flooding (the inundation of land areas along the coast caused by sea water above normal tidal actions) is related to three main drivers (excluding wind waves). These are: mean sea level; changes in the astronomical tides throughout the year; and storm surges in low-pressure systems. When one or more of these components are at an extreme level, they can temporarily push water levels across a threshold or cause the failure and breach of coastal defences, resulting in significant coastal inundation. This can cause loss of life, damage to homes, business and infrastructure, such as the worst natural disaster in the UK's recent history in terms of loss of life, the 1953 flood in the North Sea (McRobie *et al.*, 2005). In some cases the effects of storm surges are exacerbated by coincident extremes in land precipitation, whereby swollen rivers discharging at the coast meet the storm surge. Tsunamis are another cause of coastal flooding but are considered in a separate chapter (Chapter 10).

As a consequence, many research studies are beginning to explore the use of models for the prediction of coastal flooding which include a probabilistic assessment (Zerger *et al.*, 2002; Bates *et al.*, 2005; Brown *et al.*, 2007; Purvis *et al.*, 2008). One accepted scientific methodology is to employ a mass and momentum conserving inundation model, forced by a storm tide height from an estimated extreme water level (EWL) return curve for a specific return period. For example, the RASP (Risk Assessment of Flood and Coastal Defence for Strategic Planning) project-derived maps provide flood outlines for 1 in 200 years and 1 in 1000 years for coastal flooding. The main difficulty is estimating the probability of exceeding a certain EWL when often these event magnitudes have not been observed or there are very limited data on which to base these estimates.

7.2.3 Urbanisation and geological factors in flooding

There has been unprecedented worldwide growth of coastal communities and megacities in the last few decades. Moreover, in many places coastal communities are built upon areas that are subsiding due to geological processes, such as faulting or compaction of sediment. A very common situation is urban development on subsiding deltas, examples of which include New Orleans, Yangon and Karachi. Subsidence rates are comparable to, or up to more than an order of magnitude greater than, the best estimates of global sea-level rise due to climate change; for example, the Mississippi delta at New Orleans is subsiding 10–15 mm per year compared to a current global average sea-level rise of ~3 mm per year (Nicholls and Cazenave, 2010). Flood control mechanisms which reduce terrestrial sediment supply to deltas can even exacerbate the problem as sedimentation on deltas and their flood plains can counteract subsidence. Indeed, rapid subsidence at New Orleans was only recognised recently and this reflects a lack of appreciation of geological controls on coastal flood risk.

Groundwater flooding occurs as a result of water rising up from the underlying rocks or from water flowing from abnormal springs. This tends to occur after much longer periods of sustained high rainfall. Higher rainfall means more water will infiltrate into the ground and cause the water table to rise above normal levels. Groundwater tends to flow from areas where the ground level is high to areas where the ground level is low. In low-lying areas the water table is usually at shallower depths anyway, but during very wet periods, with all the additional groundwater flowing towards these areas, the water table can rise up to the surface, causing groundwater flooding (Finch *et al.*, 2004).

Groundwater flooding is most likely to occur in low-lying areas underlain by permeable rocks (aquifers). These may be extensive, regional aquifers, such as chalk or sandstone, or may be localised sands or river gravels in valley bottoms underlain by less permeable rocks. Groundwater flooding takes longer to dissipate because groundwater moves much more slowly than surface water and will take time to flow away underground. Consequently, groundwater-flooded areas are characterised by prolonged inundation which can last weeks or months and hence have a high impact on the local infrastructure affected (Morris *et al.*, 2007).

The prediction of groundwater flooding is very problematic on account of: sparse data on groundwater monitoring; events tend to be extremely rare; and the complexity of surface–groundwater interactions that sometimes make it difficult to quantify the contributions to an overall flooding event. Studies are therefore quite limited and so are not discussed further in this chapter. The reader is referred to the following articles that have dealt with significant groundwater flooding and the development of groundwater flood risk maps: Pointet *et al.* (2003); Finch *et al.* (2004); Pinault *et al.* (2005); Morris *et al.* (2007); Habets *et al.* (2010); Hughes *et al.* (2011); Macdonald *et al.* (2012).

7.3 Flood hazard/risk mapping and flood inundation

Flood mapping, as a hazard footprint (see Chapters 1–3) provides the inundation extent of inland flood waters normally associated with a specific event or related to a flood exceedance probability. Under the Floods Directive of the European Union (European Union, 2007), member states are required to produce flood hazard and risk maps; however, these maps typically do not consider estimates of epistemic and aleatory uncertainty or non-stationary internal or external factors, such as climate change. In the UK, for example, assessment of flood risk from river and coastal flooding is routine and by international standards well advanced.

7.3.1 A note on flood inundation probabilities

There is a long tradition in flood hydrology of describing flood frequencies in terms of return periods (the '100-year event', the '1000-year event'). As noted in Chapter 2, this is not to be encouraged because, although the return period has a technical meaning in flood frequency analysis (as the inverse of the annual exceedance probability), it can have an ambiguous interpretation in more common usage. Even the technical usage, in terms of an average time between exceedances of a given magnitude of event over a long enough series, can itself be considered ambiguous in the face of the potential for non-stationarity of flood frequencies into the future (Milly *et al.*, 2008).

The stationarity problem can be avoided by the use of annual exceedance probabilities as an estimate of what might happen in any one year (particularly next year) based on the statistics of past events. Since the historical record of past events might be short or certainly less than 100 or 1000 years, this will involve some assumptions about the tail behaviour of the distribution of extreme events. Although there are asymptotic arguments to favour the extreme value distributions in representing the maxima of a series of events, the short period of record means that other families of distributions might be used (IH, 1999).

The form of distribution will also affect the exceedance probabilities over any period of n years. Where n is the length of a design lifetime for flood defences, or the period for assessment of flood damages this effect might be important. Formally, if $F_q(q) = \Pr(Q \leq q)$ is the cumulative probability distribution of annual maxima flood peaks, the annual exceedance probability $\text{AEP}(q) = 1 - F_q(q)$ and the probability of exceedance over a period of n years (assuming stationarity) is

$$\Pr(q|n) = 1 - (F_q(q))^n.$$

The future stationarity of flood peaks is an important current issue. It has generally proven difficult, given the historical record, to demonstrate that there have been changes in frequency during the recent past (e.g. Robson, 2002). Change resulting from the effects of land use and management on flood characteristics have also been difficult to demonstrate except in small headwater catchments (e.g. McIntyre and Marshall, 2010), in particular because there have also been changes in rainfall characteristics (e.g. Beven *et al.*, 2008b). However, there remains much speculation that both the intensification of agriculture, urbanisation and climate change will produce significant future changes in flood frequency characteristics.

7.3.2 Inundation modelling to support flood risk mapping

Simulation of flood inundation using numerical models is a key source of flood hazard evidence to support risk mapping. There are many commercial and research codes, both in the UK and elsewhere, that utilise numerical schemes of various degrees of physical complexity. Flood inundation modelling to support hazard footprint mapping is usually concerned with the simulation of depth and inundation extents, although variables such as velocities, discharge rates and duration are important in some contexts.

In practice, models are either one- or two-dimensional. One-dimensional models calculate flow depths and discharges along a longitudinal profile at nodes. Widely used one-dimensional modelling packages such as ISIS, HEC-RAS and MIKE11 are based on the equations for shallow-water waves in open channels (the Saint-Venant equations). Two-dimensional models, such as TUFLOW, LISFLOOD-FP and JFLOW, calculate depth-averaged flow depths and velocities in two spatial dimensions on a grid or some form of unstructured mesh. Three-dimensional models are generally used only to examine very local impacts, such as bridges and obstructions to the flow. A wide variety of physical complexities and numerical solvers are available in research and industry from shallow water to rapid volume spreading models. Intercomparisons of different models have been published by Crowder *et al.* (2004), Hunter *et al.* (2008) and Neelz and Pender (2010).

The computation required by one-dimensional models is usually sufficiently low that they can be applied in some Monte Carlo sampling and real-time forecasting types of application. One-dimensional models can be used to simulate flood plain inundation, and, while they are a tried, tested and often accurate method of simulating within-bank flows, where the hydraulics can be approximated well in one dimension, model structural errors may become significant on the flood plain (especially urban flood plains). As computer power has increased, two-dimensional models have become the generally preferred framework for detailed inundation mapping.

The computation required for many two-dimensional model codes, especially in urban areas, limits the use of these models with Monte Carlo type methods, despite continued advances in computer hardware (Lamb *et al.*, 2009; Neal *et al.*, 2009d). This has led to a renewed interest in porosity-based techniques for representing sub-grid-scale features in coarse-resolution models (Guinot and Soares-Fraz, 2006; Yu and Lane, 2006a, 2006b; McMillan and Brasington, 2007), the development of volume-spreading methods (Gouldby *et al.*, 2008), inertial codes (Bates *et al.*, 2010) and an interest in emulators (Beven *et al.*, 2008c; Hall *et al.*, 2011b). The benchmarking study by Hunter *et al.* (2008) found that a representative sub-set of these codes ranging from shallow water (Liang *et al.*, 2006; Villanueva and Wright, 2006) to diffusive (Bradbrook *et al.*, 2004) and storage cell models (Bates and De Roo, 2000) simulated surprisingly similar state dynamics given the range of physical process representations and numerical solvers tested. Nevertheless, no single approach is considered to be best suited to all applications. The lack of a community inundation model under which different numerical schemes can be benchmarked also makes rigorous comparisons difficult. Simpler but faster models may be inappropriate due to large structural and parametric uncertainty, but such differences might also be dominated by epistemic errors about the spatial distribution of lateral inflows to a channel and other local boundary condition effects (e.g. Pappenberger *et al.*, 2006b). Nevertheless, benchmarking studies will continue to be essential as new codes become available, especially the rapidly evolving volume-spreading models that are being adopted by industry flood risk assessors.

7.3.3 *Sources of uncertainty in inundation modelling*

Inundation models are subject to structural errors, parameter estimation errors and boundary condition errors that contribute to simulation uncertainty (e.g. Figure 7.3). All of these sources of uncertainty might involve both aleatory and epistemic uncertainties (see Chapters 2 and 3). Inundation models are particularly sensitive to the representation of topography both in the channel (Aronica *et al.*, 1998; French and Clifford, 2000) and on the flood plain, and require accurate elevation data for both. Channel data usually take the form of surveyed cross-sections or sonar bathymetry on larger rivers, while flood plain topography is usually sampled from remote platforms using LiDAR (light detection and

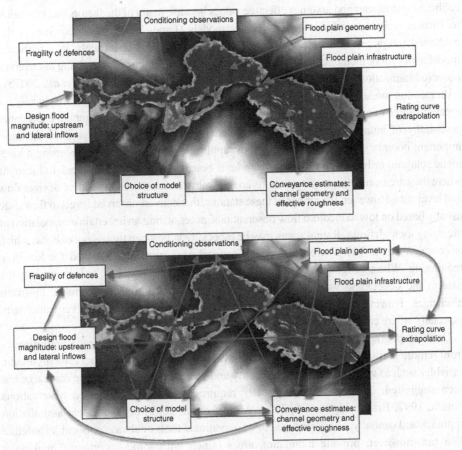

Figure 7.3 (a) Top panel: sources of uncertainty in flood hazard mapping; (b) interactions among sources of uncertainty in flood hazard mapping.

ranging), InSAR (interferometric synthetic aperture radar) or photogrammetry. Specifying flood plain geometry in urban areas can be difficult because topographic features in the region of 1–5 m in horizontal extent can influence flow pathways (Mark *et al.*, 2004). At present, deriving digital elevation models (DEMs) that represent these features, at the city scale, is only realistically possible where airborne LiDAR is available (Marks and Bates, 2000; Cobby *et al.*, 2003; Sanders, 2007), although InSAR-based products are also available (Sanders *et al.*, 2005). Recent developments in terrestrial LiDAR allow for centimetre resolution terrain modelling, but it is not yet clear how these data should be used in flood risk mapping, given that models typically run at much coarser resolutions.

Specification of surface roughness using friction coefficients can be especially problematic because it is difficult to measure the relevant land surface properties over large areas, although a number of authors have derived distributed flood plain roughness from remote sensing (Mason *et al.*, 2003; Wilson and Atkinson, 2007). Furthermore, roughness is often used as a calibration parameter and assumes effective values because many of the factors that contribute to friction resistance are not considered by models (Lane, 2005). This also means that calibrated roughness will reflect all the interacting sources of uncertainty in a particular model application, including interactions in space and time associated with fitting to uncertain observed inundation date (e.g. Romanowicz and Beven, 2003; Pappenberger *et al.*, 2007b).

Inundation models require the specification of boundary conditions, which are typically the greatest source of aleatory and epistemic uncertainty when simulating the annual exceedance probability of inundation. Flow at the upstream boundary of the river is often the most important boundary condition, although most applications will require (unless using a kinematic solution) or benefit from downstream-level boundaries (e.g. tidal reaches). In locations where they are available, gauging stations are typically the most accurate source of river flow and level data. However, the ratings at these stations, that convert observed levels to flows, are usually based on low to medium flow observations, necessitating an uncertain extrapolation of the rating for high flows. Rating errors may be especially large when flow is out of bank, which is common in the UK where the gauging network was primarily designed for low-flow monitoring. Where gauging stations are not available or spatially sparse, as is commonly the case in the developing world, hydrological models can be utilised to simulate upstream discharges. However, despite much effort, rainfall-runoff models are still very uncertain, especially where calibration/validation data are lacking.

Unlike topography and to some extent friction, discharge cannot be measured directly from remote platforms with current technology. However, the use of observable hydraulic variables such as water level, width, sinuosity and area as proxies to estimate discharge has been suggested. These methods typically require previous ground-based observations (Smith, 1997; Brakenridge *et al.*, 2003; Alsdorf *et al.*, 2007) or take a data assimilation approach and cannot yet be considered to be operational tools or always applicable. Satellite data can, however, provide inundation areas (albeit with some uncertainty) making it possible to monitor floods from space and integrate their observations into flood models, but only if their overpass trajectories coincide with flooding events.

When estimating flood risk from the hazard footprint, there is the additional uncertainty associated with the estimation of the consequences of an event. This is a factor not only of the likely damages (or other consequences) that might result from particular events, but also, in places where there are existing flood defences, of the possibility that any section of defence might fail or be overtopped, thereby enhancing the potential for damages. Evaluation of probability of the multiple potential modes of failure of defences is fraught with both epistemic and aleatory uncertainties. Procedures for doing this have been implemented into the RASP (Risk Assessment for System Planning (Sayers and Meadowcroft, 2005)) and the FLOODsite RELIABLE Software.[1]

Much of the recent research in flood inundation has been concerned with trying to reduce the potential uncertainties due to model structure, parameter values and boundary conditions. The most important factor in managing these uncertainties has been the provision of more and better

---

[1] See http://www.floodsite.net/html/partner_area/search_results3b.asp?docID=541

data, especially the use of InSAR and LIDAR to provide greatly improved representations of the flood plain topography. In fact, topographic data are now available at resolutions greater than the computationally practical model discretisations (for reasons of computational expense, particularly where many runs of the model might be required for uncertainty estimation). However, even the most detailed topographic data do not always include all the important elements of infrastructure on the flood plain that affect inundation (such as the continuity of field hedges and walls, and the accurate crest heights of flood defences).

More detailed observations of inundation areas, depths and velocities during floods are difficult to obtain since these depend on the occurrence of events that are, by definition, rare and logistically difficult to access. There are efforts being made to ensure that post-event surveys are made more often and made more accessible. This has been greatly helped by the development of differential GPS survey equipment (e.g. Neal *et al.*, 2009c). The use of such data in conditioning model uncertainties is dealt with in the next section.

7.3.4 Model calibration and uncertainty

Since models are only approximations of reality, the calibration of model parameters against some observations should, in principle, improve simulation accuracy and reduce prediction uncertainties. However, over-fitting of parameters to observations, particularly for a single flood event, may give misleading or even spurious results when simulating new events with different magnitudes. Ideally, models should be calibrated and validated against multiple events, although the available observation data are often of such limited spatial and temporal coverage that this is impractical. Past studies also suggest that different effective roughness values might be needed for different magnitudes of events or different river reaches (e.g. Romanowicz and Beven, 2003).

Distributed models are almost always over-parameterised relative to the calibration data available, as parameters must be specified at many cells, nodes or elements. To overcome this, parameters are typically lumped into spatial units or classes (e.g. river/flood plain or grassland/woodland) or fixed and thus assumed known. As a result, many quite different parameterisations of a model may result in similar simulations, leading to problems of parameter equifinality (Beven, 2006, 2009). There are many methods for calibrating deterministic models, but here the scope is limited to approaches that consider uncertainty. Recent reviews of uncertainty in flood risk mapping have been reported by Merwade *et al.* (2008) and Di Baldassarre *et al.* (2010).

New calibration methods account for parametric uncertainty (Aronica *et al.*, 2002; Beven, 2003; Bates *et al.*, 2004; Pappenberger *et al.*, 2005, 2006b, 2007a, 2007b; Romanowicz and Werner *et al.*, 2005a, 2005b), while the impact of boundary condition uncertainty on model calibration, especially from gauging stations, has also been investigated (Pappenberger *et al.*, 2006b). Calibration methods that consider structural and observational uncertainties are less common and require further attention. Hall *et al.* (2011a) take a Bayesian approach to model calibration to generate posterior parameter distributions and a Gaussian model inadequacy function to represent structural errors for a hydraulic model of a reach of the River Thames, UK. For practical purposes an emulator was used to allow many thousands of function evaluations to be used to generate the posterior distributions. An emulator (sometimes referred to as a 'surrogate model') is a statistical or data-based model which is used to provide a computationally cheap alternative to a complex computer code. The emulator is trained to replicate the code as closely as possible, using a limited number of code runs. In the statistics literature, Gaussian processes have found favour as emulators, because they possess a number of desirable properties (Oakley and O'Hagan, 2002). Young and Ratto (2011) have had considerable success in using nonlinear transfer function models for emulation of dynamical systems. The use of emulation to drastically reduce run times has also been demonstrated (Beven *et al.*, 2008c).

Many flood inundation models use uncertain boundary conditions from gauge data and are often calibrated against observations of flooding that are sparse and highly uncertain. As a consequence the uncertainty in probabilistic forecasts is often too high to sensibly guide investments in flood defence and risk management. Thus, the lack of suitable observation data, particularly of flood dynamics during flood plain wetting, is hindering advances in the simulation of flood hazard with uncertainty. More data are needed to determine minimum data requirements for robust flood inundation modelling to guide national-scale investments in data collection through the collection of benchmark ground and remotely sensed data of flooding. In urban areas, data on flood dynamics are especially limited (Hurford *et al.*, 2010; Romanowicz *et al.*, 1996), despite potential losses in these areas leading to high risk.

A particular issue in estimating boundary conditions for flood inundation models arises in the use of design events for chosen annual exceedance probabilities (AEP). At some sites there may be some gauging data available for historical floods that can be analysed to estimate the magnitude of peak expected for the design AEP. Normally this involves extrapolation in two senses: extrapolation of the rating curve to higher flows than have been measured, and extrapolation of information in a limited historical record to estimate probabilities of more extreme conditions (normally AEP 0.01 or 0.001). A third extrapolation occurs in the case of sites without discharge gauges, when the design flows must be extrapolated from information at gauging sites elsewhere; for example, using the methods provided in the *Flood Estimation Handbook* (*FEH*) (IH, 1999).

All of these extrapolations involve both epistemic and aleatory errors. They are often treated, however, as if only aleatory errors are involved (for example in the use of statistical regressions in the estimation of rating curves and frequency curves for ungauged sites, or the choice of a particular statistical distribution in the extrapolation to lower-frequency events). In the *FEH* this is normally the generalised logistic distribution for annual maximum floods or the generalised Pareto distribution for peaks over threshold series. These are, however, choices subject to epistemic error. We do not know what the true shape of the tail of the distribution of extreme events might be over the design period.

A further consideration arises when the joint distribution of flood discharges across sites needs to be assessed. This can be important in assessing damages when events affect multiple catchments (e.g. Keef *et al.*, 2009), or in correctly predicting flood inundation where multiple tributaries provide water to an area at risk of flooding. This was the case for both the Carlisle 2005 and Cockermouth 2009 events in the north-west UK (e.g. Neal *et al.*, 2011). In the latter case, the relative timing of the flood peaks can also be important as well as the joint probability of the peak discharges. This is also the case for the joint probability of extremes in river flows and tidal heights and surges for assessing the potential for flooding in sites affected by a tidal boundary condition (e.g. Heffernan and Tawn, 2004). The choices made about joint distributions will also be subject to uncertainty.

The possibility of so many potential sources of aleatory and epistemic uncertainties means that many assumptions are required to perform an analysis. Note that carrying out a deterministic analysis of the flood inundation for a chosen event or design event also depends on assumptions about the uncertainties involved. In one sense, the use of AEP values of 0.01 or 0.001 as indicators of 'rare' events is already an institutionalised way of handling uncertainty implicitly. The use of 'freeboard' in the design of flood defences is also a way of implicitly accounting for uncertainties in various factors that might affect the failure of the design (Kirby and Ash, 2000).

Taking more explicit account of the uncertainties involved will require some guidelines, such as those developed as part of the FRMRC project (Beven *et al.*, 2011b). The FRMRC guidelines are structured in the form of decisions that must be made about sources of uncertainty within a source–pathway–receptor analysis (Figure 7.4). The decisions should ideally be discussed and agreed between analyst and users. As such, they then act as a tool for the uncertainty communication process (Faulkner *et al.*, 2007) and provide an audit trail for later review and evaluation.

Methodologies have been developed to start to take account of the uncertainties in the mapping process. One result of this research is a visualisation tool where the different inundation extents and flood depths arising from uncertain estimates of either historical or design events can be investigated. Figure 7.5 shows just one possible expression of the uncertain mapping, as colour-coded probabilities of flood extent. As shown in the inset plot, each pixel within the map can also be investigated for the distribution of predicted flood depths. This also allows a link to then be made to depth-dependent damage assessments (where the uncertainty associated with the damage should also be taken into account). Other methods of visualisation are possible. In a study undertaken for the Office of Public Works, Eire (Berry, 2010), uncertainty in flood outlines was shown as a single line, but with different line types depending on the local uncertainty in the inundation extent.

7.4 Climate change impacts on flood risk

In the UK it has been estimated that one in six homes in England is at risk of flooding, according to the Environment Agency, and climate change will raise that number without better protection. The agency calculates that funding for projects that protect communities from flooding from rivers and the sea needs to double to £1 billion annually by 2035. The

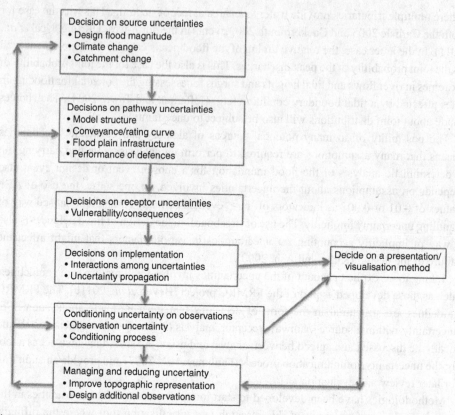

Figure 7.4 Main decision framework for incorporating uncertainty into flood hazard mapping (after Beven *et al.*, 2007).

latest UKCP09 projections (Murphy *et al.*, 2009) suggest that the UK is likely to receive more rainfall in winter periods than current conditions. Consequences of climate change will also increase the risk of coastal flooding due to rises in global sea levels. Evans *et al.* (2006) provided a discussion of the future flood risk management in the UK.

The best tools currently available to quantify climate change are global climate models (GCMs), but these lack the ability to represent local or indeed regional conditions essential for impact studies because of their coarse resolution and inability to model precipitation accurately (IPCC, 2007). Therefore, output from GCMs often cannot be used directly in local impact studies, such as runoff and flood inundation modelling, and there is a need to downscale the results in order for them to be useful (Zorita and von Storch, 1999; Wood *et al.*, 2004). Climate modelling is inherently uncertain, and although many studies have assessed uncertainty (Collins *et al.*, 2006; Deque *et al.*, 2007; Fowler *et al.*, 2007a), few studies have tried to quantify and assess rigorously the full range of uncertainties cascading through a climate impact modelling chain from GCM down to production of flood hazard maps.

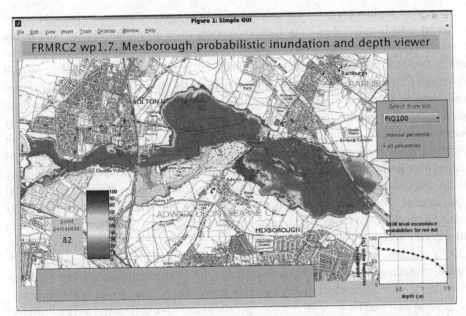

Figure 7.5 Probabilistic inundation and depth visualisation for Mexborough, Yorkshire. Colours represent the probability of inundation for the AEP 0.01 (*T* = 100 year) event as estimated (with uncertainty) from the *Flood Estimation Handbook* WinFAP analysis program. Model parameters were conditioned on post-event inundation surveys from the summer 2007 event (from Beven *et al.*, 2011b).

7.4.1 Climate change impacts on inland flooding

Using regional climate models (RCMs), Dankers and Feyen (2008) found a considerable decrease in flood hazard in the north-east of Europe, where warmer winters and a shorter snow season reduce the magnitude of the spring snowmelt peak. Also, in other rivers in central and southern Europe a decrease in extreme river flows was simulated. The results were compared with those obtained with two HIRHAM experiments at 50 km resolution for the SRES A2 and B2 scenarios. Disagreements between the various model experiments indicate that the effect of the horizontal resolution of the regional climate model is comparable in magnitude to the greenhouse gas scenario.

There are many comparison studies in climate change studies; these include the performance of downscaling techniques (e.g. Wilby and Wigley, 1997; Fowler *et al.*, 2007b); flood inundation modelling (Lane *et al.*, 2007; Veijalainen *et al.*, 2010; Karamouz *et al.*, 2011; Schneider *et al.*, 2011; Feyen *et al.*, 2012); and on the use of hydrological models in climate studies (e.g. Blazkova and Beven, 1995; Wilby *et al.*, 1999; Bell *et al.*, 2009; Cameron, 2006) and on reservoirs and lakes (e.g. Hingray *et al.*, 2007; Brekke *et al.*, 2009); on impact studies for urban flooding (e.g. Arnbjerg-Nielsen and Fleischer, 2009); and finally general changes to flooding (e.g. Milly *et al.*, 2002; Blöschl *et al.*, 2007). Few studies have tried to assess the uncertainty when modelling runoff changes in a future climate; in most cases the uncertainty is the spread in the results from the use of different models rather than a rigorous uncertainty assessment, where each contributed uncertainty is accounted for. Cameron *et al.* (2000a) used output from different scenarios run by the HadCM2 model for a catchment in the UK, and found that although the form of discharge CDF for a given return period was not greatly changed for the future scenarios compared to the uncertainty in the predictions themselves, the risk of a given flood magnitude return period does change. Wilby and Harris (2006) put forth a methodology where the different parts of the modelling framework were weighted according to their performance in a control climate. An uncertainty analysis concerning the GCM/RCM output over Europe showed that even though most uncertainties arise from the boundary GCM, the uncertainty in summer precipitation was equally due to the underlying RCM (Deque *et al.*, 2007). The uncertainty assessment took into account the spread in results arising from choice of GCM, RCM, sampling interval of the RCM and scenario. There are attempts to account for uncertainties in a more rigorous way. Serrat-Capdevila *et al.* (2007) cascaded results from 17 climate model output through downscaling to rainfall-runoff modelling on a yearly basis to estimate the long-term climate effect on water resources with an uncertainty envelope rather than a single value projection. Cameron *et al.* (2000a) analysed how sensitive parameters in a rainfall-runoff model were to different climate change scenarios and found that uncertainty within hydrologic modelling had a very large impact on the results. Wilby and Harris (2006) outlined a methodology to weight different parts of the modelling chain depending on how they performed in a control climate modelling low flows. They found that the driving GCM was the most dominant source of uncertainty.

A more complete uncertainty assessment would serve as a tool for policy-makers to make decisions about how to adapt to a future increase in flood hazard due to climate change. Ideally, a coupled climate impact modelling system should be transparent enough to follow uncertainties from specific sources through the modelling framework. The problem can be summarised in some specific questions that need more attention: (1) How can decision-makers better engage with the uncertainty of predictions? (2) What does an idealised uncertainty cascaded prediction chain look like for this problem, and what is limiting in the assessment of such a cascade? (3) Can we develop a structure for such simulations that provides a benchmark for predictions and gives guidance to the community? (4) What can help this problem from other branches of science/predictions? The current ability of RCMs to predict even current rainfall is poor and does not give confidence in future projections. Despite this, investments in flood defence of ~£500 million per year are made on the basis of such forecasts. This uncertainty urgently needs to be addressed

There are projects for the Environment Agency (EA science project code SC080004) to 'translate' UKCP09 climate projections into policy and practice, so as to help the agency understand the implications of UKCP09 for its business. In this project, an independent team (led by JBA Consulting) is providing guidance and interpretation of a very complex set of risk modelling methods and products. This could be seen as an example of an auxiliary project of a type that may be necessary more often in the future to help bridge the gap between scientific methodology and 'business users' (this is a broader population than is implied by 'policy/decision-makers').

7.4.2 Climate change impacts on coastal flooding

Factors that potentially will increase coastal flooding risk due to climate change include an increase in mean global sea level and the possibility that the frequency of storm surges will increase (Lowe *et al.*, 2009). Estimating future changes to the probability of exceeding a certain EWL can be calculated in a number of ways, including an evaluation of trends in historical data (Pirazzoli *et al.*, 2006) and the dynamic coupling of GCM/RCM models to storm surge models (and potentially flood inundation models) in a cascaded modelling framework (Lowe *et al.*, 2001). There are large uncertainties in these predictions, including understanding the responses of ice masses in the polar regions, and in the emission scenarios and indeed in the whole coupled ocean atmosphere–biosphere system that feeds back into the rates of melting of ice masses. Recent results have concluded that without making a full assessment of the uncertainty in the emissions scenarios (which may not be possible), there may result a considerable underestimation of low-probability but high-impact events. For example, Purvis *et al.* (2008) used the LISFLOOD-FP hydraulic model to map the probability of inundation for a 1 in 200-year event along a 32 km section of the north Somerset coast in the UK given uncertainty over possible future sea-level rise as described in the IPCC Third Assessment Report (see Figure 7.6).

The results of this study indicated that undertaking a future risk assessment using the most plausible sea-level rise value may significantly underestimate monetary losses as it fails to account for large but unlikely amounts of sea-level rise. These have a disproportionately damaging effect as they more readily overtop current defence systems and flood urban areas. There is therefore a need to understand how the different types of uncertainty in future predictions can be quantified by deconvoluting the various sources and understanding the cascaded uncertainties in a complex modelling chain. One way of approaching this problem is suggested in Chapter 3 of this volume in terms of the exceedance probability of risk estimates.

7.5 New technologies for improving flood risk prediction

At the catchment scale, our understanding of river form, process and function is largely based on locally intensive mapping of river reaches, or on spatially extensive but low-density data scattered throughout a catchment (e.g. the cross-sections of Marcus and Fonstad, 2010). In other words, streams have most often been treated as discontinuous systems. Recent advances in remote sensing of rivers, from the optical to the microwave spectra, indicate that it should now be possible to generate accurate and continuous datasets

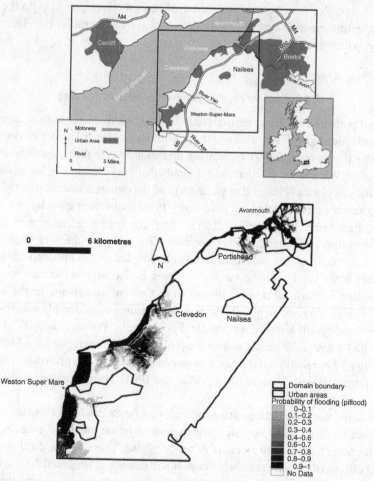

Figure 7.6 Probability of inundation predicted by the LISFLOOD-FP model for a 1 in 200-year recurrence interval event given the uncertainty in sea-level rise by 2100 stated in the IPCC Third Assessment Report.

in the form of thematic maps of flood edge and flooded area, maps of flow gradient, direction and velocity field, and spatially distributed data on storage fluxes and water-level changes. Even maps of in-stream habitats, depths, algae, wood, stream power and other features at sub-metre resolutions across entire watersheds might be possible where the water is clear and the aerial view is unobstructed (Marcus and Fonstad, 2010).

It is expected that the free availability of such data would transform river science and management by providing improved data, better models and explanation and enhanced applications. New techniques and technologies in remote sensing would enable a shift from a data-poor to a data-rich monitoring and modelling environment in flood risk studies (Bates, 2004).

7.5.1 *Remotely sensed topographic datasets*

7.5.1.1 *Photogrammetry and airborne LiDAR*

The photogrammetric technique defines shape, size and position of objects using aerial images taken from different views by either analogue or digital camera. In general, using the metric properties of an image alongside ground control points (GCP), the geometry and the position of an object can be described in a defined reference frame (Fabris and Pesci, 2005). For terrestrial applications where the camera-to-object distances are short (typically >100 m), DEM accuracy may range from millimetres to a few centimetres. In the case of aerial surveys where such distances are much greater, accuracies may decrease rapidly from a few centimetres to several meters (Fabris and Pesci, 2005).

Airborne laser altimetry (Figure 7.7a) or LiDAR has provided high-resolution, high-accuracy height data since the mid-1990s. A widespread use of LiDAR height points is the generation of digital elevation or terrain models (DEMs, DTMs). These topographic datasets have revolutionised the development of flood inundation models, particularly 2D models, as they represent the most important boundary condition for such models. Vertical accuracies range between 10–20 cm and the spatial resolution can be as fine as 0.5 m. There is also increasing use of mobile LiDAR mapping to generate new topographic datasets. LiDAR data can be acquired from a moving vehicle at over 100 000 measurements per second with a 360° field of view (FOV), while maintaining a vertical accuracy greater than 5 cm.

7.5.1.2 *Radar mapping of topography*

The development of InSAR in recent years has enabled the direct measurement of the elevation of scattering elements within a ground resolution cell. Interferometry requires two SAR images

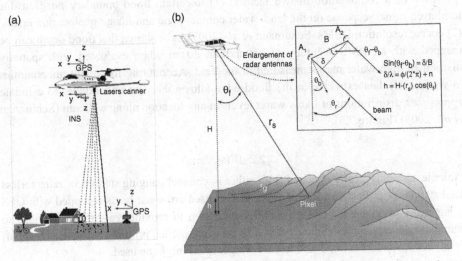

Figure 7.7 (a) Illustration of LiDAR DEM acquisition; (b) geometry of SAR interferometry (adapted from Mercer, 2004).

from slightly different viewing geometries. Measurement of surface elevation is also controlled by baseline effects, which are very difficult to determine accurately. Thus, large uncertainties and inaccuracies may be introduced to InSAR-derived DEMs, which can be in the order of (many) metres – see SRTM DEM validation studies by Farr et al. (2007). For the global DEM produced from elevation data acquired during the Shuttle Radar Topography Mission (SRTM-DEM) at 3 arc second (~90 m) spatial resolution, average global height accuracies vary between 5 m and 9 m (Farr et al., 2007). This vertical error has been shown to be correlated with topographic relief with large errors and data voids over high-relief terrain, while in the low-relief sites errors are smaller but still affected by hilly terrain (Falomi et al., 2005).

7.5.2 Work on remotely sensed flood extent and area and water level

7.5.2.1 Flood extent and area

There have been major advances in the field of flood mapping based on remote sensing technologies, especially synthetic aperture radar (SAR) (e.g. Aronica et al., 2002; Horritt et al., 2003; Bonn and Dixon, 2005). The penetration of clouds by microwaves and the ability to acquire data during day and night are the major strengths of space-borne radar imaging of floods. SAR image processing techniques can evaluate flood area or extent. Apart from direct measuring techniques from space (Alsdorf et al., 2007), river stage can be estimated at the land–water interface using high spatial resolution satellite or airborne imagery in combination with a DEM (Smith, 1997). Given that high-resolution DEMs are becoming more readily available, if flood boundaries can be adequately extracted from SAR, it is possible to map not only flood extent but also derive water stage at the shoreline (e.g. Raclot, 2006; Brakenridge et al., 2007; Mason et al., 2007). Inaccuracy is largely the result of a combination of two factors: (1) uncertain flood boundary position due to blurred signal response on the land–water contact zone and image geolocation errors; (2) coarse resolution DEMs. Schumann et al. (2007) have shown that flood depth can be mapped with an RMS accuracy of better than 20 cm when evaluated with spatially distributed high water marks measured in the field. Accounting for height uncertainties in water-level data extracted at the flood edge allows the user or modeller to estimate appropriate distributions of likely water levels at any location along a stream (Schumann et al., 2008) (Figure 7.8).

7.5.2.2 Water levels

Water levels are typically obtained from either in-channel gauging stations or reflectorless total stations (e.g. Pasternack et al., 2008). Also used are sonar boats equipped with GPS using real-time kinematic (RTK) satellite navigation to provide real-time corrections to centimetre-level accuracy (Shields et al., 2003); even satellite profiling altimeters (Frappart et al., 2006; Alsdorf et al., 2007) to cover larger areas might be used.

Surveys of water marks on objects such as buildings and lines of debris deposited at the flood edge, which are often referred to as trash lines or wrack marks, can be useful indicators

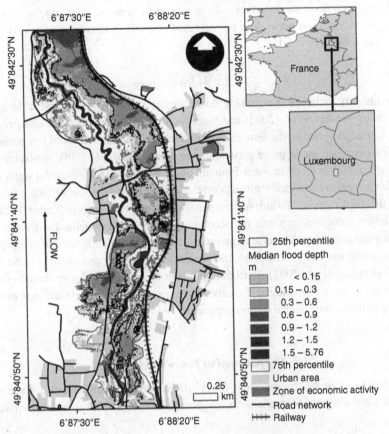

Figure 7.8 Flood depth and extent map conditioned on multiple cross-sectional water stages derived from a SAR flood extent of an event that occurred on the Alzette River (Grand Duchy of Luxembourg) in early January 2003 (modified after Schumann *et al.*, 2009).

of maximum flood extent and water surface elevation. These deposits can be surveyed after an event with a total station or differential GPS system and used to evaluate hydraulic models (e.g. Mignot *et al.*, 2006; Tayefi *et al.*, 2007; Neal *et al.*, 2009a; Horritt *et al.*, 2010) or simply map flood extents. Networks of pervasive sensors (fixed or floating) and video imagery represent promising new technologies for observing river hydraulics.

7.5.2.3 Networked pervasive sensors

Networked pervasive sensors of water depth represent a promising technology for providing flood predictions at a high spatial resolution. The data collected from such sensor networks have been successfully used for constraining the uncertainty in hydrodynamic models (Hughes *et al.*, 2008). The methodology is based upon complementing existing observational networks, with an adaptive system consisting of nodes communicating using wireless

technology. Each node can interface with one or more sensors, allowing the spatial resolution of an observation network to be increased in areas of interest at relatively low cost
(Smith *et al.*, 2009).

7.5.2.4 Video imagery

A video imagery technique for making flow measurements in streams and waterways has
been used by Bradley *et al.* (2002) and Smith *et al.* (2005) to estimate discharge. A video
camera is used to visualise the flow seeded with tracers. Measurements of free surface flow
velocities were then made using particle image velocimetry (PIV). PIV application determines surface flow velocity in rivers from oblique digital video records of the water surface.
The video images are analysed with autocorrelation algorithms that use surface water texture
to track the displacement of water surface particles between subsequent frames (Figure 7.9).
The resulting velocities in pixels per second are then ortho-normalised with total station
GCPs to yield velocities. The discharge estimate using the video technique compared well
with a conventional current meter discharge measurement with an estimated standard error
of 6.4% (Bradley *et al.*, 2002). Potential applications of video imagery include measurements of flood flows at ungauged sites, river monitoring from a remote site, and estimation
of two- and three-dimensional flow components.

7.5.3 *The potential of future satellite missions*

In addition to more systematic radar observations, other developments are required to
make possible a more comprehensive integration of models and remotely sensed flood

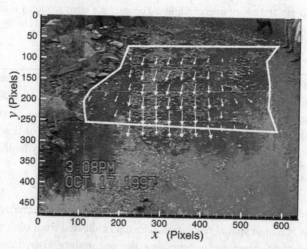

Figure 7.9 Average pixel displacement vectors (in pixels per second) overlaid on one of the digitised
images. The average is based on 60 images taken at a 1 second interval (taken from Bradley *et al.*,
2002).

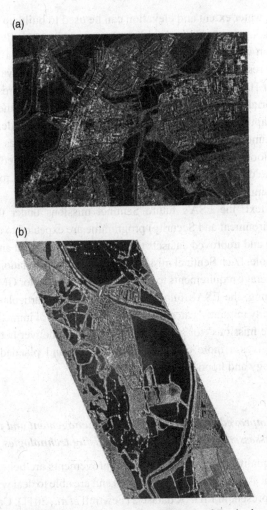

Figure 7.10 (a) A 3 m TerraSAR-X image of the flooded town of Tewkesbury © DLR; (b) a 1.2 m airborne SAR image of a rural reach of the River Severn near the town of Upton © QinetiQ Ltd (taken from Bates *et al.*, 2006).

extent and water height data. First, in many places flood plain terrain data are either insufficiently accurate or not consistently available to support modelling. Despite the progress in using LiDAR-derived DEMs for flood modelling, only SRTM data have near-global coverage and the potential to be used in some flood modelling studies (Fusco *et al.*, 2005), despite typical vertical noise of about 5 m. The DLR (German Aerospace Centre) planned the TanDEM-X mission – a unique twin-satellite constellation – which will allow the generation of global DEMs at an unprecedented accuracy (>1 mm error), coverage and quality (Figure 7.10), generated within three years after launch. Here, too, the proposed surface water ocean topography (SWOT) satellite mission may help, as over time repeated imaging of surface water extent and elevation can be used to build up detailed flood plain topography maps using the 'waterline' method (Mason *et al.*, 2001). Potentially for launch in 2019, the mission will provide a 'water mask' able to resolve 100 m rivers and 1 km^2 lakes, wetlands or reservoirs. Associated with this mask will be water-level elevations with an accuracy of 10 cm and a slope accuracy of 1 cm/1 km. Time resolutions will allow for monthly equatorial sampling and sub-monthly Arctic sampling; therefore the data will opportunistically capture more transient events (e.g. floods) for model validation. With a maximum repeat time of around ten days this potentially equates to the retrieval of approximately 75 flood plain topographic contours over the scheduled three-year mission lifetime, thereby yielding a wealth of detail concerning flood plain topographic features that can be used in inundation models.

In a similar context, the ESA's future Sentinel missions under the GMES (Global Monitoring for Environment and Security) programme are expected to contribute a significant amount of new and improved datasets to flood risk management and studies of hydrological risks as a whole. Each Sentinel mission is based on a constellation of two satellites to fulfil revisit and coverage requirements to provide robust datasets for GMES services. In the same line of reasoning, the ESA's future Explorer missions, particularly those under the lower cost 'opportunity missions' category which address areas of immediate environmental concern, or the 'core missions' for scientific research might deliver better support for risk studies. Table 7.1 gives a more complete list of recent and planned satellite missions dedicated to hydrology and flood monitoring.

7.5.4 *Improved flood risk monitoring, management and model assessment using new remote sensing technologies*

Owing to the high quality of the LiDAR data, improvements are being made to develop flood models that run at similar spatial resolutions and are able to deal with the high microtopographic details present in LiDAR datasets (Fewtrell *et al.*, 2011). Considerable efforts have been made to decrease the computational time of flood models conditioned on high-quality topographic data in order to still allow a high number of model runs to be performed – for example, within a probabilistic framework. For more details on this, see Section 7.3. At a larger scale, especially globally, elevation data of suitable resolution and vertical accuracy for flood risk assessment are not available. Current near-global DEMs include the Advanced Spaceborne Thermal Emission and Reflection Radiometer (ASTER) DEM and the Shuttle Radar Topography Mission (SRTM) DEM; however, neither are suitable for many flood risk management applications due to insufficient vertical accuracy and spatial resolution. The upcoming DEM from TanDEM-X may help with this (Moreira *et al.*, 2004).

In addition to topographic data, remote sensing can also be used for model calibration and evaluation. By using a combination of space-borne radar and aerial imagery, Schumann *et al.* (2011) were able to show that remotely sensed imagery, particularly from the new

Table 7.1 *Recent and planned satellite missions dedicated to hydrology*

Mission/Satellite	Launch year	Band (GHz)	Revisit time at equator (days)	Spatial resolution (m)	Primary use for flood hydrology
ALOS	2006	L (1.3)	46	10–100	Flood mapping; soil moisture at small scale
TerraSAR-X (TerraSAR-X 2 planned)*	2007	X (9.6)	11	1–18	High-resolution flood mapping
RADARSAT-2 (RADARSAT constellation mission – RCM – planned)**	2007	C (5.3)	24	3–100	High-resolution flood mapping; soil moisture at small scale
COSMO-SkyMed	2007	X (9.6)	16 (4 to >1, with 4 satellites in constellation)	1–100	High-resolution flood mapping
SMOS	2009	L (1.4) (passive)	2–3	30–50 km	Large-scale flood mapping; soil moisture at large scale
Cryosat-2	2010	Ku (13.575) Interferometric altimeter	369	~250 m imaging	Water level measurement; topography
TanDEM-X	2010	X (9.6) (Interferometry)	11 (in tandem orbit with TerraSAR-X)	3 m	Water-level dynamics; topography
Sentinel-1	2013	C (5.3)	12 (6, with 2 satellites in constellation)	5–100	High-resolution flood mapping; soil moisture at small scale
TanDEM-L (potential)	2015	L (1–2) (Interferometry)	8	5–100	Water-level dynamics; topography
SWOT	2019	Ka (35) Interferometric altimeter	22	>100 m imaging	Water-level measurement; topography

* Source: http://www.infoterra.de

** Source: http://www.eoportal.org

TerraSAR-X radar (Figure 7.10a), can reproduce dynamics adequately and support flood modelling in urban areas. They illustrated that image data from different remote sensing platforms reveal sufficient information to distinguish between models with varying spatial resolution, particularly towards the end of the recession phase of the event. Findings also suggest that TerraSAR-X is able to compete with aerial photography accuracies and can point to structural model errors by revealing important hydraulic process characteristics which might be missing in models; however, as Schumann *et al.* (2011) note, this has only been demonstrated for a particular urban setting and different results should be anticipated for different types of urban area.

On a 16 km reach of the River Severn, west-central England, Bates *et al.* (2006) use airborne synthetic aperture radar to map river flood inundation synoptically at fine spatial resolution (1.2 m, Figure 7.10b). Images were obtained four times through a large flood event between 8 November and 17 November 2000, and processed using a statistical active contour algorithm (Horritt, 1999) to yield the flood shoreline at each time. Intersection of these data with a high vertical accuracy survey of flood plain topography obtained from LiDAR permitted the calculation of dynamic changes in inundated area, total reach storage and rates of reach dewatering. In addition, comparison of the data to gauged flow rates, the measured flood plain topography and map data giving the location of embankments and drainage channels on the flood plain revealed more detailed under-standing of the factors controlling the development of inundation patterns at a variety of scales. Finally, the data were used to assess the performance of a simple two-dimensional flood inundation model, LISFLOOD-FP (Bates and De Roo, 2000), and allowed, for the first time, to validate the dynamic performance of the model. This process is shown to give new information into structural weaknesses of the model and suggests possible future developments, including the incorporation of a better description of flood plain hydro-logical processes in the hydraulic model to represent more accurately the dewatering of the flood plain.

Satellite imagery with a coarse spatial but high temporal resolution of a flood may provide useful information and support for timely flood inundation modelling (Di Baldassarre *et al.*, 2009a). Recently, it has also been illustrated how different flood extraction techniques can be fused to augment the information content obtained from a remotely sensed flood edge (Schumann *et al.*, 2009) and thus increase the confidence in subsequent flood model calibration (Di Baldassarre *et al.*, 2009b). It is expected that more recent and future satellite missions, such as TerraSAR-X and RADARSAT-2, or satellite constellations such as the DMC or COSMO-SkyMed, will provide more accurate image data to enhance some of the existing and indeed promising techniques. For instance, Mason *et al.* (2010) recently showed that flooding can now be detected with relatively high accuracy inside urban areas with the new TerraSAR-X sensor delivering spatial resolutions of up to 1 m. This eliminates most of the deteriorating effects on sensor saturation due to the signal bouncing off sharp-cornered objects and returning to the instrument.

7.5.5 *Remote sensing and data assimilation for flood risk management*

Studies in the field of assimilation of remote sensing data in flood forecasting systems are at present only very few in number, as there are a number of considerable challenges to be faced, such as generating forecast ensembles with adequate error statistics. Furthermore, for small rivers a simple re-initialisation of models with distributed water stages obtained from remote sensing does not lead to significant improvement because of the dominating effect of the forcing terms on the modelling results (Matgen *et al.*, 2007). In this respect, it is often necessary to include an error forecasting model for the forcing terms to get the desired forecast lead times (Madsen and Skotner, 2005; Andreadis *et al.*, 2007; Neal *et al.*, 2007). Montanari *et al.* (2009) and Giustarini *et al.* (2011) have demonstrated that spatially distributed water levels retrieved at a remote sensing-derived flood edge from SAR imagery can lead to improved calibration of a coupled hydrological-hydraulic model. In the context of using remote sensing to estimate discharge, Neal *et al.* (2009b) have illustrated how such water-level data, albeit with many inherent uncertainties, can improve discharge estimations from a hydrological model conditioned on precipitation forecasts through Ensemble Kalman Filter-based assimilation in a hydraulic model. Meanwhile, Andreadis *et al.* (2007) and Biancamaria *et al.* (2011) demonstrated the estimation of discharge from synthetic swath altimetry measurements using variants of the Ensemble Kalman Filter.

In a recent review on integration of remote sensing data and flood models, Schumann *et al.* (2009) argue that in order to ensure data availability is compatible with near-real-time remote sensing-based data assimilation for flood risk studies, new processing chains need to be developed. A possible way forward is provided by the new SWOT mission that uses radar altimetry and SRTM-type technology for timely water-level recording at a global scale. Also, the European Space Agency's (ESA) new FAIRE (Fast Access to Imagery for Rapid Exploitation) tool which is integrated into a grid-based environment (grid processing on demand, G-POD (2009)) is promising. An alternative is to use the forecasting capability of a flood forecasting system to order satellite imagery of the predicted event in advance (Aplin *et al.*, 1999). A comprehensive remote sensing data assimilation framework has the potential to become a critical component in flood forecasting.

7.6 Future challenges in flood risk

7.6.1 *How do we approach the problem of quantifying non-stationary epistemic uncertainties and the information content of conditioning data?*

This discussion shows that generic epistemic errors in the natural hazards modelling process can lead to non-stationary residual series. The influence of individual sources of error may persist over the relaxation times of the system, which can commonly be long relative to an individual event. Complex autocorrelation and covariance structures in residuals can be expected in space and time, with non-stationary statistical properties. In some cases, where the autocorrelation is simple and the non-stationarity can be treated as a simple hetero-scedasticity, then it may be possible to represent the residuals as if they were aleatory by a standard statistical model. The information content of the residual series will then also be defined. While some authors have suggested that this is the only way to assess uncertainty (e.g. Mantovan and Todini, 2006), and others have tried to transform more complex errors into a simpler statistical framework (e.g. Montanari, 2005), there is a real need to assess the effect of non-stationary epistemic uncertainties on the real information content of calibration data, testing competing models as hypotheses about system function and model inference in prediction. Beven (2006), for example, has suggested an alternative approach based on defining limits of acceptability for model performance before running a set of models in calibration (see, for example, Liu *et al.*, 2009). While it is difficult to allow for unknown input errors in setting these limits of acceptability, such an approach requires the user to think about what the appropriate limits should be and what might be considered fit-for-purpose in prediction.

7.6.2 How do uncertainties in data (input and discharge data) affect the outcome of flood risk?

Understanding the effect of data uncertainties in the modelling process is a critical aspect of any method that is quantifying flood hazard and flood risk. Various authors have tried to address specific data uncertainties for flooding problems, including: rainfall uncertainty and variability (Kuczera and Williams, 1992; Arnaud *et al.*, 2002; Vivoni *et al.*, 2007); flood level and/or discharge (Werner, 2004); boundary conditions (Pappenberger *et al.*, 2006b); on a combination of different uncertainties (Blazkova and Beven, 2009a); and the development of uncertainty cascaded approaches (McMillan and Brasington, 2008). The specific problem of extrapolating these uncertainties, necessary for defining risks associated with extreme event behaviour, has also been considered (e.g. Hall and Anderson, 2002). These studies demonstrate that significant changes to the calculated flood risk are caused by taking data uncertainties into account.

In hydrology, numerous studies have tried to quantify uncertainty in rainfall-runoff processes with high relevance to the quantification of flood risk. The BATEA methodology (Kavetski *et al.*, 2002) is an example of compensating for the epistemic model uncertainties by characterising input uncertainty error.

7.6.3 How can we effectively quantify uncertainty and risk and embed them in flood risk management decisions?

Flood risk management decisions to reduce risk require assessment of structural and non-structural methods over a range of spatial and temporal scales (Hall *et al.*, 2003). The modern flood risk decision-making process has to consider a wide variety of outcomes, not just associated with the building of new infrastructure (e.g. new defences), but also the possible effects of other mitigation measures for flood risk reduction (e.g. land-use change, planning controls, etc.). There is also a need to consider future changes to flood risk.

A framework for the inclusion of uncertainty analysis in decision-making has been suggested by Hall and Solomatine (2008), focusing on flood risk management plans and strategies rather than how to apply such risk frameworks in flood warning and emergency response. Important points concern various types of uncertainty in the decision-making process and the scope of factors involved to fully quantify the risk. They introduce a nine-point plan for an uncertainty analysis process, some points being similar to other discussions (for example, Beven, 2006). Some uncertainties are going to be very uncertain or severe, implying that it might be difficult to optimise the solution or decision space. Different approaches to conventional applied decision theory are therefore required, such as the info-gap theory of Ben-Haim (2006). Here, the risk manager determines if a given decision achieves a satisfactory outcome for a range of possible plausible future scenarios. Although overall risk analysis methods are reasonably well established, gaps remain in translating the outputs, including uncertainty, through the flood risk management decision-making process. The new decision-making methods under severe uncertainty are controversial (e.g. Hall and Solomatine, 2008). A key question is: are these research studies relevant to the management of risk? Research endeavours deal with a highly complex set of uncertain impacts and seek to provide the techniques and the computational resources to quantify risk and uncertainty. Funders and practitioners need to be convinced this is all worth it for making more informed decisions. Research is needed to find out if there is any consistency in the dominant risk factors that are important to the final decision process and so reduce the overall decision space that needs to be analysed. Taking this further, the detailed quantification of decisions within a risk-based framework may vary considerably, depending on which stakeholders are affected and what importance they might place on the specified uncertainties. This diversity of possible outcomes and requirements highlights the difficulty of making flood risk decisions.

The 'translation' of scientific knowledge into decision-making tools has been attempted in projects related to coastal management (e.g. the Wadden Sea – see http://www.modelkey. org/) and such approaches should be further developed.

7.6.4 The quantification of risk and understanding the characterisation of risk in the end product

There is concern about the dependability of the numbers and maps that are coming out of flood risk analysis. This is most acute at the national scale with, for example, the NaFRA analysis. The challenge with complex risk models like NaFRA, which combine analysis of water levels, flood defence failure, inundation and flood damage, is the number of potentially uncertain variables. Uncertainty and sensitivity analysis of a number of parameters for selected locations has been conducted. However, also of concern is the potential for gross errors, such as mis-classification of flood defences or omissions from the property database. Recently the EA has funded a research project on the handling of uncertainty in NaFRA, but this is a very recent development and the approach has not yet been fully formulated. The challenge is to do this comprehensively, given all the errors throughout the modelling process. Furthermore, underlying assumptions of an uncertainty analysis need to be explicitly stated and also communicated effectively so that they can be assessed by users of the outputs. There is a danger of incorporating all sources of uncertainty into procedures that only appear as a 'black box' to many users, which could then lead to users overlooking how important risks and uncertainties were quantified. Analyses of risk and uncertainty may need some 'translation' to be properly understood by users (Faulkner *et al.*, 2007). Research also needs to continue to look at how to express and cope with risk information that is not quantified or integrated into a probability distribution (see the discussion in Chapter 3). As noted earlier, one framework for discussing and communicating assumptions about different sources of uncertainty to users is the decision-tree approach suggested by Beven *et al.* (2011b).

A disadvantage commonly found in current flood risk quantitative analysis is that many studies are performed on relatively small scales, and for short simulation times. Certain stakeholders (e.g. the insurance sector) are particularly interested in large-scale assessments. Long simulation times are needed to estimate larger return period statistics. Potentially this means certain stakeholders often have to take a quite different approach to modelling than academia, namely less computationally intensive approaches that can assess risk over larger spatial and temporal domains. There is a need for academia to turn its attention to such methods. Assessment of extreme events has involved the development of methods that produce realistic sequences to evaluate larger return period statistics through techniques such as continuous simulation, using simple modelling structures and hundreds of years of simulations driven by an appropriately calibrated rainfall generator (e.g. Cameron *et al.*, 1999, 2000b). Thus far such studies have been focused on small catchment scale studies (though see Blazkova and Beven, 2009a for an application in a larger catchment).

7.6.5 *Closing the gap with other disciplines and stakeholders*

Researchers involved in developing risk and uncertainty methods should consider very carefully the wider context in which their work takes place, and the language used, in order to help in developing guidance for users. Information derived from risk assessment methodologies may in some cases play a smaller part in decision-making than is appreciated by researchers. Users need to understand why different approaches to risk assessment may give different answers. It is particularly important with methods that rely on complex algorithms that critical assumptions are not 'lost' in the explanation. Those providing the scientific evidence need to understand the needs of consumers of the evidence, although there is a case for seeing this as a cyclical process, where science provides evidence to meet the needs but may also drive new thinking on what these needs are.

A key approach is to support seminars and workshops to bring a wider group of flood risk stakeholders and scientists together in the same room. An exemplar of this approach has been the Flood Risk Management Consortium (FRMRC and FRMRC2), which has been funded over the last five years. The consortium has produced significant outputs (papers and reports) and web-based guidance, such as the WIKI tree (http://www.floodrisknet.org.uk/methods) for deciding the appropriate uncertainty analysis method given the stated problem for assessment (see Pappenberger *et al.*, 2006a).

Acknowledgements

This chapter was compiled with the kind cooperation of a number of academics and stakeholders. Thanks especially are to Dr Andrew Challinor, Ruth Doherty, Dr Arno Hilberts, Dr Ben Gouldby, Dr Dawei Han, Dr Neil Hunter, Dr John Huthance, Dr Rob Lamb, Dr Dave Leedal, Dr Simon McCarthy, Dr Iris Moller, Dr Mike Steel and Dr Jonathan Webb. Thanks are also directed at the participants of the SAPPUR workshop for their involvement in this process and to the main team members of SAPPUR. Finally we thank Neil McIntyre on his insightful review of the manuscript.

References

Al-Futaisi, A. and Stedinger, J. R. (1999) Hydrologic and economic uncertainties and flood-risk project design. *Journal of Water Resources Planning and Management–ASCE* **125** (6): 314–324.

Alsdorf, D. E., Rodriguez, E. and Lettenmaier, D. P. (2007) Measuring surface water from space. *Reviews of Geophysics* **45** (2): n.p.

Andreadis, K. M., Clark, E. A., Lettenmaier, D. P., *et al.* (2007) Prospects for river discharge and depth estimation through assimilation of swath-altimetry into a raster-based hydro-dynamics model. *Geophysical Research Letters* **34** (10): n.p.

Andréassian, V., Lerat, J., Loumagne, C., *et al.* (2007) What is really undermining hydro-logic science today? *Hydrological Processes* **21** (20): 2819–2822.

Apel, H., Thieken, A. H., Merz, B., *et al.* (2004) Flood risk assessment and associated uncertainty. *Natural Hazards and Earth System Sciences* **4** (2): 295–308.

Arnaud, P., Bouvier, C., Cisneros, L., *et al.* (2002) Influence of rainfall spatial variability on flood prediction. *Journal of Hydrology* **260** (1–4): 216–230.

Arnbjerg-Nielsen, K. and Fleischer, H. S. (2009) Feasible adaptation strategies for increased risk of flooding in cities due to climate change. *Water Science and Technology* **60** (2): 273–281.

Aronica, G., Hankin, B. and Beven, K. (1998) Uncertainty and equifinality in calibrating distributed roughness coefficients in a flood propagation model with limited data. *Advances in Water Resources* **22** (4): 349–365.

Aronica, G., Bates, P. D. and Horritt, M. S. (2002) Assessing the uncertainty in distributed model predictions using observed binary pattern information within GLUE. *Hydrological Processes* **16** (10): 2001–2016.

Bates, B. C. and Townley, L. R. (1988) Nonlinear, discrete flood event models: 3. Analysis of prediction uncertainty. *Journal of Hydrology* **99** (1–2): 91–101.

Bates, P. D. and De Roo, A. P. J. (2000) A simple raster-based model for flood inundation simulation. *Journal of Hydrology* **236** (1–2): 54–77.

Bates, P. D., Horritt, M. S., Aronica, G., *et al.* (2004) Bayesian updating of flood inundation likelihoods conditioned on flood extent data. *Hydrological Processes* **18** (17): 3347–3370.

Bates, P. D., Dawson, R. J., Hall, J. W., *et al.* (2005) Simplified two-dimensional numerical modelling of coastal flooding and example applications. *Coastal Engineering* **52** (9): 793–810.

Bates, P. D., Wilson, M. D., Horritt, M. S., *et al.* (2006) Reach scale floodplain inundation dynamics observed using airborne synthetic aperture radar imagery: data analysis and modelling. *Journal of Hydrology* **328** (1–2): 306–318.

Bates, P. D., Horritt, M. S. and Fewtrell, T. J. (2010) A simple inertial formulation of the shallow water equations for efficient two-dimensional flood inundation modelling. *Journal of Hydrology* **387** (1–2): 33–45.

Bell, V. A., Kay, A. L., Jones, R. G., *et al.* (2009) Use of soil data in a grid-based hydrological model to estimate spatial variation in changing flood risk across the UK. *Journal of Hydrology* **377** (3–4): 335–350.

Ben-Haim, Y. (2001) *Information-gap Decision Theory: Decisions Under Severe Uncertainty*, San Diego, CA: Academic Press.

Ben-Haim, Y. (2006) *Information-gap Decision Theory: Decisions Under Severe Uncertainty*, San Diego, CA: Academic Press.

Bernardara, P., Schertzer, D., Sauquet, E., *et al.* (2008) The flood probability distribution tail: how heavy is it? *Stochastic Environmental Research and Risk Assessment* **22**: 107–122.

Berry, G. R. (2010) The year of the flood. *Organization and Environment* **23** (2): 248–250.

Beven, K. J. (1993) Prophecy, reality and uncertainty in distributed hydrological modelling. *Advances in Water Resources* **16** (1): 41–51.

Beven, K. J. (2006) A manifesto for the equifinality thesis. *Journal of Hydrology* **320** (1–2): 18–36.

Beven, K. J. (2009) *Environmental Modelling: An Uncertain Future?*, London: Routledge.

Beven, K. J. (2011) Distributed models and uncertainty in flood risk management. In *Flood Risk Science and Management*, ed. G. Pender, G. and H. Faulkner, Chichester: Wiley-Blackwell, pp. 291–312.

Beven, K. J., Smith, P. and Freer, J. (2007) Comment on 'Hydrological forecasting uncertainty assessment: incoherence of the GLUE methodology' by Pietro Mantovan and Ezio Todini. *Journal of Hydrology* **338** (3–4): 315–318.

Beven, K. J., Smith, P. J. and Freer, J. E. (2008a) So just why would a modeller choose to be incoherent? *Journal of Hydrology* **354** (1–4): 15–32.

Beven, K. J., Young, P., Romanowicz, R., *et al.* (2008b) Analysis of historical data sets to look for impacts of land use and management change on flood generation. *FD2120 Project Report*, London: Defra.

Beven, K. J., Young, P. C., Leedal, D. T., *et al.* (2008c) Computationally efficient flood water level prediction (with uncertainty). In *Flood Risk Management: Research and Practice*, ed. P. Samuels, S. Huntingdon, W. Allsop, *et al.*, Boca Raton, FL: CRC Press/Balkema, p. 56.

Beven, K., Smith, P. J. and Wood, A. (2011a) On the colour and spin of epistemic error (and what we might do about it). *Hydrology and Earth System Sciences* **15** (10): 3123–3133.

Beven, K. J., Leedal, D. T. and McCarthy, S. (2011b) Framework for uncertainty estimation in flood risk mapping. *FRMRC Research Report SWP1* **7**. Available at: http://www.floodrisk.org.uk.

Biancamaria, S., Durand, M., Andreadis, K. M., *et al.* (2011) Assimilation of virtual wide swath altimetry to improve Arctic river modeling. *Remote Sensing of Environment* **115** (2): 373–381.

Blazkova, S. and Beven, K. J. (1995) Frequency version of TOPMODEL as a tool for assessing the impact of climate variability on flow sources and flood peaks. *Journal of Hydrological Hydromechanics* **43**: 392–411.

Blazkova, S. and Beven, K. (2009a) A limits of acceptability approach to model evaluation and uncertainty estimation in flood frequency estimation by continuous simulation: Skalka catchment, Czech Republic. *Water Resources Research* **45**: n.p.

Blazkova, S. and Beven, K. (2009b) Uncertainty in flood estimation. *Structure and Infrastructure Engineering* **5** (4): 325–332.

Bloschl, G., Ardoin-Bardin, S., Bonell, M., *et al.* (2007) At what scales do climate variability and land cover change impact on flooding and low flows? *Hydrological Processes* **21** (9): 1241–1247.

Bonn, F. and Dixon, R. (2005) Monitoring flood extent and forecasting excess runoff risk with RADARSAT-1 data. *Natural Hazards* **35** (3): 377–393.

Bradbrook, K. F., Lane, S. N., Waller, S. G., *et al.* (2004) Two dimensional diffusion wave modelling of flood inundation using a simplified channel representation. *International Journal of River Basin Management* **2** (3): 1–13.

Bradley, A. A., Kruger, A., Meselhe, E. A., *et al.* (2002) Flow measurement in streams using video imagery. *Water Resources Research* **38** (12): n.p.

Bradshaw, C. J. A., Sodhi, N. S., Peh, K. S. H., *et al.* (2007) Global evidence that deforestation amplifies flood risk and severity in the developing world. *global Change Biology* **13** (11): 2379–2395.

Brakenridge, G. R., Anderson, E., Nghiem, S. V., *et al.* (2003) Flood warnings, flood disaster assessments, and flood hazard reduction: the roles of orbital remote sensing. In *Proceedings of the 30th International Symposium on Remote Sensing of the Environment*, Honolulu, Hawaii: International Center for Remote Sensing of Environment, p. 4.

Brakenridge, G. R., Nghiem, S. V., Anderson, E., *et al.* (2007) Orbital microwave measurement of river discharge and ice status. *Water Resources Research* **43** (4): n.p.

Brekke, L. D., Maurer, E. P., Anderson, J. D., *et al.* (2009) Assessing reservoir operations risk under climate change. *Water Resources Research* **45**: n.p.

Brown, J. D., Spencer, T. and Moeller, I. (2007) Modeling storm surge flooding of an urban area with particular reference to modeling uncertainties: a case study of Canvey Island, United Kingdom. *Water Resources Research* **43** (6): n.p.

Cameron, D. (2006) An application of the UKCIP02 climate change scenarios to flood estimation by continuous simulation for a gauged catchment in the northeast of Scotland, UK (with uncertainty). *Journal of Hydrology* **328** (1–2): 212–226.

Cameron, D. S., Beven, K. J., Tawn, J., *et al.* (1999) Flood frequency estimation by continuous simulation for a gauged upland catchment (with uncertainty). *Journal of Hydrology* **219** (3–4): 169–187.

Cameron, D., Beven, K. and Naden, P. (2000a) Flood frequency estimation by continuous simulation under climate change (with uncertainty). *Hydrology and Earth System Sciences* **4** (3): 393–405.

Cameron, D., Beven, K., Tawn, J., *et al.* (2000b) Flood frequency estimation by continuous simulation (with likelihood based uncertainty estimation). *Hydrology and Earth System Sciences* **4** (1): 23–34.

Cobby, D. M., Mason, D. C., Horritt, M. S., *et al.* (2003) Two-dimensional hydraulic flood modelling using a finite-element mesh decomposed according to vegetation and topographic features derived from airborne scanning laser altimetry. *Hydrological Processes* **17** (10): 1979–2000.

Collins, M., Booth, B. B. B., Harris, G. R., *et al.* (2006) Towards quantifying uncertainty in transient climate change. *Climate Dynamics* **27** (2–3): 127–147.

Crowder, R. A., Pepper, A., Whitlow, C., *et al.* (2004) *Benchmarking of Hydraulic River Modelling Software Packages*, Bristol: Environment Agency.

Dankers, R. and Feyen, L. (2008) Climate change impact on flood hazard in Europe: an assessment based on high-resolution climate simulations. *Journal of Geophysical Research–Atmospheres* **113** (D19): n.p.

Defra (2001) *Flood and Coastal Defence Project Appraisal Guidance: Overview (Including General Guidance)*, London: Defra.

Deque, M., Rowell, D. P., Luthi, D., *et al.* (2007) An intercomparison of regional climate simulations for Europe: assessing uncertainties in model projections. *Climatic Change* **81**: 53–70.

Di Baldassarre, G., Schumann, G. and Bates, P. (2009a) Near real time satellite imagery to support and verify timely flood modelling. *Hydrological Processes* **23** (5): 799–803.

Di Baldassarre, G., Schumann, G. and Bates, P. D. (2009b) A technique for the calibration of hydraulic models using uncertain satellite observations of flood extent. *Journal of Hydrology* **367** (3–4): 276–282.

Di Baldassarre, G., Schumann, G., Bates, P. D., *et al.* (2010) Flood-plain mapping: a critical discussion of deterministic and probabilistic approaches. *Hydrological Sciences Journal–Journal Des Sciences Hydrologiques* **55** (3): 364–376.

Environment Agency (2009) *Flooding in England: A National Assessment of Flood Risk*, Bristol: Environment Agency.

Evans, E., Hall, J., Penning-Rowsell, E., *et al.* (2006) Future flood risk management in the UK. *Proceedings of the Institution of Civil Engineers–Water Management* **159** (1): 53–61.

Fabris, M. and Pesci, A. (2005) Automated DEM extraction in digital aerial photogrammetry: precisions and validation for mass movement monitoring. *Annals of Geophysics–Italy* **48** (6): 973–988.

Falorni, G., Teles, V., Vivoni, E. R., *et al.* (2005) Analysis and characterization of the vertical accuracy of digital elevation models from the Shuttle Radar Topography Mission. *Journal of Geophysical Research–Earth* **110** (F2): n.p.

Farr, T. G., Rosen, P. A., Caro, E., *et al.* (2007) The shuttle radar topography mission. *Reviews of Geophysics* **45** (2): n.p.

Faulkner, H., Parker, D., Green, C., *et al.* (2007) Developing a translational discourse to communicate uncertainty in flood risk between science and the practitioner. *Ambio* **36** (8): 692–703.

Fewtrell, T. J., Duncan, A., Sampson, C. C., *et al.* (2011) Benchmarking urban flood models of varying complexity and scale using high resolution terrestrial LiDAR data. *Physics and Chemistry of the Earth* **36** (7–8): 281–291.

Feyen, L., Dankers, R., Bodis, K., *et al.* (2012) Fluvial flood risk in Europe in present and future climates. *Climatic Change* **112** (1): 47–62.

Finch, J. W., Bradford, R. B. and Hudson, J. A. (2004) The spatial distribution of groundwater flooding in a chalk catchment in southern England. *Hydrological Processes* **18** (5): 959–971.

Fowler, H. J., Ekstrom, M., Blenkinsop, S., *et al.* (2007a) Estimating change in extreme European precipitation using a multimodel ensemble. *Journal of Geophysical Research–Atmospheres* **112** (D18): n.p.

Fowler, H. J., Kilsby, C. G. and Stunell, J. (2007b) Modelling the impacts of projected future climate change on water resources in north-west England. *Hydrology and Earth System Sciences* **11** (3): 1115–1124.

Frappart, F., Do Minh, K., L'Hermitte, J., *et al.* (2006) Water volume change in the lower Mekong from satellite altimetry and imagery data. *Geophysical Journal International* **167** (2): 570–584.

French, J. R. and Clifford, N. J. (2000) Hydrodynamic modelling as a basis for explaining estuarine environmental dynamics: some computational and methodological issues. *Hydrological Processes* **14** (11–12): 2089–2108.

Fusco, L., Guidetti, V. and van Bemmelen, J. (2005) e-Collaboration and Grid-on-Demand computing for Earth Science at ESA. http://www.ercim.eu/publication/Ercim_News/enw61/fusco.html

Giustarini, L., Matgen, P., Hostache, R., *et al.* (2011) Assimilating SAR-derived water level data into a hydraulic model: a case study. *Hydrology and Earth System Sciences* **15** (7): 2349–2365.

Gouldby, B., Sayers, P., Mulet-Marti, J., *et al.* (2008) A methodology for regional-scale flood risk assessment. *Proceedings of the Institution of Civil Engineers–Water Management* **161** (WM3): 169–182.

Guinot, V. and Soares-Fraz, O. S. (2006) Flux and source term discretization in two-dimensional shallow water models with porosity on unstructured grids. *International Journal for Numerical Methods in Fluids* **50**: 309–345.

Habets, F., Gascoin, S., Korkmaz, S., *et al.* (2010) Multi-model comparison of a major flood in the groundwater-fed basin of the Somme River (France). *Hydrology and Earth System Sciences* **14** (1): 99–117.

Hall, J. and Anderson, M. (2002) Handling uncertainty in extreme or unrepeatable hydrological processes: the need for an alternative paradigm. *Hydrological Processes* **16** (9): 1867–1870.

Hall, J. W. and Harvey, D. P. (2009) Decision making under severe uncertainty for flood risk management: a case study of info-gap robustness analysis, *Proceedings of the International Conference on Science and Information Technologies for Sustainable Management of Aquatic Ecosystems*, Concepcion, Chile: IWA.

Hall, J. W. and Solomatine, D. (2008) A framework for uncertainty analysis in flood risk management decisions. *Journal of River Basin Management* **6** (2): 85–98.

Hall, J., O'Connell, E. and Ewen, J. (2007) On not undermining the science: coherence, validation and expertise. Discussion of Invited Commentary by Keith Beven Hydrological Processes 20, 3141–3146 (2006). *Hydrological Processes*, **21** (7): 985–988.

Hall, J. W., Meadowcroft, I., Sayers, P. B., *et al.* (2003) Integrated flood risk management in England and Wales. *Natural Hazards Review* **4** (3): 126–135.

Hall, J. W., Harvey, H. and Tarrant, O. (2011a) Case study: uncertainty and sensitivity analysis of a system model of flood risk in the Thames Estuary. In *Applied Uncertainty Estimation in Flood Risk Management*, ed. K. J. Beven and J. W. Hall, Imperial College Press, inpress.

Hall, J. W., Manning, L. J. and Hankin, R. K. S. (2011b) Bayesian calibration of a flood inundation model using spatial data. *Water Resources Research* **47**: n.p.

Harvey, D. P., Peppé, R. and Hall, J. W. (2008) Reframe: a framework supporting flood risk analysis. *Journal of River Basin Management* **6** (2): 163–174.

Heffernan, J. E. and Tawn, J. A. (2004) A conditional approach for multivariate extreme values. *Journal of the Royal Statistical Society Series B–Statistical Methodology* **66**: 497–530.

Hingray, B., Mouhous, N., Mezghani, A., *et al.* (2007) Accounting for global-mean warming and scaling uncertainties in climate change impact studies: application to a regulated lake system. *Hydrology and Earth System Sciences* **11** (3): 1207–1226.

Horritt, M. S. (1999) A statistical active contour model for SAR image segmentation. *Image and Vision Computing* **17** (3–4): 213–224.

Horritt, M. S., Mason, D. C., Cobby, D. M., *et al.* (2003) Waterline mapping in flooded vegetation from airborne SAR imagery. *Remote Sensing of Environment* **85** (3): 271–281.

Horritt, M. S., Bates, P. D., Fewtrell, T. J., *et al.* (2010) Modelling the hydraulics of the Carlisle 2005 flood event. *Proceedings of the Institution of Civil Engineers–Water Management* **163** (6): 273–281.

Hughes, A. G., Vounaki, T., Peach, D. W., *et al.* (2011) Flood risk from groundwater: examples from a Chalk catchment in southern England. *Journal of Flood Risk Management* **4** (3): 143–155.

Hughes, D., Greenwood, P., Blair, G., *et al.* (2008) An experiment with reflective middleware to support grid-based flood monitoring. *Concurrency and Computation–Practice & Experience* **20** (11): 1303–1316.

Hulme, M. and Dessai, S. (2008) Negotiating future climates for public policy: a critical assessment of the development of climate scenarios for the UK. *Environmental Science & Policy* **11** (1): 54–70.

Hunter, N. M., Bates, P. D., Neelz, S., *et al.* (2008) Benchmarking 2D hydraulic models for urban flooding. *Proceedings of the Institution of Civil Engineers–Water Management* **161** (1): 13–30.

Huntington, T. G. (2006) Evidence for intensification of the global water cycle: review and synthesis. *Journal of Hydrology* **319** (1–4): 83–95.

Hurford, A. P., Maksimovic, C. and Leitao, J. P. (2010) Urban pluvial flooding in Jakarta: applying state-of-the-art technology in a data scarce environment. *Water Science and Technology* **62** (10): 2246–2255.

Institute of Hydrology (1999) *Flood Estimation Handbook*, Wallingford: Institute of Hydrology.

IPCC (2007) *Fourth Assessment Report*, Cambridge: Cambridge University Press.

Jonkman, S. N. (2005) Global perspectives on loss of human life caused by floods. *Natural Hazards* **34** (2): 151–175.

Karamouz, M., Noori, N., Moridi, A., *et al.* (2011) Evaluation of floodplain variability considering impacts of climate change. *Hydrological Processes* **25** (1): 90–103.

Kavetski, D., Franks, S. and Kuczera, G. (2002) Confronting input uncertainty in environmental modelling. In *Calibration of Watershed Models*, ed. Q. Duan, H. V. Gupta, S. Sorooshian, *et al*, Washington, DC: American Geophysical Union, pp. 49–68.

Keef, C., Svensson, C. and Tawn, J. A. (2009) Spatial dependence in extreme river flows and precipitation for Great Britain. *Journal of Hydrology* **378** (3–4): 240–252.

Kirby, A. M. and Ash, J. R. V. (2000) *Fluvial Freeboard Guidance Note*, Bristol: Environment Agency.

Kuczera, G. and Williams, B. J. (1992) Effect of rainfall errors on accuracy of design flood estimates. *Water Resources Research* **28** (4): 1145–1153.

Lamb, R., Crossley, A. and Waller, S. (2009) A fast 2D floodplain inundation model. *Proceedings of the Institution of Civil Engineers–Water Management* **162** (6): 363–370.

Lane, S. N. (2005) Roughness: time for a re-evaluation? *Earth Surface Processes and Landforms* **30** (2): 251–253.

Lane, S. N., Tayefi, V., Reid, S. C., *et al.* (2007) Interactions between sediment delivery, channel change, climate change and flood risk in a temperate upland environment. *Earth Surface Processes and Landforms* **32** (3): 429–446.

Liang, D., Falconer, R. A. and Lin, B. L. (2006) Comparison between TVD-MacCormack and ADI-type solvers of the shallow water equations. *Advances in Water Resources* **29** (12): 1833–1845.

Liu, Y. L., Freer, J., Beven, K., *et al.* (2009) Towards a limits of acceptability approach to the calibration of hydrological models: extending observation error. *Journal of Hydrology* **367** (1–2): 93–103.

Lowe, J. A., Gregory, J. M. and Flather, R. A. (2001) Changes in the occurrence of storm surges around the United Kingdom under a future climate scenario using a dynamic storm surge model driven by the Hadley Centre climate models. *Climate Dynamics* **18** (3–4): 179–188.

Lowe, J. A., Howard, T., Pardaens, A., *et al.* (2009) *UK, Climate Projections Science Report: Marine and Coastal Projections*, Exeter: Met Office Hadley Centre.

Macdonald, D., Dixon, A., Newell, A., *et al.* (2012) Groundwater flooding within an urbanised flood plain. *Journal of Flood Risk Management* **5** (1): 68–80.

Madsen, H. and Skotner, C. (2005) Adaptive state updating in real-time river flow forecasting: a combined filtering and error forecasting procedure. *Journal of Hydrology* **308** (1–4): 302–312.

Mantovan, P. and Todini, E. (2006) Hydrological forecasting uncertainty assessment: incoherence of the GLUE methodology. *Journal of Hydrology* **330** (1–2): 368–381.

Mantovan, P., Todini, E. and Martina, M. L. V. (2007) Reply to comment by Keith Beven, Paul Smith and Jim Freer on 'Hydrological forecasting uncertainty assessment: Incoherence of the GLUE methodology'. *Journal of Hydrology* **338** (3–4): 319–324.

Marcus, W. A. and Fonstad, M. A. (2010) Remote sensing of rivers: the emergence of a subdiscipline in the river sciences. *Earth Surface Processes and Landforms* **35** (15): 1867–1872.

Mark, O., Weesakul, S., Apirumanekul, C., *et al.* (2004) Potential and limitations of 1D modelling of urban flooding. *Journal of Hydrology* **299** (3–4): 284–299.

Marks, K. and Bates, P. (2000) Integration of high-resolution topographic data with flood-plain flow models. *Hydrological Processes* **14** (11–12): 2109–2122.

Mason, D. C., Davenport, I. J., Flather, R. A., *et al.* (2001) A sensitivity analysis of the waterline method of constructing a digital elevation model for intertidal areas in ERS SAR scene of eastern England. *Estuarine and Coastal Shelf Science* **53** (6): 759–778.

Mason, D. C., Cobby, D. M., Horritt, M. S., *et al.* (2003) Floodplain friction parameterization in two-dimensional river flood models using vegetation heights derived from airborne scanning laser altimetry. *Hydrological Processes* **17** (9): 1711–1732.

Mason, D. C., Horritt, M. S., Hunter, N. M., *et al.* (2007) Use of fused airborne scanning laser altimetry and digital map data for urban flood modelling. *Hydrological Processes* **21** (11): 1436–1447.

Mason, D. C., Speck, R., Devereux, B., *et al.* (2010) Flood detection in urban areas using TerraSAR-X. *IEEE Transactions on Geoscience and Remote Sensing* **48** (2): 882–894.

Matgen, P., Schumann, G., Pappenberger, F., *et al.* (2007) Sequential assimilation of remotely sensed water stages in flood inundation models. In *Symposium HS3007 at IUGG2007: Remote Sensing for Environmental Monitoring and Change Detection*. Perugia: IAHS, pp. 78–88.

McIntyre, N. and Marshall, M. (2010) Identification of rural land management signals in runoff response. *Hydrological Processes* 24 (24): 3521–3534.

McMillan, H. K. and Brasington, J. (2007) Reduced complexity strategies for modelling urban floodplain inundation. *Geomorphology* 90: 226–243.

McMillan, H. K. and Brasington, J. (2008) End-to-end flood risk assessment: a coupled model cascade with uncertainty estimation. *Water Resources Research* 44 (3): n.p.

McRobie, A., Spencer, T. and Gerritsen, H. (2005) The big flood: North Sea storm surge. *Philosophical Transactions of the Royal Society A–Mathematical, Physical and Engineering Sciences* 363 (1831): 1263–1270.

Mercer, B. (2004) DEMs created from airborne IFSAR: an update. http://www.isprs.org/ proceedings/XXXV/congress/comm2/papers/242.pdf.

Merwade, V., Olivera, F., Arabi, M., *et al.* (2008) Uncertainty in flood inundation mapping: current issues and future directions. *Journal of Hydrologic Engineering* 13 (7): 608–620.

Merz, B. and Thieken, A. H. (2005) Separating natural and epistemic uncertainty in flood frequency analysis. *Journal of Hydrology* 309 (1–4): 114–132.

Mignot, E., Paquier, A. and Haider, S. (2006) Modeling floods in a dense urban area using 2D shallow water equations. *Journal of Hydrology* 327 (1–2): 186–199.

Milly, P. C. D., Wetherald, R. T., Dunne, K. A., *et al.* (2002) Increasing risk of great floods in a changing climate. *Nature* 415 (6871): 514–517.

Milly, P. C. D., Betancourt, J., Falkenmark, M., *et al.* (2008) Climate change: stationarity is dead–whither water management? *Science* 319 (5863): 573–574.

Montanari, A. (2005) Large sample behaviors of the generalized likelihood uncertainty estimation (GLUE) in assessing the uncertainty of rainfall-runoff simulations. *Water Resources Research* 41 (8): n.p.

Montanari, A. (2007) What do we mean by 'uncertainty'? The need for a consistent wording about uncertainty assessment in hydrology. *Hydrological Processes* 21 (6): 841–845.

Montanari, M., Hostache, R., Matgen, P., *et al.* (2009) Calibration and sequential updating of a coupled hydrologic-hydraulic model using remote sensing-derived water stages. *Hydrology and Earth System Sciences* 13 (3): 367–380.

Moreira, A., Krieger, G., Hajnsek, I., *et al.* (2004) TanDEM-X: a TerraSAR-X add-on satellite for single-pass SAR interferometry. *Igarss 2004: IEEE International Geoscience and Remote Sensing Symposium Proceedings, Vols 1–7: Science for Society: Exploring and Managing a Changing Planet*, Piscataway, NJ: IEEE, pp. 1000–1003.

Morris, S. E., Cobby, D. and Parkes, A. (2007) Towards groundwater flood risk mapping. *Quarterly Journal of Engineering Geology and Hydrogeology* 40: 203–211.

Murphy, J. M., Sexton, D. M. H., Jenkins, G. J., *et al.* (2009) *UK Climate Projections Science Report: Climate Change Projections*, Exeter: Met Office Hadley Centre.

Neal, J. C., Atkinson, P. M. and Hutton, C. W. (2007) Flood inundation model updating using an ensemble Kalman filter and spatially distributed measurements. *Journal of Hydrology* 336 (3–4): 401–415.

Neal, J., Bates, P., Fewtrell, T., *et al.* (2009a) Hydrodynamic modelling of the Carlisle 2005 urban flood event and comparison with validation data. *Journal of Hydrology*.

Neal, J., Schumann, G., Bates, P., *et al.* (2009b) A data assimilation approach to discharge estimation from space. *Hydrological Processes* 23 (25): 3641–3649.

Neal, J. C., Bates, P. D., Fewtrell, T. J., *et al.* (2009c) Distributed whole city water level measurements from the Carlisle 2005 urban flood event and comparison with hydraulic model simulations. *Journal of Hydrology* 368 (1–4): 42–55.

Neal, J. C., Fewtrell, T. J. and Trigg, M. A. (2009d) Parallelisation of storage cell flood models using OpenMP. *Environmental Modelling and Software* 24 (7): 872–877.

Neal, J., Schumann, G., Fewtrell, T., *et al.* (2011) Evaluating a new LISFLOOD-FP formulation with data from the summer 2007 floods in Tewkesbury, UK. *Journal of Flood Risk Management* 4 (2): 88–95.

Neelz, S. and Pender, G. (2010) *Benchmarking of 2D Hydraulic Modelling Packages*, Bristol: Environment Agency.

Nicholls, R. J. and Cazenave, A. (2010) Sea-level rise and its impact on coastal zones. *Science* 328 (5985): 1517–1520.

Oakley, J. and O'Hagan, A. (2002) Bayesian inference for the uncertainty distribution of computer model outputs. *Biometrika* 89 (4): 769–784.

Pappenberger, F. and Beven, K. J. (2006) Ignorance is bliss: or seven reasons not to use uncertainty analysis. *Water Resources Research* 42 (5): n.p.

Pappenberger, F., Beven, K., Horritt, M., *et al.* (2005) Uncertainty in the calibration of effective roughness parameters in HEC-RAS using inundation and downstream level observations. *Journal of Hydrology* 302 (1–4): 46–69.

Pappenberger, F., Harvey, H., Beven, K., *et al.* (2006a) Decision tree for choosing an uncertainty analysis methodology: a wiki experiment. *Hydrological Processes* 20 (17): 3793–3798.

Pappenberger, F., Matgen, P., Beven, K. J., *et al.* (2006b) Influence of uncertain boundary conditions and model structure on flood inundation predictions. *Advances in Water Resources* 29 (10): 1430–1449.

Pappenberger, F., Beven, K., Frodsham, K., *et al.* (2007a) Grasping the unavoidable subjectivity in calibration of flood inundation models: a vulnerability weighted approach. *Journal of Hydrology* 333 (2–4): 275–287.

Pappenberger, F., Frodsham, K., Beven, K., *et al.* (2007b) Fuzzy set approach to calibrating distributed flood inundation models using remote sensing observations. *Hydrology and Earth System Sciences* 11 (2): 739–752.

Pasternack, G. B., Bounrisavong, M. K. and Parikh, K. K. (2008) Backwater control on riffle-pool hydraulics, fish habitat quality, and sediment transport regime in gravel-bed rivers. *Journal of Hydrology* 357 (1–2): 125–139.

Pate-Cornell, M. E. (1996) Uncertainties in risk analysis, six levels of treatment. *Reliability Engineering and System Safety* 54: 95–111.

Pinault, J. L., Amraoui, N. and Golaz, C. (2005) Groundwater-induced flooding in macropore-dominated hydrological system in the context of climate changes. *Water Resources Research* 41 (5): n.p.

Pirazzoli, P. A., Costa, S., Dornbusch, U., *et al.* (2006) Recent evolution of surge-related events and assessment of coastal flooding risk on the eastern coasts of the English Channel. *Ocean Dynamics* 56 (5–6): 498–512.

Pitt, M. (2008) *Learning Lessons from the 2007 Floods: An Independent Review by Sir Michael Pitt. Final Report*, London: Cabinet Office.

Pointet, T., Amraoui, N., Golaz, C., *et al.* (2003) The contribution of groundwaters to the exceptional flood of the Somme River in 2001: observations, assumptions, modelling. *Houille Blanche-Revue Internationale De L Eau* 6: 112–122.

Purvis, M. J., Bates, P. D. and Hayes, C. M. (2008) A probabilistic methodology to estimate future coastal flood risk due to sea level rise. *Coastal Engineering* 55 (12): 1062–1073.

Raclot, D. (2006) Remote sensing of water levels on floodplains: a spatial approach guided by hydraulic functioning. *International Journal of Remote Sensing* 27 (12): 2553–2574.

Robson, A. J. (2002) Evidence for trends in UK flooding. *Philosophical Transactions of the Royal Society of London Series A–Mathematical Physical and Engineering Sciences* **360** (1796): 1327–1343.

Romanowicz, R. and Beven, K. (2003) Estimation of flood inundation probabilities as conditioned on event inundation maps. *Water Resources Research* **39** (3): n.p.

Romanowicz, R., Beven, K. J. and Tawn, J. (1996) Bayesian calibration of flood inundation models. In *Flood Plain Processes*, ed. M. G. Anderson, D. E. Walling and P. D. Bates, Chichester: John Wiley and Sons, pp. 181–196.

Sanders, B. F. (2007) Evaluation of on-line DEMs for flood inundation modeling. *Advances in Water Resources* **30** (8): 1831–1843.

Sanders, R., Shaw, F., MacKay, H., *et al.* (2005) National flood modelling for insurance purposes: using IFSAR for flood risk estimation in Europe. *Hydrology and Earth System Sciences* **9** (4): 449–456.

Sayers, P. B. and Meadowcroft, I. (2005) RASP: a hierarchy of risk-based methods and their application. Paper presented at the Defra Flood and Coastal Management Conference, York, UK.

Schneider, C., Florke, M., Geerling, G., *et al.* (2011) The future of European floodplain under a changing climate. *Journal of Water and Climatic Change* **2** (2–3): 106–122.

Schumann, G., Matgen, P., Hoffmann, L., *et al.* (2007) Deriving distributed roughness values from satellite radar data for flood inundation modelling. *Journal of Hydrology* **344** (1–2): 96–111.

Schumann, G., Matgen, P. and Pappenberger, F. (2008) Conditioning water stages from satellite imagery on uncertain data points. *IEEE Geoscience and Remote Sensing Letters* **5** (4): 810–813.

Schumann, G., Bates, P. D., Horritt, M. S., *et al.* (2009) Progress in integration of remote sensing-derived flood extent and stage data and hydraulic models. *Reviews in Geophysics* **47**: n.p.

Schumann, G. J. P., Neal, J. C., Mason, D. C., *et al.* (2011) The accuracy of sequential aerial photography and SAR data for observing urban flood dynamics, a case study of the UK summer 2007 floods. *Remote Sensing of Environment* **115** (10): 2536–2546.

Serrat-Capdevila, A., Valdes, J. B., Perez, J. G., *et al.* (2007) Modeling climate change impacts and uncertainty on the hydrology of a riparian system: The San Pedro Basin (Arizona/Sonora). *Journal of Hydrology* **347** (1–2): 48–66.

Shields, F. D., Knight, S. S., Testa, S., *et al.* (2003) Use of acoustic Doppler current profilers to describe velocity distributions at the reach scale. *Journal of the American Water Resources Association* **39** (6): 1397–1408.

Sivakumar, B. (2008) Undermining the science or undermining Nature? *Hydrological Processes* **22** (6): 893–897.

Smith, J. C., Bérubé, F. and Bergeron, N. E. (2005) A field application of particle image velocimetry (PIV) for the measurement of surface flow velocities in aquatic habitat studies. In *Proceedings of the 26th Canadian Symposium on Remote Sensing*, Vancouver: Canadian Remote Sensing Society.

Smith, L. C. (1997) Satellite remote sensing of river inundation area, stage, and discharge: a review. *Hydrological Processes* **11** (10): 1427–1439.

Smith, P. J., Hughes, D., Beven, K. J., *et al.* (2009) Towards the provision of site specific flood warnings using wireless sensor networks. *Meteorological Applications* **16** (1): 57–64.

Tayefi, V., Lane, S. N., Hardy, R. J., *et al.* (2007) A comparison of one- and two-dimensional approaches to modelling flood inundation over complex upland floodplains. *Hydrological Processes* **21** (23): 3190–3202.

UNESCO (2007) *International Flood Initiative*. Paris: International Centre for Water Hazard and Risk Management (ICHARM).

USWRC (1983) *Economic and Environmental Principles and Guidelines for Water Related Land Resources Implementation Studies*, Washington, DC: US Water Resources Council.

Veijalainen, N., Lotsari, E., Alho, P., *et al.* (2010) National scale assessment of climate change impacts on flooding in Finland. *Journal of Hydrology* **391** (3–4): 333–350.

Villanueva, I. and Wright, N. G. (2006) Linking Riemann and storage cell models for flood prediction: Proceedings of the Institution of Civil Engineers. *Journal of Water Management* **159** (1): 27–33.

Vivoni, E. R., Entekhabi, D. and Hoffman, R. N. (2007) Error propagation of radar rainfall nowcasting fields through a fully distributed flood forecasting model. *Journal of Applied Meteorology and Climatology* **46** (6): 932–940.

Werner, M. (2004) A comparison of flood extent modelling approaches through constraining uncertainties on gauge data. *Hydrology and Earth System Sciences* **8** (6): 1141–1152.

Werner, M., Blazkova, S. and Petr, J. (2005a) Spatially distributed observations in constraining inundation modelling uncertainties. *Hydrological Processes* **19** (16): 3081–3096.

Werner, M. G. F., Hunter, N. M. and Bates, P. D. (2005b) Identifiability of distributed floodplain roughness values in flood extent estimation. *Journal of Hydrology* **314** (1–4): 139–157.

Wilby, R. L. and Harris, I. (2006) A framework for assessing uncertainties in climate change impacts: low-flow scenarios for the River Thames, UK. *Water Resources Research* **42** (2): n.p.

Wilby, R. L. and Wigley, T. M. L. (1997) Downscaling general circulation model output: a review of methods and limitations. *Progress in Physical Geography* **21** (4): 530–548.

Wilby, R. L., Hay, L. E. and Leavesley, G. H. (1999) A comparison of downscaled and raw GCM output: implications for climate change scenarios in the San Juan River basin, Colorado. *Journal of Hydrology* **225** (1–2): 67–91.

Wilson, M. D. and Atkinson, P. M. (2007) The use of remotely sensed land cover to derive floodplain friction coefficients for flood inundation modelling. *Hydrological Processes* **21** (26): 3576–3586.

Wood, A. W., Leung, L. R., Sridhar, V., *et al.* (2004) Hydrologic implications of dynamical and statistical approaches to downscaling climate model outputs. *Climatic Change* **62** (1–3): 189–216.

Young, P. C. and Ratto, M. (2011) Statistical emulation of large linear dynamic models. *Technometrics* **53** (1): 29–43.

Yu, D. and Lane, S. N. (2006a) Urban fluvial flood modelling using a two-dimensional diffusion-wave treatment, part 1: mesh resolution effects. *Hydrological Processes* **20** (7): 1541–1565.

Yu, D. and Lane, S. N. (2006b) Urban fluvial flood modelling using a two-dimensional diffusion-wave treatment, part 2: development of a sub-grid-scale treatment. *Hydrological Processes* **20** (7): 1567–1583.

Zerger, A., Smith, D. I., Hunter, G. J., *et al.* (2002) Riding the storm: a comparison of uncertainty modelling techniques for storm surge risk management. *Applied Geography* **22** (3): 307–330.

Zorita, E. and von Storch, H. (1999) The analog method as a simple statistical downscaling technique: comparison with more complicated methods. *Journal of Climate* **12** (8): 2474–2489.

8

Scientific uncertainties: a perspective from probabilistic seismic hazard assessments for low-seismicity areas

W. P. ASPINALL

8.1 Introduction

Earthquakes have claimed several million human lives over the last two millennia (Dunbar *et al.*, 1992), as well as unquantifiable numbers of casualties and limitless grief. A large, and increasing, proportion of the world's population nowadays resides in poorly constructed dwellings that are high susceptibility to seismic damage; building collapses account for more than 75% of human casualties in earthquakes (Coburn and Spence, 2002). On top of this, increasing urbanisation means that half of the world's population now lives in cities (United Nations, 2002), and a significant number of these cities are located in earthquake-prone regions (Bilham, 2004). In recent times, the average global fatality count from earthquakes has been about 100 000 per year, a rate that is determined mainly by the larger but less frequent disasters. Given the inexorable growth in the global population, an increasing worldwide earthquake fatality rate might be expected, but recent urban expansion has seen a decline in the fatality rate when expressed as a percentage of the instantaneous population (Bilham, 2004). There are two possible explanations for this. One is that the application of earthquake codes and better construction techniques – coupled to improved hazard assessments – are engendering real beneficial outcomes; while this may be true in some countries, it is more questionable for many cities in the developing world (Ambraseys and Bilham, 2011), especially mega-cities. An alternative reading is that the apparent decline in risk has arisen simply from a short-term fluctuation in large-magnitude earthquake activity. Major earthquakes recur, on average, at intervals of the order of decades to centuries, whereas the world population has dramatically increased only in the last few decades. Just one or two extreme earthquake disasters in the near future – affecting some of the world's mega-cities – could easily reverse the current downturn in apparent risk. Several of these mega-cities are close to active tectonic plate boundaries; others may be further away, but these too are not completely immune to threats from major intraplate events, albeit of much rarer occurrence.

Before the twentieth century, earthquake information came not from instruments, but from documented accounts of the severity of damage and felt effects in regions surrounding an earthquake epicentre, with local accounts serving to distinguish different grades of earthquake intensity. Such historical information is still an essential ingredient for assessing

Risk and Uncertainty Assessment for Natural Hazards, ed. Jonathan Rougier, Steve Sparks and Lisa Hill. Published by Cambridge University Press. © Cambridge University Press 2013.

future activity levels, and the treatment, interpretation and statistical characterisation of such data – for hazard assessment purposes – remains a major challenge. The majority of the sources of these damaging earthquakes lie in wait, concealed within or below the Earth's crust. This said, measuring the nature and size of seismogenic faults from surface features has long been one important element of geological science's attempts to assess potential earthquake effects, especially in California, where conditions encountered on the San Andreas Fault system – if not with the blind thrusts – suit this approach.

However, California is a unique place and elsewhere the Earth generates earthquakes in many varied ways – this much has been learned from instrumental detection of earthquakes around the world following the construction of the first seismometers at the end of the nineteenth century. Several decades elapsed after these first seismological insights were gained before Charles Richter devised a practical earthquake magnitude scale in the 1930s – intended for communicating earthquake 'size' to the press in California – that gave a quantitative reference frame with which to rank different events for scientific and engineering contexts. However, surprises, whether in terms of magnitude, location or wave-forms of an earthquake, or unanticipated levels of shaking from a hidden thrust, for example, still occur surprisingly often, even in California. These literal and metaphorical shocks have a tendency to undermine hazard maps that are based solely on the local historical record of earthquake occurrence or inferences limited to knowledge of surface geology, and thus pose a significant challenge for anyone seeking to assess possible future earthquake actions, for whatever purpose.

This chapter concerns itself with a perspective on scientific uncertainties and their role in seismic hazard assessment from experience in low seismicity areas; it does not venture very far into issues specific to regions with frequent earthquakes or into topics relating to seismic risks, which is a substantial and highly technical topic in its own right. It may appear paradoxical at first sight, but substantial advances in seismic hazard assessment methods, and the use of probabilistic concepts in this context in particular, have been made in countries where seismicity is minimal or just moderate. The reason for this is that where data are very sparse and of questionable reliability, and major safety-critical facilities are a concern, it is necessary to ensure that related statistical analyses are rigorous and defensible, and that results and implications are robust against almost any future event, however seemingly implausible at the time an assessment is undertaken.

Thus, the present discussion starts by outlining the various seismic hazards that may need to be assessed in Section 8.2, then identifies some basic seismological concepts and associated uncertainty issues in relation to seismic hazard, in Section 8.3. The main part of the discussion, concerned with seismic hazard assessment, is found in Section 8.4, where an overview is given on the way uncertainties are assessed, treated and evaluated, and implications of these uncertainties are surveyed. This is followed by Section 8.5, in which certain methodological tenets for quantitative hazard assessment in the face of uncertainty are briefly aired, based on experience with seismic hazard estimation; and then onto Section 8.6, where the topic is communication of earthquake hazards and risks. Extending the conversation, because it represents a topic complementary to hazard assessment,

Section 8.7 touches on the tricky – but still related – subject of earthquake prediction and the practicality of earthquake warning systems. Uncertainty-related topics in hazard assessment that merit further research are highlighted in Section 8.8, while Section 8.9 – in which some very brief concluding remarks are offered – closes this chapter. The content here should, however, be read in association with several other chapters in the present volume, which develop many connected themes in more detail.

8.2 Seismic hazards

The main direct hazards due to earthquakes are seismic shaking and ground rupture. If not too extreme, the former can be mitigated to some extent by simple and inexpensive engineering measures; the latter is much less easy to cope with. Ground rupture is permanent ground displacement or offset localised to the fault trace, which can have critical implications for life-lines, for instance. In some cases, ground rupture can represent neotectonic displacements in areas with little or no geological or morphological evidence for earthquake precedents (e.g. the 1988 Tennant Creek, Australia sequence – Choy and Bowman, 1990). In addition, surface fault rupture may produce extreme near-fault effects, such as 'fling', a special, and violent, case of ground movement that is not truly vibratory in the usual sense and therefore has to be modelled differently. On the other hand, the forceful disturbances of ground shaking due to strong earthquake wave motions that can cause structural damage and general societal mayhem easily extend over hundreds of square kilometres in the biggest earthquakes.

The same probabilistic methods that are used for estimating seismic ground shaking hazard, discussed below, can be deployed for calculating ground rupture hazard likelihoods, although, as a general rule, the data, relationships and associated uncertainties are all much less well-constrained than for the shaking problem. In addition to the direct threats from earthquakes, there are a number of possible secondary seismic hazards: liquefaction, groundwater seismo-hydraulic effects, as well as landslides (Chapter 9), tsunami (Chapter 10) and fires. The likelihoods of these can all be assessed probabilistically, conditional on earthquake occurrence, magnitude, location, duration, and so on, but these hazards are not embraced here.

A newly recognised secondary hazard is large-scale co-seismic ground uplift or displacement, such as occurred in the magnitude 6.6 2007 Niigataken-Chūetsu-Oki earthquake, that affected the Kashiwazaki-Kariwa nuclear power plant in Japan and resulted in up to 24 cm of permanent uplift. Even more extreme were the permanent displacements – both horizontal and vertical – experienced along the north-east coast of Japan as a result of the devastating magnitude 9 Tohoku-Oki earthquake of 11 March 2011. Thus far, no approach has been developed for quantifying these special co-seismic hazards in full probabilistic terms.

8.3 Seismological uncertainty issues

Certain seismological issues and their associated uncertainties impinge significantly on any attempt to evaluate the level of seismic hazard at a given place and within a given timeframe.

8.3.1 Magnitudes

At the heart of modern seismic hazard assessment is the long-established but often inexact characterisation of earthquake 'size' in terms of magnitude. Integrating over the kinetic energy density of seismic waves produced by an earthquake, it is found that earthquake magnitude is proportional to the logarithm of the seismic energy released at the source. From Richter's simple concept of local magnitude (Richter, 1958), different forms of magnitude scale have evolved, using alternative choices of earthquake phase on a seismogram to make measurements: other than local Richter magnitude (M_L), commonly used variants are body wave magnitude (m_b), surface wave magnitude (M_S) and, increasingly, moment magnitude (M_w). The latter, in its proper form, relies on measurements of the low-frequency spectral radiation from the source, but there is a questionable tendency among many seismologists simply to take other magnitude values (especially older estimates, registered on other scales) and convert these to M_w by crude numerical substitution.

Transitioning from M_L or M_S to M_w as the fundamental measure of earthquake size constitutes a recent and major change in methodology for seismic hazard assessment in many countries. The rationale for M_w is clear in the United States: in many parts of that country, earthquakes can occur that are large enough to saturate the M_L scale. Furthermore, while historical data are sparse in the United States, neotectonic evidence is, relatively speaking, quite abundant. Such data may allow the dimensions of past fault movements to be assessed, which can then be expressed in terms of M_w. Given this rationale for seismic source characterisation, the practice has sensibly developed for US ground-motion attenuation relations to be based on M_w. However, for many other countries the situation is reversed: seismotectonic evidence is lacking, and historical data are more abundant. The latter are often recorded as felt-intensity information from which equivalent or proxy instrumental magnitudes (commonly M_L or M_S values) can be inferred. Switching to the M_w scale for hazard assessment therefore requires that M_w assignments be made for all -significant events in a historical earthquake catalogue and, because such magnitude conversions are replete with uncertainty, this entails further analysis to gauge the sensitivity of hazard results to such conversions.

For instance, for a chosen M_L to M_w correspondence, the magnitude correlations with different felt-intensity areas need to be re-established from primary data. This can be done by plugging in M_w values for the specific earthquake event set upon which the M_L–macroseismic data link was originally derived. But, there is a significant degree of dispersion displayed in the M_L–M_w conversion: for a given M_L value, the corresponding M_w value may vary by ±0.5 units, which represents a major source of potential uncertainty. Accordingly, it is to be expected that there will be a distinct difference in statistical M_L and M_w felt area correlations, due to this variability. In low to moderate seismicity areas in particular, the reassessment into M_w terms of a few individual hazard-controlling event magnitudes can lead to a substantial change in computed hazard levels.

In short, inter-scale magnitude conversion is a fraught exercise, and handling a host of different magnitude scales is a challenge for coherent data fusion and for uncertainty estimation.

8.3.2 Magnitude–frequency relations

With accumulating magnitude data, furnished by size estimates on individual earthquakes, the Gutenberg and Richter (1954) magnitude–frequency relation is universally (and sometimes inappropriately) applied in seismology and in seismic hazard estimation although, in essence, it remains just one phenomenological representation of a large body of data. The relation takes the form:

$$\log(N_M) = a - b*(M - M_{MIN}),$$

where N is the cumulative number of earthquakes having magnitude equal to or greater than M, M_{MIN} is a minimum threshold magnitude, defined for the given dataset, and a and b are constants that depend upon those data, representing the annual activity rate of events of M_{MIN} or greater and the relative number of large magnitude events to small, respectively. While it should be no surprise that mixing magnitude values, based on different scales in a dataset or catalogue, will compromise estimation of Gutenberg–Richter relationship parameters, this happens a lot in practice. The situation is frequently exacerbated, too, by ignoring or neglecting uncertainties on individual magnitude estimates in the dataset, which themselves can be substantial.

The physical basis for such linear power-law scaling relations in complex dynamical systems is not well understood (see e.g. Barabási, 2009), and a convincing explanation for the behaviour of earthquakes and geological faulting in this regard has proved elusive. Dynamical models of faults and tectonic blocks can manifest some of the principal characteristics of earthquakes arising from crustal deformation driven by tectonic plate motions, but local, small-scale geographical and temporal details are more difficult to capture. At this scale, transfer of stresses can cause a system of faults to become dynamically unstable, and trigger a chain reaction of earthquake failures. Models of such processes can be tuned to replicate the general logarithmic relation between magnitude and the energy released in a real earthquake, and the Gutenberg–Richter relation for magnitude and frequency. But such standard data transformations can hide deviations from uniform scaling at the local fault system scale because, as with many things in seismology and the wider Earth sciences, such transformations are often simplifying constructs that originated when computational limitations necessitated expression in a tractable form. All too often, important and often unstated assumptions are imposed a priori, but subsequent results are not so qualified or communicated in a clear and explicit way.

Past seismological experience is the main basis for assessing future seismic hazards, with that knowledge often extended by combining it with statistical models that impose regularity in space and time. These statistical models undoubtedly have their limitations. For example, the assumption that earthquake arrivals reflect a random Poisson-type process may be an adequate representation of average global or regional activity over the long run, but deviations exist locally or over shorter time intervals (e.g. the conventional 50 years adopted for design, or less for human and political outlooks). At these shorter timescales and within more restricted geographical areas, earthquake occurrences may be non-stationary and depart from uniform randomness; evidence for this is starting to accumulate from careful, targeted studies in certain localities. The net effect is that related uncertainties may be inappropriately specified for hazard assessment.

New work (Clauset *et al.*, 2009) on the recognition and characterisation of power laws indicates the problem is complicated by fluctuations that can occur in the tail of a data distribution – i.e. the part of the distribution representing large but rare events – and by the difficulty of identifying the range over which power-law behaviour holds. Commonly used methods for analysing power-law data, such as least-squares fitting, can produce substantially inaccurate estimates of parameters for power-law distributions. Even in cases where such methods return accurate answers, they may still be unsatisfactory because there is no confirmation of whether the data obey a power law at all. Clauset *et al.* provide analytical comparisons of several established approaches using real-world datasets from a range of different disciplines, each of which has been conjectured to follow a power-law distribution. In the case of the Gutenberg–Richter relationship for Californian earthquakes they find the power-law assumption is not substantiated, at least for the dataset they chose.

If this is confirmed elsewhere, the implications for seismic hazard assessment will be profound, especially now that the most extreme earthquakes are an important hazard assessment consideration globally, following the great 2011 Tōhoku-Oki earthquake.

8.3.3 Hypocentral location

In addition to uncertainty about how temporal traits of earthquake occurrence should best be characterised, there is also significant uncertainty about precisely where individual earthquakes have happened. This is understandable for historic earthquakes, for which the only indication of an event's location is information on where it was felt and how strongly, but even with good-quality modern seismometry, important uncertainties persist in hypocentre location (e.g. Billings *et al.*, 1994).

These limitations can impinge significantly on ambitions to associate individual earthquakes with mapped faults (see Section 8.4.5.6), and correct quantification of numerical location uncertainties is not always assured in many seismic hazard assessments. The main reason for this is that earthquake location algorithms work by fitting observers' readings of wave arrival times at different stations to an assumed velocity structure in the Earth (usually one-dimensional, in depth only). Almost all location programs provide estimates of solution numerical misfit *to the chosen model*; they do not provide any way of judging the locational accuracy of the individual earthquake in real-world geographical terms – a different velocity model can produce a location many kilometres away. Indeed, 'errors' reported by earthquake location programs are often very small, especially with sparse observations, and often cite positions to better than ±1 km in latitude and longitude. (It may be pointed out that the source dimensions of large earthquakes are measured in tens or hundreds of kilometres – an instrumental 'hypocentre' is an approximation to the point where the rupture is believed to commence.) Matters are even more unsure when arrival times are routinely determined by automatic phase-picking algorithms.

The location picture is even less satisfactory when it comes to hypocentral depth, a potentially influential factor when it comes to seismic shaking effects and hence hazard assessment. Poor control on depth is due to two causes: frequently there is no seismograph station directly above the hypocentre to constrain depth from travel-time, and the location algorithm cannot easily prevent a trade-off occurring between the origin time of the event (a variable in its own right) and the depth value, during the minimisation process. (Depth control can be ameliorated by using timings for certain near-source surface-reflected waves called pP, but these are not always present in a seismogram and are not read routinely in ordinary seismological data processing).

With regard to gauging earthquake location accuracy, some efforts have been made to introduce Bayesian thinking into the hypocentre location problem (e.g. Matsu'ura, 1984; Engell-Sorensen, 1991a, 1991b; Fagan *et al.*, 2009). However, this type of approach is yet to make serious inroads into mainstream seismology – it is a topic that clearly merits more research for seismic hazard assessment purposes and, in particular, for testing associations of individual earthquakes with mapped faults.

As noted, earthquake depth can be a significant influence on the level of ground motions at the surface – for instance, events with deep origins in subduction zones can produce quite different – lower – intensities of shaking compared to upper crustal earthquakes of the same magnitude in the same area. However, major efforts in developing ground motion attenuation relations have been rooted in California, where deep earthquakes do not exist, so focal depth is not included as an explanatory variable in many regressions from that region (see Section 8.4.5.4). Another reason for this omission is that the hypocentre denotes the point of start of rupture, but the peak level of shaking at a particular point on the surface can depend more upon how close the propagating rupture comes to that place; thus, attenuation relations have adopted a variety of fault-to-receiver distance definitions to represent this effect, but specific depth of focus has not been incorporated directly into most relations and hence its influence on uncertainty in ground motion assessment is obscured. Furthermore, cross-comparison of hazard results derived using different distance definitions can be a thorny issue.

8.3.4 The ergodic assumption

Another statistical area of concern in seismic hazard assessment is the so-called ergodic assumption (Anderson and Brune, 1999), which currently underpins the estimation of magnitude–distance attenuation. An attenuation relationship is (generally, but not exclusively) an empirical regression representing the strength and variability of ground motions (e.g. peak and frequency-dependent spectral accelerations, velocities and displacements) at a particular site as a consequence of a variety of spatially distributed earthquake sources. For this, the conventional assumption is that sample measures of ground motion at multiple sites in multiple real earthquakes can be used to characterise what can happens in future earthquakes at a single site of interest – reassignment from the several to the specific being the essence of the 'ergodic' assumption.

Thus, the statistical distribution of observations of a particular ground-motion parameter – ensuing from earthquakes over time in a particular locality – is presumed to be the same as the statistical distribution of that same parameter from earthquakes recorded at different sites in different areas, and at different times. Although this presumption is almost certainly invalid, it is difficult to avoid relying on it when developing an attenuation relation for low and moderate seismicity areas, and even more so for several very active areas in the developing world. However, attempting to remove its effect may risk misplacing some genuine epistemic uncertainty in the process – see, e.g. Atkinson (2006), Strasser *et al.* (2009) and Chapter 3 in this volume.

8.4 Seismic hazard assessment uncertainties: general remarks

The basics of seismic hazard assessment were set out by Reiter (1991), from the perspective of a seismologist with many years experience in reviewing, sponsoring and carrying out studies in connection with nuclear power plant siting, construction, licensing and operation within regulatory and legal frameworks. The nuclear power world has motivated many of the advances in seismic hazard assessment made in recent decades, with a focus mainly on site-specific hazard analysis – one of two ways of viewing and characterising earthquake hazard patterns spatially.

8.4.1 Background

The genealogy of seismic hazard assessment has been traced (Muir-Wood, 1993) through five generations of methodology: (1) historical and geological determinism; (2) historical probabilism; (3) seismotectonic probabilism; (4) non-Poissonian probabilism; (5) and earthquake prediction.[1] Deterministic approaches, although favoured in many countries for defining structural design requirements until very recently, present special challenges when it comes to the treatment of uncertainty. This particular problem is not considered here in detail, but Chapter 13 includes discussion of the 2011 Tōhoku-Oki earthquake and tsunami disaster in Japan in the context of the legacy there of deterministic design of technological facilities against natural hazards.

While non-Poissonian probabilism is gaining ground quite rapidly (e.g. Beauval *et al.*, 2006a), the most common of the methodologies in terms of currency, motivated especially by the need to mitigate seismic hazards at safety-critical facilities, is seismotectonic probabilism. That is the variant that will form the basis of the present discussion concerning uncertainties in seismic hazard assessment; *earthquake prediction* gets a mention later, as a rider.

---

[1] The inclusion of earthquake prediction in this taxonomy is debatable: even if an earthquake could be predicted perfectly, say, one month ahead, this would be little use to engineers who have to design a building to last for 50 years. In that regard, seismic hazard assessment will not evolve into and be replaced by prediction. For this point, I am grateful to RMW Musson, pers. comm.

As noted above, there are two different ways of representing seismic hazards. The first, more generic approach uses mapping to delineate areas with similar levels of exposure. The exposures charted in hazard maps are mostly depicted in terms of the probability of exceedance of a ground-motion parameter (e.g. peak acceleration) within some defined time interval, usually equivalent to return periods of hundreds of years, at most, and such maps are the basis for many design codes for 'conventional' buildings and structures (e.g. Eurocode8[2]). There have been major recent initiatives in global seismic hazard mapping (e.g. Global Seismic Hazard Assessment Program, GSHAP, and now the Global Earthquake Model GEM[3]), but thus far these have not been completely successful or convincing, throwing up apparent anomalies at local and sub-regional scales, where engineers and designers need specific information.

For instance, there is ongoing massive and fabulous property development on the coast of the Arabian Gulf in the United Arab Emirates (UAE), in an area where, historically, earthquakes have been exceedingly rare. This did not preclude GSHAP – in order to leave no blank areas on their global map – from publishing a regional seismic hazard map which lumped the UAE in with the very active belt of Zagros seismicity in southern Iran: 'The hazard was mapped by simulating the attenuated effect of the seismic activity in the Dead Sea fault area (Near East) and in the Zagros province of Iran' (Grünthal et al., 1999). In other words, the GSHAP hazard level for the UAE was attributed uncritically from those in distant surrounding regions, without taking local seismicity or seismological history into account. While this is an extreme example of the global hazard mapping approach failing at the local scale, others can be found elsewhere in the world, where local seismotectonic understanding has not been utilised in modelling seismic hazard and risk (e.g. Smolka et al., 2000; Musson, 2012a).

The second approach is site-specific. This seeks to quantify exceedance probabilities at a single place on the Earth's surface and, quite often, these involve the determination of design parameters with very low probabilities of exceedance for regulatory requirements: annual probabilities of exceedance of 10^{-4}, or even lower, are typical for safety-critical installations and strategically important national assets. In the UK, for instance, society, through regulators and the political directorate, insist that such hazard estimates are derived in such a way as to 'achieve assured safety' (e.g. Bye et al., 1993; Coatsworth, 2005). This edict obliges a distinctive philosophical, rather than scientific, approach be taken with respect to the consideration and treatment of uncertainty in a safety-critical site-specific hazard assessment.

8.4.2 Generic uncertainty

Uncertainty is present as a ubiquitous accompaniment to all the stages of the process by which any hazard assessment is carried out. This situation arises because: (1) all mechanistic representations (i.e. models) of the process, or any part of the process, are, inevitably,

---

[2] http://www.eurocodes.co.uk/EurocodeDetail.aspx?Eurocode=8 (accessed 30 December 2011).
[3] http://www.globalquakemodel.org/ (accessed 30 December 2011).

simplifications of only tentative validity; and (2) there is always some degree of uncertainty associated with parameter values used in such representations.

In facing up to this problem, it can be helpful to consider overall uncertainty as being made up of two components – see Chapters 2 and 3, and Woo (2011: ch. 4):

(1) 'aleatory' uncertainty (i.e. corresponding in simple terms to true randomness); and

(2) 'epistemic' uncertainty (i.e. corresponding in essence to deficiency in understanding).

In studying most natural physical processes, some element of stochastic randomness is encountered. With earthquakes, this scatter, which is due to intrinsic variability in the rupture process itself and in the subsequent wave propagation to a site, is commonly termed 'aleatoric' uncertainty.

In practice, when observations are fitted to a model, the total scatter that is present (which at any given instant has to be treated as though it were all due to randomness) also includes some uncertainty that results from imperfect understanding through the use of inadequate (e.g. oversimplified) models, or from inadequate or uncertain data. This informational or knowledge-based component of the overall uncertainty, which can be reduced by improvements in modelling techniques or data quality, is usually termed 'epistemic' uncertainty (see Chapter 3).

While the use of these two concepts can be very helpful in planning analysis strategies and developing improved methodologies, it would be delusionary to infer that, in any given situation (i.e. at any given state of understanding), the overall observed scatter can be split readily and unequivocally into these two elements, one of which is fundamental and irreducible, and the other ephemeral and easily removed. For example, while the acquisition of more and more data may provide the possibility of improved understanding, it is also liable to increase the extremes of scatter (as exemplified by the Christchurch, New Zealand earthquake of 26 February 2011, where extreme single-component ground shaking of 2.2 g acceleration was recorded in a relatively modest magnitude 6.2 M_w event – Holden, 2011).

In terms of present-day practice, the removal of at least some of the existing epistemic uncertainty is most desirable in the derivation of strong motion attenuation relations, as this has a big influence on hazard level estimation (see Section 8.5).

8.4.3 *Avoiding unnecessary uncertainty*

Given the fact that there are so many potential sources of unavoidable uncertainty that can be encountered in carrying out a seismic hazard assessment, it seems self-evident that every effort needs to be made not to introduce additional sources of uncertainty and to be diligent in characterising uncertainties as thoroughly as possible.

The best way of avoiding such problems is to work, wherever possible, with directly relevant local data: each time a dataset, or a parametric relationship, or even understanding, is imported from elsewhere, additional uncertainty is potentially introduced. This risk is recognised in the principle behind the wording by the IAEA (2010: para. 4.6), which says, in the particular context of constructing a seismotectonic model: 'The construction of the model should be primarily data driven, and the data should not be interpreted in a manner that supports an individual's preconception.'

The problem in a moderate or low seismicity environment, like that of the UK, where there is only limited information, is that there is often no alternative to importing data or relationships or understanding. With very few strong motion records, for example, how else is it possible to derive useable attenuation relationships or characteristic spectral shapes? Therefore, it is crucial to minimise uncertainties that are introduced by the process of importation, and every effort has to be made to particularise and select the imported material so that it merges as directly and coherently as possible with the local data that are available. In this context, Bayesian updating suggests itself as a viable and rational approach.

Ideally, imported material (whether data, relationships or understanding) should provide a surrogate which can reliably be used, until it can be replaced by local information. Deciding what constitutes 'reliably' in this regard is non-trivial, however.

This objective imposes a number of constraints on the selection process, namely the need to:

- import only data that closely match local expectations;
- import only parametric relationships that have ranges of validity which are demonstrably appropriate to local conditions;
- select and import only parametric relationships that are couched in terms which allow them to be used directly, without any need for further manipulation (such as magnitude scale conversion, etc.);
- select and import understandings that are relevant to local conditions.

8.4.4 Addressing uncertainty

8.4.4.1 Conservatism

In cases where safety-critical issues are involved, the appropriate response to irreducible uncertainty is to employ a commensurate degree of conservatism – although invoking conservatism should not be presumed an adequate substitute for a coherent analysis of uncertainty. Inevitably, this means the exercise of expert judgement in formulating the hazard modelling basis and taking into account the safety demands of the intended application. For critical industrial installations, such as nuclear power plants, it is necessary to ensure that judgements are made soundly and fairly, but in a manner which cannot be reasonably interpreted as unconservative.

The use of a logic-tree formulation (Kulkarni et al., 1984), enumerating probability distributions for seismological parameters rather than single best-estimate values, provides a platform for expressions of belief in widely dispersed values. Freedom to choose a weighted set of values permits a spectrum of alternative viewpoints to be represented in a balanced manner. This does much to diminish the natural tendency towards over-conservatism which is common in those seismic hazard studies where only a single value can be attributed to an uncertain parameter. Where even the quantification of parameter variability is open to substantial doubt, and thereby significant uncertainty, a gesture to conservatism in the parameter distributions may be warranted. This may entail ensuring some skewness towards high-hazard parameter values.

8.4.4.2 Expert judgement

Where a decision has to be made on the basis of uncertain evidence, the only practical recourse is to make use of expert judgement (see also Chapter 4). Recognising this to be the case, the use of expert judgement in the practice of seismic hazard assessment is, nowadays, an acknowledged feature of such studies (see, for example, IAEA, 2010: para 2.9).

In the context of probabilistic risk assessment, it is important to distinguish expert judgement from engineering judgement as the differences between them can be sufficient to affect system safety in tangible way (Skipp and Woo, 1993). Expert judgements, properly elicited, are not arbitrary, but should satisfy various fundamental principles: in particular, they should be coherent with the axioms of probability theory which provide systematic rules for the synthesis of subjective judgements.

By contrast, engineering judgement, in standard civil engineering practice, for instance, is much less tangible and less well-defined, but rather is used in a holistic fashion. Thus, engineering judgement may be captured in colloquial statements of the form 'To do it that way will get you into difficulties', or 'If you do it this way things will be alright.'

The recursive use of engineering judgement in estimating appropriate factors of safety defies quantification, except in the binary sense that inadequacy is ultimately exposed by failure. The main characteristics of engineering judgement are that it:

- is not constrained by formal logic;
- cannot be easily calibrated;
- may incorporate conservatism;
- tends to be heuristic and holistic.

Differences of opinion over safety issues will always be present, but terminological confusions should be avoidable. Where an 'engineering judgement' is unwittingly offered in place of an 'expert judgement', or vice versa, problems can be expected. As Bayesian methods become more widely accepted as a method of weighing evidence (in litigation, for example) there will be an increasing need to clarify the nature of the judgement being offered or used.

In the case of expert judgement in seismic hazard assessments, more is required than some general awareness of the topic: judgements need to be made by experts with extensive experience of site-specific hazard assessments in the relevant region. A rigorous mathematical check on the consistency of expert judgements with the fundamental laws of probability, in terms of the axioms expounded by Kolmogorov (1933), can be afforded by comparative hazard assessments for a range of different sites across a region.

In practice, it often seems that the ramifications of the use of expert judgement are not fully appreciated. Two pertinent comments can be made in this regard:

(1) use of expert judgement means far more than simply adopting *ipso facto* a particular point of view which happens to have been expressed by an 'expert'; and
(2) even where formalised procedures are used for making certain decisions within a hazard assessment framework, other, similarly significant, decisions are often afforded no special attention at all.

These points are both of considerable importance and, in safety cases, the way in which expert judgement is utilised on issues involving uncertainty is a matter which has long been of considerable concern to regulators. Nowadays, such authorities expect not only the process by which expert judgement is elicited and different opinions resolved to be satisfactory, they also require appropriate attention to be paid to the reporting of that process in terms of an auditable 'trail', with uncertainties and their treatments accounted for, in particular.

In the absence of a universally accepted theory of earthquake recurrence and ground-motion attenuation, the existing observational database is insufficient, in both quality and quantity, to differentiate among a range of alternative models which might all be credited as being plausible. These options can engender a legitimate broad divergence of opinion and, as a consequence, the ensuing lack of scientific consensus has had repercussions on regulatory decision-making in relation to seismic hazard modelling (Kelly *et al.*, 1991).

This has motivated the search for formal methods of eliciting expert judgement, and aggregating individual opinions, with the issue being highlighted in a 1991 lecture by the late eminent engineering risk analyst C. Allin Cornell. He characterised it as potentially (then) the most important outstanding problem in seismic hazard assessment – and it can be argued that this has not changed in the intervening 20 years, despite some attempts to develop more structured approaches to expert elicitation (a topic of current significance in its own right, discussed in Chapter 4).

The role of expert opinion is central to the formalism that should underlie any hazard assessment – uncertainty cannot be adequately represented, consistency cannot be maintained and any conclusions may be vitiated by a casual and informal attitude towards the involvement of experts and the elicitation of their judgements.

8.4.5 Significant uncertainties in seismic hazard modelling

Even for those aspects of seismic hazard assessments where there is general recognition that expert judgement has to be called upon, the priorities are sometimes misunderstood. This section summarises some of the issues that are recognised as important in making such judgements.

8.4.5.1 Magnitude completeness thresholds

The magnitude completeness thresholds assigned to an earthquake catalogue are supposed to define, through combinations of magnitudes and dates, the portions of that catalogue which form complete datasets. Without such completeness thresholds, reliable statistical characterisation of seismicity is compromised (for the Gutenberg–Richter relationship, both activity rate and *b*-value parameters depend on them). Also, the construction of zonal seismic source models needs to be founded on the locations of only those earthquakes that belong to complete and homogeneous datasets.

The main principle when assigning magnitude completeness thresholds to an earthquake catalogue – whether across a whole map for the purposes of constructing a zonation or to a single zone – is that those thresholds should reflect conditions in that part of the area of interest where the circumstances have been *least* propitious for recording earthquakes. Thus, some data from certain areas, recorded in a historical (e.g. Musson, 2004) or instrumental catalogue, may have to be disregarded as part of the analysis in order to ensure statistical homogeneity across the whole piece (Musson, 2012a); in future, the application of Bayesian techniques may allow that information to be utilised.

A related issue is what value to use for the 'minimum magnitude' in a probabilistic hazard assessment model – noting this will be unlikely to correspond directly with any reporting completeness threshold. While minimum magnitude is a factor that strongly influences hazards results, it is not in practice a real physical parameter but a bounding construct introduced into a hazard model for various reasons – the main ones being tractability in terms of data handling and analysis, and the argument that earthquakes below a certain magnitude (usually $5\,M_w$) do not cause damage in properly engineered structures.[4] The latter reasoning is, of course, more risk-informed than physics-based and thus assessment results are strongly conditioned on the chosen cut-off. The upshot of this is that results from one assessment cannot be compared directly to results from another analysis of the same situation when the latter uses a different minimum magnitude setting – a state of affairs seemingly not grasped by many engineers or seismologists!

The effect of asserting different minimum magnitude thresholds in an assessment for a low-seismicity setting can be illustrated by reducing it in steps from, say, $4.0\,M_S$ down to $2.5\,M_S$ (at which point the very concept of surface wave magnitude becomes meaningless). For the average UK seismicity model (activity rate at $4.0\,M_S = 0.0055$ per $10\,000\ \text{km}^2$ per year, *b*-value $= 1.28$, etc.), the expected annual exceedance probabilities for a *pga* of 5 %g as a function of minimum magnitude only are:

Minimum magnitude	Annual probability of exceedance of 5 %g *pga*
$4.0\,M_S$	0.20×10^{-2}
$3.5\,M_S$	0.49×10^{-2}
$3.0\,M_S$	0.12×10^{-1}
$2.5\,M_S$	0.27×10^{-1}

[4] This is one perspective; a wider, total-system safety viewpoint is that other parts of a complex plant – such as relays – may be susceptible to short-duration high-frequency shocks and that arbitrarily fixing a minimum magnitude cut-off for a hazard assessment is putting the cart before the horse.

Thus, even for 5 %g, which is not an extreme peak acceleration by any means, there is potentially a near 15-fold apparent increase in calculated hazard exposure when the minimum magnitude is reduced from $4.0\,M_S$ to $2.5\,M_S$. In large measure, however, this effect is likely to be due to the use, in calculating the annual exceedance probabilities, of a constant b-value – argument can be made that b-value and $4\,M_S$ activity rate are not independent variables and that b-value should also be varied in such a test.

Judging the extent of the influence of minimum magnitude selection on hazard estimation in other pathways of seismic hazard assessment, for instance when it violates the magnitude validity of an attenuation relation, is more challenging, but indications of significant potential effects are afforded by the work of Bommer et al. (2007).

8.4.5.2 Activity rates

The activity rate[5] assigned to a seismic source directly affects local hazard levels and is, therefore, of primary significance. Clearly, for any area, an absence of earthquakes from its complete dataset means either that there have not been any earthquakes or that there has been no system sensitive enough to record their occurrence. Given such uncertainty, a good deal of work has been done on the most appropriate ways of modelling activity rates for attempting to judge low (typically, 10^{-4} p.a.) probability of exceedance hazard over, say, the next 50 years, faced with an earthquake record which may only be complete for the last 200 or 300 years, at best.

If it is presumed that earthquake generation in a moderate seismicity environment is a Poisson process that can be described by a Gutenberg–Richter model (which is the usual presumption), there are a whole range of possible long-term average activity rates which could have produced even that part of the known historical and instrumental earthquake record which is considered to be complete. The confidence limits which can be placed on apparent mean annual activity rates vary markedly with the number of events that have occurred. As an illustration of the variability in confidence levels, consider two situations, each suggesting a mean annual activity rate of 0.05 for magnitude 4 events:

(1) For a zone with 50 magnitude 4 events observed in 1000 years, the 95% confidence limits for the average annual activity rate at magnitude 4 are 0.035 and 0.063.
(2) For a zone with five magnitude 4 events observed in 100 years, the 95% confidence limits for the average annual activity rate at magnitude 4 are 0.013 and 0.105.

The implication of this is that, for any given seismic hazard assessment, judgement has to be exercised to decide how much reliance is placed on the historical record. It may be perfectly reasonable to use the historical record directly when addressing high probabilities and short

---

[5] In seismic hazard assessment, *activity rate* refers to the annual frequency at which earthquakes *at and above* a set minimum magnitude occur, and usually anchors the low-magnitude end of a linear Gutenberg–Richter relationship. Typically this threshold magnitude is an arbitrary chosen value from magnitude 4–5 on whichever scale is employed, and the corresponding rate may be normalised to so many quakes per year per $10\,000\,\mathrm{km}^2$.

lifetimes (although even in such circumstances there may be problems). At the other extreme, for very low probabilities and very long facility lifetimes (e.g. for nuclear waste repositories), the historical earthquake record obviously becomes less and less relatable, and more reliance has to be placed on other, usually geological, information.

In the middle of this range are facilities where the requirement is for ground motions having probabilities of exceedance around 10^{-4} p.a. for lifetimes of about 50 years. For such projects, the objective in deriving activity rates is not simply to obtain the closest possible fit to the seismicity which has occurred in the recent past. When dealing with the uncertainties which attend such probabilities, it cannot be presumed that the parameters of the time-invariant Poisson process are fully revealed by the available earthquake catalogue: even that part of the historical record which is complete needs to be regarded as being only a sample of earthquake occurrence.

Not to allow for the possibility that the historical record is insufficient to reveal all the possible fluctuations in earthquake occurrence is to ignore a fundamental aspect of the physics of seismogenicity that could well be significant at this level of probability of occurrence.

8.4.5.3 Zonation

Although earthquakes occur on pre-existing faults, there are many problems in identifying which faults are active and in associating known earthquakes with particular structures. These problems arise because, even in the more active areas, displacements occur so infrequently on some faults that, for one reason or another (e.g. the effects of glaciation, the limited duration of the historical record, etc.), the available database contains no evidence of this movement. An example of long intervals between significant displacements, and the consequent challenges of estimating a punctual hazard rate, is provided by the Meers Fault, in Oklahoma (see, e.g. Stein *et al.*, 2009).

Matters are not helped by the reality that faulted zones in many seismically active regions, e.g. Japan (Apted *et al.*, 2004) and New Zealand (Berryman and Beanland, 1991), do not present as obvious continuous systems when mapped, and that 'new' active faults or fault segments are revealed only when an unexpected earthquake announces them. Actively growing fault systems may, in detail, comprise a complex geometry of several soft-linked and kinematically connected segments (e.g. Walsh and Watterson, 1991; Walsh *et al.*, 2003). In some cases, all the segments could rupture during a single event because they are coupled at depth, having evolved to this criticality state due to crustal processes associated with discrete faulting operating over geological timescales (see, e.g. Mandl, 1987; Childs *et al.*, 1995; Walsh *et al.*, 2003; Nicol *et al.*, 2005). However, such a situation can be very difficult to discriminate on the basis of small-scale fault segmentation characteristics, discerned mainly by mapping at the surface, and so field data on faults sometimes convey only part of the story.

Potential shortcomings of this kind in datasets and their implications are extremely important: even if every detected earthquake could be associated with an actual individual fault, this will not ensure every active fault is identified. Thus, nowhere in the world can it be assumed that all future seismicity will take place on known faults, and it is this conclusion that gives rise to the need, at least in regulator-facing seismic hazard assessments, for so-called 'zones of diffuse seismicity' (IAEA, 2010). In this particular context, the concept of zones of diffuse seismicity grew out of an earlier notion of 'seismotectonic provinces' (IAEA, 1979).

The change in terminology is meaningful, reflecting the fact that such zones are simply constructs for hazard modelling, catering for uncertainties resulting from deficiencies of knowledge. The basic premise is that these zones will try to encompass:

(1) past earthquakes which cannot be attributed to known seismogenic structures; and
(2) future earthquakes on faults that have not yet evinced seismogenicity.

Therefore, as a device for future hazard estimation, zones of diffuse seismicity can only be abandoned when unforeseen earthquakes stop occurring.

From seismotectonic considerations, it should be possible to express how much of the total seismicity budget ought to be attributed to faults, but it is important not to overemphasise the relative importance of fault sources, especially in low- and moderate-seismicity regions. Guidance on this matter can be gained from generic sensitivity studies, such as those presented by Mallard and Woo (1991), which examine the sensitivity of the hazard at a site, situated locally within a typical diffuse seismicity zone, to the encroachment of a fault source with its seismicity constrained by the regional strain-rate and by the local evidence (both positive and negative) on earthquake occurrence. The effect of such a nearby fault source on site-specific hazard can be great, but begs the question of whether other, unknown, regional candidate faults may exist that can accommodate some of the strain (see Section 8.4.5.6).

Thus far, this particular discussion has been concerned with just one of the roles of zonal sources, i.e. representing the known earthquakes which cannot be attributed to faults. The other role, to cater for the active fault which cannot be identified from the available database, is addressed by defining the maximum magnitude associated with the zonal seismicity – another model parameter for which enumeration of uncertainty is challenging. As such, it is equivalent to the need, in a deterministic methodology, to define the size of the so-called 'floating earthquake'. In effect, therefore, a zone of diffuse seismicity in a model is no more than a simple representation of a volume of the Earth's crust in which the (uncertain) area density of earthquake epicentres, and their (uncertain) depth profile, can both, on the basis of existing information, be presumed to be spatially homogeneous.

In the case of site-specific hazard assessments, the boundaries of such a zone, although based on available data, should be related primarily to the site of interest, and should be constructed to be secure against future, new data. Such boundaries are not necessarily delineated for scientific seismotectonic posterity, but can vary from one site model to another, even when sites are close together. There may be instances where hard and fast seismotectonic boundaries, such as changes in geological structure, in earthquake focal mechanisms, in stress state and so on, are apparent. Even where this is the case, however, two other factors need to be considered:

(1) The boundaries should mark differences in the statistics of earthquake occurrence: this is of major importance since a zone is characterised primarily by the earthquakes it has produced. In this regard, boundaries should respect limitations of knowledge: for example, a zone should not be extended a long way offshore where there are no recorded earthquakes.

(2) The boundaries should reflect an appropriate degree of conservatism. In some cases, it may be prudent to adopt a boundary that just includes an uncertainly located large earthquake which cannot be attributed to a fault source. How close a zone ought to get to achieving the maximum seismicity density by just enclosing outlying earthquakes is a judgement call.

Generally, the primary concern is with the boundaries of the zone which surrounds the site itself and, where appropriate, with the boundaries of nearby zones of higher seismicity. However, where source models are used to predict ground motion at various frequencies, and not just the peak ground acceleration, the significance of more distant boundaries may be increased, especially for low-frequency ground motions. When the locations of zone boundaries have been decided, and the zonal seismicity has been parameterised, sensitivity analyses should be carried out to demonstrate just where the balance between conservatism and uncertainty appears to fall.

In practice, the basic requirement in constructing a zonal seismic source model is to use a methodology which respects the recommendations of the IAEA (2010) but which, in detail, reflects the local circumstances – most obviously the seismotectonic environment, but also the strengths and weaknesses of the available database. Uncertainties will, inevitably, be encountered and, as with a number of other issues, it is important that the methodology used and the decision-making process achieve coherent seismic source models in those local circumstances. In terms of seismic hazard zonation methodology and exploring seismological data and uncertainty influences, one comparison which increasingly commends itself is that between hazard results given by the conventional Cornell–McGuire approach, using zonal model sources, and those given by the more recently introduced *zone-free method*, based on a kernel modelling algorithm (Woo, 1996).

The zone-free method amalgamates statistical consistency with an empirical knowledge basis (earthquake catalogue), and incorporates parameters describing the structured character of the earthquake distribution. Then, a statistical kernel technique is used to compute probability density functions for the size and location of future events. Linear trends based on geological or other seismic information, and uncertainties in magnitude and epicentral location for each earthquake, are fully incorporated into the statistical estimates.

From the limited applications of this latter technique to-date, notwithstanding its rationality, the zone-free kernel function method appears to give results systematically lower than those produced by the more conventional methodology (see, e.g. Molina *et al.*, 2001). This suggests there could be some measure of relative conservatism in the Cornell–McGuire method and, more specifically, it was found (Molina *et al.*, 2001) that the difference between the two methods increases with increasing deviation in the catalogue from the self-similarity assumption implied by the Gutenberg–Richter relationship (see Section 8.3).

A fuller quantitative examination of the extent and nature of the variations between the two approaches has been carried out by Beauval *et al.* (2006b). Their results show that the two approaches lead to similar hazard estimates in very low-seismicity regions but, in regions of increased seismic activity, the kernel smoothing approach yields systematically lower estimates than the Cornell–McGuire zoning method, corroborating Molina *et al.* (2001). In the light of this, the zone-free approach may thus be considered a lower-bound estimator for seismic hazard, and can assist decision-making in moderate-seismicity regions where source zone definition and estimation of maximum possible magnitudes could lead to a wide variety of estimates due to lack of knowledge.

The two approaches lead, however, to very different so-called 'controlling earthquake' scenarios. Beauval *et al.* undertook disaggregation studies at a representative number of sites which showed that if the distributions of contributions are comparable between the two approaches in terms of source-site distance, then the distributions of contributions according to magnitude differ, reflecting the very different seismicity models used. The zone-free method leads to scenarios dominated by intermediate magnitudes events (magnitude 5–5.5), while the zoning method leads to scenarios with magnitudes that increase from the minimum to the maximum magnitudes considered, as the exceedance probability level is reduced.

These trends demonstrate that the choice of seismicity model is fundamental to the determination of hazard level and the definition of controlling scenarios in low- to moderate-seismicity areas; resolving which is the most appropriate model to use in any given circumstance, and understanding how seismological uncertainties play through the different models remain important open issues.

8.4.5.4 *Strong motion attenuation*

Attenuation relations describe the decay of the severity of earthquake ground motion with distance from the earthquake focus. They can be framed in terms of macroseismic intensity or in terms of one or more of the measured parameters of ground motion (most commonly, to-date, peak horizontal ground acceleration or, in Japan, peak velocity). With the increase in the use of uniform risk spectra (URS; sometimes UHS – uniform hazard spectra), and a movement towards expressing seismic hazard as the spectral acceleration at specific natural periods of oscillation (e.g. 1.0 s and 0.2 s), rather than as absolute peak acceleration, a number of frequency-specific attenuation relations may be employed.

It is important that the emphasis should be put on using attenuation relations and spectral shapes that are locally valid. Preferably, they should be derived using local data but, where this is impossible, every effort should be made to confirm the suitability of imported relations or, where new relations are being determined, the appropriateness of imported data. Also, they should require as little manipulation as possible, such as magnitude scale conversion, which, as has been noted earlier, can introduce significant uncertainties into the assessment.

The care needed in assembling a suitable instrumental database for analysing such an attenuation relationship has been recognised for a long time and various selection criteria have been proposed to eliminate instrumental inconsistencies, unintentional biases or unmanageable random scatter. These have been found to arise, respectively, from differences in recording instrument response characteristics and processing techniques, from multiple recordings from one event, and from the varieties of tectonic regime and instrument site geology encountered in practice. The bias introduced by multiple recordings from a single earthquake has been treated by statistical weighting schemes (e.g. Campbell, 1981), and modest reductions in scatter have been obtained latterly by grouping data by site response type, by focal mechanism and by magnitude interval (e.g. Abrahamson and Silva, 2008; Campbell and Bozorgnia, 2008).

This last point is of considerable significance as calculated hazard levels are strongly influenced by the scatter (σ values) associated with attenuation relations used: the larger the σ value, the higher the hazard – see Figure 8.1. Even where proper care has been taken in selecting the database, it is important to recognise that elements of conservatism are often present in conventionally derived attenuation relations. Very little attention is paid, for example, to negative evidence, such as might be provided by a particular strong motion recorder which did not trigger during an earthquake. Also, many relations are derived using only the higher of the two horizontal components of ground motion.

Several procedures for determining the constraints of attenuation relationships have been adopted. Because the distribution of the dependent strong motion parameter to be predicted (e.g. peak ground acceleration *pga*) is apparently lognormal (or nearly so), regressions have been made conventionally on the logarithm of this parameter using least-squares techniques. This approach is justified if the chosen model is linear with respect to the constants. If the model proves to be nonlinear, then the coefficients can be derived empirically using Monte Carlo techniques.

Of all the procedures adopted for handling the inevitable biases in any dataset, the most traditional statistical procedure has been weighted regression, where weights are assigned on the basis of data quality. The purpose of doing this is to reduce bias associated with uneven distributions of recordings with respect to individual earthquakes, magnitude and distance, but because this involves several variables simultaneously, a strong element of judgement enters the exercise.

In any event, it is important to recognise the significance of the very real scatter that is present in any strong motion dataset. In setting up an attenuation relationship for probabilistic seismic hazard assessment, an ergodic assumption is commonly adopted in the regression analysis used to obtain a mean curve predicting ground motion as a function of magnitude and distance (and, sometimes, other parameters). In this context (Anderson and Brune, 1999), an ergodic process is a process in which the distribution of a random variable in space is the same as the distribution of that same random variable at a single point when sampled through time. The standard deviation of this kind of ground motion regression is determined mainly by the misfit between observations and the corresponding predicted ground motions at multiple stations for a small number of well-recorded earthquakes. Thus,

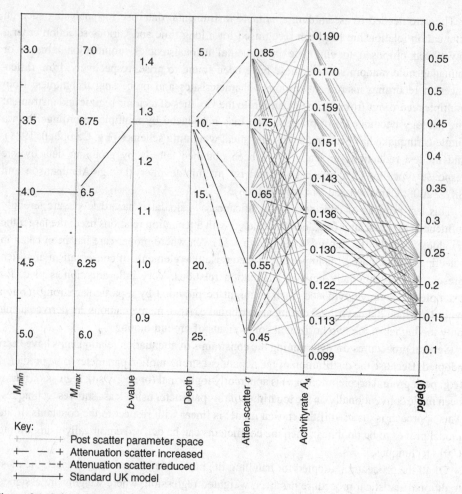

Figure 8.1 Seismic hazard model parameter cobweb plot for a typical UK site showing strong influence of attenuation scatter on peak ground acceleration (*pga*) hazard level. The solid line ('Standard UK model') a shows pathway through model for central values of each parameter distribution (n.b. variables M_{MIN} to depth are all held fixed at central values, for clarity), with dashed lines showing influence of different attenuation scatter values on hazard level: *pga* accelerations span from 0.15 *g* to nearly 0.3 *g*, depending on scatter (note: the chart does not indicate probabilities of exceedance associated with *pga* values).

the standard deviations of most published ground motion regressions are predominantly related to the statistics of the spatial variability of the ground motions, and some authors (e.g. Anderson and Brune, 1999; Woo, 1999) have drawn attention to the inflationary effect this will have on probabilistic hazard estimates.

Several computational algorithms have been proposed in the seismological literature for attenuation models with multiple variance components, including the earthquake-to-earthquake and site-to-site components, along with a remaining component of variability after the former components have been accounted for. Joyner and Boore (1981) proposed a two-stage estimation method which, although computationally simple, and widely used, may lack formal statistical meaning and, hence, justification for the resulting estimates (Chen and Tsai, 2002).

To counter this, Chen and Tsai (2002) proposed an extended approach to the derivation of strong motion attenuation relations with multiple variance components. They adopt a general attenuation functional form as a 'mixed' model, where the regression coefficients are treated as 'fixed effects' and the components of deviations are treated as 'random effects'. Both the fixed and random effects are estimated in a unified framework via an expectation-maximisation (EM) algorithm, with a modification when necessary to accommodate nonlinear attenuation models. Explicit formulae are provided for the variance estimates of the estimated model parameters. Two types of strong motion prediction are discussed:

(1) the 'unconditional prediction', in which the site-specific deviation is not accounted for; and
(2) the 'conditional prediction', which does incorporate that deviation.

A simulation study showed that the proposed procedure yields estimates with smaller biases compared with the EM procedure of Brillinger and Preisler (1985), or with the commonly used two-stage method of Joyner and Boore (1993).

The Chen and Tsai method has been applied to a large dataset of ground motions from Taiwan, which included 'remarkable' site-to-site variability. The total deviation in the regression for their data is estimated numerically to be 0.318. What is particularly relevant to the present discussion is that the estimated site-to-site uncertainty, ~0.16, represents a significant fraction of the total deviation. Hence, Chen and Tsai conclude that a probabilistic seismic hazard analysis based on an attenuation relationship which relies on the ergodic assumption (where the spatial variability of ground motions is ignored) may overestimate the *pga* hazard in Taiwan, in line with the theoretical views of Anderson and Brune (1999).

The most promising path to reducing attenuation scatter is, therefore, to differentiate clearly between the epistemic and the aleatory uncertainties involved (Chapter 3). In general, the site-to-site component of variance can be reduced if additional factors, such as the site classification (e.g. rock or soil) or the basin depth, are incorporated into the attenuation model. Also, the remaining component of variance can be reduced if information on both site properties and path effects is available (Lee *et al.*, 1998). According to Toro *et al.* (1997) and Anderson and Brune (1999), these two components of variance should, ideally, be treated as epistemic. Although such additional information is not available for their dataset, Chen and Tsai assert that it is still possible to use a random site-effect term in attenuation models like theirs to estimate the size of the site-to-site variability and thus reduce the uncertainty in predictions of strong motion at a given site. They conclude that the 'conditional prediction' has been shown to be much more accurate than the 'unconditional prediction' for their circumstances and data.

Taking up the issue of attenuation scatter determination more recently, Arroyo and Ordaz (2010a, 2010b) break new ground by applying multivariate Bayesian regression analysis to ground motion prediction equations using a set of actual records. Based on seismological arguments and a suitable functional form for the equation, Arroyo and Ordaz discuss how the prior information, required for the model, can be defined; they then compare results obtained from Bayes regression with those obtained through the one-stage maximum-likelihood and the constrained maximum-likelihood methods. Potential advantages of the Bayesian approach over traditional regression techniques are discussed.

Another emerging topic of investigation is the characterisation, via the lognormal distribution, of scatter (aleatory variability) in recorded ground motion parameters. This distribution, when unbounded, can yield non-zero probabilities for what appear to be unrealistically high ground motion values at low annual exceedance probabilities, so Huyse *et al.* (2010) question the appropriateness of the assumption by examining the tail behaviour of *pga* recordings from the newly assembled Pacific Earthquake Engineering Research – Next Generation Attenuation of Ground Motions (PEER NGA) database and from *pga* residuals using Abrahamson and Silva's (2008) NGA ground-motion relations. Their analyses show that regression residuals for *pga* data in the tail do not always follow a lognormal distribution and the tail is often better characterised by the generalised Pareto distribution (GPD).

They propose using a composite distribution model (CDM) that consists of a lognormal distribution (up to a threshold value of ground motion residual) combined with GPD for the tail region, and go on to demonstrate implications of the CDM in PSHA using a simple example. At low annual exceedance probabilities, the CDM yields considerably lower *pga* values than the unbounded lognormal distribution. It also produces smoother hazard curves than a truncated lognormal distribution because the *pga* increases asymptotically with a decreasing probability level.

It is not just intrinsic ground motion variability, but also inhomogeneous data quality which has given rise to the practice of using strong motion records in statistical ensembles. The uncertainties associated with the stochastic irregularity of seismic ground motions, together with errors in empirical observations are, of course, further compounded by uncertainties in estimation of the basic seismological descriptors of an event (location, magnitude and moment, as discussed earlier), on which all attenuation regressions rely.

Matters can be aggravated further by arbitrary modelling decisions. As just noted, it has been found in some seismic hazard assessments that implausibly large or physically incredible accelerations may be computed for high-magnitude events with very low probabilities of exceedance. The pragmatic response to this often is to truncate, deliberately, the extent of lognormal scatter on attenuation, sometimes to a spread less than three standard deviations beyond the mean, thus precluding extreme high-acceleration values from emerging (see, e.g. Strasser *et al.*, 2008).

However, care has to be taken to guard against unconservative, unintended consequences when making decisions like this to bound parameter distributions in a logic-tree model: truncating attenuation scatter, for instance, means that while extreme peak accelerations are excluded for the highest magnitude events, they are also inhibited in analysis at lower magnitudes.[6]

In summary, given all the intrinsic uncertainties that are involved, the formulation, selection and model application of a suitable attenuation relationship is one of the most challenging and most critical elements in any seismic hazard assessment.

8.4.5.5 Ground characterisation

The acquisition of further strong motion data will enhance understanding of the characteristics of seismic vibrations, and improve scientific knowledge of ground motion parameters and of the effects of near-surface geology on shaking levels and frequency content. But major gaps in this knowledge will persist: this is because such data cannot be replicated reliably in a laboratory, but only painstakingly acquired in field observations of real earthquakes, when these oblige.

The topics of seismic ground category and dynamic soil response are substantial and complicated matters in their own rights, and an in-depth review of attendant uncertainties is beyond the scope of the present chapter. Suffice it here to say that the average velocity of shear waves in the top 30 m of soil, V_{S30}, has become the ground category parameter used by many engineering design codes and most recent published attenuation equations used to estimate strong ground motion.[7]

Yet, according to Lee and Trifunac (2010), few studies exist to substantiate V_{S30} as a meaningful parameter to use, and whether ground motion estimates based on it are reliable. One major criticism is that taking only the top 30 m of a soil profile into account fails to represent, for many sites, the influence of deeper soil layers, that may extend to considerably greater depths, sometimes hundreds of metres or more. Some previous, older ground category parameter definitions incorporated less proscribed measures of layer thicknesses, and hence allowed greater scope for characterising velocities, densities and other relevant geotechnical properties over greater depths for dynamic response analysis.

Whatever seismic ground category definition is chosen, however, the difficulties of obtaining representative and reliable shearwave velocity measurements (and compressional wave velocities, or Poisson's ratio) in the field are substantial, and associated uncertainties can be very large and difficult to quantify exactly. Moreover, uncertainties can be further compounded where records are made on topographic irregularities, or in proximity to major geological or crustal structures, for example.

---

[6] Thus a probabilistic seismic hazard assessment conducted for Christchurch, New Zealand, with strongly truncated attenuation scatter would likely have ascribed zero probability to the occurrence of accelerations of 1.8 g and 2.2 g, which were recorded during the magnitude 6.2 M_w earthquake in February 2011 (Holden, 2011). In a similar vein, there is a danger of not incorporating fully the uncertainty range on a parameter in a logic-tree calculation when that distribution, for instance a Gaussian, is expressed by, say, just three quantile values, with weights corresponding to the mean and mean ± 1 standard deviation levels; more extreme values – higher and lower – are not sampled.

[7] For instance, the US National Earthquake Hazard Reduction Program (NHERP) building code assigns one of six soil-profile types to a site, from hard rock (type A) to soft soils (types E or F), based on the estimated value of V_{S30} for a particular site (BSSC, 1997). These soil profile categories are also part of the International Building Code adopted in 2001 (IBC, 2002), and a similar approach is adopted by Eurocode8.

The chaotic unpredictability of earthquake sources and attenuation with distance are the most obvious causes of irreducible uncertainty intrinsic to seismic phenomena, but these effects are compounded by lack of sufficiently detailed geological and geotechnical measurements in respect of recording sites, where data are collected, and of target sites, where hazard levels are to be determined. So uncertainty in characterising local ground conditions is also an important factor when it comes to evaluating what expected site-specific seismic ground motions will be in some future earthquake.

8.4.5.6 Status of geological faults

One of the main problems in seismic hazard assessment in regions of relatively modest seismicity is to discern which faults are active (in Britain, for example, there are many faults but relatively few earthquakes); this is why emphasis needs to be placed on having a robust rationale for adequately treating and representing faults in any assessment methodology.

Any approach to this problem has to address the two concerns raised by known faults, namely:

(1) whether they should appear in a ground motion hazard model as a distinct source of seismicity; and
(2) in the case of surface faults at, or close to, the site of interest, whether they pose a hazard in terms of surface rupture and, therefore, effectively exclude development.

Although it seems obvious that the methodology adopted should address these two concerns in a coherent fashion, experience shows that this has not always been the case.[8]

Because knowledge of the precise geometry of faults in three dimensions at seismogenic depths is usually extremely rare, and because there is always some (and often considerable) uncertainty in the location of earthquake hypocentres, the most positive diagnostic of an active fault is provided by unequivocal evidence of neotectonic displacement.[9] This, of course, requires a definition of 'neotectonic displacement' and, around the world, various time frames have been used as the defining criteria in this regard.

In the case of ground motion hazard, with perfect data and, therefore, full seismotectonic understanding, seismic source models for estimating ground motion hazard would consist only of fault sources. However, as has already been discussed, even in California or Japan, this situation is unlikely ever to be achieved (see also Mallard, 1993). Therefore, everywhere, an appropriate degree of emphasis has to be accorded to fault sources. In an area with relatively few proven active faults, the deterministic method may well overemphasise the importance of those faults: timescales can be such that the first faults to be identified as being active are not necessarily the most active. This overemphasis can work in both directions:

---

[8] One example of such inconsistency, which was corrected through the intervention of the IAEA, allowed for magnitude 6 earthquakes to occur on an on-site fault when assessing the ground motion hazard, but ignored the possibility of ground rupture on the same fault.

[9] Implying that, as well as being active, the fault is also 'capable' in regulatory terminology, i.e. it is capable of surface rupture – see IAEA 2010.

the hazard can be kept low by constraining earthquakes to occur on a fault at a distance from the site, or the hazard can be unnecessarily heightened where that fault is close to the site.

Thus, in taking proper account of a fault that needs to be regarded as being active, a major problem is deciding how much of the known local seismicity should be attributed to that fault, given the nature of the local seismogenic crust. In this regard, again, the probabilistic method has advantages in that weights can be apportioned as to whether a particular earthquake was on the fault or in the zone that surrounds it.

The challenge of finding a way to make this decision in the face of locational uncertainties in both earthquake location and fault position has been tackled by Wesson *et al.* (2003). They invoke Bayesian inference to provide a method to use seismic intensity data or instrumental locations, together with geologic and seismologic data, to make quantitative estimates of the probabilities that specific past earthquakes are associated with specific faults. Probability density functions are constructed for the location of each earthquake, and these are combined with prior probabilities through Bayes' theorem to estimate the probability that an earthquake is associated with a specific fault. Results using this method are presented by Wesson *et al.* for large historical earthquakes and for recent earthquakes with instrumental locations in the San Francisco Bay region. The probabilities for individual earthquakes can be summed to construct a probabilistic frequency–magnitude relationship for a fault segment.

Other applications of the technique include the estimation of the probability of background earthquakes, that is, earthquakes not associated with known or considered faults, and the estimation of the fraction of the total seismic moment associated with earthquakes less than the characteristic magnitude. Comparisons of earthquake locations and the surface traces of active faults as determined from geological data show significant disparities, indicating that a complete understanding of the relationship between earthquakes and faults remains elusive.

It may be for the latter reason – or perhaps because Bayesian concepts are intellectually more demanding – that the approach espoused by Wesson *et al.* (2003) is yet to gain more universal traction in seismic hazard assessment methodology.

8.4.5.7 Uncertainties and model sensitivity tests

Conventional sensitivity analyses, focusing on components of a hazard model, can, in theory, provide quantitative indications of the effects of elements of model uncertainty on the results given by that model (see Chapter 3).

While formal sensitivity testing can address the influence on the hazard estimate of individual parameters, one effective way of indicating the range of joint effects produced by parameter uncertainties is to present all possible combinations in seismological uncertainty space as a 'cobweb' plot, of the type shown in Figure 8.2. On that chart, which is for low-seismicity conditions in the UK, none of the individual parameter values are entirely implausible, and it would be challenging, scientifically, to justify values well outside the ranges shown.

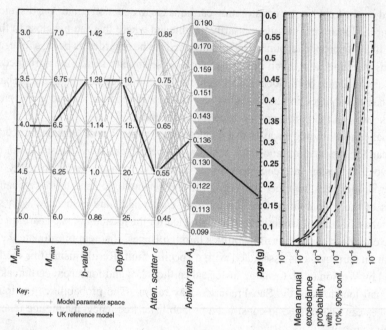

Figure 8.2 Cobweb plot showing seismic hazard model parameter space for typical low-seismicity (e.g. UK) conditions. In an equivalent logic tree, the corresponding branches would each carry distinct weights, with individual parameter weights summing to unity. As in Figure 8.1, the solid line shows a pathway through the model for central values of each parameter distribution, and the resulting peak horizontal acceleration hazard level. The right-hand panel shows a typical hazard exceedance curve from such a model, together with model-based 10% and 90% confidence bounds, determined from contributing uncertainty distributions. Note: some practitioners regard depth as aleatory, and would exclude it from a logic tree model; also, some might consider magnitude 4 activity rate (A_4) and b-value dependent variables, not independent, as shown here. However, arguments can be constructed either way in each case, depending on the type of data used, forms of relevant relationships and the probabilistic formulation underpinning the hazard assessment model. For instance, there is a special class of probabilistic models called pseudo-independent (PI) models, where a set of marginally independent domain variables shows a special type of collective dependency – see, e.g. Lee and Xiang (2010).

However, the uppermost and lowermost hazard results, which emerge from extreme combinations of parameter values, may not be seismologically reasonable for low-seismicity conditions, and care would be needed on this score when constructing a logic-tree model for low-probability hazard estimation, in particular.

Ideally, it should be possible to refine these partial insights on quantified overall uncertainties by drawing comparisons between results produced by different groups of specialists. In reality, this is not easily done (although, conceivably, it might be possible if two hazard studies for the same site were available that were equally satisfactory in terms of the expert judgements involved).

While there is potential for significantly reducing hazard level estimates through further investigation of epistemic uncertainty in any particular attenuation relation, as discussed in Section 8.4.5.4, some recent high-profile seismic hazard assessments have embraced the proliferation of such relations, warts and all (see also Section 4.3.1). This leads to such massive logic trees, with exponentially manifold branches, that it becomes unfeasible to undertake sensitivity tests in a comprehensive – or comprehensible – manner. Disentangling the contributions of a multitude of various, often competing, factors through millions of pathways is close to impossible. These trees can grow huge – a total of 10^{15} or more operative branches is not unknown – because of the contemporary obsession with 'capturing all epistemic uncertainty'; every possible or semi-plausible by-product of this notion has to be roped into the model. In particular, multiple published attenuation relationships are invoked, together with unconstrained estimates of uncertainties associated with parameter distributions that purport to characterise different seismotectonic and faulting hypotheses through variant forms of expert elicitation (Chapter 4).

It is clear that if seismic hazard modelling computations and model understanding are to remain tractable, and practicable for many less well-resourced organisations or countries, a less complex methodological approach should be formulated and adopted. This will need to provide basic central tendency hazard results, with complementary sensitivity tests providing the means for checking acute response to parameterisation uncertainties, or the existence of so-called cliff-edge effects in the basic model.

8.4.5.8 *Reliability of hazard results*

Given the uncertainties involved, it is only natural to ask whether there is any possibility of judging how reliable (or unreliable) the results of probabilistic hazard studies are likely to be. The most basic approach to judging the reliability of a hazard assessment entails gauging the reliability of its constituent decisions – considering subjectively the scientific or technical legitimacy of each of those decisions. To attempt to go beyond this, in pursuit of some form of objectivity, there is a risk that terms such as 'validation' or 'verification' are brought into the debate; these are often being used incorrectly, or as though they were interchangeable. In fact, it is important to recognise that there can be no possibility of either validating or verifying a numerical model of a natural system: indeed, it is impossible to demonstrate the truth of any proposition except in a closed system (Oreskes *et al.*, 1994).

Nevertheless, it is desirable that such empirical evidence as is available should be used to check that hazard study results are not inappropriate or invalid. In pursuit of this objective, two types of empirical evidence are, theoretically, in play:

(1) the record of macroseismic intensities that have been experienced at the site in question; and
(2) the record of instrumental strong ground motions that have been recorded at the site of interest.

Because of the much longer timescale involved, the first of these sources of information should obviously be accorded precedence, and recommended practice (IAEA, 2010) is to present comparisons drawn on this basis in any site-specific hazard study report.

In low-seismicity areas – and, indeed, in many high-activity areas – it is not so easy to establish comparisons between estimated hazard levels and experience data in the form of strong-motion records. Because initiatives to install the necessary instrumentation are relatively recent and because they still provide information from only a few locations, there is inevitably a massive difference in the overall effectiveness of the strong-motion recording system as compared with the macroseismic history accumulated over centuries.

In principle, there is a limit to the degree to which spatial data can be swapped – reliably – for historical data, and this limit arises not only because of spatial variations in activity (see below), but also because of temporal variations. As discussed above, in a hazard assessment the confidence limits which can be placed on apparent mean annual activity rates vary markedly with the number of events that have actually occurred. The number of occurrences that are required in order that one can have a high degree of confidence as to the long-term average recurrence interval is a moot point, but establishing reliably the event which has a genuine average return period of 100 years, for example, is likely to require much more than 1000 years of data.

If, in addition to all these issues, one takes into account the conditions that obtain in reality, the true complexity surrounding the meaningfulness of a short period of strong-motion data observation becomes apparent. The factors which affect the real situation include the very strong likelihood that seismicity varies from one part of a country or region to another (which affects not only the 'local', i.e. zone-specific record, but also produces complications for instruments close to zonal boundaries), and the fact that, although the instruments themselves may be set to have the same recording thresholds, various factors can conspire to introduce large variations in their true operational sensitivity.

Musson (2012b) has broached the matter of objectively validating probabilistic seismic hazard assessments with observational data, and proposed an approach.

8.4.5.9 Summing up

Drawing this part of the discussion on seismic hazard assessment uncertainties to a close, it is appropriate to take stock of what experience with many site-specific studies and their outcomes tell us. As indicated, there are many uncertainties involved in modelling seismic hazard: six of the most significant – but not necessarily in the order given – are:

(1) determination of magnitude completeness thresholds;
(2) enumeration of activity rates;
(3) delineation of zonations;
(4) the active status of geological faults;
(5) strong motion attenuation scatter;
(6) ground motion characterisation.

In an assessment, decisions which are made concerning every one of these factors can have a major effect on the resulting hazard estimates. However, the effects of these, and other, uncertainties can reliably – and fairly easily – be modelled and understood in combinations, through the logic-tree formalism; however, this may not be necessarily so straightforward for some other natural hazards (e.g. volcanic hazards – see Chapter 11) or, indeed, in other parts of the seismic risk and anti-seismic design process.

Thus, a probabilistic methodology should be configured in such a way that it provides robust hazard estimates which cater for all the well-understood process uncertainties in the problem. Within the limits of the evidence that is available for earthquakes, such estimates can be shown to be usually not inconsistent with experience data, whether macroseismic or instrumental; if, in a particular case, there is serious deviation then a flawed analysis should be suspected.

Having examined the main possibilities, it is concluded that the single area of modelling where an attempt to reduce present conservatisms through uncertainty reduction is best justified is in the scatter on strong-motion attenuation relations. While that uncertainty is easily shown to be the one that has the greatest effect on hazard levels, it is an uncertainty which, at least in part, exists because of the decision to adopt *ab initio* the ergodic assumption in selecting and using data for the construction of such relations. Discarding this would represent a major break with precedent and practice.

8.5 Uncertainties and methodological precepts

Within the context of seismic hazards and risks, an effective, robust and evolved methodology for earthquake hazards assessment has emerged in recent decades, thanks mainly to extensive work for the nuclear power industry in many countries. The lessons afforded by all this effort and by elements of good practice, engendered by multiple applications worldwide, offer several insights for assessment methodology in other natural hazards.

As a general principle, a hazard evaluation should account for all relevant uncertainties, both aleatory and epistemic (Chapter 3). In a deterministic hazard analysis, where uncertainties are identified, these are usually incorporated by assuming a conservative process at each step of the evaluation, with the consequence that outcome results tend to be overstated, numerically. In contrast, a probabilistic hazard analysis, when implemented in a logical and balanced manner, can seek to incorporate relevant uncertainties explicitly into the analysis and provide an unbiased and realistic assessment. For this reason, the latter approach is increasingly favoured, internationally, for safety-critical plants and high-value assets, such as national interest infrastructure facilities (see also Chapter 13). Besides, it is not possible to make rational cross-hazard comparisons when only deterministic estimates of different individual risks are available, just when each assessment is uniformly probability-based.

In cases where both deterministic and probabilistic results are present, however, deterministic assessments can be used as a check against probabilistic assessments in terms of the reasonableness of those results, particularly when low annual frequencies of exceedance are being considered. In exchange, probabilistic assessment results allow deterministic values to be evaluated within a probability framework so that the equivalent expected annual frequency of exceedance of a deterministic parameter value may be estimable.

Numerical techniques and methods for quantifying uncertainties continue to evolve, especially within the probabilistic hazard assessment paradigm, and good applications respect certain elementary principles for the appropriate treatment of the inevitable uncertainties that are encountered. In seismic hazard assessment, for instance, good practice also takes into account the emerging understanding of the physical basis for earthquake processes in the round – for example, insights provided by power-law scaling relations that are characteristic of complex dynamical systems – even though, *pace* Clauset *et al.* (2009), every detail is not yet well understood and a complete explanation for the behaviour of earthquakes and geological faulting is proving very elusive. These elements of uncertainty, which are essentially epistemic and due to our limited understanding of the processes involved, are unlikely to be reduced much in the short-term – but they can be accommodated in some measure in a quantitative hazard assessment by utilising different dynamical models, constrained to match the principal empirical characteristics manifest by earthquakes.

Thus the best general approach to hazard evaluation should be directed towards appropriately incorporating and constraining (and, where possible, reducing) the uncertainties at all the various stages of the evaluation process, with the purpose of deriving reliable results constrained, to the maximum extent possible, by data (rather than relying too much on computer models or conceptual hypotheses). Experience in earthquake hazard assessment shows that the most effective way of achieving this is first to collect the maximum amount of reliable and relevant data, to make that data as uniform and coherent as possible, and then decide if these data – on their own – are sufficient for purpose. If so, then state-of-the-art historiographic and probabilistic analysis techniques can be applied to them, such as peak-over-threshold and extreme value distribution fitting and time-series analysis, supplemented where necessary by formalised expert judgement elicitation.

An important corollary to searching for a sufficient foundation for hazard assessment (and event forecasting) is the need for better and more formal ways of testing the quality and reliability of such assessments. Basic decision theory tells us that, in the presence of uncertainty, it is possible to generate probabilistic forecasts which, while somewhat disposed to produce false positives, can still be used to make better choices. In premium-quality earthquake hazard assessments, formalised methods for evaluating the weight to give uncertain scientific evidence are utilised to cope with multiple strands of data and evidence from different sources.

However, it should be remembered that all probabilistic hazard assessments are model-based. The results obtained therefore depend completely on the many assumptions, constructs and simplifications that are used to make models and calculations tractable – the computation of a hazard level does not assert absolute scientific truth. And, as often as not, direct comparability of results generated by differing analyses is unwarranted.

With so many related uncertainties, inevitably imperfect characterisations of natural phenomena make for difficulties in communicating the scientific conviction that does exist about hazards and their potential risks.

8.6 Communication of earthquake hazards and risks

Where the safety of people or facilities is involved, the results of a hazard analysis feed naturally into an assessment of risk, with the latter entailing further uncertainties in relation to vulnerability of populations and fragility of systems and structures. Notwithstanding the significant inherent uncertainties involved in both domains – hazard and risk – most scientists who possess insights about the causes of disasters have a keen desire to share their insights with people who are potentially in harm's way, and with public officials who could implement policies that would reduce vulnerability. Indeed, scientists with such understanding surely have an ethical responsibility to raise awareness in their communities and to provide insights that contribute to reduction of future losses – although how to do this while simultaneously articulating the associated scientific uncertainties is hugely challenging.

As noted at the start of this chapter, the importance of such efforts for seismic risk mitigation grows every year as expanding populations occupy larger areas in earthquake-prone regions. New multidisciplinary technological developments (e.g. the US FEMA loss-estimation model, HAZUS[10]) enable future earthquake losses to be quantified regionally in order to assist policy-makers and the public to appreciate the extent of their exposure. Given the uncertainties in PSHA, a corollary question is what should be the appropriate engineering response when it comes to design?

Performance-based seismic design is being used increasingly in the United States as a way to determine and justify seismic ground motion levels for design (McGuire, 2011). Application of the performance-based seismic design concept requires three determinations to be made:

(1) discrete performance states for a structure or facility affected by uncertain future loads have to be established;
(2) acceptable levels of risk (annual probabilities), beyond which performance states will not be achieved, have to be decided; and
(3) design loads that will achieve a given performance state from (1), with an acceptable level of risk of failure from (2).

For example, for (1) it might be decided that a commercial building should not suffer partial or complete collapse during an expected 50-year lifetime. For (2), a 1% probability of partial or complete collapse in 50 years might be deemed bounding the limit of acceptability. Then, for (3), the appropriate design loads need to be derived, using probabilistic seismic hazard analysis combined with appropriate building-specific collapse fragility functions, such that

---

[10] http://www.fema.gov/plan/prevent/hazus/ (accessed 29 December 2011).

effects will not exceed the acceptable risk set out in (2). The acceptable risk in step (2) is not an abstract determination, but often is referenced to past design practice and, more crucially, to risks implied by experience. For instance, application to current design requirements for new nuclear power plants indicate that this approach leads to seismic core damage frequencies that are lower (safer) than those of existing plants by a substantial margin. Applications to the seismic design of commercial buildings indicate that a target collapse probability no worse than 1% in 50 years is achievable. This said, probabilities of lesser levels of seismic damage to commercial buildings may be relatively high and some additional benefit could accrue from explicit evaluation and management in the design process.

In such policy dialogues, Earth scientists should not feel constrained to simply characterise the hazards, but should collaborate proactively with engineers, economists and social scientists in order to make estimates of future risks and losses as robust as possible, in the face of all the intrinsic uncertainties (see Chapter 15). Internationally, there are many similarities in hazard response and preferred public policy from country to country, but there are also important contrasts in population toleration of risks and hence their behaviour and responsiveness. And there can be significant differences in the factors associated with these behaviours: cultural contexts are fundamental, and scientific uncertainty can amplify differences. Thus this is not solely a technical domain problem; the role of corruption as a major factor in earthquake disasters has been highlighted by Ambraseys and Bilham (2011). Governance structures need to be developed to prevent bad building, and building codes and legislation need to be enforced. Social scientists have a role here (see Chapter 16 on human behaviour), and further critiques of what has worked and not worked in previous earthquake-risk communication efforts would be helpful. This includes accessing insights that have been developed by risk-communication specialists.

There are two main challenges confronting communicators: presenting hazard and risk information in an intelligible way to lay people, and making the recipients and policy directors aware of ways that they can significantly reduce risk. For earthquake hazards and risk communication, there are real needs, and real opportunities, for collective research contributions from specialists in statistics, numerical data analysis and data visualisation, as well as Earth scientists and earthquake engineers – every one of these domains gives rise to or entails significant scientific or technical uncertainty, and bridging that whole, multi-dimensional uncertainty space is challenging.

Eventually, similar efforts of uncertainty understanding will be needed for seismic hazard assessment's putative consort: earthquake prediction.

8.7 Earthquake prediction and warning systems

Our current inability to predict earthquakes routinely (i.e. to specify jointly, with defined precisions, the time, place and size of a future event) can be attributed to the inadequacy of basic seismological data for underpinning a satisfactory general theory of earthquake occurrence. Some successful predictions have been made under very special circumstances, e.g. when swarm sequences produce evident foreshocks, and deviations from the usual Gutenberg–Richter linear relation between magnitude and frequency are manifest (e.g. Morgan *et al.*, 1988).

However, in most earthquake sequences, the mainshock is the first event. It is because of this time asymmetry in most major earthquake occurrences that earthquake forecasting is much more difficult than weather event forecasting (Woo, 1991). Even when there are putative foreshocks, and a major earthquake might seem to be imminent, the complex and erratic ways that these minor ruptures propagate and arrest, and modulate stresses and strains locally, can affect when a mainshock might happen. The mainshock can be delayed for so long that useful prediction or forecasting on a suitable mitigation timescale is negated, or it may be forestalled altogether.

An important corollary to searching for wide-ranging earthquake forecasting techniques is the need to develop better and more formal ways of testing the quality and reliability of such forecasts. Until now, most seismologists have sought only for a single precursory signal (or pattern of signals) that would provide a totally reliable prediction capability, and have rejected anything that amounts to less than a scientific truth or law. However, basic decision theory tells us that, in the presence of uncertainty, it is possible to generate probabilistic forecasts (*tipcasts* – temporary increases in probability) which, while liable to produce false positives, can still assist us to make better choices. For instance, detection of a moderate increase in the probability of a damaging earthquake in a particular locality could be used to put emergency services on alert, even if a decision to evacuate the public cannot be justified.

The methods for evaluating the weight to give to uncertain scientific evidence in such circumstances can be extended to cope with multiple strands of data and evidence from different sources. It should be possible to combine spatial patterns and temporal trends in seismicity data with GPS deformation measurements, strain signals, radon emissions, and so on, into a multi-parameter prognostic – the devastating 2009 Abruzzo earthquake in Italy and tragic deaths in the town of L'Aquila is a case in point, where several separate precursory signs may have been present, albeit each very weak on its own, in a predictive analytic sense. But the Earth sciences community is not yet proficient at applying probabilistic analysis to decision support (e.g. weighing elements of evidence by Bayes factors, Bayesian Belief Networks, etc.). This said, a seminal application to volcanic hazards management illustrates some of the concepts (Aspinall *et al.*, 2003; see also Chapter 11), and more recent work has seen Bayesian Belief Networks being suggested as an evidentiary reasoning tool to support tsunami warnings (Blaser *et al.*, 2011). Impetus to apply these techniques to difficult seismic hazard problems, such as multi-parameter synthesis for earthquake occurrence forecasting, would be an important scientific investment and a worthwhile objective to pursue in the context of hazard mitigation.[11]

---

[11] For certain special circumstances, instant earthquake warning systems are being developed that could help provide additional margins of safety and protection against strong seismic shaking effects: e.g. the Japanese 'bullet train', Mexico City alerting for distant great earthquakes, Kashiwazaki-Kariwa NPP scram shutdowns, Athens gas supply protection. These are, essentially, seismometric response systems. The monitoring and event-detection techniques needed for effecting useful warnings – given a

In any early warning or alerting system that relies on event detection monitoring it will be important to fully incorporate estimates of true and false positive rates *and* true and false negative rates in any conditional probability decision matrix. In the wider context of maximising the value of scientific evidence and insights, these rate assessments will involve the use of expert judgement (see Chapter 4), as well as observational data and modelling.

8.8 Future research and needs

Given the inventory of problem areas outlined above (and these were just selected examples), clearly there are significant opportunities to leverage advanced academic research in seismology into better practice in earthquake hazard assessment and risk mitigation. In relation to seismology as a science, contemporary academic research is thriving in terms of the development of physical and statistical-physical models of earthquake processes, and in advanced statistical approaches to hazard assessment and treatment of uncertainty. For instance, notable examples of work in the UK that are extending global geosciences research frontiers (as well as offering the prospect of enabling improved hazard evaluation) are the advances in understanding of the complex dynamics of fault interactions, seismicity patterns and earthquake triggering in work by Main, Steacy, McCloskey and colleagues (e.g. Hainzl *et al.*, 2010; Main and Naylor, 2010; McCloskey, 2011).

There is also an important role opening up for palaeoseismology and palaeogeodesy, which are increasingly becoming key parts of seismic hazard assessment worldwide. Good examples include trenching work by Sieh and colleagues on the San Andreas Fault (Liu-Zeng *et al.*, 2006) and, more recently, reconstructions of the tectonic history on the Sumatran fault system through studies of uplifted corals (Sieh *et al.*, 2008). The combination of a good understanding of fault geology and geochronology – applied to suitable geological materials – can yield information on stress build-up rates and strain relief in earthquakes, and longer time-series of events, as well as average fault slip rates.

Work by Stein and colleagues in the Tokyo Bay area on uplifted terraces (Toda *et al.*, 2008) is another example of understanding fault systems and recognising patterns of fault system movements in space and time. Another approach is the mapping of strain from GPS and InSAR data to identify areas of high (and low) strain and to look for mismatches between datasets: the latest data from the New Madrid seismic zone, for instance, appears to show negligible contemporary regional strain, and this observation is throwing into doubt previous conceptions of continuous albeit low deformation in mid-continental areas (Calais and Stein, 2009). Even so, such work is also starting to help identify hidden, previously unrecognised, faults – an important capability to incorporate into seismic hazard assessment methodology.

---

significant earthquake has started to rupture along a fault – are still very novel and the subject of active research. Integration into a probabilistic early warning system would be beneficial, especially if it is desirable to minimise (expensive) false alarms and (potentially dangerous) false negatives.

New models for these complex system network processes are likely to give fresh insights into aspects of the past earthquake history that have not yet been recognised. Even testable predictions may be possible for the possible evolution of the whole seismogenic system, including where strain accumulation might be expected precursory to release in future earthquakes. In principle, such model predictions can be tested using strain observations and deformation data and – if successful – could be incorporated as improvements in earthquake hazard models and used to refine event occurrence expectations.

A way to unify many of the varied elements of the seismic hazard assessment problem, touched on in preceding sections, and to improve our understanding and handling of their related uncertainties, is to explore and develop the Bayesian paradigm and what that has to offer with regard to weighing and combining disparate strands of evidence. It is desirable, for several reasons, to moderate some of the oversimplistic assumptions and oversimplified relationships that have been proposed in the recent past, and to put the treatment of uncertainty on a more rigorous basis.

Advances like these would be crucial for the development of more cost-effective seismic hazard mitigation policies, especially in areas of moderate or low seismicity, and for the improvement, for instance, of insurance portfolio risk management on a global scale. Some convergence of indefinite probabilistic hazard assessment with workable prediction may be closer than many scientists suspect.

8.9 Concluding remarks

Scientific uncertainty is ubiquitous in earthquake hazard assessment at any level of concern, and its quantification and treatment become ever more critical as non-exceedance at lower and lower levels of probability are demanded for safety. There is frequently a failure to recognise that all probabilistic hazard assessments are *model-based*, and that the results obtained therefore depend on the many assumptions, constructs and simplifications that are used to make calculations tractable. In determining a hazard level there is no absolute scientific truth to be found – the analysis is done usually for pragmatic reasons, and not for scientific posterity.

In the face of so many uncertainties, any hazard model's inevitably less-than-perfect characterisation of natural processes makes for difficulties in communicating scientific conviction relating to the associated hazards and their potential risks. This said, probabilistic seismic hazard assessment methodology is perhaps the most developed of all hazard assessment methods for low-probability hazardous natural events, thanks mainly to an acute need for scientific guidance by the civil nuclear power industry worldwide and consequent investment in detailed studies. There is clear potential for the many methodological insights gained in this domain – in terms of tenets and concepts for the treatment and representation of uncertainties, principles and subtleties in data analysis, and fallacies and pitfalls in judgement and interpretation – to be transferred to risk assessment approaches for other natural hazards.

Acknowledgements

In large measure, the thinking in this chapter has been extensively influenced by the knowledge, expertise and advice generously shared by many colleagues including, but not limited to, David Mallard, Gordon Woo, Bryan Skipp and Steve Sparks, and other colleagues over the years on the UK Seismic Hazard Working Party. However, the views expressed in this contribution, and any errors, belong to the author; he also recognises, apologetically, that the research literature has moved on during the time it took to finalise this manuscript. Roger Musson generously provided an exhaustive and constructive review of a later version of the chapter, and suggested many valuable improvements – the author is greatly indebted to him. WPA was supported in part by a European Research Council grant to Prof. R. S. J. Sparks FRS: VOLDIES Dynamics of Volcanoes and their Impact on the Environment and Society.

References

Abrahamson, N. and Silva, W. (2008) Summary of the Abrahamson & Silva NGA ground-motion relations. *Earthquake Spectra* **24**: 67–97.

Ambraseys, N. and Bilham, R. (2011) Corruption kills. *Nature* **469**: 153–155.

Anderson, J. G. and Brune, J. N. (1999) Probabilistic seismic hazard analysis without the ergodic assumption. *Seismological Research Letters* **70**: 19–28.

Apted, M., Berryman, K., Chapman, N., *et al.* (2004) Locating a radioactive waste repository in the ring of fire. *EOS, Transactions American Geophysical Union* **85**: 465.

Arroyo, D. and Ordaz, M. (2010a) Multivariate Bayesian regression analysis applied to ground-motion prediction equations, part 1: theory and synthetic example. *Bulletin of the Seismological Society of America* **100**: 1551–1567.

Arroyo, D. and Ordaz, M. (2010b) Multivariate Bayesian regression analysis applied to ground-motion prediction equations, part 2: numerical example with actual data. *Bulletin of the Seismological Society of America* **100**: 1568–1577.

Aspinall, W. P., Woo, G., Baxter, P. J., *et al.* (2003) Evidence-based volcanology: application to eruption crises. *Journal of Volcanology & Geothermal Research* **128**: 273–285.

Atkinson, G. M. (2006) Single-station sigma. *Bulletin of the Seismological Society of America* **96**: 446–455.

Barabási, A. -L. (2009) Scale-free networks: a decade and beyond. *Science* **325**: 412–413.

Beauval, C., Hainzl, S. and Scherbaum, F. (2006a) Probabilistic seismic hazard estimation in low-seismicity regions considering non-Poissonian seismic occurrence. *Geophysical Journal International* **164**: 543–550.

Beauval, C., Scotti, O. and Bonilla, F. (2006b) The role of seismicity models in probabilistic seismic hazard estimation: comparison of a zoning and a smoothing approach. *Geophysical Journal International* **165**: 584–595.

Berryman, K. R. and Beanland, S. (1991) Variation in fault behaviour in different tectonic provinces of New Zealand. *Journal of Structural Geology* **13**: 177–189.

Bilham, R. (2004) Urban earthquake fatalities: a safer world, or worse to come? *Seismological Research Letters* **75**: 706–712.

Billings, S. D., Sambridge, M. S. and Kennett, B. L. N. (1994) Errors in hypocenter location: picking, model, and magnitude dependence. *Bulletin of the Seismological Society of America* **84**: 1978–1990.

Blaser, L., Ohrnberger, M., Riggelsen, C., *et al.* (2011) Bayesian networks for tsunami early warning. *Geophysical Journal International* **185**: 1431–1443.

Bommer, J. J., Stafford, P. J., Alarcón, J. E., *et al.* (2007) The influence of magnitude range on empirical ground-motion prediction. *Bulletin of the Seismological Society of America* **97**: 2152–2170.

Brillinger, D. R. and Preisler, H. K. (1985) Further analysis of the Joyner–Boore attenuation data. *Bulletin of the Seismological Society of America* **75**: 611–614.

BSSC (Building Seismic Safety Council) (1997) NEHRP, recommended provisions for seismic regulations for new buildings, part 1: provisions (FEMA Federal Emergency Management Agency 302).

Bye, R. D., Inkester, J. E. and Patchett, C. M. (1993) A regulatory view of uncertainty and conservatism in the seismic design of nuclear power and chemical plant. *Nuclear Energy* **32** (4): 235–240.

Calais, E. and Stein, S. (2009) Time-variable deformation in the New Madrid Seismic Zone. *Science* **323**: 1442.

Campbell, K. W. (1981) Near-source attenuation of peak horizontal acceleration. *Bulletin of the Seismological Society of America* **71**: 2039–2070.

Campbell, K. W. and Bozorgnia, Y. (2008) NGA ground motion model for the geometric mean horizontal component of PGA, PGV, PGD and 5% damped linear elastic response spectra for periods ranging from 0.01 to 10s. *Earthquake Spectra* **24**: 139–171.

Chen, Y.-H. and Tsai, C.-C. P. (2002) A new method for estimation of the attenuation relationship with variance components. *Bulletin of the Seismological Society of America* **92**: 1984–1991.

Childs, C., Watterson, J. and Walsh, J. J. (1995) Fault overlap zones within developing normal fault systems. *Journal of the Geological Society of London* **152**: 535–549.

Choy, G. L. and Bowman, J. R. (1990) Rupture process of a multiple main shock sequence: analysis of teleseismic, local, and field observations of the Tennant Creek, Australia, earthquakes of January 22, 1988. *Journal of Geophysical Research* **95** (B5): 6867–6882.

Clauset, A., Shalizi, C. R. and Newman, M. E. J. (2009) Power-law distributions in empirical data. *SIAM Review* **51**: 661–703.

Coatsworth, A. M. (2005) Earthquake resistant nuclear structures: a regulatory view. *The Structural Engineer* **83**: 30–37.

Coburn, A. W. and Spence, R. J. S. (2002) *Earthquake Protection*, Chichester: John Wiley and Sons.

Dunbar, P. K., Lockridge, P. A. and Lowell, S. (1992) *Catalog of significant earthquakes, 2150 B.C.–1991 A.D., including quantitative casualties and damage*. Boulder, CO: US Dept. of Commerce.

Engell-Sorensen, L. (1991a) Inversion of arrival times of microearthquake sources in the North Sea using a 3-D velocity structure and prior information: part I. Method. *Bulletin of the Seismological Society of America* **81**: 1183–1194.

Engell-Sorensen, L. (1991b) Inversion of arrival times of microearthquake sources in the North Sea using a 3-D velocity structure and prior information: part II. Stability, uncertainty analyses, and applications. *Bulletin of the Seismological Society of America* **81**: 1195–1215.

Fagan, D., Taylor, S., Schult, F., *et al.* (2009) Using ancillary information to improve hypocenter estimation: Bayesian single event location (BSEL). *Pure and Applied Geophysics* **166**: 521–545.

Grünthal, G., Bosse, C., Sellami, D., *et al.* (1999) Compilation of the GSHAP regional seismic hazard map for Europe, Africa and the Middle East. *Annali di Geofisica* **42**: 1215–1223.

Gutenberg, B. and Richter, C. F. (1954) *Seismicity of the Earth and Associated Phenomena*, 2nd edn, Princeton, NJ: Princeton University Press.

Hainzl, S., Steacy, S. and Marsan, D. (2010) Seismicity models based on Coulomb stress calculations. Community Online Resource for Statistical Seismicity Analysis. http://www.corssa.org/articles/themev/hainzl_et_al/hainzl_et_al.pdf.

Holden, C. (2011) Kinematic source model of the 22 February 2011 Mw 6.2 Christchurch earthquake using strong motion data. *Seismological Research Letters* **82**: 783–788.

Huyse, L., Chen, R. and Stamtakos, J. A. (2010) Application of generalized Pareto distribution to constrain uncertainty in peak ground accelerations. *Bulletin of the Seismological Society of America* **100**: 87–101.

IAEA (International Atomic Energy Agency) (1979) *Earthquakes and Associated Topics in Relation to Nuclear Power Plant Siting*. Vienna: International Atomic Energy Agency.

IAEA (International Atomic Energy Agency) (2010) *Seismic Hazards in Site Evaluation for Nuclear Installations*. Vienna: International Atomic Energy Agency.

IBC (2002) *International Building Code 2003*, Country Club Hills, IL: International Building Council.

Joyner, W. B. and Boore, D. M. (1981) Peak horizontal acceleration and velocity from strong-motion records including records from the 1979 Imperial Valley, California, earthquake. *Bulletin of the Seismological Society of America* **71**: 2011–2038.

Joyner, W. B. and Boore, D. M. (1993) Methods for regression analysis of strong-motion data. *Bulletin of the Seismological Society of America* **83**: 469–487.

Kelly, G. B., Kenneally, R. M. and Chokshi, N. C. (1991) Consideration of seismic events in severe accidents. In *Proceedings of the Conference on Probabilistic Safety Assessment and Management PSAM2*, New York, NY: Elsevier, pp. 871–876.

Kolmogorov, A. N. (1933) Foundations of the theory of probability. *Erg. der Math* **2**: n.p.

Kulkarni, R. B., Youngs, R. R. and Coppersmith, K. (1984) Assessment of confidence intervals for results of seismic hazard analysis. In *Proceedings of the 8th World Conference on Earthquake Engineering*, Englewood Cliffs, NJ: Prentice-Hall, pp. 263–270.

Lee, J. and Xiang, Y. (2010) Complexity measurement of fundamental pseudo-independent models. *International Journal of Approximate Reasoning* **46**: 346–365.

Lee, V. W. and Trifunac, M. D. (2010) Should average shear-wave velocity in the top 30 m of soil be used to describe seismic amplification? *Soil Dynamics and Earthquake Engineering* **30**: 1250–1258.

Lee, Y., Zeng, Y. and Anderson, J. G. (1998) A simple strategy to examine the sources of errors in attenuation relations. *Bulletin of the Seismological Society of America* **88**: 291–296.

Liu-Zeng, J., Klinger, Y., Sieh, K., *et al.* (2006) Serial ruptures of the San Andreas fault, Carrizo Plain, California, revealed by three-dimensional excavations. *Journal of Geophysical Research* **111**: B02306.

Main, I. G. and Naylor, M. (2010) Entropy production and self-organized (sub)criticality in earthquake dynamics. *Philosophical Transactions of the Royal Society, Series A* **368**: 131–144.

Mallard, D. J. (1993) Harmonising seismic hazard assessments for nuclear power plants. In *Proceedings of the Institution of Mechanical Engineers Conference 'NPP Safety Standards: Towards International Harmonisation'*, London: Institution of Mechanical Engineers, pp. 203–209.

Mallard, D. J. and Woo, G. (1991) The expression of faults in UK seismic hazard assessments. *Quarterly Journal of Engineering Geology* **24**: 347–354.

Mandl, G. (1987) Discontinuous fault zones. *Journal of Structural Geology* **9**: 105–110.

Matsu'ura, M. (1984) Bayesian estimation of hypocenter with origin time eliminated. *Journal of Physics of the Earth* **32**: 469–483.

McCloskey, J. (2011) Focus on known active faults. *Nature Geoscience* **4**: 494.

McGuire, R. K. (2011) Performance-based seismic design in the United States. Paper presented at the Ninth Pacific Conference on Earthquake Engineering: Building an Earthquake-Resilient Society, 14–16 April, Auckland, New Zealand.

Molina, S., Lindholm, C. and Bungum, H. (2001) Probabilistic seismic hazard analysis: zoning free versus zoning methodology. *Bollettino di Geofisica Teorica ed Applicata* **42**: 19–39.

Morgan, D., Wadge, G., Latchman, J., *et al.* (1988) The earthquake hazard alert of September 1982 in southern Tobago. *Bulletin of the Seismological Society of America* **78**: 1550–1562.

Muir-Wood, R. (1993) From global seismotectonics to global seismic hazard. *Annali di Geofisica* **XXXVI** (3–4): 153–168.

Musson, R. M. W. (2004) A critical history of British earthquakes. *Annals of Geophysics* **47**: 597–609.

Musson, R. M. W. (2012a) Interpreting intraplate tectonics for seismic hazard: a UK historical perspective. *Journal of Seismology* **16**: 261–273.

Musson, R. M. W. (2012b) PSHA validated by quasi observational means. *Seismological Research Letters* **83**: 130–134.

Nicol, A., Walsh, J., Berryman, K., *et al.* (2005) Growth of a normal fault by the accumulation of slip over millions of years. *Journal of Structural Geology* **27**: 327–342.

Oreskes, N., Schrader-Frechette, K. and Belitz, K. (1994) Verification, validation, and confirmation of numerical models in the Earth Sciences. *Science* **263**: 641–646.

Reiter, L. (1991) *Earthquake Hazard Analysis: Issues and Insights*, New York, NY: Columbia University Press.

Richter, C. F. (1958) *Elementary Seismology.* San Francisco, CA: W. H. Freeman and Company.

Sieh, K., Natawidjaja, D. H., Meltzner, A. J., *et al.* (2008) Earthquake supercycles inferred from sea-level changes recorded in the corals of West Sumatra. *Science* **322**: 1674–1678.

Skipp, B. O. and Woo, G. (1993) A question of judgement: expert or engineering? In *Proceedings of the ICE Conference 'Risk and Reliability in Ground Engineering'*, London: Thomas Telford, pp. 29–39.

Smolka, A., Allman, A. and Ehrlicher, S. (2000) An Absolute Earthquake Risk Index for insurance purposes. In *Proceedings of 12th World Conference of Earthquake Engineering*, Upper Hutt, New Zealand: New Zealand Society for Earthquake Engineering.

Stein, S., Liu, M., Calais, E., *et al.* (2009) Mid-continent earthquakes as a complex system. *Seismological Research Letters* **80**: 551–553.

Strasser, F. O., Bommer, J. J. and Abrahamson, N. A. (2008) Truncation of the distribution of ground-motion residuals. *Journal of Seismology* **12**: 79–105.

Strasser, F. O., Abrahamson, N. A. and Bommer, J. J. (2009) Sigma: issues, insights and challenges. *Seismological Research Letters* **80**: 32–49.

Toda, S., Stein, R. S., Kirby, S. H., *et al.* (2008) A slab fragment wedged under Tokyo and its tectonic and seismic implications. *Nature Geoscience* **1**: 771–776.

Toro, G. W., Abrahamson, N. A. and Schneider, J. F. (1997) Model of strong ground motions from earthquakes in central and eastern North America: best estimates and uncertainties. *Seismological Research Letters* **68**: 41–57.

United Nations (2002) *World Population Prospects*, Geneva: United Nations.

Walsh, J. J. and Watterson, J. (1991) Geometric and kinematic coherence and scale effects in normal fault systems. In *The Geometry of Normal Faults*, ed. A. M. Roberts, G. Yielding, and B. Freeman, London: Geological Society of London, pp. 193–203.

Walsh, J. J., Bailey, W. R., Childs, C., *et al.* (2003) Formation of segmented normal faults: a 3-D perspective. *Journal of Structural Geology* **25**: 1251–1262.

Wesson, R. L., Bakun, W. H. and Perkins, D. M. (2003) Association of earthquakes and faults in the San Francisco Bay Area using Bayesian inference. *Bulletin of the Seismological Society of America* **93**: 1306–1332.

Woo, G. (1991) *The Mathematics of Natural Catastrophes*. River Edge, NJ: Imperial College Press.

Woo, G. (1996) Kernel estimation methods for seismic hazard area source modeling. *Bulletin of the Seismological Society of America* **86**: 353–362.

Woo, G. (2011) *Calculating Catastrophe*. River Edge, NJ: Imperial College Press.

9

Landslide and avalanche hazards

T. K. HINCKS, W. P. ASPINALL, R. S. J. SPARKS,
E. A. HOLCOMBE AND M. KERN

9.1 Introduction

The incidence and impact of landslides, avalanches and associated debris flows is increasing in many parts of the world (Aleotti and Chowdhury, 1999; Dai et al., 2002; Petley et al., 2007; Remondo et al., 2008). The development of robust landslide risk reduction policies and decision-making frameworks 'relies crucially on a better understanding and on greater sophistication, transparency and rigour in the application of science' (Dai et al., 2002: 82). This call has been responded to and complemented by the parallel emergence, since the 1980s, of a risk-based approach to landslides within the geoscience and engineering communities. This chapter provides an overview of the types and causes of landslide hazards, current advances in landslide hazard and risk assessment, with focus on issues of uncertainty, and concludes with some related challenges and future research aims. The chapter includes consideration of snow avalanches, which share many of the same physical attributes in terms of triggering and emplacement dynamics.

Section 9.2 defines landslide hazard and outlines three major areas of landslide research within the wider context of landslide risk assessment and management. Different landslide types and processes are identified and critical aspects of research into the physical mechanisms of triggering and emplacement are described. Section 9.3 considers: the assessment of landslide hazard using remote-sensing data and geographical information system (GIS) software; methods for assessing landslide consequences; and issues relating to data acquisition and the handling of uncertainty. Section 9.4 describes landslide risk management, providing examples of regional landslide risk assessment and decision-support systems, and selected new research developments. Section 9.5 identifies current research challenges and opportunities to advance the science.

9.2 Background

9.2.1 Hazard description

A landslide is broadly defined as 'the movement of a mass of rock, debris, or earth down a slope' (Cruden, 1991: 28). The term 'landslide' is used to describe a wide range of phenomena, varying widely in scale, constituent materials, rate and type of movement

Risk and Uncertainty Assessment for Natural Hazards, ed. Jonathan Rougier, Steve Sparks and Lisa Hill. Published by Cambridge University Press. © Cambridge University Press 2013.

(see Section 9.2.3). Avalanche is almost a synonym, although this is typically used to describe debris or rock falls and, of course, snow avalanches. A landslide can form a debris flow if the rock or soil debris mixes with sufficient water. Landslides vary in scale over many orders of magnitude, from the minor failure of a slope on a railway embankment or road cutting, to the gigantic ocean floor debris avalanches that form due to collapses of large parts of ocean islands and continental slopes and which can extend several hundreds of kilometres.

Risk is a product of two components: the hazard, which can be defined as the probability of occurrence of a specific event within a given time period; and the consequence, typically measured by the cost of damage or loss of life (for example, see Crozier and Glade, 2005). Several environmental and human factors contribute to landslide hazard, principally by determining the susceptibility of a slope or geological system to instability. These *preparatory factors* include slope geometry, hydrology (surface- and groundwater), soil type, underlying geology, vegetation and slope loading. As instability is approached, some process or combination of processes can trigger failure. Landslide *triggering factors* include rainfall, earthquakes (Figure 9.1), volcanic activity, changes in groundwater,

Figure 9.1 Landslides caused by the 2010 Haiti earthquake: (a) coastal landslide near Nan Diamant on the south coast of Haiti. Photograph by R. Jibson, USGS (2011a); (b) lakes formed by landslide dams blocking the flow of a tributary stream, northwest of Jacmel (USGS, 2011a); (c) rock slides triggered by the ground shaking, partially blocking the mountain road to the town of Dufort. Photograph by W. Mooney, USGS (2011a).

erosion or undercutting and human activities (e.g. mining, construction or deforestation). Hydrological processes can exert a significant and dynamic influence on slope stability, as both a landslide preparatory factor, in terms of pore pressure and groundwater levels within the slope, and as a landslide trigger as a result of rainfall infiltration or pore pressure changes. Depending on the hydrological properties of the slope material, an increase in moisture content and pore-water pressure can reduce the mechanical strength of the slope to the point of failure. Landslide consequences depend on the type and magnitude of the landslide; the process of emplacement or runout, which determines the rate, travel distance and volume of debris (the hazard footprint); and the exposure and vulnerability of elements within the area of impact. Landslides can destroy or bury buildings, infrastructure and crops, cut off transport links, water and power supplies, and result in significant direct and indirect economic losses. Fast-moving landslides and derivative debris flows have the potential to cause large numbers of fatalities. Unless the occurrence of a landslide is anticipated (e.g. by observation of precursory ground movements or potential triggers such as heavy rainfall) the onset of such flows are typically too rapid to allow warning or evacuation, and escape is unlikely. Landslides also impact on the natural environment, altering landscapes and temporarily damaging habitats and ecosystems (Schuster and Highland, 2007). Secondary hazards associated with landslides include flooding resulting from debris damming streams and rivers, and tsunamis generated by slides into or under water.

9.2.2 An overview of landslide hazard and risk research

Here we consider research in the context of landslide risk management. Risk management options typically include:

(1) risk avoidance, for example prohibiting development or access in areas exposed to landslide hazards;
(2) risk reduction – this can include non-intrusive safety measures such as warning systems, and engineering interventions to reduce the probability of slope failure (e.g. drainage, stabilisation) or to minimise the extent or impact of the landslide (e.g. barriers). This is discussed in more detail in Section 9.4.4;
(3) risk transfer, distributing the burden of loss – e.g. through insurance or legislation.

A certain level of risk may be deemed tolerable, if the costs associated with mitigating the risk are far greater than the potential loss. The suitability and effectiveness of each of these options for any given situation will depend on the extent of the hazard, value at risk and cost–benefit evaluation of any mitigation actions. In landslide risk assessment this generally requires one or more of the following activities (see Figure 9.2):

(1) Identifying zones of landslide *susceptibility*, typically based on assessment of preparatory factors and/or analysis of inventory of previous landslides.

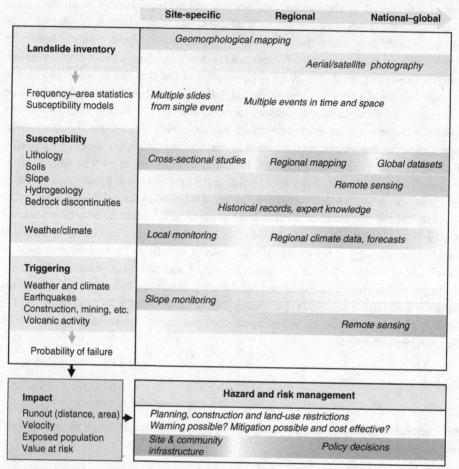

	Site-specific	Regional	National–global
Landslide inventory	Geomorphological mapping		
			Aerial/satellite photography
Frequency–area statistics Susceptibility models	Multiple slides from single event	Multiple events in time and space	
Susceptibility			
Lithology Soils	Cross-sectional studies	Regional mapping	Global datasets
Slope Hydrogeology			Remote sensing
Bedrock discontinuities	Historical records, expert knowledge		
Weather/climate	Local monitoring	Regional climate data, forecasts	
Triggering			
Weather and climate Earthquakes Construction, mining, etc. Volcanic activity	Slope monitoring		
			Remote sensing
Probability of failure			

Impact	**Hazard and risk management**	
Runout (distance, area) Velocity	Planning, construction and land-use restrictions Warning possible? Mitigation possible and cost effective?	
Exposed population Value at risk	Site & community infrastructure	Policy decisions

Figure 9.2 Schematic showing the various components of landslide hazard and risk evaluation and management.

(2) Understanding the triggering processes and predicting landslide events spatially and temporally.

(3) Assessing the probable consequences based on the exposure and vulnerability of elements at risk with respect to the landslide hazard and its emplacement dynamics.

Data relating to landslide preparatory factors and a reliable inventory of past landslides are required to identify zones of landslide susceptibility (Section 9.2.4). Depending on the spatial extent of the investigation, these data may be in map form, providing general information on features such as slope, ground cover, soils and underlying geology, or more detailed descriptions of individual terrain units (slope elements). Information is drawn primarily from geological, remotely sensed and historical data, including records and studies of past landslides. Potentially unstable slopes are identified and, in some cases, pre-landslide slope movements are monitored (see Sections 9.3.2 and 9.4.6). Such studies are augmented by theoretical and empirical models of slope stability, and by laboratory measurements of geotechnical properties.

The effect and probability of potential triggering events can sometimes be assessed at this stage. The effect of the triggering event can be assessed based on empirical evidence (e.g. a particular rainfall 'threshold' or ground-shaking intensity which could be expected to trigger landslides in a given region), or by detailed modelling of dynamic hydrological and mechanical processes occurring within a specific slope or rock outcrop. The return period or frequency of the triggering event thus translates to the probability of the landslide occurring within some specified future time interval. An objective is to anticipate landslide occurrences, and to carry out post-event investigations that improve understanding of the factors leading to landslides.

The topic of emplacement aims to understand the dynamics of landslides and debris avalanches (and potential impact), and is discussed in Section 9.2.6. From a hazards perspective, the volume of debris involved, the rheology of the constituent material, the distance and rate of travel and the areas inundated are of principal interest. The main approaches to understanding emplacement include studies of previous landslide deposits, generation of empirical relationships between different landslide geometries and their runout parameters, models of the flows based on concepts of dynamic materials properties (rheology), laboratory-scale studies of multiphase flows (mixtures of solids, air and fluids) and laboratory measurements of material properties.

9.2.3 Landslide classification

The first step in assessing landslide hazard and risk is to identify the specific type of landslide posing the threat. The term 'landslide' covers a range of phenomena involving the movement of diverse materials, varying widely in scale and rate of movement. Landslides and avalanches are typically classified according to the type of movement or mode of failure, and the material involved. Various classification schemes have been developed. The most widely used are shown by Varnes (1978 (see Table 9.1); see also Highland and Johnson, 2004; Highland and Bobrowsky, 2008) and Cruden and Varnes (1996), adopted by the British Geological Survey and the UNESCO Working Party on Landslide Inventory (WP/WLI, 1993), and Hutchinson's classification of mass movements on slopes (Hutchinson, 1988). Other variants include the EPOCH classification scheme for European landslides (Casale *et al.*, 1994; Dikau *et al.*, 1996). The wide range of materials in flows generated by landslides encompasses a very wide range of rheological and mechanical properties.

Movement can either be described as: a *fall*, resulting in a bouncing, rolling motion; a *topple*, where an entire column of rock collapses with an outward rotational movement; a *slide* (e.g. as in Figure 9.1c), where blocks remain largely intact during emplacement; or

Table 9.1 *Varnes' (1978) classification of slope movements (see also Giani, 1992). Fine material is defined as a composition of >80% sand and finer (particles <2 mm diameter)*

			Material		
				Engineering soils	
Movement			Bedrock	Coarse	Fine
Falls			Rock fall	Debris fall	Earth fall
Topples			Rock topple	Debris topple	Earth topple
Slides	Rotational	Few units	Rock slump, block slide, slide	Debris slump, block slide, slide	Earth slump, block slide, slide
	Translational	Many units			
Lateral spreads			Rock spread	Debris spread	Earth spread
Flows			Rock flow (deep creep)	Debris flow	Earth flow
				Soil creep	
Complex			Combination of two or more principle types of movement		

a *flow*, where material behaves in a fluid-like manner, commonly with complex rheology. Water can be a key component in many types of landslides. For example, rainfall and surface water infiltration is the most common trigger of slides and flows in weathered materials in the humid tropics (Lumb, 1975; Costa, 2004). *Creep* is a slow gravitational spreading. *Lateral spread* is unusual in that it occurs on almost flat or gently sloping ground. In soils this type of failure is caused by *liquefaction* – where loose, saturated material assumes a liquefied state, typically triggered by strong ground shaking, e.g. due to an earthquake or construction activities (Highland and Johnson, 2004).

Small- to medium-sized landslides commonly involve failure of the veneer of soil, superficial deposits (such as volcanic ash) and weathered rock over steep slopes. Deeper soils can experience rotational sliding and slumping movements along a circular or irregular shear zone within the soil mass. Examples include the earthquake-triggered Las Colinas landslide, El Savador in 2001, involving a volume of 200 000 m^3 of soil (Konagi *et al.*, 2002; see Figure 9.3) and the Pozzano landslide on the Sorrento Peninsula on 10 January 1997, caused by heavy rain on steep mountain slopes covered by volcanic pumice and ash layers from Vesuvius (Calceterra and Santo, 2004). At usually larger scales, a rock mass can fail along a geological weakness to generate a rock avalanche. Suitably oriented bedding planes, metamorphic foliations, faults and boundaries between major geological units with contrasted strength are examples of surfaces of weakness that can develop into failure surfaces. A large deep-seated weakness is sometimes called a decollement surface. Large landslides are commonly described as rock or debris avalanches (e.g. see Figure 9.4), which have large volumes (typically >10^6 m^3), high velocities (>30 m s^{-1}) and large runouts (Hutchinson, 2008). Well-documented examples include the Hope landslide in British

Figure 9.3 Las Colinas landslide in Santa Ana, El Salvador, January 2003. The slide was triggered by a magnitude 7.3 earthquake and caused 600 fatalities. Photograph courtesy of NGI.

Figure 9.4 The hummocky terrain of a debris avalanche deposit, due to collapse of a previous edifice of Parinacota volcano (Clavero *et al.*, 2002). The avalanche, dating from 7000 BP, extends as a sheet of rock debris to 30 km and has a volume of about 4 km^3.

Columbia in 1965, with a volume of 47 000 000 m³ (Matthews and McTaggart, 1969) and the Frank landslide in Alberta in 1903, with a volume of 30 000 000 m³ (Cruden and Krahn, 1973).

Submarine landslides are associated with unstable undersea slopes, coastal terrains and ocean islands, and can be triggered by rapid sedimentation, large-magnitude earthquakes and volcanic eruptions. Slumps and debris avalanches typically occur on continental margins, on the fronts of deltas and volcanic islands. The bathymetry of the Lesser Antilles island arc in the Caribbean shows evidence of extensive debris avalanche deposits (Figure 9.5; Boudon *et al.*, 2007). Submarine landslides produce some of the longest runout distances and largest volume collapses on earth. For example, prehistoric submarine land- slides cover at least 100 000 km² between the islands of Kauai and Hawaii in the Pacific Ocean, with some individual debris avalanches more than 200 km long and 5000 km³ in volume (Schuster and Highland, 2001). Off north-west Africa, debris avalanches associated with the failure of continental margin extend as much as 700 km across the adjacent Atlantic sea floor (e.g. the Saharan slide, Embley, 1982; Weaver, 2003). Large submarine landslides (e.g. the Storegga slide, see Bondevik *et al.*, 2003; Chapter 10 in this volume) can also result in major tsunamis. Submarine landslides are a significant threat to offshore commercial activities and sea floor installations, notably in the oil and gas industry, with production facilities located in deep-water environments.

The landslide velocity scale (Figure 9.6; see Cruden and Varnes, 1996) provides a system for classifying landslides by their velocity and resulting destructive potential. The scale ranges from velocity class 1, extremely slow creeping slope failures with movement of the order of several millimetres per year; to class 7, extremely rapid ($>5\,\mathrm{m\,s}^{-1}$) flows which have enormous destructive potential both in terms of damage to infrastructure and human fatalities. For landslides up to velocity class 3 it may be possible to intervene (e.g. to halt or divert the flow), whereas for faster flows (up to class 5), forecasting is the primary tool to reduce risk.

9.2.4 Landslide susceptibility and triggering mechanisms

Landslides occur when the shear strength of a slope material is exceeded by the shear stresses acting upon it (see Crozier, 1986 for a discussion of causes of instability). Rock slopes and weathered or fractured material are susceptible to landslides along zones of relative weakness, such as joints and geological discontinuities. Weak layers usually play a key role as the initial slide surfaces. Layer orientation is critical and the slide surface will have a strong tendency to be oriented in the direction most favourable to shear failure. The Mohr–Coloumb failure criterion or more complex variants provide the standard theoretical framework for understanding failure (e.g. see Jiang and Xie, 2011; Yang and Yin, 2004; Laouafa and Darve, 2002). The moisture content of the slope material is commonly an important factor, and many slides occur in saturated slopes, where groundwater levels have risen or where surface water has infiltrated, thus increasing pore fluid pressures and reducing

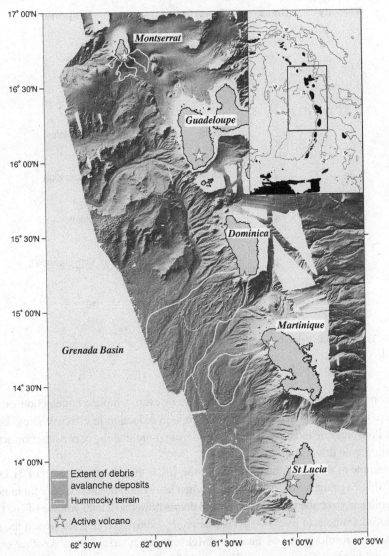

Figure 9.5 Bathymetry of the Lesser Antilles island arc in the Caribbean, from Montserrat to St Lucia, showing the extent of mapped submarine avalanche deposits off the volcanic islands (from Boudon *et al.*, 2007).

strength, especially in weak layers. Human activities can increase susceptibility and potentially trigger landslides by increasing the loading on slope, removing vegetation, excavating slope material (mining, road cuttings, construction) and altering the hydrology (changing drainage patterns or irrigating the slope). Physical triggers of landslides include intense or prolonged rainfall, rapid snowmelt, freeze–thaw weathering, earthquakes, volcanic activity

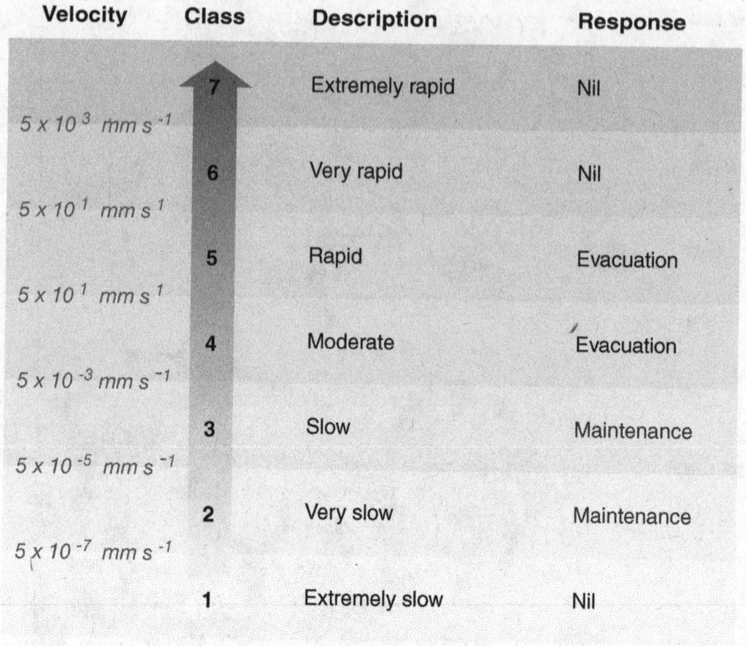

Velocity	Class	Description	Response
	7	Extremely rapid	Nil
5×10^{3} mm s^{-1}			
	6	Very rapid	Nil
5×10^{1} mm s^{1}			
	5	Rapid	Evacuation
5×10^{1} mm s^{1}			
	4	Moderate	Evacuation
5×10^{-3} mm s^{-1}			
	3	Slow	Maintenance
5×10^{-5} mm s^{-1}			
	2	Very slow	Maintenance
5×10^{-7} mm s^{-1}			
	1	Extremely slow	Nil

Figure 9.6 The landslide velocity scale.

and erosion of the toe of the slope. Slope failure or even complete liquefaction can occur when there is an increase in pore fluid pressure with reduction in effective stress between particles. Pore pressure increases can be caused by seismic shaking or construction activities (e.g. blasting, pile driving) or intense rain.

One example of a catastrophic landslide in the UK is the Aberfan disaster of 21 October 1966, in which a primary school was buried when heavy rain destabilised a coal mine spoil tip. The collapse, described as a flow slide and debris flow, had a total volume of 107 000 m^3 and produced a runout of over 600 m (Siddle *et al.*, 1996). The slide destroyed the school and 20 houses, resulting in 144 fatalities (McLean and Johnes, 2000). Another example occurred at Limbe in Cameroon on 27 June 2001. Twenty-four people were killed, 120 houses were destroyed and 2800 people made homeless after two days of heavy rain led to multiple landslides on slopes already destabilised by deforestation and undercutting as unofficial shantytowns developed at the base of steep slopes (Ayonghe *et al.*, 2004; Zogning *et al.*, 2007). The landslides all occurred within a short space of time (minutes), suggesting that they may have been triggered by a minor earth tremor from a large landslide (Ayonghe *et al.*, 2004). These examples illustrate that human intervention in landscapes can often create conditions for landslides, and that they may also have multiple causes. The potentially catastrophic nature of landslides means that the best approach to risk reduction is to identify circumstances likely to lead to slope instability.

Figure 9.7 Head scarp of the Fatalak Landslide near Rudbar, Iran, triggered by the magnitude 7.7 Manjil earthquake in June 1990. The landslide buried two villages and caused 173 fatalities (Shoaei *et al.*, 2005). The quake also caused widespread soil liquefaction, resulting in extensive building damage up to 85 km from the fault (Yegian *et al.*, 1995). Photograph courtesy of NGI.

9.2.5 Landslides and other natural hazards

Large earthquakes often trigger landslides and avalanches (Figure 9.7). Ground shaking can directly cause collapse of unstable material, or result in an increase in pore pressure that then leads to slope failure. Furthermore, the inertial forces caused by ground acceleration increase the shear stress on potential slide surfaces for short durations during the earthquake. This means that seismic hazard information and 'ShakeMaps' (near-real-time maps of ground motion and shaking intensity, Wald *et al.*, 2005) can be used to identify areas prone to landsliding (e.g. see Kaynia *et al.*, 2010). Malamud *et al.* (2004a) estimate that there is a minimum earthquake magnitude $M = 4.3 \pm 0.4$ necessary to generate a landslide, and a major earthquake can cause tens of thousands of individual landslides. The 4 February 1976 Guatemala magnitude 7.6 earthquake is estimated to have resulted in around 50 000 individual slides, with a total volume of approximately $1.16 \times 10^8 \, \mathrm{m}^3$ (see Keefer, 1994; Malamud *et al.*, 2004a, 2004b for this and several other examples). The 12 January 2010 magnitude 7.0 Haiti earthquake caused numerous landslides, particularly in the mountain-ous regions around Port-au-Prince (see Figure 9.1; Theilen-Willige, 2010).

Hydrology and climate are also critical factors in slope failure, and many landslides are triggered by extreme rainfall, storms and hurricanes. For example, the recent severe flooding in south-eastern Brazil in January 2011 resulted in extensive and devastating mudslides. There were at least 806 fatalities (EM-DAT, 2011; regional news reports state

904 deaths – O Globo, 2011), and around 14 000 were displaced from their homes (Prada and Kinch, 2011), making the event the greatest natural disaster to have hit Brazil. As a result, the government was criticised for failing to impose restrictions on construction in high-risk areas (Prada and Kinch, 2011).

Secondary hazards associated with landslides include floods related to damming of river valleys and the subsequent dam breach (Costa and Schuster, 1988; Korup, 2002) and tsunamis (e.g. Ward, 2001; Fritz *et al.*, 2004).

Landslides and lahars (volcanic mudflows and debris flows) are common in volcanic regions, and are typically triggered by intense rainfall. Slope failure can occur due to abundant unconsolidated newly erupted material, steep, faulted or fractured slopes on the volcanic edifice and areas weakened by alteration. Eruptive activity can also directly trigger collapse – for example, due to a phreatic or magmatic explosion, or shallow magma intrusion or loading of a slope by lava. On volcanoes where there is extensive hydrothermal activity, the edifice may be pervasively fractured and weak, and prone to development of high pore pressures due to circulating hydrothermal fluids. These features can result in susceptibility to large and catastrophic collapses (e.g. Clavero *et al.*, 2004). Large volcanic debris avalanches entering the sea can cause devastating local tsunamis, as happened in Mount Unzen in 1792, when around 5000 people were killed by the volcanogenic tsunami (Tanguy *et al.*, 1998). Large submarine landslides are also common in volcanic island settings. A large flank collapse on an active volcano can in turn lead to more violent eruptive activity, due to rapid decompression of either the hydrothermal system or shallow magma bodies, or both (Cecchi *et al.*, 2004). As failure arises due to a combination of processes, it can be difficult to determine whether the primary forcing is volcanic, or due to external factors such as rainfall, groundwater or seismicity. This makes it difficult to model and forecast the hazard.

Lahars typically accompany volcanic landslides and flank collapses, as unconsolidated material enters rivers or becomes saturated and remobilised by rainfall. Such flows can be highly erosive. In 1998, intense rainfall during hurricane Mitch triggered a small ($\sim 1.6 \times 10^6 \, \text{m}^3$) flank collapse at Casita volcano, Nicaragua, causing major lahars and killing 2500 (see Figure 9.8; Kerle, 2002). The initial lahars were very wet (around 60 wt % solids).

Figure 9.8 The 1998 landslide on the flanks of Casita volcano, Nicaragua. Photographs courtesy of NGI.

They rapidly entrained additional material (sediment, debris and water) and bulked significantly, to around 2.6 times the initial collapse volume (Scott *et al.*, 2005). Down-flow bulking therefore has significant implications for estimating potential runout and risk of inundation. A large flank collapse will change the topography of the volcanic edifice, and can affect the direction and areas of impact of future pyroclastic flows and lahars, changing the future hazard footprint.

9.2.6 *Runout and emplacement dynamics*

Landslide runout is an important component of hazard and risk assessment. Much attention has been given to the excess runout problem. Most small landslides have a runout, L, that can be simply related to the potential energy as measured by the height dropped (H) from the centre of the failed mass. H/L values of 0.3 are typical in small-volume landslides that are approximately two-dimensional (i.e. they flow down a slope and are confined to a valley with limited lateral spread). H/L (the Heim coefficient) is commonly equated in a rough and ready way to the friction coefficient. However, as landslide volume increases, the tendency is for H/L to decrease and the flow to runout further (Hsu, 1975; Dade and Huppert, 1998). This has led to several different explanations for long runout, including fluidisation, high basal pore fluid pressures, sliding on weak zones or substrates, acoustic shaking and shear thinning and frictional melting. Dade and Huppert (1998) compiled data on avalanche volumes and showed that first-order runout was proportional to the ratio $A/V^{2/3}$, where A is the deposit area and V is the deposit volume. Further, they suggested that this relationship could be explained if the avalanches had an approximately constant retarding stress during emplacement. Kelfoun and Druitt (2005) showed that a model using this concept for the large Socompa volcanic debris avalanche in Chile explained most of the morphology of the deposits. However, a complete physics-based explanation of excess runout remains elusive.

Given the variety of volumes, properties of the materials involved in landslides, triggering scenarios and geological complexity, it is not surprising that there are many different approaches to modelling of emplacement and some controversy. The simplest modelling concept is the movement of a rigid block sliding down a slope under gravity, opposed by friction. While a useful heuristic concept, it is rarely an adequate description since most landslides involve internal deformation and larger landslides (and avalanches) behave like a fluid. Thus, most models nowadays assume some kind of rheology for the material. A very common approach is to assume that geological materials have Mohr–Coloumb properties characterised by a basal friction angle. Materials are then assumed to be cohesionless or cohesive, depending on the nature of the geological material. In some cases landslides are modelled as non-Newtonian fluids with a yield strength and nonlinear stress–strain rate response (e.g. see Hungr, 2009). There is an increasing tendency to develop granular flow models of rock avalanches (e.g. Mangeney-Castelnau *et al.*, 2003) with Coloumb material properties.

In general terms, landslide models are highly empirical, and controversies abound because the models do not yet adequately describe the physics of these complex multiphase flows. On the other hand, the models can be tuned by data. Thus they remain very useful and indeed the only option for practical hazards assessment. The capability of emplacement models has increased dramatically with increasing computer power and the development of much more accurate digital elevation models (DEMs). Thus the emplacement of avalanches and landslides can be reproduced extremely well by numerical models. Typically the models involve two or three free parameters (such as basal and internal friction values), which can be adjusted to fit the observed distribution of landslide and avalanche deposits (e.g. Heinrich *et al.*, 2001). The model can then be applied to simulate future events in the same area, or in areas with similar geological characteristics. If many landslides have occurred in a particular area it is also possible to investigate the uncertainty empirically.

Despite the apparent success and utility of empirical models, one challenge for research is that the conceptual models for landslides are not consistent with observations of small laboratory granular avalanches. Models are typically based on shallow-water equations, where properties of the flow are depth-averaged and a simple stopping criterion is assumed. However, simple granular flows studied in the laboratory are emplaced by transfer of particles to a growing static layer across a dynamic interface at the base of the flows (e.g. Lube *et al.*, 2007) and, further, much of their emplacement is inertial with no requirement for friction until the last stages of emplacement.

9.2.7 Snow avalanches

Snow avalanches are characterised according to the type of release (e.g. loose snow or a cohesive slab), motion (air or ground), the involvement of water and the plane of failure, which can either be a surface snow layer or the ground interface. Snow layers are highly porous, close to melting point, and have complicated and spatially variable mechanical properties. These features make modelling, hazard mapping and short-term forecasting very challenging. Failure can be triggered by ground shaking (e.g. earthquakes or explosions), changes in loading (e.g. due to skiers or snowfall) or rain (Schweizer *et al.*, 2003). Currently there are models describing the mechanics of fracturing, and competing process models simulating the effects of damage and sintering, but there are gaps in the understanding of how failure initiates and develops over time (Schweizer *et al.*, 2003).

A slab avalanche involves the release of a cohesive layer of snow over a large plane of weakness. Sliding snow slabs progressively break into smaller blocks and grains. Avalanche formation involves complex interactions between the ground, snow layers and meteorological conditions. Commonly powder snow avalanches (PSAs) develop from flowing snow avalanches (SNA) when snow is suspended by turbulence due to entrainment of air into the avalanche. Flowing avalanches have a high-density basal region ($\sim$150–500 kg m^{-1}), flow depths are typically of the order of a few meters and reach speeds of 5–50 m s^{-1} (Ancey, 2012). PSAs empirically develop when the SNA reaches velocities in excess of about

20 m s^{-1}. The fluidised and powder part of dry-mixed snow avalanches can reach speeds of 70 m s^{-1} or greater (Vilajosana *et al.*, 2006), with flow depths of the order 10–100 m and densities of 5–50 kg m^{-1} (Ancey, 2012). Speed is dependent on the initial mass of snow, frictional conditions and the path of the avalanche (height, length and angle of drop). Avalanches of wet snow typically move more slowly for an equivalent mass and drop, but their higher density can make them very destructive. There are similarities between snow avalanches and pyroclastic flows in terms of the dynamics of the flow and formation of accompanying suspension clouds. Volcanic flows are discussed more fully in Chapter 11.

9.2.7.1 Snow avalanche models

An avalanche can be modelled as a flow of continuous material by solving the equations of motion, using appropriate initial and boundary conditions to account for the avalanche release area and mass, and topography of the avalanche track. Both one-dimensional and two-/three-dimensional avalanche models are used operationally, primarily for estimating runout of extreme avalanches, generating scenarios for hazard assessment and in some cases as a tool for hazard zoning. Impact pressures derived from runout velocities may serve as a basis for hazard maps, and this approach has been widely applied in Switzerland (BFF/SLF, 1984), Austria (Sauermoser, 2006; Höller, 2007) and Canada (Jamieson *et al.*, 2002). Two-/three-dimensional models can be used to evaluate potential release areas, avalanche tracks and impact areas. Such models are also used for the design of buildings and technical avalanche protection measures such as dams and obstacles (Jóhannesson *et al.*, 2009), and estimation of potentially endangered areas on large topographic scales. These models cannot provide any information on avalanche magnitude–frequency relationships. In all models there is a high degree of structural uncertainty, and this is generally handled using large input parameter ranges. For fundamental reasons, calibration of empirical frictional parameters is problematic, and heavily dependent on expert judgement. Black-box use of guideline parameters may lead to misleading results, and in two-/three-dimensional models parameter choice is typically supported by GUIs, further increasing the risk of uninformed use. Parameters of statistical models, however, are based on direct observation.

Examples of one-dimensional operational models include the Analytical Voellmy–Salm model (Salm *et al.*, 1990), which is based on the equations of motion of a block mass on a frictional incline; and AVAL-1D (Bartelt *et al.*, 1999; Gruber *et al.*, 2002), which solves equations of motion on a one-dimensional-shaped topography, taking into account erosional processes. Two-dimensional models include AVAL-2D/RAMMS (Christen *et al.*, 2002), which employs depth-averaged hydraulic equations of motion and can simulate different flow rheologies, and SamosAT (Sampl and Zwinger, 2004), which can model both flow and suspended (powder) components of avalanches. Powder transport is modelled in three dimensional models, as a two-phase flow, coupled to the main flow component. Harbitz (1998), Naaim (2004) and Barbolini *et al.* (2008) review a range of research models.

Hazard and risk assessments are largely based on estimates of potential avalanche runout. Whereas numerical avalanche models have the advantage of being able to predict avalanche velocity and impact pressure, structural and parametric uncertainties can limit their reliability. Statistical models, which include the α–β model (Lied and Bakkehøi, 1980) and the Runout Ratio model (McClung and Lied, 1987), offer the potential for more robust treatment of uncertainties (e.g. see Raftery *et al.*, 1997). The α–β model (originally developed for Norway but now widely used) is a regression model, based on the correlation between maximum runout and a number of topographic parameters. Regression coefficients are determined from the paths of several hundred extreme avalanche events, where 'extreme' in this case is defined as a return period of the order of 100 years (Gauer *et al.*, 2010). In its simplest form $\alpha = b_0\beta + b_1$, where α is the angle marking the maximum runout of the dense component of the avalanche given some release point, β is a measure of the mean slope angle, and b_0 and b_1 are regression coefficients. The residuals represent unexplained variation between the chosen model and the data, and give an indication of model uncertainty.

Further, Gauer *et al.*'s (2010) re-analysis of runout data used to formulate α–β models raises questions about classical dynamical modelling approaches for calculating avalanche-retarding forces. They find that the mean retarding acceleration is predominantly a function of mean slope angle β, and find that flow volume and fall height do not strongly influence runout ratio.

9.2.8 Quantification of impact

It can be difficult to quantify the impact of historic landslides from the literature, as losses are often attributed to the triggering event – for example, in the case of landslides resulting from earthquake or hurricane activity. For the period 1900–2010, EM-DAT (2011) reports a total of 64 015 persons killed, and over 13 million persons affected (injured, displaced or otherwise affected) by mass movements in a total of 631 events (see Figure 9.9). This equates to a global average of around 575 fatalities per year. There are some problems with

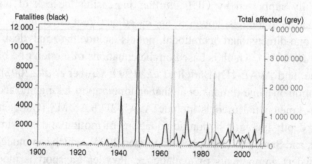

Figure 9.9 EM-DAT (2011) data for annual number of fatalities and persons affected by dry and wet mass movements since 1900. This is not an exhaustive record, and as with any catalogue there are issues of completeness and reliability. Events included in the database fulfil at least one of the following criteria: ten or more people reported killed, 100 or more people reported affected, declaration of a state of emergency, or a call for international assistance. The total affected is the sum of those made homeless, injured or otherwise affected (e.g. displaced or evacuated).

Table 9.2 *Catastrophic landslides: largest recorded by volume (USGS, 2011b)*

Year	Location/event	Volume
1980	United States (Washington) Mount St Helens rock slide/debris avalanche, caused by the eruption of Mount St Helens.	$2.8 \times 10^9 \, m^3$
1974	Peru (Huancavelica) Mayunmarca rock slide/debris avalanche. Thought to have been triggered by rainfall or river erosion.	$1.6 \times 10^9 \, m^3$
1911	Tadzhik Republic (formerly USSR) Usoy rock slide, caused by magnitude 7.4 earthquake.	$2.0 \times 10^9 \, m^3$

Table 9.3 *Catastrophic landslides: greatest number of fatalities (USGS, 2011b)*

Year	Location/event	No. fatalities
1970	Nevados Huascaran debris avalanche (volume $30–50 \times 10^6 \, m^3$) triggered by magnitude 7.7 earthquake. Town of Yungay destroyed, Ranrahirca partially destroyed. Average velocity of flow $280 \, km \, hr^{-1}$.	18 000
1949	Khait rock slide, Tadzhik Republic, triggered by magnitude 7.5 earthquake. Rock slide transformed into large loess and granite debris avalanche. 12 000–20 000 killed or missing; 33 villages destroyed.	12 000–20 000
1920	China (Ningxia) Haiyuan landslides caused by an earthquake. Many villages destroyed; 675 large loess landslides created more than 40 lakes; slide volumes uncertain.	~100 000

record accuracy; however, this is likely to be an underestimate of the true number, due to under-reporting, recording thresholds (for EM-DAT this constitutes an event with ten or more fatalities) and deaths attributed to the triggering event.

In the United States, landslides are estimated to cause losses of around US$1–3 billion (USGS, 2006; 2006 dollars), and result in 20–50 deaths per year, higher than the annual average for earthquake deaths (Schuster and Fleming, 1986; Schuster and Highland, 2001). According to Li and Wang (1992) there were at least 5000 landslide-related deaths during the period 1951–1989, an average of 130 deaths per year. Tables 9.2 and 9.3 summarise the largest and highest-impact landslides of the twentieth century.

Despite scientific advances in understanding, losses associated with landslides remain high with no sign of declining. This is attributed to population pressures driving development in landslide-prone areas (Schuster and Highland, 2001). Deforestation and land-use changes can also affect slope stability, increasing landslide susceptibility in areas with little or no historic activity. Increased regional precipitation caused by changing climatic patterns may also cause landslide activity in previously stable areas (Schuster, 1996).

9.3 Landslide hazard and risk assessment

Landslide hazard and risk research is a cross-disciplinary science. Several scientific methods and techniques typically need to be applied. These range from field- and laboratory-based work to remote sensing, from mapping and monitoring to prediction, from empirical and statistical analysis to process-based numerical modelling, and the application of both deterministic and probabilistic modelling. Areas suspected of being prone to landslides require detailed geomorphological, geological, geophysical and geotechnical investigations to ascertain the hazard, and may require instrumental monitoring if there is evidence of unstable movements, such as creep prior to a major failure (e.g. see Oppikofer *et al.*, 2009). Increasingly, automated terrain-analysis tools are being used to map susceptible slopes, employing proprietary or open source GIS (e.g. see Clerici *et al.*, 2006). Google Earth is also becoming widely used to identify regions at risk and map existing landslide scars (e.g. Teeuw *et al.*, 2009; Sato and Harp, 2009; Yang *et al.*, 2010). In places with extensive areas of high landslide susceptibility and limited field data, expert judgement of inventory or spatial extent is sometimes substituted for detailed mapping (see Section 9.3.4).

Definitions of terms such as landslide susceptibility, hazard and risk are not always used consistently in the literature. Landslide inventory maps show locations and extents of past landslides, and provide information that can be used to develop hazard or susceptibility models and determine frequency–area statistics. An inventory may constitute multiple triggered events over time, or a single triggered event resulting in multiple landslides (see Section 9.3.3). Different catalogues therefore cannot be easily compared or assimilated. A discussion of the issues associated with the production, evaluation and comparison of landslide inventory maps is presented by Galli *et al.* (2008).

Hazard maps should ideally provide information on the location, magnitude and return period, accounting for the probability of triggering events. Most landslide hazard maps indicate where failure is likely to occur, but do not address recurrence rates (Carrara *et al.*, 1995), and are therefore more accurately described as susceptibility maps. Landslide susceptibility maps indicate slope stability, based on the state and properties of the slope, including factors such as slope inclination, vegetation cover, geology, etc., but do not capture any temporal information (return periods or probability of triggering) or the potential spatial extent of any resultant landslides. By contrast, a hazard map would typically show the predicted extent or 'footprint' of the hazard and in some cases information on probabilities. Risk maps generally show the annual expected loss (cost, fatalities, etc.) for a region, accounting for the probability of the hazard event, its spatial extent and impact.

However, it is difficult to quantify the often highly uncertain elements that contribute to risk, and in some cases it can be useful to use relative or qualitative measures of assessment. Aleotti and Chowdhury (1999) list commonly used qualitative and quantitative methods of landslide hazard and susceptibility assessment. Approaches include the use of geomorphological data to make qualitative assumptions about stability, heuristic or index-based methods (rule-based combinations of indices or parameter maps), bivariate and multivariate statistical analysis (relating the spatial distribution of landslides to factors contributing to

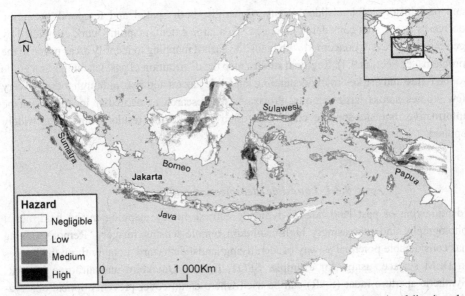

Figure 9.10 Map showing precipitation-induced landslide hazard for Indonesia, following the methodology of Nadim *et al.* (2006) (Figure 7 in Cepeda *et al.*, 2010; see also GAR, 2011). The map shows relative landslide hazard, estimated from susceptibility factors describing the slope, lithology, soil moisture, precipitation and seismicity, and can be linked to an approximate annual probability of failure (see table 7 in Nadim *et al.*, 2006). Courtesy of NGI.

instability), geotechnical engineering approaches (probabilistic and deterministic safety factor calculations) and 'black box' models using neural network analysis. Figure 9.10 is an example of a qualitative hazard map for Indonesia, developed by the NGI (Norwegian Geotechnical Institute) for GAR (2011) in a study to evaluate precipitation induced landslide risk. Risk maps (not shown here) are presented in Cepeda *et al.*, 2010 and GAR, 2011. The risk scale was developed using estimates of physical exposure and a number of socio-economic indices, namely: land cover (forest, arable); percentage population without access to clean water; percentage population without access to health facilities; Human Development Index (HDI); Gender Development Index (GDI); and Human Poverty Index (HPI). Physical exposure was calculated by weighting the landslide hazard maps (as shown in Figure 9.10) with respect to the population density at pixel level. Such an approach is useful for making broad comparisons of risk on a global scale, where regional complexities (complicated by varying measures of capacity, value at risk and risk tolerance) make quantitative risk assessment difficult. One disadvantage is the absence of uncertainty assessment.

Landslide hazard assessments range from very local studies (e.g. monitoring areas of potential slope failure and mapping landslides), to regional studies investigating the probability of failure, given such factors as slope, geology, antecedent weather conditions, regional seismicity (or probability of other 'shaking' events such as explosions), land use and past history of landslide activity. Generally, both spatial and temporal resolution decreases as the area considered increases. To a large extent, mapping work completed by geological surveys is concerned with landslide hazard mapping to identify areas prone to the problem (see Section 9.4). Regional studies map the distribution of past landslides as a guide to the future and are key to understanding landslide hazard and risk, although there are very few studies across large regions. Process-based research at particular locations can be informative, but site-specific conditions often preclude the results from being widely applied.

9.3.1 *Landslide identification and mapping*

Identification of past landslides is typically carried out by mapping in the field, aerial photography and increasingly high-resolution remote-sensing imagery. Remote sensing has considerable potential as way of identifying landslide hazard, notably through changes in DEM surface, using, for example, SPOT, Ikonos, Quickbird and airborne LIDAR. Synthetic aperture radar (SAR) can be used where cloud cover prevents optical observations, but due to shadowing effects there can be problems with data for regions with steep slopes. Areas that are very unstable may be monitored by instruments to see if the ground is creeping and whether these movements might be accelerating (see the discussion of monitoring techniques in Section 9.4.6); for example, InSAR (interferometric synthetic aperture radar) can monitor creep in slow-moving landslides and use backscatter to study differences between pre- and post-landslide images. There is also potential for manual and automated to semi-automated data analysis, including supervised and unsupervised classification. Some of the difficulties, limitations and challenges of remote-sensing techniques applied to landslides are discussed by Joyce *et al.* (2009). The International Charter 'Space and Major Disasters' provides free remote-sensing data (SAR and optical) to member countries. TerraSAR-X (launched June 2007, Infoterra) is a commercial satellite capable of providing imagery up to 1 m spatial resolution. TanDEM-X (scheduled to launch October 2009) will orbit close to TerraSAR-X, allowing the generation of high-resolution DEMs that could be used to detect elevation changes and provide better volume estimates of landslides (Joyce *et al.*, 2009).

Information on landslide hazard, typically in the form of a map, can be utilised by land-use planners and private industry. For example, such information might lead to the refusal of planning permission. Private industry can also use the information for detailed site investigations to assess slope stability. The approach is typically highly empirical, using concepts like rock quality indices, together with detailed field observations of geological structures, fracture and bedding orientation, augmented by geotechnical data. A 'factor of safety' is commonly applied where items such as slope geometry, geology, soil strength, hydrological conditions, etc., are put together to come up with an assessment of failure potential. Engineering solutions, such as grouting and pinning, are sometimes employed to stabilise the slopes and reduce the probability of failure.

9.3.2 Landslide inventories and statistical analysis

A landslide event can range from a single collapse to thousands of individual concurrent landslides – for example, when triggered by a large regional earthquake or hurricane. Landslide inventories are generally either historical, a catalogue of events that have occurred over time in a particular region or associated with a single trigger (multiple landslides in a single event). Historical landslide inventories are prone to uncertainties and reporting biases that can complicate statistical analysis. Evidence of smaller events can be lost, and it can be difficult to identify the extent of older deposits. Data can be missing where landslide events occur in remote areas (due to underreporting) or are associated with contemporaneous hazards such as tropical cyclones. Kirschbaum *et al.* (2009) provide a discussion of the limitations and uncertainties associated with global landslide reporting and cataloguing.

Landslide inventories can be characterised by magnitude–frequency relationships. Area (A) rather than volume (V) is used as a measure of magnitude because area is much easier to measure from maps and satellite images. The observation that the ratio $A / V^{2/3}$ is approximately constant in a global database (Dade and Huppert, 1998) indicates area is an approximate proxy for magnitude. The frequency–area distribution of medium to large landslides decays with the inverse power of the landslide area (Turcotte and Malamud, 2004). Such landslide magnitude–frequency curves allow for the probabilistic characterisation of regional landslide hazard. According to Guthrie *et al.* (2007), landslides exhibit self-organised criticality, including the tendency to follow a power law over part of the magnitude–frequency distribution. However, they also note that landslide distributions exhibit poor agreement with a power law at smaller sizes, with a flattening of the slope known as rollover.

9.3.3 Landslide vulnerability and risk assessment

Global hazard and risk maps for earthquake- and precipitation-induced landslides and snow avalanches were developed as part of the World Bank Hotspots Project to identify countries most at risk (Nadim *et al.*, 2006; Nadim and Kjekstad, 2009). The study assesses risk of mortality and economic loss, using a global landslide and snow avalanche hazard index map, global population data and historical loss rates from EM-DAT (2011). The landslide hazard index, used to identify high-hazard 'hotspots', is based on triggering and susceptibility characteristics including slope, soil and soil moisture conditions, precipitation, seismicity and temperature (an example for Indonesia is shown in Figure 9.10). Physical exposure for a region is defined as the product of hazard (frequency of landslides) and the total population. Nadim *et al.* (2006) use a proxy for vulnerability, computed using the ratio of past casualties and the total exposed population in a given location, with risk calculated as the product of physical exposure and the vulnerability proxy. A revised approach presented in Cepeda *et al.* (2010) and GAR (2011) uses various socio-economic indices to represent vulnerability. Nadim *et al.* (2006) note that landslide mortality data are largely incomplete, and higher-resolution data are necessary to enable more accurate local and regional assessments. In any global risk assessment there will be significant spatial variation in model uncertainty due to varying availability and quality of hazard and vulnerability data. An initial attempt to quantify and map this uncertainty would be a useful next step enabling critical information gaps to be identified, guiding further research or modelling efforts.

Other relevant work includes vulnerability functions for debris flows and torrents developed by Fuchs *et al.* (2007a, 2007b), and Bell and Glade's (2004) methodology for quantitative risk assessment for landslides, applied to Bìldudalur, north-west Iceland. The model evaluates annual individual risk to life and object risk to life for debris flows and rock falls, to determine whether the level of risk is acceptable, according to limits set by the Icelandic government (see Section 9.4.1 and Iceland Ministry for the Environment, 2000). The model of Bell and Glade (2004) is developed in ArcGIS, and combines raster data for spatial and temporal probability of impact, and the exposed population (occupants of buildings impacted by a landslide). They provide qualitative estimates of uncertainties in elements of the model and input data, but the final risk maps are deterministic.

The EU FP 6 IRASMOS project, 'Integral risk management of extremely rapid mass movements', has developed methodologies for comprehensive risk management and decision support, prevention, intervention and recovery actions, and has produced a wide range of reports on current and best practice (IRASMOS, 2008; Bischof *et al.*, 2009). Recommendations for future research include understanding potential climate change impacts on hazard and risk, development of existing hazard maps into risk maps to better inform land-use planning and the development of dynamic risk maps to model short-term changes in hazard and exposure. The latter would be useful during an incident to aid response and decision-making.

9.3.4 *Probabilistic approaches and handling uncertainty*

Probabilistic approaches are gaining traction as issues are recognised with more traditional approaches to hazard estimation, due to limited availability and resolution of data, misclassification, catalogue bias and incompleteness. Vulnerability is often particularly poorly constrained (again largely due to data limitations), highly variable and better represented probabilistically. The last 20 years have seen an increased acceptance of probabilistic landslide hazard assessment and susceptibility mapping, but there are very few fully probabilistic *risk* models that attempt to quantify and propagate uncertainties in both the hazard and vulnerability elements of the model. Making steps in this direction, Bründl (2008) outlines various approaches for handling uncertainties in risk assessment, including the use of scenarios and expert judgement to introduce uncertainty in vulnerability functions for damage and mortality. A further example of a site-specific probabilistic risk model is given by Cassidy *et al.* (2008), who present a retrospective quantitative risk assessment for the 1996 Finneidfjord submarine landslide, Norway. Their probabilistic model for slope failure accounts for uncertainty in soil properties, slope geometry and model bias; and in their consequence analysis provide quantitative estimates of both epistemic and aleatory uncertainties in exposure and vulnerability parameters (road and rail traffic, building occupancy, etc.). Although a structured treatment of uncertainty as proposed by Cassidy *et al.* (2008) is desirable for informed decision-making and identifying model and data limitations, there are difficulties associated with interpreting probabilistic risk estimates in terms of acceptable or tolerable risk criteria (see Section 9.4.4).

Various methods for evaluating and propagating uncertainty are employed in landslide hazard assessments and the development of susceptibility maps. Carrara *et al.* (1992, 1995) estimate uncertainty in mapping by comparing data obtained by different surveying techniques and by surveyors of varying experience. There are significant uncertainties associated with identifying geomorphological features (in particular boundaries) of landslide deposits, and these uncertainties can be hard to quantify, and difficult to represent on a map. Ardizzone *et al.* (2002) remark that 'inventory maps can be misleading when they are transferred to the local administrators who may assume that the landslide boundaries are certain, well-defined entities', and suggest that statistical hazard models that incorporate geological, geomorphological and other environmental characteristics of the terrain provide more reliable information for decision-making.

There is also potential to add value through the use of expert judgement – for example, to quantify parameter or (estimate) model uncertainties. Van den Eeckhaut *et al.* (2005) present a study evaluating the utility of expert judgement in landslide classification. Experts were asked to identify areas that they suspected to be possible landslides on a hillshade map, and the results were compared to detailed field survey data to evaluate the accuracy of the experts' estimates. Generally, the expert-derived inventory was of poor quality compared to the field data. However, combining both sources of information could prove valuable, for example to identify old, deep-seated landslides in vegetated areas where aerial photography and satellite imaging is less accurate, or for large-scale regional assessments where detailed field surveys would be costly.

There is growing interest in the application of data-mining techniques to investigate factors that lead to development of landslides, with some examples being given by Wang *et al.* (2005), Saito *et al.* (2009) and Wang and Niu (2009). Wang *et al.* (2005) propose a conceptual model for incorporating data mining in a GIS-based framework for hazard zonation, but do not extend this to a case study. Saito *et al.* (2009) apply data mining techniques to analyse landslide susceptibility using the open source WEKA data-mining software (University of Waikato, 2008; Hall *et al.*, 2009). They perform ensemble learning to generate a decision tree structure using a database of catchment characteristics and ten-year training dataset for landslide occurrence and reactivation, then validate the model using data for the following two years. The structure of the decision tree (and associated conditional probabilities) provides information about the relationship between landslide occurrence and contributing factors such as topography and geology. Wang and Niu (2009) similarly use decision tree algorithms to mine a database of spatial landslide criteria and forecast landslide activity. Ercanoglu and Gokceoglu (2002) use a set of simplified characteristics (binned values for slope angle, land use, groundwater conditions, etc.) and fuzzy conditional statements to generate a landslide susceptibility map.

9.4 Landslide risk management

9.4.1 Approaches and initiatives in the UK, Europe and United States

The British Geological Survey (BGS) maintains the UK National Landslide Database, a catalogue of landslides recording the location, date, morphology, extent, trigger mechanism and impact, together with associated bibliographic and research material (Foster *et al.*, 2008). This is now a freely accessible online inventory documenting over 16 000 known landslides. The BGS Landslide Response team works regionally to map the spatial distribution of landslides, as a way of understanding their causal factors and possible triggers. This mapping feeds into the National Landslide Database and assessments of landslide susceptibility (GeoSure, see BGS, 2011a). GeoSure is a digital mapping facility for assessing susceptibility to various geohazards (including landslides) to assist planners and engineers in decision-making. Landslide susceptibility is currently evaluated heuristically, but Foster (2010) is exploring statistical techniques (bivariate statistical analysis) to develop a new national landslide susceptibility map. BGS has also undertaken (for around ten years) the systematic monitoring of coastal erosion at a number of sites around the UK. Techniques for using terrestrial laser scanning (LiDAR) have been developed to study coastal erosion rates and processes. Risk assessments are typically site- or area-specific, or for safety-critical operations, e.g. high-pressure gas pipelines.

Within Europe, the European Commission Joint Research Centre (JRC) Institute for Environment and Sustainability have established a landslide experts group to provide scientific and technical support for EU policy and research (see Hervás *et al.*, 2010). JRC also provides links to a number of landslide inventories for Europe. The EC FP7 project SafeLand (coordinated by the NGI) aims to produce generic quantitative hazard and risk assessment and management tools for landslides at local, regional and national scales. A key component involves the study of future climate change scenarios, considering the impact on risk due to changes in demography and extreme precipitation (Nadim and Solheim, 2009; Lacasse *et al.*, 2010).

The USGS Landslide Hazards Program promotes research on the causes of ground failure and the development of appropriate mitigation strategies. They carry out extensive monitoring and provide online monitoring data for several US sites, and software for evaluating time-dependent slope stability in response to rainfall infiltration (TRIGRS, described by Savage *et al.*, 2003, 2004; Baum *et al.*, 2008). Another USGS initiative, the US National Landslide Hazards Mitigation Strategy is a programme of research to develop a predictive understanding of landslide processes and triggering mechanisms (Spiker and Gori, 2000; NRC, 2004). The programme promotes a national approach to mitigation of landslide hazards, based on partnerships between federal, state, local and non-governmental organisations (NRC, 2004).

9.4.2 Snow avalanche hazard and risk management

Snow avalanches have threatened human lives and property since the beginning of human settlement in mountains. In the Alps, risk has increased with the evolution of permanent settlements and increasing population pressure, which is reflected by numerous, sometimes catastrophic avalanche events (Ammann *et al.*, 1997). As a consequence of the catastrophic avalanche winters of 1951, 1954 and especially 1968, systematic hazard zoning and land-use planning became mandatory for endangered areas in the most affected countries of Austria and Switzerland. Other countries followed with similar measures, such as Norway, where, in the avalanche winters of 1993, 1994 and 1997, large avalanches with extreme runout lengths destroyed numerous buildings. A similar evolution took place in Iceland after catastrophic avalanches in January and October of 1995. Jóhannesson *et al.* (2009) give a short account of official regulations for avalanche protection measures in different countries. The estimation of avalanche hazard that underlies technical and land-use planning protection measures is based on expected runout distances of avalanches with defined return periods. For a complete risk assessment, additional information about the intensity of the expected events (frequently given in terms of impact pressures) and about the vulnerability of endangered objects is necessary. Estimates of expected avalanche velocities are an important input parameter for the design of technical protection measures.

9.4.3 Tolerable risk criteria for landslides and avalanches

A number of countries have defined tolerable risk criteria for landslides and snow avalanches. The tolerable risk criterion is often defined in terms of the annual probability of being killed by a landslide event. In 2000 the Icelandic government issued a regulation on hazard zonation and acceptable risk due to snow avalanches and landslides (Iceland Ministry for the Environment, 2000). The regulation sets an upper limit of annual risk of 3×10^{-5} for occupants of residential buildings and public buildings such as schools, 10^{-4} for commercial buildings and 5×10^{-5} for recreational homes (holiday homes). The Hong Kong Special Administrative Region Government (ERM, 1998) has set a tolerable annual risk limit for the population most at risk of 10^{-4} for existing slopes and 10^{-5} for new slopes. The Australian Geomechanics Society Guidelines for landslide risk management (AGS, 2000) give the same tolerability thresholds as Hong Kong.

9.4.4 Risk reduction

Risk can be reduced by reducing the probability of occurrence or magnitude of the physical hazard; limiting exposure to the hazard and reducing vulnerability to damage. Landslide mitigation measures can be classified as passive or active. Passive measures include the use of land-use planning and regulation of development in landslide-prone areas. In some circumstances it can be necessary (and cost-effective) to relocate critical infrastructure. Active measures include engineering interventions to stabilise the slope or to reduce the impact of debris in the potential debris emplacement zone (protection engineering), thus reducing or altering the footprint of the hazard. Emplacement-zone measures are typically employed for debris flows and snow avalanche hazards – e.g. barriers to dissipate avalanche energy or divert the flows (see Section 9.4.4.2). Evacuation can be used to temporarily reduce exposure if there is evidence of an imminent threat in populated areas.

There are a variety of methods for slope stabilisation, and the approach taken will depend on the specific site and cost effectiveness. Three general classes of engineering approach can be identified: modification of the slope geometry; reinforcement of the slope; and drainage. Excavation of the slope crest, or the addition of material at the base of the slope, can be effective for small sites where the failure surface is deep-seated. However, this method is less effective for large translational slides or where the topography is complex (Dai *et al.*, 2002). Common methods for increasing the resistance of slopes to destabilising forces include introducing structural supports (buttresses or retaining walls); grouting, meshing or pinning potentially unstable rock surfaces; the use of piles or anchors to prevent movement of unstable layers; mesh reinforcements in soils; biotechnical stabilisation using a combination of man-made structures and vegetation (see Highland and Bobrowsky, 2008).

If the driving mechanism for instability is surface water infiltration or groundwater, drainage can be an effective way of increasing slope stability. Contour drains can be used to intercept runoff and reduce surface water infiltration into the zone of unstable slope material. Simply increasing vegetation cover can also help to capture rainfall and reduce infiltration, although different plants may have different effects on slope stability and may actually create enhanced flow into the slope (Wilkinson *et al.*, 2002). Other methods for reducing infiltration include the use of man-made slope coverings such as shotcrete (a cement skim) or impermeable membranes. Sub-surface drainage can be used to de-water slopes and locally reduce the water table, again improving the slope stability by reducing positive pore water pressures. Although engineering measures are usually prohibitively expensive for developing countries, surface drainage can be a cost-effective way of reducing landslide hazard if the driving mechanism for the instability is rainfall infiltration (as is often the case in the humid tropics). The MoSSaiC project (Management of Slope Stability in Communities) has been developed for just such countries in order that engineers and vulnerable urban communities can identify and construct appropriate surface drainage measures to reduce landslide hazard (Anderson *et al.*, 2008, 2011; Holcombe *et al.*, 2011).

Engineering measures are often used in combination. Sub-surface and surface drainage systems are used to control groundwater, and anchors and large concrete piles up to several meters in diameter can stabilise slopes (see, e.g. Poulos, 1995). Channel linings to rivers and streams can prevent undercutting by stream erosion, and reduce runout and impact of debris flows in the channel. In Japan, Sabo dams and other hillside structures are employed with the intention of channelling debris flows in a controlled way (Misuyama, 2008). However, they do not always succeed and can be problematic in terms of their effects on other hazards. For example, during the 2010 eruption of Mount Merapi, a sabo dam upstream of the village of Kaliadem appears to have acted to divert pyroclastic flows into the village (see Lube *et al.*,

2011). Such structures effectively reduce the channel capacity, causing the flow to overtop or jump the barrier.

9.4.4.1 Snow avalanche mitigation

Active control involves triggering a controlled avalanche (often using explosives) to clear or stabilise the snow slope. Passive control measures include engineered structures to divert the flow away from vulnerable areas or provide protection – for example, avalanche barriers, deflection or retention dams, snow sheds or flexible structures such as snow nets (see Nicot *et al.*, 2002; Phillips, 2006; Margreth, 2007; Jóhannesson *et al.*, 2009). Other mitigation measures include construction of elevated roadways, catchments to stop the flow and measures to prevent initiation (fences and reforestation). Research to understand the dynamics of snow avalanches and their interaction with topography and engineered protective structures has played an important role in mitigation strategies (Hákonardóttir *et al.*, 2003; Jóhannesson *et al.*, 2009; Baillifard, 2007).

9.4.4.2 Cost–benefit analysis of mitigation measures

Cost–benefit analysis is a widely used tool for decision-making and is commonly associated with planning based on specific scenarios and assessment of alternative mitigation strategies. A good example is given by Kong (2002), who presents a cost–benefit analysis for mitigating shallow, rainfall-induced landslides in Hong Kong, using an event tree framework. Kong (2002) evaluates individual probability of death estimates for different exposure scenarios (e.g. outdoors, in a vehicle or indoors) given different impact scenarios based on previous records of landslide fatalities. An event tree is used to model the sequence of events, accounting for the probability of triggering (rainfall), probability of failure, the elements at risk and the resulting impact. Each branch in the tree has an associated probability of occurrence, but uncertainties are not captured in the probability (or risk) estimates. Alternative mitigation strategies are compared by calculating the 'implied cost of averting a fatality' (ICAF), where ICAF = cost to implement measure / (reduction in probability of loss of life × lifetime of measure). A further example is given by Crosta *et al.* (2005), who present a cost–benefit analysis for landslide hazard mitigation for Cortenova, Italy. The study evaluates the probability of damage and loss of life for a single scenario of a 1 500 000 m^3 landslide. They calculate costs and benefits for various mitigation strategies, including continuous monitoring with an alarm system, a defensive barrier, total relocation and non-intervention, applying societal acceptable risk criteria to rule out unacceptable mitigation strategies.

9.4.5 Remote sensing, monitoring and early warning systems

If slope instability is recognised from appearance of fractures and evidence of slow creeping movements, then it can be monitored. Communities that might be affected can then be relocated, evacuated or protected by an early warning system. Monitoring devices providing real-time data on conditions at the site, providing input to physical models of the hazard process, form the basis for hazard assessment and warning. Early warning systems are often costly, involving arrays of sensors and specialised (typically black-box) software. Multi-parameter sensor networks can include inclinometers, dilatometers, accelerometers, pressure, temperature, acoustic sensors and use of LiDAR and long-range laser scanning. Displacement transducers and GPS are particularly suitable for detecting small movements. The Åknes rockslide is a good recent example of a carefully monitored area of instability located on the western side of the Sunnylvsfjord. Its volume has been estimated to be more than $35 \times 10^6 \, \mathrm{m}^3$ (Blikra *et al.*, 2005) and its failure might result in a catastrophic tsunami in the fjord. The area is sliding at 6–7 cm year^{-1} (see Oppikofer *et al.*, 2009).

ETH Zurich (the Swiss Federal Institute of Technology) has been monitoring progressive rockslope failure at Randa in Switzerland since 2002 to understand the processes of failure and fracturing in hard rock slopes, studying the effects of seasonal temperature changes, glacial erosion and earthquake shaking. Monitoring techniques include seismics, radar, photogrammetry, ground-based differential InSAR (DInSAR), geodetic surveying and a range of *in situ* sensors to measure crack propagation, tilt, strain, temperature and meteorological data (see, e.g. Willenberg *et al.*, 2008a, 2008b). In the UK the BGS are monitoring an active inland landslide in North Yorkshire using LiDAR, geophysical instruments, a weather station and inclinometers/piezometers. The aim of the project (Landslide ALERT) is to develop a four-dimensional landslide monitoring system to explore the sub-surface structure, investigate what triggers movement, the effects of rainfall and how water moves through the landslide. Several other UK coastal locations (including sites prone to landsliding at Aldbrough and Beachy Head) are regularly monitored using terrestrial LiDAR.

Infrared reflectance (in particular shortwave IR) can be used to measure soil moisture (Lobell and Ansner, 2002). The BGS and University of Portsmouth are developing techniques to use infrared spectroscopy to look at the mechanical properties of soils and rocks and calculate a soil moisture index, to assess if a slope is saturated and close to failure. This approach could be used to provide an early warning system for large debris flows (BGS, 2011b). Soil moisture and precipitation can also be monitored using satellite-borne microwave sensors to identify conditions that could lead to landsliding (e.g. TRMM and AMSR-E passive microwave radiometer; Ray and Jacobs, 2007).

There are currently good systems in place for early warning of snow avalanches, and good public education and communication of avalanche risk, although different organisations apply different decision-making criteria. In Switzerland, the SLF (WSL Institute for Snow and Avalanche Research) regularly issue national and regional avalanche bulletins and hazard maps (SLF, 2011). This information is based on current and historical meteorological data (snowfall, wind, precipitation) from automatic weather stations, augmented by daily observations from local stations, reporting the state and depth of the snow. Avalanche warnings are issued at a local level and are based on expert knowledge and experience of the local conditions. As such, there is no standardised procedure for issuing an alert or evacuating an area. A danger scale ranging from 1 to 5 is used. Scale value 1 represents low hazard, where the snowpack is generally well-bonded and stable, and triggering unlikely,

Danger scale		Snowpack stability	Avalanche triggering probability
	Very high	The snowpack is poorly bonded and largely unstable in general	Numerous large and very large natural avalanches can be expected, even in moderately steep terrain
Red	High	The snowpack is poorly bonded on most steep slopes	Triggering is likely, even from low additional loads on many steep slopes. In some cases, numerous medium and often large natural avalanches can be expected
Orange	Considerable	The snowpack is moderate to poorly bonded on many steep slopes	Triggering is possible, even from low additional loads particularly on indicated steep slopes. In some cases medium, in isolated cases large natural avalanches are possible
Yellow	Moderate	The snowpack is only moderately well bonded on some steep slopes, otherwise well bonded in general	Triggering is possible primarily from high additional loads, particularly on indicated steep slopes. Large natural avalanches are unlikely
Green	Low	The snowpack is well bonded and stable in general	Triggering is generally possible only from high additional loads in isolated areas of very steep, extreme terrain. Only sluffs and small-sized natural avalanches are possible

Figure 9.11 The European Avalanche Danger Scale. Avalanche-prone locations are described in greater detail in the avalanche bulletin (altitude, slope aspect, type of terrain). A low additional load is defined as an individual skier or snowboarder riding softly, a snowshoer or group with good spacing (minimum 10 m) keeping distances. A high additional load is defined as two or more skiers or snowboarders without good spacing (or without intervals), snow-machine, explosives, a single hiker or climber.

and 5 corresponds a very high hazard area, where the snow is extremely unstable with a very high probability of natural avalanches occurring (see Figure 9.11).

9.4.6 Risk perception and communication

Identifying differences in risk perception is the starting point for improving both the policy and practice of landslide risk management. Landslide experts, decision-makers and those at risk will have different perceptions of the level of risk, the risk drivers and the potential management options. It is known that in developing countries the most socio-economically vulnerable may discount the risks associated with lower frequency events (such as land-slides) in order to address more pressing risks such as unemployment – for example, by locating their houses on landslide-prone slopes near cities in order to be close to potential employment opportunities. For political reasons decision-makers may adopt a short-term approach and assume that a disaster will not happen in their term of office (Kunreuther and Useem, 2009). Against a backdrop of different risk perceptions it becomes even more important to communicate risk clearly, especially when it comes to uncertainty. Malamud and Petley (2009) note that increasing model complexity, the necessary incorporation of uncertainty in model predictions and risk assessments, and the subsequent reporting of such information can lead to greater confusion among politicians and the public. Publicly differing expert opinion can also lead to conflict and confusion. People may learn from one expert that there is little to be concerned about, and from another that the very same risk is of major significance (Kunreuther and Useem, 2009).

Once landslide risk is on the agenda, raising the awareness of potential risk management solutions is the next hurdle, as risk reduction is often viewed as 'ranking low on the tractability dimension' (Prater and Londell, 2000). In order to invest in landslide risk management, decision-makers need the scientific and cost–benefit evidence to justify their investment. To provide such a basis, researchers need to translate their knowledge into practical risk management solutions (Malamud and Petley, 2009).

Within Europe, Wagner (2007) carried out a study to investigate perceptions of flash floods and landslides in four communities in the Bavarian Alps, and found that the public had a relatively poor understanding of the physical processes that contribute to landslides. This was attributed to two main factors: the relative infrequency of landslides (resulting in a population with limited experience of the hazard), and the fact that the processes and conditions that lead to slope failure are complex and (with the exception of weather conditions) largely invisible. There is therefore a need to promote public understanding of risk – for landslides and other natural hazards. One such initiative is the USGS and Geological Society of Canada's free online publication *The Landslide Handbook: A Guide to Understanding Landslides* (Highland and Bobrowsky, 2008). The book was developed for communities affected by landslides, emergency managers and decision-makers, and explains landslide processes, hazards and mitigation strategies. They provide suggestions for local government outreach for landslide hazard, and give examples of basic safety information and actions homeowners and communities can take to reduce risk.

9.5 Research developments and challenges

Over the last two decades, the literature on landslide hazard assessment has demonstrated significant advances in the numerical modelling of preparatory and triggering processes at the scale of individual slopes. More recently, sophisticated physics-based approaches have been developed in order to model the emplacement dynamics of landslides at this localised scale. Thus, *given sufficient site data*, it is currently feasible for scientists and engineers to understand and model the specific physical processes occurring at a particular site and quantify the landslide hazard posed to 'point' elements such as individual buildings, and linear elements such as roads and pipelines (van Westen *et al.*, 2006). In many cases the limiting component in deterministic landslide prediction is the acquisition of adequate data to parameterise the model. Recent probabilistic approaches aim to account for the uncertainties associated with the lack of data and to incorporate the natural variability of slope parameters.

The increasing availability of remotely sensed data and digital maps for large areas of the world has allowed the science of landslide hazard assessment to be approached at a regional level (the scale of 1:10 000–1:50 000) – in line with the demand from planners, risk managers and decision-makers for hazard and risk zonation maps. In most cases, zones of relative landslide susceptibility can be determined based on spatially approximated preparatory factors to deliver a *qualitative* landslide susceptibility map (sometimes based on expert judgement). More data-intensive studies allow the prediction of the expected number of landslides triggered by rainfall or seismic events. However, there is still a significant 'gap' between the scale of highly localised of landslide processes, and the level of detail that can be obtained from medium-scale maps. Modelling both the exact location and frequency of landslides, and the emplacement dynamics for multiple landslide initiation areas in a given region represents a particular challenge. This makes assessment of the consequences of landslides in specific areas very difficult. However, developments in remote sensing are leading to significant improvements in landslide identification, with high enough resolution to identify 'small' slides of several square metres. It is also possible to identify potential sites of mass failure using LiDAR, microwave and infrared sensors (discussed in Section 9.4.5).

This review of the state-of-the-art in landslide hazard and risk assessment reflects a challenging problem of various phenomena with wide diversities of scale and type. The variability, heterogeneity and complexity of environments that generate landslides and avalanches are large and this means that most assessments of practical value are likely to be very site-specific. The factors that influence slope stability and the processes involved in triggering collapse are complex, and failure can be difficult to predict. Different types of evidence need to be combined: site measurements, remote sensing data, evidence of past events and model simulations. Maps need to be updated regularly to take into account changing land use and occurrence of new landslides.

The science remains highly empirical due to the formidable challenges of understanding complex multiphase earth materials in motion. Many operational landslide and avalanche models that have been developed and used extensively in the last years are hardly more than crude approximations. These models are lacking statistically robust calibrations as well as formal assessments of their validity in terms of structural errors. As this might have safety-relevant consequences there is pressing need for further research on model calibration, uncertainties and applicability.

The US National Landslide Hazards Mitigation Strategy (NRC, 2004) identified a number of basic science questions that need to be addressed for different types of landslide. For debris flows, they highlight the importance of initiation, runout characteristics and magnitude–frequency relationships. For bedrock slides key research areas include spatial distribution, influence of climate events, controls on velocity and stability. Progress in understanding submarine slides will require more extensive mapping, investigation of geotechnical characteristics and the role of gas hydrates in slope stability. Van Westen *et al.* (2006) remark that advances are needed in the use of very detailed topographic data, the generation of event-based landslide inventory maps, the development of spatial-temporal probabilistic modelling and the use of land-use and climate change scenarios in deterministic models.

The occurrence of a landslide results in changes in terrain and factors such as slope angle (van Westen *et al.*, 2006), and may ultimately affect the scale or incidence of future landslide events in that area. Once there has been a landslide at a particular place, the probability of a repeat event in the same location may be reduced for some time into the future due to the removal of the unstable material or the attainment of a stable slope geometry. This has been termed 'event resistance' (Glade and Crozier, 2005). Conversely, the probability of a further landslide may actually subsequently increase after a landslide event due to the over-steepening of the slope at the landslide crest, or through the reactivation of the failed materials remaining on the slope, which will be at their residual strength. This potential time dependence is of importance for landslide hazard assessment, which traditionally assumes that the preparatory and triggering events that resulted in landslides in the past will produce the same effect in the future (Zêzere *et al.*, 2008).

As is the case for many natural hazards, most research focuses on the hazard process, and quantification of risk and uncertainty remains a major challenge. Vulnerability is a critical component, as there are limited data on landslide impacts (e.g. mortality, building damage), and the high degree of uncertainty in vulnerability is rarely captured in risk assessments. Further research is also needed to assess the effectiveness and cost–benefit of early warning systems and mitigation measures. Carrara and Pike (2008) remark that the reliability of landslide-hazard predictions varies significantly, largely due to different assessment scales and methodologies, and that it is too early in the evolution of GIS-based landslide-hazard and risk modelling to identify any 'best practice' approach or set of techniques. They stress that most regional-scale landslide studies are carried out by academic research groups, with limited overlap or interaction with civil bodies involved in land-use planning, public safety, etc. Carrara and Pike (2008) also note that poor-quality or sparse data are a major hindrance, and that there would be significant benefits in advancing collaborations between academic, public and private bodies to improve availability of data and promote sharing of scientific knowledge and tools for hazard risk and assessment. Social science collaboration is key for addressing issues such as vulnerability, public education and response. Cross-hazards collaboration is also important, as methodologies for managing different types of hazard have fundamental similarities and common problems.

Better methods of communication are needed to convey hazards information in a manner that people might use or take notice of. The work of the Geotechnical Engineering Office in Hong Kong is an excellent example of how this should be done in landslide-prone communities (see HKSS, 2011). They provide extensive public information on all aspects of landslide hazard and risk, covering landslip warnings, personal safety measures, slope maintenance, causes and control of landslides, as well as listing slopes undergoing or scheduled for preventative works. Much can be learned from other types of hazard infor-mation (for example, weather forecasts and extreme weather warnings). There are currently good systems in place for communicating early warnings of snow avalanches, but there is a need for a more formal approach to incorporate expert judgement and assess uncertainty. There is also a recognised need for improved technical measures for protection and mitigation. Many dam structures, intended to halt avalanches, are designed based on a simple energy model, yet much more sophisticated physical models of avalanche dynamics are available. The WSL (2011) are currently developing software to simulate different types of mass movement (avalanches, debris flows and rockfalls), and their interaction with natural and man-made barriers. Snow avalanches are simulated by solution of depth-averaged equations governing avalanche flow, and debris flows are represented by a two-phase, two-dimensional model based on the shallow-water equations (Christen *et al.*, 2008).

There is similar significant scope for improvements in impact assessment. Little is known about the interaction of avalanches with buildings (although the capability exists), and how such interactions might affect the total runout and damage caused by the flow. Such developments will lead to more robust cost–benefit analyses and better risk-informed decision-making.

Acknowledgements

The authors would like to thank C. Foster and M. Culshaw (British Geological Survey), B. Malamud (King's College London), M. Bründl (WSL Institute for Snow and Avalanche Research SLF), C. Graf (WSL Swiss Federal Research Institute) and D. N. Petley (University of Durham) for their valuable input and comments; and F. Nadim and P. Gauer for providing a very helpful and constructive review.

References

AGS (2000) Landslide risk management concepts and guidelines. Australian Geomechanics Society, Sub-Committee on Landslide Risk Management. *Australian Geomechanics* **35**: 49–92.

Aleotti, P. and Chowdhury, R. (1999) Landslide hazard assessment: summary review and new perspectives. *Bulletin of Engineering Geology and the Environment* **58** (1): 21–44.

Ammann, W., Buser, O. and Vollenwyder, U. (1997) *Lawinen*, Birkhäuser Verlag: Basel.

Ancey, C. (2012) Gravity flow on a steep slope. In *Buoyancy-Driven Flows*, ed. E. P. Chassignet, C. Cenedese and J. Verron, Cambridge: Cambridge University Press.

Anderson, M. G., Holcombe, E. A., Flory, R., *et al.* (2008) Implementing low-cost landslide risk reduction: a pilot study in unplanned housing areas of the Caribbean. *Natural Hazards* **47** (3): 297–315.

Anderson, M. G., Holcombe, E. A., Blake, J., *et al.* (2011) Reducing landslide risk in communities: evidence from the Eastern Caribbean. *Applied Geography* **31** (2): 590–599.

Ardizzone, F., Cardinali, M., Carrara, A., *et al.* (2002) Impact of mapping errors on the reliability of landslide hazard maps. *Natural Hazards and Earth Systems Science* **2**: 3–14.

Ayonghe, S. N., Ntasin, E. B., Samalang, P., *et al.* (2004) The June 27, 2001 landslide on volcanic cones in Limbe, Mount Cameroon, West Africa. *Journal of African Earth Sciences* **39** (3–5): 435–439.

Baillifard, M.-A. (2007) Interaction between snow avalanches and catching dams. Dissertation, Eidgenössische Technische Hochschule ETH Zürich, Nr. 17301, 2007. Swiss Federal Institute of Technology Zürich.

Barbolini, M., Cappabianca, F., Korup, O., *et al.* (2008) D3.1: Technical evaluation report of current methods of hazard mapping of debris flows, rock avalanches, and snow avalanches. IRASMOS. European Commission 6th Framework Programme, project 018412. http://irasmos.slf.ch/results_wp3.htm (accessed 2 August 2011).

Bartelt, P., Salm, B. and Gruber, U. (1999) Calculating dense-snow avalanche runout using a Voellmy-fluid model with active/passive longitudinal straining. *Journal of Glaciology* **45** (150): 242–254.

Baum, R. L., Savage, W. Z. and Godt, J. W. (2008) TRIGRS: A Fortran program for transient rainfall infiltration and grid-based regional slope-stability analysis, version 2.0. US Geological Survey Open-File Report 2008–1159.

Bell, R. and Glade, T. (2004) Quantitative risk analysis for landslides: examples from Bildudalur, NW-Iceland. *Natural Hazards and Earth System Sciences* **4** (1): 117–131.

BFF/SLF (1984) Richtlinien zur Berücksichtigung der Lawinengefahr bei raumwirksamen Tätigkeiten, Mitteilungen des Bundesamt für Forstwesen und Eidgenössischen Instituts für Schnee- und Lawinenforschung, EDMZ, Bern. http://www.bafu.admin. ch/publikationen/publikation/00778/index.html (accessed 17 June 2011).

BGS (2011a) BGS National Landslide Database. *British Geological Survey.* http://www. bgs.ac.uk/landslides/nld.html (accessed 4 August 2011).

BGS (2011b) Infrared measurement of soils. British Geological Survey. http://www.bgs.ac. uk/science/landUseAndDevelopment/new_technologies/infra_red.html (accessed 23 June 2011).

Bischof, N., Romang, H. and Bründl, M. (2009) Integral risk management of natural hazards: a system analysis of operational application to rapid mass movements. In *Safety, Reliability and Risk Analysis: Theory, Methods and Applications*, ed. S. Martorell, C. G. Soares, J. Barnett, Boca Raton, FL: CRC Press/Balkema, pp. 2789–2795.

Blikra, L. H., Longva, O., Harbitz, C., *et al.* (2005) Quantification of rock-avalanche and tsunami hazard in Storfjorden, western Norway. In *Landslides and Avalanches*, ed. K. Senneset, K. Flaate, J. O. Larsen, London: Taylor and Francis, pp. 57–64.

Bondevik, S., Mangerud, J., Dawson, S., *et al.* (2003) Record-breaking height for 8000-year old tsunami in the North Atlantic. *EOS* **84**: 289–293.

Boudon, G., Le Friant, A., Komorowski, J.-C., *et al.* (2007) Volcano flank instability in the Lesser Antilles Arc: diversity of scale, processes, and temporal recurrence. *Journal of Geophysical Research* **112**: B08205–B08232.

Bründl, M. (ed.) (2008) D5.2: Technical report highlighting the reliability, needs, and short-comings of risk quantification procedures, incl. consequences of risk under- and over-estimation for acceptable risk levels. IRASMOS. European Commission 6th Framework Programme, project 018412. http://irasmos.slf.ch/results_wp5.htm (accessed 23 June 2011).

Calceterra, D. and Santo, A. (2004) The January 10, 1997 Pozzano landslide, Sorrento Peninsula, Italy. *Engineering Geology* **75**: 181–200.

Carrara, A. and Pike, R. J. (2008) GIS technology and models for assessing landslide hazard and risk. *Geomorphology* **94** (3–4): 257–260.

Carrara, A., Cardinali, M. and Guzzetti, F. (1992) Uncertainty in assessing landslide hazard and risk. *ITC Journal, The Netherlands* **2**: 172–183.

Carrara, A., Cardinali, M., Guzzetti, F., *et al.* (1995) GIS technology in mapping land-slide hazard. *Geographical Information Systems in Assessing Natural Hazards* **5**: 135–175.

Casale, R., Fantechi, R. and Flageollet, J.-C. (eds) (1994) Temporal occurrence and fore-casting of landslides in the European Community: Final Report. EPOCH programme, EU FP2, Contract EPOC0025.

Cassidy, M. J., Uzielli, M. and Lacasse, S. (2008) Probability risk assessment of landslides: a case study at Finneidfjord. *Canadian Geotechnical Journal* **45** (9): 1250–1267.

Cecchi, E., van Wyk de Vries, B. and Lavest, J. M. (2004) Flank spreading and collapse of weak-cored volcanoes. *Bulletin of Volcanology* **67** (1): 72–91.

Cepeda, J., Smebye, H., Vangelsten, B., *et al.* (2010) Landslide risk in Indonesia. GAR 11 Contributing paper, prepared by the International Centre for Geohazards, Norwegian Geotechnical Institute. http://www.preventionweb.net/english/hyogo/gar/2011/en/home/annexes.html (accessed 22 June 2011).

Christen, M., Bartelt, P. and Gruber, U. (2002) AVAL-1D: an Avalanche dynamics program for the practice. Paper presented at the International Congress Interpraevent 2002 in the Pacific Rim, Matsumo, Japan.

Christen, M., Bartelt, P., Kowalski, J., *et al.* (2008) Calculation of dense snow avalanches in three-dimensional terrain with the numerical simulation programm RAMMS. Paper presented at the International Snow Science Workshop 2008.

Clavero, J., Sparks, R. S. J., Huppert, H. E., *et al.* (2002) Geological constraints on the emplacement mechanism of the Parinacota debris avalanche, northern Chile. *Bulletin of Volcanology* **64**: 3–20.

Clavero, J., Sparks, R. S. J., Pringle, M. S., *et al.* (2004) Evolution and volcanic hazards of Taapaca Volcanic Complex, Central Andes of Northern Chile. *Journal of the Geological Society* **161**: 1–17.

Clerici, A., Perego, S., Tellini, C., *et al.* (2006) A GIS-based automated procedure for landslide susceptibility mapping by the Conditional Analysis method: the Baganza valley case study (Italian Northern Apennines). *Environmental Geology* **50** (7): 941–961.

Costa, J. E. and Schuster, R. L. (1988) The formation and failure of natural dams. *Geological Society of America Bulletin* **100** (7): 1054–1068.

Crosta, G. B. (2004) Introduction to the special issue on rainfall-triggered landslides and debris flows. *Engineering Geology* **73**: 191–192.

Crosta, G. B., Frattini, P., Fugazza, F., and Caluzzi, L. (2005) Cost–benefit analysis for debris avalanche risk management. In *Landslide Risk Management: Proceedings of the International Conference on Landslide Risk Management*, ed. O. Hungr, R. Fell, R. Counture, *et al.* Philadelphia, PA: Taylor and Francis, pp. 367–374.

Crozier, M. J. (1986) *Landslides: Causes, Consequences and Environment.* London: Croom Helm.

Crozier, M. J. and Glade, T. (2005) Landslide hazard and risk: issues, concepts and approach. In *Landslide Hazard and Risk*, ed. T. Glade, M. Anderson and M. Crozier, Chichester: John Wiley and Sons, pp. 1–40.

Cruden, D. M. (1991) A simple definition of a landslide. *Bulletin of Engineering Geology and the Environment* **43** (1): 27–29.

Cruden, D. M. and Krahn, J. (1973) A re-examination of the geology of the Frank Slide. *Canadian Geotechnical Journal* **10**: 581–591.

Cruden, D. M. and Varnes, D. J. (1996) Landslide types and processes. In *Landslides: Investigation and Mitigation*, ed. A. K. Turner and R. L. Schuster, Washington, DC: National Research Council.

Dade, W. B. and Huppert, H. E. (1998) Long-runout rockfalls. *Geology* **26**: 803–806.

Dai, F. C., Lee, C. F. and Ngai, Y. Y. (2002) Landslide risk assessment and management: an overview. *Engineering Geology* **64** (1): 65–87.

Dikau, R., Brunsden, D., Schrott, L., *et al.* (eds) (1996) *Landslide Recognition: Identification, Movement and Causes*, Chichester: John Wiley and Sons.

Embley, R. W. (1982) Anatomy of some Atlantic margin sediment slides and some comments on ages and mechanisms. In *Marine Slides and Other Mass Movements*, ed. S. Saxov and J. K. Nieuwenhuis, New York, NY: Plenum, pp. 189–213.

EM-DAT (2011) EM-DAT: The OFDA/CRED International Disaster Database: www.emdat.net. Université Catholique de Louvain, Brussels, Belgium.

Ercanoglu, M. and Gokceoglu, C. (2002) Assessment of landslide susceptibility for a landslide-prone area (north of Yenice, NW Turkey) by fuzzy approach. *Environmental Geology* **41** (6): 720–730.

ERM (1998) *Landslides and Boulder Falls from Natural Terrain: Interim Risk Guidelines*, Hong Kong: Geotechnical Engineering Office.

Foster, C. (2010) Landslide susceptibility mapping: the GB perspective. 2nd Experts Meeting on Harmonised Landslide Susceptibility Mapping for Europe. JRC, Ispra, 26–27 May 2010. http://eusoils.jrc.ec.europa.eu/library/themes/LandSlides/wg.html (accessed 23 June 2011).

Foster, C., Gibson, A. and Wildman, G. (2008) The new national landslide database and landslide hazard assessment of Great Britain. First World Landslide Forum, Tokyo, Japan, 18–21 November 2008, pp. 203–206.

Fritz, H. M., Hager, W. H. and Minor, H. E. (2004) Near field characteristics of landslide generated impulse waves. *Journal of Waterway Port Coastal and Ocean Engineering – ASCE* **130** (6): 287–302.

Fuchs, S., Heiss, K. and Hübl, J. (2007a) Towards an empirical vulnerability function for use in debris flow risk assessment. *Natural Hazards and Earth System Sciences* **7** (5): 495–506.

Fuchs, S., Heiss, K. and Hübl, J. (2007b) Vulnerability due to torrent events: a case study from Austria. In *Geomorphology for the Future: Conference Proceedings*, ed. A. Kellerer-Pirklbauer, M. Keiler, C. Embleton-Hamann, *et al.*, Innsbruck: Innsbruck University Press, pp. 97–104.

Galli, M., Ardizzone, F., Cardinali, M., *et al.* (2008) Comparing landslide inventory maps. *Geomorphology* **94** (3–4): 268–289.

GAR (2011) *2011 Global Assessment Report on Disaster Risk Reduction*. http://www. preventionweb.net/english/hyogo/gar/2011 (accessed 4 August 2011).

Gauer, P., Kronholm, K., Lied, K., *et al.* (2010) Can we learn more from the data underlying the statistical α–β model with respect to the dynamical behavior of avalanches? *Cold Regions Science and Technology* **62**: 42–54.

Giani, G. P. (1992) *Rock Slope Stability Analysis*. London: Taylor and Francis.

Glade, T. and Crozier, M. J. (2005) A review of scale dependency in landslide hazard and risk analysis. In *Landslide Hazard and Risk*, ed. T. Glade, M. G. Anderson and M. J. Crozier, Chichester: John Wiley and Sons, pp. 75–138.

Gruber, U., Bartelt, P. and Margreth, S. (2002) *Kursdokumentation Teil III: Anleitung zurBerechnung von Fliesslawinen*. Davos Dorf: Swiss Federal Institute for Snow and Avalanche Research.

Guthrie, R. H., Deadman, P. J., Cabrera, A. R., *et al.* (2007) Exploring the magnitude–frequency distribution: a cellular automata model for landslides. *Landslides* **5** (1): 151–159.

Hákonardóttir, K. M., Hogg, A. J., Batey, J., *et al.* (2003) Flying avalanches. *Geophysical Research Letters* **30** (23): n.p..

Hall, M., Frank, E., Holmes, G., *et al.* (2009) The WEKA data mining software: an update. *SIGKDD Explorations* **11** (1): 10–18.

Harbitz, C. (1998) *A Survey of Computational Models for Snow Avalanche Motion*. Oslo: SAME Collaboration/Norwegian Geotechnical Institute.

Heinrich, P., Boudon, G., Komorowski, J. C., *et al.* (2001) Numerical simulation of the December 1997 debris avalanche in Montserrat, Lesser Antilles. *Geophysical Research Letters* **28**: 2529–2532.

Hervás, J., Günther, A., Reichenbach, P., *et al.* (2010) Harmonised approaches for landslide susceptibility mapping in Europe. In *Proceedings of the International Conference Mountain Risks: Bringing Science to Society*, ed. J.-P. Malet, T. Glade and N. Casagli, Strasbourg: CERG Editions, pp. 501–505.

Highland, L. and Bobrowsky, P. (2008) *The Landslide Handbook: A Guide to Understanding Landslides*. Reston, VA: US Geological Survey Circular.

Highland, L. and Johnson, M. (2004) Landslide types and processes. US Department of the Interior, US Geological Survey Fact Sheet 2004–3072, July.http://pubs.usgs.gov/fs/2004/3072/ (accessed 15 June 2011).

HKSS (2011) Website of the Civil Engineering and Development Department of the Government of the Hong Kong Special Administrative Region. http://hkss.cedd.gov.hk (accessed 5 September 2011).

Holcombe, E. A., Smith, S., Wright, E., *et al.* (2011) An integrated approach for evaluating the effectiveness of landslide hazard reduction in vulnerable communities in the Caribbean. *Natural Hazards* **61** (2): 351–385.

Höller, P. (2007) Avalanche hazards and mitigation in Austria: a review. *Natural Hazards* **43** (1): 81–101.

Hsu, K. J. (1975) Catastrophic debris streams (Sturzstroms) generated by rockfalls. *Geological Society of America Bulletin* **86** (1): 129–140.

Hungr, O. (2009) Numerical modelling of the motion of rapid, flow-like landslides for hazard assessment. *KSCE Journal of Civil Engineering* **13** (4): 281–287.

Hutchinson, J. N. (1988) General report: morphological and geotechnical parameters of landslides in relation to geology and hydrogeology. In *Proceedings, Fifth International Symposium on Landslides*, ed. C. Bonnard, Rotterdam: Balkema, pp. 3–35.

Hutchinson, J. (2008) Rock slide hazards: detection, assessment and warning. Presentation at the 33rd International Geological Congress, Oslo.

Iceland Ministry for the Environment (2000) Regulation on hazard zoning due to snow- and landslides classification and utilization of hazard zones and preparation of provisional hazard zoning (505/2000) Icelandic Meteorological Office. http://en.vedur.is/avalanches/articles/ (accessed 4 August 2011).

IRASMOS (2008) Integral Risk Management of Extremely Rapid Mass Movements. European Commission 6th Framework Programme, project 018412. http://irasmos.slf.ch/ (accessed 23 July 2011).

Jamieson, B., Stethem, C., Schaerer, P., *et al.* (2002) *Land Managers Guide to Snow Avalanche Hazards in Canada*, Revelstoke, BC: Canadian Avalanche Association, http://www.avalanche.ca (accessed 17 June 2011).

Jiang, H. and Xie, Y. (2011) A note on the Mohr–Coulomb and Drucker–Prager strength criteria. *Mechanics Research Communications* **38** (4): 309–314.

Jóhannesson, T., Gauer, P., Issler, D., *et al.* (2009) The design of avalanche protection dams: recent practical and theoretical developments. Technical report, European Commission, Directorate-General for Research. No. EUR 23339, Climate Change and Natural Hazard Research Series 2.

Joyce, K. E., Belliss, S. E., Samsonov, S. V., *et al.* (2009) A review of the status of satellite remote sensing and image processing techniques for mapping natural hazards and disasters. *Progress in Physical Geography* **33** (2): 183–207.

Kaynia, A., Skurtveit, E. and Saygili, G. (2010) Real-time mapping of earthquake-induced landslides. *Bulletin of Earthquake Engineering* **9** (4): 955–973.

Keefer, D. K. (1994) The importance of earthquake-induced landslides to long-term slope erosion and slope-failure hazards in seismically active regions. *Geomorphology* **10** (1–4): 265–284.

Kelfoun, K. and Druitt, T. H. (2005) Numerical modeling of the emplacement of Socompa rock avalanche, Chile. *Journal of Geophysical Research* **110**: B12202.

Kerle, N. (2002) Volume estimation of the 1998 flank collapse at Casita volcano, Nicaragua: a comparison of photogrammetric and conventional techniques. *Earth Surface Processes and Landforms* **27**: 759–772.

Kirschbaum, D. B., Adler, R., Hong, Y., *et al.* (2009) A global landslide catalog for hazard applications: method, results, and limitations. *Natural Hazards* **52**: 561–575.

Konagi, K., Johansson, J., Mayorca, P., *et al.* (2002) Las Colinas landslide caused by the January 13, 2001 earthquake off the coast of El Salvador. *Journal of Japan Association for Earthquake Engineering* **2**: 1–15.

Kong, W. K. (2002) Risk assessment of slopes. *Quarterly Journal of Engineering Geology and Hydrogeology* **35**: 213–222.

Korup, O. (2002) Recent research on landslide dams: a literature review with special attention to New Zealand. *Progress in Physical Geography* **26** (2): 206–235.

Kunreuther, H. and Useem, M. (2009) Principles and challenges for reducing risks from disasters. In *Learning from Catastrophes: Strategies for Reaction and Response*, Upper Saddle River, NJ: Prentice Hall.

Lacasse, S., Nadim, F. and Kalsnes, B. (2010) Living with landslide risk. *Geotechnical Engineering Journal of the SEAGS & AGSSEA* **41** (4): n.p.

Laouafa, F. and Darve, F. (2002) Modelling of slope failure by a material instability mechanism. *Computers and Geotechnics* **29** (4): 301–325.

Li, T. and Wang, S. (1992) *Landslide Hazards and their Mitigation in China*, Beijing: Science Press.

Lied, K. and Bakkehøi, S. (1980) Empirical calculations of snow-avalanche run-out distance based on topographic parameters. *Journal of Glaciology* **26** (94): 165–177.

Lobell, D. B. and Asner, G. P. (2002) Moisture effects on soil reflectance. *Soil Science Society of America Journal* **66** (3): 722–727.

Lube, G., Huppert, H. E., Sparks, R. S. J., *et al.* (2007) Static and flowing regions in granular collapses down slopes. *Physics of Fluids* **19**: 1–9.

Lube, G., Cronin, S. J., Thouret, J.-C., *et al.* (2011) Kinematic characteristics of pyroclastic density currents at Merapi and controls on their avulsion from natural and engineered channels. *Geological Society of America Bulletin* **123** (5–6): 1127–1140.

Lumb, P. (1975) Slope failures in Hong Kong. *Quarterly Journal of Engineering Geology* **8**: 31–65.

Malamud, B. D. and Petley, D. (2009) Lost in translation. *Public Service Review: Science and Technology* **2**: 164–169.

Malamud, B. D., Turcotte, D. L., Guzzetti, F., *et al.* (2004a) Landslides, earthquakes, and erosion. *Earth and Planetary Science Letters* **229** (1–2): 45–59.

Malamud, B. D., Turcotte, D. L., Guzzetti, F., *et al.* (2004b) Landslide inventories and their statistical properties. *Earth Surface Processes and Landforms* **29** (6): 687–711.

Mangeney-Castelnau, A., Vilotte, J. P., Bristeau, M. O., *et al.* (2003) Numerical modeling of avalanches based on Saint Venant equations using a kinetic scheme. *Journal of Geophysical Research* **108** (B11): 2527.

Margreth, S. (2007) *Defense Structures in Avalanche Starting Zones: Technical Guideline as an Aid to Enforcement*. Bern: WSL Swiss Federal Institute for Snow and Avalanche Research.

Matthews, W. H. and McTaggart, K. C. (1969) The Hope landslide, British Columbia. *Proceedings of the Geological Association of Canada* **20**: 65–75.

McClung, D. and Lied, K. (1987) Statistical and geometrical definition of snow avalanche runout. *Cold Regions Science and Technology* **13** (2): 107–119.

McLean, I. and Johnes, M. (2000) *Aberfan: Government and Disasters*, Cardiff: Welsh Academic Press.

Misuyama, T. (2008) Sediment hazards and SABO works in Japan. *International Journal of Erosion Control Engineering* **1**: 1–4.

Naaim, M. (ed.) (2004) Final Report of project CADZIE (catastrophic avalanches, defense structures and zoning in Europe) CEMAGREF, Grenoble EVG1-199-00009.

Nadim, F. and Kjekstad, O. (2009) Assessment of global high-risk landslide disaster hotspots. In *Landslides: Disaster Risk Reduction*, ed. K. Sassa and P. Canuti, Berlin: Springer-Verlag, pp. 213–221.

Nadim, F. and Solheim, A. (2009) Effects of climate change on landslide hazard in Europe. AGU Fall Meeting 2009, abstract NH52A-03.

Nadim, F., Kjekstad, O., Peduzzi, P., *et al.* (2006) Global landslide and avalanche hotspots. *Landslides* **3**: 159–173.

Nicot, F., Gay, M. and Tacnet, J. M. (2002) Interaction between a snow mantel and a flexible structure: a new method to design avalanche nets. *Cold Regions Science and Technology* **34** (2): 67–84.

NRC (2004) *Partnerships for Reducing Landslide Risk: Assessment of the National Landslide Hazards Mitigation Strategy*, Washington, DC: National Academies Press.

O Globo (2011) Número de mortos na Região Serrana já passa de 900 após chuvas de janeiro. News report, 16 February. http://oglobo.globo.com/rio (accessed 17 June 2011).

Oppikofer, T., Jaboyedoff, M., Blikra, L., *et al.* (2009) Characterization and monitoring of the Aknes rockslide using terrestrial laser scanning. *Natural Hazards and Earth System Sciences* **9** (3): 1003–1019.

Petley, D. N., Hearn, G. J., Hart, A., *et al.* (2007) Trends in landslide occurrence in Nepal. *Natural Hazards* **43** (1): 23–44.

Phillips, M. (2006) Avalanche defence strategies and monitoring of two sites in mountain permafrost terrain, Pontresina, Eastern Swiss Alps. *Natural Hazards* **39** (3): 353–379.

Poulos, H. G. (1995) Design of reinforcing piles to increase slope stability. *Canadian Geotechnical Journal* **32** (5): 808–818.

Prada, P. and Kinch, D. (2011) Brazil landslide and flood toll reaches 665. *Wall Street Journal*, Latin America News, 17 January. http://online.wsj.com/article/SB1000 1424052748704029704576087893783940786.html (accessed 20 December 2011).

Prater, C. S. and Londell, M. K. (2000) Politics of hazard mitigation. *Natural Hazards Review* **1** (2): 73–82.

Raftery, A. E., Madigan, D. and Hoeting, J. A. (1997) Bayesian model averaging for linear regression models. *Journal of the American Statistical Association* **92** (437): 179–191.

Ray, R. L. and Jacobs, J. M. (2007) Relationships among remotely sensed soil moisture, precipitation and landslide events. *Natural Hazards* **43** (2): 211–222.

Remondo, J., Bonachea, J. and Cendrero, A. (2008) Quantitative landslide risk assessment and mapping on the basis of recent occurrences. *Geomorphology* **94** (3–4): 496–507.

Saito, H., Nakayama, D. and Matsuyama, H. (2009) Comparison of landslide susceptibility based on a decision-tree model and actual landslide occurrence: the Akaishi Mountains, Japan. *Geomorphology* **109** (3–4): 108–121.

Salm, B., Burkard, A. and Gubler, H. (1990) *Berechnung von Fliesslawinen, eine Anleitung für Praktiker mit Beispielen. Mitteilungen des Eidgenössischen Institutes für Schnee und Lawinenforschung*, Switzerland: Davos.

Sampl, P. and Zwinger, T. (2004) Avalanche simulation with SAMOS. *Annals of Glaciology* **38**: 393–398.

Sato, H. P. and Harp, E. L. (2009) Interpretation of earthquake-induced landslides triggered by the 12 May 2008, M7.9 Wenchuan earthquake in the Beichuan area, Sichuan Province, China using satellite imagery and Google Earth. *Landslides* **6** (2): 153–159.

Sauermoser, S. (2006) Avalanche hazard mapping: 30 years' experience in Austria. *Proceedings of the 2006 International Snow Science Workshop in Telluride*, 1–6 October, Telluride, CO: International Snow Science Workshop, pp. 314–321.

Savage, W. Z., Godt, J. W. and Baum, R. L. (2003) A model for spatially and temporally distributed shallow landslide initiation by rainfall infiltration. In *Debris-Flow Hazards Mitigation: Mechanics, Prediction and Assessment*, ed. D. Rickenmann and C. Chen, Rotterdam: Millpress, pp. 179–187.

Savage, W. Z., Godt, J. W. and Baum, R. L. (2004) Modeling time-dependent aerial slope stability. In *Landslides: Evaluation and Stabilization. Proceedings of the 9th International Symposium on Landslides*, ed. W. A. Lacerda, M. Erlich, S. A. B. Fontoura, *et al.*, London: AA Balkema, pp. 23–36.

Schuster, R. L. (1996) Socioeconomic significance of landslides. In *Landslides: Investigation and Mitigation*, ed. A. K. Turner and R. L. Schuster, Washington, DC: National Academy Press, pp. 12–35.

Schuster, R. L. and Fleming, R. W. (1986) Economic losses and fatalities due to landslides. *Bulletin of the Association of Engineering Geologists* **23**: 11–28.

Schuster, R. L. and Highland, L. (2001) Socioeconomic and environmental impacts of landslides in the Western Hemisphere. US Geological Survey open-file report; 01–0276. US Dept. of the Interior. http://purl.access.gpo.gov/GPO/LPS14705 (accessed 4 August 2011).

Schuster, R. L. and Highland, L. M. (2007) Overview of the effects of mass wasting on the natural environment. *Environmental & Engineering Geoscience* **13** (1): 25–44.

Schweizer, J., Jamieson, J. B. and Schneebeli, M. (2003) Snow avalanche formation. *Reviews of Geophysics* **41** (4): n.p.

Scott, K. M., Vallance, J. M., Kerle, N., *et al.* (2005) Catastrophic precipitation-triggered lahar at Casita volcano, Nicaragua: occurrence, bulking and transformation. *Earth Surface Processes and Landforms* **30**: 59–79.

Shoaei, Z., Shoaei, G. and Emamjomeh, R. (2005) *Interpretation of the Mechanism of Motion and Suggestion of Remedial Measures Using GPS Continuous Monitoring Data*. Berlin: Springer-Verlag.

Siddle, H. J., Wright, M. D. and Hutchinson, J. N. (1996) Rapid failures of colliery spoil heaps in the South Wales Coalfield. *Quarterly Journal of Engineering Geology* **29**: 103–132.

SLF (2011) WSL Institute for Snow and Avalanche Research SLF website (providing avalanche bulletins, reports and hazard map). http://www.slf.ch (accessed 24 June 2011).

Spiker, E. C. and Gori, P. L. (2000) *National Landslide Hazards Mitigation Strategy: A Framework for Loss Reduction*. US Geological Survey Open-File Report 00-450.

Tanguy, J. C., Ribiere, C., Scarth, A., *et al.* (1998) Victims from volcanic eruptions: a revised database. *Bulletin of Volcanology* **60** (2): 137–144.

Teeuw, R., Rust, D., Solana, C., *et al.* (2009) Large coastal landslides and tsunami hazard in the Caribbean. *Eos, Transactions, American Geophysical Union* **90** (10): 80–81.

Theilen-Willige, B. (2010) Detection of local site conditions influencing earthquake shaking and secondary effects in Southwest-Haiti using remote sensing and GIS-methods. *Natural Hazards and Earth System Sciences* **10** (6): 1183–1196.

Turcotte, D. L. and Malamud, B. D. (2004) Landslides, forest fires, and earthquakes: examples of self-organized critical behaviour. *Physica A: Statistical Mechanics and its Applications* **340** (4): 580–589.

University of Waikato (2008) WEKA data mining software. http://www.cs.waikato.ac.nz/ml/weka/ (accessed 4 August 2011).

USGS (2006) Landslide hazards program 5-year plan (2006–2010): Appendix B. Partnerships for reducing landslide risk. US Department of the Interior, US Geological Survey. http://landslides.usgs.gov/nlic/LHP_2006_Plan_Appendix_B.pdf (accessed 21 January 2011).

USGS (2011a) USGS Multimedia Gallery. http://gallery.usgs.gov/ (accessed 4 August 2011).

USGS (2011b) Worldwide overview of large landslides of the 20th and 21st centuries. http://landslides.usgs.gov/learning/majorls.php (accessed 17 June 2011).

van den Eeckhaut, M., Poesen, J., Verstraeten, G., *et al.* (2005) The effectiveness of hillshade maps and expert knowledge in mapping old deep-seated landslides. *Geomorphology* **67** (3–4): 351–363.

van Westen, C. J., van Asch, T. W. J. and Soeters, R. (2006) Landslide hazard and risk zonation: why is it still so difficult? *Bulletin of Engineering Geology and the Environment* **65**: 167–184.

Varnes, D. J. (1978) Slope movement types and processes. In *Landslides: Analysis and Control*, ed. R. L. Schuster and R. J. Krizek, Washington, DC: National Research Council, pp. 11–33.

Vilajosana, I., Khazaradze, G., Suriñach, E., *et al.* (2006) Snow avalanche speed determination using seismic methods. *Cold Regions Science and Technology* **49** (1): 2–10.

Wagner, K. (2007) Mental models of flash floods and landslides. *Risk Analysis* **27** (3): 671–682.

Wald, D. J., Worden, B. C., Quitoriano, V., *et al.* (2005) *ShakeMap Manual: Technical Manual, User's Guide, and Software Guide*, Reston, VA: US Geological Survey.

Wang, H. B., Liu, G. J., Xu, W. Y., *et al.* (2005) GIS-based landslide hazard assessment: an overview. *Progress in Physical Geography* **29** (4): 548–567.

Wang, X. M. and Niu, R. Q. (2009) Spatial forecast of landslides in Three Gorges based on spatial data mining. *Sensors* **9** (3): 2035–2061.

Ward, S. N. (2001) Landslide tsunami. *Journal of Geophysical Research – Solid Earth* **106** (B6): 11201–11215.

Weaver, P. P. E. (2003) Northwest African continental margin: history of sediment accumulation, landslide deposits, and hiatuses as revealed by drilling the Madeira Abyssal Plain. *Paleoceanography* **18** (1): 1009.

Wilkinson, P. L., Anderson, M. G., Lloyd, D. M., *et al.* (2002) Landslide hazard and bioengineering: towards providing improved decision support through integrated model development. *Environmental Modelling and Software* **7**: 333–344.

Willenberg, H., Loew, S., Eberhardt, E., *et al.* (2008a) Internal structure and deformation of an unstable crystalline rock mass above Randa (Switzerland): Part I. Internal structure from integrated geological and geophysical investigations. *Engineering Geology* **101**: 1–14.

Willenberg, H., Evans, K. F., Eberhardt, E., *et al.* (2008b) Internal structure and deformation of an unstable crystalline rock mass above Randa (Switzerland): Part II. Three-dimensional deformation patterns. *Engineering Geology* **101**: 15–32.

WP/WLI (1993) A suggested method for describing the activity of a landslide. *Bulletin of Engineering Geology and the Environment* **47** (1): 53–57.

WSL (2011) Swiss Federal Institute for Forest, Snow and Landscape Research. http://www.wsl.ch (accessed 24 June 2011).

Yang, X. L. and Yin, J. H. (2004) Slope stability analysis with nonlinear failure criterion. *Journal of Engineering Mechanics – ASCE* **130** (3): 267–273.

Yang, H., Robert, F. A., Dalia, B., *et al.* (2010) Capacity building for disaster prevention in vulnerable regions of the world: development of a prototype global flood/landslide prediction system. *Disaster Advances* **3** (3): 14–19.

Yegian, M. K., Ghahraman, V. G., Nogolesadat, M. A. A., *et al.* (1995) Liquefaction during the 1990 Manjil, Iran, Earthquake: 1. Case-history data. *Bulletin of the Seismological Society of America* **85** (1): 66–82.

Zêzere, J. L., Garcia, R. A. C., Oliveira, S. C., *et al.* (2008) Probabilistic landslide risk analysis considering direct costs in the area north of Lisbon (Portugal). *Geomorphology* **94**: 467–495.

Zogning, A., Ngouanet, C. and Tiafack, O. (2007) The catastrophic geomorphological processes in humid tropical Africa: a case study of the recent landslide disasters in Cameroon. *Sedimentary Geology* **199** (1–2): 13–27.

10

Tsunami hazard and risk

T. K. HINCKS, R. S. J. SPARKS AND W. P. ASPINALL

10.1 Introduction

Tsunamis have been given enhanced attention since the Sumatran earthquake of 26 December 2004, one of the greatest disasters of recent decades, with a death toll of at least 230 000. This event was triggered by the largest global earthquake for 40 years (magnitude 9.2) and highlighted the impact of extreme natural events. Tsunami waves hit 12 countries bordering the Indian Ocean region (Figures 10.1 and 10.2), and tourists and nationals from many more countries were affected; for example, it is considered the greatest natural disaster in Sweden's history due to the large number of Swedish tourists killed (estimated total of 543 fatalities – Kivikuru and Nord, 2009). Between 1900 and 1990 the NGDC Global Tsunami Database reports a total of 33 500 fatalities due to tsunamis (NOAA/NGDC, 2010), while in the period 1990 to 2004 there were in excess of 230 000 deaths, largely related to this single event. As a consequence, tsunami hazard has been elevated towards the top of the natural hazards agenda. The tsunami that affected American Samoa in September 2009, the Chilean earthquake tsunami of February 2010 (Figure 10.3) and the Indonesian tsunami of October 2010 that resulted in over 400 deaths have re-emphasised the importance of tsunami hazard to coastal communities. The unanticipated scale of the tsunami impact on the Fukushima nuclear power plant in March 2011 further highlighted the need to advance the science of risk assessment for safety-critical facilities, and to develop more robust frameworks for assessing uncertainty (see Chapters 2 and 3).

This chapter provides an overview of tsunami hazards, the current science of hazard and risk assessment, existing operational systems for monitoring and early warning, challenges and future research needs. For an in-depth account of current tsunami research, the volume by Bernard and Robinson (2009) is recommended reading.

10.2 Background

10.2.1 Hazard description

The word tsunami derives from Japanese, *tsu* meaning harbour and *nami*, a wave, but has the wider meaning of a compressional wave anywhere in the ocean. A tsunami is a set of ocean

Risk and Uncertainty Assessment for Natural Hazards, ed. Jonathan Rougier, Steve Sparks and Lisa Hill. Published by Cambridge University Press. © Cambridge University Press 2013.

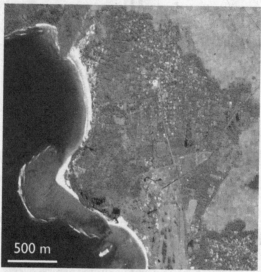

500 m

10 January 2003

29 December 2004

Figure 10.1 Satellite images from before and after the 2004 Sumatran earthquake tsunami, showing inundation of the town of Lhoknga, on the west coast of Sumatra near Banda Aceh, where the most severe damage was reported. Ikonos Satellite Images © Centre for Remote Imaging, Sensing and Processing, National University of Singapore (2004).

waves generated by a large, rapid disturbance of the sea surface or connected body of water (Bernard *et al.*, 2006). Tinti *et al.* (2004) define a large-scale tsunamigenic source as being capable of generating long water waves with periods of the order of tens of seconds or more, although periods are more typically tens of minutes. Most tsunamis are generated by marine

Figure 10.2 Remains of a house on M. Kolhufushi in the Maldives, showing the severe damage caused by the wave. Photograph by Bruce Richmond, USGS (Richmond and Gibbons, 2005).

Figure 10.3 The magnitude 8.8 earthquake and tsunami on 27 February 2010 caused extensive damage in the Chilean town of Dichato and fishing villages along the coastline. Photographs by Walter D. Mooney (USGS, 2011).

or coastal earthquakes, but can also be triggered by landslides, submarine gravitational collapse, volcanic activity or meteorological events (e.g. fast-moving meteorological fronts). The scale and shape of the source disturbance determines the wavelength, period and amplitude of the tsunami, and ultimately its propagation.

Tsunami waves are characterised by long wavelengths, typically in the range 10 to 500 km, with wave periods of tens of minutes. Tsunamis can be treated as shallow-water waves as the wavelength λ is large compared to the water depth d. For the simple case where $\lambda \gg d$ the group and phase speed of the wave are approximately equal and proportional to the square root of the depth (see Figure 10.4). Thus, in deep oceans (with typical average depth ~4 km) tsunamis can travel at speeds of several hundred kilometres an hour, and can cross a major ocean in a few hours. Major transoceanic tsunamis typically comprise several waves that reach coasts around an ocean basin at intervals of between 20 and 90 minutes.

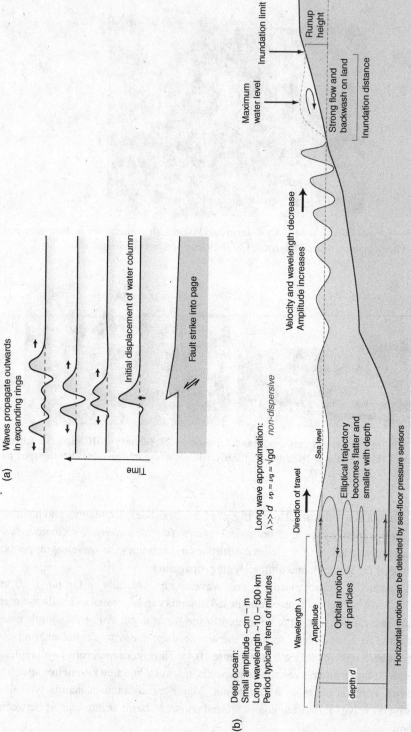

Figure 10.4 (a) Typical mode of tsunami generation (earthquake); (b) tsunami propagation and runup. The initial waveform is determined by the source geometry and dynamics (e.g. vertical displacement, rate and area of slip) and water depth at the source; ocean bathymetry influences propagation. High-resolution coastal bathymetric and topographic data are necessary to accurately predict runup and inundation. Runup is the height above sea level at the maximum inundation distance, although in some instances is used to denote to the maximum-recorded wave height either measured on land and or offshore by tide gauge. Coastal features (bays, estuaries, rivers and valleys) can amplify runup locally.

In the deep ocean, tsunami wave amplitudes typically range from a few centimetres to tens of centimetres. As the wave approaches the coastline and the water depth decreases, wave height increases by a process known as shoaling. Wave velocity and wavelength decrease with decreasing depth, which by conservation of energy results in an increase in amplitude as kinetic energy is converted to potential energy. This can result in huge wave heights at the coast, transporting enormous water masses inland. For a major tsunami, wave heights at the coast can locally be tens of metres, depending on the source, energy, propagation path and local conditions (e.g. tides, coastal bathymetry, resonant wave periods). Maximum-recorded tsunami wave heights are reported in the NOAA/WDC Historical Tsunami Database (NOAA/NGDC, 2010; see Section 10.2.3).

In low-lying regions waves of a few meters in height at the coast can cause flooding hundreds of meters inland, and transoceanic tsunamis can impact large areas, leading to widespread devastation. The 2004 Sumatran earthquake tsunami caused damage and a total of approximately 230 000 fatalities in 12 countries, including 13 deaths in Tanzania, some 6000 km from the source (see NOAA/NGDC, 2010). Synolakis and Kong (2006) provide a detailed review of field surveys carried out in the aftermath of the event. The greatest runup height associated with the event (50.9 m) was recorded in Labuhan, Indonesia, approximately 250 km from the source NOAA/NGDC, 2010). Indonesia also suffered extensive flooding, with inundation up to 5 km inland (associated runup 15 m). On Penang Island, Malaysia, where the water reached a maximum height of 2.6 m the flooding extended 3 km inland. A tsunami generated by a magnitude 6.8 earthquake in the Banda Sea, Indonesia, in 1964 produced a runup of 100 m locally on Ambon Island (NOAA/NGDC, 2010).

Tsunamis in enclosed water bodies can result in even more extreme wave heights: the highest recorded tsunami wave reached an elevation of 525 m above sea level, generated by an earthquake-induced landslide at the head of Lituya Bay, south-eastern Alaska in 1958. Due to the method of generation – a landslide in shallow waters of the bay, the tsunami only had local effect (limited propagation) and fortunately resulted in only two fatalities. Previous landslides in Lituya Bay in 1853 and 1936 are reported to have produced runups of 120 m and 149 m, respectively.

The initial shape of the wave affects runup and drawdown behaviour at the coast (Tadepalli and Synolakis, 1994), and is determined by the generation mechanism. In normal faults the disturbance is due to the initial downward motion of the sea bed, giving rise to a dilatational first motion. Thrust-faulting earthquakes result in a compressional first motion (Fowler, 2004). Drawdown (where the sea appears to retreat before the wave arrives and breaks) occurs if the trough of the first wave reaches the coast before the peak. The period between successive waves is typically in the range 20–90 minutes, so the unwary may be attracted back after the first wave only to be impacted or killed by the next waves.

10.2.2 Triggering mechanisms

Most major tsunamis typically result from large (greater than magnitude 7) shallow-focus earthquakes, generally at depths less than 30 km (Bernard *et al.*, 2006), which produce a large vertical or sub-vertical movement on the fault plane. The majority of such large, devastating tsunamigenic earthquakes occur in subduction zone settings (Moore *et al.*, 2007). Earthquakes (even those well below magnitude 7, and strike-slip faults which do not produce vertical displacement) can also trigger underwater landslides, which can in turn cause devastating tsunamis. For example, a magnitude 7 earthquake off the coast of Aitape, Papua New Guinea, in 1998 triggered a submarine slump of $\sim$4 km^3, which produced a locally devastating tsunami with a peak wave height of $\sim$15 m, and caused at least 2200 fatalities (Okal and Synolakis, 2004).

Major tsunamis can also be produced by very large landslides from the submarine slopes of continental margins and from flank collapse of volcanic islands. A good example of an extreme event is the Storegga slide, a sequence of massive submarine landslides off the west coast of Norway. The first Storegga slide is estimated to have occurred between 30 000–50 000 years BP, and displaced 3880 km^3 of sediment, producing a 34 000 km^2 slide scar (Bugge *et al.*, 1987, 1988). Two further slides occurred around 8200 years ago, with a total volume of approximately 1700 km^3 and a runout of 800 km (see Section 10.2.5). Reconstruction of the resulting tsunami from models and deposits indicate the wave was 10–15 m high along the Norwegian coasts and a few metres high along the eastern Scottish coast. In this case the cause of the slide seems to have been the rapid accumulation of glacial sediments on top of marine muds with elevated pore pressures. Ocean islands, mostly volcanic in origin, commonly show evidence of major landslides that are likely to have been tsunamigenic (e.g. Deplus *et al.*, 2001). Here, a likely trigger is volcanic activity, but volcanic islands are inherently unstable so earthquakes and pore pressure effects may also cause major landslides. Examples include the Canary Islands off western Africa, and Hawaii (Lockridge, 1998). The Clarke slide, a submarine slide off the Hawaiian island of Lanai around 150 ka (Moore *et al.*, 1989), is thought to have produced a major tsunami, and Young and Bryant (1992) propose that the erosive effects of the wave can be observed in south-eastern Australia. However, interpretation of palaeo-tsunami deposits and erosive features is challenging and a source of great uncertainty. The magnitude and dynamics of tsunamis from very large landslides into the ocean are also topics of controversy (see Løvholt *et al.*, 2008).

Volcanic eruptions can trigger tsunamis by at least three mechanisms, namely explosive volcanic eruptions in subaqueous settings (e.g. Krakatoa, 1883), pyroclastic flow or volcanic landslides into the ocean (e.g. Mount Unzen, Japan in 1792) and caldera collapse. Other triggering mechanisms include meteorite/asteroid impacts into the ocean and meteorological effects (see Rabinovich and Monserrat, 1996). In one unusual case, the explosion of a munitions ship in Halifax Harbour, Nova Scotia, in 1917 caused a strong local tsunami. However, explosion-generated tsunamis typically have short periods (tens of seconds to minutes) resulting in more rapid dissipation of energy, and are only likely to have a very localised impact (Clague *et al.*, 2003).

Analysis of the NGDC Historical Tsunami Database by Dunbar and Stroker (2009) finds that 81% of all historic tsunamis were triggered by earthquakes; 6% were associated with volcanic eruptions; 3% due to landslides; 1% of meteorological origin; and the remaining 9% of unknown origin. According to the NGDC database, earthquake-induced tsunamis have caused the greatest number of fatalities per event, and generally cause damage on a larger scale and at much greater distances from the source (Dunbar and Stroker, 2009). The process of earthquake tsunami generation is discussed in more detail in Section 10.3.3.1. Volcanogenic tsunamis account for a significant proportion of volcano fatalities: estimates range from 16.9% (Tanguy *et al.*, 1998) to 21% (Simkin *et al.*, 2001), making tsunamis a major volcanic hazard.

Meteorological tsunamis are generated by atmospheric disturbances producing waves in the open ocean, which become amplified near the coast. These waves are much less energetic than seismic tsunamis and only have local effects. Monserrat *et al.* (2006) report that 'destructive meteotsunamis are always the result of a combination of several resonant factors' and are therefore location dependent.

10.2.3 Tsunami catalogues

The NGDC provides a searchable global database of tsunami records from 2000 BCE to the present (see Dunbar *et al.*, 2008; NOAA/NGDC, 2010), compiled from a number of sources and comprising over 2600 recorded events. The database contains information on the source, runup locations and impacts, and includes an estimate of event validity and intensity (Figure 10.5). The NGDC also provide an online bibliographic database of tsunami deposits and proxies for tsunami events. The NGDC database is a widely used resource; however, it is known to contain some errors, and several catalogued estimates of tsunami intensities are not documented in the refereed literature and may be questionable (Synolakis, pers. comm.). Work is ongoing to verify data against original source documents, evaluate accuracy and improve and expand the database.

The Novosibirsk Tsunami Laboratory maintains a second (and in part overlapping) historical global tsunami database (Gusiakov, 2001; NTL, 2011).

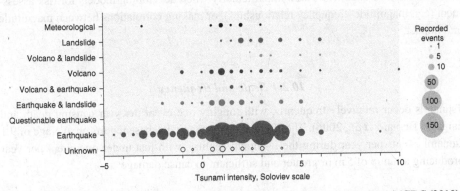

Figure 10.5 Plot showing tsunami intensity by trigger, using source data from NOAA/NGDC (2010). Intensity is defined by Soloviev and Go (1974) as $i_s = \log_2 \sqrt{2}h$, where h is the maximum runup height. Circle size is proportional to the number of events recorded, to give an indication of the relative frequency of each event type.

The first Europe-wide catalogue of tsunamis was compiled under the EU GITEC and GITEC-TWO projects (Genesis and Impact of Tsunamis on the European Coasts). GITEC developed standards for recording tsunami events, data quality and reliability, using a database structure derived from the Italian tsunami catalogue of Tinti and Maramai (1999) and Tinti *et al.* (2004). The database has been subsequently developed into the European/ NEAMTWS Tsunami Catalogue, hosted on the Intergovernmental Oceanographic Commission Tsunami Programme website (IOC, 2011). An important feature of the database is the representation of data reliability, using a scale from 0 to 4 ('very improbable' to 'definite') or 'no tsunami' to represent the validity of the event based on assessment of available evidence. The reliability index is determined by a combination of parameters describing the information available on the trigger, the tsunami itself and the source location.

Ambraseys and Synolakis (2010) provide an assessment of the reliability of tsunami catalogues for the eastern Mediterranean, and remark that in many cases, estimates of tsunami intensity are subject to significant uncertainty. For events with very limited information it can be inappropriate to even attempt to quantify magnitude or intensity, and additional sedimentological evidence may be necessary to enable statistical analysis of return periods, particularly for large, infrequent events. Runup varies substantially and can also be an unreliable indicator of magnitude. Local topographic and bathymetric features have a major effect on runup locally, and methods of surveying and recording are not always consistent between locations or events. Similarly, historical reports of damage and fatality numbers may be subject to historical or geographical bias. There is a clear need to quantify these uncertainties in a more robust and consistent manner, to enable reliable assessments of future hazard and risk. Although the NGDC and Novosibirsk databases both use validity indices as a proxy for uncertainty, it is not clear that the methodologies are consistent, compounded by the fact that both are compiled from earlier independent databases with potentially different reporting criteria. There is also a risk of human error, with information becoming corrupted in translation from one source to another. Such databases must therefore be used with caution, particularly when developing models for risk assessment (e.g. magnitude–frequency relationships), or making correlations between magnitude and impact.

10.2.4 Scale and frequency

Tsunamis occur relatively frequently, with roughly one event per year causing death and damage (Bernard *et al.*, 2006). The Novosibirsk tsunami database reports an average of 9.1 tsunami events per year during the twentieth century, with just under one event per year producing a runup of 5 m or greater and sufficient to cause damage.

10.2.4.1 Quantifying tsunamis: magnitude and intensity scales

Over the last 80 years there have been various attempts to quantify tsunami magnitude and intensity. A true intensity scale is a measure of the macroscale effects of the phenomena, and is independent of the physical parameters associated with the event (e.g. wave height or period). Intensity scales are typically descriptive, classifying the impact on buildings, humans or the natural environment. Conversely, magnitude is the measure of a physical characteristic of the phenomenon (typically energy), and is independent of its impact, although these definitions are not strictly adhered to in the literature.

Sieberg (1927) initially proposed a six-point intensity scale, analogous to earthquake intensity and based on the level of destruction. This was modified by Ambraseys (1962) to describe further effects such as visual observations of the waves, casualties and effects on ships at sea. The first tsunami magnitude scale was devised by Imamura and Iida, and was developed from an extensive catalogue of Japanese tsunamis dating back to 1700 (Imamura, 1942; Iida, 1956). Imamura–Iida magnitude is defined as $m_i = \log_2 H_{max}$, where H_{max} is the maximum wave height measured at the coast or by tide gauge (see Papadopolous, 2003). Shuto (1993) analysed damage sustained by different structures over a range of wave heights to develop a damage scale corresponding to the magnitude scale of Imamura–Iida. Soloviev (1970) and Soloviev and Go (1974) propose an alternative formulation for tsunami intensity, defined as $i_s = \log_2 \sqrt{2}h$, where h is the mean runup height at the coast or maximum runup according to NOAA/NGDC (2010) (e.g. see Figure 10.5). As i_s derives from runup and is therefore in some respects related to impact, Soloviev defined this measure as intensity, although strictly it is a measure of magnitude with similarities to the Imamura–Iida scale. Papadopolus and Imamura (2001) devised a 12-point intensity scale to enable more precise classification, although this is not widely used.

The most widely applied tsunami magnitude scale for seismogenic tsunamis (M_t) is that of Abe (see Abe, 1979, 1981, 1989), which has the general form $M_t = a \log H + b \log \Delta + D$, where H is the maximum amplitude (in metres) measured by tide gauge, Δ is the distance in kilometres from the earthquake epicentre along the shortest oceanic path, and a, b and D are constants. Murty and Loomis (1980) propose a magnitude scale based on energy, defining the Murty–Loomis tsunami magnitude $ML = 2 (\log E^{-19})$, where E is the potential energy of the tsunami (ergs). Potential energy E is estimated from $E = \frac{1}{2} \rho g h^2 A$, where ρ is the density of water (g cm^{-3}), g is acceleration due to gravity (cm s^{-1}), h is the elevation or subsidence at the source (cm) and A is the area of the ocean bottom affected by the uplift/subsidence (cm^2). The constant is derived from observations that tsunami potential energy must be of the order of 10^{19} ergs to be detectable ocean-wide, thus giving a tsunami of this threshold energy a magnitude of zero. Negative-magnitudes therefore correspond to tsunamis that will not propagate far and only have the potential to cause damage locally.

In an era when field measurements of runup exist, measures of intensity have become less useful, particularly as it is very difficult to assign values on 6- or even 12-point scales accurately and consistently in the field (Ambraseys and Synolakis, 2010). Lack of consistency in measurement is a particular problem in tsunami databases, and is a major source of uncertainty in any statistical analysis of past events. Runup is now much more widely used in hazard assessment and estimation of magnitude–frequency relationships (see Section 10.2.4.2).

10.2.4.2 Inter-event times and magnitude–frequency relationships

As the majority of tsunamis are earthquake-generated, inter-event times are largely governed by earthquake recurrence rates. Earthquake magnitude–frequency is typically modelled using the Gutenberg–Richter power law. However, rates of seismicity vary regionally and magnitude–frequency relationships for earthquakes determined from relatively short timescale datasets do not necessarily apply to return periods associated with extreme events (e.g. see Sornette and Knopoff, 1997). A further complication arises from the fact that other characteristics of the source (e.g. rupture geometry, slip distribution) are critical in determining tsunami generation and resulting wave properties, therefore earthquake magnitude alone is not a good predictor of tsunami size. The amplitude of far-field earthquake-generated tsunamis can be estimated reasonably well from earthquake moment magnitude. However, runup is much more variable for near-field tsunamis (see Geist, 1999 and references therein). Geist (2002) recommend probabilistic modelling to account for uncertainties in the rupture mechanism. Okal and Synolakis (2004) find that for near-field tsunamis the maximum runup on a smooth beach is limited by the slip on the fault (for earthquake-induced tsunamis), and for submarine landslides by the amplitude of the original depression.

Temporally, tsunamigenic sources are generally uncorrelated and as such inter-event times are generally treated as independent, identically distributed random variables with their distribution typically modelled as exponential, associated with a Poisson process (Geist *et al.*, 2009). As for earthquakes, the relationship between tsunami size (usually defined by runup) and frequency at a given location is typically modelled as a power law distribution, although the limit and shape of the tail of the distribution is not well characterised (Geist *et al.*, 2009). Using several decades of tsunami runup data for Japan, Burroughs and Tebbens (2005) found that a power law or upper-truncated power law distribution best describes the frequency–runup relationship. Truncation was only apparent at certain locations, and may be due to a relatively short time-series of observations, with very large events lacking. Alternatively, it could reflect a true limiting effect of the source (cf. the modified Gutenberg–Richter distribution for large earthquakes – Geist *et al.*, 2009). Van Dorn (1965) found that runup heights at numerous coastal locations on the Hawaiian Islands in the period 1946–1957 followed a lognormal distribution. This has been recognised for several other tsunami events (listed in Choi *et al.*, 2002), and is also a feature of some simple numerical simulations. Choi *et al.* (2002) propose that the lognormal distribution arises from random variations in the bathymetry near the coast.

Empirical distributions are useful for estimating scale and frequency of future events for risk assessment, although there are problems associated with extrapolating to extreme events. It is necessary to account for data uncertainty due to censoring and catalogue incompleteness. In addition, there is a high degree of measurement uncertainty in catalogue data, because parameters such as runup and magnitude are not always well defined or measured consistently.

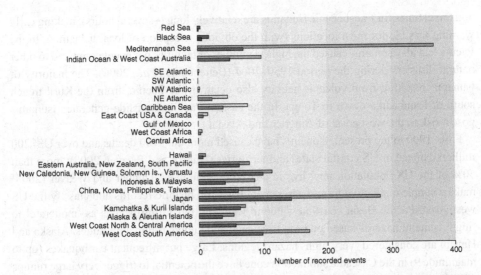

Shaded bars denote events since 1900. Only events with a validity of 2 (questionable tsunami) or greater are included.

Figure 10.6 Number of recorded tsunamis by region. The darker shaded bars denote events since 1900. Data are from the NOAA/WDC Historical Tsunami Database (NOAA/NGDC, 2010). Only events with a validity of 2 (questionable tsunami) or greater are counted here. In many areas there are few records prior to 1900, and prior to 1992 (and the advent of more extensive monitoring) there is a notable bias in historical records leading to under-reporting of small events (Ambraseys and Synolakis, 2010).

10.2.5 Areas affected

Approximately 59% of historical tsunamis have occurred in the Pacific Ocean, due to high levels of earthquake activity associated with the Pacific Ring of Fire. A further 25% of tsunamis occur in the Mediterranean Sea, 12% in the Atlantic Ocean and 4% in the Indian Ocean (PTWC, 2011). Figure 10.6 shows the global distribution of tsunamis as recorded in the NOAA/WDC Historical Tsunami Database. Geographical as well as temporal variation in data completeness and reliability is a major problem with the use of such databases to evaluate future hazard and risk. For example, the Historical Tsunami Database includes very detailed studies of tsunamis in parts of Europe and the Mediterranean region, dating back to 2000 BCE (Syria) and the ~1600 BCE volcanic eruption of Thera (Santorini). Records elsewhere, including higher-risk regions of the Pacific, are much sparser. However, Ambraseys and Synolakis (2010) caution that almost 75% of the entries in the existing Mediterranean tsunami catalogues may be spurious.

10.2.5.1 Pacific Ocean

Tsunamis occur most frequently in the Pacific Ocean and are predominantly earthquake generated due to high levels of seismicity associated with plate subduction around the Pacific Rim. Due to the size of the ocean, boundary reflections are not generally significant and travel times for transoceanic tsunamis are relatively long (~several hours), making early warning feasible for most locations (with the obvious exception of local tsunamis). In the Pacific Islands, tsunamis caused the highest number of fatalities per event compared to other natural disasters during the period 1950–2004 (Bettencourt *et al.*, 2006). The majority of tsunamis resulting from volcanic activity also occur in the Pacific, from the Kuril trench south of Kamchatka down to Tonga. In the eastern Pacific, landslide-generated tsunamis pose a risk to the west coast of America and Alaska.

From 1900 to the present, tsunamis have caused more than 700 deaths and over US$200 million damage in US coastal states and territories (Dunbar and Weaver, 2008); more than 50% of the US population now live in coastal communities. Dunbar and Weaver (2008) make a simple qualitative hazard assessment based on historic records and classify the US west coast, Pacific Basin coastlines, Puerto Rico and the Virgin Islands as 'moderate' to 'high' tsunami hazards based on both frequency and known runup amplitudes. Alaska and Hawaii are considered 'very high' hazard regions. Large but infrequent earthquakes (up to magnitude 9) in the Cascadia subduction zone have the potential to trigger very large runups affecting Oregon, Washington, and northern California.

The South China Sea (western Pacific) is effectively a closed basin, with the narrow straits connecting to the Pacific and Indian Oceans and adjoining basins limiting the propagation of tsunami waves. However, Okal *et al.* (2011) remark that tsunami hazard in this region may be underestimated and identify several locations where large-magnitude tsunamigenic earthquakes could occur, including the Luzon Straits, Luzon Trench and the Taiwan Straits. Okal *et al.* (2011) also note the possibility of tsunamis triggered by large underwater landslides (e.g. a reactivation of the Brunei slide).

10.2.5.2 Indian Ocean and Indonesia

Major tsunamigenic earthquakes are infrequent in the Indian Ocean, and prior to 2004 were a relatively neglected hazard. Most tsunamis originate in the tectonically and volcanically active eastern Indian Ocean, the Indonesian archipelago and the Bay of Bengal in the north. According to the Smithsonian Global Volcanism Program database (GVP, 2011), there are 78 historically active volcanoes in Indonesia and seven in the Indian Ocean (mainly in the east, with the exception of Karthala and La Réunion in the west). Cummins (2007) notes that the subduction zone off the coast of Myanmar (northern Bay of Bengal) is an environment that could generate giant megathrust earthquakes and therefore poses a significant tsunami hazard. In this region warning times are likely to be short due to short distances from source to coast, and in many regions it might not be possible to communicate warnings in time. Education and preparedness is therefore key to reducing risk. Simulations of seismic sources along the Sumatran–Andeman rupture (the location of the 2004 tsunami) by Okal and Synolakis (2008) show that source direction-ality and resulting refraction and focusing due to the bathymetry are critical factors affecting wave amplitude, making accurate, location-specific forecasting for the region very difficult.

10.2.5.3 North Atlantic

Tsunamis are much less frequent in the North Atlantic Ocean, as large earthquakes are relatively rare. Barkan *et al.* (2009: 109) note 'the Azores–Gibraltar plate boundary is the source of the largest earthquakes and tsunamis in the north Atlantic basin'. A notable historic event was the 1755 Lisbon tsunami, caused by a magnitude 8.5 earthquake. The tsunami produced runup heights of 5–15 m along the Portuguese and Moroccan coasts (Barkan *et al.*, 2009), and resulted in an estimated 60 000 fatalities (NOAA/NGDC, 2010). However, Baptista *et al.* (1998) remark that in Portugal most casualties were due to the earthquake itself or the resulting fires, so the number of fatalities resulting directly from the tsunami might be lower (for example, Baptista *et al.* give an estimate of 1000 fatalities in Portugal). The hypocentral location of the 1755 earthquake, and hence its causative fault source, is a matter of ongoing debate and analysis (e.g. Carvalho *et al.*, 2004 and references therein).

The principal tsunami risk in northern Europe is due to submarine landsliding of thick sediments deposited during the last ice age (see the Tsunami Risks Project – Dawson, 2000). Slope failure can be triggered by low-magnitude earthquakes, or the release of methane gas (clathrates) contained within the sediments. One of the largest known submarine landslides, the Storegga slide, caused a major tsunami that affected the coasts of Norway, Shetland, Scotland and the Faroe Islands approximately 8200 years ago. The slide occurred off the Norwegian continental shelf, and is identified as the last megaslide in the region (Bryn *et al.*, 2005). The landslide and turbidite deposits extend for more than 800 km from the edge of the continental shelf down to the abyssal floor. The landslide itself travelled approximately 300 km, continuing as a turbidity current for a further 500 km (Bugge *et al.*, 1988). Slides in the Storegga area are understood to have occurred with an average interval of approximately 100 000 years since the onset of continental shelf glaciations at 0.5 Ma. Seismic stratigraphy indicates that sliding typically occurs at the end of a glaciation or shortly after deglaciation, as rapid loading from glacial deposits gives rise to increases in pore pressures and a resulting decrease in shear strength (Bryn *et al.*, 2005). In addition, glacio-isostatic rebound is thought to have led to elevated earthquake activity in the region (including earthquakes of magnitude 7 and greater). The Storegga slide is conjectured to have resulted from a combination of unstable sediments and a strong earthquake trigger. Bugge *et al.* (1988) also remark that the presence of gas hydrates might have contributed to slope failure. Analysis of sediment records show that the resulting tsunami produced runup heights of 20–30 m on Shetland, 10–12 m on the west coast of Norway, more than 10 m on the Faroe Islands and 3–6 m on the north-east coast of Scotland (Bondevik *et al.*, 2003, 2005).

Two reports commissioned for Defra address tsunami risk for the UK and Ireland (Defra, 2005, 2006). Defra (2005) provides a summary of potential tsunamigenic sources, hazard scenarios and impacts (including effects on groundwater), and presents a framework for cost–benefit assessment of tsunami detection and warning schemes for the UK. Defra (2006) presents results of more detailed tsunami generation and runup simulations, considering both a North Sea source and a 1755 Lisbon-type earthquake in the Atlantic.

Tsunami impact at the coast is evaluated in terms of water height and flow velocity, and compared to extreme sea level and storm surge risk. The study concludes that wave heights at the coast are unlikely to exceed those expected for major storm surges, and for a Lisbon-type event travel times should be sufficiently long to effectively communicate warnings (~4.5 hours).

10.2.5.4 *Mediterranean*

Approximately 84% of Mediterranean tsunamis are earthquake-triggered, with only around 4% due to volcanic activity (NOAA/NGDC, 2010, events with validity 2 or greater). Earthquake source locations in the Mediterranean basin are typically close to coastlines, therefore a relatively small event could have a significant local impact. Tinti *et al.* (2005) simulate earthquake-triggered tsunamis associated with active fault zones in the Mediterranean basin, using estimates for maximum earthquake magnitude based on historic activity. Results indicate that the wave will reach the nearest coastline within 15 minutes, and cross the basin within one hour; however, they do not attempt to estimate event probability or return period.

There is potential for tsunami hazards from volcanism in the Mediterranean, with several submarine and active volcanic islands, notably in the Eolian and Aegean volcanic arcs. Two landslides from Stromboli volcano on 30 December 2002 generated tsunamis with wave heights of several metres, which damaged buildings and injured several people in villages of Stromboli and Ginostra. One of the largest explosive eruptions of the Holocene formed the caldera of Santorini, and geological evidence indicates that major tsunamis affected the coasts of the eastern and central Mediterranean (e.g. Bruins *et al.*, 2009).

10.2.5.5 *Caribbean*

Tsunamis also pose a significant hazard in the Caribbean region. For example, a tsunami struck Hispaniola and the Virgin Islands in 1842, killing 4000–5000 people (Lander *et al.*, 2002). The most recent destructive tsunami in 1946 was triggered by a magnitude 8.1 earthquake in the Dominican Republic, killing around 1800 people (Grindlay *et al.*, 2005). The NGDC database reports 77 tsunami events (validity 2 or greater) in the Caribbean; the majority (64 events) were seismic in origin and ten events (~13%) were associated with volcanic activity. Many of the volcanic islands in the Lesser Antilles island arc show large sector collapse scars and extensive submarine debris avalanche deposits (Deplus *et al.*, 2001).

10.2.6 *Impacts and quantification of harm and damage*

Tsunamis can cause damage to buildings, roads, power and water supplies, and loss of human life, health and livelihood. The extent and magnitude of the impact will depend on wave height and coastal topography. Low-lying islands and atolls are particularly vulnerable, and in some locations it can be difficult or impossible to evacuate to high ground.

Where it is possible to provide a timely warning and evacuate, direct loss of life can be prevented. However, subsequent impacts on food and water supplies, and reduced access to emergency or basic health services can cause significant and long-lasting problems for affected communities.

Displaced persons living in temporary shelters are highly vulnerable to disease and mental health problems. Infection can be a major problem (Uckay *et al.*, 2008). In the aftermath of a disaster many public services are limited or unavailable, including waste disposal, education, transport and medical care. Survivors can suffer severe psychological trauma and acute shock as a result of their losses and experience of the event (van Griensven *et al.*, 2006).

The 2004 Indian Ocean tsunami resulted in water shortages in many areas as saltwater infiltrated shallow coastal sand aquifers, and wells became polluted with hazardous waste, debris and sea water. This was a particular problem in Sri Lanka, where around 80% of the rural domestic water supply is provided by shallow wells (Leclerc *et al.*, 2008). In the low-lying atoll islands of the Maldives aquifers became polluted, and damage to septic tanks and inundation of rubbish dumps caused extensive pollution and public health risks (Chatterjee, 2006). The Food and Agriculture Organization of the United Nations (FAO) estimates that in Indonesia a total of 61 816 ha of agricultural land was damaged, leading to major crop losses and deterioration of land fertility (FAO, 2005). Depending on local conditions (e.g. rainfall) impacts on crops and drinking water can persist months to years after the inundation event.

10.2.6.1 Human impact

For the period 1900–2009 the EM-DAT database (CRED, 2009) reports a total of 240 959 fatalities and a further 2.48 million persons affected by tsunamis, although this is likely to be an underestimate of the total loss. The Indian Ocean tsunami on 26 December 2004 alone resulted in around 230 000 fatalities, over one million displaced persons, and financial losses of around US\$ 10 billion (RMS, 2006). According to a preliminary damage and loss assessment by BAPPENAS and the World Bank (BAPPENAS/World Bank, 2005), Indonesia suffered losses of roughly Rp 41.4 trillion (US\$ 4.5 billion), 66% constituting direct damage to property and infrastructure, and the remaining 34% the estimated income lost to the economy. Insurance penetration was low and only 12% of the losses were insured (estimate from RMS, 2006).

Prior to the 2004 Indian Ocean tsunami, historical tsunami fatality numbers had been quite low relative to other natural hazards, giving rise to a false perception (and severe underestimation) of risk (see Table 10.1). The 2004 disaster resulted in a major re-evaluation of tsunami risk and risk management strategies, and the potential impact of extreme events. On 29 September 2009, a magnitude 8 earthquake near Samoa and Tonga in the South Pacific triggered a tsunami that caused 181 fatalities and displaced more than 5000 people. This was the first tsunami to cause significant damage and fatalities on US territory (American Samoa) in more than 40 years (Okal *et al.*, 2010).

Table 10.1 *Major tsunami disasters*

Date	Location/event	Fatalities
26 December 2004	Earthquake off the west coast of Sumatra, Indian Ocean	Approximately 230 000
12 December 1992	Earthquake 50 km north of Maumere, capital of Flores Island, Indonesia (see Shi *et al.*, 1995)	Approximately 2500 (Marano *et al.*, 2010)
17 July 1998	Earthquake off the coast of northern Papua New Guinea	2000
16 August 1976	Earthquake in the Moro Gulf, Philippines	>5000 (possibly 8000)
28 March 1964	Good Friday earthquake, Prince William Sound, Alaska	122 (Alaska, Oregon and California)
22 May 1960	Earthquake off the coast of South Central Chile; Tsunami hits Chile, Hawaii, the Philippines, and Japan	1000 in Chile, 61 in Hawaii
1 April 1946	Earthquake in the Aleutian Trench, Alaska	5 in North Cape, 159 in Hawaii
31 January 1906	Offshore earthquake, Colombia; impacts Colombian and Ecuadorian coastlines	500–1500
15 June 1896	Earthquake off the coast of Sanriku, Japan	26 000
27 August 1883	Eruption of Krakatau; tsunami impacts Java and Sumatra	36 000

Note: Data from BAPPENAS/World Bank (2005) and NOAA/NGDC (2010) unless otherwise stated. There is some uncertainty regarding fatality numbers, as historical reports do not always clearly discriminate between deaths from concurrent hazard events (earthquakes, landsliding, etc.).

10.2.6.2 *Impact on buildings and infrastructure*

Tsunami waves cause damage to structures due to hydrodynamic pressure, buoyancy, uplift, scour and impact by debris carried by the water (Ghobarah *et al.*, 2006). The first wave will destroy most structures (except reinforced concrete), generating large quantities of debris. Subsequent waves entrain this debris, causing further damage and impacting stronger structures. Ghobarah *et al.* (2006: 312) discuss the impact of the 2004 Indian Ocean earthquake and tsunami on building structures and infrastructure. They report that 'failure of critical infrastructure such as bridges, harbour docks, hospitals and communication systems delayed search and rescue operations and relief efforts', exacerbating the impact.

Many nuclear power plants (NPPs) are located on the coast and may be vulnerable to tsunami impact. The magnitude 9 Tohoku earthquake off the east coast of Japan on 11 March 2011 triggered a major tsunami which caused severe flooding at the Fukushima Dai-ichi NPP. The tsunami wave (estimated to be around 14 m high, exceeding design standards for the plant) breached the sea defences, causing damage to the cooling systems of three reactors, and ultimately led to release of radionuclides to the sea and atmosphere (Wakeford, 2011). In a similar but less catastrophic event, the Kalpakkam power plant (on the coast in Tamil Nadu, India) was safely shut down following flooding from the 2004 Indian Ocean tsunami. The NRC Standard Review Plan (USNRC, 2007) defines procedures and acceptance criteria for assessing the probable maximum tsunami (PMT) hazard for NPPs and establishing inundation limits for site selection and design. This, and the subsequent NRC report by Prasad *et al.* (2009) provide guidelines for various aspects of hazard assessment, including the application of numerical models for tsunami propagation and inundation. However, the Fukushima incident highlights the importance of reassessing risk for older facilities as science advances and new data become available.

To further improve response to such threats, the International Seismic Safety Centre of the International Atomic Energy Agency (IAEA) are continually working to assess the hazards generated by external events, and provide extensive guidance on preparedness in cases of emergencies. In March 2010, the IAEA concluded a project on the protection of NPPs against tsunamis and post-earthquake considerations in the external zone (TiPEEZ), which was undertaken with contributions from the Japan Nuclear Energy Safety Organisation (JNES) and the US NRC (see IAEA, 2010 and Ebisawa *et al.*, 2008 for a description of the original PEEZ system for earthquake protection). The TiPEEZ system is an emergency response information management system that aids evaluation of the post-tsunami state of an NPP, and also helps evaluate offsite damage and potential impacts on safe plant operation – e.g. possible damage to bridges and roads in the vicinity of installations, and effects on evacuation routes, location of shelters and arrangements of vehicles for evacuation. The project has involved the development of an online tsunami occurrence notification system, specifically configured for informing nuclear sites worldwide as close to real-time as possible.

10.3 Hazard and risk assessment

Tsunami arrival time forecasts have been routine since the late 1960s, and the NOAA National Geophysical Data Center now routinely provide inundation estimates for many US Pacific cities, and real-time tsunami height forecasts for the Pacific and Indian Oceans and the Caribbean. Because inundation on land is difficult to predict (particularly in real time), pre-event tsunami hazard information for other parts of the globe typically consists of travel-time maps and probability estimates for wave height at the coast. Real-time forecasts for arrival time and wave height are computed from rapid first estimates of the seismic source parameters (see Section 10.3.3.4 on operational forecasting models). Travel time is mapped in two ways: either as a source-based map, giving estimates of the minimum travel time from the source to locations around the globe; or a location-based map showing the minimum time a wave would take to reach a specific location from different parts of the ocean. Maps are deterministic, typically giving contours of one hour. NOAA provides minimum travel-time maps for selected coastal locations and first-arrival travel-time maps for historic tsunami sources, computed using Tsunami Travel Time software (Wessel, 2011). Figure 10.7 shows an example travel-time map for San Francisco.

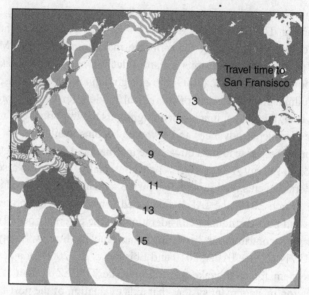

Figure 10.7 Estimated minimum travel time (in hours) to San Francisco from a tsunamigenic source in the Pacific Ocean. Figure developed from the online mapping facility on the NOAA/NGDC website (NGDC, 2011).

10.3.1 Challenges in predictability

It is as yet impossible to predict precisely when and where a tsunami will occur, although areas susceptible to tsunamis (e.g. earthquake-prone plate margins and continental margins) can be identified. Once generated and measured by deep-ocean sensors, it is possible to forecast the tsunami arrival time and area of impact accurately, although warning times are typically short (hours at most) (Bernard *et al.*, 2006). To limit loss of life it is necessary to provide rapid detection and forecasting of the tsunami impact areas, followed by fast and effective communication of warnings and alerts. Suitable early warning systems must be in place with communication systems that enable warnings to be disseminated rapidly to populations – and these systems must be working, and populations at risk must be ready to respond and evacuate. Any breakdown or excessive delay in any part of this chain will result in failure to mitigate the dangers.

10.3.1.1 Cross-hazard implications: assessment and forecasting of triggering events

There is movement towards great earthquake forecasting (in a broad sense) using fault stress transfer models (e.g. in Sumatra). There may be operational utility in such models for tsunami hazard assessment and forecasting too, where there is a high probability of a tsunami being associated with the anticipated seismic event. For example, McCloskey *et al.* (2008) present an assessment of tsunami hazard in the Indian Ocean resulting from a future megathrust earthquake west of Sumatra. They simulate a range of possible complex earthquakes, and calculate the resulting sea-floor displacements and tsunami wave-height distributions to give an estimate of the potential hazard. About 20% of their simulations produce tsunami wave heights in excess of 5 m for the southern Sumatran cities of Padang and Bengkulu. Such assessments provide valuable information for developing tsunami preparedness strategies and understanding long-term risk. A second line of potential forecasting is with volcanic eruptions, which may well have periods of unrest sufficiently long for related tsunami hazards to be assessed. A third line of assessment is modelling worst-case tsunami scenarios to identify potential flooding areas and design evacuation routes to extract the coastal population from the tsunami hazard.

10.3.2 Assessment using geological deposits and historical evidence

10.3.2.1 Tide gauge data and historical records

Before the nineteenth century, only tsunamis that resulted in casualties or significant damage were likely to have been recorded (Scheffers and Kelletat, 2003). In terms of observational data, there are very few tide gauge records prior to the twentieth century, and most records that do exist are from the Northern Hemisphere. The NGDC provide a comprehensive online bibliographic database of tsunami deposits and proxies for tsunami events (see Section 10.2.3 on tsunami catalogues; also see NOAA/NGDC, 2010).

10.3.2.2 Tsunami deposits

Tsunami deposits from modern, historical and late Quaternary times are well documented in the literature, but it becomes difficult to identify evidence further back in the geological record (Dawson and Stuart, 2007; Bourgeois, 2009). Even more recent records are incomplete and highly biased towards large events. Most studies focus on fine sediments, although these can be difficult to identify and the means of deposition may be ambiguous (Scheffers and Kelletat, 2003). There is a particular issue of distinguishing tsunami deposits from storm surge deposits (e.g. Switzer *et al.*, 2005). Tsunamis typically impact environments subject to frequent reworking: flood plains, low-lying coastal areas, shallow seas and submarine canyons; as such, deposits are poorly (and rarely) preserved, or may be indistinguishable from storm wave deposits. Lowe and de Lange (2000) suggest that an inundation height of 5 m or more is necessary to produce an onshore deposit that could be identified in the sedimentary record. Dawson and Stuart (2007) suggest that (for earthquake-generated tsunamis) an earthquake of magnitude 8 or greater is required to leave a recognisable tsunami deposit. However, they note that not all such events will produce a discernible trace, and other features of the earthquake (e.g. speed of seismic rupture) affect the energy transferred to the tsunami wave.

The processes involved and sedimentary features associated with tsunami deposits remain contentious, and distinguishing between tsunami and storm surge deposits remains a challenge; see discussions by Bridge (2008), Jaffe *et al.* (2008), Kelletat (2008) and Dawson *et al.* (2008). Compared to storm waves, tsunamis generally deposit sediment sheets over relatively wide areas and considerable distances (often several kilometres) inland (Dawson and Shi, 2000). Storm deposits have been observed to exhibit better sorting than the tsunami sediments (Nanayamaya *et al.*, 2000; Switzer *et al.*, 2005), as the backwash results in deposition of larger clasts and more poorly sorted material. Storm waves typically only show evidence of one direction of palaeocurrent, from offshore to onshore (Dawson and Shi, 2000).

Dawson and Stuart (2007) present a review of tsunami deposits in the geological record, discussing the relationships between the processes of tsunami generation, propagation, runup and associated sedimentary responses. They remark that 'boulder complexes deposited by tsunami may represent the most promising area of scientific enquiry', and that offshore deposits could also provide valuable information, if 'sedimentary "signatures" of these tsunami-related bottom currents can be distinguished from the turbidity and debris-flow sediment sequences' (p. 180). On land, tsunami sediment sheets also contain distinctive macro- and microfauna (including fish remains, shell debris, diatoms and foraminifera).

The Late Bronze Age explosive volcanic eruption of Thera (Santorini), around 1628–1525 BCE, is believed to have generated multiple major tsunamis, resulting from caldera collapse, pyroclastic flows and debris flows entering the sea, and possibly pre-eruption seismicity. There is a large body of literature (from Marinatos, 1939, to more recent publications by Cita and Aloisi, 2000; McCoy and Heiken, 2000; Minoura *et al.*, 2000), and some controversy over the scale and impact of the event. Re-analysis by Dominey-Howes (2004) suggests there is no strong evidence on land for a major distal tsunami, and offshore deposits can prove difficult to interpret or relate to a specific event. However, tsunami deposits have been identified on the coast of Crete (Bruins *et al.*, 2008, 2009).

There is a need for further research into the physical processes of sediment erosion, transport and deposition that occur during tsunami propagation and runup. Pritchard and Dickinson (2008) remark that it is very difficult to reconstruct wave hydrodynamics from sedimentary records using physical models. However, numerical simulations (e.g. Pareschi *et al.*, 2006), in conjunction with observations from sea-floor instruments and coastal monitoring networks, can provide valuable insights into the dynamics of tsunami waves, and forces and effects on the sea floor and coastline.

10.3.3 *Assessment using models*

Tsunami models aim to provide fast and reliable forecasts of tsunami propagation through the ocean, estimating the time to arrival and potential impact on coastal communities. Most operational models for tsunami warning or hazard and risk assessment describe tsunamis generated by offshore earthquakes. When an earthquake of sufficient magnitude has been detected, seismic source data and any subsequent observations of tsunami wave propagation (e.g. tsunameter/DART data, see Section 10.4.3) are used to constrain model parameters using data assimilation techniques. The modelling process can be broken down into three main stages: generation, propagation and inundation. In the United States the NOAA Center for Tsunami Research (NCTR) provides forecast modelling software for the NOAA tsunami warning centres, and inundation models to assist coastal communities in their efforts to assess the risk, and mitigate the potential of tsunami hazard.

10.3.3.1 Tsunami generation

The tsunami-generation process is complex and relatively poorly understood. In earthquake-triggered tsunamis, the surface perturbation is typically assumed to be the same as the vertical displacement on the sea floor. Estimates of this vertical displacement are highly uncertain as they are estimated from seismic moment, which is in turn some function of the average displacement on the fault. Some models (e.g. Tanioka and Satake, 1996) also consider the effects of horizontal displacement. It can also be necessary to model aftershocks (Song *et al.*, 2005). Papers by Hwang and Divoky (1970), Ward (1982) and Pelayo and Wiens (1992) describe the mechanisms of earthquake-generated tsunamis. Song *et al.* (2005: n.p.) note 'the slip function is the most important condition controlling the tsunami strength, while the geometry and the rupture velocity of the tectonic plane determine the spatial patterns of the tsunami'. It is believed that splay and 'megasplay' faults (branching faults at the extremity of a major fault) may be important in tsunami generation, accumulating convergent stresses and transferring displacement near the surface (Moore *et al.*, 2007).

Liu *et al.* (2005) and Lynett and Liu (2005) use laboratory and numerical simulations to determine correlations between landslide characteristics and maximum runup on adjacent beaches. In studies of both subaerial and submerged slides, they find the key parameters controlling runup are initial elevation and the specific weight of the slide. Following the methodology of Pelinofsky and Poplavsky (1996), Grilli and Watts (2005) and Watts *et al.* (2005) use numerical simulations to develop predictive empirical equations describing tsunami generation by submarine mass failure. Modelling a two-dimensional slide with a Gaussian profile, they find that tsunami amplitude and runup depend strongly on the initial depth of the failed mass, and follows a power law distribution (although this is highly simplified geometry, and this relationship is not widely used in practice).

The generation of tsunamis from large flank collapses of volcanic islands has proved to be highly controversial. Several of the volcanic islands in the Canary Islands archipelago have had major collapses and it has been inferred (Ward and Day, 2001) that they generated mega-tsunamis that would have devastated the whole North Atlantic coastal region. These models have been criticised for oversimplifying the process of debris flow emplacement and interaction with the water (Masson *et al.*, 2006; Løvholt *et al.*, 2008).

10.3.3.2 Propagation

Tsunami wavelengths are several times greater than the depth of the ocean so the wave motion involves the whole column of water from the sea bed to the surface. Propagation is modelled using the shallow-water equations (e.g. using a finite element method – Kawahara *et al.*, 1978), accounting for bottom friction and dispersion. Dispersion occurs as different frequencies propagate at different speeds, resulting in changes in wave shape. To model propagation over large distances it is also necessary to account for the effect of the Coriolis force and curvature of the earth. Titov and Synolakis (1998) describe a method of splitting the propagation equations to solve for onshore and longshore propagation separately.

10.3.3.3 Runup and inundation

Inundation models typically aim to predict wave arrival times, maximum wave height and maximum current speed along the coastline, and estimate the locations of maximum inundation. This information can be used to manage evacuations, protect critical infrastructure and prepare for deployment of emergency services. Inundation models are also used to simulate future tsunami impact scenarios, e.g. long-term risk assessments for infrastructure or land-use planning, and flood simulations for evacuation planning and training (González *et al.*, 2005).

Tsunami inundation is a highly nonlinear process (Tang *et al.*, 2009). Small-scale variations in local bathymetry and coastal morphology, tidal effects and hydrodynamic processes, such as wave breaking and wave–structure interaction, can have significant local effects. Titov and González (1997: 5) remark that 'runup of a tsunami onto dry land is probably the most underdeveloped part of any tsunami simulation model'. This is largely due to a lack of high-resolution bathymetric and topographic grids needed to accurately reproduce wave dynamics during inundation, and high-quality field data for model validation. Quality and availability of data are problems in many parts of the world, making it difficult to forecast runup height or perform risk assessments. The NGDC maintains the US national archive for bathymetric data, and from 2006 the NGDC has been collating high-resolution (10 m) digital elevation models (DEMs) for a number of US coastal locations (both onshore and offshore topography), for tsunami forecasting and inundation modelling. These bathymetric and topographic datasets require ongoing updating, as better data become available and coastal changes occur.

10.3.3.4 Operational forecasting models

The main objective of a forecast model is to provide an estimate of wave arrival time, wave height and inundation area immediately after tsunami generation. Tsunami forecast models are run in real time while a tsunami is propagating in the open ocean; consequently, they are designed to perform under very stringent time limitations. Following recommendations made by the Intergovernmental Coordination Group for the Indian Ocean Tsunami Warning and Mitigation System, the US Pacific Marine Environmental Laboratory (PMEL) and partner organisations, including the Australian Bureau of Meteorology, have developed and are supporting the use of ComMIT, a 'Community Modelling Interface for Tsunamis'. This is a computational suite based on the MOST model (see below) that allows for online development of inundation maps using remote hardware (Titov *et al.*, 2011). ComMIT aims to widen access to modelling tools and improve capability for forecast and hazard assessment in the Indian Ocean. To date (August 2011), the tool is being used for operational inundation mapping in 17 countries.

MOST (method of splitting tsunamis) MOST is an integrated suite of simulation codes for modelling tsunami generation, propagation and inundation, and is the standard operational model used by the NOAA Center for Tsunami Research (NOAA/PMEL, 2011). The tsunami-generation process is based on elastic deformation theory, and the method of computing inundation is a refinement of the model described by Titov (1997) and Titov and Synolakis (1998). MOST can be used in real time to forecast tsunami arrival time and inundation, or to run multiple scenario simulations for risk assessment. Real-time forecasting is a two-stage process. The first step is data assimilation: observations are compared with a database of pre-computed model runs to identify the most likely simulation parameters. The first pass just uses immediately available earthquake source data, and parameters are subsequently refined as more information becomes available from ocean buoys. Sensitivity analysis of the MOST model (Titov *et al.*, 1999, 2001) shows that source magnitude and location are the most critical parameters for inversion. The second stage is forecasting: best-fit model parameters are used to extend the simulation to locations where data are not available, and make forecasts ahead of the wave. Local forecasts are generated using a high-resolution two-dimensional inundation model (Titov *et al.*, 2003), which runs in minutes. Parallel or distributed computing is typically used to produce forecasts for many localities. The MOST model uses nested computational grids with resolution increasing towards the coastline of interest, to maximise accuracy and minimise computational time. Relatively coarse grids are able to resolve the wave in deep water where wavelengths are long, but in shallow water the wavelength shortens (and the amplitude rises), requiring more node points. With current technology there is generally sufficient time to forecast first-wave impacts of far-field earthquake-generated tsunami. Statistical methods are used to forecast the maximum height of secondary tsunami waves that can threaten rescue and recovery operations (Mofjeld *et al.*, 2000).

SIFT (short-term inundation forecasting for tsunamis) SIFT is a forecasting system currently being developed by the US Pacific Marine Environmental Laboratory for NOAA Tsunami Warning Centers (see Titov *et al.*, 2003; NOAA/PMEL, 2011). An extension of the MOST forecasting approach, SIFT uses an inundation model constrained by real-time data from DART deep-ocean buoys (deep-ocean assessment and reporting of tsunamis, see Section 10.4.3.2). Forecast products include estimates of tsunami amplitudes, flow velocities and arrival times for offshore, coastal and inundation areas. Stand-by inundation models (SIMs) using high-resolution bathymetric and topographic data are being developed for a number of coastal locations, to provide rapid, real-time forecasts before the first wave arrives.

T1 T1 is a forecasting system developed by the Australian Bureau of Meteorology (Greenslade *et al.*, 2007). Like SIFT, it is based on the MOST model, and uses a tsunami scenario database. Greenslade and Titov (2008) remark that although there are major differences in implementation of the MOST model and construction of scenarios, differences in the resulting SIFT and T1 forecasts are small, and for two case study events (Tonga 2006 and Sumatra 2007), results from both systems compared very well with observations. Greenslade *et al.* (2007) note that the model produces accurate forecasts when good real-time data are available, and performs particularly well for locations far from the source. However, for complex triggers (for example, the 26 December 2004 Sumatra earthquake), model performance is less accurate.

TREMORS TREMORS (Tsunami risk evaluation through seismic moment from a real-time system) is a tsunami warning system developed by the Centre Polynésien de Prévention des Tsunamis (CPPT, see Talandier, 1993), part of the French Commissariat à l'Énergie Atomique CEA. TREMORS is an automated system which derives estimates of potential tsunami amplitude from seismic moment. (Reymond *et al.*, 1996). Preliminary focal mechanism determination (PDFM) is used to obtain a quick first estimate of the seismic moment and focal depth (Reymond and Okal, 2000). TREMORS software is operational in French Polynesia (IOC-UNESCO, 2009), Chile (Gutierrez, 1998) and Europe. CEA, in partnership with the Euro-Mediterranean Seismological Centre, is also responsible for seismic monitoring in Europe.

10.3.3.5 *Scenario modelling for long-term hazard and risk estimation*

Simulations of tsunami scenarios (hypothetical future events) are widely used to understand long-term hazard and risk, and can be particularly useful in regions where tsunamis are relatively infrequent or there are limited historic data. For example, Lorito *et al.* (2008) argue the need for a more comprehensive database of scenarios for the Mediterranean. They present a methodology for determining source parameters using detailed tectonic information and earthquake datasets (to obtain the 'maximum credible earthquake'), and perform simulations to evaluate the potential impact at the coast of a tsunami associated with such an earthquake. Scenario modelling is used in the development of inundation maps used for evacuation planning and guidance during actual emergencies. In the United States, the National Tsunami Hazard Mitigation Program (NTHMP), launched in 1999 in the aftermath of the Papua New Guinea event, has strived to produce inundation maps of the Pacific coastline of the United States, and since 2004 for the Atlantic coast. California is the first state to have completed its mapping effort for all coastal communities (Barberopoulou *et al.*, 2011).

10.3.4 *Use of remote sensing data*

Remote sensing can be used to detect, monitor and assess tsunami propagation, impact and inundation extent. Satellite observations of tsunami propagation in the open ocean can provide valuable information for forecasting and early warning. Images from SPOT and IKONOS satellites have been used to track wave propagation and impact in coastal regions. Aitkenhead *et al.* (2007) describe an automated procedure using neural networks to analyse

IKONOS satellite data (pre-processed data at 1 m and 4 m resolution) and identify inundation extent. Aitkenhead *et al.* (2007) also develop an algorithm to identify urban areas that have been cut off and can no longer be reached by road, providing valuable information for emergency services. This method gives a measure of classification uncertainty. MODIS (moderate resolution imaging spectroradiometer) data have also been used to map flood extent, sediment load and impact (Wang and Li, 2008). Section 10.4.3.2 discusses the use of radar for detecting and tracking tsunamis.

10.3.5 Probabilistic models and treatment of uncertainty

Although operational models are largely deterministic (in part due to the need for speed and simplicity), the need for probabilistic hazard information is becoming increasingly recognised. One of the first probabilistic tsunami hazard assessments is that of Rikitake and Aida (1988), who estimated probability of exceedance for wave height at coastal locations in Japan, accounting for both local and distant earthquake sources. Earthquake probability is determined by observed recurrence rates and predicted accumulation of near-shore crustal strain. In regions with high seismicity they apply a Weibull distribution to account for increased probability immediately after an event. In the United States, the Tsunami Pilot Study Working Group (2006) have developed probabilistic inundation maps based on the principles of flood risk assessment. The project was part of a US Federal Emergency Management Agency (FEMA) initiative to improve their flood insurance rate maps, and resulted in the first ever probabilistic map for the town of Seaside, Oregon (González *et al.*, 2009). The analysis incorporated estimates of return periods from sediment studies and inferred earthquake return periods. The model accounts for multiple tsunami sources, using estimates for the maximum earthquake magnitude that can occur along a particular subduction zone (or segment) and applying the Gutenberg–Richter law to estimate mean frequency. Multiple simulations are performed to estimate the 100-year and 500-year probability of exceedance inundations (referenced to conventional hazard levels in flood risk assessment).

Grilli *et al.* (2009) present a probabilistic impact analysis for the upper north-east coast of the United States, for tsunami events triggered by earthquake-induced submarine landslides on the nearby continental slope. Probability distributions are defined for all key parameters (seismicity; sediment properties; type, location, volume, and dimensions of slide; water depth, etc.), and used to perform a Monte Carlo simulation. Probability of slope failure is assessed for a range of seismic ground acceleration values, each with an associated return period. For unstable slopes, the characteristic height of the tsunami source is estimated using empirical equations, and used to determine the corresponding breaking wave height and runup on the nearest coastline. Results are given in terms of expected runup for a given return period. They conclude that the overall tsunami hazard for the study area (for 100- and 500-year seismic events) is quite low at most locations, compared to the storm hazard for a similar return period (a 100-year storm event).

10.3.5.1 Probabilistic forecasts using Bayesian Networks

Woo and Aspinall (2005) suggest that the Bayesian Belief Network (BBN) provides a good framework for a risk-informed tsunami warning system, to capture uncertainties in the forecasting model (see Section 10.4.2 on systems for decision support). Blaser *et al.* (2009) develop this idea, and describe the use of a BBN to handle uncertain seismic source information, and provide probabilistic forecasts for tsunami warnings.

For tsunami events where the triggering earthquake occurs relatively close to the coast, time to impact can be of the order of 20 minutes or less. The first seismic source information is typically available after five minutes, but at such an early stage the data are highly uncertain. Existing operational warning systems are rule based, and do not account for uncertainties (or missing data) in seismic source information. The BBN of Blaser *et al.* (2009) captures the uncertainty in relationships between earthquake parameters (location, magnitude, rupture geometry, focal depth and mechanism) and tsunami runup at the coast. A synthetic database of tsunami events is used as input data, and structure and parameter learning is performed to generate a BBN. They demonstrate application of this approach to evaluate tsunami impact for the city of Padang, based on simulations of earthquake activity on the Sumatran subduction zone and the resulting tsunami generation, propagation and runup. Given some initial seismic data, the BBN produces a probabilistic forecast of tsunami size at the coast. The forecast can be updated as more or better-constrained earthquake data become available, and ultimately used to inform evacuation decisions.

10.3.5.2 Logic trees

Annaka *et al.* (2007) use a logic-tree approach (see Woo, 1999) to produce probability of exceedance curves for tsunami runup height. They present an example hazard assessment for Yamada, considering local tsunami sources around Japan and distant tsunami sources along the South American subduction zones. The model accounts for variability in tsunami source zones; size, frequency and recurrence rates of tsunamigenic earthquakes; alternative fault models; and resulting tsunami height at the coastline. Weightings for branches of the logic tree that represent alternative hypotheses, or where data are unavailable (e.g. the probability of a tsunami-generating earthquake occurring in a region where none have occurred in the past) were determined by means of a survey of tsunami and earthquake experts. Weightings for branches that represent natural variability of measurable quantities were obtained from historical data.

10.4 Risk management

Tsunami risk management requires an integrated approach, involving hazard assessment (mapping tsunami inundation potential), real-time monitoring and warning systems and an ongoing programme of public education to improve public awareness and response (Bernard, 1997).

10.4.1 Tsunami early warning: systems and organisations

The Pacific Tsunami Warning Center (PTWC) in Hawaii was established after the 1946 Alaska tsunami killed 173 people in Hawaii (Bernard *et al.*, 2006), and became the operations centre for the Pacific Basin. In 1960, following another major tsunami incident off the Chilean coast, the Tsunami Commission (a science organisation), and the International Coordinating Group for Tsunami Warnings in the Pacific (ITSU, a government organisation), were formed. The Intergovernmental Coordination Group for the Pacific Tsunami Warning and Mitigation System (ICG/PTWS, formerly ITSU) now acts to coordinate international tsunami warning and mitigation activities. Independent systems operate in Australia and Japan (see Section 10.4.1.1). All international tsunami warning centres cooperate under arrangements coordinated by the UNESCO Intergovernmental Oceanographic Commission (IOC). Technology currently exists to issue tsunami warnings within five minutes for earthquakes occurring along US coastlines, and within three minutes for local tsunamis off the coast of Japan. Section 10.3.3.4 discusses operational forecasting models employed by the various warning centres.

10.4.1.1 Areas of responsibility for tsunami warning

The PTWC issues tsunami watches, warnings and information bulletins for all American territories in the Pacific, and 25 other member countries in the Pacific Ocean Basin (see PTWC, 2011). The PTWC is also responsible for providing rapid warnings for local tsunamis off the coast of Hawaii. The PTWC provides interim warnings for countries in the Caribbean Sea, the South China Sea (China, Macao, Hong Kong, Taiwan, the Philippines, Malaysia, Brunei, Indonesia, Singapore, Thailand, Cambodia, Vietnam), and also acts as the interim warning centre for the Indian Ocean Tsunami Warning System. The West Coast/Alaska Tsunami Warning Center (WCATWC, 2011) issues warnings and information bulletins for Canadian coastal regions, Puerto Rico, the Virgin Islands, and the ocean coasts of all US states except Hawaii.

The Japanese tsunami warning system (part of the Japan Meteorological Agency – JMA, 2011) consists of a network of approximately 300 seismic sensors and 80 sea bed pressure sensors (Wang and Li, 2008). Local earthquake-generated tsunamis pose the greatest risk to the Japanese coastline, requiring a rapid response. Seismic intensity is typically obtained within two minutes of an earthquake, enabling an initial tsunami warning or advisory to be issued after three minutes. For a local tsunami this gives around ten minutes of evacuation time. Further information on the location and magnitude of the earthquake can be obtained after about three minutes, enabling more detailed tsunami forecasts (including arrival time and height of the first wave) to be issued for each of the 66 regions around the coast. To enable such rapid forecasting, the system uses a database of pre-computed simulations, including around 100 000 local earthquakes and a number of distant source events, and searches using seismic source information (magnitude and location) to find the closest simulation.

The Joint Australian Tsunami Warning Centre (JATWC, 2011) is operated by the Australian Government Bureau of Meteorology, and provides independent, 24-hour capability to detect and monitor tsunamis and issue warnings. The JATWC receives data from 50 seismic stations in Australia, more than 120 international seismic stations, a network of coastal sea-level gauges and DART buoys. Tsunami warnings form part of the standard set of hazard information issued by the Bureau of Meteorology, disseminated to the public via the web, local TV and radio, and communicated to the emergency services, local councils, port authorities, and police.

Following the 2004 Indian Ocean tsunami, UNESCO formed a number of intergovernmental coordination groups (ICGs) to support global cooperation and improve coverage in regions beyond the Pacific Ocean. Groups include ICG IOTWS for the Indian Ocean, ICG CARIBE for the Caribbean and neighbouring regions and ICG NEAMTWS for the northeastern Atlantic and Mediterranean. These organisations are in varying stages of development, but have systems in place for receiving and communicating alerts from the primary tsunami warning centres, and sharing data between member countries. The ICG NEAMTWS aims to establish an effective tsunami warning system by the end of 2011 (UNESCO, 2007). Although monitoring capabilities in the Atlantic are currently very limited, there are a significant number of existing offshore stations in Europe that could potentially be used for tsunami detection, including meteorological-oceanographic buoys and ocean-bottom seismic sensors.

10.4.2 *Formal mechanisms for decision support*

10.4.2.1 *Tsunami watch and warning*

There are four main levels of tsunami alert currently issued by the primary tsunami warning centres. *Warnings* are given when there is an imminent threat of a tsunami from a large undersea earthquake, or following confirmation that a potentially destructive tsunami is underway. Warnings recommend appropriate mitigation actions (e.g. to evacuate low-lying coastal areas, or move boats and ships out of harbours to deep water) and are updated at least hourly as conditions warrant. Tsunami *watches* are based on seismic information only, and are issued before confirmation that a destructive tsunami is underway, with the aim of providing a rapid advance alert to areas at risk. Watches are updated at least hourly and may be upgraded to a warning, or cancelled if a tsunami has not been generated or if the wave is small and non-destructive. Tsunami *advisories* are issued to coastal populations in areas not currently in warning or watch status if a tsunami warning has been issued for another region of the same ocean. A tsunami *information bulletin* reports the potential for generation of a tsunami as soon as an earthquake occurs. In most cases there will be no threat, and the aim of the bulletin will be to prevent unnecessary evacuations in coastal areas where the earthquake may have been felt. At each level of alert, monitoring will continue and the tsunami warning centres will review the alert status until the threat has passed. The uncertainties associated with initial seismic data analysis, observations, propagation models and simulation databases used to predict runup are not currently communicated in hazard information issued by the tsunami warning centres.

10.4.2.2 Evidence-based criteria for tsunami alerts

Initially, tsunami warning centres had good seismic network data, but were unable to accurately detect and confirm actual tsunami generation. This initially led to a very high false-alarm rate. In the period 1949–2000, 75% of warnings were false alarms (Bernard *et al.*, 2006), resulting in high evacuation costs and loss of credibility. The state of Hawaii estimate that a single false alarm would result in an average loss of around US$68 million (estimate for 2003 – González *et al.*, 2005). To minimise false alarms, early warning systems now incorporate real-time monitoring data. Earthquakes that could trigger tsunamis occur frequently in the Pacific Ocean (annually) so the warning system uses additional data from pressure sensors, tide gauges and satellite imagery to identify if a tsunami has been generated. Outside the Pacific, where monitoring capabilities are limited and warning time might be short, it is impractical to insist on certainty before issuing a warning. Woo and Aspinall (2005: 457) suggest that the 'alternative approach is to establish evidence-based criteria for raising different levels of tsunami alert. This would reduce the chance of complete alert failure, at a cost of some additional false alerts.' Woo and Aspinall propose developing a BBN to provide a framework for a risk-informed warning system and capture uncertainties in the forecasting model (see Section 10.3.5 on probabilistic models). This may be of particular value for developing nations, where high population densities in vulnerable coastal regions make the cost–benefit of a warning (even with increased probability of a false alarm) likely to outweigh the cost of a false alarm and evacuation.

10.4.3 Monitoring networks

10.4.3.1 Tide gauges

Water level or tide gauges provide information on sea level and tsunami propagation, and are a key component of the Pacific tsunami warning system. The majority of tide gauge stations in the Pacific network are designed for tsunami detection, providing fast sampling and near real-time reporting (Merrifield *et al.*, 2005). Traditional instruments (e.g. mechanical tide gauges and pneumatic pressure gauges – see Lee *et al.*, 2003) were originally designed to measure ocean tides, and as such do not perform as well in the tsunami frequency band. Newer installations (post-1990) typically use acoustic gauges with sounding tubes. These devices measure the travel time of acoustic pulses reflected vertically from the air–sea interface, and typically also transmit other environmental data (wind, air temperature, sea temperature, atmospheric pressure) as well as water level. Acoustic gauges form the basis of the US and Australian (SEAFRAME) tide gauge networks. Coastal tide gauges are typically installed inside harbours, and as a result suffer from harbour response (Titov *et al.*, 2005). Sources of measurement error include tidal effects, interference due to harbour geometry, harbour resonance and instrument response.

10.4.3.2 Tsunameters

Tsunameters consist of a bottom pressure recorder with an acoustic link to a surface buoy with satellite telecommunications. Tsunameters are located in the deep ocean, and provide early detection and real-time measurement of tsunamis. Data are used to update forecasts and limit false alarms by preventing or cancelling unnecessary warnings. During the 2003 Rat Islands tsunami, Alaska, tsunameter data were used to generate a forecast before the first wave reached most coastlines. The wave reached the first buoy after 80 minutes, and rapid data assimilation enabled generation of a high-resolution inundation forecast (Titov *et al.*, 2005; Bernard *et al.*, 2006). Tsunameter networks are expanding and currently provide good coverage for the northern Pacific, but less so for other coastlines. Larger arrays would increase forecast accuracy and lead-time (Titov *et al.*, 2003); however, they are costly systems to install and maintain. Ocean buoys are at risk of damage from fishermen, storms and boat collisions, resulting in loss of data and impaired forecasting ability. Ongoing servicing and maintenance of tsunameter networks is essential. According to Lubick (2009), in September 2009, 16 of the 55 buoys in the Tropical Atmosphere Ocean (TAO) array in the Pacific were out of action (missing or damaged). Lines of TAO buoys have a six-month maintenance schedule. In 2005 the US National Weather Service reported that two DART buoys were out of action for around 14 months (Knight, 2009).

DART DART (Deep-ocean Assessment and Reporting of Tsunamis) is a tsunameter system developed by the Pacific Marine Environmental Laboratory (PMEL) in 1986. The DART system consists of a sea-floor bottom pressure recorder (BPR) capable of detecting tsunamis as small as 1 cm, and a moored surface buoy for real-time data transmission via satellite. First-generation DART devices featured automatic detection and reporting, triggered by a threshold wave height. Second-generation devices (DART II) have two-way communications, enabling tsunami data transmission on demand (for example, following a seismic event). This allows measurement and reporting of tsunamis with amplitudes below the trigger threshold. In 2010 a third-generation, easy-to-deploy DART system became commercially available which allows the buoy to be deployed off small boats, reducing the costs of installing and maintaining DART arrays (Chwaszczewski, 2010). This cost-saving development will allow more countries to participate in the global tsunami warning system advocated by the IOC.

10.4.3.3 Future developments: radar

High-frequency coastal radar can detect changes in surface currents when a tsunami wave reaches a continental shelf (Heron *et al.*, 2008), and could be used to give warnings with lead times of the order of 40 minutes to two hours (depending on the width of the continental shelf). Heron *et al.* (2008) estimate that the limit of detection would be an open ocean wave height of around 2.5 cm, making it possible to detect relatively small tsunamis.

Godin (2004) suggests that changes in ocean surface roughness may be used to detect the propagation of tsunami waves, noting that extended darker strips on the ocean surface were observed along a front of a weak tsunami off Oahu, Hawaii (a 'tsunami shadow'). Godin (2004) remarks that this method may have potential value for early warning, but that further research is necessary to quantify changes in surface roughness under variable atmospheric conditions, and to evaluate the operational suitability of space-borne radars.

10.4.3.4 Multi-parameter ocean observatories

There are a large number of global ocean and earth monitoring initiatives, providing multiple data streams (physical, chemical and biological) for a range of scientific applications. Ocean monitoring data typically include temperature, depth, salinity, current flow, turbidity, chlorophyll, and oxygen and nitrate levels. These networks provide high-resolution spatial and temporal data, often in real time, and can be valuable for development and improvement of tsunami warnings, forecasting and inundation models.

United States The Ocean Observatories Initiative (OOI, 2011) is a long-term (25–30 years) project funded by the US National Science Foundation to provide coastal-, regional- and global-scale ocean monitoring. At the global scale, satellite-linked, moored instruments will operate at four sites in the North Pacific, South Atlantic, south-west of Chile in the South Pacific and south-east of Greenland in the North Atlantic. Regional-scale monitoring will comprise oceanic plate-scale studies of the water column, sea floor and sub-sea floor using cabled instruments (providing higher bandwidth). Coastal-scale monitoring (using buoys, underwater vehicles, etc.) will provide long-term, high-resolution spatial data. Data will be streamed live over the internet, providing real-time and adaptive capabilities.

Japan In June 2011, the Japan Agency for Marine-Earth Science and Technology (JAMSTEC) completed deployment of DONET (the Dense Oceanfloor Network System for Earthquakes and Tsunamis – Kaneda, 2010). The cabled network comprises 20 sets of instruments (including high-precision pressure sensors for tsunami monitoring) deployed at 15–20 km intervals across the sea floor, enabling detailed analysis of seismic activity, and supporting a tsunami early warning system.

Europe The European Seas Observatory Network (ESONET – Person *et al.*, 2008) is an EC initiative to promote the implementation and management of a network of long-term multidisciplinary ocean observatories around Europe, and the development of best practice and standards for knowledge and data sharing (Puillat, 2008). ESONET will provide real- or near-real-time deep-sea data as part of the Global Earth Observation System of Systems (GEOSS). A major science objective is improving the understanding of the relationship between earthquakes, tsunami generation and submarine slope failures. The Hellenic Center for Marine Research (HCMR) operates a multi-parameter cable observatory off Pylos, Greece, including a BPR which acts as a tsunamograph.

Canada The NEPTUNE-Canada cabled sea floor observatory is located off the west coast of Vancouver Island, British Columbia, across the Juan de Fuca plate. Instrumentation includes temperature and conductivity sensors and a number of high-precision, rapid (one second) sampling BPRs capable of measuring tsunami wave height in the open ocean (see Thomson *et al.*, 2011).

10.4.4 Preparedness and communication

Given very short warning and forecast lead times, preparedness is critical to tsunami hazard mitigation. Local tsunamis (triggered up to a few hundred kilometres offshore) are highly destructive and can reach the coastline in several minutes, often before a warning is possible. As such, populations at risk need to be aware of the hazards and how to respond to signs that a tsunami could be imminent: official warnings, strong ground shaking (which may be a precursor to a local tsunami) and unusual behaviour of the sea (e.g. water rapidly receding as the trough of the wave arrives). Evidence is emerging that indigenous knowledge based on societal memory of past events can increase community resilience considerably. An example of this was documented for the 2007 Solomon Islands tsunami where the proportion of casualties among immigrants was much higher than local people because they lacked the indigenous knowledge of the local people (Fritz and Kalligeris, 2008; McAdoo *et al.*, 2009).

Bettencourt *et al.* (2006) note that the 2004 Indian Ocean tsunami highlighted the importance of public awareness and education in a situation where timely forecasting and warning was (at the time) operationally impossible. For example, teaching children to run to high ground or a safe, elevated structure. It is also important to make sure that the public (both local and tourist populations) are aware of vulnerable areas and safe escape routes. Tsunami response and evacuation plans need to be based on reliable models of projected inundation, and estimates of probability of occurrence. It is also essential to consider tsunami risk when enforcing building codes, land-use planning and locations of services and infrastructure, and maintain natural wave breaks that can moderate the impact of a tsunami (e.g. coral reefs, mangroves and sand banks).

The US National Weather Service has developed the TsunamiReady programme (NWS, 2011) to engage and inform communities and emergency managers about tsunami risk, and promote planning, education and preparedness. TsunamiReady defines a set of criteria for communities to meet to officially join the programme. Requirements include establishing a 24-hour warning and emergency operations centre, promoting public preparedness and education and carrying out emergency training exercises. Jonientz-Trisler *et al.* (2005) describe the work of the US National Tsunami Hazard Mitigation Program (NTHMP), current mitigation strategies and ongoing programmes of action aimed at building tsunami-resilient communities.

Tsunami warning systems require regular testing and review. Simulations of hazard scenarios or evacuations and disaster management exercises are effective ways to improve preparedness and identify problems with existing procedures. One such example of a recent training exercise is Exercise Pacific Wave 08. This event involved response to a simulated magnitude 9.2 earthquake off north-west Japan. The exercise was not conducted through to community level, but was intended to test operational lines of communication within the Pacific tsunami warning system member states (IOC, 2008).

10.4.5 *Disaster response and recovery*

During the Asia tsunami and in Pakistan, the entire disaster response system was found lacking in commitment to evidence-based response, transparency, and data sharing. In such well-funded operations, a lack of funding is not a credible explanation for shortcomings in data management.

(Amin and Goldstein (2008: 40))

The Tsunami Evaluation Coalition (TEC, part of the UK Overseas Development Institute think-tank) was established in 2005 in response to the 2004 Indian Ocean tsunami. The TEC operates as an independent learning and accountability initiative in the humanitarian sector. The TEC carried out a number of studies to gauge the international response to the 26 December 2004 tsunami (Telford *et al.*, 2006; Telford and Cosgrave, 2007). They report that the relief phase was effective in ensuring that immediate survival needs were met, but that relief efforts were not well coordinated, leading to an excess of some interventions such as medical teams, but with shortages in less accessible areas or less popular sectors such as water supply (Cosgrave, 2005). A further conclusion of the TEC study was that local capacity building is key to saving lives, but this capacity is often underestimated and undervalued by the international aid community. In addition, recovery proved a far bigger challenge than relief; due in part to the greater complexity of recovery needs and increased demands on aid agencies.

Richmond *et al.* (2009) carried out interviews with first responders and survivors of the 29 September 2009 South Pacific tsunami, and conclude that hazard education played a key role in saving lives during the disaster.

10.4.5.1 *Social networking for disaster response*

Electronic networking tools are being increasing adopted by communities, organisations and individuals to share information and assist disaster response efforts. During the 2004 Indian Ocean tsunami, blogs (such as SEA-EAT, 2004) were used to identify needs and coordinate damage assessment, relief and volunteers. Open-source geo-referencing and communication tools, from Google Earth/Google Maps to text messages (SMS) were used during Hurricane Katrina to track down missing people, coordinate shelters and distribute aid (Butler, 2009).

InSTEDD (Innovative Support to Emergencies Diseases and Disasters) is a non-profit organisation developing open-source geo-referencing and communication tools for use in disaster situations and the management of public health threats (InSTEDD, 2011). Products include tools for data sharing (GeoChat) and mobile communication in the field that can be accessed and updated via SMS, twitter, email and web browser. In a GeoChat group, the locations of individuals, aid teams, resources and any additional information (news feeds, sensor data, imagery, needs) are geo-referenced, mapped and shared via a web interface.

10.5 Good practice and future research needs

10.5.1 Standards and good practice

There are no explicit, globally adopted standards for tsunami modelling and forecasting tools, although increasing cooperation and data sharing between tsunami warning centres is resulting in a more unified approach to the provision of alerts. Synolakis *et al.* (2007) note that without defined standards, forecasts may be produced using old or untested methods, potentially giving poor or misleading results. If inundation extent is underestimated lives may put be at risk. On the other hand excessive false alarms can result in unnecessary evacuation (at cost to the local economy), and loss of confidence in the alert system. Papers by Synolakis *et al.* (2007, 2008) discuss standards and 'best-practice' procedures for evaluation and application of tsunami models, and stress the importance of ongoing model validation and verification, ongoing scientific scrutiny by peer review of documentation and the need to maintain operational functionality – models must perform under constraints of time and limited data.

For effective global tsunami risk management and early warning, multi-parameter monitoring networks require common standards and tools governing instrumentation, data collection and processing, and user interfaces (see, e.g. Bernard *et al.*, 2006; Dunbar, 2009). For example, standards for DART have been developed by the IOC International Tsunameter Partnership.

Despite best efforts, bathymetric and topographic data used for tsunami modelling and forecasting are rarely of uniform resolution and quality. In some parts of the world data are very limited, leading to significant model uncertainty. Data can be obtained by a number of methods (with varying accuracy and different sources of uncertainty) in different terrestrial environments, and at various scales and resolutions. Changes to the region subsequent to data collection (both geomorphologic and anthropogenic change) can also result in inaccuracies. The NGDC have a standard procedure for data processing to maintain consistency in their modelling efforts. Topographic and bathymetric DEMs used in the MOST model are assessed for quality and accuracy both within each dataset and between datasets to ensure consistency and gradual topographic transitioning along the edges of datasets.

In terms of historical data and catalogues, the NGDC are working to ensure data sources in their tsunami hazards database are fully verified (NOAA/NGDC, 2010), and have developed standardised reporting formats.

10.5.2 Future needs

There is a need for further development of methodologies and software tools for assessing tsunami risk. Most existing research focuses on hazard processes, and this appears to be common across the natural hazards. Another common need is a more formalised approach to handling uncertainty. Current operational tsunami forecasting and alert systems are rule-based, and use databases of pre-computed simulations. As such, there are few means of

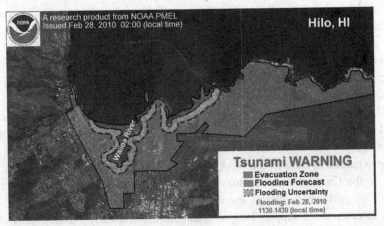

Figure 10.8 Proposed future tsunami warning product that indicates tsunami evacuation zones, overlain with tsunami flooding forecast including uncertainty and time of flooding for Hilo, Hawaii. Such information could be delivered direct to smartphones. Image courtesy of NOAA/Pacific Marine Environmental Laboratory (NOAA/PMEL, 2011).

accounting for uncertainties in the model structure, limitations associated with the simulated scenarios and input parameters, variation in quality or accuracy of bathymetric data, or uncertainties in the observations used for data assimilation. However, the United States has established a method for measuring skill of tsunami forecast models by comparing forecasted time-series with observed tsunami time-series at coastal measurement sites. After comparing forecast and observed tsunamis for 14 measureable tsunamis, skill level is about 90% (Titov, 2009). A 10% uncertainty could be incorporated into forecast products as illustrated in Figure 10.8, with dissemination to mobile devices (Bernard, pers. comm.).

Tsunami-generation mechanisms remain poorly understood (González *et al.*, 2005). For earthquake-triggered tsunamis, the slip function, fault geometry and rupture velocity are important factors governing tsunami energy and dynamics, but in existing forecasting models the earthquake process is highly simplified. Improved understanding of the source term and generation process would enable better forecasting and provision of more accurate and timely warnings for local tsunamis.

Continuing model validation and verification is essential, as reinforced by Synolakis *et al.* (2008: 2197):

In the aftermath of the 26 December, 2004 tsunami, several quantitative predictions of inundation for historic events were presented at international meetings differing substantially from the corresponding well-established paleotsunami measurements. These significant differences attracted press attention, reducing the credibility of all inundation modeling efforts. Without exception, the predictions were made using models that had not been benchmarked.

Local coastal measurements and runup data are critical for inundation model verification, but are often sparse and of poor quality. In the aftermath of a tsunami, data collection is crucial for improving future forecasts and more reliable risk assessments. Tide gauge data can be difficult to analyse due to the complex nature of the interaction of the wave with the oceanic shelf, coastline and harbour, and instruments may be damaged by the tsunami (González *et al.*, 2005). Deep-ocean buoys (e.g. DART tsunameters) provide the most unambiguous data for quantifying tsunami source and propagation, and give high-resolution temporal data, although there is still insufficient coverage in many areas.

Global tsunami monitoring and warning systems are continually expanding in scope and quality, but there is still significant room for improvement (e.g. Bernard *et al.*, 2007), and lack of mitigation funding is still the largest impediment to progress. It is sobering to recall that prior to the Asian tsunami the NOAA tsunami research programme was under funding pressure in the United States and may even have been closed down (E. Bernard, pers. comm.). This problem highlights the need for continued vigilance to persuade funding agencies to continue to support long-term monitoring and early warning networks even when there has not been a headline-grabbing event for some time. For example, there is currently no system in place to alert south-west Pacific countries of local tsunamis, which can be highly destructive and reach the coastline in only a few minutes. There are also no capabilities to warn remote communities. Tsunamis can be highly directional (González *et al.*, 2005), making it essential to increase the number of operational tsunameters and network coverage. It is important to develop systems that are affordable and sustainable, and provide rapid analysis of seismic data and communication of warnings to coastal populations and officials. The high costs of installation and ongoing maintenance of advanced monitoring networks will make such systems prohibitively expensive for many developing countries, and may not be justifiable in regions where tsunamis are infrequent. As a result, there is a need for evidence-based approaches to hazard forecasting and early warning, developing systems that can account for uncertainties in both models and observations and provide the best available risk information in a timely manner (e.g. as proposed by Woo and Aspinall, 2005 and Blaser *et al.*, 2009).

Given very short warning and forecast lead times, preparedness and rapid response is essential. There has been relatively little research into human response to tsunamis, although such information is essential for development of effective education programmes and evacuation management. However, relevant research is beginning (Mercer *et al.*, 2007; McAdoo *et al.*, 2009), which emphasises the importance of integrating Western scientific with indigenous knowledge in creating resilient communities. There is also potential for development of simulations (e.g. agent-based models) to train emergency response teams and develop evacuation strategies.

There is a clear need for further research to better understand tsunami impacts. For example, modelling the forces and interactions with structures to evaluate potential damage, and assess the effectiveness of protective structures (both man-made and natural). This would enable more reliable and comprehensive risk assessments, and lead to better and more cost-effective land-use planning and mitigation decisions.

There is great potential for more advanced use of real-time (or near-real-time) remote sensing data for damage evaluation to identify flooded roads or populated areas that have been cut off, and guide emergency response teams. Voigt *et al.* (2007) discuss the benefits of using satellite mapping to inform and assist disaster relief efforts, and describe the efforts of the German Aerospace Centre's Centre for Satellite-based Crisis Information (ZKI) in instigating rapid satellite mapping campaigns for a number of recent disasters. Examples include an impact assessment for the 2004 Indian Ocean tsunami, mapping hotspot extent during the forest fires in Portugal in 2005, a damage assessment for the 2005 Kashmir earthquake, and 2006 Philippine landslides. Rapidly acquired satellite information can be used to plan logistics, evaluate the extent of the impact and assess the potential locations and numbers of affected people.

Lastly, there are huge scientific challenges to be confronted in respect of finding ways to assess the probabilities of tsunami-causing events where these are very rare but where population densities, infrastructure and societal assets are highly concentrated, and to characterise the potential risk impacts in such areas.

Acknowledgements

The authors would like to thank C. Synolakis and E. Bernard for providing detailed and constructive reviews.

References

Abe, K. (1979) Size of great earthquakes of 1837–1974 inferred from tsunami data. *Journal of Geophysical Research* **84** (4): 1561–1568.

Abe, K. (1981) Physical size of tsunamigenic earthquakes of the northwestern Pacific. *Physics of the Earth and Planetary Interiors* **27** (3): 194–205.

Abe, K. (1989) Quantification of tsunamigenic earthquakes by the Mt scale. *Tectonophysics* **166** (1–3): 27–34.

Aitkenhead, M. J., Lumsdon, P. and Miller, D. R. (2007) Remote sensing-based neural network mapping of tsunami damage in Aceh, Indonesia. *Disasters* **31** (3): 217–226.

Ambraseys, N. (1962) Data for the investigation of the seismic sea-waves in the Eastern Mediterranean. *Bulletin of the Seismological Society of America* **52** (4): 895–913.

Ambraseys, N. and Synolakis, C. (2010) Tsunami catalogs for the eastern Mediterranean, revisited. *Journal of Earthquake Engineering* **14** (3): 309–330.

Amin, S. and Goldstein, M. (eds) (2008) *Data against Natural Disasters.* Washington, DC: World Bank.

Annaka, T., Satake, K., Sakakiyama, T., *et al.* (2007) Logic-tree approach for probabilistic tsunami hazard analysis and its applications to the Japanese coasts. *Pure and Applied Geophysics* **164** (2–3): 577–592.

BAPPENAS/World Bank (2005) Indonesia: preliminary damage and loss assessment. The December 26, 2004 natural disaster. Technical Report no. 31380, 19 January.

Baptista, M. A., Heitor, S., Miranda, J. M., *et al.* (1998) The 1755 Lisbon tsunami: evaluation of the tsunami parameters. *Journal of Geodynamics* **25** (1–2): 143–157.

Barberopoulou, A., Legg, M. R., Uslu, B., *et al.* (2011) Reassessing the tsunami risk in major ports and harbors of California I: San Diego. *Natural Hazards* **58** (1): 479–496.

Barkan, R., ten Brink, U. S., Lin, J. (2009) Far field tsunami simulations of the 1755 Lisbon earthquake: implications for tsunami hazard to the US East Coast and the Caribbean. *Marine Geology* **264** (1–2): 109–122.

Bernard, E. N. (1997) Reducing tsunami hazards along U.S. coastlines. In *Perspectives on Tsunami Hazards Reduction: Observations, Theory and Planning, Advances in Natural and Technological Hazards Research*, ed. G. Hebenstreit, Dordrecht: Kluwer Academic.

Bernard, E. N. and Robinson, A. R. (eds) (2009) *The Sea, Volume 15: Tsunamis. Ideas and Observations on Progress in the Study of the Seas*, Cambridge, MA: Harvard University Press.

Bernard, E. and Titov, V. (2006) Improving tsunami forecast skill using deep ocean observations. *Marine Technology Society Journal* **40** (4): 86–89.

Bernard, E. N., Mofjeld, H. O., Titov, V., *et al.* (2006) Tsunami: scientific frontiers, mitigation, forecasting and policy implications. *Philosophical Transactions of the Royal Society A: Mathematical, Physical and Engineering Sciences* **364** (1845): 1989–2006.

Bernard, E. N., Dengler, L. A. and Yim, S. C. (2007) National Tsunami Research Plan: report of a workshop sponsored by NSF/NOAA. NOAA Technical Memorandum OAR PMEL-133. Pacific Marine Environmental Laboratory. Seattle, WA. March.

Bettencourt, S., Croad, R., Freeman, P., *et al.* (2006) Not if but when: adapting to natural hazards in the Pacific Islands region. A policy note. World Bank, East Asia and Pacific region, Pacific Islands Country Management Unit.

Blaser, L., Ohrnberger, M., Riggelsen, C., *et al.* (2009) Bayesian belief network for tsunami warning decision support. *Symbolic and Quantitative Approaches to Reasoning with Uncertainty, Proceedings* **5590**: 757–768.

Bondevik, S., Mangerud, J., Dawson, S., *et al.* (2003) Record-breaking height for 8000-year old tsunami in the North Atlantic. *EOS* **84**: 289–293.

Bondevik, S., Lovholt, F., Harbitz, C., *et al.* (2005) The Storegga Slide tsunami: comparing field observations with numerical simulations. *Marine and Petroleum Geology* **22** (1–2): 195–208.

Bourgeois, J. (2009) Geologic effects and records of tsunamis. In *The Sea*, ed. E. N. Bernard and A. R. Robinson, Cambridge, MA: Harvard University Press, pp. 53–91.

Bridge, J. S. (2008) Discussion of articles in 'Sedimentary features of tsunami deposits' discussion. *Sedimentary Geology* **211** (3–4): 94.

Bruins, H. J., MacGillivray, J. A., Synolakis, C. E., *et al.* (2008) Geoarchaeological tsunami deposits at Palaikastro (Crete) and the Late Minoan IA eruption of Santorini. *Journal of Archaeological Science* **35** (1): 191–212.

Bruins, H. J., van der Plicht, J. and MacGillivray, A. (2009) The Minoan Santorini eruption and tsunami deposits in Crete (Palaikastro): dating by geology, archaeology, ^{14}C, and Egyptian chronology. *Radiocarbon* **51** (2): 397–411.

Bryn, P., Berg, K., Forsberg, C. F., *et al.* (2005) Explaining the Storegga Slide. *Marine and Petroleum Geology* **22** (1–2): 11–19.

Bugge, T., Befring, S., Belderson, R. H., *et al.* (1987) A giant three-stage submarine slide off Norway. *Geo-Marine Letters* **7**: 191–198.

Bugge, T., Belderson, R. H. and Kenyon, N. H. (1988) The Storegga Slide. *Philosophical Transactions of the Royal Society of London Series A – Mathematical, Physical and Engineering Sciences* **325** (1586): 357–388.

Burroughs, S. M. and Tebbens, S. F. (2005) Power-law scaling and probabilistic forecasting of tsunami runup heights. *Pure and Applied Geophysics* **162** (2): 331–342.

Butler, D. (2009) Networking out of natural disasters. *Nature*. http://www.nature.com/news/2009/090325/full/news.2009.187.html.

Carvalho, A., Campos Costa, A. and Oliveira, C. S. (2004) A stochastic finite: fault modeling for the 1755 Lisbon earthquake. Paper No. 2194, Thirteenth World Conference on Earthquake Engineering, Vancouver, BC, Canada, 1–6 August.

Chatterjee, P. (2006) Maldives struggle to reinstate tsunami-hit water supplies. *The Lancet* **367** (9504): 18.

Choi, B. H., Pelinovsky, E., Ryabov, I., *et al.* (2002) Distribution functions of tsunami wave heights. *Natural Hazards* **25** (1): 1–21.

Chwaszczewski, R. S. (2010) Bringing ocean technologies from research to operations. *Ocean News and Technology* **6** (7): 42–43.

Cita, M. B. and Aloisi, G. (2000) Deep-sea tsunami deposits triggered by the explosion of Santorini (3500 y BP) eastern Mediterranean. *Sedimentary Geology* **135** (1–4): 181–203.

Clague, J. J., Munro, A. and Murty, T. (2003) Tsunami hazard and risk in Canada. *Natural Hazards* **28** (2–3): 433–461.

Cosgrave, J. (2005) *Tsunami Evaluation Coalition: Initial Findings*. London: Tsunami Evaluation Coalition.

CRED (2009) EM-DAT: The OFDA/CRED International Disaster Database. Centre for Research on the Epidemiology of Disasters. Université Catholique de Louvain, Brussels, Belgium. www.emdat.be (accessed 15 December 2009).

Cummins, P. R. (2007) The potential for giant tsunamigenic earthquakes in the northern Bay of Bengal. *Nature* **449** (7158): 75–78.

Dawson, A. G. (2000) Tsunami risk in the northeast Atlantic region. Natural Environment Research Council, London. www.nerc-bas.ac.uk/tsunami-risks/html (accessed 7 August 2012).

Dawson, A. G. and Shi, S. Z. (2000) Tsunami deposits. *Pure and Applied Geophysics* **157** (6–8): 875–897.

Dawson, A. G. and Stewart, I. (2007) Tsunami deposits in the geological record. *Sedimentary Geology* **200** (3–4): 166–183.

Dawson, A. G., Stewart, I., Morton, R. A., *et al.* (2008) Reply to Comments by Kelletat (2008) comments to Dawson, AG and Stewart, I. (2007) tsunami deposits in the geological record, Sedimentary Geology, 200, 166–183. Discussion. *Sedimentary Geology* **211** (3–4): 92–93.

Defra (2005) The threat posed by tsunami to the UK. Study commissioned by Defra Flood Management. Department for Environment, Food and Rural Affairs. http://www.defra.gov.uk/environment/flooding/documents/risk/tsunami05.pdf (accessed 4 January 2011).

Defra (2006) Tsunamis: assessing the hazard for the UK and Irish coasts. Study commissioned by the Defra Flood Management Division, the Health and Safety Executive and the Geological Survey of Ireland. http://www.defra.gov.uk/environment/flooding/documents/risk/tsunami06.pdf (accessed 4 January 2011).

Deplus, C., Le Friant, A., Boudon, G., *et al.* (2001) Submarine evidence for large-scale debris avalanches in the Lesser Antilles Arc. *Earth and Planetary Sciences Letters* **192** (2): 145–157.

Dominey-Howes, D. (2004) A re-analysis of the Late Bronze Age eruption and tsunami of Santorini, Greece, and the implications for the volcano-tsunami hazard. *Journal of Volcanology and Geothermal Research* **130** (1–2): 107–132.

Dunbar, P. (2009) Integrated tsunami data for better hazard assessments. *EOS, Transactions, American Geophysical Union* **90** (22): 189–190.

Dunbar, P. and Stroker, K. (2009) Statistics of global historical tsunamis. Conference presentation at the 24th International Tsunami Symposium of the IUGG Tsunami Commission, Novosibirsk, Russia, July.

Dunbar, P. K. and Weaver, C. S. (2008) US States and Territories national tsunami hazard assessment: historical record and sources for waves. US Geological Survey, National Oceanic and Atmospheric Administration. Report prepared for the National Tsunami Hazard Mitigation Program. http://nthmp.tsunami.gov/documents/Tsunami_Assessment_Final.pdf (accessed 7 August 2012).

Dunbar, P. K., Stroker, K. J., Brocko, V. R., *et al.* (2008) Long-term tsunami data archive supports tsunami forecast, warning, research, and mitigation. *Pure and Applied Geophysics* **165** (11–12): 2275–2291.

Ebisawa, K., Yamada, H., Taoka, H., *et al.* (2008) Development of information system for post-earthquake plant evaluation and evacuation support. Paper presented at the Fourteenth World Conference on Earthquake Engineering, 12–17 October, Beijing, China.

FAO (2005) Report of the regional workshop on salt-affected soils from sea water intrusion: strategies for rehabilitation and management.

Fowler, C. M. R. (2004) *The Solid Earth: An Introduction to Global Geophysics*, Cambridge: Cambridge University Press.

Fritz, H. M. and Kalligeris, N. (2008) Ancestral heritage saves tribes during 1 April 2007 Solomon Islands tsunami. *Geophysical Research Letters* **35** (1): n.p.

Geist, E. L. (1999) Local tsunamis and earthquake source parameters. *Advances in Geophysics*, **39**: 117–209.

Geist, E. L. (2002) Complex earthquake rupture and local tsunamis. *Journal of Geophysical Research – Solid Earth.* **107** (B5): n.p.

Geist, E. L., Parsons, T., ten Brink, U. S., *et al.* (2009) Tsunami probability. In *The Sea*, ed. E. N. Bernard and A. R. Robinson, Cambridge, MA: Harvard University Press. pp. 93–135.

Ghobarah, A., Saatcioglu, M. and Nistor, I. (2006) The impact of the 26 December 2004 earthquake and tsunami on structures and infrastructure. *Engineering Structures* **28** (2): 312–326.

Godin, O. A. (2004) Air–sea interaction and feasibility of tsunami detection in the open ocean. *Journal of Geophysical Research – Oceans* **109** (C5): n.p.

González, F. I., Bernard, E. N., Meinig, C., *et al.* (2005) The NTHMP tsunameter network. *Natural Hazards* **35** (1): 25–39.

González, F. I., Geist, E. L., Jaffe, B., *et al.* (2009) Probabilistic tsunami hazard assessment at Seaside, Oregon, for near- and far-field seismic sources. *Journal of Geophysical Research – Oceans* **114**: n.p.

Greenslade, D. J. M. and Titov, V. V. (2008) A comparison study of two numerical tsunami forecasting systems. *Pure and Applied Geophysics* **165** (11–12): 1991–2001.

Greenslade, D. J. M., Simanjuntak, M. A., Burbidge, D., *et al.* (2007) A first-generation real-time tsunami forecasting system for the Australian Region. BMRC research report no. 126. Australian Government Bureau of Meteorology, Melbourne. http://cawcr.gov.au/bmrc/pubs/researchreports/researchreports.htm (accessed 7 August 2012).

Grilli, S. T. and Watts, P. (2005) Tsunami generation by submarine mass failure: I. Modeling experimental validation, and sensitivity analyses. *Journal of Waterway Port Coastal and Ocean Engineering – ASCE* **131** (6): 283–297.

Grilli, S. T., Taylor, O. D. S., Baxter, C. D. P., *et al.* (2009) A probabilistic approach for determining submarine landslide tsunami hazard along the upper east coast of the United States. *Marine Geology* **264** (1–2): 74–97.

Grindlay, N. R., Hearne, M. and Mann, P. (2005) High risk of tsunami in the northern Caribbean. *EOS, Transactions, American Geophysical Union* **86** (12): 121.

Gusiakov, V. K. (2001) Basic Pacific tsunami catalogs and database, 47 BC–2000 AD: results of the first stage of the project. In *Proceedings of the International Tsunami Symposium*, Seattle, WA: PMEL/NOAA, pp. 263–272.

Gutierrez, D. R. (1998) Advances in the Chilean Tsunami Warning System and application of the TIME project to the Chilean coast. In *Conference on Early Warning Systems for the Reduction of Natural Disasters*, ed. J. Zschau and A. N. Kuppers, Berlin: Springer-Verlag, pp. 543–547.

GVP (2011) Global Volcanism Program. Smithsonian National Museum of Natural History. www.volcano.si.edu (accessed 26 August 2011).

Heron, M. L., Prytz, A., Heron, S. F., *et al.* (2008) Tsunami observations by coastal ocean radar. *International Journal of Remote Sensing* **29** (21): 6347–6359.

Hwang, L. and Divoky, D. (1970) Tsunami generation. *Journal of Geophysical Research* **75** (33): 6802–6817.

IAEA (2010) Braving the waves: IAEA works to protect nuclear power plants against tsunami and flooding hazards. Staff report. http://www.iaea.org/newscenter/news/2010/bravingwaves.html (accessed 29 June 2011).

Iida, K. (1956) Earthquakes accompanied by tsunamis occurring under the sea off the islands of Japan. *Journal of Earth Sciences, Nagoya University, Japan* **4**: 1–53.

Imamura, A. (1942) History of Japanese tsunamis. *Kayo-No-Kagaku (Oceanography)* **2**: 74–80 (Japanese).

InSTEDD (2011) Innovative support to emergencies, diseases and disasters. http://instedd.org (accessed 30 August 2011).

IOC (2008) *Exercise Pacific Wave 08: A Pacific-wide Tsunami Warning and Communication Exercise, 28–30 October*, Paris: UNESCO.

IOC (2011) Intergovernmental Oceanographic Commission Tsunami Programme. www.ioc-tsunami.org (accessed 16 August 2011).

IOC-UNESCO (2009) National Report Submitted by France (French Polynesia). http://ioc-unesco.org/components/com_oe/oe.php?task=download&id=5613&version=1.0&lang=1&format=1 (accessed 20 November 2009).

Jaffe, B. E., Morton, R. A., Kortekaas, S., *et al.* (2008) Reply to Bridge (2008) Discussion of articles in 'Sedimentary features of tsunami deposits' Discussion. *Sedimentary Geology* **211** (3–4): 95–97.

JATWC (2011) Joint Australian Tsunami Warning Centre, Commonwealth of Australia, Bureau of Meteorology. http://www.bom.gov.au/tsunami (accessed 26 August 2011).

JMA (2011) Japan Meteorological Agency. http://www.jma.go.jp (accessed 26 August 2011).

Jonientz-Trisler, C., Simmons, R. S., Yanagi, B. S., *et al.* (2005) Planning for tsunami-resilient communities. *Natural Hazards* **35** (1): 121–139.

Kaneda, Y. (2010) The advanced ocean floor real time monitoring system for mega thrust earthquakes and tsunamis: application of DONET and DONET2 data to seismological research and disaster mitigation. Paper presented at the Oceans 2010 Conference, 20–23 September.

Kawahara, M., Takeuchi, N. and Yoshida, T. (1978) 2 step explicit finite-element method for tsunami wave-propagation analysis. *International Journal for Numerical Methods in Engineering* **12** (2): 331–351.

Kelletat, D. (2008) Comments to Dawson, AG and Stewart, I. (2007) Tsunami deposits in the geological record. – Sedimentary Geology 200, 166–183 Discussion. *Sedimentary Geology* **211** (3–4): 87–91.

Kivikuru, U. and Nord, L. (eds) (2009) *After the Tsunami: Crisis Communication in Finland and Sweden*, Göteborg: Nordicom.

Knight, W. (2009) Two Pacific tsunami-warning buoys out of action. *New Scientist*, 19 January. http://www.newscientist.com/article/dn6901.

Lander, J. F., Whiteside, L. S., Lockridge, P. A. (2002) A brief history of tsunamis in the Caribbean Sea. *Science of Tsunami Hazards* **20** (1): 57–94.

Leclerc, J. P., Berger, C., Foulon, A., *et al.* (2008) Tsunami impact on shallow groundwater in the Ampara district in Eastern Sri Lanka: conductivity measurements and qualitative interpretations. *Desalination* **219** (1–3): 126–136.

Lee, W. H. K., Kanamori, H., Jennings, P., *et al.* (eds) (2003) *International Handbook of Earthquake & Engineering Seismology*, London: Elsevier.

Liu, P. L. F., Wu, T. R., Raichlen, F., *et al.* (2005) Runup and rundown generated by three-dimensional sliding masses. *Journal of Fluid Mechanics* **536**: 107–144.

Lockridge, P. (1998) Potential for landslide-generated tsunamis in Hawaii. *Science of Tsunami Hazards* **16** (1): 8.

Lorito, S., Tiberti, M. M., Basili, R., *et al.* (2008) Earthquake-generated tsunamis in the Mediterranean sea: scenarios of potential threats to Southern Italy. *Journal of Geophysical Research – Solid Earth* **113** (B1): n.p.

Løvholt, F., Pedersen, G. and Gisler, G. (2008) Oceanic propagation of a potential tsunami from the La Palma Island. *Journal of Geophysical Research* **113**: C09026.

Lowe, D. J. and de Lange, W. P. (2000) Volcano-meteorological tsunamis, the c. AD 200 Taupo eruption (New Zealand) and the possibility of a global tsunami. *The Holocene* **10** (3): 401–407.

Lubick, N. (2009) Buoy damage blurs El Niño forecasts. *Nature* **461** (7263): 455.

Lynett, P. and Liu, P. L. F. (2005) A numerical study of the run-up generated by three-dimensional landslides. *Journal of Geophysical Research – Oceans* **110** (C3): n.p.

Marano, K. D., Wald, D. J. and Allen, T. I. (2010) Global earthquake casualties due to secondary effects: a quantitative analysis for improving rapid loss analyses. *Natural Hazards* **52** (2): 319–328.

Marinatos, S. (1939) The volcanic destruction of Minoan Crete. *Antiquity* **13**: 425–439.

Masson, D. G., Harbitz, C. B., Wynn, R. B., *et al.* (2006) Submarine landslides: processes, triggers and hazard prediction. *Proceedings of the Royal Society A* **364** (1845): 2009–2039.

McAdoo, B. G., Moore, A. and Baumwell, J. (2009) Indigenous knowledge and the near field population response during the 2007 Solomon Islands tsunami. *Natural Hazards* **48**: 73–82.

McCloskey, J., Antonioli, A., Piatanesi, A., *et al.* (2008) Tsunami threat in the Indian Ocean from a future megathrust earthquake west of Sumatra. *Earth and Planetary Science Letters* **265** (1–2): 61–81.

McCoy, F. W. and Heiken, G. (2000) Tsunami generated by the Late Bronze Age eruption of Thera (Santorini) Greece. *Pure and Applied Geophysics* **157** (6–8): 1227–1256.

Mercer, J., Dominy-Howes, D., Kelman, I., *et al.* (2007) The potential for combining indigenous and western knowledge in reducing vulnerability to environmental hazards in small island developing states. *Natural Hazards* **7**: 245–256.

Merrifield, M. A., Firing, Y. L., Aarup, T., *et al.* (2005) Tide gauge observations of the Indian Ocean tsunami, December 26, 2004. *Geophysical Research Letters* **32** (9): n.p.

Minoura, K., Imamura, F., Kuran, U., *et al.* (2000) Discovery of Minoan tsunami deposits. *Geology* **28** (1): 59–62.

Mofjeld, H. O., Gonzalez, F. I., Bernard, E. N., *et al.* (2000) Forecasting the heights of later waves in Pacific-wide tsunamis. *Natural Hazards* **22** (1): 71–89.

Monserrat, S., Vilibic, I. and Rabinovich, A. B. (2006) Meteotsunamis: atmospherically induced destructive ocean waves in the tsunami frequency band. *Natural Hazards and Earth System Sciences* **6** (6): 1035–1051.

Moore, G. F., Bangs, N. L., Taira, A., *et al.* (2007) Three-dimensional splay fault geometry and implications for tsunami generation. *Science* **318** (5853): 1128–1131.

Moore, J. G., Clague, D. A., Holcomb, R. T., *et al.* (1989) Prodigious submarine landslides on the Hawaiian ridge. *Journal of Geophysical Research – Solid Earth and Planets* **94** (B12): 17465–17484.

Murty, T. S. and Loomis, H. G. (1980) A new objective tsunami magnitude scale. *Marine Geodesy* **4** (3): 267–282.

Nanayama, F., Shigeno, K., Satake, K., *et al.* (2000) Sedimentary differences between the 1993 Hokkaido-nansei-oki tsunami and the 1959 Miyakojima typhoon at Taisei, southwestern Hokkaido, northern Japan. *Sedimentary Geology* **135** (1–4): 255–264.

NGDC (2011) *Tsunami Travel Time Maps*. National Geophysical Data Center. www.ngdc. noaa.gov/hazard/tsu_travel_time.shtml (accessed 29 June 2011).

NOAA/NGDC (2010) National Geophysical Data Center Historical Tsunami Database. http://www.ngdc.noaa.gov/hazard/tsu_db.shtml (accessed 15 December 2012).

NOAA/PMEL (2011) NOAA Center for Tsunami Research, US Department of Commerce. nctr.pmel.noaa.gov (accessed 29 June 2011).

NTL (2011) Historical Tsunami Databases for the World Ocean. Novosibirsk Tsunami Laboratory, Institute of Computational Mathematics and Mathematical Geophysics, Novosibirsk, Russia. http://tsun.sscc.ru/ (accessed 26 August 2011).

NWS (2011) TsunamiReady. US National Weather Service. http://www.tsunamiready.noaa. gov (accessed 30 August 2011).

Okal, E. A. and Synolakis, C. E. (2004) Source discriminants for near-field tsunamis. *Geophysical Journal International.* **158** (3): 899–912.

Okal, E. A. and Synolakis, C. E. (2008) Far-field tsunami hazard from mega-thrust earth-quakes in the Indian Ocean. *Geophysical Journal International* **172** (3): 995–1015.

Okal, E. A., Fritz, H. M., Synolakis, C. E., *et al.* (2010) Field survey of the Samoa tsunami of 29 September 2009. *Seismological Research Letters* **81** (4): 577–591.

Okal, E. A., Synolakis, C. E. and Kalligeris, N. (2011) Tsunami simulations for regional sources in the South China and adjoining seas. *Pure and Applied Geophysics* **168** (6–7): 1153–1173.

OOI (2011) Ocean Observatories Initiative. http://www.oceanobservatories.org (accessed 21 December 2011).

Papadopoulos, G. A. (2003) Quantification of tsunamis: a review. In *Submarine Landslides and Tsunamis, Proceedings of the NATO Advanced Research Workshop, Istanbul, Turkey, 23–26 May 2001*, ed. A. C. Yalçiner, E. N. Pelinovsky, E. Okal, *et al.*, Dordrecht: Kluwer, pp. 285–292.

Papadopoulus, G. and Imamura, F. (2001) A proposal for a new tsunami intensity scale. International Tsunami Symposium Proceedings, Seattle, Session 5, Number 5-1, pp. 569–577.

Pareschi, M. T., Favalli, M. and Boschi, E. (2006) Impact of the Minoan tsunami of Santorini: simulated scenarios in the eastern Mediterranean. *Geophysical Research Letters* **33** (18): n.p.

Pelayo, A. M. and Wiens, D. A. (1992) Tsunami earthquakes: slow thrust-faulting events in the accretionary wedge. *Journal of Geophysical Research – Solid Earth* **97** (B11): 15321–15337.

Pelinovsky, E. and Poplavsky, A. (1996) Simplified model of tsunami generation by submarine landslides. *Physics and Chemistry of the Earth* **21** (1–2): 13–17.

Person, R., Beranzoli, L., Berndt, C., *et al.* (2008) ESONET: a European sea observatory initiative. In *International Conference OCEANS 2008 and MTS/IEEE Kobe Techno-Ocean*, Japan, Arlington, VA: Compass Publications, pp. 1215–1220.

Prasad, R., Cunningham, E. and Bagchi, G. (2009) Tsunami Hazard Assessment at Nuclear Power Plant Sites in the United States of America – Final Report. United States Nuclear Regulatory Commission. Pacific Northwest National Laboratory, March 2009. http://www.nrc.gov/reading-rm/doc-collections/nuregs/contract/cr6966 (accessed 7 August 2012).

Pritchard, D. and Dickinson, L. (2008) Modelling the sedimentary signature of long waves on coasts: implications for tsunami reconstruction. *Sedimentary Geology* **206** (1–4): 42–57.

PTWC (2011) Pacific Tsunami Warning Center, US Dept of Commerce/NOAA/NWS. http://ptwc.weather.gov/ (accessed 26 August 2011).

Puillat, I., Person, R., Leveque, C., *et al.* (2008) Standardization prospective in ESONET NoE and a possible implementation on the ANTARES Site. In *3rd International Workshop on a Very Large Volume Neutrino Telescope for the Mediterranean Sea.* Toulon, France, Amsterdam: Elsevier, pp. 240–245.

Rabinovich, A. B. and Monserrat, S. (1996) Meteorological tsunamis near the Balearic and Kuril Islands: descriptive and statistical analysis. *Natural Hazards* **13** (1): 55–90.

Reymond, D. and Okal, E. A. (2000) Preliminary determination of focal mechanisms from the inversion of spectral amplitudes of mantle waves. *Physics of the Earth and Planetary Interiors* **121** (3–4): 249–271.

Reymond, D., Robert, S., Thomas, Y., *et al.* (1996) An automatic tsunami warning system: TREMORS application in Europe. *Physics and Chemistry of the Earth* **21** (1–2): 75–81.

Richmond, B. and Gibbons, H. (2005) Assessing tsunami impacts in the Republic of Maldives. *USGS Sound Waves Monthly Newsletter*, April.http://soundwaves.usgs.gov/2005/04/fieldwork3.html (accessed 29 September 2011).

Richmond, B. M., Dudley, W. C., Buckley, M. L., *et al.* (2009) Survivor interviews from the Sept. 29, 2009 tsunami on Samoa and American Samoa. *EOS* **90** (52): n.p.

Rikitake, T. and Aida, I. (1988) Tsunami hazard probability in Japan. *Bulletin of the Seismological Society of America* **78** (3): 1268–1278.

RMS (2006) Managing tsunami risk in the aftermath of the 2004 Indian Ocean earthquake & tsunami. Risk Management Solutions, Reconnaissance Report. http://www.rms.com/Reports/IndianOceanTsunamiReport.pdf (accessed 18 August 2011).

Scheffers, A. and Kelletat, D. (2003) Sedimentologic and geomorphologic tsunami imprints worldwide: a review. *Earth-Science Reviews* **63** (1–2): 83–92.

SEA-EAT (2004) The South-East Asia Earthquake and Tsunami Blog. http://tsunamihelp.blogspot.com (accessed 29 June 2011).

Shi, S. Z., Dawson, A. G. and Smith, D. E. (1995) Coastal sedimentation associated with the December 12th, 1992 tsunami in Flores, Indonesia. *Pure and Applied Geophysics* **144** (3–4): 525–536.

Shuto, N. (1993) Tsunami intensity and disasters. In *Tsunamis in the World*, ed. S. Tinti, Boston, MA: Kluwer, pp. 197–216.

Sieberg, A. (1927) *Geologische Einführung in die Geophysik.* Jena: G. Fischer.

Simkin, T., Siebert, L. and Blong, R. (2001) Policy forum: disasters – volcano fatalities – lessons from the historical record. *Science* **291** (5502): 255–255.

Soloviev, S. L. (1970) Recurrence of tsunamis in the Pacific. In *Tsunamis in the Pacific Ocean,* ed. W. M. Adams, Honolulu: East–West Center Press, pp. 149–163.

Soloviev, S. L. and Go, C. N. (1974) *A Catalogue of Tsunamis on the Western Shore of the Pacific Ocean,* Moscow: Nauka Publishing House. Translation available from the Canada Institute for Scientific and Technical Information, National Research Council, Ottawa.

Song, Y. T., Ji, C., Fu, L. L., *et al.* (2005) The 26 December 2004 tsunami source estimated from satellite radar altimetry and seismic waves. *Geophysical Research Letters* **32** (20): n.p.

Sornette, D. and Knopoff, L. (1997) The paradox of the expected time until the next earthquake. *Bulletin of the Seismological Society of America* **87** (4): 789–798.

Switzer, A. D., Pucillo, K., Haredy, R. A., *et al.* (2005) Sea level, storm, or tsunami: enigmatic sand sheet deposits in a sheltered coastal embayment from southeastern New South Wales, Australia. *Journal of Coastal Research* **21** (4): 655–663.

Synolakis, C. E. and Kong, L. (2006) Runup measurements of the December 2004 Indian Ocean Tsunami. *Earthquake Spectra* **22**: S67–S91.

Synolakis, C. E., Bernard, E. N., Titov, V. V., *et al.* (2007) Standards, criteria, and procedures for NOAA evaluation of tsunami numerical models. NOAA Technical Memorandum OAR PMEL-135. Pacific Marine Environmental Laboratory, Seattle, WA, May.

Synolakis, C. E., Bernard, E. N., Titov, V. V., *et al.* (2008) Validation and verification of tsunami numerical models. *Pure and Applied Geophysics* **165** (11–12): 2197–2228.

Tadepalli, S. and Synolakis, C. E. (1994) The run-up of n-waves on sloping beaches. *Proceedings of the Royal Society of London Series A – Mathematical, Physical and Engineering Sciences* **445** (1923): 99–112.

Talandier, J. (1993) French Polynesia Tsunami Warning Center. *Natural Hazards* **7** (3): 237–256.

Tang, L., Titov, V. V. and Chamberlin, C. D. (2009) Development, testing, and applications of site-specific tsunami inundation models for real-time forecasting. *Journal of Geophysical Research – Oceans* **114**: n.p.

Tanguy, J. C., Ribiere, C., Scarth, A., *et al.* (1998) Victims from volcanic eruptions: a revised database. *Bulletin of Volcanology* **60** (2): 137–144.

Tanioka, Y. and Satake, K. (1996) Tsunami generation by horizontal displacement of ocean bottom. *Geophysical Research Letters* **23** (8): 861–864.

Telford, J. and Cosgrave, J. (2007) The international humanitarian system and the 2004 Indian Ocean earthquake and tsunamis. *Disasters* **31** (1): 1–28.

Telford, J., Cosgrave, J. and Houghton, R. (2006) Joint evaluation of the international response to the Indian Ocean tsunami: Synthesis Report. Tsunami Evaluation Coalition. http://www.alnap.org/initiatives/tec (accessed 26 August 2011).

Thomson, R., Fine, I., Rabinovich, A., *et al.* (2011) Observation of the 2009 Samoa tsunami by the NEPTUNE-Canada cabled observatory: test data for an operational regional tsunami forecast model. *Geophysical Research Letters* **38**: L11701.

Tinti, S. and Maramai, A. (1999) Large tsunamis and tsunami hazard from the new Italian tsunami catalog. *Physics and Chemistry of the Earth Part A – Solid Earth and Geodesy* **24** (2): 151–156.

Tinti, S., Maramai, A. and Graziani, L. (2004) The new catalogue of Italian tsunamis. *Natural Hazards* **33** (3): 439–465.

Tinti, S., Armigliato, A., Pagnoni, G., *et al.* (2005) Scenarios of giant tsunamis of tectonic origin in the Mediterranean. *ISET Journal of Earthquake Technology*, **42** (4): 171–188.

Titov, V. V. (1997) Numerical modeling of long wave runup. PhD thesis, University of Southern California.

Titov, V. V. (2009) Tsunami forecasting. In *The Sea*, ed. E. N. Bernard, and A. R. Robinson, Cambridge, MA: Harvard University Press, pp. 371–400.

Titov, V. V. and González, F. I. (1997) Implementation and testing of the Method of Splitting Tsunami (MOST) model NOAA Technical Memorandum ERL PMEL-112.

Titov, V. V. and Synolakis, C. E. (1998) Numerical modeling of tidal wave runup. *Journal of Waterway Port Coastal and Ocean Engineering – ASCE* **124** (4): 157–171.

Titov, V. V., Mofjeld, H. O., González, F. I., *et al.* (1999) Offshore forecasting of Alaska–Aleutian Subduction Zone tsunamis in Hawaii. NOAA Tech. Memo ERL PMEL-114, NOAA/Pacific Marine Environmental Laboratory, Seattle, WA.

Titov, V. V., Mofjeld, H. O., González, F. I., *et al.* (2001) Offshore forecasting of Alaskan tsunamis in Hawaii. In *Tsunami Research at the End of a Critical Decade*, ed. G. T. Hebenstreit, Dordrecht: Kluwer, pp. 75–90.

Titov, V. V., González, F. I., Mofjeld, H. O., *et al.* (2003) Short-term inundation forecasting for tsunamis. *Submarine Landslides and Tsunamis* **21**: 277–284.

Titov, V. V., González, F. I., Bernard, E. N., *et al.* (2005) Real-time tsunami forecasting: challenges and solutions. *Natural Hazards* **35** (1): 41–58.

Titov, V., Moore, C., Greenslade, D., *et al.* (2011) A new tool for inundation modeling: community modeling interface for tsunamis (ComMIT). *Pure and Applied Geophysics* **168** (11): 2121–2131.

Tsunami Pilot Study Working Group (2006) *Seaside, Oregon Tsunami Pilot Study: Modernization of FEMA Flood Hazard Maps*. US Department of the Interior, US Geological Survey Open-File Report 2006-1234. http://pubs.usgs.gov/of/2006/1234/ (accessed 4 January 2011).

Uckay, I., Sax, H., Harbarth, S., *et al.* (2008) Multi-resistant infections in repatriated patients after natural disasters: lessons learned from the 2004 tsunami for hospital infection control. *Journal of Hospital Infection* **68** (1): 1–8.

UNESCO (2007) North-East Atlantic, the Mediterranean and Connected Seas Tsunami Warning and Mitigation System, NEAMTWS, Implementation Plan. Third Session of the Intergovernmental Coordination Group for NEAMTWS, IOC Technical Series No. 73. USGS (2011) USGS Multimedia Gallery.http://gallery.usgs.gov/ (accessed 4 August 2011).

USNRC (2007) Standard review plan for the review of safety analysis reports for nuclear power plants. Section 2.4.6: Probable maximum tsunami hazards. US Nuclear Regulatory Commission. http://www.nrc.gov/reading-rm/basic-ref/srp-review-stand ards.html (accessed 18 August 2011).

Van Dorn, W. G. (1965) Tsunamis. In *Advances in Hydroscience*, ed. V. T. Chow, New York: Elsevier, pp. 1–48.

van Griensven, F., Chakkraband, M. L. S., Thienkrua, W., *et al.* (2006) Mental health problems among adults in tsunami-affected areas in southern Thailand. *Journal of the American Medical Association* **296** (5): 537–548.

Voigt, S., Kemper, T., Riedlinger, T., *et al.* (2007) Satellite image analysis for disaster and crisis-management support. *IEEE Transactions on Geoscience and Remote Sensing* **45** (6): 1520–1528.

Wakeford, R. (2011) And now, Fukushima. *Journal of Radiological Protection* **31** (2): 167–176.

Wang, J. F. and Li, L. F. (2008) Improving tsunami warning systems with remote sensing and geographical information system input. *Risk Analysis* **28** (6): 1653–1668.

Ward, S. N. (1982) Earthquake mechanisms and tsunami generation: the Kurile Islands event of 13 October 1963. *Bulletin of the Seismological Society of America* **72** (3): 759–777.

Ward, S. N. and Day, S. (2001) Cumbre Vieja Volcano: potential collapse and tsunami at La Palma, Canary Islands. *Geophysical Research Letters* **28** (1717): 3397–3400.

Watts, P., Grilli, S. T., Tappin, D. R., *et al.* (2005) Tsunami generation by submarine mass failure: II. Predictive equations and case studies. *Journal of Waterway Port Coastal and Ocean Engineering – ASCE* **131** (6): 298–310.

WCATWC (2011) West Coast/Alaska Tsunami Warning Center, US Dept of Commerce, National Oceanic and Atmospheric Administration. http://wcatwc.arh.noaa.gov (accessed 26 August 2011).

Wessel, P. (2011) TTT (Tsunami Travel Times) Software, version 3.2. http://www.geoware-online.com/tsunami.html.

Woo, G. (1999) *The Mathematics of Natural Catastrophes*, River Edge, NJ: Imperial College Press.

Woo, G. and Aspinall, W. (2005) Need for a risk-informed tsunami alert system. *Nature* **433** (7025): 457–457.

Young, R. W. and Bryant, E. A. (1992) Catastrophic wave erosion on the south-eastern coast of Australia: impact of the Lanai tsunamis CA-105-KA. *Geology* **20** (3): 199–202.

11

Risk and uncertainty assessment of volcanic hazards

R. S. J. SPARKS, W. P. ASPINALL, H. S. CROSWELLER
AND T. K. HINCKS

11.1 Introduction

Over 600 million people live close enough to active volcanoes to have their lives disrupted if there are signs of unrest that might lead to eruption, and threatened when an eruption does occur. This estimate is an update of the analysis of Tilling and Peterson (1993) and Small and Naumann (2001) using 2009 World Bank population data. In comparison to other natural hazards, such as earthquakes and floods, the historic death toll from volcanoes is quite small. Since about 1500 CE the Smithsonian Global Volcanism Program database records about 200 000 in total, and slightly below 100 000 since 1900 CE (Simkin et al., 2001; Witham, 2005; Siebert et al., 2011). However, there have been a small number of infamous historic eruptions with horrific casualty figures, such as the destruction of St Pierre in 1902 by Mont Pelée volcano, Martinique, with 29 000 fatalities (Heilprin, 1903) and the burial of the town of Armero, Colombia by a volcanic mudflow in 1985, with 25 000 fatalities due to an eruption of Nevado del Ruiz (Voight, 1990).

Although volcano deaths and economic losses are historically small compared to earthquakes, floods and droughts, this is misleading. In the 1991 eruption of Mount Pinatubo in the Philippines, a mega-disaster was narrowly averted when tens of thousands of people were evacuated from the flanks of the volcano just in time; as a consequence, the largest explosive eruption of the twentieth century only caused a few hundred casualties (Newhall and Punongbayan, 1996a, 1996b). In the decade that followed, more than 200 000 people had to be permanently evacuated because their towns were buried by lahars. Many lahar warnings were issued, both long-term and immediate. Roughly 400 people were killed by lahars, but this is also a relatively small number compared to the 200 000.

Disruption to societies and associated economic costs due to volcanic activity has been considerable, although not so easily quantified as deaths. During the first few years of the volcanic emergency on the small island of Montserrat (caused by the eruption of the Soufrière Hills volcano that started in 1995) over 8000 people were evacuated from the island or left voluntarily, two-thirds of the total population, and the estimated economic losses by 1999 were estimated at about US$1 billion (Clay et al., 1999). In 1976 what turned out to be a quite limited phreatic eruption of La Soufrière volcano in Guadeloupe led to the evacuation of 73 000 people for over three months, major economic costs for the island

Risk and Uncertainty Assessment for Natural Hazards, ed. Jonathan Rougier, Steve Sparks and Lisa Hill. Published by Cambridge University Press. © Cambridge University Press 2013.

(some persisting to the present day) and significant turmoil among scientists and politicians in France (Fiske, 1984). The economic impact of volcanism was again more recently highlighted, in 2010, when a relatively minor eruption of the Eyjafjallajökull volcano in Iceland resulted in disruption to flights in European air space, severe inconvenience for millions of passengers and impacts on business and losses that are estimated to be US$1.7 billion for the aviation industry (IATA, 2011).

Nowadays, there are many big cities and megacities growing close to highly dangerous volcanoes, thereby increasing vulnerability. Naples in Italy is perhaps the best known, but there are many more – another example is Jakarta (Indonesia), which is built in part on deposits of landslides and debris flows from Salak and possibly Pangrango volcanoes, eruptions of which were triggered in 1699 by a strong regional earthquake. Volcanism adversely affects such countries disproportionately. This is in part due to the large number of active volcanoes in low-income countries, but is also due to limited resources (e.g. volcano monitoring, emergency management), limited resilience and a limited capacity for recovery. The possibility of massive and unprecedented numbers of casualties being suffered in some city lying close to an erupting volcano (a super-eruption) is real, and growing.

To top this, a giant volcanic eruption is also the only natural hazard apart from meteor impact that is capable of creating a truly global catastrophe (Rampino, 2002; Sparks *et al.*, 2005; Bryan *et al.*, 2010). In terms of geological time, advanced human societies are very recent, and volcanological and archaeological evidence indicates that very large magnitude eruptions can have environmental consequences devastating to human populations. A counter view is provided by Grattan (2006).

Living with a volcano is part of everyday existence for many communities, which in general are becoming more vulnerable as populations expand, reliance on technology increases and the effects of other kinds of environmental stress mount. On the other hand, advances in volcanology are helping with the provision of early warning and to improve the management of volcanic emergencies. The analysis, assessment and communication of risk and uncertainty are central to all aspects of living with volcanoes and through avoidance of disaster and minimising deaths and losses when they erupt.

This chapter summarises the current state of knowledge on risk assessment for volcanism and discusses future developments and needs. Section 11.2 outlines the main kinds of volcanic hazard. Section 11.3 synthesizes the state-of-the-art in prediction, forecasting and early warning. Section 11.4 discusses approaches to hazard and risk assessment, including hazards zonation maps. Section 11.5 considers risk management and the communication of hazard and risk during volcanic emergencies. Section 11.6 concludes with a discussion of future outlooks and challenges.

11.2 Volcanic hazards

There are an estimated 550 historically active volcanoes and 1300 active volcanoes in the world, the latter defined somewhat arbitrarily as those having evidence of eruptions during

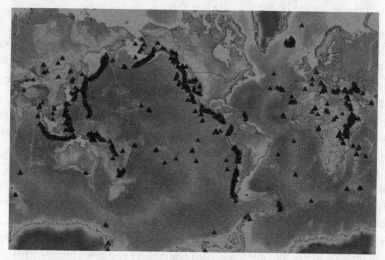

Figure 11.1 Distribution of the world's Holocene volcanoes.

the Holocene (last 10 000 years). About 20 volcanoes are erupting on average at any one time (see Siebert *et al.*, 2011 for more authoritative information from the Smithsonian Institution's Global Volcanism Program). There are many more Quaternary volcanoes (defined as the last 2.58 million years) and it is likely that a significant number of these are dormant rather than extinct. Most volcanoes are located along plate tectonic boundaries and so coincide with places prone to large earthquakes (Figure 11.1).

Volcanoes come in all sorts of shapes and sizes and display a rich diversity of eruptive phenomena, so that hazard assessments have to consider several very different potential physical phenomena. Furthermore, each of these phenomena can range very widely in scale and intensity over several orders of magnitude and the footprint of a particular hazardous phenomenon is also highly variable as a consequence. Hazardous volcanic flows may be strongly dependent on topography, while ash hazards depend on meteorological factors. The different phenomena can occur simultaneously or can be causally linked, making volcanic hazards a rich and diverse subject that is quintessentially multi-hazard in character. Past activity is often a good guide to the future, but this is not always the case since volcanoes evolve and their eruptions can develop into new and sometimes unexpected eruptive regimes.

Eruptions span several orders of magnitude in terms of size, intensity and duration. Size is conventionally measured by the mass or volume of erupted material, which is defined as the magnitude (M). Like earthquakes and floods, the frequency of eruptions decreases markedly with magnitude, but this relationship is not yet very well characterised, partly because volume is often poorly constrained. As an example, the eruption of Mount Pinatubo in 1991 is an M = 6.5 event, where $M = \log[m] - 7$ and m is the erupted mass in kilograms. Figure 11.2 shows a generalised magnitude versus return period relationship for explosive eruptions based on the studies of Mason *et al.* (2004) and Deligne *et al.* (2010). Intensity is a very useful indicator of the violence of an eruption and is measured in kilograms of magma

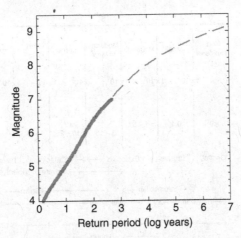

Figure 11.2 Global magnitude versus return period for explosive eruptions of magnitude 4 and greater from the analysis of Deligne *et al.* (2010) (solid curve) and an extrapolation (dashed curve) based on an upper threshold of $M = 9.3$ for the Earth and the analysis of Mason *et al.* (2004) of eruptions with $M > 8$.

erupted per second. Since intensity commonly fluctuates greatly during many eruptions, peak intensity is often a useful parameter to estimate. The volcanic explosivity index (VEI) introduced by Newhall and Self (1982) is widely used as a metric of eruption scale. The VEI (Figure 11.3) is assigned according to both quantitative estimates of volume and intensity and more qualitative indicators. Durations can range from less than a day, sometimes with extreme intensity ($5 \, km^3$ of magma erupted from Mount Pinatubo on 15 June 1991), to persistently active volcanoes that erupt almost continuously with many small eruptions (Stromboli volcano in Italy has been erupting since Roman times or before). Rare extreme eruptions ($M > 8$, i.e. up to 300 times the magnitude of Pinatubo) are the only natural phenomena apart from meteor impact that can have a worldwide impact through global effects and consequent extreme short-term climate change, sometimes described as volcanic winter, when cooling of several degrees may last for several years (Robock, 2000).

The main hazardous phenomena are now summarised. A very good source of up-to-date information on volcanic hazards and volcanoes is the US Geological Survey (http://volca-noes.usgs.gov/). More detailed accounts of volcanic processes and hazards can be found in Blong (1984), Tilling (1989) and Sparks *et al.* (1997).

Explosive eruptions are the most important primary volcanic hazard. There are two basic phenomena (Figure 11.4). First, explosive eruptions form high volcanic plumes in the atmosphere (Figure 11.4a), which disperse volcanic fragments ranging in size from a few metres to dust. Much of these ejecta are centimetres to millimetres to a few tens of microns in size and are collectively known both as tephra and as pyroclastic particles. Volcanic ash is defined as particles of less than 4 mm across. Volcanic plumes range from a few kilometres to a few tens of kilometres in height (Sparks *et al.*, 1997) and are dispersed by wind patterns (Figure 11.5a). Particles fall out of the plume to form accumulations on the ground known as

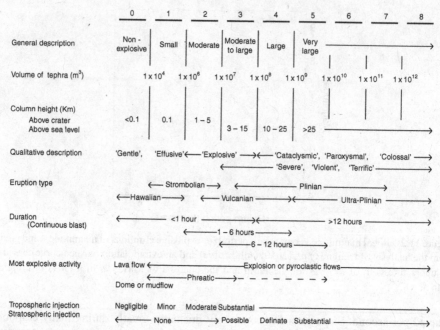

Figure 11.3 Description of the volcanic explosivity index (VEI), after Newhall and Self (1982).

tephra fall deposits. Near the volcano, typically within a few kilometres, large ejecta, commonly known as ballistics, can break roofs and cause fires, and fragments larger than 10 cm are likely to cause death or serious injury (Baxter, 1990). Tephra fall deposits that are sufficiently thick can cause roofs to collapse and death or injury to those inside (Blong, 1984; Spence *et al.*, 2005). Volcanic ash can have major environmental impacts (Durant *et al.*, 2010). Tephra commonly carries toxic chemical components and acids such as H_2SO_4 (sulphuric acid), metals and fluorine. Ash can prevent photosynthesis so cause crop failure, poison livestock, pollute water supplies and cause health hazards such as dental and skeletal fluorosis. Thus areas affected by tephra fall can threaten food security, and famines have followed a number of very large eruptions. For example, about one-third of the population of Iceland died due to the famine caused by the Laki volcano eruption in 1783, and there were severe environmental and health effects in Europe (Grattan and Charman, 1994). Very fine ash is a health hazard (Baxter, 1990; Hansell *et al.*, 2006) and can compromise the operation of technological facilities such as nuclear power stations and electrical infrastructure (Bebbington *et al.*, 2008) and aircraft, as widely experienced in Europe during April and May 2010 as a result of the eruption of the Eyjafjallajökull volcano in Iceland.

Second, explosive eruptions can generate hot, rapidly moving turbulent flows of lava fragments and ash (known as pyroclastic flows) by a phenomenon called column collapse (Figures 11.4b and 11.5b), where the erupted mixture is too dense to keep rising and collapses from height above the vent. Such pyroclastic flows can spread across terrain at

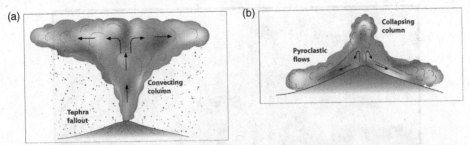

Figure 11.4 (a) Schematic diagram to illustrate the structure and dynamics of a convection eruption column with tephra fallout. Here an intense explosive discharge of pyroclastic ejecta and gas at the vent mixes with the atmosphere to form a buoyant column in the atmosphere, which then spreads out laterally around a height of neutral buoyancy to form an umbrella cloud. Tephra (volcanic fragments of all sizes) fall out of the column. (b) Schematic diagram to show the formation of pyroclastic flows during explosive eruptions. Here the intense vertical discharge of volcanic ejecta and gas mixes with the atmosphere but runs out of kinetic energy while still denser than the surrounding air. The column collapses as a fountain around the vent and forms flows of hot pyroclastic particles.

Figure 11.5 (a) Small volcanic plume generated by explosive eruption at Lascar volcano, Chile showing the effects of the wind as the plume levels out at the height of neutral buoyancy. (b) Pyroclastic flow formed by column collapse during the 1980 eruption of Mount St Helens (photograph from US Geological Survey).

speeds of tens to hundreds of kilometres per hour, at temperatures of several hundred degrees. Nothing in their path can survive and they are historically a major killer (Baxter, 1990). Almost the entire population of St Pierre (29 000) in Martinique was killed by a laterally directed explosion, essentially a particularly energetic variety of pyroclastic flow (Figure 11.6).

Figure 11.6 Destruction in the town of St Pierre in Martinique (8 May 1902) from a violent pyroclastic flow or volcanic blast from Mont Pelée volcano.

Very viscous lavas form unstable domes that can accumulate on a volcano and then generate dome collapse pyroclastic flows or high-speed lateral explosions when the lava dome reaches a critical internal pressure. Dome collapse pyroclastic flows can be very energetic and destructive. A recent example was the death of 20 people on Montserrat due to a collapse of the andesite lava dome at the Soufrière Hills volcano on 25 June 1997 (Loughlin *et al.*, 2002a). Pyroclastic flows vary greatly in scale and intensity. Most of them share the characteristic of having a dense concentrated basal flow and an overriding dilute hot and turbulent cloud of fine ash that usually extends to greater heights. This upper part, sometimes described as a surge, can spill out of valleys and move into unexpected areas. This dual character of the flow has led in the recent past to tragedy. In 1991 a pyroclastic flow formed by the collapse of a lava flow at Mount Unzen, Japan, moved down a deep valley on the volcano's flanks (Ui *et al.*, 1999). Forty-three people situated on a ridge well above the valley were killed by the dilute surge cloud.

Lava flows, while usually not life-threatening, can also be very destructive, bulldozing and burying whole villages and sometimes setting towns on fire. The town of Catania in Sicily was destroyed by lava in 1699. A recent example is the January 2002 eruption of Nyiragongo volcano in the Democratic Republic of Congo, where lava destroyed 15% of the town and caused fires with many burn injuries (Komorowski *et al.*, 2003). An explosion of a gas station overrun by lava led to over 470 people with burns and gas intoxication. Viscous lava domes are very hazardous due to their potential to generate pyroclastic flows, as described above.

There are a number of secondary hazards associated with eruptions that are also highly destructive. Volcanic edifices are often unconsolidated and fractured, and potentially unstable, so landslides are common. Large volcanic landslides, known as debris avalanches, may involve collapse of a large proportion of the volcanic edifice forming an avalanche that can travel tens of kilometres in more extreme events. Such large slope failures can be triggered by an eruption when magma is intruded into the edifice and the collapse can also trigger a very violent lateral explosion. This happened at Mount St Helens on 18 May 1980, with the blast devastating $600 \, km^2$ in four minutes (Voight *et al.*, 1981). There were only 57 fatalities because there was an exclusion zone and the eruption took place early on a Sunday morning, when loggers were not working and hikers were few in number. A variety of causes have been identified as triggers for major slope failure on volcanoes. These include loading of the edifice by growth of lava domes; rise in pore pressure due to fluid movement perhaps activated by magma rise and related unrest; earthquakes; and intense rainfall. In the latter two cases the collapse may be unrelated to volcanic processes. For volcanoes in or close to the sea or lakes, tsunamis are a major hazard. About 15 000 people were killed by the tsunami caused by a debris avalanche due to the collapse of a lava dome at Mount Unzen, Japan, in 1792.

Another major volcanic hazard is flows of volcanic debris and water, known as lahars (an Indonesian word). The most important factor is heavy rain, which can remobilise large amounts of loose debris generated by eruptions. A good example is the generation of lahars after the 1991 eruption of Mount Pinatubo. In this case the valleys cutting the flanks of the volcano were choked with huge amounts of unconsolidated and easily eroded deposits generated by pyroclastic flows during the eruption. For almost two decades since the eruption lahars have been episodically generated around Pinatubo by intense tropical rain-storms (van Westen and Daag, 2005) and associated fatalities are estimated to be about 700. Another common mechanism is where pyroclastic flows enter into water bodies, such as a river, and then transform to a lahar. Lahars can also be generated by rapid melting of ice. In 1985, 25 000 people were killed in the town of Armero, Colombia, when an explosive eruption of Nevado del Ruiz melted the icecap on the volcano to form lahars, which buried the town 70 km away (Voight, 1990). In some cases lahars are generated by muddy water being directly extruded from the ground, likely as a consequence of disturbance of the groundwater systems by magma ascent (Sparks, 2003). Mudslides can be generated by intense rainfall on dormant volcanic edifices, a good example being Casita volcano, Nicaragua, in 1998 due to the passage of Hurricane Mitch (Kerle *et al.*, 2003).

Other hazards associated with volcanoes can include lightning due to charge generation and separation of particles in the plume (Mather and Harrison, 2006), fire caused by hot ejecta (and lightning); shockwaves caused by powerful explosions (e.g. damage to buildings 3 km from Kirishima volcano, Japan, February 2011); pollution of water supplies and the poisoning and asphyxiating effects of gases. Of current interest is the hazard to aviation of suspended fine ash in the atmosphere.

The diversity and, in many cases, strong interdependencies of these various hazards poses special problems for risk assessment. The different kinds of phenomena can occur simultaneously, or one may follow from another. For example a pyroclastic flow eruption or heavy ash fall leads to conditions favourable to lahar generation. The hazards footprint can also be very different between the hazards, making the preparation and presentation of hazard zoning maps quite complicated.

11.3 Forecasting, prediction and early warning

Volcanic eruptions occur when magma rises to the earth's surface. The ascent and underground movement of magma and associated geothermal fluids gives rise to several precursory phenomena, such as numerous small earthquakes, ground deformation, heating of groundwater, chemical changes in springs and fumaroles and release of volcanic gases at the surface in advance of the magma (Figure 11.7). Such phenomena may be detected prior to an eruption to allow early warnings to be given. Likewise, geophysical and geochemical data generated by monitoring support the management of volcanic crises as they unfold.

The technology to monitor volcanoes is improving rapidly, as is the ability to deal with large amounts of data using the ever-increasing power and speed of computers. For those volcanoes that are well monitored there have been some notable successes in forecasting (Sparks, 2003). Although precise prediction is rarely possible, there are now examples of volcanic eruptions where the assessment of precursory signals made it possible for the scientific teams to recognise that the volcano was in a dangerous or critical state, leading to timely evacuation based on this advice. The best example of a successful evacuation is the case of the 1991 eruption of Mount Pinatubo (Newhall and Punongbayan, 1996b), where as many as 20 000 lives were likely saved. As well as the technological advances, the skill of experienced volcanologists and their appreciation of the power of eruptions are critical to effective early warning and crisis management because large amounts of diverse data and information have to be integrated together within the framework of conceptual understanding of volcanic processes.

Current improvements in volcano forecasting and early warning are largely driven by technology linked into improved understanding of the physics of volcanic processes and advances in data-led conceptual models of volcanic systems and processes (Sparks, 2003; Sparks and Aspinall, 2004). Monitoring of volcanic earthquakes by seismic networks remains the most tried and trusted approach for almost all active volcanoes (McNutt, 2005). Magma and fluid movements typically cause myriad small earthquakes and their detailed examination can allow distinctions to be made between various kinds of subterranean phenomena and increasingly quite well-defined identification of the magma conduits, fluid pathways and fracture systems that are activated during volcanic unrest and eruption. The dimensions, shapes of fractures, imposed external stress systems and internal pressures can be inferred from seismic observations. Small earthquakes related to breaking rock in shear can be distinguished from movement of fluids and gases along fractures. Volcano seismologists are getting better at distinguishing source and pathway effects. Over the last decade or more the widespread deployment of broadband three-component seismometers, which detect motions over a wide frequency range in three dimensions, has revolutionised the ability to assess the processes acting on and the state of stress in volcanic systems.

Ground deformation accompanies most eruptions and their precursory events (Dzurisin, 2003), and these movements reflect fluctuations in pressure of magma and fluid bodies in the crust. Such movements can be detected by several different complementary techniques,

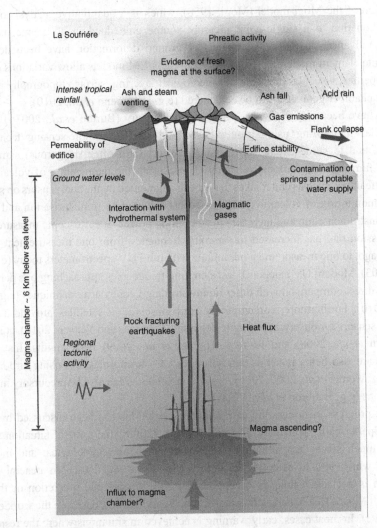

Figure 11.7 Schematic diagram depicting the main effects of magma rising up a conduit to supply an eruption.

including precise levelling, electronic distance measurement (EDM), tilt meters, GPS, synthetic aperture radar interferometry (inSAR) and borehole strain meters. Mass changes can also be detected by gravity measurements. InSAR has become particularly important (Burgmann *et al.*, 2000) in detecting volcano state changes in remote parts of the world (e.g. Pritchard and Simons, 2004). The recent introduction of L-band (25 cm wavelength) devices is enabling useful data to be retrieved from areas where vegetation has previously been a problem. These change detection methods act as indicators of pressure variations in volcanic systems. The interpretation of ground deformation models has been dominated by application of the Mogi model (Mogi, 1958), which is a quantitative analysis of surface deformation due to a single-point pressure source embedded in an elastic half-space. Recently advances in numerical models of ground deformation have been developed, using finite element modelling (FEM) for example. FEM models allow variations of elastic rock properties, viscoelastic rheologies, geological layering, surface topography and more complex chamber geometries to be considered (e.g. Hautmann *et al.*, 2010).

There have been major advances in gas geochemistry (Burton *et al.*, 2007; Edmonds, 2008). Gas monitoring using ground-based and satellite remote-sensing technologies augment, and to some extent supersede, the direct (and often dangerous) sampling of volcanic fumaroles. Traditional methods of measuring the temperature of volcanoes and volcanic features, such as crater lakes, have been augmented by thermal sensors on satellites. Remote monitoring of volcanic gases has been facilitated by miniaturisation and reduced costs of instrumentation and increases in computing power. Spectroscopic measurement of SO_2 emission rate has increased in sampling frequency from one measurement per day, 20 years ago, to one measurement per minute with mini UV spectrometers today (e.g. Galle *et al.*, 2003). Modern UV cameras boast sampling frequencies approaching one per second, allowing direct comparison with other rapid-sampling geophysical techniques (e.g. Dalton *et al.*, 2010). Furthermore, contemporary earth observation satellites provide a suite of potential sensors for observing volcanic degassing (Thomas and Watson, 2010), at both UV (e.g. Carn *et al.*, 2008) and IR (e.g. Thomas *et al.*, 2009) wavelengths. Instrumental advances are also being made in monitoring of CO_2 from volcanoes using airborne platforms (e.g. Werner *et al.*, 2000, 2006). Soil gas emanations can also be precursory indicators of eruptions (e.g. Baubron *et al.*, 1991).

Recent progress in forecasting and predicting eruptions has been discussed by Sparks (2003), Sparks and Aspinall (2004) and McNutt (2005); the general situation has not changed much since then, although there continue to be technological and modelling advances which add to scientific capability (Sparks *et al.*, 2012). In general, precise prediction is not achievable, an exception being the remarkable prediction of the 2000 eruption of Mount Hekla using strain meter data (Sparks, 2003), based on the work of Linde *et al.* (1993). In most cases, early warning is achieved in situations where the responsible experts in an observatory make a judgement call, taking account of all the scientific evidence and the associated uncertainties. Sometimes these judgements have to be made rapidly because the build up from low background activity, with little cause for concern, to a significant eruption can be as short as a few tens of minutes.

There are permanent observatories on many active volcanoes dedicated first to recognising the signals of an impending eruption and then to monitoring eruptions. Worldwide there are 80 such observatories (the World Organisation of Volcano Observatories, or WOVO; http://www.wovo.org/) and also regional networks of seismometers and ground deformation measurements (such as GPS). However, many of the world's volcanoes remain unmonitored with little or no baseline information. Major volcanic crises in the developing world have often necessitated calls on outside scientific assistance. The Volcano Disaster Assistance

Programme (VDAP) of the US Geological Survey has played a particularly prominent role in responding to such emergencies (http://volcanoes.usgs.gov/vhp/vdap.php).

Signals precursory to volcanic eruptions take place on a wide range of timescales, from many years to a few tens of minutes (Sparks, 2003). Precise prediction about the onset time of an eruption is rarely possible, but there are many examples where eruptions have been successfully forecast hours, days or weeks ahead of the actual event. A major problem is that active volcanoes often show unrest, such as earthquakes, ground deformation, gas release and steam explosions, but this unrest does not always lead to eruption. Indeed, there are many more cases of periods of unrest without subsequent eruptions than there are eruptions (Newhall and Dzurisin, 1988; Biggs *et al.*, 2009; Poland, 2010; Moran *et al.*, 2011). Volcanic unrest may come about because of movement of fluids underground, tectonic processes and the intrusion of magma at depth or a combination of these factors. Failed eruptions can lead to so-called 'false alarms', although these are better termed unfulfilled alarms (the cause for alarm is usually real enough!). Thus, when there is volcanic unrest above background levels, the outcome is highly uncertain.

The difficulty of interpreting volcanic unrest and predicting eruptions has serious consequences for risk assessment and crisis management. One example is the Soufrière Hills volcano, Montserrat, where periods of unrest without eruption occurred in 1896–1897, 1933–1937 and 1966–1967. When new unrest began in 1992, people on Montserrat were not anticipating that an eruption would follow, so were ill-prepared when a major eruption started in July 1995 (Robertson *et al.*, 2000). Another example is that of Miyakejima volcano, Japan, where over 90% of shallow dyke intrusions that take place do not connect to the surface as eruptions (Geshi *et al.*, 2010). Evacuations are quite commonly called during periods of strong volcanic unrest, exemplified by the evacuation of the town of Basse Terre on Guadeloupe in 1976 (Feuillard *et al.*, 1983; Fiske, 1984; Section 11.5.2). One of the consequences of so-called false alarms is that populations may perceive that an evacuation is called unnecessarily, with loss of trust in the scientists and a reluctance to respond when another period of unrest starts.

Once an eruption has started there continue to be great challenges in judging the course of the eruption. It is typically very difficult to forecast the exact style, size and duration of eruptive activity, and volcanologists are always faced with significant uncertainties that have to be communicated to decision-makers. For volcanoes like Vesuvius, which are close to large populations, evacuation plans are typically based as much on logistics as they are on the ability of scientists to make reliable forecasts of an eruption. In the case of Naples, the evolving evacuation plan still assumes there will be several days of advance warning, but many in the scientific community are sceptical that a confident and reliable forecast that an eruption is likely to take place can be given on that timescale. Likewise, evacuations can create major problems when there have been unfulfilled alarms. Managing the realities of the large uncertainties in forecasting and giving reliable early warnings, and matching these realities to the requirements of decision-makers and expectations of the public, are very common challenges in volcano crisis management.

11.4 Hazard and risk assessment

The classical basis for forecasting the nature of a future eruption is the historical record or geological studies of past eruptions. Since the 1970s the primary communication tool in applied volcanology has been the hazards map. A typical study involves charting out young volcanic deposits to generate maps for each type of hazard, reflecting areas that have been affected by past volcanic events. Tilling (1989) provides a comprehensive account of this classical approach. Increasingly, such studies are augmented by modelling of the processes involved. Here, models are run under the range of conditions thought to be plausible for the particular volcano and commonly calibrated to observed deposit distributions. Examples of modelling investigations for the Soufrière Hills volcano, Montserrat include distribution of pyroclastic flows (Wadge, 2009) and probabilistic distribution of ash fall deposits (Bonadonna *et al.*, 2002). Mapping is complemented by chrono-lithostratigraphic studies that seek to characterise the frequency of eruptions of different size and diagnose the eruption style.

Typically the outcome of such geological studies is a zonation map. The area around a volcano is typically divided into zones of decreasing hazard that are used to identify communities at high risk, to help in the management of volcanic crises and for planning purposes by the authorities. A very common type of map will have a red zone of high hazard, orange or yellow zones of intermediate hazard (often both), and a green zone of low hazard. Implicitly, such maps also depict the risk to people occupying a zone. Boundaries between zones are typically marked initially by lines on maps based on judgement by scientists about hazard levels. The precise positioning of such boundaries in published versions of these maps, however, may be modified to take account of administrative issues and practical matters, such as evacuation routes, as determined by civilian or political authorities. Figure 11.8 shows the 2010 hazard map for the Soufrière Hills volcano, Montserrat (http://www.mvo.ms/). Here the hazard map is linked to a hazard-level scheme that depends on the activity of the volcano. The map zones change their status in terms of hazard level as the activity of the volcano changes. As the hazard level of each zone changes, restrictions on the allowed activities are increased or decreased and actions by administrative authorities specified for each zone are changed.

The position of hazard zone boundaries is implicitly probabilistic, but it is only recently that more rigorous approaches to locating such boundaries have been developed. Hazard and derivative risk maps of volcanoes are produced by a variety of organisations. Geological surveys or government institutions typically have official responsibility for providing scientific information and advice to civilian, political or military authorities, who have the responsibility to make policy or decisions such as whether to evacuate. Academic groups and insurance companies also generate maps, so there is the opportunity for serious, and unhelpful, contention if any of these do not appear to agree with hazard or risk maps from an official source.

For disaster mitigation purposes, volcanic risk is usually defined in terms of loss of life and the main strategy in most volcanic crises is to move people out of harm's way by evacuation. It is useful in this context to distinguish between exposure and vulnerability in calculating risk. An exposed population may be living in normal circumstances in an area that has potentially high

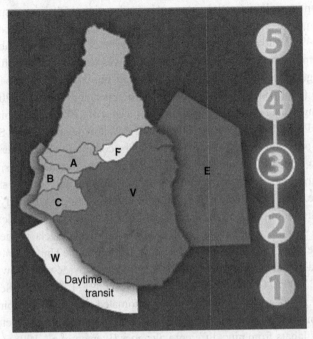

Figure 11.8 A hazard map for Soufrière Hills volcano, Montserrat at alert level 3. The letters on the map designate regions which are colour coded to imply their status in terms of access. For example, area A is the red exclusion zone under alert level 3.

hazard when the volcano becomes active. Their risk will not only depend on their exposure but on their vulnerability, which is dependent on many physical, economic, cultural and social factors (see Chapter 16), as well as the existence of a well-thought-out plan for evacuation.

In periods of dormancy or low activity, reduction of risk can be promulgated through land-use planning, evacuation planning and raising awareness of the hazards and attendant risks in potentially affected communities. However, in many parts of the world the application of hazard and risk assessment to preparedness, resilience and mitigation is limited. Volcanic risk should also be defined in terms of economic loss or potential destruction of key facilities. Such analysis can be a useful guide to inform urban planning and location of major technological facilities such as oil refineries, large dams and nuclear power stations. Damage to critical infrastructure could impact a large number of people and their economy long after an eruption. The only arena where such hazard and risk assessments have been developed substantially is in connection with the possible locating of sites for radioactive waste repositories (Connor *et al.*, 2009). In a related context, the International Atomic Energy Authority (IAEA) will be publishing guidelines in 2011 for volcanic hazards and risk assessment for nuclear installations (Hill *et al.*, 2012).

In the last 15 years or so, there has been a strong move to develop more robust and structured approaches to volcanic hazard and risk assessment and the related task of volcano crisis management. The methods being developed have much in common with those being developed for other natural hazards. For example, ash fall hazards and risks are increasingly assessed by eruption column models, atmospheric advection-diffusion transport models and statistical information on the atmospheric wind systems (e.g. Bonadonna *et al.*, 2002; Costa *et al.*, 2006, 2009). Validation and calibration is done by comparison with observations, but systematic comparison between the models themselves has not yet been fully developed. There is a range of models including sophisticated numerical physics-based models of volcanic flows (e.g. Todesco *et al.*, 2002; Charbonnier and Gertisser, 2009), ensemble approaches using Monte Carlo techniques (e.g. Hincks *et al.*, 2006) and empirical simplified models such as PYROFLOW (Wadge, 2009) or PFz (Widiwijayanti, 2009). So far, the more sophisticated models are largely research tools and the community has not reached accord as to the extent to which these models should be applied in hazards assessments and risk analysis. One argument is that they are the best we have got and so they should be used, while others see them as too deficient in terms of understanding the underlying physics so they could be misleading. To a large extent it is the empirical simplified models that are most widely used in hazards assessments. An example of such models is the TITAN code developed at the State University of New York at Buffalo in the United States for granular avalanches, and which is now being widely applied to pyroclastic flow hazards (e.g. Saucedo *et al.*, 2005; Charbonnier and Gertisser, 2009). LAHARZ (Schilling, 1998; Schneider *et al.*, 2008) is an example of an automated and highly empirical model of lahar inundation widely used for hazard assessment. The model parameters have been calibrated by 27 lahars from nine different volcanoes (Iverson *et al.*, 1998).

Thus, it is only in the last decade or so that risk and uncertainty analyses have entered into common discourse in volcanology. Logic and event trees, with associated analysis of uncertainties, have been introduced (Newhall and Hoblitt, 2002; Marzocchi *et al.*, 2004). The event tree has been used extensively, particularly by the USGS VDAP team, in several volcanic emergencies (C. Newhall, pers. comm.), but there are only a few examples of published descriptions of the application of the methodology, including Pinatubo (Punongbayan *et al.*, 1996) and Montserrat (Sparks and Aspinall, 2004). The Bayesian event tree (BET) model is a flexible tool to provide eruption forecast and volcanic hazard and risk assessment (the INGV code can be downloaded free at http://bet.bo.ingv.it/). The BET is a graphical representation of events in which individual branches are alternative steps from a general prior event, state or condition, through increasingly specific subsequent events (intermediate outcomes) to final outcomes. Such a scheme shows all relevant possible outcomes of volcanic unrest at progressively higher degrees of detail. The probability of each outcome is obtained by combining opportunely the probabilities at each node of the tree. The Bayesian approach applied to the event or logic tree brings two key advantages. First, the probabilities at the nodes are described by distributions, instead of by single values; this allows appropriate estimates of the uncertainty related to the probability evaluations to be incorporated into the calculations formally. Second, the Bayesian approach makes easier the merging of all the relevant available information such as theoretical models, prior beliefs, monitoring measures and past data.

The BET model has been formulated in different ways, depending on intended use. The application to eruption forecasting, BET_EF (Marzocchi *et al.*, 2008), is focused on the first part of the event tree, i.e. from the onset of the unrest up to eruption occurrence of a specific size. The input data are mostly monitoring measures; depending on what the monitoring observations indicate and how they change, BET_EF updates the evaluation of an eruption forecast in near real time. The primary purpose of this code is to assist decision-makers in managing a phase of unrest. The code has been set up for different volcanoes and used in two recent experiments that simulate the reactivation of Vesuvius (MESIMEX experiment: Marzocchi and Woo, 2007; Marzocchi *et al.*, 2008) and the Auckland Volcanic Field (Ruaumoko: Lindsay *et al.*, 2010). It has also been applied retrospectively to the pre-eruptive phases of the 1631 eruption of Vesuvius by using the information reported in contemporary chronicles (Sandri *et al.*, 2009).

The BET_VH model can be applied for long-term volcanic hazard estimation (Marzocchi *et al.*, 2010), focusing on the impact of different hazardous phenomena in areas surrounding the subject volcano. As inputs, BET_VH incorporates results from numerical models simulating the impact of hazardous volcanic phenomena area by area, and data from the eruptive history of the volcano of concern. For output, the code provides a wide and exhaustive set of spatio-temporal probabilities of different events. This version of the BET code has been used in a recent application to long-term tephra fallout hazard assessment at Campi Flegrei, Italy (Selva *et al.*, 2010).

However, such sophisticated probabilistic hazard assessments, with expressions of scientific uncertainty embedded, pose a significant challenge for decision-makers to digest, since they have to take decisions under their own constraints, which are mostly non-scientific. A rational use of the probabilistic BET model, or any probabilistic assessment model in general, requires the definition of a volcanic risk metric (VRM). Marzocchi and Woo (2009) propose a strategy based on coupling probabilistic volcanic hazard assessment and eruption forecasting with cost–benefit analysis via the VRM concept. This strategy, then, has the potential to rationalise decision-making across a broad spectrum of volcanological questions. When should the call for evacuation be made? What early preparations should be made for a volcano crisis? Is it worthwhile waiting longer? What areas should be covered by an emergency plan? During unrest, what areas of a large volcanic field or caldera should be evacuated, and when? The VRM strategy has the paramount advantage of providing a set of quantitative and transparent rules that can be established well in advance of a crisis, optimising and clarifying decision-making procedures and responsibilities. It enables volcanologists to apply all their scientific knowledge and observational information to assist authorities in quantifying the positive and negative risk implications of any decision.

The eruption of the Soufrière Hills volcano, Montserrat, was the first time that structured expert elicitation techniques have been applied in a real crisis (Aspinall and Cooke, 1998; Sparks and Aspinall, 2004). During a volcanic eruption the approach is to identify all the possible outcomes over some practical risk management time period, such as a year or a few weeks. Various outcomes may be benign or lethal, volcanic risk so far being almost exclusively defined in such situations in terms of the annualised probability for loss of life, there being little that can be done to mitigate property damage on an immediate basis.

Typically, any particular hazard is represented in this structured approach by an inter-linked series of disaggregated events or processes. Commonly, the ultimate risk level depends on conditional probabilities in the chain of events that lead to risk. As an example, the size and likelihood of a dome collapse pyroclastic flow at the Soufrière Hills volcano depends on the volume of the dome, its direction of growth and the rate of growth, and can be modulated by the occurrence of intense rainfall that promotes destabilisation of the dome. The chance of a pyroclastic flow reaching a particular place, such as a village, depends on the initial volume and momentum of the flow, the rheological properties of the flow and topography. Some, such as dome volume, may be accurately known. Others, like the long-term occurrence of intense rainfall events, may require regional data that are treated statistically. Yet others may require an empirical model (e.g. PYROFLOW: Wadge, 2009) based, for example, on observed runouts of pyroclastic flows of different volumes or laboratory data on rheology. All of these components have epistemic and aleatory uncertainties. Where data are scarce or understanding of the processes is poor, expert judgement can be used. To go from hazard to risk requires additional information on vulnerability and exposure, adding further complexity and uncertainty. Variation in vulnerability can be modelled statistically; for example, by applying Monte Carlo re-sampling methods the overall risk level can be evaluated using different population distributions to show how risk can be mitigated by relocating groups of people away from the hazards.

In line with this thinking, during the eruption of the Soufrière Hills volcano in Montserrat, a scientific advisory committee (SAC) has met approximately every six months since 1997, applying the methods described above. The eruption is still ongoing at the time of going to press (September 2012) and the SAC remains operational. The SAC typically consists of several scientists representing a range of expertise, and interfaces with staff from the Montserrat Volcano Observatory (MVO), who are responsible for the day-to-day monitoring, hazard assessment and provision of advice to the government authorities. The SAC looks at the long-term trends and outlook for the volcano to facilitate planning and management. A period of six months was chosen by the SAC as a benchmark, but both longer and shorter periods have been considered, depending on circumstances. Occasionally the SAC has convened at short notice to support the MVO and provide advice in periods of elevated activity, when evacuations may be necessary.

Figures 11.9, 11.10 and 11.11 show examples of some key products of these formal hazard and risk assessments. The event tree (Figure 11.9) shows an example of a range of potential future hazardous events in a dome-building eruption, with branches that depict alternative outcomes. Probabilities and uncertainties in these probabilities are assessed for each branch on the tree through integration of all pertinent evidence and models using expert elicitation. Such event trees develop and grow as an eruption proceeds; the skills of the science team can be assessed by comparing the actual outcome with the events that are assessed to be most likely. Curves of societal risk in Figure 11.10 consist of plotting the probability of a certain number of casualties being exceeded versus the number of casualties. Each curve is based on a certain assumed distribution of population and, to be cautious, also assumes there is no warning. In the case of Montserrat, the island was divided into several

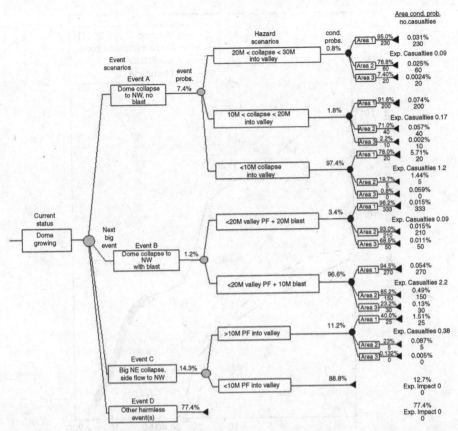

Figure 11.9 A simplified event tree for four pyroclastic flow and blast event scenarios in a dome-building eruption, and potential casualty risks in three populated areas (based on a typical case for the Belham Valley and Soufrière Hills volcano, Montserrat). Event D denotes any eruption scenario that does not impact the Belham Valley and therefore presents no risk to residents there; events A, B and C are three different ways in which a dangerous flow or blast could impact the valley and adjacent areas. The hazard scenarios are rudimentary classifications, scaled by volume(s) of material involved – e.g. '20M' means 20 million cubic metres of lava. Branching probabilities are usually obtained by elicitation; potential numbers of casualties are based on global experience and are functions of the total number of people in each area and the nature of the hazard. In a quantitative risk analysis based on such an event tree, all values would be characterised by appropriate statistical distribution to represent uncertainties. For most eruption situations, a comprehensive risk tree like this can easily extend to hundreds or thousands of branches. Expected casualties indicate the statistical values for the particular scenario. These expectation values show it is the higher probability/less intense events that provide the greatest risks, because the more extreme events may kill more but have much lower probabilities proportionately.

areas, with population estimates for each zone. If people are evacuated from a high-risk area then the risk curve moves downwards. Decision-makers can then see the reduction in risk due to selective timely evacuation. Such curves can also be used to monitor how volcanic risk varies due to fluctuations in the scientists' appraisals of volcanic activity, as well as

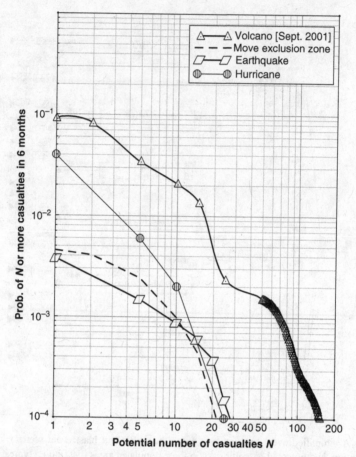

Figure 11.10 Example of probability curves for societal risk in Montserrat. Each curve shows the probability plotted against number of casualties over a six-month period, and is the mean of thousands of simulations using Monte Carlo re-sampling from uncertainty distributions on the parameters that influence risk. The upper curve (solid line) is the exposure within the Belham Valley area populated before the evacuation, and the lower curve (dashed line) shows the reduction in risk with evacuation. Regional risk curves for hurricanes and tectonic earthquakes are shown for comparison. Note that for each curve uncertainties at the 5% and 95% levels were calculated but for clarity are not shown.

compare the volcanic risk with the risk from other familiar hazards such as hurricanes and earthquakes.

While Figure 11.10 shows overall societal risk levels for the whole population, usually there are also concerns about levels of exposure to the volcano for individuals, and what is acceptable or tolerable in this context. In health and safety terms, this measure of risk is commonly expressed as the individual risk per annum (IRPA) of death and, in the case of Montserrat, a typical individual risk ladder is shown in Figure 11.11. The ladder shows a resident's relative risk exposure due to the proximity of the volcano when living full-time in

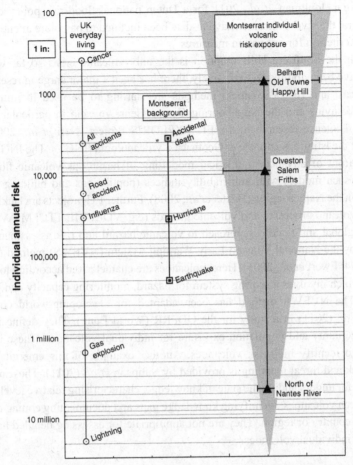

Figure 11.11 Example of an individual risk ladder for Montserrat residents living near the Belham Valley, showing an individual's volcanic risk exposure as a function of locality in March 2010 (right-hand ordinate). The adjoining ladders on the chart show: background risk levels for an individual from natural hazards on island and accidental death (excluding road accidents) and some everyday living risks for the UK (adapted from SAC 14 Report – available on the MVO website).

certain areas. One of the main purposes of such a chart is to help convey and communicate the extent of any additional risk from living on the island, over and above the long-term background accidental death risk and due to dangers from other natural hazards. Because of the multiplicity of possible volcanic hazards, relative to other sources of risk, the evaluation of such risk levels is complex; a nonlinear relationship exists between the estimates of IRPA shown in Figure 11.11 and the societal curves in Figure 11.10, making it difficult to connect values from one to the other uniquely across all circumstances. This dichotomy adds to the challenges faced by decision-makers and, as yet, no satisfactory way of utilising both forms of risk measure has been found for policy setting in Montserrat or, indeed, further afield (see e.g. Jonkman *et al.*, 2011 for a Dutch national flood-risk policy discussion). In Montserrat the civil authorities have to-date been inclined to pay more attention to IRPA values as a criterion for mitigation measures.

Although the Soufrière Hills eruption is the only volcanic crisis so far where these methods have been applied so extensively, there has been a major surge in research in this arena. Hazard and risk assessment methods are starting to be used in mitigation and planning. Vesuvius and the neighbouring Campi Flegri volcano in particular have been the focus of intense research in the EU EXPLORIS Project (Neri *et al.*, 2008) and by scientists at the Istituto Nazionale di Geofisica e Vulcanologia (INGV). The INGV has been at the forefront of advanced numerical modelling of hazardous volcanic flows and in integrating such models with vulnerability indices (populations and building quality) to assess risk in the Naples region (Baxter *et al.*, 2008). Further EU projects are tackling related aspects of volcano processes and volcanic hazards (e.g. VOLUME TTC; MIAVITA).

A more global and regional approach to volcanic hazard and risk assessment has been pioneered by the National Volcano Early Warning System (NVEWS) method (Ewert and Harpel, 2004; Ewert *et al.*, 2005). Here, volcanoes are characterised according to their level of threat, which involves a scoring system for hazard, monitoring capacity and population exposure. The NVEWS method has been adapted for developing-world countries by Aspinall *et al.* (2011) in a study for the GDFRR (World Bank). They define indices for hazard, uncertainty and population exposure for individual volcanoes. These indices are then used to identify high-risk volcanoes. Another example of this emerging arena of volcanic risk and threat mapping is provided by Simpson *et al.* (2011). These approaches are suitable for understanding the state of knowledge, characterising relative levels of risk or threat between volcanoes, identifying knowledge gaps and documenting coping capacity in a particular country or region. They are not appropriate for assessing detailed hazards and risk around individual volcanoes.

11.5 Risk management and communication

Effective risk management during periods of volcanic unrest and eruption is essential to saving lives and minimising losses. Here, science provides critical information for decision-making and responses, such as evacuations. In many cases decisions have to be made at very short notice and good communications between scientists, responsible authorities and affected citizens are pivotal. This section explores some of the main issues and what can be learned from some past volcanic emergencies.

11.5.1 Risk management

Risk management around active volcanoes is required right through what is sometimes called the disaster cycle. In periods of dormancy, risk management activities include assessing the hazards, improving monitoring and early warning systems, land-use planning, raising awareness of volcanic hazards and risk among citizens and development of evacuation plans in the event of an impending eruption. In periods of volcanic unrest that might lead to an eruption there is heightened awareness and emergency planning among authorities, emergency services, scientific institutions (commonly associated with a volcano observatory) and the public. These activities become even more prominent once an eruption starts. Commonly, decisions have to be made very quickly – for example, on whether to evacuate. For persistently active volcanoes the management will be a long-term requirement to enable the society to live safely with the volcano, with emergency situations being declared when the activity becomes heightened. In the recovery period after a crisis lessons learned can be applied to improve future responses, an activity that is probably likely to be more effective if recriminations are avoided. A critical issue is at what stage evacuated people are allowed to return home and rebuilding of disrupted lives and economies can begin. In some cases discussions will be necessary to decide whether it is wise to reoccupy areas of high hazard and relocation of people may become necessary.

In practice, risk management around active volcanoes can become immensely complicated due to many factors unrelated to the physical hazards. Conflicting views and tensions can arise between authorities, scientists and communities for many different reasons. A volcanic eruption is a traumatic and highly disruptive event in which attitudes to and perceptions of risk can vary widely between individuals, institutions, authorities and other elements of civil society (Johnston *et al.*, 1999; Paton *et al.*, 2001; Davis *et al.*, 2005). Authorities taking on emergency powers can come into conflict with individuals who feel their human rights are being infringed. Responses of individuals, communities, societal organisations and institutions and governing authorities include denial, being risk averse and being a risk taker based on perceived benefits from taking risks. In all cases there will be unique cultural, political and economic factors that influence how risk is managed, or not, in a particular circumstance. The Soufrière Hills eruption on Montserrat (1995 to the present) illustrates many of these complexities, and readers are referred to Haynes (2006) for a synthesis of that case history.

One of the major problems for volcanic risk management relates to the ability of scientists to make confident forecasts of impending hazardous activity and the authorities to respond by taking action such as evacuation.

As already discussed in detail earlier, time periods for precursory unrest and low-level activity to build to hazardous eruptions can vary greatly from many years to only a few hours. More often than not, increasing unrest does *not* lead to an eruption, but when the volcano is on an inevitable path to eruption the period of time during which unrest causes meaningful concern may be rather short.

While increasingly supported by better scientific data and improved models, judgement remains central to scientific assessment in such circumstances. On the other hand, decision-making and implementation of these decisions takes time and it is an unfortunate fact that some volcanoes ramp up to dangerous eruptions in times that can be much shorter than the ability of authorities to make and implement mitigation decisions. A future eruption of Vesuvius is an example of a severe issue because evacuation of over 600 000 people from around the flanks of the volcano will take many days, even under the most optimistic assumptions, whereas it is quite likely that the build up to the eruption will be shorter. The common mismatch in timescales leads inevitably to what are sometimes called false alarms, as many government authorities are likely to take a precautionary approach and instigate evacuation orders well before there is complete certainty that an eruption will take place. The 1976 evacuation of the town of Basse Terre in Guadeloupe is the best-known case, when 73 000 people were moved away for three months due to strong seismic unrest and vigorous explosive phreatic activity at the nearby La Soufrière de la Guadeloupe volcano. The fear was a major explosive eruption with pyroclastic flows. At the time there was much controversy and a widespread view in the communities and among some scientists that the evacuation had been unnecessary. In retrospect, the evacuation was a sensible and precautionary decision, but there was much loss of credibility and trust for scientists, which still resonates on the island. Ironically, this 'non-event' may make the communities on the island more vulnerable to a future eruption of La Soufrière.

Because evacuation is the only effective response to dangerous volcanic activity, evacuation modelling (Marzocchi and Woo, 2007; Woo, 2008) is a new area of research for volcanology, with the potential of providing key input into planning for volcanic emergencies. Marzocchi and Woo (2007) have developed a probabilistic scheme that integrates eruption forecasting and cost–benefit analysis of evacuation options. The method incorporates available knowledge on hazards into a quantitative decision-analysis framework using transparent rules. The approach enables prior scrutiny of any scientific input into the model and so may help to reduce the stress on scientists during an emergency phase. Modelling of this kind, however, poses some significant research challenges, as predicting what happens on the ground after an evacuation decision has been made is very difficult because it will likely depend critically on human behaviour. The extent to which human behaviour and economic disruption can be incorporated into evacuation models is problematic and likely contentious.

Evacuation is always traumatic as people have to abandon their homes, belongings and livelihoods at short notice, often under conditions of fear, anxiety and chaos. In general, evacuees are placed in emergency accommodation of some kind and it is often not possible for them, or for the volcanologists, to be sure when they may be able to return, if at all. Prior relocation is becoming increasingly used as a mitigation policy by some governments. Usually this policy is developed during or after an event, when it becomes clear that the evacuated area is going to remain too dangerous to reoccupy. In the case of Galeras in Colombia the government has passed a law to relocate communities living in areas deemed to be at high risk. Whether the policy is evacuation, reoccupation or relocation, the decisions are rarely straightforward and can become contentious and politically charged: scientific uncertainty is an ever-present factor.

11.5.2 Communication

The previous discussion of risk management implies a critical role for communication between key actors before, during and after a volcanic emergency. Scientists are inevitably at the centre of communication networks in that they have specialist knowledge about the potential hazards, the information needed for giving early warning, and informed experience necessary for assessing hazards and risk. Some of the major volcanic disasters reflect communications failure. Analysis of the 1985 Nevado del Ruiz tragedy by Voight *et al.* (2012), when over 23 000 people were killed in the town of Armero, Colombia by a lahar, suggests that factors that contributed were: delays in producing a hazard map; inadequately prepared local authorities; an unprepared populace; and a refusal to accept false alarms. Voight *et al.* (2013) identify the lessons learned at Nevado del Ruiz as: scientists and authorities having to accept the responsibility to communicate the risk to the public; the need to plan critical decisions in advance; to test warning systems in advance; to anticipate technological problems with communication systems; and to develop effective relationships with the media. Similar lessons relating to the importance of effective communication systems were drawn from the 1994–1998 eruptions of Merapi, Indonesia (Voight *et al.*, 2000a).

However, volcanic eruptions at most volcanoes are infrequent and very large eruptions extremely rare, such that they are commonly outside the experience of most official decision-makers and populations. Even with recent efforts to inform people with videos and documentary programmes, it is still difficult to convey the full extent, subtleties and dangers of the hazards that volcanoes can produce. Without previous experience of an eruption, at-risk communities can only make inferences from events elsewhere as to what may happen when their volcano erupts and this may be vastly different. This is not to say that a population with recent experience of volcanic activity necessarily has a more accurate perception of future hazard. If the previous hazard experience was relatively benign, people can experience a 'normalisation bias' (Mileti and O'Brien, 1993) whereby this becomes the archetypal eruption, even if there is a strong likelihood of the volcano erupting more violently in the future. A recent example of this situation is the October–November 2010 eruption of Merapi volcano in Indonesia, where quite frequent episodes of pyroclastic flows in the last 100 years had led to hazards zonation and evacuation plans that extended to 10 km from the volcano. Past history, though, suggested that Merapi can have much more energetic and larger magnitude eruptions (Voight *et al.*, 2000b), the last major one prior to 2010 being in the 1800s. The 2010 eruption produced pyroclastic flows, one of which ran out to 17 km, and hazard zonations had to be extended urgently to 20 km during the crisis.

Volcanic eruptive processes are intrinsically unpredictable, and therefore scientists are limited in what they can say with confidence about any future volcanic activity. This uncertainty is sometimes not understood and can be seen by some as incompetence. Conversely, some people can be overly reliant on scientists to provide an adequate warning. For example, in relation to earthquake hazards, investigations conducted by Valery (1995) after the Kobe, Japan, earthquake found that many citizens knew about the risks and how to prepare but believed that the science was so advanced that they would receive a warning before the earthquake struck, so there was no real need to prepare. By contrast, people may become so used to volcanic activity that they become overconfident in their own ability to judge the threat. This seems to have happened in the case of the eruption of the Soufrière Hills volcano on 25 June 1997, when 20 people were killed in the exclusion zone; interviews with survivors (Loughlin *et al.*, 2002b) indicate that most understood the risks but were prepared to accept them for the benefits of looking after their farms or property. A similar situation has been found for Etna, where surveys of citizens indicate an objective and informed perspective concerning the volcanic hazard (Davis *et al.*, 2005). In contrast, people living in the highest risk areas at Vesuvius showed high levels of fear and perceived risk, but low levels of perceived ability to protect themselves from the effects of an eruption, as well as low levels of awareness concerning evacuation plans and confidence in the success of such plans (Davis *et al.*, 2005). This combination of scientific uncertainty, limited hazard experience of the local population and mismatches between public and scientific evaluations of risk represent challenges for effective communication.

Compounded with this are official and political concerns about the best way to provide advice to a threatened population. Officials are sometimes wary of giving vivid descriptions of worst-case scenarios to the public, fearing widespread panic. Officials routinely may expect the public to panic or, at best, to misinterpret orderly efforts to mitigate disaster. Thus, being anxious not to create panic can lead officials to make over-reassuring statements, to suppress information, sometimes contradicting scientific advice, and to belittle those who are anxious as irrational (see also Chapter 16). On the other hand, public officials can also be afraid of being blamed in the case of casualties and therefore can take a precautionary, risk-averse approach. A balance needs to be struck between reassuring the population that any future crisis will be dealt with and arousing enough interest in the subject that complacency does not set in. This is particularly pertinent for volcanic hazards where there can be very long periods of quiescence between eruptions. Unfortunately there are few published analyses of these issues for volcanic crises, but useful insights can be gained from health fears (Leventhal, 1971).

It is imperative that the goals of communication are properly defined before attempting to communicate risks relating to volcanic hazards, or any other hazard for that matter. Is the aim to simply inform about the hazards and potential mitigative actions that can be taken, or is it to effect behaviour change? Moreover, who will be responsible for such communication? It has long been believed that scientists should be responsible for monitoring and prediction, whereas the government or emergency management officials should be those who communicate the necessary information, along with whatever policy decisions they draw from this information, to the public (Peterson and Tilling, 1993; Peterson, 1996). However, scientists are becoming more and more involved in the communication of at least the hazard information, if not risk; for example, at Mount St Helens (Driedger *et al.*, 2008; Frenzen and Matarrese, 2008). There is no single model for risk communication that will work in all situations, as both the hazard and the political situations vary from one location to another, and so the communication methods will also need to adapt.

That being said, one of the principal challenges for good risk communication is to instil trust in those who are responsible for public safety and risk communication, particularly when personal experience is lacking (Renn and Levine, 1991). Research in this area conducted on three separate islands in the Lesser Antilles, each in different stages of unrest, had similar findings with regard to trust in scientists; they were found to be one of the most trustworthy sources, ranked significantly higher than the local governments on each of the islands (see Haynes *et al.*, 2008; Crosweller, 2009). Much of this can be explained through the fact that scientists are often seen as being 'value neutral'. Trust, however, is fragile and is much easier to lose than it is to gain, and so every effort must be made to communicate clearly about the uncertainties and limitations of scientific predictions. It should be made clear that unfulfilled alarms will be likely in any crisis management. This issue of unfulfilled warnings is also an important point for the scientific community to grapple with: scientists, by inclination, will seek to find very reliable prediction and forecast methods, because that is the basis for scientific progress, almost to the point of requiring nothing less than a physical law before making predictions. However, total reliability of predictions or forecasts is never going to be achievable, and good risk communication and management needs to accommodate uncertainties in the best ways possible, and this means taking advantage of emerging 'evidence science' theories and methods (Aspinall *et al.*, 2003), which are finding application in many safety-critical fields where scientific uncertainty is a present and key factor. It also requires a readiness on the part of officials and the public to accept some 'false alarms' if they require a high degree of safety. In effect, social contracts between scientists, officials and those at risk are needed.

We live in a world of rapid changes in communication with mobile phones, the internet and social networking augmenting, and to some extent replacing, traditional ways of communication. There are certainly opportunities for innovation as well as new challenges. In recent volcanic emergencies the 'new media' played a key role in the dissemination of information and, unfortunately, misinformation. New forms of communication, such as the blogosphere, Twitter and mobile phones provide opportunities to disseminate scientific advice, early warnings and raise awareness. On the other hand, they also allow the spreading of unfounded rumours, the promotion of antagonistic scepticism against mainstream science and other contrarian views, and enable conspiracy theories to proliferate at lightning speed. Working more closely with the media and with these new forms of communication is going to be essential, but challenging.

11.6 Future outlook and challenges

Advances in enabling technologies and understanding of volcanic processes over the last decade have greatly enhanced the ability of volcanologists to anticipate the behaviour of volcanoes and to assess their hazards. However, application of these advances to forecasting and emergency management during eruptions is contingent on having suitable resources in place, and this is not necessarily the case in developing countries, where many volcanoes remain poorly monitored and other priorities exist for limited resources. Reduction in costs of electronic equipment and increasing availability of remotely sensed observations are helping to improve the monitoring status of many volcanoes. Methodologies to assess hazards footprints in probabilistic terms have developed together with quantitative approaches to risk assessment. The importance of evaluating uncertainties is becoming ever more apparent.

This said, there is a significant gulf between quite advanced academic research and routine practice in 'applied volcanology'. Academic research is thriving in terms of the development of physics-based models of hazardous volcanic processes and advanced statistical approaches to risk assessment and treatment of uncertainty. Most volcanic crises are still handled using the traditional tools of hazards mapping and forecasting based on monitoring. Observations and knowledge of past eruptive activity are valuable, but interpretation of the underlying processes and future behaviour is difficult and uncertain and requires an approach that uses all the advanced science available and more robust statistical treatment. Risk and uncertainty are often dealt with qualitatively and there remains a conservative element among practitioners and some observatory scientists who do not see the value in trying to quantify either. Examples in the UK (with Montserrat), New Zealand and Italy show this situation is going to change as more practitioners are trained in research groups in the expertise and skills to assess risk quantitatively. However, much still needs to be done to train young scientists and educate the older generation of scientists in the methods of analysis and communication of risk and related uncertainties.

Volcanologists are increasingly being asked to go beyond traditional roles of monitoring, hazard assessment, giving early warning and providing scientific advice, and are increasingly being called upon to participate in risk assessment applied to decision-making. These pressures reflect a broader trend for science to be more actively engaged in addressing societal problems but will bring scientists into new and more complex arenas and roles. There is certainly some discomfort for many. Volcanologists have the expertise to quantify hazard and risk as well as attendant uncertainties, but as many social scientists would remark, risk is essentially a human construct. Volcanologists must work with engineers, other technical experts and social scientists to make full assessments of volcanic risk. There are also limits to the benefits of such quantification and indeed the ability to quantify some facets of risk. Volcanologists can estimate the risk and uncertainties of loss of life, given assumptions about the hazard and the location of people, but what is much harder to quantify is how people will respond to, for example, evacuation orders. The meaning of 'risk', which is intrinsically multifaceted, needs to be articulated, not least because there is some concern that attempts to quantify risk in support of public safety may also open up its practice to litigation if things go 'wrong'.

Risk management for volcanoes has largely been viewed in terms of emergency response, with the goal of avoiding loss of life. The priority accorded to this perspective is not likely to change soon. However, there is increasing attention being paid to better planning so that economic losses and loss of life are reduced. In the arena of development, reducing the impact of natural disasters on sustainable livelihoods is becoming a higher priority, and so approaches that increase the resilience of communities is now the *lingua franca* of discussions. This has provoked, rightly, many questions about what is the best approach to volcanic risk mitigation and the political or economic impact of volcanic disasters. A repeated narrative for natural disasters is that society has hitherto tended to be reactive rather than proactive. In the future it seems likely that the role of scientists will become more complex, and challenging, as they are drawn deeper into risk assessment and forecasting the future.

Acknowledgements

We thank SAPPUR colleagues for their support. The manuscript was improved with reviews by Gill Jolly, Patty Mothes and Chris Newhall. The authors were supported by the European Research Council Advanced Grant and by the Natural Environment Research Council.

References

Aspinall, W. P. and Cooke, R. (1998) Expert judgement and the Montserrat volcano eruption. In *Proceedings of the 4th International Conference on Probabilistic Safety Assessment and Management PSAM4, 13–14 September 1998*, ed. A. Mosleh. and R. A. Bari, New York: Springer.

Aspinall, W. P., Woo, G., Voight, B., *et al.* (2003) Evidence-based volcanology: application to eruption crises. *Journal of Volcanology and Geothermal Research* **128**: 273–285.

Aspinall, W. P., Auker, M., Hincks, T., *et al.* (2011) Volcanic hazard and risk in GDRFF priority countries. NGI report 20100806, May. http://www.globalvolcanomodel.org.

Baubron, J.-C., Allard, P., Sabroux, J-C., *et al.* (1991) Soil gas emanations as precursory indicators of volcanic eruptions. *Journal of the Geological Society* **148**: 571–576.

Baxter, P. J. (1990) Medical effects of volcanic eruptions. *Bulletin of Volcanology* **52**: 532–544.

Baxter, P. J., Aspinall, W. P., Neri, A., *et al.* (2008) Emergency planning and mitigation at Vesuvius: a new evidence-based approach. *Journal of Volcanology and Geothermal Research* **178**: 454–473.

Bebbington, M., Cronin, S. J., Chapman, I., *et al.* (2008) Quantifying ash fall hazard to electrical infrastructure. *Journal of Volcanology and Geothermal Research* **177**: 1055–1062.

Biggs, J., Amelung, F., Gourmelen, M., *et al.* (2009) InSAR observations of 2007 Tanzania rifting episode reveal mixed fault and dyke extension in an immature continental rift. *Geophysical Journal International* **179**: 549–558.

Blong, R. J. (1984) *Volcanic Hazards: A Source Book on the Effects of Eruptions*, Orlando, FL: Academic Press Inc.

Bonadonna, C., Macedonio, G. and Sparks, R. S. J. (2002) Numerical modelling of tephra fallout associated with dome collapses and Vulcanian explosions: application to hazard assessment on Montserrat. In *The Eruption of the Soufrière Hills Volcano, Montserrat 1995 to 1999*, ed. T. H. Druitt and B. P. Kokelaar, London: Geological Society of London, pp. 517–538.

Bryan, S. E., Peate, I. U., Peate, D. W., *et al.* (2010) The largest volcanic eruptions on Earth. *Earth-Science Reviews* **102**: 207–229.

Burgmann, R., Rosen, P. A. and Fielding, E. J. (2000) Synthetic aperture radar interferometry to measure Earth's surface topography and its deformation. *Annual Review of Earth and Planetary Sciences* **28**: 169–209.

Burton, M. R., Allard, P., Murè, F., *et al.* (2007) Magmatic gas composition reveals the source of Strombolian explosive activity. *Science* **371**: 227–230.

Carn, S. A., Krueger, A. J., Krotkov, N. A., *et al.* (2008) Daily monitoring of Ecuadorian volcanic degassing from space. *Journal of Volcanology and Geothermal Research* **176**: 141–150.

Charbonnier, S. J. and Gertisser, R. (2009) Numerical simulations of block-and-ash flows using the Titan2D flow model: examples from the 2006 eruption of Merapi Volcano, Java, Indonesia. *Bulletin of Volcanology* **71**: 953–959.

Clay, E., Barrow, C., Benson, C., *et al.* (1999) *An Evaluation of HMG's Response to the Montserrat Volcanic Emergency*, London: DFID.

Connor, C. B., Chapman, N. and Connor, L. J. (eds) (2009) *Volcanism, Tectonism and the Siting of Nuclear Facilities*, Cambridge: Cambridge University Press.

Costa, A., Macedonio, G. and Folch, A. (2006) A three-dimensional Eulerian model for transport and deposition of volcanic ashes. *Earth Planetary Science Letters* **241**: 634–647.

Costa, A., Dell'Erba, F., Di Vito, M., *et al.* (2009) Tephra fallout hazard assessment at the Campi Flegrei caldera (Italy). *Bulletin of Volcanology* **71**: 259–273.

Crosweller, H. S. (2009) An analysis of factors influencing volcanic risk on two islands in the Lesser Antilles. PhD thesis, University of East Anglia.

Dalton, M. P., Waite, G. P., Watson, I. M., *et al.* (2010) Multiparameter quantification of gas release during weak Strombolian eruptions at Pacaya Volcano, Guatemala. *Geophysical Research Letters* **37**: L09303.

Davis, M. S., Ricci, T. and Mitchell, L. M. (2005) Perceptions of risk for volcanic hazards at Etna and Vesuvio. *The Australian Journal of Disaster and Trauma Studies.* http://www.massey.ac.nz/~trauma/issues/2005-1/davis.htm

Deligne, N. I., Coles, S. G. and Sparks, R. S. J. (2010) Recurrence rates of large explosive volcanic eruptions. *Journal of Geophysical Research* **115**: B06203.

Driedger, C. L., Neal, C. A., Knappenberger, T. H., *et al.* (2008) Hazard information management during the autumn 2004 reawakening of Mount St. Helens Volcano, Washington. In *A Volcano Rekindled: The Renewed Eruption of Mount St. Helens, 2004–2006*, US Geological Survey Professional Paper 1750, ed. D. R. Sherrod, W. E. Scott and P. H. Stauffer, Chapter 24.

Durant, A. J., Bonadonna, C. and Horwell, C. J. (2010) Atmospheric and environmental impacts of volcanic particulates. *Episodes* **6**: 235–240.

Dzurisin, D. (2003) A comprehensive approach to monitoring volcano deformation as a window on the eruption cycle. *Reviews of Geophysics* **41**: 1001.

Edmonds, M. (2008) New geochemical insights into volcanic degassing. *Philosophical Transactions of the Royal Society A* **366**: 4559–4579.

Ewert, J. W. and Harpel, G. J. (2004) In harm's way: population and volcanic risk. *Geotimes* **49**: 14–17.

Ewert, J. W., Guffanti, M. and Murray, T. L. (2005) An assessment of volcanic threat and monitoring capabilities in the United States: framework for a National Volcano Early Warning System (NVEWS). USGS Open-File Report 2005-1164.

Feuillard, M., Allegré, C. J., Brandeis, G., *et al.* (1983) The 1975–1977 crisis of La Soufrière de Guadeloupe (F.W.I.): a still-born magmatic eruption. *Journal of Volcanology and Geothermal Research* **16**: 317–334.

Fiske, R. (1984) Volcanologists, journalists, and the concerned local public: a tale of two crises in the eastern Caribbean. In *Explosive Volcanism: Inception, Evolution and Hazards. Studies in Geophysics*, ed. F. Boyd, Washington, DC: National Academy Press, pp. 170–176.

Frenzen, P. M. and Matarrese, M. T. (2008) Managing public and media response to a reawakening volcano: lessons from the 2004 eruptive activity of Mount St. Helens.

A Volcano Rekindled: The Renewed Eruption of Mount St. Helens, 2004–2006, US Geological Survey Professional Paper 1750, ed. D. R. Sherrod, W. E. Scott and P. H. Stauffer, Chapter 23.

Galle, B., Oppenheimer, C., Geyer, A., *et al.* (2003) A miniaturised ultraviolet spectrometer for remote sensing of SO_2 fluxes: a new tool for volcano surveillance. *Journal of Volcanology and Geothermal Research* **119**: 241–254.

Geshi, N., Kusumoto, S. and Gudmundsson, A. (2010) Geometric difference between non-feeder and feeder dikes. *Geology* **38**: 195–198.

Grattan, J. P. (2006) Aspects of Armageddon: an exploration of the role of volcanic eruptions in human history and civilization. *Quaternary International* **151**: 10–18.

Grattan, J. P. and Charman, D. J. (1994) Non-climatic factors and the environmental impact of volcanic volatiles: implications of the Laki Fissure eruption of AD 1783. *The Holocene* **4**: 101–106.

Hansell, A. L., Horwell, C. J. and Oppenheimer, C. (2006) The health hazards of volcanoes and geothermal areas. *Occupational and Environmental Medicine* **63**: 149–156.

Hautmann, S., Gottsmann, J., Sparks. R. S. J., *et al.* (2010) The effect of mechanical heterogeneity in arc crust on volcano deformation with application to Soufrière Hills Volcano, Montserrat (W.I.). *Journal of Geophysical Research* **115**: B09203.

Haynes, K. A. (2006) Volcanic island in crisis: investigating environmental uncertainty and the complexities it brings. *The Australian Journal of Emergency Management* **21** (4): 21–28.

Haynes, K. A., Barclay, J. and Pidgeon, N. (2008) The issue of trust and its influence on risk communication during a volcanic crisis. *Bulletin of Volcanology* **70**: 605–621.

Heilprin, A. (1903) *Mont Pelée and the Tragedy of Martinique: A Study of the Great Catastrophes of 1902, with Observations and Experiences in the Field*, Philadelphia, PA: J.B. Lippincott Company.

Hill, B. E., Aspinall, W. P., Connor, C. B., *et al.* (2012) Recommendations for assessing volcanic hazards at sites of nuclear installations. In *Volcanic and Tectonic Hazard Assessment for Nuclear Facilities*, ed. C. B. Connor, N. A. Chapman and L. J. Connor, Cambridge: Cambridge University Press, pp. 566–592.

Hincks, T. K., Aspinall, W. P., Baxter, P. J., *et al.* (2006) Long-term exposure to respirable volcanic ash on Montserrat: a time series simulation. *Bulletin of Volcanology* **68**: 266–284.

IATA (International Air Transport Association) (2011) *Annual Report 2011*, http://www.iata.org/pressroom/Documents/annual-report-2011.pdf.

Iverson, R. M., Schilling, S. P. and Vallance, J. W. (1998) Objective delineation of lahar-inundation zones. *Geological Society of America Bulletin* **110**: 972–984.

Johnston, D. M., Bebbington, M. S., Lai, C.-D., *et al.* (1999) Volcanic hazard perceptions: comparative shifts in knowledge and risk. *Disaster Prevention and Management* **8**: 118–126.

Jonkman, S. N., Jongejan, R. and Maaskant, B. (2011) The use of individual and societal risk criteria within the Dutch Flood Safety Policy: nationwide estimates of societal risk and policy applications. *Risk Analysis* **31**: 282–300.

Kerle, N., van Wyk De Vries, B. and Oppenheimer, C. (2003) New insight into the factors leading to the 1998 flank collapse and lahar disaster at Casita Volcano, Nicaragua. *Bulletin of Volcanology* **65**: 331–345.

Komorowski, J.-C., Tedesco, D., Kasereka, M., *et al.* (2003) The January 2002 flank eruption of Nyiragongo volcano (Democratic Republic of Congo): chronology, evidence for a tectonic rift trigger, and impact of lava flows on the city of Goma. *Acta Vulcanologica* **15**: 27–61.

Leventhal, H. (1971) Fear appeals and persuasion: the differentiation of a motivational construct. *American Journal of Public Health* **61**: 1208–1224.

Linde, A. T., Agustsson, K., Sacks, I. S., *et al.* (1993) Mechanism of the 1991 eruption of Hekla from continuous borehole strain monitoring. *Nature* **365**: 737–740.

Lindsay, J., Marzocchi, W., Jolly, G., *et al.* (2010) Towards real-time eruption forecasting in the Auckland Volcanic Field: application of BET_EF during the New Zealand national disaster exercise 'Ruaumoko'. *Bulletin of Volcanology* **72**: 185–204.

Loughlin, S. C., Calder, E. S., Clarke, A., *et al.* (2002a) Pyroclastic flows and surges generated by the 25 June 1997 dome collapse, Soufriere Hills Volcano, Montserrat. In *The Eruption of Soufrière Hills Volcano, Montserrat, from 1995 to 1999*, ed. T. H. Druitt and B. P. Kokelaar, London: Geological Society of London, pp. 191–209.

Loughlin, S. C., Baxter, P. J., Aspinall, W. P., *et al.* (2002b) Eyewitness accounts of the 25 June 1997 pyroclastic flows and surges at Soufrière Hills Volcano, Montserrat, and implications for disaster mitigation. In *The Eruption of Soufrière Hills Volcano, Montserrat, from 1995 to 1999*, ed. T. H. Druitt and B. P. Kokelaar, London: Geological Society of London, pp. 211–230.

Marzocchi, W. and Woo, G. (2007) Probabilistic eruption forecasting and the call for an evacuation. *Geophysical Research Letters* **34**: L22310.

Marzocchi, W. and Woo, G. (2009) Principles of volcanic risk metrics: theory and the case study of Mount Vesuvius and Campi Flegrei, Italy. *Journal of Geophysical Research* **114**: n.p.

Marzocchi, W., Sandri, L., Gasparini, P., *et al.* (2004) Quantifying probabilities of volcanic events: the example of volcanic hazard at Mt. Vesuvius. *Journal of Geophysical Research* **109**: B11201.

Marzocchi, W., Sandri, L. and Selva, J. (2008) BET_EF: a probabilistic tool for long- and short-term eruption forecasting. *Bulletin of Volcanology* **70**: 623–632.

Marzocchi, W., Sandri, L. and Selva, J. (2010) BET_VH: a probabilistic tool for long-term volcanic hazard assessment. *Bulletin of Volcanology* **72**: 705–716.

Mason, B. G., Pyle, D. M. and Oppenheimer, C. (2004) The size and frequency of the largest explosive eruptions. *Bulletin of Volcanology* **66**: 735–748.

Mather, T. A. and Harrison, R. G. (2006) Electrification of volcanic plumes. *Surveys in Geophysics* **27**: 387–432.

McNutt, S. R. (2005) Volcano seismology. *Annual Reviews of Earth and Planetary Sciences* **32**: 461–491.

Mileti, D. S. and O'Brien, P. W. (1993) Public response to aftershock warnings. US Geological Survey Professional Paper 1553-B, pp. 31–42.

Mogi, K. (1958) Relations between the eruptions of various volcanoes and the deformations of the ground surfaces around them. *Bulletin of Earthquake Research Institute University of Tokyo* **36**: 99–134.

Moran, S., Newhall, C. G. and Roman, D. (2011) Failed eruptions; late stage cessation of magma ascent. *Bulletin of Volcanology* **73**: 115–122.

Neri, A., Aspinall, W. P., Cioni, R., *et al.* (2008) Developing an event tree for probabilistic hazard and risk assessment at Vesuvius. *Journal of Volcanology and Geothermal Research* **178** (3): 397–415.

Newhall, C. G. and Dzurisin, D. (1988) *Historical Unrest at Large Calderas of the World*, Washington, DC: US Government Print Office.

Newhall, C. G. and Hoblitt, R. P. (2002) Constructing event trees for volcanic crises. *Bulletin of Volcanology* **64**: 3–20.

Newhall, C. G. and Punongbayan, R. S. (1996a) The narrow margin of successful volcanic-risk mitigation. In *Monitoring and Mitigation of Volcanic Hazards*, ed. R. Scarpa and R. I. Tilling, Berlin: Springer-Verlag, pp. 807–838.

Newhall, C. G. and Punongbayan, R. (eds) (1996b) *Fire and Mud: Eruptions and Lahars of Mount Pinatubo, Philippines*. Quezon City and Seattle, WA: Philippine Institute of Volcanology and Seismology and University of Washington Press.

Newhall, C. G. and Self, S. (1982) The volcanic explosivity index (VEI): an estimate of explosive magnitude for historical volcanism. *Journal of Geophysical Research* **87**: 1231–1238.

Paton, D., Johnston, D. and Houghton, B. (2001) Direct and vicarious experience of volcanic hazards: implications for risk perception and adjustment adoption. *Australian Journal of Emergency Management* **15**: 58–63.

Peterson, D. W. (1996) Mitigation measures and preparedness plans for volcanic emergencies. In *Monitoring and Mitigation of Volcano Hazards*, ed. R. Scarpa and R. I. Tilling, Berlin: Springer-Verlag, pp. 701–718.

Peterson, D. W. and Tilling, R. I. (1993) Interactions between scientists, civil authorities and the public at hazardous volcanoes. *Active Lavas*, ed. C. R. J. Kilburn and G. Luongo, London: UCL Press, pp. 339–365.

Poland, M. (2010) Learning to recognise volcanic non-eruptions. *Geology* **38**: 287–288.

Pritchard, M. E. and Simons, M. (2004) An InSAR-based survey of volcanic deformation in the central Andes. *Geochemistry Geophysics Geosystems* **5**: n.p.

Punongbayan, R. S., Newhall, C. G., Bautista, L. P., *et al.* (1996) Eruption hazard assessments and warnings. In *Fire and Mud: Eruptions and Lahars of Mount Pinatubo, Philippines*, ed. C. G. Newhall and R. Punongbayan, Quezon City and Seattle, WA: Philippine Institute of Volcanology and Seismology and University of Washington Press.

Rampino, M. (2002) Supereruptions as a threat to civilizations on Earth-like planets. *Icarus* **156**: 562–569.

Renn, O. and Levine, D. (1991) Credibility and trust in risk communication. In *Communicating Risks to the Public: International Perspectives*, ed. R. E. Kasperson and P. J. M. Stallen, Dordrecht: Kluwer Academic Publishers, pp. 175–218.

Robertson, R. E. A., Aspinall, W. P., Herd, R. A., *et al.* (2000) The 1995–98 eruption of the Soufriere Hills volcano, Montserrat. *Philosophical Transactions of the Royal Society* **358**: 1619–1637.

Robock, A. (2000) Volcanic eruptions and climate. *Reviews of Geophysics* **38**: 191–219.

Sandri, L., Guidoboni, E., Marzocchi, W., *et al.* (2009) Bayesian Event Tree (BET) for eruption forecasting at Vesuvius, Italy: a retrospective forward application to 1631 eruption. *Bulletin of Volcanology* **71**: 729–745.

Saucedo, R., Macias, J. L., Sheridan, M. F., *et al.* (2005) Modelling of pyroclastic flows of Colima Volcano, Mexico: implications for hazard assessment. *Journal of Volcanology and Geothermal Research* **139**: 103–115.

Schilling, S. (1998) LAHARZ: GIS Programs for Automated Mapping of Lahar-inundation Hazard Zones: U.S. Geological Survey Open-File Report 98–638.

Schneider, D., Delgado Granados, H., Huggel, C., *et al.* (2008) Assessing lahrz from ice-capped volcanoes using ASTER satellite data, the SRTM DTM and two different flow models: case study on Iztaccihuatl (Central Mexico). *Natural Hazards Earth System Science* **8**: 559–571.

Selva, J., Costa, A., Marzocchi, W., *et al.* (2010) BET_VH: exploring the influence of natural uncertainties on long-term hazard from tephra fallout at Campi Flegrei (Italy). *Bulletin of Volcanology* **72**: 717–733.

Siebert, L., Simkin, T. and Kimberly, P. (2011) *Volcanoes of the World*, 3rd edn, Washington, DC: University of California Press.

Simkin, T., Siebert, L. and Blong, R. (2001) Volcano fatalities: lessons from the historical record. *Science* **291**: 255.

Simpson, A., Johnson, W. R. and Cummins, P. (2011) Volcanic threat in developing countries of the Asia-Pacific region: probabilistic hazard assessment, population risks, and information gaps. *Natural Hazards* **57**: 151–165.

Small, C. and Naumann, T. (2001) The global distribution of human population and recent volcanism. *Environmental Hazards* **3**: 93–109.

Sparks, R. S. J. (2003) Forecasting volcanic eruptions. *Earth and Planetary Science Letters, Frontiers in Earth Science Series* **210**: 1–15.

Sparks, R. S. J. and Aspinall, W. P. (2004) Volcanic Activity: Frontiers and Challenges in Forecasting, Prediction, and Risk Assessment, Washington, DC: American Geophysical Union.

Sparks, R. S. J., Bursik, M. I., Carey, S. N., *et al.* (1997) *Volcanic Plumes* Chichester: John Wiley and Sons.

Sparks, R. S. J., Self, S., Grattan, J., *et al.* (2005) Super volcanoes: global effects and future threats. Geological Society of London, Report of a Working Group, June.

Sparks, R. S. J., Biggs, J. and Neuberg, J. W. (2012) Monitoring volcanoes. *Science* **335**: 1310–1311.

Spence, R., Kelman, I., Baxter, P., *et al.* (2005) Residential building and occupancy vulnerability to tephra fall. *Natural Hazards and Earth Systems Science* **5**: 477–494.

Thomas, H. E. and Watson, I. M. (2010) Observations of volcanic emissions from space: current and future perspectives. *Natural Hazards* **54**: 323–354.

Thomas, H. E., Watson, I. M., Kearney, C. K., *et al.* (2009) A multi-sensor comparison of sulphur dioxide emissions from the 2005 eruption of Sierra Negra volcano, Galápagos Island. *Remote Sensing of Environment* **113**: 1331–1342.

Tilling, R. I. (1989) Volcanic hazards and their mitigation: progress and problems. *Reviews of Geophysics* **27**: 237–269.

Tilling, R. I. and Peterson, L. W. (1993) Lessons in reducing volcanic risk. *Nature* **364**: 277–279.

Todesco, M., Neri, A., Esposti Ongaro, T., *et al.* (2002) Pyroclastic flow hazard assessment at Vesuvius (Italy) by using numerical modeling: I. Large-scale dynamics. *Bulletin of Volcanology* **64**: 155–177.

Ui, T., Matsuwo, N., Sumita, M. and Fujinawa, A. (1999) Generation of block and ash flows during the 1990–1995 eruption of Unzen Volcano, Japan. *Journal of Volcanology and Geothermal Research* **89**: 123–137.

Valery, N. (1995) Fear of trembling: a survey of earthquake engineering. *The Economist* **22**: 3–12.

Van Westen, C. J. and Daag, A. S. (2005) Analysing the relation between rainfall characteristics and lahar activity at Mount Pinatubo, Philippines. *Earth Surface Processes and Landforms* **30**: 1663–1674.

Voight, B. (1990) The 1985 Nevado del Ruiz volcano catastrophe: anatomy and retrospection. *Journal of Volcanology and Geothermal Research* **1**: 349–386.

Voight, B., Glicken, H., Janda, R. J., *et al.* (1981) Catastrophic rock-slide avalanche of May 18. In *The 1980 Eruption of Mount St Helens*, US Geological Survey Professional Paper 1250, ed. P. W. Lipman and D. R. Mullineaux, pp. 347–378.

Voight, B., Constantine, E. K., Siswowidjoyo, S., *et al.* (2000a) Historical eruptions of Merapi Volcano, Central Java, Indonesia, 1768–1998. *Journal of Volcanology and Geothermal Research* **100**: 69–138.

Voight, B., Young, K. D., Hidayat, D., *et al.* (2000b) Deformation and seismic precursors to dome-collapse and fountain-collapse nuées ardentes at Merapi Volcano, Java, Indonesia, 1994–1998. *Journal of Volcanology and Geothermal Research* **100**: 261–287.

Voight, B., Calvache, M. L., Minard, V., Hall, L. and Monsalve, M. L. (2013) The tragic 13 November 1985 eruption of Nevado del Ruiz Volcano, Colombia; the worst can happen. *Encyclopaedia of Natural Hazards* Bobrowsky, P. (ed.) *(in press). Springer Verlag.* ISBN 978-90-481-8699-0.

Wadge, G. (2009) Assessing the pyroclastic flow hazards from dome collapse at Soufriere Hills Volcano, Montserrat. In *Studies in Volcanology: The Legacy of George Walker*, ed. T. Thordarson, S. Self, G. Larsen, *et al.*, London: Geological Society, pp. 211–224.

Werner, C. A., Brantley, S. L. and Boomer, K. (2000) CO2 emissions related to the Yellowstone volcanic system 2: statistical sampling, total degassing, and transport mechanisms. *Journal of Geophysical Research* **105**: 10831–10846.

Werner, C., Christenson, B. W., Hagerty, M., *et al.* (2006) Variability of volcanic gas emissions during a crater lake heating cycle at Ruapehu Volcano, New Zealand. *Journal of Volcanology and Geothermal Research* **154**: 291–302.

Widiwijayanti, C., Voight, B., Hidayat, D., *et al.* (2009) Objective rapid delineation of areas at risk from block-and-ash pyroclastic flows and surges. *Bulletin of Volcanology* **71**: 687–703.

Witham, C. (2005) Volcanic disasters and incidents: a new database. *Journal of Volcanology and Geothermal Research* **148**: 191–233.

Woo, G. (2008) Probabilistic criteria for volcano evacuation decisions. *Natural Hazards* **45**: 87–97.

12

Risk assessment and management of wildfires

T. K. HINCKS, B. D. MALAMUD, R. S. J. SPARKS,
M. J. WOOSTER AND T. J. LYNHAM

12.1 Background

12.1.1 Introduction

Wildfires are dramatic and can be highly destructive, yet palaeoenvironmental records show that fire has been a part of the earth's natural environment for hundreds of millions of years (Bowman *et al.*, 2009). Some ecological systems in drought-prone climates have adapted to frequent wildfires. However, as human systems have gradually taken over much of the planet, wildfires are increasingly seen as a threat. Wildfire frequency is likely increasing as a consequence of human activities, either inadvertently, intentionally or maliciously. Complex dynamics and atmospheric feedbacks make fire behaviour very hard to predict. Whether started intentionally or naturally, wildfires can quickly get out of control and spread at alarming speeds. Major wildfires have been documented at speeds of spreading up to $10-20 \, km \, hr^{-1}$ (Noble, 1991; Rasmussen and Fogarty, 1997), and can cover areas of tens, hundreds or in some cases tens of thousands of square kilometres (see Azuma *et al.*, 2004; Gill and Allan, 2008; VBRC, 2009: appendix A). Embers can blow hundreds of metres or even kilometres ahead of the fire front and can catch vulnerable structures ablaze in seconds (Wang, 2006). Air temperatures in a major wildfire can reach many hundreds of degrees Celsius, making survival or escape difficult or even impossible. Impacts include destruction of crops and buildings, damage to ecosystems, huge economic losses and societal disruption and the release of smoke that represents a flux of carbon and greenhouse gas to the atmosphere and which can be detrimental to visibility and human health.

Wildfires are uncontrolled fires in the natural environment, typically burning in areas of grassland, woodland, shrubland, forest or peatland. They include fires triggered by natural causes, fires resulting from accidental or malicious ignition and managed burns that get out of control (e.g. for land clearance or forest management). Wildfires occur on all continents except Antarctica; anywhere there is a source of fuel and conditions for combustion. The frequency, timing (e.g. seasonality), extent and intensity of fire in an area are collectively referred to as the fire regime, which can be described either qualitatively (e.g. Boer *et al.*, 2008; Chuvieco *et al.*, 2008) or quantitatively (e.g. Millington *et al.*, 2006).

Wildfire hazard has become increasingly prominent in recent years. Wildfires are an expensive natural hazard and costs appear to be increasing, largely due to increasing

Risk and Uncertainty Assessment for Natural Hazards, ed. Jonathan Rougier, Steve Sparks and Lisa Hill. Published by Cambridge University Press. © Cambridge University Press 2013.

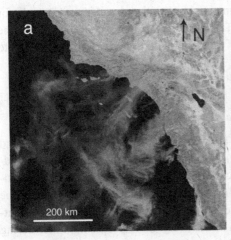
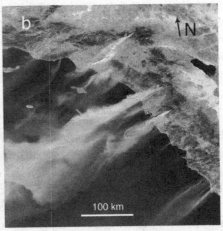

Figure 12.1 Two examples of smoke plume spread from Southern California wildfires. (a) A combination of visible and infrared electromagnetic observations from the Moderate Resolution Imaging Spectroradiometer (MODIS) on the Terra satellite from 26 October 2003. In the original false-colour image (see NASA, 2011a) it is possible to identify burned areas (brick red), smoke (blue) and the flaming fire fronts (bright pink spots) on the perimeter of the fires. Image courtesy of Jacques Descloitres, MODIS Rapid Response Team at NASA GSFC (NASA, 2011a). (b) European Space Agency ENVISAT image showing smoke plumes from the Southern California wildfires on 22 October 2007 (ESA, 2011a). Fires are an annual occurrence in California, driven by the seasonal Santa Ana winds (see Section 12.1.3). Scales are approximate.

vulnerability, but climate change may be having a role too (discussed in Section 12.1.3). For example, in January and February 2009 wildfires in Victoria, Australia were estimated to have caused at least AU$4 billion of damage, including AU$1.2 billion in insurance claims, and resulted in 173 deaths and destruction of 2133 houses (VBRC, 2009). Major wildfires also occur frequently in California; the autumn 2007 firestorms (Figure 12.1) are estimated to have cost US$1.8 billion in property damage (Karter, 2008), and the 2008 fire season resulted in losses of over US$2 billion (Munich Re, 2009). Fire service costs for suppression and preparedness activities also appear to be increasing (see Section 12.3).

This chapter discusses wildfires from the perspective of the assessment and effective management of hazard and risk, with a particular focus on remote-sensing techniques. The chapter starts with a summary of their causes, phenomenology and spreading mechanisms. Section 12.2 describes the basic types of data (and associated uncertainties) used to evaluate fire frequency and extent, and presents an overview of global wildfire incidence. Section 12.3 discusses the various impacts of wildfires on humans and the ecosystems. Section 12.4 describes current research challenges and tools used in the assessment of wildfire hazard and risk, including physical and empirically based models (12.4.1) and statistical models (12.4.2). Section 12.5 provides an overview of historical and current approaches to wildfire risk management (12.5.1), formal mechanisms for decision support, including ratings and indices for quantifying fire hazard or risk (Section 12.5.2) and public communication (12.5.4). Section 12.5.3 summarises operational and research tools for remote sensing and monitoring of wildfires. The chapter concludes with a summary of the primary research needs, and highlights the need for multidisciplinary collaboration to improve risk management capability.

12.1.2 Types, causes and drivers of wildfire

The major factors influencing fire behaviour are weather and climate, fuel properties and topography (Johnson, 1992; Flannigan 2005). Human activity is also significant, providing causes of ignition, fire suppression and changes in land use that can affect susceptibility or vulnerability to fire. Initiation of wildfires is more frequent during the daytime due to generally lower humidity, increased temperatures, increased wind speeds and heightened opportunities for ignition due to human activity (e.g. more people light fires during the day).

The majority of wildfires (particularly grassland fires) are started by human ignition: this includes accidental, malicious and managed fires that escape control. According to US National Interagency Fire Center data (NIFC, 2010a), 80–89% of wildfires in the United States from 2001–2009 were ignited by humans. However, 'natural' ignition sources such as lightening are important in sparsely populated ecosystems, where unchecked growth can occur. In boreal forests the most common cause of ignition is lightening strike, and this is significant as forest fire fuel consumption can be ten times higher than in grassland fires. For comparison, Ito and Penner (2004: table 8) provide estimates of dry mass fuel loads for grassland (giving an average of $0.37\,\mathrm{kg\,m^{-2}}$) and forest ($\sim 12\,\mathrm{kg\,m^{-2}}$). Analysis of experimental fires and wildfires in Canadian boreal forests gives an average value for forest floor fuel consumption of $\sim 3\,\mathrm{kg\,m^{-2}}$ (de Groot *et al.*, 2009: table 7). In boreal regions soil is a key fuel component and source of emissions (van der Werf *et al.*, 2010), and the forest floor constitutes a large proportion of stand biomass (de Groot *et al.*, 2009).

There are three primary types of wildfire:

(1) *Ground fires* (Figures 12.2a–12.2b), effectively 'below surface' fires, involve the combustion of roots or other underground organic matter, and can burn for days to months. These typically occur during periods of drought, when the organic soil layer is sufficiently dry. For example, large expanses of Indonesia's peatlands have been logged and drained, and are now affected by frequent fires. In Indonesia, most peat fires are deliberate burns ignited for land clearance, but an unusually long El Niño dry season in 1997 resulted in numerous fires in the Central Kalimantan region getting out of control and burning extensively (see Page *et al.*, 2002; Wooster and Strub, 2002; and further discussion in Section 12.3.2). Peat fires are very difficult to manage, as the fire spreads underground and can blaze up unexpectedly. Areas of major peatland burning where there is a high natural fire hazard as well as extensive managed burning include Borneo, Sumatra and West Papua. Drained wetlands are particularly vulnerable to ground fire (Figure 12.2a).

Figure 12.2 Types of fire. (a) ESA Envisat medium resolution imaging spectrometer image showing smoke plumes from peat and forest fires to the east of Moscow, Russia, on 29 July 2010 (ESA, 2011a). Extensive areas of drained peatland burned across Central Russia and around Moscow, following record high summer temperatures (over 35 °C) and drought (RIA Novosti, 2010). (b) Peat fire in a black spruce stand in interior Alaska, burned in 2004. The photo depicts *Sphagnum* moss hummocks that escaped combustion due to high soil moisture levels. Image (taken in 2005) courtesy of Merritt Turetsky, University of Guelph. (c) An understorey burn (surface fire) at Okefenokee National Wildlife Refuge, Georgia, United States, in 2003 (FWS, 2011). This was a controlled burn to clear vegetation. (d) Another controlled surface fire in the Florida Panther National Wildlife Refuge, Southwest Florida (FWS, 2011). (e) Trees torching, Tetlin National Wildlife Refuge, Alaska, United States (FWS, 2011). This is an example of a passive crown fire, primarily spreading as a surface fire but with trees occasionally torching. (f) A crown or canopy fire. Photo of the King County Creek Wildfire on the Kenai Peninsula, Alaska, United States, on 26 June 2005 (FWS, 2011). The fire produced a convection column about 2400 m high and burned approximately 4 ha before being extinguished (*Firehouse*, 2005; KPB, 2010). 1 km^2 = 100 ha.

(2) *Crawling or surface fires* involve the burning of any low-lying vegetation on or near the ground (e.g. timber, dropped leaves, grass; Figures 12.2c–12.2d). Surface fire models typically include all surface fuels (including small trees and bushes) less than 2 m high (Pastor *et al.*, 2003).

(3) *Crown or canopy fires* involve burning of aerial vegetation. These are the most severe type of forest fire and difficult to control because of their intensity (Figures 12.2e–12.2f). A surface fire may make the transition to crown fire depending on the surface fire intensity and crown characteristics (Van Wagner, 1977, 1993). If the fire front propagates at canopy height in addition to surface spreading, it is known as an *active crown fire*. In a *passive crown fire*, fire spreads largely as a surface fire with occasional torching of individual trees (Pastor *et al.*, 2003).

The term *ladder fire* is sometimes used to describe the transition from surface to crown fire, typically caused by burning of small trees, low and broken branches and vines.

12.1.2.1 Fuel

Wildfires require sufficient fuel (in a flammable condition), oxygen and an ignition source. The *total fuel load* is defined as the quantity of fuel (mass per unit area) that could potentially burn in a fire of the highest intensity and under the driest conditions (Byram, 1959; Scott and Reinhardt, 2001). *Available fuel* is defined as the amount of fuel actually consumed in a fire – this varies significantly and depends on factors such as fuel moisture and fire intensity. Weather conditions most conducive to fire are high temperatures, strong winds, low humidity and lack of rain. Rainfall is important as fires are unlikely to occur immediately after a few centimetres of rain even if it is hot, dry and windy; however, fuel dries at a negative exponential rate (Van Wagner, 1979) and a fire might occur in these same conditions a day or more later. For example, in the south-eastern United States, fires readily occur in high-temperature and high-humidity conditions following thunderstorms. Strong positive feedback effects can exacerbate the spread of wildfire. Hot air rising due to convection can produce intense winds at the fire front, drawing in more oxygen and fuelling the burn. Atmospheric feedbacks are therefore an important component of wildfire simulation.

The *rate of fire spread* is the rate of increase in area or linear dimension of the fire (Byram, 1959). Rate of spread generally increases with increasing wind speed, ground slope and amount of fine fuels. *Fine fuels* are defined as fast-drying dead vegetation (such as grass, leaves, needles and twigs), with a high surface area:volume ratio, and dimensions less than about 6 mm (Rothermel, 1972; Pyne *et al.*, 1996). This material provides the main fuel for surface fires. For a given fuel load (kg m^{-2}), the rate of burning increases with increasing fuel surface area, assuming there is sufficient oxygen supply (Byram, 1959). Quick-drying materials with high surface area:volume ratios (needles, leaves, twigs, etc.) typically provide the main sources of fuel in forest fires. As such, mature forests generally have larger fuel loads than younger forests. Some plant species ignite and spread fire more rapidly. Fires of sufficient intensity can pre-heat the fuels in advance of the flame front, increasing rates of spread. This is one of the reasons why fires burning upslope generally spread more rapidly. Certain species depend on fire to disperse seed and clear ground for new seedlings (e.g. eucalyptus), and promote fire spread, producing combustible oil, copious litter and long shreds of hanging bark or other 'ladder fuels' (Williams, 2002; see also Section 12.3.1 on the ecological benefits of wildfire). Weather and climate also affect the nature and availability of fuels (Section 12.1.3).

Regular fires prevent the excessive build up of fuel, removing brush and other ground vegetation that might otherwise accumulate and can increase the size, intensity and duration of subsequent fires (GAO, 2008). As a result, it is widely believed that excessive fire suppression can lead to fuel build up, and many forest management agencies use prescribed (controlled) burning to counteract this (see Section 12.5.1 on wildfire management). However, there is debate as to whether long-term extensive fire suppression necessarily leads to larger confla-grations in all types of wildland. Keeley *et al.* (1999) present an analysis of the California Statewide Fire History Database, comprising records from 1910, plus some earlier reports, and find no evidence for fuel build-up due to fire suppression resulting in larger and more destructive fires in brush-covered terrain. However, the effect is recognised in the coniferous forests of the western United States, and Keeley *et al.* (1999) conclude that effective fire management strategies will be different for different types of forest or wildland. Westerling *et al.* (2006) find evidence that on interannual to decadal scales, climate is a more significant driver of wildfire risk than land use and forest management.

12.1.3 Effect of weather and climate

Wildfires are influenced by climate, and are typically associated with prolonged periods of drought and/or heat waves (see, e.g. Westerling *et al.*, 2003; McKenzie *et al.*, 2004). For example, the extremely unusual April 2011 wildfires seen throughout many areas of the UK were associated with a period of very low rainfall, and persisted for over a month in many regions of the country. On a wider scale, climatic oscillations such as the El Niño Southern Oscillation (ENSO) events can have dramatic effects on rainfall in large parts of the tropics, which can lead to extensive wildfires. During the 1997 ENSO event, fires in Southeast Asia's tropical forests resulted in economic costs of around US$9 billion, and in the same episode of drought, more than 20 million hectares were destroyed by fire in Latin America, with a loss of US$10–15 billion (Bowman *et al.*, 2009). There is therefore a growing perception that wildfires are linked to changing regional climates and extremes in weather, perhaps exacerbated by climate change that is expected to alter the fire regime of many areas (e.g. Flannigan *et al.*, 2000). Bowman *et al.* (2009) use sedimentary charcoal records to comment on potential longer-term links between climate and fire activity in the earth's past.

Climate controls the type and distribution of vegetation, and affects fire activity across a wide range of spatial and temporal scales (Trouet, 2009). Timescales of hours to days relate to variations in precipitation and wind speed that affect fire intensity and spread on a local scale. Inter-annual and seasonal variations affect growth and flammability of vegetation over wide areas (e.g. episodes of drought). High temperature and low humidity affect fuel moisture and increase the chance of ignition and spread (Pyne *et al.*, 1996). Local topography affects both the microclimate and the ability of wildfires to spread. The effect of synoptic-scale circulation weather patterns related to atmospheric phenomena such as the ENSO and the Pacific Decadal Oscillation (PDO) can be identified in fire datasets and tree-ring records (see, e.g. Westerling *et al.*, 2006; Trouet, 2009). Swetnam and Betancourt (1990) and others have shown that ENSO affects the frequency and extent of wildfires, with, for example, dry spring weather being associated with the high-SO phase (la Niña), resulting in increased fires in the south-western US states. Conversely, Wooster *et al.* (2011) find that the low-SO phase (el Niño) is associated with drought in areas of Southeast Asia, which, when accompanied by human ignition sources, results in severe forest and peatland burning. Balling *et al.* (1992) identify a positive correlation between the Palmer Drought Severity Index and fire season severity.

Wind is a major factor affecting fire behaviour (Beer, 1991). It drives the fire (the *fire behaviour impact*), increasing the intensity and governing the speed and direction of spread. Hot embers carried by the wind can cause the fire to 'jump' across natural or engineered firebreaks; this is a particular problem in crown fires and is known as 'spotting'. Catastrophic wildfires in brush-covered regions such as California are often driven by high winds (Figure 12.1). The Santa Ana winds are strong, warm and dry adiabatic winds that blow from deserts east of the Sierra Nevada to the coast of Southern California. Wind speeds can exceed $100\,km\,hr^{-1}$, and under such conditions fire spread is much less dependent on fuel structure (type and age), and unlikely to be mitigated by fuel reduction strategies such as prescribed burning or thinning (Keeley *et al.*, 1999). Wind also acts to dry the fuel (the *fire danger impact*), increasing the likelihood of conditions favourable to fire occurrence and spread, and is an important component of certain agencies' fire indices (e.g. the Canadian Forest Service Fire Weather Index – see Section 12.5.2).

Fire-atmosphere interactions are also crucial in governing fire behaviour. Buoyancy effects can generate strong local winds as air is drawn into the convecting column, which can interact with the prevailing wind (Beer, 1991; CSIRO, 2008). Rotational updrafts can produce vortices of smoke or flame (see Figure 12.3; Umscheid *et al.*, 2006). Dold *et al.* (2005) and Weber and Dold (2006) report the observation of fire-generated cumulonimbus clouds (or pyro-cumulonimbus – Fromm *et al.*, 2005) and whirlwinds in a bushfire in Canberra, Australia on 18 January 2003. One particularly strong whirlwind travelled over 16 km, felling trees in its path. The interaction between two nearby fires can also have a major influence on the fire behaviour (Finney and McAllister, 2011).

12.2 Scale, frequency and areas affected

12.2.1 Data sources

Knowledge of past wildfires is largely based on the following sources of information:

(1) *Detailed studies of directly observed, discrete events*. This can include instrumentation on the ground, airborne or remote sensing and direct visual observation, all of which provide different levels of completeness, resolution and accuracy. Ground-based observations are

Figure 12.3 A fire whirl or 'fire tornado', a vortex of flame produced by strong rotation in the convective column (e.g. see Umscheid *et al.*, 2006). This photograph was taken during prescribed burning of fields in Tule Lake NWR, California (FWS, 2011).

relatively sparse, requiring greater human power and resources. Remote sensing enables wider coverage, but can suffer from limitations related to resolution (spatial and temporal) and classification accuracy. Frequently measured parameters include wildfire burned area, hotspot (active fire) location and ecological effects such as vegetation mortality and changes in soil characteristics (e.g. see Certini, 2005).

(2) *Inferences about fire activity based on indirect observations or measurements of combustion products.* Trace gases and aerosols are released into the atmosphere in copious quantities by wildfires (van der Werf *et al.*, 2006). The main trace gas released in wildfires is carbon dioxide, which has a long lifetime and a relatively high atmospheric abundance. Smoke contains dozens of other lower concentration species (Andreae and Merlet, 2001), and as with CO_2, for some of these species wildfires represent one of the major global sources. Direct sampling or remote sensing is generally used to identify the presence of such species. Very long-range transport of smoke is a noted phenomenon, particularly if the material has been injected into the stratosphere through extreme convection. For example, see Fromm *et al.* (2005) and Jost *et al.* (2004), who identified carbon monoxide and particulate matter from forest fires up to altitudes of 15.8 km.

(3) *Wildfire inventories or databases* that consist of many records of historical wildfires in a given area and over a given time period (by aerial photography, remote sensing and again direct observation). Examples include the European Space Agency's ATSR World

Fire Atlas (ESA, 2011b), which provides near-real-time hotspot data and monthly global fire maps from 1995 to the present; and the Canadian Large Fire Database (CLFDB), which comprises forest fire data for fires over 200 ha, used for spatial and temporal analyses of large-scale fire impacts (see, e.g. Stocks *et al.*, 2002). Russell Smith *et al.* (1997) use Landsat Multispectral Scanner (MSS) imagery to reconstruct a 15-year fire history (albeit with uncertainties) for the Kakadu National Park, northern Australia.

(4) *Wildfire activity as interpreted from palaeo records*, including dendrochronology (Heyerdahl *et al.*, 1995) and studies of charcoal deposits (Power *et al.*, 2008; Scott, 2009).

To a large degree, much of the current reporting of wildfire scale, frequency and area affected comes from inventories and databases of historical wildfires compiled by local, regional, national and international agencies. Depending on the inventory or database, there are many different thresholds for inclusion, and different attributes of wildfires that might be recorded. Records typically include: ignition date and time; extinction date and time; ignition spatial location; fire cause or causes; total burned area, energy released or pixels affected (from remote-sensing data); spatial outline of the final burned area, including in some cases: unburned areas; predominant vegetation types; land-use type and ownership; meteorological conditions; effectiveness of fire management strategies; and fire type (single, re-ignition). However, the completeness of wildfire inventories can vary greatly not only across different countries, but also within a country or region and over different time periods. Any given region might only record fires above a given size, leaving out the many small fires that occur, or only record those fires that have a direct monetary loss or cause fatalities. For example, some US government agencies report any fire with burned area <1 acre (<0.00405 km^2) as '0 acres', whereas others might report a small fire as '0.1 acre' or '1 acre' (see Brown *et al.*, 2002; Malamud *et al.*, 2005). In addition, small-scale fires in remote or unpopulated areas are likely to be undetected and unreported. The completeness of inventories can also change in any given region as a function of changes in technology or the staffing of government agencies. For example, in a study of US wildland fire records, Brown *et al.* (2002) found that between 71% of US Forest Service and 90.7% of US Department of the Interior records were deemed complete or 'useable'. In this section, we report the scale and frequency of fire globally, in Europe and in the UK, but recognising that there are some limitations inherent in the inventories used to compile the broader conclusions reached. In Section 12.4.2 we discuss more fully the frequency–size statistics of wildfires.

12.2.1.1 Palaeo and geological records

Past fire history derived from geological, palaeo-ecological and archaeological records provides a major information resource. Such records commonly involve study of macro-scopic and mesoscopic charcoal assemblages, and application of a range of modelling techniques with differing limitations and uncertainties. A detailed review of methods for reconstructing past fire regimes is given by Conedera (2009). Many of these palaeo- and geologic records provide a long-term perspective on fire frequency and size in different kinds of environment, especially for the more extreme events that have a greater chance of being preserved in stratigraphic records.

Palaeo-ecological methods are used to determine time-series of fire history and recurrence rates in the recent past. Techniques include dendrochronology to determine forest age and identify fire scars on trees, and identification of charcoal deposits in lakes, bogs, soils and alluvial sediments (e.g. Clark, 1990; Power *et al.*, 2008). The majority of chronologies based on tree fire scars typically extend back 500 years or so, but are biased towards surface fires that damage but do not kill trees (Whitlock, 2004; Whitlock *et al.*, 2004). However, in some long-lived tree species it is possible to extract chronologies extending over 1000 years. For example, Heyerdahl *et al.* (1995) constructed a fire history database of the western United States based on over 100 authors' published and unpublished tree-ring reconstructions for fire regimes in forested areas, 400 BCE to 1900 CE. Using this study, Malamud *et al.* (1998) provide an analysis of the frequency–size statistics of burned areas for the period 1150–1960 CE, finding an inverse power-law distribution in agreement with shorter-term but more complete recent records from the last century (see Section 12.4.2). In more recent work, Brown and Schoettle (2008) present a history based on fire-scar and tree-recruitment chronologies in Limber pine and Rocky Mountain Bristlecone pines in Colorado, with the earliest specimen dating from 780 CE. The oldest recorded Rocky Mountain Bristlecone pine dates from 442 BCE, and Great Basin Bristlecone pines are reported to achieve ages of ~4000 years (Brunstein and Yamaguchi, 1992). Charcoal records from lake sediments can extend back several millennia, but these studies typically focus on montane and subalpine forest (Whitlock, 2004). Fire regimes are also studied in deep time before the involvement of man (Scott, 2000, 2009).

12.2.2 Global extent

Biomass burning, which includes all wildfires and deliberate, controlled burns (e.g. for land clearance), occurs widely across almost all continents on an annual basis, including savannah, peatland, tropical and boreal forests (van der Werf *et al.*, 2010). Using multiple remote-sensing datasets, Giglio *et al.* (2010) estimate that global burned area for the years 1997–2008 varied between 330 and 430 million hectares (3.3–4.3 million square kilometres). The most frequent wildfires and the largest areas burned are in Africa and Australia. Van der Werf *et al.* (2010) estimate that during the period 1997–2004, globally, 41% of the total area burned was in Northern-Hemisphere Africa, 23% in Southern-Hemisphere Africa and 14% in Australia. However, uncertainties in area estimates vary between landcover classes, and are often greatest in regions of dense forest and where peat fires occur, largely due to problems in mapping these types of burned area from satellite imagery (e.g. the burns can be very small, and sometimes occur below a closed forest canopy; Roy *et al.*, 2008). Fire return intervals vary for different areas (Chuvieco *et al.*, 2008), and depend, for example, on the plant species, landscape, climate and ignition frequency

Table 12.1 *Number of fires and burned area in Portugal, Spain, France, Italy and Greece (1980–2009)*

Number of fires[1]	Portugal	Spain	France	Italy	Greece[2]	Total
Average 1980–1989	7381	9515	4910	11 575	1264	34 645
Average 1990–1999	22 250	18 152	5538	11 164	1748	58 851
Average 2000–2009	24 949	18 337	4406	7259	1569	56 645
Average 1980–2009	18 194	15 335	4951	9999	1569	50 047
Total (1980–2009)	545 805	452 848	148 531	299 977	47 058	1 501 409
Burned areas (ha)						
1980–1989	73 484	244 788	39 157	147 150	52 417	556 995
1990–1999	102 203	161 319	22 735	118 573	44 108	448 938
2000–2009	150 101	125 239	22 342	83 878	49 238	430 798
1980–2009	108 956	177 115	28 078	116 534	48 587	78 910
Total (1980–2009)	3 257 886	5 313 457	842 332	3 496 005	1 457 624	14 367 304

Notes
[1] Data given by the JRC (2010) do not indicate the minimum fire size considered by each country, so number of fires should be considered with care.
[2] Provisional data for 2009.
Source: JRC (2010).

Taking the United States as an example, an average of 2.8 million hectares per year are burned by wildfires, with an average of 78 522 recorded fires per year (data for 2000–2008 – NIFC, 2010b). However, the total number of fires reported is problematic as a reliable number, as it is greatly influenced by the completeness of the record and the minimum fire size recorded in the database (see Sections 12.2.1 and 12.4.2). Within Europe, wildfires are most frequent and extensive (in terms of total area burned) in the Mediterranean region. The annual average area burned in this region for the countries of Portugal, Spain, France, Italy and Greece was approximately 480 000 ha for 1980 to 2009 (Table 12.1; JRC, 2010), although there is significant yearly variation due to climate. Roughly 36% of the total recorded fires occur in Portugal and 30% in Spain (JRC, 2010). The wildfires in Portugal in 2005 were associated with the worst drought in the region for several decades and provide a prominent example: 300 000 ha were burned, 20 people died as a direct consequence (including 12 firefighters) and 100 people were arrested on suspicion of arson (Viegas, 2006).

Data sources for wildfires include the European Commission annual reports on forest fires (see Camia *et al.*, 2008; JRC, 2010) and the United Nations Economic Commission for Europe statistics on forest fires in the UNECE region (incorporating Europe, North America and Commonwealth of Independent States countries – see UNECE, 2011). The European Commission Joint Research Centre also provides online tools for extracting

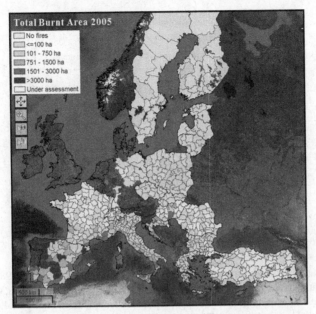

Figure 12.4 Example of the European Forest Fire Information System (EFFIS, 2011) online tools for examining fire statistics for selected European countries. Shown here is the total wildfire burned area as a function of administrative province, for the year 2005. A close-up figure for Spain and Portugal is shown in Figure 12.5, including the statistics by year for the north-east administrative region of Portugal.

wildfire statistics spatially and temporally for selected countries, from 1980 to the present (EFFIS, 2011). An example showing burned area as a function of each country's administrative area is shown in Figure 12.4 for the year 2005, where the large wildfires in Portugal and Spain are clearly evident in the lower left-hand corner. Figure 12.5 shows the annual burned areas for the north-east administrative region of Portugal, Tras-os-Montes, 1980–2008, showing a strong annual variability, similar to most other regions where wildfires occur.

The UK normally has a relatively low fire risk due to its maritime climate, relatively high rainfall and no real 'dry season'. Nonetheless, Gazzard (2009a) report that there are an estimated 95 000 wildfire incidents per year in the UK. Most moorland fires in the UK are caused by humans, and lightning strike ignition is rare (Albertson *et al.*, 2009). Burning of forest and woodlands, in addition to UK heathlands that do already occasionally burn (Scott *et al.*, 2000), is likely to have significant economic and health consequences.

The UK Department of Communities and Local Government provide data on fire incidents, including wildfires (DCLG, 2011). However, McMorrow and Legg (2009) and Gazzard (2009b) remark that current UK data on vegetation fires are poor, with limited information on the nature of the fire (e.g. type of vegetation, extent of burn), impact, response and recovery and operational effectiveness. As a result of the lack

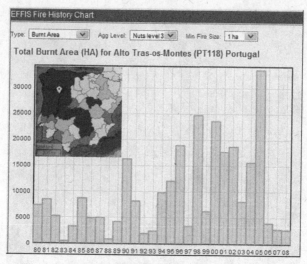

Figure 12.5 Variation of total burned area in hectares (*y*-axis) as a function of year, 1980–2008 (*x*-axis) for the administrative area of Tras-os-Montes, Portugal, for all fires greater than 1 ha, using the EFFIS (2011) online tools. The inset figure in the upper left-hand corner is from Figure 12.4, burned area for the year 2005, with the same legend as Figure 12.4. The location of Tras-os-Montes is indicated by the four-sided polygon in the inset figure.

of coherence in data collection and reporting, the Forestry Commission have produced the UK Vegetation Fire Standard to provide a framework for data capture (Gazzárd, 2009b).

12.3 Impact and cost

The major impacts of wildfire are loss of life, injury, air pollution with related health consequences, destruction of natural environments, financial losses and costs of fire-fighting and prevention services. The diversion of resources away from normal fire-fighting operations is also a significant impact. However, wildfires can have benefits for some ecological systems (see Section 12.3.2), such as by promoting species diversity. Economic impacts include destruction of property, crops, loss of livestock and disruption of facilities, the costs of fighting the wildfire and the cost of reconstruction and recovery. There are also political impacts, as exemplified by recent wildfires in Australia (Commonwealth of Australia, 2010) and Greece (see Xanthopoulos, 2004; and the news report by Attewill, 2007), where perceptions have developed that governments have not invested in sufficient prevention measures and the adequacy of the emergency responses have been questioned. Wildfires also have potentially large implications for the carbon cycle (Section 12.3.3).

12.3.1 Economic and societal impact

In the UK, heath, moor and forest fires have implications for water quality, as reservoirs are small and vulnerable to catchment disturbance. However, forest and heath fires are not currently part of water resource management strategies. Effects include increased erosion and transport of nutrients that can result in eutrophication in downstream aquatic systems, toxic algal blooms, fish kills and pollution of potable water (Blake, 2009; Blake *et al.*, 2009). Wildfire can also affect environmentally sensitive areas. In July 2006 a fire destroyed much of the Thursley Nature Reserve in Surrey, a home to unique wildlife, and smoke impacted upon local and regional transport routes (A. Scott, pers. comm.). Unlike countries such as Australia and the United States, the UK does not currently cost the local economic impacts of wildfire (FIRES, 2009). For example, McLennan and Birch (2006) estimate that between 1966 and 1999 the cost of Australian wildfires corresponded to about 10 per cent of the total cost of all natural disasters for Australia.

In the United States the annual costs of fire suppression are increasing (GAO, 2009; see Figure 12.6), with a general trend over the period 1996–2007 from US$1 billion to US$3 billion (adjusted to 2007 dollars). Prior to 1994 annual expenditure (net of inflation) did not show any significant increase (Schuster *et al.*, 1997). However, fire loss data (like incidence and area burned) are also subject to uncertainties associated with record

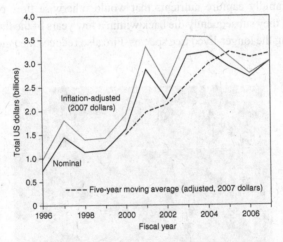

Figure 12.6 US Wildland fire appropriations (allocated funds) for 1996–2007. Values represent the total budgeted spending on wildfire preparedness and suppression activities for the US Forest Service and Department of the Interior (Bureau of Land Management, National Park Service, Bureau of Indian Affairs and US Fish and Wildlife Service). The dark black line indicates the actual amount budgeted each year, and the thinner grey line the equivalent values adjusted for 2007 dollars accounting for inflation. The five-year moving average (dashed line, also adjusted to 2007 dollars) shows an increase in average annual appropriations for all wildfire-related activities (fire suppression and preparedness) from US$1.2 billion for the period 1996–2000, rising to just over US$3 billion for 2003–2007. Data from US General Accounting Office Statement GAO-09–877 (GAO, 2009: table 1).

incompleteness and variation in level and accuracy of reporting. Procedures for recording fire expenditure have changed over time, making it difficult to compare records over long time periods, and different agencies employ different approaches.

12.3.2 Ecological benefits of wildfire

Natural wildfires have ecological benefits. They can support habitat diversity, controlling the spread of insects and disease, recycling nutrients and promoting new and sustainable growth of fire-dependent and resistant species. A number of species have developed adaptations to survive fire, or need fire to propagate (see Williams, 2002; Bond and Keeley, 2005). For example, the ponderosa pine has thick bark and tall crowns to resist fire, and re-establishes quickly post-fire. Certain species of pine have serotinous cones, and need fire to melt the resin and disperse seed, e.g. the lodgepole pine. Cones remain on the tree and require temperatures in excess of 45 °C to release seed. Lodgepole pine forests require fires to occur at least every 200–300 years to prevent succession by other species (Perry and Lotan, 1979). In other species seed germination is prompted by the effects of fire (Fenner, 1992). Some plants regenerate by sprouting from roots protected from heat underground, or take advantage of the fire clearance to establish seedlings (Figure 12.7). Many early successional plants, whose germination is prompted by fire, play an important role because they are ephemeral (e.g. see Wright, 1974; Nilsson and Wardle, 2005). The early coloniser plants rapidly capture nutrients that would otherwise flow off the site due to rainfall, and when they subsequently die back within a few years of the fire, the nutrients are released, benefiting the longer-lived tree species. Fire also reduces the amount and vigour of

Figure 12.7 'Fireweed' (*Epilobium angustifolium* or Rosebay Willowherb) re-colonising ground after a fire, Kenai Peninsula, Alaska (FWS, 2011). Fireweed (found extensively in the temperate Northern Hemisphere) can regenerate from buried rhizomes reaching tens of centimetres deep into the soil, and spreads rapidly due to wind dispersal of light, fluffy seeds (Foster, 1985; Landhäusser and Lieffers, 1997). Seeds can also remain viable in the soil until conditions are suitable for germination (prompted by light).

broad-leaf saplings during the regeneration phase. This is important to the regeneration of conifer species, creating space while maintaining protection for the conifer seedlings. The protection comes in the form of lower surface temperatures and reduced evaporation during times of heat and drought. Since the broad-leaf trees are shorter-lived, over time they also become a source of nutrients for the conifers. Thick-bark tree species (such as Douglas fir) are more resistant to fire, and mature trees often survive wildfires. Conflicts with human society arise where development takes place in areas prone to natural wildfires (see Guyette *et al.*, 2002).

12.3.3 Carbon emission

Fire has consequences for the global carbon cycle and atmospheric chemistry (Kasischke *et al.*, 1995; Cofer *et al.*, 1997; Simmonds *et al.*, 2005; Power *et al.*, 2008). Van der Werf *et al.* (2010) estimate that globally an average of approximately 2.0 gigatonnes of carbon is released annually from biomass burning, although this estimate includes controlled burns (largely for land clearance) as well as wildfires. They also note large interannual variability in the carbon release from fires (of more than $1\,Gt\,yr^{-1}$) due to climate effects (see Section 12.1.3). An ENSO-related drought resulted in a prolonged and extremely severe fire season in Southeast Asia in 1997–1998, with extensive wildfires resulting from deliberate burns getting out of control. Page *et al.* (2002) estimate that in Indonesia in 1997 between 0.81 and 2.57 gigatonnes of carbon were released to the atmosphere as a result of burning peat and vegetation (deliberate and uncontrolled burns), equivalent to 13–40% of the mean annual global carbon emissions from all fossil fuel combustion. Wooster *et al.* (2011) present fire maps derived from a series of AVHRR Global Area Coverage datasets (processed using the methodology of Wooster and Strub, 2002), and show that fires were concentrated in the low-lying areas of East, Central and South Kalimantan, rather than being equally spread across the whole island.

Wildfire processes form an important component of dynamic global vegetation models (Krinner *et al.*, 2005; Prentice *et al.*, 2007), and are significant in terms of vegetation–climate–atmosphere feedbacks in global climate models. An analysis of the total effects of gas, aerosols, carbon deposition and albedo changes due to boreal forest fires is given by Randerson *et al.* (2006). They conclude that in that particular case, the long-term net effect is a decrease in global radiative forcing, indicating that if this was true across all such events future increases in boreal fire would not accelerate 'global warming'.

12.3.4 Human health risks

Burns, suffocation and building collapse can result in injury or death. However, other impacts of wildfires on human health have only more recently been assessed. At the fire front, where air temperatures can exceed 300 °C, radiant heat is the greatest threat. Dehydration and heat exhaustion are well-recognised health risks for firefighters

(Johnston, 2009). Johnston *et al.* (2007) identify an increase in hospital admissions for respiratory conditions in areas frequently affected by smoke from wildfires in Darwin, Australia. Exposure to smoke can cause major health problems, with effects ranging from eye and respiratory tract irritation to more serious disorders, including reduced lung function, bronchitis, exacerbation of asthma and premature death (OEHHA, 2008). Smoke particles are typically small, in the range 0.4–0.7 μm, and are therefore largely within the fine particle ($PM_{2.5}$) fraction that can be inhaled into the deep lung (OEHHA, 2008). Carbon monoxide is also produced, with relative concentrations being highest during the smouldering stages of a fire. The composition of smoke is variable, and depends on a number of factors (fuel type, age, moisture, weather conditions, etc.); however, respiratory irritants such as acrolein and formaldehyde and the carcinogen benzene are typically present – albeit in low concentrations. The impact of the smoke plume (area affected, exposure time, etc.) will be highly dependent on the interaction of winds and the terrain, as well as fire size, intensity and duration.

Heavy metals such as mercury are also a problem as fire aids their release into the atmosphere. Since the industrial revolution, atmospheric transport of mercury from localities such as the American Midwest has led to contamination of peatlands and forests to the north (e.g. see Givelet *et al.*, 2003), although complex biogeochemistry makes this difficult to quantify (Fitzgerald *et al.*, 1998). These cold, wet and frozen regions have largely acted as sinks for heavy metals; however, changes in climate and water levels has led to an increased incidence of fires, releasing pulses of methyl mercury to the atmosphere. Sigler *et al.* (2003) estimate that annual mercury emissions released by burning Canadian boreal forests could be as high as 30% of industrial emissions, and much greater in severe fire years; a significant contribution to the global mercury budget (see also Turetsky *et al.*, 2006).

12.3.5 Across hazards: triggers and implications

Secondary effects of wildfires include propensity for erosion, landslides and debris flows, flooding, introduction of invasive species and changes in water quality (USGS, 2006). Post-fire erosion is related to the size of the wildfire, the intensity and severity of the burn and post-fire weather conditions (Moody and Martin, 2001; Robischaud and Elenbeer, 2001; Graham, 2003; Shakesby and Doerr, 2006; Moody and Martin, 2009a), and is typically more severe in wetter climates (Moody and Martin, 2009b). Sediment transport can lead to excessive nutrient loads, causing eutrophication and affecting water quality. Post-fire flooding can also result in pollution of reservoirs, as was shown in the Hayman Fire of 2002 in Colorado, where the water supply to Denver was affected (Graham, 2003). Forestry and environmental planners need to be able to predict and understand these effects. Shakesby and Doerr (2006) provide a review of hydrological and geomorphological impacts of wildfire, and emphasise the need for a fire severity index based on critical soil changes, rather than the standard approach measuring biomass destruction. Further short-term impacts include destruction of timber, forage and wildlife habitats. Longer-term effects include destruction of buildings and infrastructure, reduced access to recreational areas and resulting impacts on tourism and loss of cultural, environmental and economic resources (USGS, 2006).

12.4 Hazard and risk assessment

Wildfire is a complex process on all scales, from the scale of the combustion process where fuel chemistry and fluid dynamics dominate, to larger scales where convection, weather conditions, landscape, topography, fuel properties and human intervention control behaviour and extent of the burn. All these factors are highly variable and pose difficult challenges in forecasting future fire risk and behaviour.

Flannigan *et al.* (2005) emphasise the need to understand the complex feedbacks between fire, climate, biosphere and human activity in order to understand risk. The record dry summer of 1988 resulted in 248 fires in the Yellowstone region (50 in the National Park), leaving a total of 486 000 ha scorched and 321 000 ha burned (NPS, 2009). The cost of firefighting was around $US120 million. Similarly the 2005 wildfires in southern Europe that particularly affected Portugal and southern France occurred during the most severe summer drought and heat wave for decades. Future fire risk is likely to be strongly affected by climate change (see, e.g. Marlon *et al.*, 2009). In a globally warmer climate, many areas are predicted to see more severe dry weather, potentially resulting in greater areas burned, more fires and longer fire seasons (Flannigan *et al.*, 2000).

12.4.1 Assessment using empirical and physics-based models

Fire is a multi-scale, multi-physics problem. The complex combustion chemistry, highly variable nature of the fuels and the vagaries of weather conditions together make wildfire simulation modelling computationally challenging. Operational wildfire models used to predict fire spread and behaviour are largely empirically based. Such models are employed by fire services and land management agencies to inform planning and decision-making on issues such as fire suppression, risk management, issuing of warnings and evacuation. Empirical models rely on data from real fires, and are therefore constrained by the data used to construct the model, so their applicability will be limited with extrapolations and generalisations quite likely to be misleading. Semi-empirical models apply conservation of energy, but generally don't distinguish between conductive, convective and radiative heat transfer. Anthropogenic and natural changes in climate and other factors that influence fire behaviour limit the utility of operational models based on historic data. There is therefore a need to further develop physical and quasi-physical models for operational applications, and to provide more robust, accurate and practical forecasts of future fire activity (Sullivan, 2009).

Most empirically based fire spread forecasting models assume quasi-steady-state behaviour, and predict the medium- to long-term (0.5–6 hr) mean behaviour and forward rate of spread, given mean estimates of fuel and weather conditions (Sullivan, 2009). On these timescales, predictions are useful for planning and public warning. The key inputs (the most significant being fuel and wind) have high spatial and temporal variability. Even when good-quality observational data are available, fire behaviour can deviate significantly from predictions, particularly on short timescales and over small spatial scales. As a result, such predictions, although they do have their uses, must be carefully assessed and tensioned against the complexity and reality by firefighters on the ground.

Two-dimensional fire growth models such as FARSITE (used by the US National Park Service and Forest Service) and Prometheus (the Canadian wildland firegrowth model – see Nunes *et al.*, 2008; Tymstra *et al.*, 2010) use the Huygens principle of wave motion to model the propagation of the fire front. Some empirical fire spread models incorporate information from numerical weather prediction models to provide inputs such as wind velocity. The Missoula Fire Sciences Laboratory Fire Models website (MFSL, 2011) provides compre-hensive information on operational fire models used by US authorities (including FARSITE, fire weather and danger indices), and a number of research tools that are under development to simulate varying wind fields and the effects of topography.

Physically based models capture the physics, and in some cases also the chemistry of fire spread, and generally involve complex, high-resolution fluid dynamics. One example is the physics-based, three-dimensional coupled fire–atmosphere model for grassland fires of Mell *et al.* (2007), developed by the US National Institute of Standards and Technology. This 'Fire Dynamics Simulator' (FDS, 2011) uses Navier–Stokes equations appropriate for low-speed, thermally driven flow, to simulate smoke and heat transport. Such models can be used to investigate fire behaviour that cannot be captured by empirical models; for example, to evaluate the effectiveness of fuel treatments, the mechanisms and conditions that lead to blow-up (rapid increase in intensity), and fire spread through heterogeneous fuels. A model with similar characteristics to FDS is FIRETEC, a three-dimensional coupled atmosphere–fire model developed by Los Alamos National Laboratory for improved forecasting of fire behaviour under variable meteorological conditions and on complex terrain (Linn *et al.*, 2002; Cunningham and Linn, 2007). Similarly, NCAR has developed a coupled atmos-phere–wildland fire–environment model that employs the Clark–Hall atmospheric predic-tion model coupled to a wildfire model. As the fire releases heat and moisture into the atmosphere, strong winds develop near the fire front, which in turn influences the rate and direction of spread. Changing atmospheric conditions affect the shape of the fire front, including sudden outbursts of flame and formation of vortices. NCAR are currently devel-oping a wildfire model component for their weather research and forecasting model (see Grell *et al.*, 2005; WRF, 2011). Sun *et al.* (2006) attempt to validate both the FDS and the Clark coupled wildfire model (Clark *et al.*, 1996) by comparing model simulations of fire plume behaviour with experimental data.

A wildfire is a complex system that involves transitional behaviour (e.g. from a surface to crown fire or across firebreaks) and both horizontal and vertical variation in critical parameters such as fuel structure and composition. Various models have been developed to simulate different aspects of fire behaviour, and a review is given by Pastor *et al.* (2003).

Butler *et al.* (2004) present a physically based crown fire spread rate model that simulates steady-state fire spread dominated by radiative heat transfer. The model currently captures the relative response of fire spread to fuel and environment variables, but over-predicts the magnitude of the spread rate. Sullivan and Knight (2008) have developed a 'hybrid' two-dimensional cellular automata model that attempts to capture the effects of convection and wind. Cell-based rules for the spread of fire through a fuel layer are combined with a semi-physical model of convection above the fire, and used to model the fire perimeter. The authors believe that with development, this type of model could eventually be used to give insight into wildfire behaviour, e.g. coalescence and response to extreme fire weather conditions. Mandel *et al.* (2007, 2008) present a methodology for using data assimilation with a simple, partial differential equation-based model for wildland fire, to enable prediction in 'faster than real time' for operational use. They remark that the next challenge is to incorporate coupling with an atmospheric model to better represent important feedbacks.

12.4.2 Assessment using statistical models

12.4.2.1 Gaussian and heavy-tailed distributions

Here the statistics of wildfires based on fire inventories are considered. Frequency–size distributions are fundamental for a better understanding of models, data and processes when studying any natural hazard, particularly for a better understanding of risk, but also as a quantitative characterisation of the wildfire regime. Frequency–size distributions characterise the probability of a given size event (e.g. fire burned area, fire released energy) occurring, and are commonly based on data from natural hazard inventories, but can also be based on or related to empirical equations. The frequency–size statistics give an idea of how many small vs. medium vs. large events might occur in a given region and over a given period of time.

A standard approach for frequency–size analyses in much of the social and physical sciences has been to assume that the frequency–size distribution of all values in a given time-series approaches a Gaussian (normal) distribution, i.e. that the central limit theorem applies. Often this approach is appropriate and provides correct statistical distributions. However, in many other cases, particularly for wildfires, it is necessary to consider a broader range of statistical distributions, such as log-normal, Gumbel, Weibull or power-law (Stedinger *et al.*, 1993; Evans *et al.*, 2000; White *et al.*, 2008). An example of two strongly contrasting distributions, the Gaussian and the power-law, are shown in Figure 12.8.

12.4.2.2 Wildfire frequency–size statistics

The frequency–size statistics of wildfires have been shown to follow inverse heavy-tailed, non-exponential distributions, such as power-laws (Millington *et al.*, 2006; Cui and Perera, 2008). Studies of wildfires in many countries have found non-cumulative frequency–size distributions that are inverse power-laws over several orders of burned area, with power-law

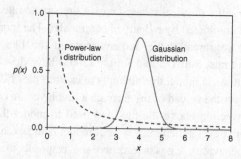

Figure 12.8 Example of the probability of an event occurring $p(x)$ as a function of the size of the event x. Two contrasting non-cumulative frequency–size distributions are shown. A Gaussian distribution (mean 4.0, standard deviation 0.5) and an inverse power-law distribution (exponent 1.2). The frequency–size statistics of wildfire sizes are strongly non-Gaussian, following power-law or other heavy-tailed distributions.

exponents ranging from 1.1 to 2.1 (e.g. Malamud *et al.*, 1998, 2005; Ricotta *et al.*, 1999, 2001; Niklasson and Granstrom, 2000; Song *et al.*, 2001; Ward *et al.*, 2001; Zhang *et al.*, 2003; Turcotte and Malamud, 2004; Jiang *et al.*, 2009). Others (e.g. Reed and McKelvey, 2002; Schoenberg *et al.*, 2003) have found different types of strongly non-Gaussian distributions, such as truncated power-laws and Weibull distributions. What is common to all of these frequency–size distributions is that there are few large wildfires, more medium ones, and many small ones.

An example using $N_T = 7422$ wildfires originating on United States Forest Service (USFS) lands in California for the 37-year period 1970–2006 and with burned areas $A_F \geq 0.012 \, \text{km}^2$ is given in Figure 12.9. Figures 12.9a (linear–linear axes) and 12.9b (log–linear axes) display burned area of each wildfire as a function of time; an annual periodicity in the wildfire season is evident, with wildfires ranging in size from $0.012 \, \text{km}^2$ (the smallest values considered here) to $1130 \, \text{km}^2$. Figures 12.9c (linear–linear axes) and 12.9d (log–log axes) show the frequency densities $f(A_F)$ of the wildfires as a function of burned area, A_F. The frequency density $f(A_F) = n/\delta$, with n the number of values in a given bin $A_F \pm 0.5\delta$, and δ the width of the bin. An inverse power-law distribution with exponent 1.46 provides a fairly robust fit to the frequency densities $f(A_F)$ as a function of burned area A_F. There is a slight 'roll-over' at the largest wildfire burned areas. In this example, typical of most wildfire distributions, the largest 75 wildfires (just 1% of those considered) account for 57% of the total burned area. If we had considered smaller wildfires in the inventory, then there would be more wildfires in total, and the largest 1% of the fires would therefore account for a larger percentage of the total burned area.

The total number of wildfires N_T will get larger as the increasing completeness of a fire database accounts for smaller and smaller fires. The total number of wildfires reported is very sensitive to the completeness of the inventory. For instance, for a region characterised by a frequency–size distribution for burned areas that is an inverse power-law with an exponent of 1.5, then reporting all wildfires with an area greater than 0.1 ha will give

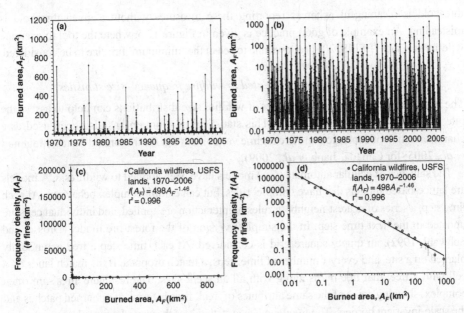

Figure 12.9 Time-series (a–b) and frequency–area wildfire statistics (c–d) of 7422 wildfires originating on USFS lands in California for the 37-year period 1970–2006 and with burned areas $A_F \geq 0.012 \, km^2$. (a) On linear–linear axes and (b) on log–linear axes show the burned area of each wildfire A_F as a function of time. (c) Linear–linear axes and (d) log–log axes are the frequency densities $f(A_F)$ (the number of wildfires per 'unit bin' of 1 km², circles) as a function of burned area, A_F. Also shown in (c) and (d) is the best least-squares power-law fit to the values (solid line) and associated equation, with coefficient of determination, r^2; vertical error bars represent two standard deviations of the frequency densities $f(A_F)$. Data from T. J. Brown, Desert Research Institute (updated dataset described in Brown *et al.*, 2002).

3.2 times the total number of wildfires if only wildfires above 1 ha are reported. This three-fold increase is large, particularly as reporting policies in many countries have systematically started to include fires that are smaller and smaller, sometime two orders of magnitude smaller now compared to two decades ago. However, although the total number of wildfires is sensitive to the completeness of the inventory, the total area burned is not. This is because the largest wildfires account for the majority of the burned fire area or energy released in any given inventory (e.g. Strauss *et al.*, 1989; Cui and Perera, 2008). For example, Stocks *et al.* (2002) showed that in Canada from 1959–1997, the largest 3% of those fires that were reported were responsible for 97% of the burned area.

As the total number of fires in the inventory is very sensitive to the completeness of the inventory, so too is the average area burned (calculated by taking the total burned area and dividing by the total number of fires) a commonly reported figure by local and national authorities. Therefore, care must be taken when either reporting total number of fires or average burned area of fires, particularly when comparing different inventories, or even the same inventory over time. Reporting the number of fires or average area burned above a given threshold is meaningful, whereas reporting these measures without a given threshold is misleading. An example of good practice is given in Figure 12.5, where the tool to examine wildfire statistics (EFFIS, 2011) allows one to select the 'minimum' fire size to be considered.

12.4.2.3 Other research related to wildfire frequency–size statistics

The heavy-tailed (power-law) character of wildfire size distributions can help to assess the rate of occurrence of large wildfires. This statistical distribution has also been used as a characterisation of ecoregion and fire regime variability (e.g. for the United States, Malamud *et al.*, 2005; for Canada, Jiang *et al.*, 2009).

The statistics of cellular automata (CA) models have been applied to wildfires. CA models are lattice-based models that have 'simple' rules, but can exhibit complex behaviour. At each time step, a series of nearest-neighbour rules of interaction are applied, and individual cells are updated in the next time step. In the simplest version of the forest-fire model (Drossel and Schwabl, 1992), an empty square grid is considered. At each time step a tree is randomly planted on a site, and every *x* number of time steps, a match dropped. If the match lands on a cell with a tree, that tree burns along with all adjacent sites. The resultant 'fires' are often complex, showing some of the same attributes of 'real' fires in terms of unburned patches and the scale-invariant borders. The frequency–size statistics of these model fires show power-law distributions over many orders of magnitude (for reviews, see Turcotte, 1999; Malamud and Turcotte, 2000). Millington *et al.* (2005) and Krenn and Hergarten (2009) review studies that compare 'real' wildfires and CA models, as both follow power-law statistics; many of these studies then argue that both wildfires and these 'simple' toy models can be considered as part of broader 'universal' models (self-organising systems).

Some questions that arise with wildfire frequency–size distributions include:

(1) If there is power-law or other heavy-tailed behaviour, over how many orders of magnitude is it applicable? For example, physical constraints on fire size (e.g. due to land use, fire separation, etc.) can vary globally and with topography or vegetation type. There are similarities with the problem of estimating maximum magnitude for earthquakes (e.g. see Kijko, 2004; Chapter 8 in this volume).
(2) What is the smallest scale at which the method can be applied, either due to data limitations or due to changes in the physical system?
(3) What are the uncertainties associated with the best-fit distribution (and implications for frequency of large-magnitude events)?
(4) How do the statistical distributions vary due to different factors (e.g. human activities, bio-physical factors)?
(5) What are the implications for government reporting and risk?

12.4.3 Uncertainty in wildfire hazard and risk research

As for all natural hazards, there are many sources of uncertainty inherent in wildfire hazard and risk research. There are two primary types of uncertainty: aleatory uncertainty is irreducible, and pertains to natural variability in the system; epistemic uncertainty, due to incomplete knowledge or lack of data, can be reduced. In risk assessment it is important to attempt to quantify these uncertainties (or at least consider their implications) to enable informed decision-making and make risk managers aware of the limitations of the science. A generic statistical methodology for handling both aspects of uncertainty is given in Chapters 2 and 3 of this volume, with Chapter 2 focusing on aleatory uncertainty and Chapter 3 describing the sources and implications of epistemic uncertainty.

A primary source of epistemic uncertainty is in the measurements made on the fires themselves. The scale at which measurements are made can lead to 'binning' effects, such as when a wildfire manager reports area burned rounded to the nearest 1, 2, 5 or 10 acres. In remote sensing, uncertainty in area burned is influenced by the resolution (e.g. aerial photographs or remote-sensing equipment) and choice of method applied. For example, having measured the perimeter of a fire, do we consider that everything inside is burned, or take into consideration the fact that there is almost always some unburned vegetation? Such assumptions or inconsistencies in measurement can have significant implications for wildfire frequency–size statistics.

For process-based wildfire models, uncertainties arise due to the accuracy and resolution (both temporal and spatial) of measurements of independent variables, e.g. a rainfall time-series subject to measurement uncertainty or made at a different geographical location, or topography used to model the slope and dynamics of the fire. Model structural uncertainty is also critical here – how accurately are the physical and chemical processes of combustion represented? This is something that is very difficult to quantify. There are also uncertainties associated with model parameterisation (e.g. the combustion temperature of a given type of vegetation), and methods of data assimilation. Over the last 5–10 years, development of probabilistic models to account for parameter variability has become more widespread, for example in modelling fire spread or intensity (see Thompson and Calkin, 2011 for a review).

Thompson and Calkin (2011) remark that wildfire management and decision-making is additionally hampered by a limited scientific understanding of the impact of fire management treatments (e.g. suppression and fuel reduction) and the ecological response to fire. There are also difficulties in valuing resources at risk. To address some of the problems associated with operational wildfire management and decision-making, Haight and Fried (2007) develop a scenario-based model to optimise daily deployment and dispatch of fire-fighting resources accounting for uncertainties (e.g. the number, location and intensity of fires). The California Fire Economics Simulator (Fried and Gilless, 1999) is used to model the initial attack and estimate the probability of fires escaping control. Brillinger *et al.* (2009) present a probabilistic wildfire risk model for the San Diego wildland–urban interface using data from the 2003 Cedar Fire. They investigate probability of destruction (and associated uncertainty) for houses in different locations, using a number of explanatory variables such as vegetation class, land area associated with the building, value and house size. The result is a simple model for estimating probability of loss. Brillinger *et al.* (2009) also remark on the limited scope and suitability of price-based models for evaluating risks to non-market resources (such as recreation, air and water quality and cultural heritage). To this end, Thompson *et al.* (2011) have developed a quantitative, geospatial model for decision support to evaluate human and ecological values at risk from wildfire for the continental United States. They use ratio-scale preference models to derive weightings based on the relative 'importance' of various resources, including wildlife habitats, national trails, watersheds and infrastructure.

12.5 Risk management

12.5.1 *Wildfire management and research*

From the advent of the US Forest Service in 1905, suppression of all fires was policy for federal land management agencies, until the late 1960s when the National Park Service officially recognised fire as a natural and necessary process (van Wagtendonk, 2007). Excessive fire suppression leads to the build up of fuel and can exacerbate risk. 'Prescribed natural fire zones' were subsequently defined, and criteria used to determine whether fires should be extinguished, contained or allowed to burn. Montiel and Kraus (2010) remark that policies for fire suppression in Europe may have led to increasingly large fires (or 'megafires'), for example in Catalonia in 1986, 1994 and 1998 (Badia *et al.*, 2002); Galacia in 1994 and 2005 and Greece in 2001 and 2006. Suppression remains the main course of action in the UK (A. Scott, pers. comm.).

A primary action in management of wildfires is fuel reduction (GAO, 2008), although the ecological need for fire is also recognised in order to encourage fire-adapted species and maintain species diversity. In the United States, the Federal Wildland Fire Management Policy recognises fire as a critical natural process requiring *appropriate management response*. Depending on the circumstances and the potential ecological, social and legal consequences, naturally occurring fires may be permitted to burn under controlled conditions. This is known as wildland fire use, and is recognised as an essential fire risk management strategy in forests for fuel reduction (see NIFC, 2009). Thinning of trees and vegetation, and *prescribed fire* (deliberate burning) are also effective means of reducing fuel build-up to minimise the impact (and extent) of natural fires (see Agee and Skinner, 2005; Figure 12.10).

Variation in wind conditions and other factors influencing fire spread are very difficult to predict, and this unpredictability can lead to problems in fire management. The Cerro Grande Fire in 2000 began as a prescribed fire, ignited by National Park fire managers. Within a day, wind changes led to the fire escaping the prescribed boundaries and it was declared a wildfire. The towns of Los Alamos and White rock were evacuated. In total 19 400 hectares were burned and around 280 houses destroyed (Hill, 2000).

Online fire mapping and forecasting tools such as EFFIS (EFFIS, 2011) GeoMAC (DoI/ USGS, 2011) and the USDA Forest Service Fire Detection Maps (USDA, 2011) are widely used by fire managers to identify and track wildfire activity (see Section 12.5.3). These platforms are used to assimilate and communicate information from a range of sources, from

Figure 12.10 Prescribed burning to reduce fuel build up (dead trees, fallen branches and low-lying vegetation) in the Bear Valley National Wildlife Refuge, Oregon (FWS, 2011).

incident reports on the ground to remote-sensing observations from earth observation satellites and aircraft.

Fire risk can be particularly difficult to manage at the wildland–urban interface, where there is a conflict of human and environmental needs, and mitigation actions (e.g. creation of buffer zones) can have unwanted ecological impacts (Keeley *et al.*, 1999). An increasing global population is driving development in wildland areas. For example, in the United States approximately 44.8 million housing units (39%) are classified as within the wildland–urban interface, including 5.1 million in California (Radeloff *et al.*, 2005). However, Schoennagel *et al.* (2009) note that much of the US wildland–urban interface and bordering land is privately owned, limiting federal agencies' capacity to implement fire risk reduction measures.

There are also several issues related to fire-fighting. Existing operational air tankers cannot carry enough suppressant to control and extinguish very large fire fronts. Airborne fire fighting cannot currently be undertaken at night, and current fire-fighting systems (including foam or water pumps, and aerial drops) have limited effectiveness in high winds. In some instances prescribed fire can be used actively for suppression, to direct or slow down the spread of the original wildfire or reduce fire intensity (see Castellnou *et al.*, 2010). For example 'burnout' involves using deliberate ignition to clear fuel near the edge of the wildfire. However, use of suppression fires is technically challenging due to difficulties in predicting fire behaviour and constantly changing conditions.

12.5.1.1 Evaluating fire intensity and severity

Fire intensity and fire severity are two key measures of wildfires, but there are no standardised definitions, and as such defining any relationship between them can be problematic. Keeley (2009) attempts to provide more formal definitions, but these have not yet been fully accepted by the fire community, and fire intensity is often used interchangeably with fire severity to describe the effects of fire on an ecosystem (Omi, 2005).

Fire intensity has generally been defined as the rate of energy release measured by flame length in the field (technically the rate of heat release), calculated using the rate of spread (calculated by the number of metres burned per second), and the energy flux defined by kilowatts generated per metre burned (Lentile *et al.*, 2006). The latter term is relatable to the fuel consumption and the fuel heat yield. The term 'fire line intensity' is often used (Byram, 1959).

Fire severity is typically defined as the degree to which fire affects vegetation and site characteristics – for example the soil or the percentage of basal tree area killed (Chappell and Agee, 1996). A number of attempts have been made to assess fire severity levels; from a points-scoring technique to broad descriptors such as light, moderate or severe (Chafer *et al.*, 2004; Raybould and Roberts, 2006). These schemes include a range of observations, mainly from the field but also from analysis of satellite images (Cocke *et al.*, 2005).

Fire intensity and severity are typically evaluated using remote sensing and/or ground observational data. However, linkage between interpretations or quantitative examination of remote-sensing data products and ground observational data of wildfires has often proved difficult (Pyne *et al.*, 1996; Hammill and Bradstock, 2006; Lentile *et al.*, 2006). Although quantitative relationships can be determined for certain parameters (e.g. thermal conductivity of the soil and changes in soil temperature and water content beneath surface fires) such relationships have not yet been fully validated in the field (DeBano *et al.*, 1998).

The lack of consistent terminology can lead to confusion, and potential misuse of information. As stated by Lentile *et al.* (2006), land managers and scientists ideally require unambiguous remote-sensing products for fire management. An important research goal is therefore to relate satellite and ground data, and investigate new approaches to enable more consistent and robust measurement. Developing techniques include the use of charcoal residues to infer temperature post-fire (Scott, 2009). Charcoal reflectance, which is measured from polished blocks under oil in reflected light, has been shown to increase with temperature, and could therefore be used to provide a minimum fire temperature (see Scott, 2000; Scott and Glasspool, 2007). Other useful measures include fire radiative power (FRP), which represents the rate of radiant energy release of the fire, and fire radiative energy (FRE) which is the time-integral of FRP and relates to the amount of fuel burned (Wooster *et al.*, 2005). FRP is typically measured using airborne or satellite-derived middle-infrared measurements (see Section 12.5.3), and can be used to estimate 'radiative' fire line intensity (Smith and Wooster, 2005).

12.5.2 Formal mechanisms for decision support

12.5.2.1 Fire danger ratings and indices

A fire danger rating is a measure of fire potential, and is typically based on recent and forecast weather and other factors leading to an increased probability of fire, such as topography and fuel. Fire danger ratings are used to inform fire management activities, availability of fire-fighting personnel, setting prescribed fires and keeping the public alert to fire danger. There are various systems in operation globally, developed for different vegetation types and conditions, although the basic model structures are similar (see Figure 12.11). Matthews (2009) provides

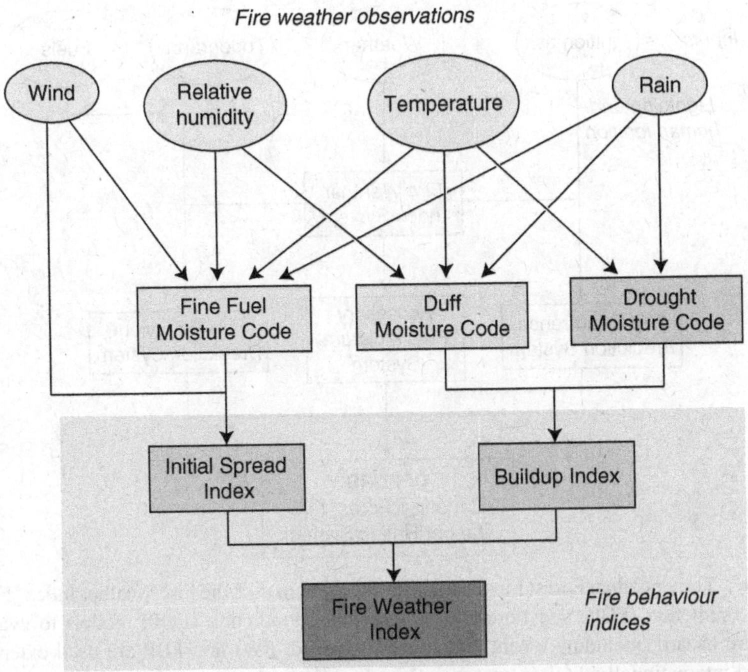

Figure 12.11 Schematic describing calculation of the Fire Weather Index, using weather observations and suite of codes to evaluate fuel properties, the build-up of dry, combustible fuels and potential for fire spread. Adapted from NRC (2011a).

an overview and comparison of three key models (the Forest Fire Danger Index and the Forest Fire Behaviour Tables developed for use in Australian forests, and the Canadian Forest Fire Weather Index – FWI), and finds they agree reasonably well.

The Canadian Forest Fire Danger Rating System (CFFDRS, Figure 12.12) is a broad umbrella system of rating fire danger that includes many elements and accounts for the probability of initiation, weather and fire behaviour (Stocks *et al.*, 1989; Taylor and Alexander, 2003). One element of the CFFDRS is the Canadian Forest Fire Weather Index (Figure 12.11), which assesses the impact of temperature, fuel moisture (for a range of fuels, from loose, fine material to deep, compact organic layers), wind and precipitation on fire danger (NRC, 2011a). The FWI was initially developed for Canadian pine forests and is a general index of fire danger for all Canadian forested areas. Elements of the CFFDRS, including the FWI, are widely applied in countries outside of Canada, including New Zealand, Fiji, several US states, Mexico, Southeast Asia, and extensively in Europe (Taylor and Alexander, 2003; Matthews, 2009).

The US National Fire Danger Rating System (NFDRS – Schlobohm and Brain, 2002) characterises the expected upper limit for fire danger in a given region for a 24-hour period, and gives an indication of potential fire growth and behaviour should a wildfire

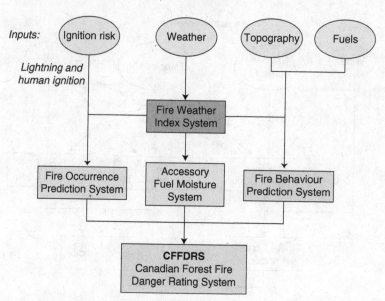

Figure 12.12 The Canadian Forest Fire Danger Rating System uses the Fire Weather Index (FWI), Fire Behaviour Prediction (FBP) and Forest Fire Occurrence Prediction (FOP) models to evaluate the potential fire hazard, including extent and speed of spread. FWI and FBP are used extensively in Canada and internationally. FOP is operational in several regional schemes, and the Accessory Fuel Moisture System is in development. Adapted from NRC (2011a).

start. Inputs include 24-hour weather forecasts for lightning activity level, temperature, humidity and precipitation; current observations of humidity, temperature, cloud cover, wind speed; live and dead fuel inputs; topographic factors; and risk of human initiation. The model generates indices for fire occurrence and burning, which are used to estimate the fire load index or *fire danger* (Table 12.2). Fire danger rating maps are updated daily on the internet, and local conditions are communicated to the public via signs in national parks and forests.

The Fire Potential Index (FPI) is an additional measure used to identify areas susceptible to ignition, and can be expressed in terms of the number of expected ignitions in a given area (e.g. Eidenshink and Klaver, 2005), or as a simple scale (low–high) indicating probability of ignition. FPI maps use satellite information to assess the effects of weather conditions and vegetation on fire danger (see Preisler *et al.*, 2009). FPI is based on the Soil Moisture Index, which in turn is derived from temperature and vegetation measurements taken by NOAA's satellite-based Advanced Very High Resolution Radiometer (AVHRR). Preisler *et al.* (2009) present a statistical model for assessing the skill of fire danger indices and forecasting the distribution of the expected numbers of large fires.

The Normalised Difference Vegetation Index (NDVI) is a commonly used satellite observation-derived metric sensitive to the presence of photosynthetically active vegetation. From AVHRR, NDVI data are derived at a 1.1 km spatial resolution at nadir, and a

Table 12.2 *Summary of the US National Fire Danger Rating System (Schlobohm and Brain, 2002). Fire danger is determined by the burning index, based on available fuel, moisture, weather conditions and potential spread, and an ignition component, accounting for human or lightning initiation*

Fire danger	Ignition	Spread	Spotting	Control
Low (green)	Fuels do not ignite readily from small firebrands. Lightning may start fires in dry, rotted wood or organic matter.	Fires in open grasslands may burn freely, wood fires spread slowly.	Little danger of spotting.	Easy
Moderate (blue)	Fires can start from most accidental causes, but with the exception of lightning fires in some areas, the number of starts is generally low.	Fires in open grasslands will spread rapidly on windy days. Timber fires spread slowly to moderately fast. The average fire is of moderate intensity.	Short-distance spotting may occur, not persistent.	Fires are not likely to become serious and control is relatively easy.
High (yellow)	All fine dead fuels ignite readily and fires start easily from most causes. Unattended brush and campfires are likely to escape.	Fires spread rapidly. High-intensity burning may develop on slopes or in concentrations of fine fuels.	Short-distance spotting is common.	Fires may become serious and control difficult unless they are attacked while small.
Very high (orange)	Fires start easily from all causes.	Fires immediately spread rapidly and increase quickly in intensity. Fires may quickly develop high-intensity characteristics such as long-distance spotting and fire whirlwinds.	Spot fires are a constant danger; long-distance spotting likely.	
Extreme (red)	Fires start quickly and burn intensely. All fires are potentially serious.	Furious spread likely, along with intense burning.	Spot fires are a constant danger; long-distance spotting occurs easily.	Direct attack rarely possible and may be dangerous except immediately after ignition.

daily temporal scale. Multi-temporal compositing is generally used to obtain cloud-free coverage of the area of interest. In certain ecosystems, NDVI can be combined with other indices derived from satellite data to identify areas at high risk of fire. However, studies by Gabban *et al.* (2006, 2008) identify limitations in the use of NDVI, and find that the FWI (Figure 12.11) performs better as an indicator of fire risk. Gabban *et al.* (2006, 2008) studied a ten-year dataset of satellite imagery and fire incidence for Spain, and found little differentiation in NDVI between areas where fires took place and areas where fire did not occur.

12.5.3 *Remote sensing, monitoring and early warning systems*

Advances in satellite earth observation are having a significant impact on (1) forecasting conditions for wildfires; (2) monitoring fire occurrence, spread and areas burned; (3) observing or calculating emissions; and (4) post-fire assessment. Primary techniques include infrared imaging to identify actively burning fire 'hotspots' (e.g. Stearns *et al.*, 1986) and to measure radiative power output (which relates to the fire intensity and rate of fuel consumption; see Section 12.5.1); optical imaging to map burned areas and identify locations for subsequent post-fire assessment; and atmospheric remote sensing of smoke plumes and chemical species (e.g. Lamarque *et al.*, 2003).

The IR imaging sensors onboard environmental earth observation satellites typically have spatial resolutions of the order of hundreds to thousands of metres. At these resolutions, flaming and smouldering zones are generally significantly smaller than the pixel, and in many cases less than 1% of the pixel area. Nevertheless, hotspots are readily detected due to the intense energy emissions in certain spectral regions resulting from high-temperature combustion (Robinson, 1991; Zhukov *et al.*, 2008). High temporal resolution thermal imagery (e.g. from the BIRD, MODIS or geostationary-type sensors) can be used to identify active fires, in some cases in close to real time, with information available on open-access web databases (e.g. King's College London, 2011; University of Maryland, 2011). IR systems can therefore be used for early warning of new fires, for monitoring the development of large fire events, and potentially as an input into models aimed at near real-time smoke forecasting (e.g. MACC, 2011). However, the method can fail to detect fires in areas with frequent cloud cover, and can also miss rapidly burning, small and/or low-intensity fires (e.g. Schroeder *et al.*, 2008a, 2008b) – leading to discrepancies within and between datasets (e.g. Hawbaker *et al.*, 2008). Quantitative analysis of IR and other data can provide information related to active fire temperature, aerial coverage and rate of radiant energy release, as shown in Figure 12.13 (see, e.g. Giglio and Kendall, 2003; Wooster *et al.*, 2003; Zhukov *et al.*, 2006; Eckmann *et al.*, 2009; Xu *et al.*, 2010).

In terms of burned area detection based on optical satellite imagery, Joyce *et al.* (2009) provide an extensive summary of remote-sensing data and image-processing techniques for natural hazard monitoring and identification, including techniques related to wildfire.

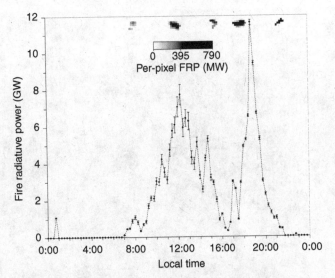

Figure 12.13 Fire radiative power (FRP) time-series of a savannah fire burning in northern Botswana on 4 September 2003, calculated from 15-minute temporal resolution Meteosat SEVIRI infrared imagery. For each image, the FRP value in megawatts for each active fire 'hotspot' pixel is calculated, and summed to give a 'total fire' value. The pixel maps show the per-pixel FRP values, while the graph denotes the total. The fire is seen to be most active, i.e. emitting most energy per unit time, during the daylight period. Error bars denote the uncertainty in derived FRP. Figure adapted from Wooster *et al.* (2005).

Burned areas (sometimes called 'fire scars') are typically identified using multi-spectral high to moderate spatial resolution data (e.g. IKONOS to MODIS class systems). Identification of burns based on spectral classification is relatively rapid in many cases, and can be applied over wide areas, although results can be highly sensitive to the algorithm used, and non-unique spectral responses can require subsequent manual processing. Figure 12.14 illustrates how the dark 'char' of a typical burn scar contrasts with the non-burned surroundings.

More recently, fully automated algorithms have been developed for mapping global burned area using moderate spatial resolution sensors such as MODIS (e.g. Roy *et al.*, 2005), providing daily data on global burned area. Fire M3 (NRC, 2011b) is a similar operational system (part of the Canadian Wildland Fire Information System), which uses low-resolution satellite imagery to identify and locate actively burning fires; assess burned area; and model fire behaviour, biomass consumption and carbon emissions. Daily fire hotspot data are made publicly available online (see NRC, 2011a).

12.5.4 Communication to the public and others

Online fire mapping tools such as EFFIS, the European Forest Fire Information System (EFFIS, 2011), have been established for communicating fire data, much of which are

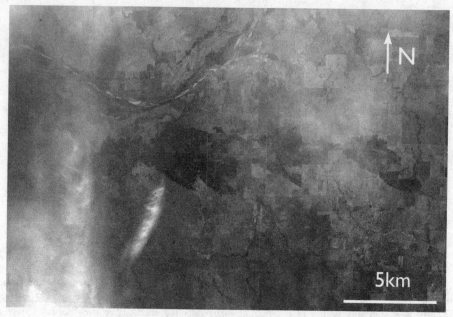

Figure 12.14 ASTER false-colour composite image of a grass-fire burn (dark) close to the Oklahoma and Texas border during a period of exceptional drought. NASA image created by Jesse Allen (NASA, 2011b) using data provided by the NASA/GSFC/METI/ERSDAC/JAROS and US/Japan ASTER Science Team. Scale is approximate.

derived from remotely sensed information, to land managers, policy-makers and the public. EFFIS provides a web GIS interface to display historical fire data, meteorological fire danger maps and forecasts, daily satellite images and maps showing the latest hot spots and fire perimeters for European wildfires. GeoMAC is a similar tool for the United States (DoI/USGS, 2011). The website allows users in remote locations to manipulate map information displays, view and download fire information at various scales and levels of detail. In the United States, the national Firewise Communities programme is a multi-agency effort designed to reach beyond the fire service by involving homeowners, community leaders and developers to promote effective emergency response and encourage individual responsibility for mitigating fire hazard (see NFPA, 2011). In the UK, the Forestry Commission website provides information on forest fires and the level of risk for the New Forest (Forestry Commission, 2011). On a global scale, the Global Fire Information Management System (GFIRMS – FAO, 2011) integrates remote-sensing and GIS technologies to rapidly deliver MODIS-derived active fire hotspot and burned area information to users around the globe, while the prototype European GMES Global Atmospheric Monitoring Service (MACC, 2011) uses polar orbiting and geostationary active fire observations to derive information on fire radiative power aerial density and thus smoke emissions (Kaiser *et al.*, 2010; Figure 12.15).

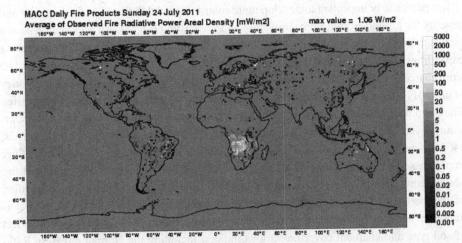

Figure 12.15 Global fire radiative power (FRP) area density data record for 24 July 2011 from MACC (2011). Data represent the observed average FRP from all active fires detected on this day in 125 km grid cells and expressed in units of FRP divided by grid-cell area [MW/m^2]. The thermal radiation flux is believed to be related to the rate at which fuel is consumed and smoke produced, and therefore can be used to estimate global open vegetation fire trace gas and particulate emissions to the atmosphere (Kaiser *et al.*, 2010, 2011). Note that this was originally a colour image, and the majority of the globe (shown in mid-grey) has no fire activity.

12.6 Summary and recommendations

Climate change will affect vegetation type and distribution, and climate drivers of wildfire (increasing temperature, decreasing precipitation and humidity). A warmer, dryer climate will potentially lead to more ignitions, larger and more intense fires and increased risk of fire at higher latitudes and altitudes. Wildfire management policy in the UK and globally is largely based on historic fire regimes and data, and there is currently little recognition that the changing climate may lead to more fires. Many funding agencies expect to see benefits within a five-year period, yet it is likely to be decades before it will be possible to see the impact of climate change on wildfire frequency in the UK (A. Scott, pers. comm.). To move forward it is essential to assess the potential climatic impacts on future risk.

Thompson and Calkin (2011) remark that recent developments in simulation and resource mapping are enabling more robust risk assessments, but that there is a need to further explore non-market and societal values of resources at risk (see, e.g. Brillinger *et al.*, 2009; Thompson *et al.*, 2011). In terms of understanding financial risk, Chen and Goodwin (2011) provide an analysis of regional (county-scale) wildfire risk based on records of annual wildfire frequency, burned area, fire per acre and burned ratio, to develop an actuarially fair wildfire insurance scheme for the US forest sector. As well as the more obvious drivers such as weather and climate, they find that socio-economic conditions have a significant impact on wildfire risk.

It is particularly important to develop more robust, formal definitions of fire intensity and severity for ground-truthing fire simulation and forecasting models. An important research goal is to link satellite and ground observational data, to make better use of remote-sensing capabilities, and develop new techniques for obtaining data (e.g. the use of temperature data from charcoal residues – Scott, 2009). Recent advances in this area include work by Disney *et al.* (2011), who propose a model for evaluating fire severity based on radiative transfer. Their approach involves constructing a three-dimensional model of overstorey and understorey layers based on detailed field observations of structural and radiometric properties, used to simulate pre- and post-fire reflectance. This will enable testing and validation of fire impact assessments in locations where it is difficult to obtain both pre- and post-fire observations. Smith *et al.* (2010) remark that fire severity is crucial to our understanding of the long-term biogeochemical, ecological and societal impacts of wildfire, and it is important to quantify uncertainties in the methods used. Smith *et al.* (2010) study the effect of soil type and charcoal cover on various indices, including the Normalised Burn Ratio (NBR), widely used for mapping fire severity; the Normalised Difference Vegetation Index (NDVI, see Section 12.5.2), typically used to evaluate post-fire vegetation recovery; and the Optimised Soil-Adjusted Vegetation Index (OSAVI), and suggest that NDVI or OSAVI may be more suitable for regional-scale fire severity mapping.

There is also a need to bridge the gap between the forest fire, fire safety and combustion science communities. The ESRC/NERC-sponsored 2008–2009 FIRES seminar series was initiated to promote interdisciplinary research and actively involve fire and rescue services in academic debate on UK wildfires, and such collaborations need to be supported. The Mediterranean Combustion Symposium Workshop on Wildfire Research (MCS, 2009) reported that although specialist foresters, biologists, ecologists and geoscientists are well represented in the forest fire research community, there are relatively few teams involved in mathematical, chemical or physical science research.

It is recognised that wildfires can cause significant economic loss, loss of life, damage to property and infrastructure and impact on the ecosystem. However, there are few studies that attempt to quantify the risk (rather than the hazard) to evaluate the cost-effectiveness of fire-fighting and mitigation measures, or assess potential change in risk under changing land use or climate regimes. Thompson and Calkin (2011) provide a comprehensive review of strategies for quantifying and managing uncertainty, with a focus on wildfire risk management and decision support. They remark that existing decision support efforts are typically directed toward active fire management, and that characterisation of resources at risk remains a key challenge, including consideration of non-market values and social preference. Decisions regarding allocation of resources for fuel reduction are not necessarily risk-informed and cost-effective (GAO, 2008), and funding is often based on professional judgement and historical expenditure. Better use of resources requires allocation of funding based on up-to-date wildland fire risk assessments and evaluations of fuel treatment effectiveness, accounting for change and uncertainty. There is also a clear need for a more coordinated and multidisciplinary approach to wildfire risk assessment and management.

Acknowledgements

The authors would like to thank Professors Andrew Scott and Keith Beven for their valuable input, and Drs Alistair Smith and Kevin Tansey for detailed and constructive reviews of this chapter.

References

Agee J. K. and Skinner C. N. (2005) Basic principles of forest fuel reduction treatments. *Forest Ecology and Management* **211** (1–2): 83–96.

Albertson, K., Aylen, J., Cavan, G., *et al.* (2009) Forecasting the outbreak of moorland wildfires in the English Peak District. *Journal of Environmental Management* **90** (8): 2642–2651.

Andreae, M. O. and Merlet, P. (2001) Emission of trace gases and aerosols from biomass burning. *Global Biogeochemical Cycles* **15** (4): 955–966.

Attewill, F. (2007) Anger at Greek government's wildfire response. *Guardian*, 28 August. www.guardian.co.uk/world/2007/aug/28/naturaldisasters.features11 (accessed 13 April 2011).

Azuma, D. L., Donnegan, J. and Gedney, D. (2004) Southwest Oregon Biscuit Fire: an analysis of forest resources and fire severity. *US Forest Service Pacific Northwest Research Station Research Paper PNW-RP* **560** (May): 1–32.

Badia, A., Saurí, D., Cerdan, R., *et al.* (2002) Causality and management of forest fires in Mediterranean environments: an example from Catalonia. *Global Environmental Change Part B: Environmental Hazards.* **4** (1): 23–32.

Balling, R. C., Meyer, G. A. and Wells, S. G. (1992) Relation of surface climate and burned area in Yellowstone National Park. *Agricultural and Forest Meteorology* **60** (3–4): 285–293.

Beer, T. (1991) The interaction of wind and fire. *Boundary-Layer Meteorology* **54** (3): 287–308.

Blake, W. (2009) Impacts of wildfire on downstream water quality. Presentation at the Wildfire 2009 Conference, Lyndhurst, New Hampshire. www.wildfire2009.org.uk.

Blake, W. H., Wallbrink, P. J., Wilkinson, S., *et al.* (2009) Deriving hillslope sediment budgets in wildfire-affected forests using fallout radionuclide tracers. *Geomorphology* **104**: 105–116.

Boer, M., Sadler, R. and Grierson, P. (2008) Objective characterisation of fire regimes for science-based management of fire-prone landscapes. The International Bushfire Research Conference 2008, incorporating the 15th annual AFAC Conference, Adelaide, Australia.

Bond, W. J. and Keeley, J. E. (2005) Fire as a global 'herbivore': the ecology and evolution of flammable ecosystems. *Trends in Ecology & Evolution* **20** (7): 387–394.

Bowman, D. M. J. S., Balch, J. K., Artaxo, P., *et al.* (2009) Fire in the Earth system. *Science* **324**: 481–484.

Brillinger, R. D., Autrey, B. S. and Cattaneo, M. D. (2009) Probabilistic risk modeling at the wildland urban interface: the 2003 Cedar Fire. *Environmetrics* **20**: 607–620.

Brown, P. M. and Schoettle, A. W. (2008) Fire and stand history in two limber pine (*Pinus flexilis*) and Rocky Mountain bristlecone pine (*Pinus aristata*) stands in Colorado. *International Journal of Wildland Fire* **17** (3): 339–347.

Brown, T. J., Hall, B. L., Mohrle, C. R., *et al.* (2002) Coarse assessment of federal wildland fire occurrence data. Report for the National Wildfire Coordinating Group, Program CEFA (Climate, Ecosystem and Fire Applications) Report 02-04.

Brunstein, F. C. and Yamaguchi, D. K. (1992) The oldest known Rocky Mountain bristle-cone pines (*Pinus aristata* Engelm.). *Arctic and Alpine Research* **24**: 253–256.

Butler, B. W., Finney, M. A., Andrews, P. L., *et al.* (2004) A radiation-driven model for crown fire spread. *Canadian Journal of Forest Research–Revue Canadienne De Recherche Forestiere* **34** (8): 1588–1599.

Byram, G. M. (1959) Combustion of forest fuels. In *Forest Fire: Control and Use*, ed. K. P. Davis, New York: McGraw Hill, pp. 61–89.

Camia, A., San-Miguel-Ayanz, J., Oehler, F., *et al.* (2008) *Forest Fires in Europe 2008*. Luxembourg: Office for Official Publications of the European Communities.

Castellnou, M., Kraus, D. and Miralles, M. (2010) Prescribed burning and suppression fire techniques: from fuel to landscape management. In *Best Practices of Fire Use: Prescribed Burning and Supression Fire Programmes in Selected Case-Study Regions in Europe*, ed. H. Hasenauer and M. Korhonen, Joensuu: European Forest Institute.

Certini, G. (2005) Effects of fire on properties of forest soils: a review. *Oecologia* **143** (1): 1–10.

Chafer, C. J., Noonan, M. and Macnaught, E. (2004) The post-fire measurement of fire severity and intensity in the Christmas 2001 Sydney wildfires. *International Journal of Wildland Fire* **13** (2): 227–240.

Chappell, C. B. and Agee, J. K. (1996) Fire severity and tree seedling establishment in *Abies magnifica* forests, Southern Cascades, Oregon. *Ecological Applications* **6** (2): 628–640.

Chen, X. and Goodwin, B. K. (2011) Spatio-temporal modeling of wildfire risks in the U.S. forest sector. Paper presented at the Agricultural and Applied Economics Association 2011 Annual Meeting. http://purl.umn.edu/103628.

Chuvieco, E., Giglio, L. and Justice, C. (2008) Global characterization of fire activity: toward defining fire regimes from Earth observation data. *Global Change Biology* **14** (7): 1488–1502.

Clark, J. S. (1990) Fire and climate change during the last 750 yr in Northwestern Minnesota. *Ecological Monographs* **60** (2): 135–159.

Clark, T. L., Jenkins, M. A., Coen, J., *et al.* (1996) A coupled atmosphere–fire model: convective feedback on fire-line dynamics. *Journal of Applied Meteorology* **35** (6): 875–901.

Cocke, A. E., Fule, P. Z. and Crouse, J. E. (2005) Comparison of burn severity assessments using Differenced Normalized Burn Ratio and ground data. *International Journal of Wildland Fire* **14** (2): 189–198.

Cofer, W. R., Koutzenogii, K. P. and Kokorin, A., *et al.* (1997) Biomass burning emissions and the atmosphere. In *Sediment Records of Biomass Burning and Global Change*, ed. J. S. Clark, H. Cachier, J. G. Goldammer, *et al.*, Berlin: Springer, pp. 189–206.

Commonwealth of Australia (2010) *The Incidence and Severity of Bushfires across Australia*, Canberra: The Senate Select Committee on Agricultural and Related Industries.

Conedera, M., Tinner, W., Neff, C., *et al.* (2009) Reconstructing past fire regimes: methods, applications, and relevance to fire management and conservation. *Quaternary Science Reviews* **28** (5–6): 555–576.

CSIRO (2008) Fire-generated wind. Commonwealth Scientific and Industrial Research Organisation (CSIRO) Fact sheet. www.csiro.au/resources/FireGeneratedWind.html (accessed 11 April 2011).

Cui, W. and Perera, A. H. (2008) What do we know about forest fire size distribution, and why is this knowledge useful for forest management? *International Journal of Wildland Fire* **17**: 234–244.

Cunningham, P. and Linn, R. R. (2007) Numerical simulations of grass fires using a coupled atmosphere–fire model: dynamics of fire spread. *Journal of Geophysical Research – Atmospheres* **112** (D5): n.p.

DCLG (2011) Fire and resiliance. UK Department for Communities and Local Government. www.communities.gov.uk/fire/ (accessed 12 May 2011).

DeBano, L. F., Neary, D. G. and Ffolliott, P. F. (1998) *Fire's Effects on Ecosystems*. New York: John Wiley and Sons.

de Groot W. J., Pritchard, J. M. and Lynham, T. J. (2009) Forest floor fuel consumption and carbon emissions in Canadian boreal forest fires. *Canadian Journal of Forest Research – Revue Canadienne De Recherche Forestiere* **39** (2): 367–382.

Disney, M. I., Lewis, P., Gomez-Dans, J., *et al.* (2011) 3D radiative transfer modelling of fire impacts on a two-layer savanna system. *Remote Sensing of Environment* **115** (8): 1866–1881.

DoI/USGS (2011) GeoMAC Wildland Fire Support (internet-based mapping application). US Department of the Interior and US Department of Agriculture. www.geomac.gov (accessed 12 May 2011).

Dold, J. W., Weber, R. O., Gill, M., *et al.* (2005) Unusual phenomena in an extreme bushfire. In *Proceedings of 5th Asia-Pacific Conference on Combustion 2005*, ed. G. J. Nathan, B. B. Dally, M. Kalt, *et al.*, Adelaide: University of Adelaide, pp. 309–312.

Drossel, B. and Schwabl, F. (1992) Self-organized critical forest fire model. *Physical Review Letters* **69**: 1629–1632.

Eckmann, T. C., Roberts, D. A. and Still, C. J. (2009) Estimating subpixel fire sizes and temperatures from ASTER using multiple endmember spectral mixture analysis. *International Journal of Remote Sensing* **30** (22): 5851–5864.

EFFIS (European Forest Fire Information System) (2011) Fire history online tools. effis.jrc.ec.europa.eu/fire-history (accessed 1 April 2011).

Eidenshink, J. and Klaver, J. (2005) USGS fire science: fire danger monitoring and forecasting. US Department of the Interior, US Geological Survey, Fact Sheet 2005-3066. egsc.usgs.gov/isb/pubs/factsheets/FS-2005-3066.pdf (accessed 13 May 2011).

ESA (2011a) European Space Agency. Multimedia Gallery. www.esa.int/esa-mmg/mmghome.pl (accessed 12 May 2011).

ESA (2011b) ATSR world fire atlas. European Space Agency, data user element. http://due.esrin.esa.int/wfa/ (accessed 2 September 2011).

Evans, M., Hastings, N. and Peacock, B. (2000) *Statistical Distributions*, New York: Wiley-Interscience.

FAO (2011) Global fire information management system (GFIMS). Food and Agriculture Organization of the United Nations. www.fao.org/nr/gfims/en (accessed 21 July 2011).

FDS (2011) Fire Dynamics Simulator and Smokeview (FDS-SMV). National Institute of Standards and Technology (NIST). www.fire.nist.gov/fds/ (accessed 22 February 2011).

Fenner, M. (1992) *Seeds: The Ecology of Regeneration*. Wallingford: CABI Publishing.

Finney, M. A. and McAllister, S. S. (2011) A review of fire interactions and mass fires. *Journal of Combustion*, doi: 10.1155/2011/548328.

Firehouse (2005) King County Creek, Alaska fire continues to grow, spreading smoke. *Firehouse* news archives, 30 June. www.firehouse.com (accessed 10 April 2011).

FIRES (2009) Economic impacts of wildfires and adaptive land management to reduce wildfire risk and impact. FIRES Seminar 4. Fire Interdisciplinary Research on Ecosystem Services: Fire and Climate Change in UK Moorlands and Heaths. 14 May. www.fires-seminars.org.uk (accessed 21 July 2011).

Fitzgerald, W. F., Engstrom, D. R., Mason, R. P., *et al.* (1998) The case for atmospheric mercury contamination in remote areas. *Environmental Science & Technology* **32** (1): 1–7.

Flannigan, M. D., Stocks, B. J. and Wotton, B. M. (2000) Climate change and forest fires. *Science of the Total Environment* **262** (3): 221–229.

Flannigan, M. D., Amiro, B. D., Logan, K. A., *et al.* (2005) Forest fires and climate change in the 21st Century. *Mitigation and Adaptation Strategies for Global Change* **11**: 847–859.

Forestry Commission (2011) The Forestry Commission. www.forestry.gov.uk (accessed 13 May 2011).

Foster, D. R. (1985) Vegetation development following fire in Picea Mariana (black Spruce) – Pleurozium forests of south-eastern Labrador, Canada. *Journal of Ecology* **73** (2): 517–534.

Fried, J. S. and Gilless, J. K. (1999) *CFES2: The California Fire Economics Simulator Version 2 User's Guide*, Oakland: University of California.

Fromm, M., Bevilacqua, R., Servranckx, R., *et al.* (2005) Pyro-cumulonimbus injection of smoke to the stratosphere: observations and impact of a super blowup in northwestern Canada on 3–4 August 1998. *Journal of Geophysical Research – Atmospheres* **110** (D8): n.p.

FWS (2011) US Fish and Wildlife Service National Digital Library. digitalmedia.fws.gov/ (accessed 12 May 2011).

Gabban, A., San-Miguel-Ayanz, J. and Viegas, D. X. (2006) On the suitability of the use of normalized difference vegetation index for forest fire risk assessment. *International Journal of Remote Sensing* **27** (22): 5095–5102.

Gabban, A., San-Miguel-Ayanz, J. and Viegas, D. X. (2008) A comparative analysis of the use of NOAA-AVHRR NDVI and FWI data for forest fire risk assessment. *International Journal of Remote Sensing* **29** (19): 5677–5687.

GAO (2008) Wildland fire management: federal agencies lack key long- and short-term management strategies for using program funds effectively. United States Government Accountability Office.

GAO (2009) Wildland fire management: federal agencies have taken important steps forward, but additional, strategic action is needed to capitalize on those steps. United States Government Accountability Office Report to Congressional Addressees.

Gazzard, R. (2009a) UK vegetation fire standard. Presentation at the Wildfire 2009 Conference, Lyndhurst, New Hampshire.

Gazzard, R. (2009b) UK vegetation fire standard: data fields and terminology for wildfire incidents and prescribed burning operations within Great Britain and Northern Ireland. www.forestry.gov.uk/website/forestresearch.nsf/ByUnique/INFD-7WKJDJ.

Giglio, L., Kendall, J. D. and Mack, R. (2003) A multi-year active fire dataset for the tropics derived from the TRMM VIRS. *International Journal of Remote Sensing* **24** (22): 4505–4525.

Giglio, L., Randerson, J. T., van der Werf, G. R., *et al.* (2010) Assessing variability and long-term trends in burned area by merging multiple satellite fire products. *Biogeosciences* **7** (3): 1171–1186.

Gill, A. M. and Allan, G. (2008) Large fires, fire effects and the fire-regime concept. *International Journal of Wildland Fire* **17** (6): 688–695.

Givelet, N., Roos-Barraclough, F. and Shotyk, W. (2003) Predominant anthropogenic sources and rates of atmospheric mercury accumulation in southern Ontario recorded by peat cores from three bogs: comparison with natural 'background' values (past 8000 years). *Journal of Environmental Monitoring* **5** (6): 935–949.

Graham, R. T. (ed.) (2003) *Hayman Fire Case Study*, Ogden, UT: US Dept. of Agriculture, Forest Service, Rocky Mountain Research Station.

Grell, G. A., Peckham, S. E., Schmitz, R., *et al.* (2005) Fully coupled 'online' chemistry within the WRF model. *Atmospheric Environment* **39** (37): 6957–6975.

Guyette, R. P., Muzika, R. M. and Dey, D. C. (2002) Dynamics of an anthropogenic fire regime. *Ecosystems* **5** (5): 472–486.

Haight, R. G. and Fried, J. S. (2007) Deploying wildland fire suppression resources with a scenario-based standard response model. *INFOR* **45** (1): 31–39.

Hammill, K. A. and Bradstock, R. A. (2006) Remote sensing of fire severity in the Blue Mountains: influence of vegetation type and inferring fire intensity. *International Journal of Wildland Fire* **15**: 213–226.

Hawbaker, T. J., Radeloff, V. C., Syphard, A. D., *et al.* (2008) Detection rates of the MODIS active fire product in the United States. *Remote Sensing of Environment* **112** (5): 2656–2664.

Heyerdahl, E. K., Berry, D. and Agee, J. K. (1995) *Fire History Database of the Western United States*. Seattle, WA: US Environmental Protection Agency, USDA Forest Service, University of Washington.

Hill, B. T. (2000) *Lessons Learned from the Cerro Grande (Los Alamos) Fire*, Washington, DC: United States General Accounting Office.

Ito, A. and Penner, J. E. (2004) Global estimates of biomass burning emissions based on satellite imagery for the year 2000. *Journal of Geophysical Research – Atmospheres* **109** (D14): n.p.

Jiang, Y., Zhuang, Q., Flannigan, M. D., *et al.* (2009) Characterization of wildfire regimes in Canadian boreal terrestrial ecosystems. *International Journal of Wildland Fire* **18**: 992–1002.

Johnson, E. A. (1992) *Fire and Vegetation Dynamics: Studies from the North American Boreal Forest*, Cambridge: Cambridge University Press.

Johnston, F. H. (2009) Bushfires and human health in a changing environment. *Australian Family Physician* **38**: 720–725.

Johnston, F. H., Bailie, R. S., Pilotto, L. S., *et al.* (2007) Ambient biomass smoke and cardio-respiratory hospital admissions in Darwin, Australia. *BMC Public Health* **7**: 240.

Jost, H. J., Drdla, K., Stohl, A., *et al.* (2004) In-situ observations of mid-latitude forest fire plumes deep in the stratosphere. *Geophysical Research Letters* **31** (11): L11101.

Joyce, K. E., Belliss, S. E., Samsonov, S. V., *et al.* (2009) A review of the status of satellite remote sensing and image processing techniques for mapping natural hazards and disasters. *Progress in Physical Geography* **33** (2): 183–207.

JRC (2010) Forest fires in Europe 2009. JRC-IES/Land Management & Natural Hazards Unit. effis.jrc.ec.europa.eu/reports/fire-reports/doc/34/raw (accessed 1 April 2011).

Kaiser, J. W., Benedetti, A., Flemming, J., *et al.* (2010) From fire observations to smoke plume forecasting in the MACC services. ESA Special Publication, SP-688.

Kaiser, J. W., Heil, A., Andreae, M. O., *et al.* (2011) Biomass burning emissions estimated with a global fire assimilation system based on observed fire radiative power. *Biogeosciences Discussions* **8** (4): 7339–7398.

Karter, M. J. (2008) Fire loss in the United States 2007. National Fire Protection Association Fire Reports, Quincy, MA. www.nfpa.org/assets/files/PDF/Public Education/FireLoss2007.pdf (accessed 5 April 2011).

Kasischke, E. S., Christensen, N. L. and Stocks, B. J. (1995) Fire, global warming, and the carbon balance of boreal forests. *Ecological Applications* **5** (2): 437–451.

Keeley, J. E. (2009) Fire intensity, fire severity and burn severity: a brief review and suggested usage. *International Journal of Wildland Fire* **18**: 116–126.

Keeley, J. E., Fotheringham, C. J. and Morais, M. (1999) Reexamining fire suppression impacts on brushland fire regimes. *Science* **284** (5421): 1829–1832.

Kijko, A. (2004) Estimation of the maximum earthquake magnitude, m_{max}. *Pure and Applied Geophysics* **161**: 1655–1681.

King's College London (2011) Wildfire research at King's College London. wildfire.geog. kcl.ac.uk (accessed 1 July 2011).

KPB (2010) Wildfires (section 3) Kenai Peninsula Borough (KPB) All-Hazard Mitigation Plan. KPB, Soldotna, Alaska. www.borough.kenai.ak.us/emergency/hazmit/plan.htm (accessed 10 April 2011).

Krenn, R. and Hergarten, S. (2009) Cellular automaton modelling of lightning-induced and man made forest fires. *Natural Hazards and Earth System Sciences* **9**: 1743–1748.

Krinner, G., Viovy, N., de Noblet-Ducoudre, N., *et al.* (2005) A dynamic global vegetation model for studies of the coupled atmosphere–biosphere system. *Global Biogeochemical Cycles* **19** (1): n.p.

Lamarque, J. F., Edwards, D. P., Emmons, L. K., *et al.* (2003) Identification of CO plumes from MOPITT data: application to the August 2000 Idaho–Montana forest fires. *Geophysical Research Letters* **30** (13): n.p.

Landhäusser, S. M. and Lieffers, V. J. (1997) Seasonal changes in carbohydrate storage and regrowth in rhizomes and stems of four boreal forest shrubs: applications in *Picea glauca* understorey regeneration. *Scandinavian Journal of Forest Research* **12** (1): 27–32.

Lentile, L. B., Holden, Z. A., Smith, A. M. S., *et al.* (2006) Remote sensing techniques to assess active fire characteristics and post-fire effects. *International Journal of Wildland Fire* **15** (3): 319–345.

Linn, R., Reisner, J., Colman, J. J., *et al.* (2002) Studying wildfire behavior using FIRETEC. *International Journal of Wildland Fire* **11** (3–4): 233–246.

MACC (2011) Monitoring Atmospheric Composition and Climate Project (MACC): global fire emissions. www.gmes-atmosphere.eu/services/gac/fire/ (accessed 1 July 2011).

Malamud, B. D. and Turcotte, D. L. (2000) Cellular-automata models applied to natural hazards. *IEEE Computing in Science and Engineering* **2**: 42–51.

Malamud, B. D., Morein, G. and Turcotte, D. L. (1998) Forest fires: an example of self-organized critical behavior. *Science* **281**: 1840–1842.

Malamud, B. D., Millington, J. D. A. and Perry, G. L. W. (2005) Characterizing Wildfire Regimes in the U.S.A. *Proceedings of the National Academy of Science* **102**: 4694–4699.

Mandel, J., Beezley, J. D., Bennethum, L. S., *et al.* (2007) A dynamic data driven wildland fire model. *Computational Science–ICCS 2007, Pt 1, Proceedings* **4487**: 1042–1049.

Mandel, J., Bennethum, L. S., Beezley, J. D., *et al.* (2008) A wildland fire model with data assimilation. *Mathematics and Computers in Simulation* **79** (3): 584–606.

Marlon, J. R., Bartlein, P. J., Walsh, M. K., *et al.* (2009) Wildfire responses to abrupt climate change in North America. *Proceedings of the National Academy of Sciences of the United States of America* **106**: 2519–2524.

Matthews, S. (2009) A comparison of fire danger rating systems for use in forests. *Australian Meteorological and Oceanographic Journal* **58** (1): 41–48.

McKenzie, D., Gedalof, Z., Peterson, D. L., *et al.* (2004) Climatic change, wildfire, and conservation. *Conservation Biology* **18** (4): 890–902.

McLennan, J. and Birch, A. (2006) A potential crisis in wildfire emergency response capability? Australia's volunteer firefighters. *Global Environmental Change Part B: Environmental Hazards.* **6** (2): 101–107.

McMorrow, J. and Legg, C. (2009) Ways forward for the management of wildfire in the UK. Paper presented at Wildfire 2009, Lyndhurst, Hampshire, 16–17 June.

MCS (2009) The 6th Mediterranean Combustion Symposium, Workshop on Wildfire Research. www.ichmt.org/mcs-6/content/view/41/54/ (accessed 8 October 2009).

Mell, W., Jenkins, M. A., Gould, J., *et al.* (2007) A physics-based approach to modelling grassland fires. *International Journal of Wildland Fire* 16 (1): 1–22.

MFSL (2011) FireModels. Fire behaviour and fire danger software, Missoula Fire Sciences Laboratory. www.firemodels.org/ (accessed 12 May 2011).

Millington, J. D. A., Perry, G. L. W. and Malamud, B. D. (2006) Models, data and mechanisms: quantifying wildfire regimes. In *Fractal Analysis for Natural Hazards*, ed. G. Cello and B. D. Malamud, London: Geological Society, pp. 155–167.

Montiel, C. and Kraus, D. (eds) (2010) Best practices of fire use: prescribed burning and suppression fire programmes in selected case-study regions in Europe. European Forest Institute Research Report 24.www.efi.int/files/attachments/publications/efi_rr24.pdf (accessed 13 April 2011).

Moody, J. A. and Martin, D. A. (2001) Hydrological and sedimentologic response of two burned watersheds. US Geological Survey Water-Resources Investigation report.

Moody, J. A. and Martin, D. A. (2009a) Forest fire effects on geomorphic processes. In *Fire Effects on Soils and Restoration Strategies*, Enfield, NH: Science Publishers Inc., pp. 41–79.

Moody, J. A. and Martin, D. A. (2009b) Synthesis of sediment yields after wildland fire in different rainfall regimes in the western United States. *International Journal of Wildland Fire* 18: 96–115.

Munich Re. (2009) *Topics Geo: Natural Catastrophes 2008, Analyses, Assessments, Positions*, Munchen: Munchener Ruck.

NASA (2011a) Visible Earth (NASA image catalogue). EOS Project Science Office, NASA Goddard Space Flight Center. visibleearth.nasa.gov (accessed 7 July 2011).

NASA (2011b) Earth Observatory Images. EOS Project Science Office, NASA Goddard Space Flight Center. earthobservatory.nasa.gov (accessed 21 July 2011).

NFPA (2011) Firewise communities program. US National Fire Protection Association. www.firewise.org/ (accessed 13 May 2011).

NIFC (2009) Guidance for implementation of federal wildland fire management policy. US National Interagency Fire Center. http://www.nifc.gov/policies/policies_main.html (accessed 2 September 2011).

NIFC (2010a) Lightning and human caused fires (by geographic area): wildland fire statistics, provided by the National Interagency Coordination Center. National Interagency Fire Center. www.nifc.gov/fire_info/fire_stats.htm (accessed 3 December 2010).

NIFC (2010b) Total wildland fires and acres: wildland fire statistics, provided by the National Interagency Coordination Center. National Interagency Fire Center. www.nifc.gov/fire_info/fires_acres.htm (accessed 3 December 2010).

Niklasson, M. and Granstrom, A. (2000) Numbers and sizes of fires: long-term spatially explicit fire history in a Swedish boreal landscape. *Ecology* 81: 1484–1499.

Nilsson, M. C. and Wardle, D. A. (2005) Understory vegetation as a forest ecosystem driver: evidence from the northern Swedish boreal forest. *Frontiers in Ecology and the Environment* 3 (8): 421–428.

Noble, J. C. (1991) Behaviour of a very fast grassland wildfire on the riverine plain of southeastern Australia. *International Journal of Wildland Fire* 1 (3): 189–196.

NPS (2009) Wildland fire in Yellowstone. National Park Service, US Department of the Interior. www.nps.gov/yell/naturescience/wildlandfire.htm (accessed 2 October 2009).

NRC (2011a) Canadian wildland fire information system. Natural Resources Canada. cwfis.cfs.nrcan.gc.ca/ (accessed 6 May 2011).

NRC (2011b) Fire monitoring, mapping, and modeling (Fire M3): data sources and methods. Natural Resources Canada. cwfis.cfs.nrcan.gc.ca/en_CA/background/dsm/fm3 (accessed 13 May 2011).

Nunes, J. R. S., Soares, R. V. and Batista, A. C. (2008) Prometheus: an integrated wildfire control system. Modelling, monitoring and management of forest fires I. *Ecology and the Environment* **119**: 253–263.

OEHHA (2008) Wildfire guide: a guide for public health officials. Office of Environmental Health Hazard Assessment, California Environmental Protection Agency. www.oehha. ca.gov/air/risk_assess/wildfire.html.

Omi, P. N. (2005) *Forest Fires: A Reference Handbook*, Santa Barbara, CA: ABC-CLIO.

Page, S. E., Siegert, F., Rieley, J. O., *et al.* (2002) The amount of carbon released from peat and forest fires in Indonesia during 1997. *Nature* **420** (6911): 61–65.

Pastor, E., Zárate, L., Planas, E., *et al.* (2003) Mathematical models and calculation systems for the study of wildland fire behaviour. *Progress in Energy and Combustion Science* **29**: 139–153.

Perry, D. A. and Lotan, J. E. (1979) A model of fire selection for serotiny in lodgepole pine. *Evolution* **33** (3): 958–968.

Power, M. J., Marlon, J., Ortiz, N., *et al.* (2008) Changes in fire regimes since the Last Glacial Maximum: an assessment based on a global synthesis and analysis of charcoal data. *Climate Dynamics* **30** (7–8): 887–907.

Preisler, H. K., Burgan, R. E., Eidenshink, J. C., *et al.* (2009) Forecasting distributions of large federal-lands fires utilizing satellite and gridded weather information. *International Journal of Wildland Fire* **18** (5): 508–516.

Prentice, I. C., Bondeau, A., Cramer, W., *et al.* (2007) Dynamic global vegetation modeling: quantifying terrestrial ecosystem responses to large-scale environmental change. In *Terrestrial Ecosystems in a Changing World*, ed. J. G. Canadell, D. A. Pataki and L. Pitelka, Berlin: Springer, pp. 175–192.

Pyne, S. J., Andrews, P. L. and Laven, R. D. (1996) *Introduction to Wildland Fire*, New York: John Wiley and Sons.

Radeloff, V. C., Hammer, R. B., Stewart, S. I., *et al.* (2005) The wildland–urban interface in the United States. *Ecological Applications* **15** (3): 799–805.

Randerson, J. T., Liu, H., Flanner, M. G., *et al.* (2006) The impact of boreal forest fire on climate warming. *Science* **314** (5802): 1130–1132.

Rasmussen, J. H. and Fogarty, L. G. (1997) A case study of grassland fire behaviour and suppression: the Tikokino Fire of 31 January 1991. New Zealand Forest Research Institute, Rotorua, in association with the National Rural Fire Authority, Wellington. FRI Bulletin No. 197, Forest and Rural Fire Scientific and Technical Series, Report No. 2.

Raybould, S. and Roberts, T. (2006) A matrix approach to fire prescription writing. *Fire Management Today* **66** (1): 79–82.

Reed, W. J. and McKelvey, K. S. (2002) Power-law behaviour and parametric models for the size-distribution of forest fires. *Ecological Modelling* **150**: 239–254.

RIA Novosti (2010) Wildfires in Russia in 2010. en.rian.ru/trend/wildfires_2010/.

Ricotta, C., Avena, G. and Marchetti, M. (1999) The flaming sandpile: self-organized criticality and wildfires. *Ecological Modelling* **119**: 73–77.

Ricotta, C., Arianoutsou, M., Díaz-Delgado, R., *et al.* (2001) Self-organized criticality of wildfires ecologically revisited. *Ecological Modelling* **141**: 307–311.

Robinson, J. M. (1991) Fire from space: global fire evaluation using infrared remote sensing. *International Journal of Remote Sensing* **12** (1): 3–24.

Robischaud, P. R. and Elenbeer, H. (2001) Wildfire and superficial processes. *Hydrological Processes* **15**: 2865–3091.

Rothermel, R. C. (1972) *A Mathematical Model for Predicting Fire Spread in Wildland Fuels*. Ogden, UT: US Department of Agriculture, Forest Service, Intermountain Forest and Range Experiment Station.

Roy, D. P., Jin, Y., Lewis, P. E., *et al.* (2005) Prototyping a global algorithm for systematic fire-affected area mapping using MODIS time series data. *Remote Sensing of Environment* **97** (2): 137–162.

Roy, D. P., Boschetti, L., Justice, C. O., *et al.* (2008) The collection 5 MODIS burned area product: global evaluation by comparison with the MODIS active fire product. *Remote Sensing of Environment* **112** (9): 3690–3707.

Russell Smith, J., Ryan, P. G. and Durieu, R. (1997) A LANDSAT MSS-derived fire history of Kakadu National Park, monsoonal northern Australia, 1980–94: seasonal extent, frequency and patchiness. *Journal of Applied Ecology* **34** (3): 748–766.

Schlobohm, P. and Brain, J. (2002) Gaining an understanding of the national fire danger rating system. National Wildfire Coordinating Group Fire Danger Working Team, National Interagency Fire Center, NFES document number 2665, PMS 932.

Schoenberg, F. P., Peng, R. and Woods, J. (2003) On the distribution of wildfire sizes. *Environmetrics* **14**: 583–592.

Schoennagel, T., Nelson, C. R., Theobald, D. M., *et al.* (2009) Implementation of National Fire Plan treatments near the wildland–urban interface in the western United States. *Proceedings of the National Academy of Sciences of the United States of America* **106** (26): 10706–10711.

Schroeder, W., Csiszar, I. and Morisette, J. (2008a) Quantifying the impact of cloud obscuration on remote sensing of active fires in the Brazilian Amazon. *Remote Sensing of Environment* **112** (2): 456–470.

Schroeder, W., Prins, E., Giglio, L., *et al.* (2008b) Validation of GOES and MODIS active fire detection products using ASTER and ETM plus data. *Remote Sensing of Environment* **112** (5): 2711–2726.

Schuster, E. G., Cleaves, D. A. and Bell, E. F. (1997) Analysis of USDA Forest Service Fire-related Expenditures 1970–1995. Pacific Southwest Research Station, Forest Service, U.S. Department of Agriculture. Research Paper PSW-RP-230.

Scott, A. C. (2000) The Pre-Quaternary history of fire. *Palaeogeography, Palaeoclimatology, Palaeoecology* **164**: 281–329.

Scott, A. C. (2009) Forest fire in the fossil record. In *Fire Effects on Soils and Restoration Strategies.*, ed. A. Cerdà and P. Robichaud, Enfield, NH: Science Publishers Inc., pp. 1–37.

Scott, A. C. and Glasspool, I. J. (2007) Observations and experiments on the origin and formation of inertinite group macerals. *International Journal of Coal Geology* **70** (1–3): 53–66.

Scott, J. H. and Reinhardt, E. D. (2001) *Assessing Crown Fire Potential by Linking Models of Surface and Crown Fire Behavior*, Fort Collins, CO: US Department of Agriculture, Forest Service, Rocky Mountain Research Station.

Scott, A. C., Cripps, J., Nichols, G., *et al.* (2000) The taphonomy of charcoal following a recent heathland fire and some implications for the interpretation of fossil charcoal deposits. *Palaeogeography, Palaeoclimatology, Palaeoecology* **164**: 1–311.

Shakesby, R. A. and Doerr, S. H. (2006) Wildfire as a hydrological and geomorphological agent. *Earth Science Reviews* **74**: 269–307.

Sigler, J., Lee, X. and Munger, W. (2003) Emissions and long-range transport of gaseous mercury from a large-scale canadian boreal forest fire. *Environmental Science and Technology* 37 (19): 4343–4347.

Simmonds, P. G., Manning, A. J., Derwent, R. G., *et al.* (2005) A burning question: can recent growth rate anomalies in the greenhouse gases be attributed to large-scale biomass burning events? *Atmospheric Environment* 39 (14): 2513–2517.

Smith, A. M. S. and Wooster, M. J. (2005) Remote classification of head and backfire types from MODIS fire radiative power and smoke plume observations. *International Journal of Wildland Fire* 14 (3): 249–254.

Song, W. G., Fan, W. C., Wang, B. H., *et al.* (2001) Self-organized criticality of forest fire in China. *Ecological Modelling* 145: 61–68.

Stearns, J. R., Zahniser, M. S., Kolb, C. E., *et al.* (1986) Airborne infrared observations and analyses of a large forest fire. *Applied Optics* 25: 2554–2562.

Stedinger, J. R., Vogel, R. M., Foufoula-Georgiou, E., *et al.* (1993) Frequency analysis of extreme events. In *Handbook of Hydrology*, ed. D. R. Maidment, New York: McGraw-Hill.

Stocks, B. J., Lawson, B. D., Alexander, M. E., *et al.* (1989) The Canadian forest fire danger rating system: an overview. *Forestry Chronicle* 65 (6): 450–457.

Stocks, B. J., Mason, J. A., Todd, J. B., *et al.* (2002) Large forest fires in Canada, 1959–1997. *Journal of Geophysical Research – Atmospheres* 108 (D1): n.p.

Strauss, D., Bednar, L. and Mees, R. (1989) Do one percent of forest fires cause ninety-nine percent of the damage? *Forest Science* 35: 319–328.

Sullivan, A. (2009) Improving operational models of fire behaviour. In *18th World IMACS Congress and MODSIM09 International Congress on Modelling and Simulation, July 2009*, ed. R. S. Anderssen, R. D. Braddock, and L. T. H. Newham, Christchurch, New Zealand: Modelling and Simulation Society of Australia and New Zealand and International Association for Mathematics and Computers in Simulation, pp. 2377–2383.

Sullivan, A. L. and Knight, I. K. (2008) A hybrid cellular automata/semi-physical model of fire growth. *Complexity International* 12. www.complexity.org.au/ci/vol12/msid09/.

Sun, R., Jenkins, M. A., Krueger, S. K., *et al.* (2006) An evaluation of fire-plume properties simulated with the Fire Dynamics Simulator (FDS) and the Clark coupled wildfire model. *Canadian Journal of Forest Research* 36 (11): 2894–2908.

Swetnam, T. W. and Betancourt, J. L. (1990) Fire–Southern oscillation relations in the southwestern United States. *Science* 249 (4972): 1017–1020.

Taylor, S. W. and Alexander, M. E. (2003) Considerations in developing a national forest fire danger rating system. In *Proceedings of the XII World Forestry Congress (September 21–28, 2003, Quebec, Que.) Volume B – Forest for the Planet*, Quebec: Food and Agricultural Organization of the United Nations, Natural Resources Canada, Canadian Forest Service and Government of Quebec. warehouse.pfc.forestry.ca/pfc/23859.pdf (accessed 31 March 2011), p. 227.

Thompson, M. P. and Calkin, D. E. (2011) Uncertainty and risk in wildland fire management: a review. *Journal of Environmental Management* 92 (8): 1895–1909.

Thompson, M. P., Calkin, D. E., Finney, M. A., *et al.* (2011) Integrated national-scale assessment of wildfire risk to human and ecological values. *Stochastic Environmental Research and Risk Assessment* 25 (6): 761–780.

Trouet, V., Taylor, A., Carleton, A., *et al.* (2009) Interannual variations in fire weather, fire extent, and synoptic-scale circulation patterns in northern California and Oregon. *Theoretical and Applied Climatology* 95 (3–4): 349–360.

Turcotte, D. L. (1999) Self-organized criticality. *Reports on Progress in Physics* **62**: 1377–1429.

Turcotte, D. L. and Malamud, B. D. (2004) Landslides, forest fires, and earthquakes: examples of self-organized critical behaviour. *Physica A* **340**: 580–589.

Turetsky, M. R., Harden, J. W., Friedli, H. R., *et al.* (2006) Wildfires threaten mercury stocks in northern soils. *Geophysical Research Letters* **33** (16): n.p.

Tymstra, C., Bryce, R. W., Wotton, B. M., *et al.* (2010) *Development and Structure of Prometheus: The Canadian Wildland Fire Growth Simulation Model*, Edmonton: Natural Resources Canada, Canadian Forest Service. firegrowthmodel.ca/documentation.html (accessed 12 May 2011).

Umscheid, M. E., Monteverdi, J. P. and Davies, J. M. (2006) Photographs and analysis of an unusually large and long-lived firewhirl. *Electronic Journal of Severe Storms Meteorology* **1** (2): n.p.

UNECE (2011) Forest fire statistics. United Nations Economic Commission for Europe and FAO European Forestry Commission. www.unece.org/timber/ff-stats.html (accessed 31 March 2011).

University of Maryland (2011) Fire Information for Resource Management System. maps.geog.umd.edu/firms/ (accessed 1 July 2011).

USDA (2011) Active Fire Mapping Program. USDA Forest Service, Remote Sensing Applications Center. activefiremaps.fs.fed.us (accessed 13 May 2011).

USGS (2006) Wildfire Hazards: A National Threat. US Department of the Interior, US Geological Survey Fact Sheet 2006–3015. pubs.usgs.gov/fs/2006/3015/ (accessed 21 July 2011).

van der Werf, G. R., Randerson, J. T., Giglio, L., *et al.* (2006) Interannual variability in global biomass burning emissions from 1997 to 2004. *Atmospheric Chemistry and Physics* **6**: 3423–3441.

van der Werf, G. R., Randerson, J. T., Giglio, L., *et al.* (2010) Global fire emissions and the contribution of deforestation, savanna, forest, agricultural, and peat fires (1997–2009). *Atmospheric Chemistry and Physics* **10** (23): 11707–11735.

Van Wagner, C. E. (1977) Conditions for the start and spread of crownfire. *Canadian Journal of Forest Research* **7**: 23–24.

Van Wagner, C. E. (1979) A laboratory study of weather effects on the drying rate of jack pine litter. *Canadian Journal of Forest Research* **9** (2): 267–275.

Van Wagner, C. E. (1993) Prediction of crown fire behavior in two stands of jack pine. *Canadian Journal of Forest Research* **23**: 442–449.

van Wagtendonk, J. W. (2007) The history and evolution of wildland fire use. *Fire Ecology* **3** (2): 3–17.

VBRC (2009) Victorian Bushfires Royal Commission Final Report. State Government of Victoria. www.royalcommission.vic.gov.au (accessed 3 December 2010).

Viegas, D. X. (2006) Forest fires in Portugal in 2005: an overview. *International Forest Fire News* **34**: n.p.

Wang, H. (2006) Ember attack: its role in the destruction of houses during ACT Bushfire in 2003. Paper presented at the Bushfire Conference 2006, Life in a Fire-Prone Environment: Translating Science into Practice. Brisbane, 6–9 June. www.griffith.edu.au/conference/bushfire2006/ (accessed 6 July 2011).

Ward, P. C., Tithecott, A. G. and Wotton, B. M. (2001) Reply: a re-examination of the effects of fire suppression in the boreal forest. *Canadian Journal of Forest Research* **31**: 1467–1480.

Weber, R. O. and Dold, J. W. (2006) Linking landscape fires and local meteorology: a short review. *JSME International Journal Series B–Fluids and Thermal Engineering* **49** (3): 590–593.

Westerling, A. L., Gershunov, A., Brown, T. J., *et al.* (2003) Climate and wildfire in the western United States. *Bulletin of the American Meteorological Society* **84** (5): 595–604.

Westerling, A. L., Hidalgo, H. G., Cayan, D. R., *et al.* (2006) Warming and earlier spring increase western US forest wildfire activity. *Science* **313** (5789): 940–943.

White, E., Enquist, B. and Green, J. (2008) On estimating the exponent of power-law frequency distributions. *Ecology* **89**: 905–912.

Whitlock, C. (2004) Land management: forests, fires and climate. *Nature* **432**: 28–29.

Whitlock, C., Skinner, C. N., Bartlein, P. J., *et al.* (2004) Comparison of charcoal and tree-ring records of recent fires in the eastern Klamath Mountains, California, USA. *Canadian Journal of Forest Research–Revue Canadienne De Recherche Forestiere* **34** (10): 2110–2121.

Williams, T. (2002) America's largest weed. *Audubon Magazine*, January. www.audubon-magazine.org/incite/incite0201.html (accessed 21 July 2011).

Wooster, M. J. and Strub, N. (2002) Study of the 1997 Borneo fires: Quantitative analysis using global area coverage (GAC) satellite data. *Global Biogeochemical Cycles* **16** (1): n.p.

Wooster, M. J., Zhukov, B. and Oertel, D. (2003) Fire radiative energy for quantitative study of biomass burning: derivation from the BIRD experimental satellite and comparison to MODIS fire products. *Remote Sensing of Environment* **86** (1): 83–107.

Wooster, M. J., Roberts, G., Perry, G. L. W., *et al.* (2005) Retrieval of biomass combustion rates and totals from fire radiative power observations: FRP derivation and calibration relationships between biomass consumption and fire radiative energy release. *Journal of Geophysical Research* **110** (D24): n.p.

Wooster, M. J., Perry, G. L. W. and Zoumas, A. (2011) Fire, drought and El Niño relationships on Borneo during the pre-MODIS era (1980–2000). *Biogeosciences Discussions* **8**: 975–1013.

WRF (Weather Research and Forecasting) (2011) WRF Model. www.wrf-model.org (accessed 1 February 2011).

Wright, H. E. (1974) Landscape development, forest fires, and wilderness management. *Science* **186** (4163): 487–495.

Xanthopoulos, G. (2004) Who should be responsible for forest fires? Lessons from the Greek experience. *Proceedings of the Second International Symposium on Fire Economics, Planning and Policy: A Global View, Cordoba, Spain, 19–22 April*, Albany, CA: USDA Forest Service, Pacific Southwest Research Station, p. 128.

Xu, W., Wooster, M. J., Roberts, G., *et al.* (2010) New GOES imager algorithms for cloud and active fire detection and fire radiative power assessment across North, South and Central America. *Remote Sensing of Environment* **114** (9): 1876–1895.

Zhang, Y. H., Wooster, M. J., Tutubalina, O., *et al.* (2003) Monthly burned area and forest fire carbon emission estimates for the Russian Federation from SPOT VGT. *Remote Sensing of Environment* **87**: 1–15.

Zhukov, B., Lorenz, E., Oertel, D., *et al.* (2006) Spaceborne detection and characterization of fires during the bi-spectral infrared detection (BIRD) experimental small satellite mission (2001–2004). *Remote Sensing of Environment* **100** (1): 29–51.

13

Catastrophic impacts of natural hazards on technological facilities and infrastructure

R. S. J. SPARKS, W. P. ASPINALL, N. A. CHAPMAN, B. E. HILL,
D. J. KERRIDGE, J. POOLEY AND C. A. TAYLOR

13.1 Introduction

The term 'technological facilities' is used here to describe facilities that, were their integrity to be seriously compromised by impacts from natural events, would cause either catastrophic loss of vital services to a community or nation, significant environmental pollution, loss of life or disruption of commercial activities and supply chains that lead to economic losses – possibly all of these. Such facilities include power stations, energy supply hubs and lines, dams and reservoirs, pipelines carrying hazardous materials, electronic communications (especially the internet), plants and factories containing hazardous materials, offshore facilities such as oil rigs, storage or disposal facilities for hazardous materials, key transport infrastructure such as airports, ports, roads and railway systems, medical facilities such as hospitals, and industrial facilities such as oil refineries and an increasing dependence on space-based infrastructure for a wide variety of services (Figures 13.1–13.4). Specific examples that are discussed in more detail in this chapter are: nuclear power stations, pipelines, electricity distribution grids, hazardous chemical refineries, deep geological repositories for long-lived radioactive wastes and disruption of transport services and communication systems.

This chapter includes a survey of examples of the effects of the major natural events on technological facilities. We focus on a number of recent case studies, notably the effects of recent earthquakes on nuclear power plants in Japan and the effects of volcanic ash in closing down airports and European airspace in 2010. These cases illustrate some key generic issues in risk assessment such as the need for probabilistic as well as deterministic hazard and risk assessments, and the issues of identifying and forecasting extreme natural events, and accounting for all relevant uncertainties. We include also in this chapter a discussion of space weather, as this is a natural hazard that has not been considered elsewhere in the book, but poses major threats to communication and electricity supply systems. We note that natural hazards only become a meaningful concept when natural processes interact with and influence human activities.

13.2 Generic issues

The catastrophic impacts of natural hazard events are in essence the disruption or destruction of the services, and their associated resource flows, that enable all aspects of societal well-

Risk and Uncertainty Assessment for Natural Hazards, ed. Jonathan Rougier, Steve Sparks and Lisa Hill. Published by Cambridge University Press. © Cambridge University Press 2013.

Figure 13.1 Nuclear power plant in the Eifel region, Germany with Holocene volcanic deposits in the inset.

Figure 13.2 Significant earthquake-induced liquefaction damage to the quay walls located at the eastern edge of the Kashiwazaki Post terminal adjacent to the mouth of the U-river, Japan.

Figure 13.3 Ground failures at the site of the Kashiwazki-Kariwa nuclear power plant, Japan.

Figure 13.4 Safety-critical industrial facilities are increasingly being monitored for earthquake shaking, even in low-seismicity regions: installation of a seismic acceleration recording instrument at an above-ground pig trap on a major buried 1200 mm diameter natural gas pipeline in the UK.

being. These service and resource flows are delivered through the technological facilities and infrastructure of our modern communities. Damage to, for example, a bridge, power station, port, or telecommunication system will reduce the level of service that it provides, with knock-on consequences to the service levels of other dependent facilities and infrastructure. Thus, catastrophes are systems effects and need to be viewed and treated as such. Improving or protecting the performance of a component of the system may be necessary, but it is not always sufficient to assure the continued satisfactory performance of the system as a whole (Blockley and Godfrey, 2000). As will become apparent in this chapter, systems behaviours as a result of a natural hazard event are often complex, unexpected and uncertain. We therefore need to evolve methods for dealing with this complexity and the risks that emerge from it.

For technological facilities risk assessments provide confidence that the facility or service is robust to natural hazards, that the threat from loss of function or release of pollutants is very low, and, in the case of a critical facility such as a power station, that the consequences of its failure are recognised and steps are taken to reduce risk to acceptable levels based on national or international regulations. Commonly, technological facilities represent localised vulnerabilities, but we also include here consideration of distributed facilities such as national electrical supply grids, transport systems and civil aviation services. Toxic and radioactive wastes are examples of substances held in technological facilities where there are associated public health concerns. These concerns arise from the potential consequences if safety is compromised and from the potential for widespread impacts of varying intensity. Localised failures can lead to wider impacts. For example, a power station or electricity sub-station that ceases functioning can cause downstream risks and threats, such as disruption or failure to emergency responses or failure of safety-critical systems. Disruption to electricity supplies at a wider level can create a cascade of risks to health and well-being. Disruption to major communications centres that direct telecommunications and internet-based connections, along with the complex network systems involved, has the potential to be enormously damaging, as many national-scale (as well as facility-scale) supervisory control and data acquisition (SCADA) systems can be dependent on such communication. Their design and operation should therefore have resilience as a performance goal. In an increasingly globalised world with complex interdependent technological systems and cross-border dependencies, major natural events can cause multiple failure and disruption in cascades of causes and effects. As such, there is increasing demand for risk information that anticipates complex events and their consequences.

A more sophisticated understanding of infrastructure performance is emerging, driven strongly by the appreciation of the importance of resilience across all technological and societal systems. Performance is not a static property in general. While under normal conditions it might be at a steady state, at times of perturbation, for example during and after an earthquake, the acceptable performance requirements of a facility and the attainment of that performance might change. Some pragmatic sacrifices in normal performance expectations might have to be made in order to maintain crucial performance aspects that are deemed, by society, to be imperative or sacrosanct.

One useful view of resilience is as the pathway that a technological system follows to return to normality after a perturbation or shock, and the speed with which this is achieved (Bruneau and Reinhorn, 2004). For mild perturbations that follow well-understood and predictable patterns, such resilient recovery can be made intrinsic to a system. However, for catastrophic perturbations, such as major earthquakes, resilience generally cannot be automatic. In such cases, the actual conditions will not be known until the event happens, at which point the controlling actors must recognise failure symptoms and adapt rapidly so as to manage the system performance within acceptable levels until they re-establish normality. Scenario planning and enactment are valuable for anticipating the impacts of events, in particular for identifying the actions (and the means that need to be provided to enact them) that are required to improve the system performance. Such scenarios can never anticipate the uniqueness of the actual event. Building in the means for rapid adaptability is crucial for attaining resilience; this applies equally to the learning capacity of the individuals and organisations involved as it does to the artefacts themselves (Marashi and Davis, 2006).

A serious event can even sometimes prompt re-evaluation of what constitutes normality. At times of catastrophe, determining acceptable performance runs hand in hand with achieving it. This generally requires political will as well as good technical decision-making, both of which must be harmonised in purpose and approach if they are to be truly successful. The concept of risk implicitly underpins these decisions. Making the concept more explicit, through framing and enabling it consistently, coherently and appropriately, should make such decisions more effective and enable us to be more realistic about what performance levels can actually be achieved.

Risk assessment for technological facilities and networks related to natural hazards involves consideration of the engineering impacts (e.g. robustness to seismic shaking and tsunami inundation) and characteristics of harmful substances in relation to their interaction with the environment, in particular the biosphere. Thus, it is likely that risk assessments are going to be highly specific to the site or network, the nature of the technology and its function, and harmful substances that might be released. There is also a need for broader risk assessment and risk forecasting that looks at how natural events impact on those connected systems of technological facilities and infrastructure that support many aspects of the modern world. There are also complex questions of responsibility when it comes to the identification and management of risk. Numerous different parties may all have a role in this, depending on the technology involved, and their roles may change; this includes, for example, government, at national and local levels, regulators, international agencies, the insurance sector and private industry. There are also cross-border aspects of responsibility and liability that must be considered.

The methods of quantitative risk and uncertainty assessment are based on some generic principles, so we recommend taking an inclusive approach to this area of research (see Chapter 2). Locations affected by natural hazards are usually complex systems of populations and infrastructures, including technological facilities. This complexity involves interactions that result in knock-on effects; for example, failure of a power station may lead to failure of other critical facilities and services whose functioning relies on that power. The timescales in risk assessment are longer than those traditionally involved in natural hazards, which tend to focus on imminent dangers and evolving crises. Threats to specific facilities and technological networks typically need long-term risk assessment and siting, design and operational procedure steps that reduce the risk of deleterious events to acceptably low levels. The concept of acceptable risk is commonly framed by environmental regulations, laws and guidelines developed by national, regional (e.g. the European Union) and international agencies. However, it is a complex societal issue involving the weighing and balancing of different benefits, disbenefits and values, a multidisciplinary issue spanning social sciences, natural sciences and engineering, which has not been fully explored.

We note that the scientific literature on the effects of hazard and risk assessment concerning technological facilities and critical infrastructure is very unevenly developed, especially with regard to publications that document the evidence and describe systematic investigations and risk assessment. Much of the information is anecdotal, found within sources such as newspapers, websites and grey literature, and may not be authoritative or accurate and is rarely peer-reviewed. In major disasters, the literature is often focused on the natural hazard event, fatalities, economic losses and social consequences. Damage to technological facilities and critical infrastructure is sometimes noted in passing, but is not so commonly the focus of systematic study. It seems likely that there is abundant information, evidence and analysis, but this may arise from internal investigations in, for example, private companies and government organisations, and is hard to access or may be confidential. The research literature is growing, but remains quite sparse compared to other aspects of natural hazards and disasters.

13.3 Timescales

One of the key generic issues is the timescales of concern. A technological facility may be in existence for several decades, perhaps up to 100 years. In the case of some radioactive waste repositories (Chapman and Hooper, 2012), the assessment timescales become geological, as regulations typically require safety to be demonstrated over timescales of many thousands – up to a million – years. Over such periods (a few thousand to a couple of hundreds of thousands of years), radioactive waste decays to a low hazard potential, often to radioactivity or radiotoxicity levels equivalent to the natural ore from which nuclear fuel was first extracted. Thus, risk assessment contrasts greatly with the needs of, for example, the insurance industry, where only a few years might be considered the priority, and events with return periods of much more than 100 years cease to be of interest. For very long-term critical technological facilities, with active, operational lifetimes of 100 years or more, or passive functional lifetimes of many thousands of years, what are normally viewed as rare and incredible events may be considered credible in the lifetime of the facility. Additionally, uncertainties in the suitability and applicability of models and data to assess risk can increase significantly when extrapolated to millennia, or longer, or are used to estimate shorter-term probabilities of occurrence that are lower than 10^{-4} per annum.

Temporal variability of natural hazards occurrences is an important factor for technological risk assessment. Many modern technologies have emerged very rapidly and spread globally, on a timescale of just a few years and have become crucial factors in the way that society and its infrastructure functions. If such systems were to fail catastrophically on a global scale the consequences to human health and welfare would be enormous. The internet and mobile telephony lie at the core of this enormous reliance on new technologies that penetrates every part of our existence: failure of the internet would destroy our ability to control and manage huge areas of our transport, communications, safety and security infrastructure and bring whole areas of activity to a standstill within hours or days. Very few of the latest technologies have been subjected to the stresses that can come from intermittent major natural hazard events, which typically recur only on decadal or longer timescales or vary in intensity with long-term trends. As examples, solar activity and space weather have been subdued for several years and volcanic activity in Iceland and controlling weather patterns have until recently kept ash away from north-west Europe's dense, and still growing, civil aviation networks and infrastructure. Extending research into the longer-term temporal aspects of technological risk assessment is needed.

There is also an increasing need, in a globalised world, for longer-term risk assessments to inform major investment decisions by commerce and governments, and risk information is a rapidly growing arena of business into which the insurance industry is naturally moving. The global dependency issue has been graphically illustrated by the 2011 floods in Thailand, where technological supply chains from component manufacturers in that country have been severely disrupted, leading to major problems in car and electronic industries from Japan to the UK (Courbage *et al.*, 2012). While individual components are often environmentally qualified during their design and manufacture (for example, by being shaken with simulated earthquake loads), testing of whole distributed systems is problematic and only happens fully during the hazard event.

13.4 Effects of natural hazards on technological facilities: some examples

The need for hazard and risk assessment of industrial complexes is paramount, since their failure can cause damaging explosions and sometimes release pollutants and poisons (e.g. Chernobyl, Flixborough chemical plant, Union Carbide at Bhopal, the fires at Buncefield, UK and the Texas City refinery in 2005 and Deep Water Horizon oil platform, Gulf of Mexico in 2010). Here we are focused on failure initiated by natural hazards, while all of these examples are related to failure of the engineering or the control system, possibly compounded by human error. Technological accidents are not within the scope of this chapter, but the interested reader is referred to the European Environment Agency Report 13/2010 for a discussion. Nevertheless, there are many examples where technological facilities have been affected by natural events, examples of which are now described. Further technological accidents can be exacerbated or ameliorated by environmental conditions or coincidence of technological failures with hazardous natural events. In some cases the distinction between natural disasters and technological accidents is blurred and may even be counter-productive.

There are many instances of earthquakes causing major problems for technological facilities, highlighted by the events at the Fukushima Dai-ichi power plant in Japan as a consequence of the magnitude (M) 9 Tohoku earthquake and associated tsunami on 11 March 2011. The Chuetsu-oki earthquake in Niigata, Japan 2007 caused liquefaction and ground deformation, and some limited damage to non-safety-related systems at Kashiwazaki-Kariwa nuclear power plant complex, with public concerns resulting in its closure for two years (Kayen *et al.*, 2009). The Shagonar earthquake of 13 March 1994, the Krasnoyarsk earthquake of 27 October 2000 and the Altai earthquake of 27 September 2003 (M7.3) caused damage at the Sayano-Shushenskaya hydroelectric power station in Russia, including disruption to the foundations of the dam (Marchuk Izvestiya, 2008). Likewise, the power-generation facilities of hydropower plants were heavily damaged following a Wenchuan earthquake in May 2008 (Houqon, 2009). The Spitak earthquake (M6.9) in 1988 shut down the Metsamor nuclear power plant in Armenia for seven years, although there is no record of any damage.

Just the threat of earthquakes can cause problems if the seismic hazard is thought to be too high. The Mülheim-Kärlich nuclear power plant, Germany was taken offline after three years of operation due to problems with building permits (Wirtz, 1987; Roth, 1998). The location was moved in the early stages of construction because it was thought that the construction site lay in an earthquake-prone basin (Neuwieder Becken). Subsequently, this same site has attracted attention as a case example for volcanic hazards due to its proximity to the Holocene Eifel volcanic field. For example, this site is underlain by several metres of tephra fall and associated sedimentary deposits that resulted from blockage of the Rhein River during the Laacher See eruption about 13 000 years ago (Schmincke, 2004). New IAEA guidelines for siting of nuclear facilities in relation to volcanic hazards (IAEA, 2011) would require a new facility at this location to be assessed in detail.

Many nuclear reactors are designed to shut down automatically during an earthquake. Examples of automatic shutdown include: three reactors at the Chinshan and Kuosheng power plants in Taiwan due to a M7.6 earthquake on 21 September 1999 and three reactors in the Onegawa power plant in Tohoku, Japan due to a M7.2 earthquake on 16 August 2005 (http://world-nuclear.org/info/inf18.html; Nuclear Safety Commission, 2006). The 23 August 2011 M5.8 earthquake in Virginia triggered the first automatic shutdown of a nuclear power plant in the United States due to seismicity, causing the North Anna plant to remain offline for nearly three months while safety assessments were conducted.

Damage to buildings is the most prominent consequence of earthquakes, and some public and commercial properties constitute critical technological facilities and infrastructure. There are numerous examples of earthquake damage to hospitals that adversely affect communication and technical systems, the ability to respond to the disaster, trauma care and the provision of equipment, water and gas supplies (Myrtle *et al.*, 2005; Kirsch *et al.*, 2010; McIntyre *et al.*, 2011). Telecommunications buildings can be damaged as in Dujiangyan County, China, following the Wenchuan earthquake in 2008, when buildings suffered 8 cm subsidence due to liquefaction (Huang and Jiang, 2010). Damage to dams from earthquakes can affect water supplies and irrigation and hydroelectric schemes, and can lead to catastrophic failure and associated floods (e.g. Blazquez and Lopez-Querol, 2007; Huang and Jiang, 2010). Damage, including cracking, slope sliding, crest settlement, leakage and failure of spillways and outlets, was reported to 1803 dams used mostly for water supply and irrigation following Wenchuan earthquake, China. Earthquakes can also damage submarine telecommunication cables. For example, the December 2006 Taiwan earthquake cut submarine cables, interrupting telecommunication traffic throughout Southeast Asia and Australasia for approximately two months (Dominey-Howes and Goff, 2009).

Modern seismic design standards, for example those embodied in Eurocode 8, have been shown to be very effective in ensuring life safety during catastrophic earthquakes. A critical issue is to ensure that those applying these standards – designers, constructors and operators – learn and apply these standards properly for the benefit of the people who rely on them – owners, users and the public. Education, training and quality assurance are equally important as the technical standards. Although these standards are proving effective for protecting life, they are based often on the concept of allowing controlled damage to develop in the structure or artefact, to ride out the event; this damage may render the latter unserviceable after the event. This is prompting a move towards performance-based engineering (PBE) approaches that seek to achieve service continuity as well as life safety. PBE methods depend on reliability and risk-based approaches (http://nisee.berkeley.edu/lessons/krawinkler.html).

Volcanic hazards commonly disrupt technological facilities in a variety of ways that reflect the multi-hazards character of many volcanic eruptions (see Chapter 6). Lahars generated during the eruption of Raupehu, New Zealand in 1995 caused infilling of Rangipo Reservoir and damage to blades of Rangipo hydroelectric power station (Cronin *et al.*, 1997). Ash penetrated the water supply and electric generating plant at Whakapapa during the same eruption (Malcolm and van Rossen, 1997), while volcanic ash caused shorting of high-voltage electric power lines at the base of the volcano, affecting all of the North Island, including some hospitals (Johnston *et al.*, 2000). Sediments from the 18 May 1980 Mount St Helens eruption extended at least 8 km upstream from the point of entry in the Columbia River to the location of the Trojan nuclear power plant, Oregon (Schuster, 1981). The Trojan plant was offline for refuelling prior to the eruption, so the plant's water intake system was not affected by suspended sediments. Although sediment deposits reached 12 m thickness in the main Columbia River channel near Trojan, these sediments did not reach the plant's water intake system. The most widely known effect of volcanism on technological infrastructure is the disruptive effects of ash in the atmosphere on aviation, which is discussed in more detail as a case study below.

Landslides and concerns about slope stability are commonly a major issue. For example, the Rosone landslide damaged facilities at Turin hydroelectric power plant in the Orco River valley, western Italian Alps (Castelli and Scavia, 2008; Pisani *et al.*, 2010), while landslides in the Mailuu-Suu Valley, Kyrgyzstan, threatened exposure and contamination by mine tailings and radioactive waste dumps along the bank of the river in what is a very landslide-prone area (Havenith *et al.*, 2006). An earthquake in May 2008 at Xiaoba Reservoir in Sichuan, China affected slope stability at the dam and caused concerns for the nearby power plant (Li *et al.*, 2009).

Storms remain a major cause of technological disruption, with damage to electrical and communications systems being commonplace. The report of the European Environment Agency (2010) on natural hazards in Europe lists three storms (Anatol, Jeanett and Kyrill) in the 1998–2009 period as causing major disruption to power supplies. Kyrill caused major power losses in six countries. Hurricane Katrina in 2005 demonstrated numerous examples of disruption and destruction to critical infrastructure, including loss of internal power-generation capability and overheating at Oschner Health System in New Orleans, with intermittent losses of commercial power forcing a shift to generator power (Gamble, 2008). Damage to public structures, infrastructure and utilities (including roads, railroads, water defences, electricity network, drainage, etc.) due to Katrina was estimated at US$7 billion (Pistrika and Jonkman, 2010). Hurricane Katrina also damaged telecommunication and power infrastructure, preventing calls for aid and hampering relief efforts (Kwasinski *et al.*, 2009). Failure was due to storm surges and high winds, which caused damage to the grid and transmission and distribution system. Communication network availability was reduced by approximately 15% and up to 75% of some areas were still without utility power six weeks after the storm.

Extreme cold or heat can be another natural cause of problems for technological facilities (Lehner *et al.*, 2005). For example, the summer heat wave in Europe in 2003 resulted in some nuclear and conventional power stations having to shut down (UNEP, 2004). The nuclear reactors in France are cooled by river water and the river levels became too low in some places to allow the cooling to continue effectively, with water being returned to the natural environment after cooling being too warm to meet environmental protection regulations. The reduced capacity for electricity generation combined with greatly increased use of air-conditioning led to some power cuts, exacerbating the problems associated with the heat wave. Droughts can also cause problems for hydroelectric generation if rivers run dry. In 2003–2004, drought on the Iberian Peninsula caused a reduction of 40% production in hydroelectric plants (García-Herrera *et al.*, 2007).

Extreme precipitation and floods are immensely disruptive and cause huge economic losses. Specific, well-documented cases of damage to technological facilities and services are surprisingly sparse. The 2007 floods in the UK were reportedly responsible for the biggest civil emergency in the history of the UK since the Second World War (European Environment Agency, 2010). Half a million people lost access to their water and electricity supplies, and in Gloucestershire 350 000 people lost their water supply for 17 days. Storm floods overtopped the Alem da Fazenda gravity dam on the Flores Island in the Azores, June 1995, damaging the structure, causing heavy erosion and severe loss of confinement and leakage, and threatening the thermal and hydroelectric power station which supplies the island's power (Vasquez *et al.*, 2001). The 2011 floods in Thailand are a tragic example of the effects of extreme floods on a complex technologically based economy with cascading effects around the world. Changes in land-use, rapid urbanisation and inadequate flood defences seem to have been contributing factors to a disaster that has halved the growth forecast for Thailand. The floods resulted in major economic losses for companies like Sony and Honda, and have had adverse impacts on component supply chains, disrupting manufacturing in several parts of the world.

Space weather also creates conditions potentially hazardous to assets in space and on the ground. The effects are known as geomagnetically induced current (GIC), which can have dramatic effects on electrical systems and can induce failure. Historical accounts give some indications of the effects of space weather hazard. US-wide and Canadian reports of the famous Carrington event of 28 August–2 September 1859 described earth currents entering the telegraph wires, Morse line communication difficulties and electric shocks experienced by telegraph operators. Alternating currents evidently entered the telegraph wires as a consequence of interference due to auroral currents. The magnetic storm of 13 May 1921 disrupted transatlantic and US telegraph lines. It caused the New York central railroad signals and switching system to be put out of action, and there were electrical fires at New England Central railway station and a Swedish telephone station.

The term 'space weather' refers to conditions potentially hazardous to assets in space and on the ground, driven by a variety of solar processes. One important space weather effect is the generation of large magnetic disturbances known as magnetic storms. During magnetic storms, rapid changes in the earth's magnetic field induce electric fields in the solid earth and oceans, driving GICs into the low-resistance paths provided by grounded systems. Historical accounts show how space weather hazard first came to light as telegraphic communications began. US and Canadian reports of the Carrington event described the effects of GIC entering telegraph wires and generally disrupting communications, but allowing communications without batteries in some cases. The association of telegraph disruption with auroral displays was already established by experience, but not explained. The Carrington event firmly established the link between solar disturbances, magnetic storms and effects on the earth.

The potential for major disruption and damage to modern-day ground-based systems due to GIC became apparent during the magnetic storm of 13 March 1989, when failure of capacitors to regulate currents resulted in collapse of the whole of the Hydro-Québec power grid (9500 MW output), leaving six million people without electricity for nine hours (Bolduc, 2002). This was a cascade event in which protective components of the Hydro-Québec power system tripped in little over a minute. The direct costs of the blackout have been estimated at US$2 billion. During the same storm, a large step-up transformer at the Salem nuclear power plant in New Jersey was severely damaged and Erinmez *et al.* (2002) refer to damage to two transformers in the UK National Grid due to overheating.

As the space age began, space-based technologies became exposed to space weather hazard and measurements of the hazard were made by spacecraft. The solar storm that started on 2 August 1972 began with three powerful solar flares, and the Pioneer-9 spacecraft detected a shockwave in the solar wind the following day. The threat to humans involved in space exploration became apparent. Apollo 16 astronauts had been on the moon in April 1972, and the Apollo 17 mission followed in December 1972. If the August storm had happened during either mission, while astronauts were on the lunar surface, they would have been exposed to severe levels of radiation from high-energy protons released at the time of the flares.

The 'Halloween Storms' between 19 October and 7 November 2003, more than three years after the peak of an 'average' solar cycle, demonstrated how modern systems and services can be affected by a major space weather event. Webb and Allen (2004) list 33 satellite anomalies. The storms were implicated in the loss of the US$640 million Japanese ADEOS-II research satellite, airlines re-routed polar flights, astronauts on the International Space Station were ordered to shelter in a shielded compartment, there was a power outage in Sweden, degradation of GPS accuracy and widespread radio blackouts (NOAA, 2004). Hapgood and Thomson (2010) provide a thorough survey of impacts from GIC.

Much of human development in the last century has involved unremitting growth in coastally located infrastructure, with the result that many industrial complexes and other critical facilities are close to the ocean. Megacities and sprawling urban ribbon developments (with populations in the tens of millions) have appeared in several regions of the world that are coastal and prone to major tectonic events and ground subsidence. Commonly this is a consequence of water extraction, as well as climatically induced changes in sea level. A disaster in a megacity from natural hazards is going to affect a very complex system. One of the main hospitals in Tehran, for example, is built directly on the thrust fault system that has destroyed Tehran four times in its history. The destruction of this hospital in a major earthquake will exacerbate these problems when another large earthquake impacts this megacity of 17 million people.

The enormous sprawl of development that is metropolitan Tokyo has a population of over 32 million and is considered to be at high and imminent (within the next decades) risk of another major earthquake (Toda *et al.*, 2008). San Francisco, Mexico City, Seattle and Wellington are among other major cities at risk from seismic and volcanic hazard. Arguably, all megacities should have a comprehensive risk assessment that considers all aspects of the complex interactions associated with a natural event. Sadly this is not the case in many cities, including those in the developed world, as Hurricane Katrina illustrated in the city of New Orleans. It is actions taken in response to risk assessments that are critical.

Viewing these kind of disasters triggered by natural hazards only in terms of the immediate effects of one particular kind of hazard is necessary, but not sufficient. Instead we believe that the future lies in multi-hazard risk assessment of regions, cities and communities that takes a long-term view, which can then be fed into planning, mitigation and steps to improve resilience. Understanding and consistent framing of system performance, together with understanding the vulnerability of the society, is crucial.

13.5 Nuclear power plants

Political and societal perceptions of radiation risks affect health and safety decisions for nuclear facilities more strongly than for any other energy technology. Because of these risk perceptions, nuclear facilities have undergone highly detailed safety assessments during the last 50 years that often have included natural and engineered events with very low likelihoods of occurrence (Connor *et al.*, 2009). For example, many countries use an annual occurrence likelihood of 10^{-7} to screen external events with potentially serious radiological consequences from consideration in safety assessments at nuclear facilities (IAEA, 2002). However, with the benefit of hindsight, what was perceived at the time in some countries as the most appropriate science-based approach to hazard and risk estimation can now be shown to have understated an appropriate design basis for infrequent phenomena.

This is illustrated, for example, by the unexpectedly high peak accelerations experienced at the Kashiwazaki-Kariwa (KK) nuclear power plant (NPP) in Japan in the 2007 Niigataken Chuetsu-Oki earthquake. The M6.6 earthquake, which occurred at an intermediate crustal depth offshore of the west coast of Japan, with an epicentre about 19 km from the site, produced a horizontal peak acceleration under reactor unit 1 of about 0.69 g. This was well above the safe shutdown specification of about 0.45 g, and far in excess of the plant rapid restart threshold of 0.28 g. The four operating reactors on the KK NPP site (three others were undergoing routine maintenance) were all powered down, and although the owner, TEPCO, was prepared to restart some units within 24 hours, the civil authorities ordered operations to be halted until additional safety checks could be completed. Largely owing to these societal concerns, all seven reactors remained shutdown for a further two years, with some units not generating electricity again until 2010. During that time, thorough examinations of the KK NPP showed that the earthquake did not damage the essential safety equipment or safety functions in the plant (IAEA, 2009). Nevertheless, in the year after the 2007 earthquake, TEPCO had to depend on non-nuclear power plants for about 50–60 TW-hr annually until the plant resumed operations, and this led to carbon emissions of 24 million tons of CO_2 (approximately 3% of Japan's emissions per year). This is an example of where reaction to one risk can cause increased risk through another mechanism, namely climate change.

At the time of the 2007 earthquake, the Japanese nuclear industry was still using its traditional deterministic approach for seismic design basis and life extension criteria. The KK NPP experience accelerated the process of introducing probabilistic seismic hazard assessment methodology as a counterpart perspective. Retrospective probabilistic seismic hazard analysis for KK NPP suggests (C. Connor, pers. comm.) that the probability of one or more >M6.0, crustal depth (<25 km) earthquakes within 10 km of the site during the 40-year lifetime of the facility was 0.005 to 0.02, with 95% confidence, indicating that prior to 2007 the site would likely have been assessed, on a probabilistic basis, as lying within a region of persistent significant seismicity. An elementary peak acceleration hazard analysis for this inferred event activity rate indicates that the observed value of 0.69 g at KK NPP would have had an estimated exceedance probability of just over 10^{-4} per annum, i.e. not beyond the design basis criterion conventionally invoked for nuclear power facilities in other parts of the world.

Although this incident was much less serious than the subsequent 2011 Fukushima Dai-Ichi NPP accident, the shutdown of the KK NPP due to the external seismic hazard concerns cost billions of dollars. In Japan, public confidence in the industry was shaken as it became known that analyses by geologists in the 1980s, during site selection for this relatively new facility, had concluded there was no significant risk of such an event.

Less than four years later came the tragic 11 March 2011 M9 Tohoku-oki earthquake off the north-east coast of Japan and the resulting massive tsunami that caused devastating damage along the coast of Japan, which included the Fukushima Dai-Ichi NPP site (Norio *et al.*, 2011). At the NPP site, the tsunami had a height of 10–14 m, inundated all of the site, destroyed back-up cooling systems, took out the on-site power generators that had replaced offsite power lost during the earthquake shaking, and resulted in loss of control of the operating reactors and the release of significant quantities of radiation into the atmosphere, onto the ground and out into the ocean. In some sectors near the plant, surface contamination levels significantly exceed the standards established for safe inhabitation by the general public. A health fund of over US$1 billion was set aside for further mid- to long-term monitoring and evaluations, and the regional clean-up plans proposed will result in enormous costs for the Japanese economy.

Both the national and global impacts of the Fukushima disaster will be felt for decades. Nationally, NPPs that were shut down for regular safety inspections at the time of the accident were required to remain offline until given permission to restart. Eventually all the NPPs were out of operation and electricity supply was under great stress for many months. Globally, the disaster caused both Germany and Switzerland to make policy decisions to withdraw from nuclear power and Italy to decide not to embark on a new programme of nuclear power building. At the time of writing, Japan itself was considering a complete closure of all its NPPs by about 2040. These decisions, requiring recourse to carbon-based energy supplies and causing economic knock-on effects associated with energy supply and cost, will contribute to a changing climate with associated changes in extreme events, as well as have unknown environmental and social impacts over the coming decades, as well as creating political issues of security of supply.

The narratives of the preceding external hazard assessments for both Fukushima Dai-Ichi NPP and for KK NPP are jarringly alike. Perhaps this is unsurprising given the plant owner and the regulatory regime are the same in both cases, but clearly there are important hazard and risk assessment lessons to be learned. During siting of the Fukushima Dai-Ichi six-reactor NPP in the 1960s, tsunami hazards were assessed based on historical observations and bounded by the great 1960 Chilean earthquake, which produced an approximately 3 m tsunami near the site. Subsequent tsunami models focused on tsunamis resulting from earthquakes originating in the area immediately adjacent to the NPP site. In 2002 TEPCO evaluated the design tsunami height using numerical methods presented by the Japan Society of Civil Engineers (JSCE) and a maximum earthquake magnitude of 8.0. This analysis established a tsunami design basis of 5.7 m, which resulted in several seawater pump motors in unit 6 being raised several decimetres (Government of Japan, 2011). Following the 2004 Indian Ocean quake and tsunami, the IAEA requested that TEPCO reassess earthquake and tsunami hazards at all their coastal nuclear facilities, including Fukushima Dai-Ichi. This hazard study was completed in 2007 when it was concluded that the maximum earthquake to be considered, magnitude 8.3, would produce a tsunami less than the design basis of 5.7 m at the NPP site. In contrast, a probabilistic tsunami hazard assessment conducted by Sakai *et al.* (2006) indicated a 10% probability of exceedance during the next 50 years for a tsunami >6 m to occur along the Fukushima coastline.

It is salutary for nuclear safety specialists to contemplate how this very recent tsunami assessment can be reconciled, even in a deterministic approach, given the knowledge of the great 869 CE Sanriku M8.6 earthquake. This historical earthquake generated a tsunami that caused widespread flooding of the Sendai plain, with sand deposits being found up to 4 km from the coast. Although the implications of the 869 CE earthquake for tsunami hazards assessments were recognised in July 2009, TEPCO had not completed an analysis of this event on the design tsunami height at Fukushima prior to the March 2011 disaster (Government of Japan, 2011).

Another report from the IAEA recommended that NPP cooling systems be reinforced against tsunami effects, but apparently this request led to no action by TEPCO. Concern arose because a tsunami directly affecting NPP operations was not unprecedented: Kalpakkam-2 (unit 2 of Madras Atomic Power Station – MAPS) tripped following the Indian Ocean tsunami that hit the east coast of India on 26 December 2004. The 170 MW pressurised heavy water reactor was operating at nominal power when the wave sent sea water into its pump house, causing electrical problems and a trip that required operators to bring the unit to safe shutdown. On this part of the Coramandel Coast of Tamil Nadu, the tsunami caused 6–7 m high waves that also swept through the residential section of the Kalpakkam research centre, killing five employees, who were among a total of 30 casualties reported from that area. The power station, which is located 9.7 m above the ocean's surface, had been designed with sea protection works to withstand waves 5.2 m high. An additional three-tier system of sandbags, rocks and embankments were put in place after the 2004 tsunami. These were intended to reduce the destructive potential of wave surge over a coastal buffering zone extending 500 m from the plant's boundaries, and to be strong enough to withstand waves up to 9 m high. Operations at Kalpakkam NPP were said to be back to normal within three days.

These two events, the KK earthquake and the Fukushima Dai-Ichi disaster following the tsunami, unequivocally demonstrate an intrinsic fallibility in deterministic analyses of natural hazards. At both sites, the magnitude of credible hazardous events used to determine the design basis for these facilities was significantly underestimated. Hard-to-find active geological faults were presumed not to exist in the case of KK, and scientific hypothesising, which suggested that zones with older subducted oceanic lithosphere, such as off north-east Japan, could not support earthquakes much greater than M8 (Stern, 2002) was uncritically relied upon for Fukushima. It is in the nature of deterministic analyses that they invite, almost inevitably, the truncation of the tail of a distribution of potential events, either because it is expedient on cost grounds to ignore this tail or it is deemed that insufficient data are available to enumerate it. In both the KK earthquake and the Fukushima Dai-Ichi tsunami, the disastrous events that occurred were subjectively adjudged beyond consideration, a sure sign of a non-comprehensive hazard assessment approach that was deficient in considering uncertainties in the recurrence of potentially catastrophic events.

In addition to transient tsunami waves, there are also longer-term hazard assessment issues related to sea-level rise and climate change and the siting of new NPPs, including some in the UK. The likely re-use of existing licensed NPP sites (which are all coastal) gives the sites a currently foreseeable operational lifetime of around 100 years, which is sufficient to span the likely operational life (of perhaps 60 to 80 years) and the initial stages of decommissioning. Further re-use, or prolonged decommissioning, could double or triple this duration of activity. As noted above, such cases suggest that the UK government, its agencies and the research councils should be considering risk assessments from natural hazards over many decades and, perhaps, even centuries, in terms of strategic locations for critical facilities.

As well as looking forward, there also may be grounds for looking back in certain cases of NPP siting. For instance, the Metsamor nuclear power plant in Armenia (Figure 13.5) is the country's main and crucial source of energy, and is located in a dense region of Quaternary volcanism. Current international guidelines for NPP siting require a detailed hazard assessment if volcanic flow phenomena appear capable of affecting the site (IAEA, 2011). The presence of numerous Quaternary scoria cones and lavas adjacent to the site, and inundation of the site by a Quaternary pyroclastic flow, show that determining the likelihood of similar events occurring in the future is a crucial component of the suitability assessment for this site. Of the same vintage is a nuclear power station built at Bataan, Morong, in the Philippines in the 1970s, but never commissioned, for numerous reasons. One important objection to the plant came from the Union of Concerned Scientists, who pointed out that it is situated only 14 km from the Holocene Natib volcano, which would be considered capable of producing future activity under current IAEA guidelines (IAEA, 2011).

Research into why such situations arise would be of huge value in preventing more examples and finding more effective methods of alerting authorities to the dangers. For many sites, the annual likelihoods of natural phenomena that could significantly exceed

Figure 13.5 A nuclear power plant in Armenia with volcanic cones in the background.

traditional design bases may be very low, on the order of 10^{-4} to 10^{-7}. Such low likelihoods can be perceived as incredible or so uncertain as to warrant inattention, with focus being given to the 'credible' events that constitute a facilities design basis. Nevertheless, credibility is readily given to human-induced events or event sequences with comparable likelihoods. As the world community continues to evaluate the lessons learned from the tragic 2011 Great Tohoku-oki earthquake and tsunami, this dichotomy hopefully will be resolved. One mitigation approach, advocated notably by Perrow (1984), is to distribute facilities into smaller units, because the destructive impact of a natural hazard will be less on the community or country. However, such a decision involves many more factors (including safety and security issues) than simply natural hazard planning, and is thus an issue of national or regional strategic planning. Back-up plans and system robustness are also important, as in the example of power generation, to reduce vulnerabilities in case beyond-design-basis events actually occur.

13.6 Radioactive waste

Almost all countries with active nuclear power programmes have concluded that geological disposal deep underground (Chapman and Hooper, 2012) is the best option for high-level waste (which often includes spent nuclear fuel), although some environmental groups have a preference for indefinite surface storage. Thus, for radioactive waste repositories, some very slow geological and environmental processes and rare catastrophic events start to become the principal interest. The primary contributors to calculated future releases from the proposed Yucca Mountain repository in the United States, for example, were disruptions caused by large earthquakes and basaltic volcanic eruptions (US Department of Energy, 2008). For timescales of the order 10^5 or 10^6 years, geological processes could result in partial exhumation of a deep repository and changes to the groundwater system, and even more severe disruption in the case of a volcanic eruption. Climate affects erosion and sedimentation rates, hydrogeological systems and indirectly influences transport pathways and rates for radionuclide movement underground. For coastal facilities, the consequences of sea-level rise may need to be considered. Over such large timescales it is not sufficient to assume that site characteristics will be the same as today. Long timescales need to be considered in light of the progressively declining hazard potential of the waste, owing to radioactive decay – and the fact that the most critical period for isolation and containment of the highest activity wastes is in the first 1000 years after their disposal, when they are many orders of magnitude more hazardous. Hazards assessments nevertheless have to forecast the spatial and temporal recurrence rates of natural processes that might compromise the performance of the repository. While some of these processes are the same as those normally thought of as hazards, some are also very slow processes, such as erosion or the onset of a major glaciation.

The recent publication *Volcanism, Tectonism and the Siting of Nuclear Facilities* (Connor *et al.*, 2009) summarises the latest research on geohazards for nuclear facilities. The general approach in hazards assessment is to analyse the spacing and recurrence rates of features such as volcanoes and active faults. Note here that for waste repositories the hazard is usually expressed in terms of the probability of the volcano or fault occurring close enough to the repository to make it likely that hazardous events will disrupt the repository and compromise safety by allowing elevated releases of radionuclides. For disruptive events that have a reasonable likelihood of occurring at a value usually specified by regulations there also needs to be an analysis of the consequences. In terms of risk, the assessment is governed by international standards and national regulations concerning exposure to harmful radio-activity, the design of the repository, the design of the waste containment packages, characteristics of the waste, as well as natural processes such as chemical corrosion, disruptive events, the hydrogeological properties of the geological barrier, the pathways and the characteristics of the surface biosphere to which radioactivity is transported. This is a complex topic, since there are many different radionuclides, whose concentrations and distributions vary with time due to radioactive decay and their different environmental chemistries. All these parameters and controls vary and have intrinsic uncertainties. There is uncertainty in data, uncertainty in models and uncertainty in our knowledge about future states of the system. There is also variability in regulations from country to country and the possibility of regulations changing – as has happened recently in the United States, when the performance period was increased from 10^4 to 10^6 years.

To analyse these uncertainties and variabilities systematically and obtain a result that is useful for decision-making, a post-closure performance assessment is often used to demon-strate that exposure is not expected to exceed some threshold level determined by regu-lations. The UK approach is fairly typical. The Nuclear Decommissioning Agency (NDA) has developed a methodology for addressing uncertainty in a systematic way, based on the analysis of features, events and processes (FEPs) and development of scenarios, which are then addressed in detail in a numerical performance assessment. In this approach, a system-atic analysis of all the FEPs relevant to the performance of the facility leads to the identification of a base scenario and a number of variant scenarios that define potential evolutions of the facility. Each scenario is represented by a range of conceptual models, developed from knowledge of the FEPs operating in the scenario. Mathematical models are developed from the conceptual models for each scenario and are implemented in software. Typically, all parameters that affect risk are sampled from some probability distribution function and then studied through Monte Carlo simulation. The products are expressed in terms of average annualised probability of death to an individual as a function of time. In the UK, if this probability exceeds 10^{-6} then the threshold is exceeded.

Not surprisingly, radioactive waste disposal is controversial in all sorts of contexts. The long timescales and rarity of potential disruptive events due to natural hazards, and the difficulty of evaluating consequences of events which have never happened, provide considerable challenges for the scientific communities. Alternative conceptual models are often advocated for some FEPs, with varying degrees of model support and different types of effects in the performance assessments. Disagreements have arisen among experts and between organisations, which contribute to the uncertainty in the assessments. The organisational scene of actors and stakeholders inevitably becomes complex, as does the potential for real or perceived conflicts of interest.

While there are many differences in how countries approach the problem of waste management, the organisational complexities and difficulties that scientists face in remaining impartial are common. There are thus many issues which provide research challenges for the social sciences, including ethical issues of balancing responsible and timely management of wastes created to benefit today's society against leaving a potentially harmful legacy to future generations, tractable projections of human society into the future and public acceptability (Chapter 16). Indeed, to date several national radioactive waste disposal programmes have run into difficulties because of the opposition of the public and of anti-nuclear environmental groups. For example, the failure of the planning application by UK Nirex for an underground laboratory at Sellafield in a public enquiry (1997) demonstrated the importance of political dimensions and public perceptions about risk, as well as the science and engineering aspects. As a consequence there was a decade of stagnation in the UK programme and the creation of the Committee for Radioactive Waste Management in 2007 (http://www.corwm.org.uk/default.aspx) in which there was strong involvement from social scientists, legal experts and environmentalists, as well as scientists and technical experts. CoWRM's main outcome at this juncture was to re-affirm geological disposal as the best option and to convince the government that a volunteer approach was crucial for site selection. Internationally, it is also critical as nuclear facilities proliferate.

Similar problems in societal acceptance have affected programmes recently in Canada, Japan, Germany, Korea and the United States. Although each potential disposal site presents challenges for achieving the appropriate degree of waste isolation, the overarching technical challenges of geologic disposal are sufficiently well known to have confidence in the ability to achieve safe, permanent disposal of high-level waste in deep geologic repositories (e.g. IAEA, 2003). Nevertheless, building societal confidence in safety remains a key area in which scientists must contribute in order to implement geologic disposal in a way that is acceptable to all stakeholders (IAEA, 2007).

The Swedish geologic repository programme provides an excellent example of successful engagement between scientists and stakeholders. Throughout the decades of site characterisation, scientists from the repository developer (SKB, the Swedish Nuclear Fuel and Waste Management Company) met regularly with local residents to discuss the positive and negative results of ongoing investigations. The Swedish Radiation Safety Authority (SSM) also conducted independent meetings to discuss these investigations with stakeholders. Many stakeholders participated in meetings and tours of SKB's Äspö underground research laboratory, canister laboratory, waste transport ship M/S Sygin and the interim fuel storage facility at Clab. These visits helped demonstrate the processes that SKB was undertaking to assure safety of the potential repository. Consequently, when the Forsmark site was selected in 2009 as the candidate repository site, there was widespread understanding of the favourable characteristics that led to selection of this site and of the accompanying engineering approach that was planned for safe and secure disposal. In March 2011, SKB submitted its licence application for the Forsmark repository to SSM and the Environmental Court in Stockholm for review. A similar approach of scientific involvement and openness is being followed in Finland's repository development, which enjoys a comparable measure of stakeholder support for its candidate repository site at Onkalo.

13.7 Aviation and volcanic ash

The hazard of volcanic eruptions to aviation exemplifies how a natural hazard can affect distributed transport networks through disruption of flights and closure of airports. Here we take the view that air transport systems are distributed technological facilities. Vulnerability arises because of the complex independencies of a largely engineered system, including the planes and their engines, the airports and air traffic control. Dealing with such a complex system is a challenge for natural-hazards forecasting, the resilience of the system components and development of adequate regulations.

In April and May 2010, volcanic ash from Eyjafjallajökull volcano, in Iceland, was dispersed over north-west Europe by unfavourable and unusually strong winds for many days. The ensuing disruption to air traffic provoked several challenges for using science to support risk management decision-making, and in May 2011, when Europe had just completed its first volcanic ash crisis exercise to validate changes and improvements to volcanic ash contingency plans and procedures following the 2010 experiences, the Grimsvötn volcano erupted and airborne ash again threatened air travel and economic losses, just over a year after Eyjafjallajökull. There have been many historical incidents, which have been compiled in a US Geological Survey database (Guffanti *et al.*, 2010).

In the 2010 north-west Europe episode, the immediate economic impact was on the aviation industry. The International Air Transport Association (IATA) estimated that the Icelandic volcano crisis cost airlines more than US$1.7 billion in lost revenue in the six days after the initial eruption. For a three-day period (17–19 April), when disruptions were greatest, lost revenues reached US$400 million per day. As many as 1.2 million passengers each day, accounting for 29% of global passenger traffic, were affected, as were large numbers of freight flights. In total, around ten million passengers were affected by the Eyjafjallajökull ash cloud. Although there were no aviation accidents, many stranded passengers looked to alternative means of travel, and it seems quite likely that this led to additional deaths and injuries from road accidents.

The hazard of volcanic ash to aviation came to prominence after two high-profile near disasters. In 1982 all four engines of a British Airways flight cut out after encountering a dense ash cloud from Galungung volcano, Indonesia. Similarly, a KLM flight en route from Amsterdam to Anchorage encountered ash from an eruption of Mount Redoubt, Alaska in 1989, with all engines failing. In both cases, at first sight miraculously, some of the engines re-started when the aircraft left the ash cloud and they managed to land safely. It appears (Casadevall, 1994) that melting of ash in the high-temperature core of the engine resulted in loss of power and engine failure, but the ash then vitrified and shattered when the engines cooled, allowing some to re-start. These famous incidents and numerous others have led to a very effective international risk management system of early warning and avoidance. Nine volcanic ash advisory centres (VAACs) have been set up around the world, hosted by national or regional meteorological services. Volcano observatories alert VAACs, who then use a combination of ash dispersion models and observations, notably satellite images, to issue early warnings and forecasts to aviation authorities, which in turn issue advisories to air traffic control, airports, airlines and pilots. Flights are diverted and in some cases airports are closed to avoid the ash hazard. This system has been very effective and there have been no major accidents caused by aircraft interacting with volcanic ash since the VAACs were established.

The Eyjafjallajökull eruptions challenged this risk management system, which was found to be inadequate for dealing with the high intensity of air traffic over Europe. The strategy of flights turning around to avoid ash in such busy air space and finding another airport proved not to be viable. The dramatic closure of air space and consequent disruption quickly led to intense pressure on civil aviation authorities and governments to change the regulations. It also raised fundamental questions about whether engines could tolerate low concentrations of ash safely. In any case, within a few days the engine manufacturers pronounced that concentrations of below $2\,\mathrm{mg\,m^{-3}}$ were not a cause for concern and should be adopted as a threshold for safe flying. This regulatory change greatly reduced the no-fly zones. However, the change also raised major challenges for the VAACs, and the ability of numerical ash dispersion models to deliver reliable forecasts was questioned.

Initially, it seemed the first couple of days of shutdowns were acceptable to society as a suitable precautionary measure. When the disruption started to last longer questions were asked. Was it really necessary to restrict all air traffic in Europe? Who was determining the restrictions and on what did they base their decisions? These were questions about risk management which meant, in turn, they were questions about risk assessment and hence about hazard assessment and scientific uncertainty. It was widely perceived that airspace was being closed based upon a theoretical model, not on factual observations or measurements. Test flights by some airlines claimed that the models were erroneous in terms of actual ash concentrations that might be encountered. 'Europe needed to find a way to make decisions based on facts and risk assessment, not theories', said Giovanni Bisignani, Director General of the IATA.

The science behind the assessment of ash hazard to aviation is summarised by Prata and Tupper (2009). The basic operational approach is to combine numerical models of advection and diffusion of ash in the atmosphere integrated into a weather forecasting model with various observations, notably from satellite images and from ground observations systems such as radar and LiDAR. When the task was to simply forecast the presence of ash for avoidance this approach was adequate. However, the new requirement to calculate spatial variations in concentrations of ash has created major challenges for ash hazard forecasting, such as defining the volcanic source terms, understanding the ash transport processes, interpreting and assimilating remotely sensed data into forecast models and understanding the uncertainties in the ash dispersion models.

A key input into the dispersion models is the volcanic source term that defines the mass flux of ash into the atmosphere, the heights of ash injection and the properties of the ash such as grain size distribution and density. Established relationships allow eruption column height observations to be converted into mass flux (Sparks *et al.*, 1997). A significant complicating factor is that fine ash particles have a strong tendency to form aggregates in the atmosphere, a process that remains poorly understood, with limited empirical information (Rose and Durant, 2011). There are very large uncertainties in the source term. Another major challenge is to use satellite data to help constrain source terms, validate models and to improve forecasts through data assimilation (Prata and Tupper, 2009; Stohl *et al.*, 2011). Currently a major issue is that models are tuned to observations, reducing the ability to use observations to independently validate the models. Even the best models appear to have large uncertainties (factors of up to 30) in the ability to forecast ash concentrations (Webster *et al.*, 2012).

The 2010 volcanic eruption on Iceland involved two central VAACs: VAAC London hosted by the UK MetOffice, and VAAC Toulouse, hosted by MeteoFrance. Before 2010, the operational activities of these centres were not fully familiar to many volcanologists, who might have been able to help better constrain model source terms and input parameters for the different ash forecasting models in use. The episode has led to much closer links between meteorologists and volcanologists in Europe. It has also highlighted the complexity of decision-making when so many different actors are involved. Actors in the Icelandic ash emergency included national governments, the EU, civil aviation authorities, national air traffic control agencies, engine manufacturers, airlines, the travel industry, government science agencies that had to coordinate their work (UK Met Office, the British Geological Survey, the Icelandic Meteorological Office and MeteoFrance) and members of the academic community with specialist knowledge in volcanology and meteorology. Adding to this complexity, the crisis was across borders and truly international as hundreds of thousands of passengers were stranded from Tokyo to Los Angeles to Johannesburg. The emergency revealed major inadequacies in how modern society copes with disruptive and unexpected events.

13.8 Space weather hazards

Most of the major natural hazards included in this chapter have been described elsewhere in this book. In this section we provide a synopsis of space weather hazards and include some discussion of their effects and assessments of risk to technological facilities and networks.

13.8.1 Background

The sun is a continuous source of electromagnetic radiation over a wide spectrum, and of charged particles that stream through space forming the ever-present, though variable, solar wind. Solar flares, radio bursts, solar energetic particle (SEP) events and coronal mass ejections (CME) are occasional impulsive releases of electromagnetic energy and particles. These phenomena are often closely associated, linked through processes reorganising the solar magnetic field. The frequency of impulsive events is modulated by the solar activity cycle of around 11 years, most commonly characterised by sunspot numbers, with fewer events close to the solar minimum. Coronal holes are structures that may be present at any phase of the solar cycle, and during the declining phase they tend to persist for months, injecting fast streams of charged particles into the solar wind. These various solar emissions create conditions and drive processes in the near-Earth space environment and upper atmosphere, and the term space weather is used to describe this solar influence.

Solar ultraviolet and X-ray emissions are absorbed in the Earth's atmosphere, creating the ionosphere, which affects the transmission of radiowaves and supports the flow of electric currents which generate magnetic fields. The solar wind interacts with the earth's magnetic field, confining it to a bullet-shaped cavity in space, the magnetosphere, with its boundary at a distance of about ten Earth radii on the sunward side and extending well beyond the moon's orbit on the downwind side. The magnetosphere contains radiation belts and various electric current systems, and energy and matter are transferred to it from the solar wind at its boundary, the magnetopause. The magnetosphere is the dynamic environment occupied by operational satellites.

When CMEs reach the Earth, magnetospheric currents may intensify, leading to large and rapid magnetic field variations termed magnetic storms, and the flux of relativistic electrons in the outer radiation belt is generally enhanced after an initial decrease. The strongest increases in the energetic electron flux are associated with recurrent magnetic storms caused by fast streams from long-lived coronal holes. (The recurrence results from the approximately 27-day solar rotation period.) The magnetic field embedded in the solar wind is a crucially important control on the response of the magnetosphere. Strong coupling results when the solar wind magnetic field is directed anti-parallel to the earth's field when it encounters the magnetopause. More detailed descriptions of space weather sources and processes are provided by Eastwood (2008).

13.8.2 *Electrical power networks*

During magnetic storms GICs flow through earthed conductors, including electrical power networks and pipelines. The level of GIC is determined by the rate of change of the geomagnetic field (dB / dt). For transformers operating at 50–60 Hz the GIC are quasi-DC currents that can cause transformer cores to saturate, leading to magnetic flux leakage, overheating and potentially catastrophic damage. Transformers are expensive, not available as 'off-the-shelf' items, and replacing one is a costly exercise that can take months. A secondary effect is generation of harmonics of the operating frequency which can cause protective equipment to malfunction, switching out equipment designed to stabilise the network. This effect led to the power outages in the Hydro-Québec power grid in March 1989 and in Malmo, Sweden in October 2003.

Kappenman (2006) uses indirect evidence to suggest that dB / dt changes in the great magnetic storm of 13–15 May 1921 were up to an order of magnitude greater than observed in North America in 1989 (around $500\,\text{nT}\,\text{min}^{-1}$) when the Hydro-Québec power grid collapsed. Kappenman points out that magnetic storm effects can be continental in scale (although dB / dt will not be uniform over such a large area), and the growth of the US high-voltage power grid to more than $250\,000\,\text{km}$ with around $12\,000$ major substations has created a serious vulnerability. A report by the US National Academy of Sciences (2008) has estimated that if the May 1921 magnetic storm was repeated today, around 130 million people in the United States would lose their electricity and more than 350 transformers would be at risk of permanent damage.

The geographical distribution of magnetic storm effects is controlled by the earth's magnetic dipole axis, inclined at about 10° to the rotation axis. Geomagnetic latitude may be defined relative to the geomagnetic poles and the greatest amplitude magnetic disturbances on the ground are close to the auroral regions, usually at a geomagnetic latitude of around 70°. However, in major magnetic storms the auroral oval expands to lower latitudes. This is why auroral sightings are occasionally reported from low-latitude areas such as southern Europe and the southern states of North America. Consequently, in major magnetic storms the maximum hazard moves closer to the latitudes of vulnerable electrical transmission equipment in the developed world. It has long been assumed that GIC effects on power networks are confined to high latitudes, but this appears not to be the case. Thomson *et al.* (2010) point to evidence of GIC damage to transformers in South Africa in the months following the October–November 2003 magnetic storm.

The first magnetic observatories were established more than 170 years ago, and so there are extensive historical records of magnetic storms to form the basis for determining a statistical distribution of events. However, these are largely photographic records from which it is difficult to accurately estimate dB / dt. Digital recording has been widespread for around 30 years and Thomson *et al.* (2011) have used digital data from observatories in Europe to estimate the return periods of major magnetic storms. Their results indicate that events with dB / dt of around $2000–6000\,\text{nT}\,\text{min}^{-1}$, similar to the value estimated by Kappenman (2006) for the May 1921 storm, occur about once every 100–200 years in Europe.

The CMEs responsible for initiating magnetic storms typically take 2–3 days to reach the earth. However, in the case of the Carrington event, generally taken to be the most extreme known space weather event, the sun–earth transit time was only around 18 hours. This indicates the lead time power network operators have to plan and implement an operational response to a major event once an earth-directed CME has been confirmed. Over the longer term, engineering options for mitigation include installing capacitors to block GIC, and using devices to change the resistance to ground of transformer earth connections.

Most governments are now alert to the threat posed by space weather. In the United States the Grid Reliability and Infrastructure Defense (GRID) Act, to give the Federal Energy Regulatory Commission the authority to develop and enforce standards for power companies to protect the electric grid from geomagnetic storms and other threats, is in the legislative process (Tretkoff, 2010). In the UK, the risk posed by severe space weather has been included, for the first time, in the National Risk Register (Cabinet Office, 2012). The approach taken for the UK National Risk Register is to identify a 'reasonable worst case' representing a plausible but challenging manifestation of each risk under consideration. The reasonable worst case adopted for space weather has been the scenario presented by a repeat of the 1859 Carrington event. For most space weather hazards this involves extrapolation from measurements made in more recent events.

13.8.3 *Satellites and communications*

Satellites are susceptible to various types of disruption and damage due to severe space weather. SEP events release high-energy charged particles (>30 MeV), which threaten satellite operations by accelerating cumulative damage to the solar arrays providing power and by their effects on electronic systems. High-energy particles can penetrate chips in digital electronic systems causing single event effects (SEEs), flipping memory and changing the state of software. Recent developments in microelectronics are leading to equipment with increasing chip density and increased vulnerability to SEEs (Dyer, 2011).

Energetic electrons (>2 MeV) in the radiation belts can penetrate satellites and cause build-up of charge in insulating materials. Discharges can permanently damage electronic components or generate false signals to which the satellite may respond. Satellite anomalies are also associated with differential charging of satellite surfaces by lower-energy electrons (1–100 keV) and subsequent discharges. Choi *et al.* (2011) have analysed reports of anomalies experienced by 95 satellites in geosynchronous orbits between 1997 and 2009 and conclude that spacecraft charging is the dominant contributor to the anomalies. Satellite operators attempt to mitigate the effects of particle penetration and discharges by hardening chips against radiation and by using multiple circuits so that a malfunctioning circuit can be outvoted by ones that are operating correctly.

Global navigation satellite systems (GNSSs), such as the US Global Positioning System (GPS), provide positioning, navigation and timing (PNT) services which are increasingly important to ground-based applications. Navigational applications of GPS are now commonplace and the use of GPS-derived time has become integral in areas as diverse as scientific monitoring, telecommunications and finance. Society's increasing reliance on, and the implications of loss of, PNT services has been highlighted by the Royal Academy of Engineering (2011). Among the vulnerabilities considered in the Royal Academy of Engineering report are those due to space weather. The GPS constellation has a minimum of 24 satellites orbiting at an altitude of about 20 000 km above the earth's surface, which places them close to the centre of the outer radiation belt. Geosynchronous satellites, in equatorial orbits at an altitude of about 36 000 km, are also exposed to radiation hazard, but to a lesser degree as they are close to the edge of the outer radiation belt.

GPS receivers on the ground rely on receiving radio signals from the GPS satellites. During magnetic storms the upper atmosphere is heated as a result of the intensification of ionospheric currents. This changes the ionospheric density profile, which affects the propagation time of the radio signals, leading to positional errors. This effect can be largely overcome by the use of dual-frequency receivers. Magnetic storms also cause small-scale ionospheric irregularities, leading to scintillation, rapid changes in signal amplitude and phase, which can result in loss of signal lock by receivers. These effects degrade the quality of PNT services while the ionosphere is disturbed. There is evidence that solar radio bursts, which may last a few hours, can overwhelm GPS satellite signals, leading to loss of service (Cerruti *et al.*, 2008).

Some SEP events, through the production of secondary neutrons (>10 MeV) from collisions in the upper atmosphere, lead to increases in radiation at ground level, giving a strong likelihood of increased error rates in digital control systems. The effects are greater at aircraft altitudes and there is clear evidence of SEEs in aircraft avionics during such events (Dyer *et al.*, 2003, 2007). Because of their exposure to radiation, aircrew on long-haul flights are classified as radiation workers by the European Union and frequent flyers are also at risk (Hapgood and Thomson, 2010). The only mitigation strategies for aircraft to reduce exposure during a radiation event are to fly at lower altitude to increase atmospheric shielding, or re-route to lower latitudes.

SEP events may also affect the state of the ionosphere in the polar regions (polar cap absorptions), leading to loss of high-frequency (HF) communications for 24 hours or more. Consequently, aircraft on polar routes must be re-routed, adding considerably to flight costs. Solar flares can produce bursts of X-rays creating ionospheric layers that absorb HF signals in the sunlit hemisphere, causing communications blackouts for several hours, again a serious concern to aviation.

The adoption of optical cables for most telephone and internet communications makes them largely immune to GICs. Transoceanic cables have electronic systems to amplify signals, a potential, but relatively minor vulnerability. The ability of relatively recent and rapidly developing wireless technologies – including mobile phones, wireless internet and device controllers – to reject interference from radio bursts has not yet been established by exposure to significant events as their widespread adoption has been during a quiet period in space weather. Evidence on the resilience of these devices is expected to emerge as solar cycle 24 develops.

13.9 Emerging risk assessment issues for distributed systems

As noted earlier, the susceptibility of a major distributed technological system to natural hazards impacts is only fully tested, if ever, by an actual event, and then only rarely. Satellite systems, energy generation and supply and the civil aviation network and its infrastructure are prime examples of distributed technological systems of critical importance to modern society and life. Others include long gas or oil pipelines with multiple pumping or pressure-reduction stations, and giant energy production centres such as NPP sites with multiple reactors like KK NPP or Fukushima Dai-Ichi NPP in Japan.

Each presents a special challenge for demonstrating system-wide robustness to the multiple and time-varying loads and forces engendered by natural phenomena. For instance, the diverse separate effects of a large earthquake which simultaneously shakes many different parts of a complex plant in different ways might produce compounding failure states in subsystems that are difficult to envisage. One extreme scenario in this context could be the vulnerability of a system of dams on a major river, such as exists in British Columbia, Canada. If one of the upper dams fails, perhaps due to a major Cascadia earthquake, the ensuing torrent could cause sequentially failure of those downstream, with the resulting compound flooding and inundations causing major widespread destruction, possibly reaching far into the United States.

This issue of distributed system vulnerability is an emerging – and challenging – area for research in risk analysis: methodological developments are needed by which site-specific probabilistic risk analysis methods can be extended to the assessment of urban lifeline systems, highway networks, electric power grids and regional healthcare delivery systems, as well as the cases just mentioned. As in the dams case, spatio-temporal correlation between earthquake ground motion effects across multiple sites is complicated to characterise (Goda *et al.*, 2011), but important for determining system loading in terms of capacity, functionality or failure, and is a non-trivial probabilistic problem for hazard or risk assessment.

In trying to make risk assessment tractable, and hence useful, with increasingly large and complex interlocking technological systems, one of the main difficulties is that the problem is multifactorial, with high dimensionality – a problem well-known to mathematicians and familiar to systems engineers as a 'curse' (Bellman, 1961). Losses and damage need to be avoided, or at least minimised and mitigated. In the latter regard, providing effective catastrophe loss securitisation by insurance and reinsurance for complex technological systems is strongly coupled to the difficult – sometimes opaque and always uncertain – scientific and engineering issues that natural hazards uniquely entail. In terms of methodological approaches, the challenges cross-cut disciplines and require unified thinking to bridge existing gaps. For instance, determining exactly how to accommodate spatial and temporal damage correlations in low probability–high consequence earthquake disaster scenarios, without leaving the insurer or the insured open to calamitous losses, is starting to be tackled from the seismic engineering side (e.g. Bazzurro and Park, 2007; Goda and Ren, 2010; Goda and Yoshikawa, 2012), with technical analyses of structural capacity and response that advance into loss quantification. Similar pioneering work is being done for volcanic hazards (e.g. Spence *et al.*, 2004).

These examples illustrate a very important point in all aspects of natural hazards assessment and risk mitigation: specialists who understand the causative processes and their uncertainties most fully are, in many ways, better placed to exercise judgement on ramifications and consequences than people for whom the natural phenomena are totally unfamiliar, or who are without any experience. This commonly applies in most natural hazards domains to politicians, other professionals, the media and the public. Conversely, scientists need to improve their understanding of the political and public perceptions of these issues, so strong linkages, interactions and better communication between these communities are needed. Thus, uncertainty and its appraisal in natural hazard estimation feeds across into public safety, societal risk and insurance risk, and a coherent approach is critical for large loss mitigation or management, whether in monetary terms or numbers of lives.

Ultimately, we accommodate and minimise the impacts of natural hazards through improving the performance of the technological facilities and infrastructure that deliver the services and resources on which our communities depend. As is apparent from the earlier discussions, these technological facilities are complicated interacting systems, created to satisfy diverse societal needs. These high-level needs are mapped onto technical performance requirements, which become the focus of the design, operation and ultimately end-of-life disposal of the facility (Hall *et al.*, 2004). Needs and performance have many aspects: safety, sustainability, cost-effectiveness, commercial, legal, regulatory, environmental impact, and political, each represented by their own set of specialist actors (Marashi and Davis, 2006). The importance and priority of each aspect might change with time, the severity of an impacting event or the power and advocacy of the actor. We require the facility to sustain acceptable performance throughout its life, accommodating and adapting to the changing societal needs and environmental actions upon it and to the impacts of the inevitable degradation of the facility itself. All of these processes are complex, difficult to define and measure, and are typically characterised by much uncertainty. Yet the hierarchy of people and organisations tasked with maintaining satisfactory facility performance must seek to understand, work within and, if possible, exploit these uncertainties. Developing a tenable, pragmatic and shared understanding of risk among all these actors is the basis for harmonising well-grounded decision-making, on which satisfactory facility performance hinges. A clear, integrated, consistent and shared definition and understanding of facility performance sets the common purpose of the actors and establishes a unifying framework for their explicit and implicit risk assessment and risk-based decision-making. This in turn implies that the framing both of risk and the method of risk assessment should be in tune with the needs and sophistication of the decision-maker. A critical factor here is the recognition of the kinds of decision each actor is required to make and the relative viability and power of their consequent actions. Their risk assessment methodology, be it explicit or implicit, and the information they use in it, should be the simplest that satisfies their purpose; yet it should still be in harmony with the methodology used by co-actors of different sophistication and power. Once again, a common focus on understanding, describing and attaining system performance offers an effective unifying and integrating framework (Marashi and Davis, 2006). Common engineering practice has long established such unification at component artefact level through the use of pragmatic codes of practice. The efficacy of this approach is evident by our burgeoning technological society. However, such strides have not kept pace when it comes to whole-system performance, especially for interacting and interdependent systems. Much still needs to be learned in this area; risk assessment, risk management and risk-based decision-making need to be central to this learning.

13.10 Final remarks

Over the last decades, rapid population growth, rapid technological advances and increasing, everyday, human reliance on these advances mean that our vulnerability to all types of hazards is increasing. Climate change, urbanisation that is increasingly concentrated onto tectonically vulnerable coastal margins, environmental stresses, changes in land use (such as deforestation) and industrialisation in many emerging economies all contribute to our growing vulnerability to natural hazards. On the other hand, there has been great investment in mitigation measures in many countries. Measures, such as flood defences and improved early warning systems, have reduced the impact of hazards. Reported disasters are increasing according to a recent analysis by the World Bank (Okuyama and Sahin, 2009). According to this same study in the period 1960–2007 the relative economic impact of disaster related to natural hazards has remained about the same, but the impact is much higher in low- and middle-income countries. However, events in 2011 have been described as an *annus horribilis* in terms of economic impact of extreme events (Courbage and Stahel, 2012), and highlight the lack of anticipation and preparedness in governments and commercial systems.

The modern globalised and highly interdependent world has developed sophisticated technological facilities and networks for support all facets of modern life. Yet these advances, complexities and interdependencies make the systems more vulnerable and society less resilient to natural hazards. There is an urgent need for rational, dependable objectivity in decision-making. The purpose of risk assessments is then to identify likelihood of consequences that then prompts contingency planning leading to enhanced robustness and resilience or adaptive reaction in specific situations.

Much of natural hazards research has been developed in the context of disaster risk reduction. Saving lives has been rightly high up the agenda, and the science has been devoted to issues of mitigation, evacuation, recovery and planning to protect communities and reduce economic losses. Most attention has been on immediate responses to disasters, such as evacuations, and putting in place engineering solutions to mitigate the effects of large events, such as flood defences and development of building codes for earthquake resistance. These approaches have only been partially successful. In several recent major disasters, engineered defences and mitigation plans have been found inadequate for a variety of reasons, including underestimating the potential for extreme events and their impacts, lack of political will or resources to take the necessary steps even if they are known, and human factors such as corruption (e.g. Ambraseys and Bilham, 2011).

What is becoming an emerging phenomenon is the way that technological facilities and critical infrastructure are becoming impacted in complex, cascading event sequences. In a globalised world the effects of natural disasters are no longer largely confined to within national borders. Complicated and sometimes unforeseen and unexpected interactions take place to cause effects across borders and around the world. Flights are cancelled in Japan because of ash from Iceland. Nuclear power is questioned and re-appraised in Europe due to an earthquake in Japan. Production at the Honda car factory in the UK is compromised due to lack of critical components caused by floods in Thailand. Stock markets and national economies are tangibly affected by the bigger events. Radioactive waste programmes for geological disposal around the world are inhibited at least in part by a lack of trust that citizens have in the ability of scientists to see into the future and be reassured that natural hazards won't cause a catastrophe.

Responding to the emergent phenomena of impacts from natural hazards on technological and critical infrastructural is a major challenge. Perhaps the first admission must be that so far the response has been inadequate and arguably simply not good enough. Too many hazard and risk assessments are based on narrow, deterministic approaches, which commonly fail to anticipate extremes and to take full account of the complexity of the systems. We believe that the only way forward is through robust systematic hazard and risk assessment that is probabilistic in character and includes quantitatively as far as possible an analysis of uncertainty. Such analysis must begin with an objective identification of facilities and systems that are at risk, together with evaluation of how those risks could manifest themselves (for example, using event trees to look at how systematic failure scenarios might develop at a high level). This will allow preliminary ranking of the potential hazard to facilities and systems, allowing decisions to be made on where to focus detailed probabilistic studies of the risks themselves.

Some very promising scientific, statistical and analytical tools have been developed using numerical modelling of hazards to make forecasts informed by observations, probabilistic mapping of hazards, logic trees, expert elicitation, system engineering analysis and Bayesian belief networks. However, inadequate historical data, the complex heterogeneous nature of the risks, complex nonlinear hazard–loss relationships and emerging socio-economic trends complicate these efforts. While generic methods of risk and uncertainty assessment can be applied here, it seems doubtful that a black-box generic modelling approach can be developed. Some recognition needs to be given to the human factors in disasters and failure of technological systems, which are hard to quantify but critical to the full analysis of risk, including the assessment of the human factors that affect the safety analyses for decision-making. Each technological facility is likely to have unique features of location, environment, function and design that mean there will need to be site-specific risk assessments. Thus, there is both a need for technology or site-specific risk assessments and larger-scale risk assessment of whole, interconnected systems.

Of course, probabilistic hazard and risk assessment methods applied to natural hazards can raise many technical issues. For instance, how best can incompletely understood, complex geological processes be represented so that the recurrence rate of major earthquakes or their magnitudes are not significantly over- or understated? Are there enough data to quantitatively bound recurrence rates of rare disastrous events? These difficulties notwithstanding, the great virtue of probability distributions, whatever the limitations, is that they do have tails. Precise enumeration of extreme values may be challenging and endlessly debatable in scientific terms, but at least extreme events can be characterised to some extent for hazard assessment purposes, and not simply precluded from consideration as has happened so often and now so disastrously with a deterministic approach. The introductory chapter of Connor *et al.* (2009) on tectonic hazard to nuclear facilities (Chapman *et al.*, 2009; written before the Fukushima disaster) observed

> that humankind has had a tendency to disbelieve, discount or forget the magnitude of events that Earth can throw at us. Events such as the December 2004 Indian Ocean tsunami show that large regions of the planet, the *home of numerous potentially hazardous* facilities, can be put at risk with little or no warning. Nuclear power systems have yet *to be exposed to a really large*-magnitude earthquake or volcanic event on their doorstep.
>
> *Chapman* et al. *(2009)*

It has taken less than two years to show the truth of these observations.

Responses to natural hazards that threaten technological facilities vary according to circumstances. In an ideal world it would be best not to build such facilities in areas vulnerable to natural hazards. Multi-hazards risk assessment should always be done where such facilities are being planned, and planning regulations should prevent construction in unsuitable places. However, this is often not an option, as there is commonly no consensus on what constitutes an unsuitable location for periods of millennia or longer. Thus the practical approach is to build robust and where possible distributed facilities that will withstand the natural phenomena. Thus engineering solutions remain a major ingredient of the response, but the right engineering solutions need to be informed by probabilistic analysis of risk.

In essence, all technological disasters are man-made – if our systems were robust enough we would not have a disaster. As Woo (2011) points out, because they are constructions of one form or another, catastrophic failures of technological systems in natural disasters should be susceptible to analysis, the identification of weak spots and hence improvements in strength and resilience. However, the problems, and the associated uncertainties, are more than just technical: systemic weaknesses include latent factors relating to man's central role in technological facilities, such as slipshod working procedures, poor corporate ethos and weak safety and risk-management cultures. A single fault-line in any one of the barriers of defence against human shortcomings, in design, build or operation, exposes a complex system to failure when the stresses of a natural hazard are applied. Understanding and characterising the interactions of the myriad uncertainties across the whole piece – from the complexities of the initiating natural hazard event through to how an intricate compound system responds – and finding feasible and cost-effective ways of mitigating impacts are, arguably, a much greater challenge than simply designing the system to function in the first place. There is also the need to address the political and educational realities in many countries that create barriers to mitigation or risk reduction. In addition to the issues of uncertainty regarding hazards and vulnerabilities, and uncertainty with respect to benefits and costs of actions, there is often a lack of public demand and organised advocacy, and perhaps most important, fragmented incentives and resources and lack of political will (Henstra and McBean, 2005).

Methods for quantitative hazard and risk assessment have been described and developed in this book. However, these methods have largely been applied only in certain well-defined and specific circumstances, and have not really yet been deployed widely enough. They offer the prospect of providing much more robust and informed evidence for planning, decision-making and policy development, so that losses are greatly reduced and the world surprised far less by how easily nature seems able to destroy man's constructions.

References

Ambraseys, N. and Bilham, R. (2011) Corruption kills. *Nature* **469**: 153–155.

Bazzurro, P. and Park, J. (2007) The effects of portfolio manipulation on earthquake portfolio loss estimates. In *Proceedings of the 10th International Conference on Applications of Statistics and Probability in Civil Engineering, Tokyo, Japan.* Abingdon: Taylor and Francis.

Bellman, R. E. (1961) *Adaptive Control Processes*, Princeton, NJ: Princeton University Press.

Blazquez, R. and Lopez-Querol, S. (2007) Endochronic-based approach to the failure of the Lower San Fernando dam in 1971. *Journal of Geotechnical and Geoenvironmental Engineering* **13**: 1144–1153.

Blockley, D. I. and Godfrey, P. S. (2000) *Doing it Differently*, London: Thomas Telford.

Bolduc, L. (2002) GIC observations and studies in the Hydro-Québec power system. *Journal of Atmospheric and Solar-Terrestrial Physics* **64**: 1793–1802.

Bruneau, M. and Reinhorn, A. (2004) Seismic resilience of communities: conceptualization and operationalization. In *Proceedings, International Workshop on Performance-based Seismic-Design*, Bled,Slovenia, 28 June–1 July, Berkeley, CA: Pacific Earthquake Engineering Research Center.

Cabinet Office (UK) (2012) National risk register for civil emergencies, January.

Casadevall, T. J. (1994) Volcanic ash and aviation safety. In *Proceedings of the First International Symposium on Volcanic Ash and Aviation Safety*, Seattle, WA, July 1991, Washington, DC: United States Government Printing Office.

Castelli, M. and Scavia, C. (2008) Multidisciplinary methodology for hazard and risk assessment of rock avalanches. *Rock Mechanics and Rock Engineering* **41**: 3–36.

Cerruti, A. P., Kintner Jr., P. M., Gary, D. E., *et al.* (2008) Effect of intense December 2006 solar radio bursts on GPS receivers. *Space Weather* **6**: S10D07.

Chapman, N. A. and Hooper, A. (2012) The disposal of radioactive wastes underground. *Proceedings of the Geologists Association* **123**: 46–63.

Chapman, N. A., Tsuchi, H. and Kitayama, K. (2009) Tectonic events and nuclear facilities. In *Volcanic and Tectonic Hazard Assessment for Nuclear Facilities*, ed. C. Connor, N. A. Chapman and L. Connor, Cambridge: Cambridge University Press, pp. 1–23.

Choi, H.-S., Lee, J., Cho, K.-S., *et al.* (2011) Analysis of GEO spacecraft anomalies: space weather relationships. *Space Weather* **9**: S06001.

Connor, C. B., Chapman, N. and Connor, L. J. (eds) (2009) *Volcanism, Tectonism and the Siting of Nuclear Facilities*, Cambridge: Cambridge University Press.

Courbage, C. and Stahel, R. (2012) *The Geneva Report No 5, Extreme Events and Insurance: 2011 Annus Horribilus*, Geneva: The International Association for the Study of Insurance Economics.

Courbage, C., Orie, M. and Stahel, W. R. (2012) Thai flood and insurance. In *The Geneva Report No 5, Extreme Events and Insurance: 2011 Annus Horribilus*, ed. C. Courbage and R. Stahel, Geneva: The International Association for the Study of Insurance Economics, pp. 121–132.

Cronin, S. J., Neall, V. E. and Palmer, A. S. (1997) Lahar history and lahar hazard of the Tongariro River, northeastern Tongariro Volcanic Centre, New Zealand. *New Zealand Journal of Geology and Geophysics* **40**: 383–393.

Dominey-Howes, D. and Goff, J. (2009) Hanging on the line: on the need to assess the risk to global submarine telecommunications infrastructure. An example of the Hawaiian 'bottleneck' and Australia. *Natural Hazards and Earth Systems Science* **9**: 605–607.

Dyer, C. S. (2011) Written evidence published in the report of the UK House of Commons Science and Technology Committee on Scientific advice and evidence in emergencies. www.publications.parliament.uk/pa/cm201011/cmselect.

Dyer, C. S., Lei, F., Clucas, S. N., *et al.* (2003) Solar particle enhancements of single event effect rates at aircraft altitudes. *IEEE Transactions on Nuclear Science* **50** (6): 2038–2045.

Dyer, C. S., Lei, F., Hands, A., *et al.* (2007) Solar particle events in the QinetiQ atmospheric radiation model. *IEEE Transactions on Nuclear Science* **54** (4): 1071–1075.

Eastwood, J. P. (2008) The science of space weather. *Philosophical Transactions of the Royal Society A* **366**: 4489–4500.

Erinmez, I. A., Kappenman, J. G. and Radasky, W. A. (2002) Management of the geo-magnetically induced current risks on the national grid company's electric power transmission system. *Journal of Atmospheric and Solar-Terrestrial Physics* **64** (5–6): 743–756.

European Environment Agency (2010) *Mapping the Impacts of Natural Hazards and Technological Accidents in Europe*, Copenhagen: European Environment Agency.

Gamble, K. (2008) Weathering the storm: having a disaster recovery plan can mean the difference between scrambling for a quick IT fix and smooth sailing in the storm. *Healthcare Informatics* **25**: 36–38.

García-Herrera, R., Hernández, E., Barriopedro, D., *et al.* (2007) The outstanding 2004/05 drought in the Iberian Peninsula: associated atmospheric circulation. *Journal of Hydrometeorology* **8**: 483–498.

Goda, K. and Ren, J. (2010) Assessment of seismic loss dependence using copula. *Risk Analysis* **30**: 1076–1091.

Goda, K. and Yoshikawa, H. (2012) Earthquake insurance portfolio analysis of wood-frame houses in south-western British Columbia, Canada. *Bulletin of Earthquake Engineering* **10** (2): 615–643.

Goda, K., Atkinson, G. M. and Hong, H. P. (2011) Seismic loss estimation of wood-frame houses in southwestern British Columbia. *Structural Safety* **33**: 122–135.

Government of Japan (2011) Report of the Japanese Government to the IAEA Ministerial Conference on Nuclear Safety, June. http://www.iaea.org/newscenter/focus/fukush-ima/japan-report/

Guffanti, M., Casadevall, T. J. and Budding, K. (2010) Encounters of aircraft with volcanic ash clouds: a compilation of known incidents, 1953–2009. US Geological Survey Data Series 545, ver. 1.0: 12 p, http://pubs.usgs.gov/ds/545.

Hall, J. W., Le Masurier, J. W., Baker-Langman, E. J., *et al.* (2004) A decision-support methodology for performance-based asset management. *Civil Engineering and Environmental Systems* **21** (1): 51–75.

Hapgood, M. A. and Thomson, A. W. P. (2010) Space weather: its impact on Earth and implications for business. Lloyds 360° Risk Insight Briefing.

Havenith, H. B. Torgoev, L., Maleshko, A., *et al.* (2006) Landslides in the Mailuu-Suu Valley, Kyrgyzstan: hazards and impacts. *Landslides* **3**: 137–147.

Henstra, D. and McBean, G. A. (2005) Canadian disaster management policy: moving toward a paradigm shift? *Canadian Public Policy* **31**: 303–318.

Houqun, C. (2009) Lessons learned from Wenchuan earthquake for seismic safety of large dams. *Earthquake Engineering & Engineering Vibration* **8**: 241–249.

Huang, Y. and Jiang, X. (2010) Field-observed phenomena of seismic liquefaction and subsidence during the 2008 Wenchuan earthquake in China. *Journal of Natural Hazards* **54**: 839–850.

International Atomic Energy Agency (2002) *External Human Induced Events in Site Evaluation for Nuclear Power Plants*, Vienna, Austria: IAEA.

International Atomic Energy Agency (2003) *Scientific and Technical Basis for Geological Disposal of Radioactive Wastes*, Vienna, Austria: IAEA.

International Atomic Energy Agency (2007) *Factors Affecting Public and Political Acceptance for the Implementation of Geological Disposal*, Vienna, Austria: IAEA.

International Atomic Energy Agency (2009) *2nd Follow-up IAEA Mission in Relation to the Findings and Lessons Learned from the 16 July 2007 Earthquake at Kashiwazaki-Kariwa NPP: Report of the Engineering Safety Review Services Seismic Safety Expert Mission 1–5 December 2008*, Vienna, Austria: IAEA.

International Atomic Energy Agency (2011) Volcanic hazards in site evaluation for nuclear installations. Draft Specific Safety Guide DS405 Rev. 11.

Johnston, D. M., Houghton, B. F., Neall, V. E., *et al.* (2000) Impacts of the 1945 and 1995–1996 Ruapehu eruptions, New Zealand: an example of increasing societal vulnerability. *Geological Society of America Bulletin* **112**: 720–726.

Kappenman, J. G. (2006) Great geomagnetic and extreme impulsive geomagnetic field disturbance events: an analysis of observational evidence including the great storm of May 1921. *Advanced Space Research* **38**: 188–199.

Kayen, R., Brandenberg, S., Collins, B., *et al.* (2009) Geoengineering and seismological aspects of the Niigata-Ken Chuetsu-Oki earthquake of 16 July 2007. *Earthquake Spectra* **25**: 777–802.

Kirsch, T. D., Mitrani-Reiser, J., Bissell, R., *et al.* (2010) Impact on hospital functions following the 2010 Chilean Earthquake. *Disaster Medicine and Public Health Preparedness* **4**: 122–128.

Kwasinski, A., Weaver, W. W., Chapman, P. L., *et al.* (2009) Telecommunications power plant damage assessment for hurricane Katrina: site survey and follow-up results. *IEEE Systems Journal* **3**: 277–287.

Lehner, B. Czisch, G. and Vassolo, S. (2005) The impact of global change on the hydropower potential of Europe: a model-based analysis. *Energy Policy* **33**: 839–855.

Li, L. L., Cheng, S. G. and Xu, Q. F. (2009) Slope stability under transfusion of the upper reaches of the Xiaoba Reservoir in Sichuan. Paper presented at the 7th International Conference on Calibration and Reliability in Groundwater Modeling, Wuhan, China.

Malcolm, L. and van Rossen, A. (1997) Ashes to ashes? Rangipo rebuild. *International Water Power & Dam Construction* **49**: 22–24.

Marashi, E. and Davis, J. P. (2006) An argumentation-based method for managing complex issues in design of infrastructure systems. *Reliability Engineering and System Safety* **91** (12): 1535–1545.

Marchuk Izvestiya, N. A. (2008) The Altai Earthquake response of the measuring systems of the Sayano-Shushenskaya hydroelectric power station dam on the Yenisei River. *Physics of the Solid Earth* **44**: 226–231.

McIntyre, T., Hughes, C. D., Pauyo, T., *et al.* (2011) Emergency surgical care delivery in post-earthquake Haiti: partners in health and Zanmi Lasante experience. *World Journal of Surgery* **35**: 745–750.

Myrtle, R. C., Masri, S. E., Nigbor, R. L., *et al.* (2005) Classification and prioritization of essential systems in hospitals under extreme events. *Earthquake Spectra* **21**: 779–802.

NOAA (2004) Service assessment: intense space weather storms October 19–November 07: 2003. NOAA National Weather Service, Silver Spring, Maryland.

Norio, O., Ye, T., Kajitani, Y., *et al.* (2011) The 2011 eastern Japan earthquake disaster: overview and comments. *International Journal of Disaster Risk Science* **2**: 34–42.

Nuclear Safety Commission (2006) Regulatory guide for reviewing seismic design of nuclear power reactor facilities. http://www.nsc.go.jp/english/taishin.pdf

Okuyama, Y. and Sahin, S. (2009) Impact estimation of disasters: a global aggregate for 1960 to 2007. World Bank Policy Development Paper 4963.

Perrow, C. (1984) *Normal Accidents: Living With High Risk Technologies*, Princeton, NJ: Princeton University Press.

Pisani, G., Kastelli, M. and Scavia, C. (2010) Hydrogeological model and hydraulic behaviour of a large landslide in the Italian Western Alps. *Natural Hazards Earth System Science* **10**: 2391–2406.

Pistrika, A. and Jonkman, S. (2010) Damage to residential buildings due to flooding of New Orleans after hurricane Katrina. *Natural Hazards* **54**: 413–434.

Prata, F. and Tupper, A. (2009) Aviation hazards from volcanoes. *Natural Hazards* **51**: 239–244.

Rose, W. I. and Durant, A. J. (2011) Fate of volcanic ash: aggregation and fallout. *Geology* **39**: 895–897.

Roth, E. (1998) Mulheim-Karlich, a German tragedy. *ATW-Internationale Zeitschrift Fur Kernenergie* **43**: 162–166.

Royal Academy of Engineering (2011) *Global Navigation Space Systems: Reliance and Vulnerabilities*, London: Royal Academy of Engineering.

Sakai, T., Soraoka, H., Yanagisawa, K., *et al.* (2006) Development of a probabilistic tsunami hazard analysis in Japan. *Proceedings of ICONE14, International Conference on Nuclear Engineering 17–20 July, Miami*, New York: American Society of Mechanical Engineers.

Schmincke, H.-U. (2004) *Volcanism*, Berlin: Springer-Verlag.

Schuster, R. L. (1981). Effects of the eruptions on civil works and operations in the Pacific Northwest. In *The 1980 Eruptions of Mount St. Helens*, ed. P. W. Lipman and D. R. Mullineaux, Washington, DC: US Geological Survey, pp. 701–718.

Sparks, R. S. J., Bursik, M. I., Carey, S. N., *et al.* (1997) *Volcanic Plumes*, Chichester: John Wiley and Sons.

Spence, R. J. S., Baxter, P. J. and Zuccaro, G. (2004) Building vulnerability and human casualty estimation for a pyroclastic flow: a model and its application to Vesuvius. *Journal of Volcanology and Geothermal Research* **133**: 321–343.

Stern, R. J. (2002) Subduction zones. *Reviews of Geophysics* **40**: 1–38.

Stohl, A., Prata, A. J., Eckhardt, S., *et al.* (2011) Determination of time and height-resolved volcanic ash emissions and their use for quantitative ash dispersion modeling: the 2010 Eyjafjallajokull eruption. *Atmospheric Chemistry and. Physics* **11**: 4333–4351.

Thomson, A. W. P., Gaunt, C. T., Cilliers, P., *et al.* (2010) Present day challenges in understanding the geomagnetic hazard to national power grids. *Advances in Space Research* **45** (9) 1182–1190.

Thomson, A. W. P., Dawson, E. B. and Reay, S. J. (2011) Quantifying extreme behaviour in geomagnetic activity. *Space Weather* **9**: S10001.

Toda, S., Stein, R. S., Kirby, S. H., *et al.* (2008) A slab fragment wedged under Tokyo and its seismic and hazard implications. *Nature Geosciences* **1**: 771–776.

Tretkoff, E. (2010) Legislation seeks to protect power grid from space weather. *Space Weather* **8**: S05004.

UNEP (2004) Impacts of the summer 2003 heat wave in Europe. *Environmental Alert Bulletin* **2**: n.p.

US Department of Energy (2008) Yucca Mountain license application, safety analysis report chapter 2: repository safety after permanent closure. DOE/RW-0573, Rev. 0.

US National Academy of Sciences (2008) *Severe Space Weather Events: Understanding Societal and Economic Impacts, Workshop Report*, Washington, DC: National Academies Press. http://www.nap.edu/catalog/12507.html.

Vazquez, J., Pimenta, L. and Freitas, H. (2001) The Alem da Fazenda dam rehabilitation: technical, constructive, maintenance and exploitation issues. In *Dams in a European Context*, Lisse: Balkema, pp. 533–542.

Webb, D. F. and Allen, J. H. (2004) Spacecraft and ground anomalies related to the October–November 2003 solar activity. *Space Weather* **2**: 3–8.

Webster, H. N., Thomson, D. J., Johnson, B. T., *et al.* (2012) Operational prediction of ash concentrations in the distal volcanic cloud from the 2010 Eyjafjallajokull eruption. *Journal of Geophysical Research – Atmospheres* **17**: n.p.

Wirtz, P. (1987) The erection and commissioning of the Mülheim-Kärlich nuclear power plant. *Nuclear Engineering and Design* **100**: 297–306.

Woo, G. (2011) *Calculating Catastrophe*, London: Imperial College Press.

14

Statistical aspects of risk characterisation in ecotoxicology

G. L. HICKEY AND A. HART

14.1 Introduction

When assessing the safety of a chemical substance, whether it is a general chemical or a pesticide, an assessment must be undertaken to determine the risk and possible consequences to the environment. The assessment comprises many (possibly overlapping) components, which include human health; persistent, bio-accumulative and toxic assessment; and ecological risk assessments. Risk managers, who may also consider additional socio-economic aspects, subsequently review the final dossier. This chapter is an overview of the basic elements of ecotoxicological risk assessment for chemical products, with a primary focus on how the role of statistics has improved the process of so-called *risk characterisation* – a joint summary of hazard and exposure.

Across the broad spectrum of international regulation is a compendium of technical guidance documents (TGDs) pertaining to risk characterisation which risk assessors base risk assessments decisions on. To give some examples, in the context of general chemical regulation within the EU (pertaining also to import and export), the newly introduced REACH (Registration, Evaluation, Authorization and Restriction of Chemicals) legislation (EC, 2006) has corresponding guidance provided by the European Chemicals Agency (ECHA, 2008a). In the context of metal contamination, a TGD has been issued by the International Council on Metals and Mining (ICMM, 2007a). In the context of European pesticide legislation (EC, 1991), there are a series of TGD that are updated periodically for different categories of ecological risk (for example, the TGD for aquatic organisms is EC, 2003, and the TGD for birds and mammals is the European Food Safety Authority (EFSA, 2008). The concepts and approaches reviewed in this chapter are potentially applicable to most categories of ecological risk, as we explain later on. Obviously, other components of the risk assessment will differ according to the type of substance, and so we do not discuss these here.

The intention of the ecotoxicological element of the risk assessment report is to deduce whether the environmental exposure is likely to exceed a conservative threshold of tolerance to the toxicant in the ecosystem under consideration. Ecosystems are highly complex interacting systems with many components and variables, thus making the risk assessment process difficult and uncertain. The aforementioned TGDs have adopted a tiered risk

Risk and Uncertainty Assessment for Natural Hazards, ed. Jonathan Rougier, Steve Sparks and Lisa Hill. Published by Cambridge University Press. © Cambridge University Press 2013.

assessment strategy such that the process is relatively efficient for a large quantity of chemicals that pose little risk. Under the tiered structure framework, failure of a preceding tier will trigger a higher tier risk assessment. At lower tiers the procedure will be inexpensive and is intended to be conservative, often being based on 'worst case' assumptions. This conservatism is a reflection of the uncertainty associated with the tier. Often, however, the degree of conservatism is not well characterised for certain areas of ecotoxicological risk assessment. The assessment is refined at higher levels to reduce the uncertainty where identified as necessary. Overall failure of the risk characterisation to satisfy some regulatory accepted threshold that is set in advance will lead to outcomes such as a chemical product being refused a licence for use or the triggering of a cleanup operation.

14.2 Risk assessment principles

The foundation of an ecotoxicological risk assessment is based on the following three components:

(1) toxicity assessment;
(2) exposure assessment; and
(3) risk characterisation.

The first component deals with assessing the sensitivity of the ecosystem to the chemical in question, typically by attempting to predict a critical concentration threshold. Since it is implausible to model every ecosystem or species community that we wish to protect, models based on the effects to individual species are used as a proxy for all communities. Moreover, uncertainty arises because it is not possible, or ethical, to measure toxicity for every different species. The second component deals with the possible routes and scenarios that may lead to species being exposed to the chemical. Finally, the third component is effectively the joint integration of the first two components, allowing a risk manager to draw suitable conclusions on whether the risks are being adequately controlled, and ultimately for decision-makers to act accordingly. Components 1 and 2 are complex in their own right, hence individual TGDs specific to these aspects are available for many chemical types (e.g. metals, general chemicals).

From the viewpoint of *uncertainty analysis*, the tiers may (for the purposes of this summary) be separated as: (1) qualitative, (2) deterministic and (3) probabilistic (for example, ECHA, 2008c). In fact, higher tiers will generally involve more ecological relevance of the data collected. A fundamental (dimensionless) quantity in risk characterisation is the 'risk characterisation ratio' (RCR), defined as

$$RCR = \frac{PEC}{PNEC},$$

where PEC is the 'predicted environmental concentration' and PNEC is the 'predicted no-effect concentration'. The PNEC and PEC fall under the domains of the toxicity and exposure assessments respectively. Tier 1 would allow for the RCR to be qualitatively evaluated as being below or above 1. This may be suitable for cases where the level of risk is sufficiently self-evident that a subjective expert judgement is a sufficient basis for risk management. Tier 2 treats the PNEC and PEC as point estimates and thus the RCR is also a point estimate. If the RCR is less than unity then the risk assessment would be considered to have been adequately passed. On the other hand, if the RCR is greater than unity, then permission for use (in the context of pesticides, etc.) will not be granted until a more appropriate (higher tier) risk assessment shows that it will have no unacceptable impact (EC, 2002). Tier 3 treats both the PNEC and PEC as random variables; hence the RCR will also be a random variable. Although it is probably sufficient to treat the two random variables independently, care must be given to ensure they are meaningfully compatible. There are numerous ways a risk manager might use the RCR distribution to evaluate the risk assessment dossier; we discuss some of these later on.

A key point of modern risk assessment is that these tiers are not distinct; typically a risk assessor would start at tier 1 and progressively utilise tiers 2 and 3 along a continuum where evidence of refinement is indicated as being appropriate. This may include, for example, determining pointwise estimates from conservative probabilistic modelling prior to a full probabilistic assessment.

As mentioned earlier, the discussion of risk characterisation is effectively the same across the spectrum of substance regulations. The RCR is inherent to the newly introduced REACH TGD (ECHA, 2008a) and is denoted as 'the risk quotient (RQ)' for assessment of metals (ICMM, 2007a). For pesticide assessment the relevant quantity is called the 'toxicity exposure ratio' (TER) and is effectively the ratio of a typical (perhaps sensitive) species toxicity concentration (which can be an *effect* or *no effect* concentration) to the PEC. For all intents and purposes, a functional equivalence therefore exists between the RCR and TER, although in practice currently methodology separates them.

In Section 14.3 we discuss the 'deterministic' evaluation of the RCR. This method, as well as the fully probabilistic risk characterisation, relies on the availability of laboratory toxicity data. Therefore we provide a brief overview, which in the interest of simplicity, we limit to the scope of aquatic ecosystem risk assessment; however, the general principles remain similar for other aspects of ecotoxicological risk assessment.

14.2.1 Toxicity data

For a given species, an individual laboratory experiment is conducted to determine its sensitivity to the toxicant, measured as a concentration – for example, in micrograms per litre ($\mu g\,L^{-1}$) of water. There is no definitive measure of sensitivity used in practice and it is usually chosen depending on the context of the assessment.

One typical toxicological measure of a species' sensitivity is the median effect concentration, denoted as EC_{50}. This concentration is defined as the concentration that would affect 50% of the tested species population over a fixed period of time. The endpoint with which the 'effect' is measured would usually be the most sensitive of growth, reproduction or mortality, although other endpoints could be investigated if deemed to have ecological significance. Estimating the EC_{50} for a species requires the estimation of a concentration–response curve (also termed dose–response curves in certain fields); see Figure 14.1 for an example of a concentration–response curve of a hypothetical chemical. That is, a fitted model (usually a log-probit or log-logistic regression curve) of the scientifically recorded effects (e.g. the proportion of species individuals which survived) of different bioassays to incremental concentrations of administered chemical. The EC_{50} is then estimated as the median of this concentration–response curve. A concentration–response model and this time the estimated EC_{50} is only defined for the period of time the experiment was conducted, which is usually fixed by design according to scientific consensus for particular groups of species, e.g. fish are regularly 96-hour experiments and daphnids 48-hour experiments. In this example, the endpoint would be referred to as *acute*. The use of acute EC_{50} values would be of practical benefit where a risk assessment was interested in the short-term effects of a one-off chemical discharge that would biodegrade in a short period of time. We refer to the observed collection of species-specific toxicity tolerance values for each species as toxicity data.

Some TGDs limit the application of acute toxicity (such as the EC_{50}) to deterministic tiers of risk assessment, as for example in ECHA (2008b). This is because, where the primary goal is to prevent long-term effects, short-term effects may be of limited relevance. A standard toxicological measure of a species' sensitivity for measuring *chronic* effects is the 'no observed effect concentration' (NOEC). A NOEC is defined by the ECHA

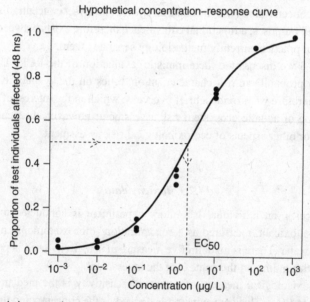

Figure 14.1 A logistic-regression concentration–response curve modelling the effect of a hypothetical substance on a species. The EC_{50} is estimated as the median of the fitted curve (solid black line), as indicated by the dashed lines.

(2008b: table R.10–1) to be the maximal test concentration at which the substance is observed to have no statistically significant effect ($P < 0.05$) when compared with the control group within a stated exposure period. This exposure time is generally much longer than those used in acute testing. Pires *et al.* (2002) criticise the NOEC as a toxicological endpoint because, although weakly reliant on statistical modelling, it is highly sensitive to experimental design, for example, the reported NOEC must be one of the chosen test concentrations. An alternative suggestion recently put forward is the 'no effect concentration' that is calculated as the threshold parameter of effect from a suitable (usually three-parameter) statistical concentration–response model (Pires *et al.*, 2002 and references therein). Notwithstanding the implied benefits to a chemical safety assessment of using chronic toxicity data, namely setting environmental standards that will be unlikely to cause adverse effects in the long term, the resources required to determine them are usually far greater than those required for acute toxicity data.

The number of tolerance values that is required for a risk assessment varies, although is generally quite small; a sample size of one is acceptable in certain regulatory contexts! While particularly low, it is not entirely unreasonable if appropriate uncertainty factors are used and, furthermore, it is also an ethically sound approach. Adequately defining an appropriate uncertainty factor is, however, non-trivial. Additional species can be tested, but the costs to chemical manufacturers and the desire to limit animal testing on ethical grounds means they are unlikely to volunteer to such requests unless higher tiers of assessment are triggered. In any quantitative risk characterisation method, uncertainty may be introduced by the typically small sample of data.

In the deterministic approach, for which a point-estimate is required, additional experimental data is either discounted or aggregated. In the former case, we mean that a risk assessment based on a collection of, say, EC_{50} values will not propagate the uncertainty about the fitted concentration–response model into the overall model. In the latter case, we mean that where multiple records for a species–chemical pair are listed (e.g. in historical literature or risk assessment reports), then the standard rule-of-thumb, subject to a systematic review of the earlier studies, is to take the geometric mean in order to determine a single concentration value. This aggregation method, referred to as 'harmonisation' by the ECHA (2008b), is considered to avoid biasing the risk assessment with certain regularly tested species. These procedures have been criticised by Duboudin *et al.* (2004) because information regarding intra-species variation and measurement error is discarded. It is, however, possible to use statistical methodology to improve the experimental design (e.g. sample size requirements) in order to control the uncertainty and develop models to incorporate repeated measurements.

14.3 PNEC specification

As stated earlier, the RCR may be evaluated with point estimates of the PNEC. Here we provide details of the two standard approaches for determining this point value under the current TGDs (e.g. ECHA, 2008b). Unlike exposure analysis, such statistical extrapolation approaches can be more widely generalised to many different assessment scenarios. The first approach is fully deterministic using fixed value extrapolation factors; the second approach uses probabilistic summary statistics. The latter approach can be considered a refinement of the former, with some of the uncertainty being quantifiable.

The PNEC can be described as the concentration of toxicant below which adverse effects are unlikely to occur in the multi-species assemblage (EFSA, 2005). This is usually based on conservative reasoning. It is not a strict protection threshold limit, in fact EC (2003: 99) state, in the context of industrial substance risk assessment (which REACH replaces): 'It is not intended to be a level below which the substance is considered safe. However, again, it is likely that an unacceptable effect will not occur.' Since it is neither ethical nor practical (and in certain cases illegal by consequence of protected species legislation) to determine the tolerance of every species type in every ecosystem that may be exposed, an extrapolation routine is suggested based on a sample of the species in a generic target community. The species upon which the risk assessment is based are usually non-randomly selected standard laboratory species. For example, in the context of pesticide effects on fish, the rainbow trout (*Oncorhynchus mykiss*) is a required test species. This is because traits and life-cycles are usually well understood for these species, as well as ease of cultivation. The EFSA (2005) and Hickey (2010) have demonstrated that this can have important consequences in the determination of PNECs, possibly resulting in overprotection.

14.3.1 Assessment factors

For a given sample of toxicity data of n distinct species: $x_1, x_2, ..., x_n$ (where each x_i is a concentration, such as an EC_{50}), the assessment factor approach defines the PNEC as

$$\text{PNEC} = \frac{\min(x_1, x_2, ..., x_n)}{\text{AF}},$$

where AF denotes the assessment factor. The assessment factor is a positive number (usually a power of 10) between 10 and 10 000 which is prescribed, albeit in many cases with limited explicit justification, by the relevant TGD subject to some criteria. These criteria are usually based on the environmental compartment (e.g. freshwater aquatic systems, or birds), the sample size and endpoint type (acute or chronic). For standard aquatic ecosystem risk assessments, this sample will typically comprise a fish species, an invertebrate and algae (Duboudin et al., 2004). As an example, three (acute) EC_{50} values would likely require AF = 1000 for freshwater compartments and AF = 10 000 for marine compartments, whereas three (chronic) NOEC values would typically require AF = 100–1000 (see ECHA, 2008b for precise specification). The difference between freshwater and marine assessment factors is partly a reflection of the presumed greater species diversity in the latter compared to the former, thus possibly requiring a larger extrapolation.

One might deduce from this that the consequence of measuring acute endpoints as opposed to chronic endpoints is describable by an increase in the assessment factor of one or two orders of magnitude. This implied portion of the assessment factor is referred to as the acute-to-chronic ratio (ACR).

There is no absolute definition of what an assessment factor represents; however, a list of some of the uncertainties believed to be accounted can be found in some TGDs. The EFSA (2005) notes that assessment factors are intended to account for all or some of, depending on the context:

- intra- and inter-laboratory variation of toxicity data;
- intra- and inter-species variation (biological variance);
- laboratory data to field impact extrapolation;
- short-term to long-term toxicity extrapolation (i.e. the ACR).

The proportion of uncertainty each assessment factor accounts for is unclear. Moreover, it is not clear whether one may treat assessment factors as combining multiplicatively. Suter (2007) suggests, however, that this is almost certainly the viewpoint taken. Consequently, the level of protection to the assemblage each assessment factor offers is undeclared. Nonetheless, recent technical guidance (ECHA, 2008b) states that the assessment factors should not be treated as fixed; where appropriate a 'weight of evidence' approach may be used to lower them. Suter (2007) suggests that they are of use primarily for screening-based assessments – a viewpoint consistent with the existing tiered framework.

The PNEC is intended to represent an estimation of the concentration of the toxicant which is below the lowest species tolerance value in the generic ecological community. Other means of summarising this data (less conservatively) have been suggested, which we discuss later. The assessment factor is the most contentious part of the risk assessment.

This deterministic approach, which assumes no model for the toxicity data, is restrictive since there is no way of quantitatively accounting for the uncertainty, but is justified by appealing to the *precautionary principle*. Forbes and Calow (2002a: 249) describe this as: 'applying controls to chemicals in advance of scientific understanding if there is a presumption that harm will be caused'. However, Suter (2007: 18) notes that, confusingly, there are at least 14 different definitions of the precautionary principle. Notwithstanding the lack of uncertainty quantification and potential over-conservatism, probabilistic models have become a natural refinement tool (see Section 14.3.2). However, it is expected that the tiers of risk assessment are coherent, such that if a low risk is indicated using fixed assessment factors, then a refined risk assessment should similarly indicate a low risk. In principle, therefore, there would be little to gain from a more refined risk assessment at the determin- istic level. Such a method is therefore considered to be robust and economically efficient.

14.3.2 Species sensitivity distributions

A refinement of the assessment factor approach uses probabilistic modelling. Moreover, such approaches are useful for situations in which the RCR is close to I; in such a situation the risk manager will find it difficult to ascertain whether permission should be granted.

The approach is based on a statistical construct, widely referred to as the 'species sensitivity distribution' (SSD), that treats tolerance values as realisations from a (defined) community representative distribution. One might view the SSD as a community-level concentration–response curve. For a general overview of SSDs, see Posthuma *et al.* (2002) and references therein. When the risk assessment is pertinent to an ecological community, the SSD represents an important source of variability, namely the interspecies variation in tolerance or, equivalently, sensitivity. The SSD concept is now scientifically accepted within the regulatory arena for intermediate and higher tier probabilistic ecotoxicological risk assessment. For example, SSDs are in use for regulatory risk assessment of: new and existing chemicals (US EPA, 1998; ECHA, 2008b), setting water quality guidelines (Stephan *et al.*, 1985; ANZECC and ARMCANZ, 2000), pesticide assessment (US EPA, 2004); and risk assessment from metal contamination (ICMM, 2007c).

The ECHA (2008b) define the PNEC to be the fifth percentile of the SSD (for example, see Figure 14.2), denoted the 'hazardous concentration' to 5% of species (HC_5) in the species assemblage, with the option for a risk manager to divide by a post-hoc assessment factor between 1 and 5 to account for additional uncertainty not captured by the model. The implied threshold defined by the HC_5 is not to be treated as an 'all or nothing' threshold. Some ecological community structures would likely sustain structure with higher proportions of loss (through ecological processes such as species adaptation and immigration), as well as possibly others that may not be sustained by this level of protection; see Newman *et al.* (2000: 508, 'List of important concerns', item 2).

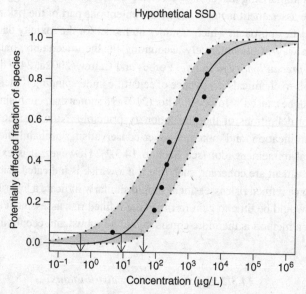

Figure 14.2 A hypothetical estimated log-normal species sensitivity distribution based on nine EC_{50} values (black points) with 90% pointwise confidence bands (grey-shaded region). The HC_5 (and 90% confidence interval) is indicated at the bottom of the plot. The HC_5 is estimated to be 8.72 µg L^{-1} with 90% confidence interval (0.48, 42.13).

There have been a number of suggestions regarding how to estimate the HC_5, with one of the most widely accepted in the EU regulatory context being the Aldenberg and Jaworska (2000) estimator (class), denoted AJ from here onwards. This is defined to be a $100\gamma\%$ one-sided lower confidence limit for the HC_5 under the assumption of a lognormal SSD. For an in-depth review, see Aldenberg and Jaworska (2000) and Hickey *et al.* (2009).

In the context of REACH, the current guidelines require the median HC_5 value to be specified with a confidence interval; however, no guidance regarding how a risk manager might interpret the confidence interval is provided. It is interesting to note that authors of the associated TGD (ECHA, 2008b) have confused confidence limits and intervals; a source of great confusion to the authors of this chapter. Within the scientific distribution literature there have been some recommendations that a one-sided lower confidence limit, namely the 95% underestimate limit (or equivalently the fifth percentile of the fifth percentile posterior from the Bayesian viewpoint) be applied in order to err on the side of caution; see the discussion by Chen (2003, and references therein). This choice of induced conservatism has since received partial empirical support for certain cases based on some comparisons with field studies (Maltby *et al.*, 2005).

The (standard) SSD approach and HC_5 are criticised in the ecotoxicological risk assessment literature. One notable deficiency of the model is the lack of species representation and non-randomness in selection, which is potentially over-protecting some groups and under-protecting others (Forbes and Calow, 2002b). This further increases ambiguity as to precisely what is being protected (Kefford *et al.*, 2005). Some species, including standard laboratory test species, have been noted as being either generally sensitive or tolerant. A statistical method for adjusting the HC_5 estimator to take account of a single species whose tolerance value is non-exchangeable with other species in the population was proposed by EFSA (2005) and followed up in Craig *et al.* (2011).

Whereas the assessment factor approach is generally based on sample sizes of between one and three species, current data requirements for SSDs are more stringent. For example, the REACH TGD states that the sample size of distinct species tested must be at least ten (preferably 15) and this must span at least eight taxonomic groups. Other suggestions have been made in the scientific literature (e.g. Stephan *et al.*, 1985). In addition, the ECHA (2008b) states that only NOECs should be used for fitting an SSD. However, due to the limited availability of chronic toxicity data, the ecotoxicology literature tends to focus on modelling EC_{50} values. A further (deterministic) acute-to-chronic extrapolation factor (cf. the ACR) would be required to determine an equivalent PNEC in this case.

Small sample sizes (e.g. ten, and below), would most likely be considered far too small for robust probabilistic modelling. A parsimonious approach for modelling the SSD is therefore preferred. The lognormal distribution is standard for European risk assessment. Other simple distributions are also considered: for example, the log-logistic distribution is frequently used in the United States and Europe, and the Burr Type III is used in Australia (for a discussion, consult Wheeler *et al.*, 2002: §4). In some risk assessments of diverse multi-taxa ecosystems there may be evidence of bimodality, which may exhibit itself from the toxicant having a particular mode of action for certain taxonomic groups. In these cases, standard distribution choices would be inappropriate. Suggestions of alternative SSD models that may account for this have been proposed by Duboudin *et al.* (2004), Grist *et al.* (2006) and Hickey *et al.* (2008); the latter two methods are based on mixture distribution SSDs. In general, model uncertainty is not considered beyond simple goodness-of-fit tests, and epistemic uncertainty about the SSD parameters is sizeable due to small sample sizes.

Whereas the assessment factor approach supposedly accounts for other uncertainties, such as laboratory uncertainty, intra-species uncertainty and laboratory-to-field uncertainty, this is not immediate from the SSD model. However, this may be captured by the post-hoc assessment factor applied to the HC_5 that can also be justified based on external data, e.g. structurally similar substances. This section has named just a few of the criticisms of the SSD and its basis for PNEC determination; for a more exhaustive list consult Newman *et al.* (2000), Forbes and Calow (2002b) and Kefford *et al.* (2005).

It is interesting to note that (1) the HC_p and (2) the strictly deterministic PNEC estimator described in Section 14.1 can be considered analogous to one another. Both estimators can be described as an assessment factor applied to a summary statistic of the observed toxicity data. The assessment factor in (2) is fixed and independent of the data, whereas in (1) it is dependent on the data. The summary statistic is generally the minimum observed tolerance value for (2) and the geometric mean for (1). An alternative estimator was proposed by the EFSA (2005), which applied fixed deterministic assessment factors to the geometric mean of the toxicity data as opposed to the minimum value. This was proposed to avoid the assessment becoming more conservative as more data are provided while maintaining mathematical tractability. Where legislation stipulates that only one species test is required, then the revised estimator provides *at least* the same (undefined) level of protection; a result which was shown to hold for a collection of different hypothetical distributional shapes by EFSA (2008: appendix 7).

There are many other methods of deriving HC_p decision rules by reconsidering the underlying behavioural models. For example, the EFSA (2005) proposed three such models, later considered by Hickey (2010), based on (1) complete independence of chemical risk assessments; (2) a hierarchical model for a defined population of chemicals; (3) scale parameter homogeneity across a population of substances. Hickey *et al.* (2009) proposed a decision-theoretic approach based on minimising loss functions as an alternative to arbitrarily selecting a conservative confidence limit. Like most aspects of ecological risk assessment, changes in standard methodology can only gain acceptance from the different stakeholders if the more sophisticated quantitative methods are transparent, the consequences effectively communicated and the statistical assumptions justified.

14.4 PEC specification

The exposure assessment requires the derivation of potential environmental concentrations for different ecological compartments. For example, pesticides sprayed on crops will affect soil compartments and possibly aquatic compartments if there is a nearby body of water. They may also enter the atmosphere. In fact, there are many aspects that must be considered to determine a PEC (or multiple PECs): for example, substance properties, exposure paths (e.g. spray drift), background exposure and temporal and spatial frequencies. Assessments will usually be based on generic reusable models at the initial tier of assessment. When refinement is required, a more site-specific exposure model will be used and evaluated by scientific experts. As for PNEC specification, PEC specification has a corresponding TGD associated with it: for example, ICMM (2007b) in the context of metal exposure.

At the lowest tier, a conservative PEC is determined using a standard model. Conservative plug-in values for the parameters of the model are chosen by risk assessors and experts and subsequently agreed by the risk manager. In some cases parameter values are pre-determined for an application, based on historical data. In accordance with the toxicity assessment, the models can be analysed probabilistically by treating the parameters as uncertain. Since the models are typically non-tractable, Monte Carlo sampling methods might be employed to approximate the exposure distribution (e.g. EUFRAM, 2006) and statistical software is available to automate these processes. In recognition of the inherent aleatory and epistemic uncertainties, two-dimensional Monte Carlo (referred to as nested Monte Carlo by Suter, 2007) methods have been suggested. At each iteration of an 'outer-loop' which accounts for epistemic uncertainty, an 'inner-loop' is conditionally iterated multiple times to account for the aleatory uncertainty. Exposure distributions can also be described by statistically fitting models to measured data.

Based on an exposure (concentration) distribution (denoted ECD), a 'worst-case' point value of the PEC is defined as an appreciably high percentile; this is usually the ninetieth (EUFRAM, 2006) or the ninety-fifth (ECHA, 2008c) percentile. An average case (fiftieth percentile) is also derived and compared to the worst-case-based RCR. It is important to recognise that the exposure distribution can have different interpretations. For example, it might represent exposures in individual ditches, or, alternatively, the proportion of ditches (in some homogeneous population of 'standard' ditches) with a certain exposure. Generally the distributions will not be simple; however, in certain cases one may use the lognormal distribution as a simple representation (Aldenberg *et al.*, 2002).

14.5 Flash Environmental Assessment Tool

SSD-based risk assessment has recently been incorporated into other risk assessment approaches, including the United Nations (UN) Flash Environmental Assessment Tool (FEAT) (see van Dijk *et al.*, 2009). FEAT is a tool designed to assist the UN Disaster Assessment and Coordination teams, responsible for being the initial field response unit, in identifying acute risks pertaining to hazardous material exposure in secondary impacts of natural hazards. As an example, a flood or earthquake could destroy or weaken the hull integrity of a chemical plant, thus posing a substantial risk to the environment. The objective is to identify those hazardous substances (i.e. toxicity) and scenarios (i.e. exposures) which pose the most immediate (i.e. temporal consideration) and greatest (i.e. spatial consideration) risk. Specialist risk assessors may be unavailable in the immediate aftermath of an incident, and so FEAT is intended as a 'carefully balanced compromise between simplicity and scientific rigor, with emphasis on usefulness', rather than a surrogate for an in-depth environmental assessment (van Dijk *et al.*, 2009).

The core concept of FEAT is embodied in the functional relationship

$$\text{impact} = F(\text{hazard}, \text{exposure}, \text{quantity}).$$

Given that FEAT is intended to be used by non-specialist assessors, the impact is measured in simple units: distance from source over which toxicant levels exceed pre-defined acceptable levels. The hazard is the underlying toxicity of the chemical, e.g. a HC_{50}. Exposure focuses on the different exposure pathways and potential, with priority given to the 'most likely' exposure scenario. The quantity is the amount of toxic material(s) stored at or near the disaster site. Because it is conservative, FEAT would consider all the toxic substance to be released unless inventory supported otherwise. Weighing all this up through a function F one obtains a measure of the magnitude of impact that is returned in units of metres. The method, while crude, is an effective parsimonious tool that requires minimal knowledge and only relies on lookup tables and basic mathematics.

14.6 Integrated probabilistic risk assessment approach

EUFRAM (2006) and Suter (2007) both note that the representation of risk should be motivated by the problem and questions of the risk manager, meaning that no single method is suitable for all risk assessments.

The pointwise evaluation of the RCR may not be appropriate for all risk assessments. Multiple RCR values may be considered corresponding to worst- and average-case scenarios pertaining to the toxicity and exposure assessments, some of which may exceed the critical threshold, and others which may not. If defensible probabilistic models of the toxicity and exposure can be determined, then a more intuitive strategy for a risk manager would be to also treat the RCR probabilistically. This circumvents the need to specify levels of protection in advance, as well as the need to make conservative assumptions a priori – for example, choosing the one-sided lower confidence limit of the fifth percentile. This is because all outcomes with their associated probabilities are captured by the overall distribution. However, toxicological endpoints must still be predefined and this has consequences for the way in which the exposure distribution should be defined in order to be compatible (Aldenberg *et al.*, 2002). For example, Verdonck *et al.* (2003) state that the time measure of the endpoints should be at least as small as the exposure distribution time intervals for the two to be compatible. It should be noted that such approaches are considered higher tier risk assessments, as they require much more judgement from risk managers. A number of suggestions have been discussed in the literature regarding how a risk manager might utilise the fully probabilistic approach as a decision-making tool; we offer an overview of some suggestions here, focusing primarily on the statistical foundations.

14.6.1 Expected risk

The most straightforward and intuitive procedure for evaluating a probabilistic RCR would be to evaluate the probability that the RCR is greater than 1. Of course, this is conditional on the RCR being meaningful, i.e. the exposure distribution being compatible with the SSD. This is in line with the deterministic decision-making criteria. Equivalently, the probability is $P[X_{SSD} < X_{ECD}]$ where $X_{SSD} \geq 0$ and $X_{ECD} \geq 0$ are random variables (concentrations) distributed according to the SSD and ECD, respectively. Conventionally, researchers and risk assessors use log (base 10) transformed random variables, i.e. log concentration. So, letting $Y_{SSD} = \log_{10}(X_{SSD})$ and $Y_{ECD} = \log_{10}(X_{ECD})$, it is clear that this probability is determined by evaluating

$$P[Y_{SSD} < Y_{ECD}] = \int_{-\infty}^{\infty} (1 - F_{ECD}(y)) dF_{SSD}(y)$$

$$\equiv \int_{-\infty}^{\infty} F_{SSD}(y) dF_{ECD}(y)],$$

where F_{SSD} and F_{ECD} are the cumulative distribution functions of Y_{SSD} and Y_{ECD}, respectively (with support on the entire real line). (The second line follows from integration by parts.) In the ecotoxicological risk assessment literature, the probability is often referred to as the *risk* (e.g. Verdonck *et al.*, 2003), *ecological risk* (e.g. van Straalen, 1990: § 4) or perhaps more suitably the *expected risk* (e.g. Warren-Hicks *et al.*, 2002). It is interesting to note there was strong opposition from the three reviewers for describing this quantity as the *average risk*, a term used in EUFRAM (2006). A less dubious description would simply be to denote the quantity as the statistical mean fraction of species whose end-points are exceeded, whereby the mean is taken with respect to the distribution of exposure outcomes. However, from a more decision-theoretical perspective, it is the expected loss under a 0–1 loss function ($0 = \{RCR < 1\}$ and $1 = \{RCR \geq 1\}$). The terminology about this quantity is somewhat confusing, but for consistency we continue with the term 'expected risk'. An extensive treatment of this risk measure is given by Aldenberg *et al.* (2002).

A standard graphical representation of this is given by what Aldenberg *et al.* (2002), EUFRAM (2006), EFSA (2005) and ECHA (2008c) describe as a 'van Straalen ecological risk plot'; in Figure 14.3 we show a hypothetical plot. The area under the solid curve (which is determined as the product of the SSD cumulative distribution and exposure distribution density function) is equal to the expected risk, which in this case is 24%.

Interpretation of the expected risk is difficult and there is limited guidance on how a risk manager should make decisions; clearly a threshold of risk would need to be proposed if decisions are to be made in a manner consistent with the RCR threshold criterion. Even without such a threshold, one example would be calculation of the risk allows a risk manager to rank and contrast different scenarios; in determining the allocation of limited resources to environmental clean-up operations of accidental contamination for competing ecological sites.

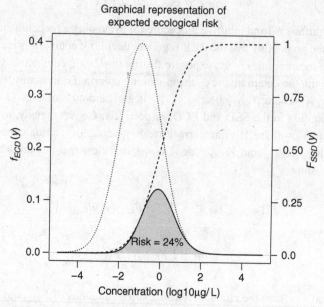

Figure 14.3 Graphical representation of the expected risk by the construction of a van Straalen ecological risk plot based on a hypothetical SSD and exposure distribution. The dotted curve indicates exposure distribution probability density function (left axis); dashed curve indicates SSD cumulative distribution function (right axis); solid curve is derived as the product of the former two curves (with respect to left axis). The shaded area is the expected risk, in this case 24%.

A weakness of the expected risk as a summary is that very different combinations of frequency and magnitude of effects can lead to the same overall value (see below). This has motivated further research.

14.6.2 *Joint probability curves*

Joint probability curves (JPCs) are parametric plots of the SSD (a probability distribution with domain over toxicant concentration) and exposure distribution (also a probability distribution with domain over toxicant concentration). A standard type of JPC is the exceedance profile plot (EPP), which parametrically plots: (1) the cumulative probability (as defined by the SSD, $F_{SSD}(\cdot)$) on the horizontal axis; and (2) the exceedance probability of exposure (equivalently, the survival function $1 - F_{ECD}(\cdot)$) on the vertical axis. The aforementioned probability is generally denoted as the 'potentially affected fraction' (PAF) of species in the ecological risk assessment literature. A discussion of the PAF is provided by Traas *et al.* (2002). Mathematically, one has a coordinate mapping of $\log_{10}$ concentration $y \in \mathbf{R}$:

$$y \mapsto (F_{SSD}(y), 1 - F_{ECD}(y)).$$

Consequently, all EPPs are guaranteed to start at $(0, 1)$ for $\log_{10}$ concentration $y = -\infty$, and conclude at $(1, 0)$ for $y = \infty$. The risk manager, when weighing up the evidence, can then examine the JPC graphical representation. Other ways of graphically representing a JPC have also been proposed, effectively through one-to-one transformations and/or reflections of the axes. Uncertainty bounds could potentially be overlaid onto the JPC, allowing for epistemic uncertainty relating to the SSD and exposure distribution to be shown.

From the EPP a risk manager could determine, by reading directly off the point ordinate, the proportion of exposures (assumed independent) that a fixed proportion of species will have their toxicological tolerances exceeded. For example, if $(0.10, 0.40)$ is a point on the EPP corresponding to some $\log_{10}$ concentration y, then 10% or more of species will have their sensitivity values exceeded for 40% of exposures. This allows the risk manager to straight-forwardly analyse impact of different exposure probabilities on an ad-hoc basis. A simple way of interpreting the risk based on an EPP is to examine how close the curve is to the axis (closer being minimal risk, further away being greater risk) (Warren-Hicks *et al.*, 2002; EUFRAM, 2006). This can be justified by considering the following thought experiment.

If most of the probability mass in the SSD is *below* the probability mass of the exposure concentration distribution, then the expected risk will be close to 1 because any random exposure concentration will almost surely exceed the tolerance value of random-selected species from the assemblage. In this case, the EPP curve will encapsulate almost the entire domain. On the other hand, if most of the probability mass in the SSD is *above* the probability mass of the exposure concentration distribution, then the expected risk will be close to 0. In this case, the EPP curve will tend to approach the axes.

Expected risk is, however, only a single (integrated) summary statistic of the risk. From this it is clear that the definition of risk in this context is problematic by noting that the expected ecological risk can be the same for two distinct risk assessments, but have very different JPCs. In other words, the method integrates over probability and magnitude of effect so that a high probability and low magnitude may lead to similar end values as a low probability and high magnitude. Consider the two EPPs, originally proposed by Verdonck *et al.* (2003), corresponding to two separate risk assessment scenarios, which we show in Figure 14.4.

The EPPs in Figures 14.4a–b suggest two very different scenarios, but yield identical expected risk. For the purposes of decision-making, the interpretation of the exposure must be clear, as the following thought experiment makes clear. Suppose that the exposure distribution was temporally defined (e.g. for a single chemical plant leakage) and the species endpoint was mortality. Then, for assessment (a) (Figure 14.4, left panel) the interpretation is that approx-imately at least 50% of species will die 100% of the time, whereas for (b) (Figure 14.4, right panel) it is that approximately 0% of the species will die 50% of the time, and approximately 100% of the species will die the in the remaining 50% of the time. Arguably, risk assessment (a) is the more desirable from a risk-management perspective since one expects to have some of the species survive all of the time (recall this is a single assemblage site). Now, consider if the exposure was spatially defined (e.g. multiple chemical plants), then the interpretation for (a) is that approximately 50% of species will die in 100% of exposures, whereas in (b) approx-imately 0% of species will die in 50% of the exposures and approximately 100% die in the

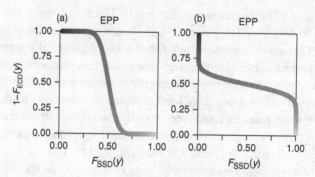

Figure 14.4 Exceedance profile plots for two risk assessments, each yielding an expected ecological risk of 50%. (a) $Y_{SSD} \sim N(0, 25)$ and $Y_{ECD} \sim N(0, 1)$. (b) $Y_{SSD} \sim N(0, 1)$ and $Y_{ECD} \sim N(0, 25)$, where $N(\mu, \sigma^2)$ denotes a normal distribution which mean and variance parameters μ and σ^2. The shading of the curves indicates increasing (log−) concentration with dark corresponding to the lowest concentration ($y = -\infty$) and light the highest ($y = \infty$).

other 50% of exposures. Hence, arguably now the risk manager might prefer outcome (b) over (a) because from a biodiversity perspective half of the locations will remain intact, whereas under assessment (a) biodiversity would be expected to be severely reduced at all locations. This example demonstrates why the expected risk measure is not fully informative.

14.6.3 Risk distributions

An alternative description of the risk has been proposed based on the potentially affected fraction of species, as specified by the SSD. In this case, one transforms the $\log_{10}$ transformed exposure concentration random variable Y_{ECD} to define a new random variable $R = F_{SSD}(Y_{ECD})$ which, for all intents and purposes, models the risk.

A number of useful properties are derivable from this. First, it is straightforward to determine that the statistical expectation of R is equivalent to the earlier definition of the expected risk. Second, the distribution function of R can be determined (which is expressible in terms of the JPC) and subsequently the probability density function. This indicates where the mass of likelihood is concentrated on the scale of fractions of species affected (e.g. for mortality) and allows for other summary measures to be conveyed to the risk manager. A library of functions to automate this style of risk assessment is now available for use with the Mathematica© software package (Wolfram Research, Inc., 2008) under the BUSY project (http://www.busymath.googlepages.com), based on a wide range of distributions.

14.7 Terminology

There is no uniformly accepted definition of risk. The International Programme on Chemical Safety (IPCS, 2004) defines risk as: 'The probability of an adverse effect in an organism, system, or (sub)population caused under the specified circumstances by exposure to an agent.' In the context of ecotoxicological risk assessment, the latter is equivalent to the probability that the RCR breaches unity. A basic – albeit misleading – risk equation, which appears frequently in the less-formal (eco)toxicological risk assessment literature (see, e.g. OPEP, 2009: ch. 9; US EPA 2007), is

$$\text{risk} = \text{toxicity} \times \text{exposure}.$$

The terminology in this field is somewhat case-specific. The 'toxicity' element of the risk assessment is described by ecotoxicologists as 'hazard assessment', and involves assessment of the intrinsic sensitivity of a (collection of) species to the toxicant. This is analogous to the concept of 'vulnerability' in natural hazards assessment. To add further confusion, exposure in ecotoxicology is analogous to the concept of 'hazard' in natural hazards assessment. Volcanic eruptions and tsunamis (see Chapters 10 and 11), for example, are natural hazards, whereas in chemical risk assessment the presence of a toxicant (which one might consider on par with the aforementioned natural hazards) would be considered an exposure (scenario). This emphasises the need for clear definitions when expressing risk and uncertainty. In particular, one should note that in this case both the hazard (toxicity) and exposure components are separate probabilistic components with associated uncertainty.

While on one hand the simple expression above implicitly suggests the need to incorporate the magnitude of effect along with the corresponding probability, it is confusing to assessors who may not have expertise in quantitative risk assessment and therefore take the multiplier literally. The definition of risk given by the Codex Alimentarius Commission procedural manual (UN FAO, 2008) is: 'A function of the probability of an adverse health effect and the severity of that effect, consequential to a hazard(s) in food.' This, although falling under the context of food standards, is considered more appropriate as it addresses both aspects. However, it must be borne in mind that there is associated uncertainty with both of these aspects which needs to be considered.

Following on from the earlier description of expected risk, we noted that we might define the summary measure in terms of expected loss:

$$E_{\downarrow}"\text{ECD}"[E_{\downarrow}"\text{SSD}"[\mathbf{1}_{\downarrow}\{Y_{\downarrow}"\text{ECD}">Y_{\downarrow}"\text{SSD}"\}|Y_{\downarrow}"\text{ECD}"]],$$

where $1_{\{A\}}$ is the indicator function for the event A: 1 if A occurs, 0 otherwise. The statistical setup mirrors mathematical definitions from other areas of risk assessment, including natural hazards (e.g. Chapters 2, 3 and 4). Whether it is appropriate or not to use 0–1 loss (the so-called 'all or nothing' measure) is questionable. However, the principal is well established for deterministic risk assessment, and for the sake of coherency it has been adopted for higher tier probabilistic risk assessments.

14.8 Conclusions

The current framework of decision-making for this single element of ecotoxicological risk assessment is relatively straightforward. A tiered approach is considered the best management strategy within the current regulatory arena, as it guides assessors to where refinement in assessment is required. Of course, we have simplified the problem and not taken account of individual case-specific points that will influence decision-makers.

As one extends into a (semi-) probabilistic approach, then subjective elements of the risk manager's cost–benefit portfolio will come into play and an 'overall risk' value may not be the definitive solution. The full probabilistic setup we described here, which is recommended by EUFRAM (2006) and others, is clearly appealing in light of the current state of the science. However, there are many underlying assumptions that require further investigation. Moreover, guidance for practitioners on how to interpret risk and the various recent proposals on integrated assessments is required.

Acknowledgements

We would like to thank Jonty Rougier (Bristol University) who reviewed this chapter more than once and made many suggestions that allowed us to improve it considerably. We also thank the reviewers Peter Craig (Durham University) and Tom Aldenberg (RIVM) for bringing our attention to ideas we had not thought of.

References

Aldenberg, T. and Jaworska, J. S. (2000) Uncertainty of the hazardous concentration and fraction affected for normal species sensitivity distributions. *Ecotoxicology and Environmental Safety* **46**: 1–18.

Aldenberg, T., Jaworska, J. S. and Traas, T. P. (2002) Normal species sensitivity distributions and probabilistic ecological risk assessment. In *Species Sensitivity Distributions in Ecotoxicology*, ed. L. Posthuma, G. W. Suter and T. P. Traas, Boca Raton, FL: Lewis Publishers, pp. 49–102.

ANZECC and ARMCANZ (Australian and New Zealand Environment and Conservation Council and Agriculture and Resource Management Council of Australia and New Zealand) (2000) Australian and New Zealand guidelines for fresh and marine water quality. National Water Quality Management Strategy Paper No. 4.

Chen, L. (2003) A conservative, non-parametric estimator for the 5th percentile of the species sensitivity distribution. *Journal of Statistical Planning and Inference* **123**: 243–258.

Craig, P. S., Hickey, G. L., Hart, A., *et al.* (2011) Species non-exchangeability in probabilistic ecotoxicological risk assessment. *Journal of the Royal Statistical Society Series A: Statistics in Society* **175** (1): 243–262.

Duboudin, C., Ciffroy, P. and Magaud, H. (2004) Effects of data manipulation and statistical methods on species sensitivity distributions. *Environmental Toxicology and Chemistry* **23**: 489–499.

EC (European Commission) (1991) Council Directive 91/414/EEC of the Council of the European Communities of 15 July 1991 concerning the placing of plant protection products on the market. *Official Journal of the European Union* L 230, p. 1.

EC (European Commission) (2002) Guidance document on aquatic ecotoxicology in the context of the Directive 91/414/EEC. SANCO/3268/2001, 4: 62.

EC (European Commission) (2003) Technical Guidance Document on Risk Assessment in Support of Commission Directive 93/67/EEC on Risk Assessment for New Notified Substances Commission Regulation (EC) No. 1488/94 on Risk Assessment for Existing Substances Directive 98/8/EC of the European Parliament and of the Council Concerning the Placing of Biocidal Products on the Market, Part II. Luxembourg: Office for Official Publications of the European Communities.

EC (European Commission) (2006) Regulation (EC) No 1907/2006 of the European Parliament and of the Council of 18 December 2006 concerning the Registration, Evaluation, Authorization and restriction of CHemicals (REACH) establishing a European Chemicals Agency, amending Directive 1999/45/EC and repealing Council Regulation (EEC) No 793/93 and Commission Regulation (EC) No 1488/94 as well as Council Directive 76/769/EEC and Commission Directives 91/155/EEC, 93/67/EEC, 93/105/EC and 2000/21/EC. *Official Journal of the European Union* L 396, p. 1.

ECHA (European Chemicals Agency) (2008a) Guidance on information requirements and chemical safety assessment Part E: risk characterization.

ECHA (European Chemicals Agency) (2008b) Guidance for the implementation of REACH: guidance on information requirements and chemical safety assessment chapter R.10: characterization of dose–response for environment.

ECHA (European Chemicals Agency) (2008c) Guidance for the implementation of REACH: guidance on information requirements and chemical safety assessment chapter R.19: uncertainty analysis.

EFSA (European Food Safety Authority) (2005) Opinion of the scientific panel on plant health, plant protection products and their residues on a request from EFSA related to the assessment of the acute and chronic risk to aquatic organisms with regard to the possibility of lowering the uncertainty factor if additional species were tested. *The EFSA Journal* **301**: 1–45.

EFSA (European Food Safety Authority) (2008) Scientific opinion of the panel on plant protection products and their residues (PPR) on the science behind the guidance document on risk assessment for birds and mammals. *The EFSA Journal* **734**: 1–181.

EUFRAM (2006) *The EUFRAM Framework: Introducing Probabilistic Methods into the Ecological Risk Assessment of Pesticides*, vol. 1, ver. 5. http://www.eufram.com (accessed 31 August 2009).

Forbes, V. E. and Calow, P. (2002a) Extrapolation in ecological risk assessment: balancing pragmatism and precaution in chemical controls legislation. *BioScience* **52**: 249–257.

Forbes, V. E. and Calow, P. (2002b) Species sensitivity distributions revisited: a critical appraisal. *Human and Ecological Risk Assessment* **8**: 1625–1640.

Grist, E. P. M., O'Hagan, A., Crane, M., *et al.* (2006) Bayesian and time-independent species sensitivity distributions for risk assessment of chemicals. *Environmental Science and Technology* **40**: 395–401.

Hickey, G. L. (2010) *Ecotoxicological risk assessment: developments in PNEC estimation.* PhD thesis, Durham University.

Hickey, G. L., Kefford, B. J., Dunlop, J. E., *et al.* (2008) Making species salinity sensitivity distributions reflective of naturally occurring communities: using rapid testing and Bayesian statistics. *Environmental Toxicology and Chemistry* **27**: 2403–2411.

Hickey, G. L., Craig, P. S. and Hart, A. (2009) On the application of loss functions in determining assessment factors for ecological risk. *Ecotoxicology and Environmental Safety* **72**: 293–300.

ICMM (International Council on Metals and Mining) (2007a) Metals environmental risk assessment guidance (MERAG) fact sheet 1: risk characterization – general aspects.

ICMM (International Council on Metals and Mining) (2007b) Metals environmental risk assessment guidance (MERAG) fact sheet 2: exposure assessment.

ICMM (International Council on Metals and Mining) (2007c) Metals environmental risk assessment guidance (MERAG) fact sheet 3: effects assessment. Data compilation, selection and derivation of PNEC values for the risk assessment of different environmental compartments (water, STP, soil, sediment).

IPCS (International Programme on Chemical Safety) (2004) *IPCS Risk Assessment Terminology*, Geneva: World Health Organization.

Kefford, B. J., Palmer, C. G., Jooste, S., *et al.* (2005) 'What is it meant by 95% of species'? An argument for the inclusion of rapid tolerance testing. *Human and Ecological Risk Assessment* 11: 1025–1046.

Luttik, R. and Aldenberg, T. (1997) Assessment factors for small samples of pesticide toxicity data: special focus on LD50 values for birds and mammals. *Environmental Toxicology and Chemistry* 16: 1785–1788.

Maltby, L., Blake, N., Brock, T. C. M., *et al.* (2005) Insecticide species sensitivity distributions: the importance of test species selection and relevance to aquatic ecosystems. *Environmental Toxicology and Chemistry* 24: 379–388.

Newman, M. C., Ownby, D. R., Mezin, L. C. A., *et al.* (2000) Applying species sensitivity distributions in ecological risk assessment: assumptions of distribution type and sufficient numbers of species. *Environmental Toxicology and Chemistry* 19: 508–515.

OPEP (Ontario Pesticide Education Program) (2009) *Grower Pesticide Safety Course Manual*. Guelph, ON: University of Guelph Ridgetown Campus.

Pires, A. M., Branco, J. A., Picado, A., *et al.* (2002) Models for the estimation of a 'no effect concentration'. *Environmetrics* 13: 15–27.

Posthuma, L., Suter, G. W. and Traas, T. P. (eds) (2002) *Species Sensitivity Distributions in Ecotoxicology*, Boca Raton, FL: Lewis Publishers.

Stephan, C. E., Mount, D. I., Hansen, D. J., *et al.* (1985) Guidelines for deriving numerical national water quality criteria for the protection of aquatic organisms and their uses. Report No. PB 85-227049. US Environmental Protection Agency, Office of Research and Development.

Suter, G. (2007) *Ecological Risk Assessment*, Boca Raton, FL: CRC Press.

Traas, T. P., van de Meent, D., Posthuma, L., *et al.* (2002) The potentially affected fraction as a measure of ecological risk. In *Species Sensitivity Distributions in Ecotoxicology*, ed. L. Posthuma, G. W. Suter and T. P. Traas, Boca Raton, FL: Lewis Publishers, pp. 315–344.

UN FAO (United Nations Food and Agriculture Organization) (2008) *Codex Alimentarius Commission Procedural Manual*, 18th edn, Rome: FAO.

US EPA (US Environmental Protection Agency) (1998) Guidelines for ecological risk assessment: notice. *Federal Register* 63: 26846–26924.

US EPA (US Environmental Protection Agency) (2004) *Overview of the Ecological Risk Assessment Process in the Office of Pesticide Programs*, Washington, DC: Office of Prevention, Pesticides and Toxic Substances.

US EPA (US Environmental Protection Agency) (2007) Assessing health risks from pesticides. Factsheet. http://www.epa.gov/opp00001/factsheets/riskassess.htm (accessed 30 May 2011).

van Dijk, S., Brand, E., de Zwart D., *et al.* (2009) *FEAT: Flash Environmental Assessment Tool to Identify Acute Environmental Risks Following Disasters. The Tool, the Explanation and a Case Study*. Bilthoven: RIVM.

van Straalen, N. M. (1990) New methodologies for estimating the ecological risk of chemicals in the environment. In *Proceedings of the 6th Congress of the International Association of Engineering Geology*, ed. D. G. Price, Rotterdam: Balkema, pp. 165–173.

Verdonck, F. A. M., Aldenberg, T., Jaworska, J., *et al.* (2003) Limitations of current risk characterization methods in probabilistic environmental risk assessment. *Environmental Toxicology and Chemistry* **22**: 2209–2213.

Warren-Hicks, W. J., Parkhurst, B. J. and Butcher, J. B. (2002) Methodology for aquatic ecological risk assessment. In *Species Sensitivity Distributions in Ecotoxicology*, ed. L. Posthuma, G. W. Suter and T. P. Traas, Boca Raton, FL: Lewis Publishers, pp. 345–382.

Wheeler, J. R., Grist, E. P. M., Leung, K. M. Y., *et al.* (2002) Species sensitivity distributions: data and model choice. *Marine Pollution Bulletin* **45**: 192–202.

Wolfram Research, Inc. (2008) Mathematica. Software, Version 7.0.

15

Social science perspectives on natural hazards risk and uncertainty

S. E. CORNELL AND M. S. JACKSON

15.1 Introduction

Many of the most pressing challenges for uncertainty analysis in the context of natural hazards relate to the human dimensions of risk. Earthquakes, volcanoes, landslides and many other hazards are themselves, *sensu stricto*, simply events in the natural world. How they come to be classed as hazardous, however, is a consequence of social factors which translate their happening as events for collective experience. Designating these event hazards as risks is a further conceptual step that invokes temporal awareness of hazards from past experience. Risks, and the uncertainties around their prediction, thus emerge from complex spatial and temporal interrelationships of natural and social worlds. Understanding the complexities of these interrelationships between natural and social domains is vital, because the analyses of risks, the processes of designing and managing responses to them and the nature and scope of their uncertainties are part and parcel of how risk *itself* comes to be constituted as a societal 'matter of concern' (Latour, 2004a).

Risks are usually understood as threats to people. Increasingly they are even seen as opportunities to be exploited. Natural hazards become risky as a function of how they are considered problematic within the systems or 'assemblages' (see Çalişkan and Callon, 2010) of which people are a part. A volcano on Earth is hazardous, and presents risks in numerous ways in many human situations, such as air travel or nearby habitation, and so on. In contrast, a volcano on Saturn's moon, Titan, is not a hazard, *unless* it becomes incorporated into a socially interested assemblage and presents a threat to a social activity, for instance, interfering with the function of a space probe. Potential risks and uncertainties presented by such interactions between humans and the natural environment are, therefore, products of how hazards are incorporated into social systems. It is widely accepted within contemporary social science that the way that modern societies negotiate and conceive risk is a function of how risk is itself a manufactured product of specific industrial, technological and scientific capacities (Beck, 1992, 1999). Risks become manifest for us in new ways precisely through the capacities afforded to us by our ever-changing social behaviours and capabilities. Risks, then, are labile functions of changing social descriptions and values. As a result, social research is of central importance to the framing and understanding of

Risk and Uncertainty Assessment for Natural Hazards, ed. Jonathan Rougier, Steve Sparks and Lisa Hill. Published by Cambridge University Press. © Cambridge University Press 2013.

contemporary risk and uncertainty. However, mainstream risk research has primarily concentrated on environmental and technical dimensions of risk, as the subject matter of this book attests.

In this chapter, we explore the emerging rationale for a more comprehensive engagement with social science expertise and techniques, in assessing risks associated with a wide range of socio-environmental or socio-technical hazards (e.g. Macnaghten *et al.*, 2005; Klinke and Renn, 2006; Demeritt, 2008; LWEC, 2008). Key features of this engagement are the opening up of both interdisciplinary and societal dialogues, and a more experimental, self-conscious and provisional approach to identifying responses, recognising that multiply optimised solutions may not exist. The recently created international programme for Integrated Research on Disaster Risk (ICSU-IRDR) is a prime example:

The Programme Plan breaks new ground in that it calls for multiple starting points: natural sciences; socio-economic sciences; engineering sciences; health sciences; and the policy-making/decision-making arena. There is need for full interaction and involvement of these groups, with each being clear what it needs from the other groups. It is also necessary to work across the interfaces, with continual re-examination as the Programme proceeds. The overall goal of contributing to a reduction in the impacts of hazards on humanity would require some relatively non-traditional research approaches.

(ICSU, 2008: 8)

As we will see, this kind of new engagement across disciplines and the channelling of academic research into policy and society present radical new challenges for the research process. Key among these is the need to integrate a greater range of expert evidence into the suite of risk calculation and response strategies. In this chapter we refer to innovative integrative studies that exemplify how varieties of traditional and non-traditional expertise (in the physical science, social scientific and non-academic communities) shape the production of new knowledges about environmental risks.

To close, we will focus briefly on the capacities for academics to be proactive in shaping society's responses, amidst demands for greater involvement by scientists in hazards policy (Royal Society, 2006; HM Government, 2011). If social scientific and non-traditional learning can be better incorporated in the interdisciplinary research design and analysis of knowledge about natural hazards, then scientific actors will also be better prepared for the demands of political application and social action.

15.1.1 *Setting the scope*

Insight from social science can and, we argue, *should* – because it already does, even if it is not always explicitly acknowledged – add to our understanding of natural hazards risks and uncertainties. Although a great deal of research into risk and uncertainty has been mobilised in the social sciences (for example, Douglas and Wildavsky, 1982; Beck, 1992; Douglas, 1992; Luhman, 1993; Giddens, 1999; Taylor-Gooby and Zinn, 2006; Callon *et al.*, 2009), arguably the crossover of insights to the parallel community in the natural and physical sciences is still wanting. Technocratic models of scientific knowledge production, risk assessment and risk communication prevail in the everyday assumptions and practices of publics, scientists, policy-makers and the media (Hulme, 2009: 102–104; Millstone, 2009; Shrader-Frechette, 2010). This is despite many long arguments about the value complexities of scientific knowledge production and the co-determinate interfaces of science and policy (see, among many examples, Jasanoff, 1990; Shrader-Frechette, 1991; Funtowicz and Ravetz, 1993; National Research Council, 1996; Etkin and Ho, 2007; Turnpenny *et al.*, 2011). While our main focus in this chapter is on the qualitative and theoretical approaches that integrate social, technical and environmental systems, as will become clear, this also implies a thorough-going critique of technocratic forms of knowledge production, including those in the social scientific arena. Technocratic models assume that objective and value-free knowledge produced by technically literate experts is both necessary and sufficient for policy decision-making. Related 'top-down' knowledge formation processes like so-called 'decisionist' models (see Weber, 1946; Hulme, 2009: 100–102; Millstone, 2009: 625) similarly privilege the role of experts in defining the terms and frames for what counts as risk-relevant knowledge. Despite surface appearances of doing otherwise, such efforts often unwittingly reproduce technocratic biases by emphasising the role of social science expertise in the communication of scientific knowledge to those affected by hazards (see, for instance, Pidgeon and Fischhoff, 2011). We argue, like Hulme (2009) and others, for a more 'co-productive' model of knowledge formation which seeks, in open forms of democratic and public dialogue with scientists and advisors, the means to embed risk-relevant values, rationalities, world-views and knowledges of those affected by hazards in the production of what counts as scientific knowledge. We even go further to suggest, by way of offering some limited examples, that how risk and hazards are understood, rationalised and communicated is a function of extremely complex human and non-human assemblages of interest. Indeed, the agencies at play in decision forecasting are distributed products of these socio-technical assemblages, and include humans, their material environments, their resonant politics, particular techniques and resultant world-views. In order to effectively understand, mitigate and possibly intervene in hazardous environments we need to take account holistically of these 'knowledge ecologies' which, at a minimum, demand that risk-relevant knowledge is co-produced from its heterogeneous contexts (Farrell, 2011). Furthermore, where once strict academic boundaries governed the non-human and human (and hence the social), increasingly such lines are becoming blurred, expanding the remit of social exploration (see Latour, 2004b; Escobar, 2008). New developments in the sciences of life and environment compel physical scientists to direct attention to the human dimensions of the issues that interest them. They also demand that social scientists themselves reconfigure their own assumptions, politics and starting points, and begin to see the social as a product of hybrid collectives beyond simply human action. We argue that such 'socio-natural' evidence and experience (to use the term of Whatmore, 2002) can provide constitutive grounds for important operational insights that help us in framing and responding to risk and uncertainty in natural hazards research.

The recognition of socio-natural co-implication and imbrication builds from previous social scientific treatments of risk. The sociologist Ulrich Beck, in his influential books *Risk*

Society (1992) and *World Risk Society* (1999), argued that, for modern social and political subjects, the risks that arise from scientific and industrial modernity constitute a new means of social change. He emphasised how contemporary social shifts away from loyalty and trust in institutions and traditional social structures, and towards science and technology and liberalised economic structures, shaped the emergence of risk as a key facet of contemporary life. As a result, risk becomes a function of greater individualism, and indeed, risk is the product of social and technical modernity itself. Risks, defined as the products of modern industrial societies and their future-oriented perspectives and capabilities, are for Beck and others (such as Giddens, 1999) key features of how modern societies understand and organise themselves. The risks that concern us today, that in Beck's words, ask us to move 'beyond a modernity of classical industrial design' (1992: 10) are what Beck calls 'manufactured uncertainties' (Beck, 1999: 19–48). They are the product of things like greater predictive capacities, increased industrialisation (pollution, stress), increased knowledge about disease and risky behaviours (smoking or over-eating, for instance), and the capacity of technology and social action, to create means – to some extent at least, if within always uncertain parameters – to assess and address these newfound risks (e.g. with climate modelling, improved geological sensing, flood course prediction, storm tracking, identification of disease and its spread). The social scientific focus since Beck's work, then, has been on analysing how such social changes impact upon individual and collective strategies or abilities to deal with what are labelled risks and uncertainties (for example, Ericson *et al.*, 2000; O'Malley, 2004; Rose, 2006). People living in what have since come to be termed 'risk societies' find their options are re-configured within complex, often market-led, structures and forces, and, as a result, they are increasingly left to manage their own safety and risk exposures. However, at the same time, the discourse of science – which is seen to objectively describe the so-called 'natural' conditions of hazard, risk and its avoidance – has become the key means by which risk is thinkable, and as such sets the context for risk and its various politics (Giddens, 1999). The critical social sciences, then, have focused largely on the politicisation of risk as a function of living in scientifically and economically mediated risk societies.

In what follows, our approach is not to survey the extensive social scientific literature on risk societies. We would not be able to do justice to that body of work in the available space, and, we suspect, we would be misaddressing the communities of interest targeted in this book. Instead, we describe some existing and emerging areas of research at the interfaces between society and the natural environment, where a social understanding of risk and uncertainty is being applied. The disciplinary divide in academia between science and social science has meant that this is still a comparatively small area, and further methodological developments are needed. We aim to flag areas where researchers can bridge the divide and attend to calls for greater and more productive collaborations around hazard risks and uncertainties. We highlight three domains where social dimensions are already recognised as important factors of uncertainty in natural hazard risk assessment and management (discussed more fully in Section 15.3).

The first and most fundamental of these domains is in the framing of the problem itself. The social sciences can shed important light on how the knowledge that pertains to hazards and risks is framed, produced and implemented (see Collins and Evans, 2002). To date, risk assessments have largely looked at the social and environmental components separately, so we need to revisit how risk and uncertainty are typically constructed and addressed. While we expect that the quantitative, applied models of complex systems that often frame approaches to natural hazards risks and uncertainties will continue to play a vital role in risk management, constraining the attempt to capture social perceptions and responses to risks simply through quantitative modelling jeopardises the capacity to understand how risks and uncertainties emerge from and are shaped by social and natural interactions (see Smil, 2003; Pilkey and Pilkey-Jarvis, 2007; Odoni and Lane, 2010). Our capacity to respond effectively to natural hazards may thus be blunted by an over-reliance on quantification. To see risk as a product of whole-system relations requires an understanding not just of the natural hazard, but also of the interactions between environmental and human systems (Figure 15.1). As we know, what is considered an acceptable or unacceptable risk (and thus, a rational or irrational response) shifts with geographical, cultural, economic and political context. Uncertainty, too, is a function of time and social expectation, translated and acted upon by the societies and cultures through which it is modulated and becomes meaningful. Thus, knowledge of the social and institutional systems operating in the context of natural hazards is required, just as an understanding of the environmental and technological drivers and constraints of human choices that lead to risk exposure is also necessary. Understanding and invoking the non-quantifiable aspects of risk and uncertainty lends a rich and necessary picture to the resonance and application of natural hazards decisions.

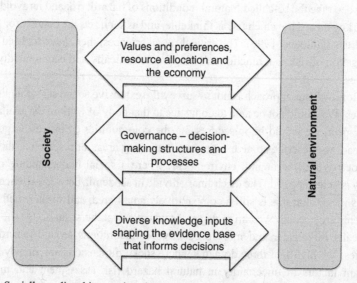

Figure 15.1 Socially mediated interactions between human society and its natural environment.

A fuller characterisation of risks and uncertainties from a coupled socio-environmental perspective necessarily involves a much wider range of epistemic inputs into the assessment process than we often currently see. In addition to the customary inputs from economics and political science, new research fields, approaches and dialogues within longer-established disciplines are shaping the framing of risk and uncertainty. For instance, within sociology, growing attention is being given to these issues in the fields of science and technology studies and actor network theory, and in geography, 'hybrid geographies' are emerging, which draw on both of these new sociological fields. Relevant research in psychology includes prospect theory and plasticity; from philosophy comes game theory and assemblage theory. Input to environmental risk also comes from ethics and law (specifically, socio-legal studies). Bringing this multidisciplinary knowledge together has become an important aspect of assessing risks and addressing social responses to natural hazards. Significantly, there are widespread moves towards the engagement of communities of interest in defining the risk problem and agreeing how they prefer to manage the uncertainties, rather than merely in accepting the proposed solutions. Their experience and knowledge makes them 'contributory experts' (Collins and Evans, 2007), whose voices are no less important than those of the scientists bringing specialist technical expertise.

This takes us to the second domain where a social science perspective plays a vital role. Like many other contemporary environmental issues, the societal risks linked to natural hazards are particularly complex problems to understand, represent and communicate. Uncertainty features prominently in the comprehensibility of environmental risks and their communication to the people who will be affected. Better interdisciplinary engagement can often help in shaping how scientific knowledge and understanding is communicated, received and acted upon in non-scientific contexts – for example, in public understanding or policy application (see Callon, 1999). Ensuring the effectiveness of science communication in the context of environmental risks is a well-documented challenge (e.g. MacDonell *et al.*, 2002; Heath *et al.*, 2007; Brulle, 2010; Sterman, 2011). However, while social research can shed useful light on how best to communicate scientific uncertainty, for us what is significant is that it can also bring social and cultural perspectives to natural hazards science via methodological processes that emphasise the 'co-production of scientific knowledge' through 'knowledge theoretic' approaches (see Odoni and Lane, 2010; Lane *et al.*, 2011). The scientific and methodological implications for people's access to, and constructions of, knowledge, and thus how risks can be understood, translated across different contexts and managed, mark one of the primary contributions of social research for natural hazards risk and uncertainty science.

Our third and final domain is less commonly appreciated. While social scientific perspectives on scientific knowledge production are increasingly shaping how physical or 'hard scientific' knowledge is produced (see, e.g. Odoni and Lane, 2010), they can also shape what 'the social' itself comes to mean within changing contexts. Like the scientific, so the social also changes and adapts through time and space. The rich and nuanced understanding that social research can bring is an important component in responding to natural hazards along the whole chronology of hazard identification, avoidance, mitigation, adaptation and post-event recovery. In this regard, we focus in this chapter on the social learning and actions at the collective and community levels; and on the systems of relations that shape how the social is constituted and involved in the production of scientific knowledge about hazard risks. There is already a considerable body of research into individuals' perceptions of risk, and psychological reasons for making choices (for example, see Kahneman *et al.*, 1982; Tversky and Kahneman, 1991; Fox and Tversky, 1995; Kahneman and Tversky, 2000). This is not reviewed in depth here, but is covered in detail in Chapter 16. Instead, we will address how understanding motivations and values influences the collective productions of, and responses to, risk and uncertainty, and how these are deployed in natural hazard risk management. Even when an understanding of the problem is 'complete', social choices – as matters of concern – regarding what to do in response to the problem are often difficult. Issues of collective understanding, choice and human agency bring a different analytical register to uncertainty, the latter of which is coming increasingly to the fore as the predictive power of the scientific dimensions of risk assessment grows.

15.1.2 The social in the science

This chapter seeks a precarious balance. On the one hand we emphasise how social scientific perspectives on hazard risks – as arising primarily from events that impact people – can contribute to effective scientific production, communication and implementation strategies. The social sciences, of course, do have a crucial role to play in the dissemination and mediation of knowledge between and across interested parties and domains. But, critically, on the other hand, social science is not a 'bolt-on' whereby so-called objective scientific truth is disseminated to affected audiences with the help of socially literate intermediaries, although it is too often regarded, unfortunately, in this way by many implicit and explicit technocratic approaches (see, for instance, Pidgeon and Fischhoff, 2011). Critical social analyses need to be woven into the fabric of research design from the outset. Contemporary social analysis, since at least the 1970s and what is known as 'the cultural turn' (Bonnell *et al.*, 1999), has been engaged in analysing how scientific discourses and practices are themselves part of the entire constitution of our physical and risk environments. As we illustrated above, many risks are *only* risks because they are products of modern social and scientific practices (see Beck, 1999). In addition, through its development of new forms of evidence and predictive capacity, scientific practice and knowledge-making is constantly producing new parameters of what risk means, and thus moulding the ever-changing landscapes of uncertainty. As a result, how evidence is constructed is a product of the various communities and interests in which it is applied and made meaningful. Van Asselt and Rotmans (2002: 77) broadly summarise the consequence of this often vexing and polarising philosophical fact with two interrelated claims: 'Science is not a purely objective, value-free activity of discovery: science is a creative process in which social and individual values interfere with observation, analysis and interpretation', and 'Knowledge is not equivalent with truth and certainty.'

We suggest that in order to better integrate natural hazards research to efficacious social outcomes, understanding how evidence and its applications are always already products of particular interested communities can go a long way to addressing scientific issues of risk and its translation, reception and application.

But, interdisciplinary alliances are not consequence-free. Balancing social and scientific demands requires significant work on *both* sides of the knowledge-production process. Social science and physical science are almost always changed, sometimes quite radically, by becoming co-implicated with one another. By bringing social scientific research, insights, and methods into natural and physical hazards research, the physical sciences must open themselves to experimental processes with new scientific actors and radical modes of knowledge production. To illustrate this requirement and its benefits, we refer in this chapter to a recent UK research project on natural hazards risk, which exemplifies just such transformative work. Understanding Environmental Knowledge Controversies (UEKC) aimed to integrate specialist physical science, social science and non-science actors within evidence generation, analytic and response processes in order to develop new approaches to flood risk forecasting and communication. UEKC produced scientific and social scientific knowledge, integrated models and policy-relevant findings from various social engagements in the co-production of environmental science. The project, for us, stands as a landmark in integrated scientific analysis, and an exemplar of the radical changes to natural science and hazard processes that social analyses can and do bring – and vice versa. As the UEKC project attests, both natural science and social science are changed by their fundamental re-engagement with one another.

In writing this chapter, we started out by entreating the physical science community to be more sensitive and receptive to the insights that the social sciences provide. Crucially, however, as also evidenced by the UEKC project, the social sciences must *also* become much more attuned to, and literate with, the significant cutting-edge developments in the physical and natural sciences. Sociology, politics, human geography and some branches of the humanities are often far too parochial and unrigorous in their approach to contemporary developments within the natural sciences. Significant time and work is urgently required by many qualitative social researchers to understand the rigours and subtleties of scientific knowledge production. Critical social scientists are often far too quick to dismiss scientific approaches as positivist, reductive and simplistic. Such reactionary judgements are, today, quite often wrong and certainly not mutually beneficial. So while physical science must open itself to the indubitable fact that science is a social, historical, political and cultural product, so too must social researchers also educate themselves in the thorough delicacies of formal and physical research. Both will benefit as a result.

15.2 The knowledge challenge

15.2.1 Integrating knowledge

Quantitative approaches are essential components in the characterisation of risks. Much research has focused on quantifying and modelling the physical and material (including economic) elements of natural hazards, as outlined in the other chapters of this book. There has also been substantial progress towards predictive approaches in many areas, often combining real-time earth observation from monitoring or satellite imaging with computer models. Yearley (1999) has explored this trend from a sociological perspective, proposing that much more consultation and participation are needed at the stage of model development, especially for models that are applied to public policy.

Yet this is more than just an issue of public understanding of science. The multiple uncertainties that pertain to human vulnerabilities to natural hazards – which include social, ethical and political elements – mean that quantitative approaches are incomplete and internally fraught in so far as they can represent and intervene effectively in the world. The main approaches for quantitative integration draw on methods for representation and assessment already described in the previous chapters, and are therefore only summarised briefly here. Key sources of uncertainty relate to: the quality or quantity of input data; the selection or definition of model parameters; different model structures and resolvability; the representation of functional relations; and the degree of model 'completeness' or abstraction from reality. We also know that quantitative models are themselves fundamentally limited. As Oreskes (2003: 13) writes, 'Models can never fully specify the systems that they describe, and therefore their predictions are always subject to uncertainties that we cannot fully specify.' But although numerical models never can provide deterministic predictions, this does not preclude adaptive responses to perceived risk. For instance, Dessai *et al.* (2009) argue that the fundamental limits to prediction should not obviate risk-attentive decision-making. Quite the opposite – they argue that 'decision-making is less likely to be constrained by epistemological limits and therefore *more likely to succeed* than an approach focused on optimal decision-making predicated on predictive accuracy' (Dessai *et al.*, 2009: 64; emphasis added).

Improving information can inform policy decisions about risk, but the availability of information or data is not *in itself* the basis for reaching or justifying those decisions, as the increasing complexity of numerical models would sometimes lead us to believe. Therefore, how this information is packaged and applied as data is significant. We move below to describe how the importance of qualitative information emerges through an awareness of the limitations of quantitative data.

15.2.2 *Qualifying quantification*

Data issues abound in quantitative social research, as in the physical sciences. Economic data are available from diverse sources on trade, industrial production, household consumption, government expenditure and so on. Typically, social data are obtained from population and household surveys (such as the annual US Survey of Income and Program Participation[1] and the UK General Household Survey[2]). Demographic and economic data, and the use of spatial analysis methods, help elucidate questions such as a population's exposure to the hazard, or a nation's dependence on sectors affected by the hazard. As in most

---

[1] www.census.gov.   [2] www.statistics.gov.uk.

data-dependent situations, uncertainty is associated with the challenges of scaling and reso-lution; the non-synchroneity of different datasets; the representativeness or otherwise of population sampling; and the necessary splitting or aggregation of datasets in order to obtain usable spatial or temporal coverage. There is a huge body of literature on social data handling and statistics, outlining refinements that are designed to reduce the uncertainty in the applica-tion of the data. For example, Davern *et al.* (2006) outline approaches for constructing synthetic variables from publicly available datasets so that estimates can be made of the standard errors of the original surveys, an important factor for the use and reuse of social data. Zintzaras *et al.* (1997) describe methods for deriving individual estimates from household-level data, address-ing another instance where the available data do not provide the desired information.

With regard to modelling, there has been a steady evolution from essentially actuarial approaches for estimating risk, which apply algorithms based on observed relationships between the hazard incidence and the societal impact (usually measured in money terms), to simulation modelling and probabilistic approaches that address the dynamics and the multiple causal factors in the human dimensions of risk. These models are generally termed 'integrated assessment models' (Table 15.1). Often they consist of linked physical or environmental and economics modules. Such models are widely used for regional and global climate impacts assessment, and have also been applied to socio-environmental

Table 15.1 *Tools in use in interdisciplinary risk and uncertainty research*

Approach and technique	Context and examples of application
Qualitative or semi-quantitative approaches to uncertainty identification:	
Uncertainty tables help in the systematic elicitation and evaluation of uncertainties, often presented as a preliminary step before 'better' quantification is undertaken.	European Food Safety Authority (2006); UK Food & Environment Research Agency (Hart *et al.*, 2010)
Multi-criteria analysis for decision support, based on multi-attribute utility theory or other forms of ranking and optimisation.	Industrial disasters (Levy and Gopalakrishnan, 2009); sustainability of water systems (Ashley *et al.*, 2005); climate risk (Brown and Corbera, 2003); air quality problems (Phillips and Stock, 2003)
Decision-mapping, where the goal is not to zero in on 'the answer' but to display (and discuss) the breadth of views on alternative actions.	Nanotechnology, genetically modified crops (Stirling and Mayer, 2001)
Combined approaches or multi-stage procedures (qualitative-quantitative synthesis):	
Integrated assessments – most are fully quantitative, such as impacts models or quantitative risk assessment consisting of hazards models coupled to econometric models.	Flood risk (e.g. the RASP model used in the UK Foresight Future Flooding project, Evans *et al.*, 2004); global climate impacts (Costanza *et al.*, 2007)

Table 15.1 (*cont.*)

Approach and technique	Context and examples of application
Mediated modelling: computational model development in ongoing dialogue between the modellers and stakeholders.	Ecosystem services assessments, catchment management (Antunes *et al.*, 2006; Prell *et al.*, 2007)
Some forms of Bayesian synthesis	Many health contexts (Roberts *et al.*, 2002; Voils *et al.*, 2009)

Other participatory deliberative techniques for assessing risk and associated uncertainty:

Expert elicitation using a range of formal methods (such as Delphi methods, multi-criteria analysis, quantitative risk assessments, and others from the list above) to take account of 'expert uncertainty'.	Geohazards, nuclear installations (Aspinall, 2010); climate impacts (Titus and Narayanan, 1996; Lenton *et al.*, 2008)
Narrative approaches: horizon-scanning, 'imagineering' (narrative approaches combined with education/training plans), foresight exercises – usually engage expert stakeholders. These methods are not intended as formally predictive, unless as enablers for a particular decision pathway.	Former DTI's Foresight programme, European Science Foundation's Forward Looks. International Institute for Sustainable Development; GECAFS food security scenarios
Community participation methods: citizens' juries and consensus conferences, community visioning exercises, backcasting – these techniques are often used for obtaining community co-development and acceptance of risk responses and action plans. Experts/scientists can be involved to varying extents.	Many examples at all scales, such as the World Bank's study by Solo *et al.* (2003)

Note: in this chapter we mention several of these approaches but do not expound on them nor critique them. This table is included as an information resource, showing some of the integrative tools that are in use that link the social and physical/natural sciences in risk management.

problems such as pollution (e.g. Hordijk, 1991; Gabbert, 2006) and coastal zone management (e.g. Mokrech *et al.*, 2009). Van Asselt and Rotmans (2002) have reviewed these models, with particular attention to the management of uncertainties associated with them. The paradox with these models is that the approximations and simplifications needed to make a manageable, usable description of the real world mean that uncertainty arising in and from the model does not have a direct relation to true uncertainty in the real world.

The result is a tension evident in the literature: quantitative approaches demand quantification of uncertainty, even when quantification is impossible. The existence of indeterminacy, non-linearity and other features of complex socio-natural systems is one area where quantification is constrained. Van der Sluijs (1996) and Batty and Torrens (2005) discuss the limitations of integrated modelling and the lack of practically usable methodologies for handling disconti-nuities and discrete random events or surprises. Quite often, the socially relevant issues of culture, values, choice and so on are explicitly highlighted in discussions about the uncertainty associated with integrated assessment (see also Eisenack *et al.*, 2005; Reichert and Borsuk, 2005 for discussions about dealing with complex socio-environmental system uncertainty in decision-making). Nevertheless, usually the proposed uncertainty management solutions are rigidly structural and assume a deterministic system: 'more consistent interpretations of uncertain parameters and inputs' and 'alternative model structures and equations' (van Asselt and Rotmans, 2002). Van der Sluijs (1996: 49) suggests that once the issues of inputs, parameters, and equations are dealt with, attention should turn to 'model completeness', where model intercomparisons and competition between modelling groups would provide the impetus for reducing the remaining uncertainty arising from model incompleteness. Yet a little earlier in the same article, he expresses his misgivings about adding more and more to models:

The solution of unknowns cannot simply be 'let's put every single detail that we know in the computer and let's hope that something shows up'. In our view, what is needed for scientific progress is to ask the right questions, not to produce an endless stream of 'what-if answers'. In that sense computers are useless, because they cannot ask questions, they only produce answers.

(van der Sluijs, 1996: 39)

In short, quantifications and model representations are necessarily reductive, simplified forms of evidential understanding. Even if these are not inherently inimical to societal application, they are at least made more complicated by social considerations. To understand socio-environmental risk and uncertainty, social considerations must be understood both to contextualise and frame any quantitative application.

Indeed, uncertainties expose the fundamental limits of quantitative models. How quanti-tative models are parameterised and applied is only ever subjective (Oreskes and Belitz, 2001). Once developed in and for one site, it is often the case that models cannot travel to other contexts (Konikow and Bredehoeft, 1992). And, given that models are necessarily simplified representations, problems with the boundary conditions around prediction are often easily exacerbated and magnified the larger or more complex the model gets (Pilkey and Pilkey-Jarvis, 2007). As Oreskes explains,

the more we strive for realism by incorporating as many as possible of the different processes and parameters that we believe to be operating in the system, the more difficult it is for us to know if our tests of the model are meaningful. A complex model may be more realistic, yet it is ironic that as we add more factors to a model, the certainty of its predictions may decrease even as our intuitive faith in the model increases.

(Oreskes, 2003: 13)

In the next section we address how *any* description, whether quantitative or qualitative, social or physical, necessarily appeals to simplified representations of the systems described, explained and predicted. Such limitation is not a failure of description or analysis as such, but it does demand – as recent social science has been at pains to explore – that knowledge production needs to become much more attuned to its own constitution and application. Doing so, and doing it well, involves subjecting the work to a continuous and always exploratory and self-reflexive process, but it produces knowledges and meanings that are much more democratically aware, transparent and responsive to their various publics.

15.2.3 Systems, models, stories

Many consider that human society, conceived as a purposive system that reacts to the knowledge it is given in pursuit of its goals – or indeed as a *purposeful* system that intentionally defines and adapts its goals (Ackoff and Emery, 1972) – cannot be considered adequately from a purely quantitative, mechanistic point of view. Concepts like community life, behaviour and the mind cannot be consistently or meaningfully quantified. However, we would argue that with the exception of carefully controlled experimental arrangements, the same holds for all systems – social, physical or hybrid. The reason lies in what 'systems' are. Any designation of a system, whether physical or social, is a designation of an observable relationship among components; it is meaningful to refer to their relations as a system, but this systematicity is an epistemic, and thus social, construct. We simplify by attributing information from the complexity of the environment to the components. As Niklas Luhmann's 'general systems theory' (1995) argues, social systems *qua* systems are inherently less complex than their environments; the virtue of a system is that it is less complex than its environment – such is the difference between representations (i.e. models, theories, etc.) and reality itself. In other words, systems are products of self-consciously selected pieces of information from within an environment. Researchers must acknowledge that a system's selection is the result of contingent propositions; the system could have been selected differently – and its behaviour would be seen to be different. After all, what information to include, and what to ignore, constitutes much of any calculation regarding natural hazard risk, and is the basis for any decision about parameterisation. Such contingency entails its own internal risk, but reducing the representation of the system further so as to fit applied quantitative constraints exacerbates this contingency, and often prevents the complexities of the environment from being explained in the system's representation. This can happen for practical reasons (e.g. calculability, resolution, etc.) or even ideological constraints, as was the case with the application of flawed financial models leading to the financial crisis of 2008 and beyond (see, for example, Crotty, 2009; Best, 2010; Haldane and May, 2011; McDowell, 2011).

Some recent social analyses go further than this, contesting any teleologically purposeful system, basing their arguments for social emergence in complexity, nonlinearity and volatile systems immanence. These argue that purely quantitative or purely qualitative means fail to capture adequately how patterns of organisation successively transform the manifold elements of the systems themselves, especially when these effects are as diffuse and nebulous as the role of perception in affect, emotional contexts, political interest, tacit coercion, and so on (Coole and Frost, 2010: 14). Thus we must accept that any explanation of social processes is only and always already partial, and can never be otherwise. This is not to say, however, that what is said cannot be meaningful or useful, but the awareness of this partiality means that careful description through multiple modalities is required to make meaningful explanation and possible intervention in social systems.

In short, systems are representational simplifications; quantified systems (sometimes in the form of numerical or computational models) are further simplifications still. While many such quantitative representations are absolutely vital to understanding natural hazards processes, attempting to argue that they are the most effective means of addressing, communicating and equivocating hazards risk assessment is a gross error. The challenge from social science to natural science lies in blending different forms of evidence and methods of representation and analysis such that complexity and heterogeneity can be preserved as much as possible in any description of a system and its interactions. Much of what has been termed, not uncontroversially, the 'post-normal science' paradigm has attempted to define just such ways and means of incorporating 'social robustness' into the quality of scientific knowledge production (Petersen *et al.*, 2011: 368; see also Turnpenny *et al.*, 2011). Indeed, as Stefan Hartmann (1999) argued, empirical adequacy and logical consistency are not the only criteria for the acceptance of a model or even an explanation about a system. The numbers that models generate need stories by which they can be justified as valuable (a kind of 'dequantification'; see also Rademaker, 1982; Randall and Wielicki, 1997). These stories guide and constrain the construction of models and explanatory systems by prescribing what should be accounted for, what should and should not be taken into account, what the relevancies are for understanding a phenomenon and its relationships among its various systemic aspects. The heterogeneities demanded by multivalent explanations (physical and social, say) of systems reveal that such stories constitute the normative frameworks by which phenomena are modelled and come to be seen to have explanatory value (Peschard, 2007: 166).

The importance of examining physical elements in close conjunction with the valuable stories told via the social lenses is increasingly identified as a major research and management priority in many risk assessment contexts. Examples from food and nanotechnology risks bear this out. First, in the context of food risks, the Royal Society/FSA (2005: 6, 7) argue that 'social scientists on risk assessment committees could help identify behavioural assumptions that need to be taken into account in the assessment process', and 'a range of social science techniques exist to support public engagement'. The report also notes that the risk assessment committees involved in the study were not well aware of social science methods that could aid in improving inclusiveness, equity and engagement. Similarly, in the context of new technologies, particularly nanotechnology, Macnaghten *et al.* (2005: 270) outlines a rationale and process for building in social science insights into societal, technical, scientific and political deliberations, based not on 'critical' engagement, but on 'constructive, collaborative interactions'. Klinke and Renn (2002, 2006) posit that a new form of risk analysis is needed for all systemic risks, namely those 'at the crossroads between natural events (partially altered and amplified by human action such as the emission of greenhouse gases), economic, social and technological developments and policy driven actions' (Klink and Renn, 2006: p. 2). They argue that systemic socio-environmental or socio-technical risks cannot be reduced to the classic risk components of hazard and probability. This new approach needs a broadened evaluation of risk that gives attention to issues of inequity, injustice, social conflict and so on. They also emphasise the importance of deliberation as a means of coping with uncertainty, complexity and ambiguity, which is a theme explored more fully below.

Zinn (2010) calls for narrative and biographical approaches to be better integrated into risk assessment, and reviews work on individual identity and experience in technological and environmental risks. He argues that this field of research potentially helps in understanding the differences in people's perceptions and responses to risk. Such work does not promise to lead automatically to better environmental risk decision-making, but highlights that 'decisions are not just a matter of weighing up present day options or values, but rather are deeply rooted in complex identity structures and social contexts, which evolve over time' (p. 6).

Integration like this – of words and numbers, both of which are used as signifiers of ideas or concepts – is a non-trivial challenge. It entails broad-scale discursive translation across theoretical, empirical and methodological domains (Table 15.1 summarises some of the interdisciplinary and integrative tools currently in development and use). Yet it is at the juxtaposition of these often-contentious translations that both misunderstandings and opportunities for transformative understanding can occur. Progressive and domain-shaping research depends on facilitating the dynamic interchange at these difficult, shifting and productive boundaries. Liverman and Roman Cuesta's (2008) simple appeal for 'a certain level of tolerance and mutual understanding' in such interdisciplinary interaction certainly needs to be borne in mind, but as we will see later on, quite radical forms of knowledge-theoretic description can emerge to change what counts as scientific explanation, evidence and social accountability (Whatmore, 2009; Odoni and Lane, 2010).

In Section 15.3 we address these insights with some attention to how social science might add to the description and explanation of natural hazards risks. Before that, we briefly turn to social conceptions of uncertainty.

15.2.4 Conceptualising uncertainty

The terminology for categories of uncertainty is – rather unhelpfully – often used in slightly different ways in the social sciences than in the physical sciences, partly reflecting the greater contingency and complexity of social phenomena. Here, we outline the three main forms of uncertainty (epistemic, aleatory and ontological) and some of their conceptual resonance within social science.

Epistemic, or systematic, uncertainty arises from a lack of knowledge, due to processes that are unknown or inadequately understood, or the poor characterisation of variability. In principle (indeed, by definition), obtaining more information about the system in question reduces epistemic uncertainty. But in the socio-natural systems of concern to us, epistemic uncertainty itself can never be eliminated, following the two arguments outlined above: first, knowledge is a processual and dynamic social product and not a fixed quantity; and, second, fundamental philosophical problems of perspectival framing and induction preclude determinist or apodictic certainty from descriptive and predictive accounts. Indeed, a central concern in many critical social science approaches today lies not with providing certainty and indubitability about complex and open social processes, but with analysing the work that specific processes do in the world, their staging and their effect. Providing certain foundations for social knowledge has long been deemed impossible. The role of epistemology now is not to provide such foundations, but to analyse how knowledge is produced and enacted. From these post-foundationalist approaches, all knowledge is thus considered to be inherently uncertain – but that does not mean that it is not rational, valuable or necessary. It simply means that we commit ourselves to the everyday messy tasks of analysing its work, application, politics and possibilities, and so see our epistemic task in less hubristic lights than perhaps we once did.

Aleatory or stochastic uncertainty is related to the inherent variability of systems, including their nonlinearity, randomness and contingency. Here, in principle, obtaining more data on the system may assist in constraining or characterising its stochastic or probabilistic elements, but it will not reduce this uncertainty. In the physical sciences, this improved characterisation of variability has emerged as a priority for environmental modelling and prediction (e.g. for climate, extreme weather events and other natural hazards). For the social sciences, the interest lies less in prediction *per se* than in how prediction is mobilised as a strategy for shaping human action and politics. The interest lies in how probabilities are communicated, understood and enacted within the various social and political contexts in terms of which they are deemed valuable or effective. Indeed, recent work on the rise of probabilistic decision-making, strategic forecasting and contingency planning, especially within deregulated market structures and the demise of the welfare state, have produced trenchant critiques of the new forms of governmental strategies that have come to dominate so much of social life (see, for instance, Beck, 1992; Dillon, 2010).

Ontological uncertainty is a term often taken as synonymous with aleatory uncertainty, especially in economics (as described by Dequech, 2004), but it holds slightly different resonance in other areas of social research, where uncertainty is not framed with the same conceptions of probability. Ontological uncertainty relates to what the nature of the real world is, not to knowledge about the real world. In social research, different ontological characterisations of reality can be employed (the issues are clearly summarised by Blaikie, 1991). For most natural and physical scientists, the nature of the real (natural and physical) world is the object of knowledge inquiry, and its reality is obviously not questioned in the normal positivist empirical work that makes up 'the scientific method'. In contrast, much social research needs to recognise the subjectivity of social reality. Whether the observer 'stands outside' the social system of interest – universally deemed an impossibility by contemporary social science – or is a participant within it, understanding the (real) nature of the social involves interpretive or deliberative processes. Some social realities are socially constructed (institutions, for instance), and they are not immutable. In such instances, it makes little sense to think of 'the inherent variability' of the system.

Several people have proposed classifications or taxonomies of uncertainty for social and socio-environmental systems (e.g. van der Sluijs, 1996; van Asselt and Rotmans, 2002; Regan *et al.*, 2002; Meijer *et al.*, 2006; Maier *et al.*, 2008; Lo and Mueller, 2010). Broadly, they all group the 'kinds' of uncertainty into two strands – epistemic and aleatory – that are coherent with use in the biophysical sciences. Morgan and Henrion (1990) and Rowe (1994), in their typologies of uncertainty, identify human choice dimensions as important, but in addressing how to deal with these issues in their articles they still propose more refined analysis and data acquisition, bringing these dimensions within the epistemic strand.

However, a consideration of human and social dimensions spotlights conceptual and operational challenges in both strands of uncertainty. Human choice interferes with randomness, making characterisation of the variability problematic. It also has implications for the knowability of future social processes. This was a pet issue for Popper (2002), both on logical and moral grounds:

No scientific predictor – whether a human scientist or a calculating machine – can possibly predict, by scientific methods, its own future results. Attempts to do so can attain their result only after the event. [...] This means that no society can predict, scientifically, its own future states of knowledge.

(p. xiii)

But is it not possible to control the human factor by science – the opposite of whim? No doubt, biology and psychology can solve, or soon will be able to solve, the 'problem of transforming man'. Yet those who attempt to do this are bound to destroy the objectivity of science, and so science itself, since these are both based upon free competition of thought. Holistic control, which must lead to the equalization not of human rights but of human minds, would mean the end of progress.

(p. 147)

Dequech (2004), in his critique of the socio-economic literature on uncertainty, argues that, for these reasons and others, the epistemic and ontological aspects of the conceptualisation of uncertainty are 'strongly entwined'. Because of human capacity to think, know and act, knowledge about the world can change the nature of the world, as, in fact, the material world agentially changes the nature, potential and capacities for human thought and politics (Bennett, 2010). Socio-environmental systems are complex systems, and human choice and agency are important factors in creating the complexity in the real world, while, crucially also being created by it. The implication of this entwinement of knowledge about the world and the actual agential nature of the material world is that obtaining more analytical information about the issue will not reduce, characterise or constrain uncertainty in domains of human choice and agency. Figure 15.2 shows an attempt to reinstate the different 'flavour' of uncertainty relating to the human world of choice and agency back into the characterisation of socio-environmental uncertainty.

It is apparent that this idea is hard to grasp. Several scholars, including very eminent thinkers, have written about the 'cascade' or 'explosion' of uncertainty, particularly in the context of the assessment of societal risks from climate impacts (Henderson-Sellers, 1993; Jones, 2000; Moss and Schneider, 2000; Mearns *et al.*, 2001; Pilkey and Pilkey-Jarvis, 2007; Wilby and Dessai, 2010). Uncertainty in the decision-making process is represented

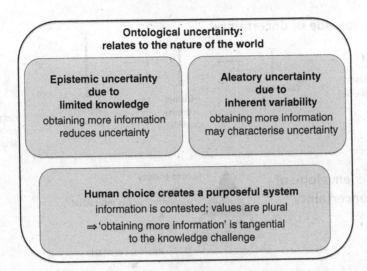

Figure 15.2 Uncertainty in socio-environmental systems.

(Figure 15.3) as expanding pyramids – implying a decision-tree analysis – or growing error bars – implying a numerical error aggregation approach. As an example of the latter, Jones (2000) states that 'ranges are multiplied to encompass a comprehensive range of future possibilities'. Wilby and Dessai (2010) take a decision-tree approach, saying 'information is cascaded from one step to the next, with the number of permutations of emission scenario, climate model, downscaling method, and so on, proliferating at each stage'. Many of these articles conclude that the escalation of these uncertainties hampers any sensible response. For example, Wilby and Dessai state 'the range (or envelope) of uncertainty expands at each step of the process to the extent that potential impacts and their implied adaptation responses span such a wide range as to be practically unhelpful'.

There are many conceptual problems with this approach. Not all of the 'uncertainty steps' in the cascade are multiplicative in nature, nor do all possible permutations of analysis and intervention take place. However, *human choices* can manifestly reduce or even remove uncertainty in some senses and contexts, and they are often given little recognition or attention. Co-productive approaches can often mitigate and translate risks and hazards in ways that address the resonance of uncertainties and their publics. It must also be noted, however, as is evidenced with the recent financial and banking crises between 2008 and the present, human choices can increase and exacerbate uncertainties, often in deeply irrational and sometimes catastrophic ways (Kahneman, 2011). Furthermore, the different kinds of uncertainty are mixed up in these representations, so there are also operational problems with the cascade idea. By bundling epistemic, aleatory and human-agency uncertainties together as these authors do, activities to assess and manage the uncertainty cannot be focused where they are needed. In particular, the uncertainty associated with human choices cannot be dealt with by routines or rational analysis – it requires critical reflexivity and

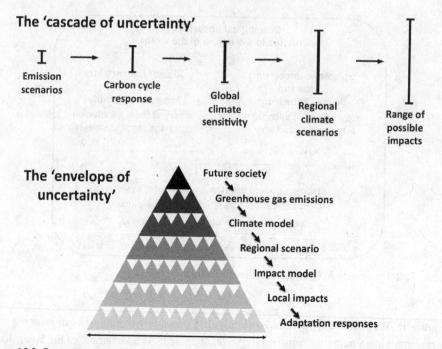

Figure 15.3 Some representations of socio-environmental uncertainty from the literature. Note that the 'cascades' treat very different kinds of uncertainty (arising from models, empirical observations, human choices) as if they were commensurate, quantitative/quantifiable, and unidirectional, although choices can reduce as well as expand uncertainty.

dialogue. A purposeful system (Checkland, 1999: 316–317) – which is what human systems are – cannot be treated as if it were deterministic (the domain covered by epistemic uncertainty), or even contingent (the domain of aleatory uncertainty). Later sections of this chapter review how such systems may be dealt with more fully.

Why might we care about these semantics? Setting appropriate margins for the stochastic uncertainty relating to the deterministic components of the risk problem, and adding information about epistemic knowledge limitations are both vital but non-trivial components of the advice mix for risk assessment. However, in responding to real-world environmental risks, the provision of knowledge is not the whole picture. The stakeholders in this mix are diverse: scientists, lobbyists, political decision-makers, media, advocacy groups and members of the public all interacting and all with their own interests to serve. This is the motivation for highlighting the human agency dimensions of uncertainty so prominently in this chapter.

15.3 Where social science matters

In this section we discuss some key areas where we feel social science matters: where improved engagement across knowledge communities would benefit environmental risk assessment and management, and where the take-up and integration of social research in recent experimental approaches has already started to shape the knowledge-production processes around natural hazards risk.

15.3.1 Relevance: defining and framing risk assessment objectives

The uncertainty implications of a lack of engagement with social issues are potentially serious for environmental risk management. A kind of 'concept-blindness' can arise, where essential elements of the hazard are not seen or perceived (the infamous unknown unknowns), adding to risk. There is also a linked problem of spurious precision, where lots of information is given about just part of the issue. In risk management, if the risk assessment objectives are framed too narrowly, and in particular, if the human and social dimensions are not considered fundamentally within the process, there is an inability to respond to the 'whole' real risk. Social science perspectives can elucidate some of these potential dangers.

Defining the risks to people arising from natural hazards seems self-evidently to require the involvement of a broad base of experts, yet in many contexts, there is still not enough engagement of interdisciplinary expertise. There is a pervasive view that risk identification and responses have been 'scientised' (Bäckstrand, 2003; Ekberg, 2004). That is, the normal methods of natural science – portrayed as a value-neutral, objective endeavour – are applied to characterise and predict environmental problems, for a separate, downstream political management response. Concerns are that this approach makes overly instrumental presumptions about how society works, and is insensitive to the potentially problematic nature of the socio-political context in which responses are implemented. Many areas of the social sciences invest a considerable amount of their time in considering the nature and implications of values and world-views in their investigations. Engaging in debates where this activity is omitted is alien to their disciplinary culture. It also asks something of knowledge production which it cannot give. Indeed, to assume that value-neutral, objective knowledge is possible is not only epistemically naive, but it is often an ideological sleight of hand that masks more than mere hubris. Knowledge is a product of communities of action and interest, and our role as social scientists is to render as transparently as possible the various inner workings of these communities, with the appreciation that our efforts too are themselves situated and give a fundamentally partial account.

There is an extensive literature on what constitutes a community of understanding (or epistemic community, as described by Haas, 1992), and significant analysis into why the earth sciences and social sciences operate so substantially in their own orbits (e.g. Demeritt, 2001; Nowotny *et al.*, 2001). But when problems that require a decision about action, such as responding to natural hazards, are framed in the limited terms of one discipline or knowledge community, difficulties are known to arise (Ricci *et al.*, 2003; Owens, 2005; Petts and Brooks, 2006). Here, we focus mainly on the interdisciplinary challenges.

Natural scientists deliver quantitative information on risk and uncertainty based on process models, observations of hazards and statistical analysis of past hazards data.

Quantitative risk assessments include some social data, such as demographic information, housing inventories and so on, in the development of tools such as risk loss curves. This detailed scientific and technical information is needed, but it is not sufficient for informed decision-making. Indeed, many factors in assessing risk and uncertainty are either unquantifiable or little more than educated guesses, and simply attempting to include them in quantitative models compounds a cascade of uncertainty and dilutes forecasting reliability (Smil, 2003: 179). While more knowledge is usually better (Cox, 1999), the value of this nuanced environmental understanding may in some cases be limited by the scientific narrowness of the framing. A risk assessment needs to be accurate (true to the real world), not just precise, and this requires an understanding of citizens' needs, of political processes and of wider public attitudes – all of which lie in the domain of 'the social sciences', but in order to be useful, cannot remain isolated there. They need to be deployed in concert with physical analysis, open to incorporation in knowledge-theoretic approaches too (see Odoni and Lane, 2010).

At a strategic level, more deliberative working for risk analysis has been strongly endorsed (Stern and Fineberg, 1996; House of Lords Select Committee on Science and Technology, 2000; HM Treasury/DTI/DfES, 2004; HM Government, 2005a[3], 2011; Royal Society/FSA, 2006). Good-practice guidance abounds on engaging in such a deliberative process (e.g. Renn *et al.*, 1993; Stirling and Mayer, 2001; Petts *et al.*, 2003; HM Government, 2005b), but good practice seems only now to be emerging. Thus, it is at the operational level that challenges still remain. What is the role of non-quantitative social research in the decision-making process? How, for instance, should insights from history be integrated? Clearly lessons can be drawn from the past (Kasperson and Kasperson, 2001; Harremoës *et al.*, 2002), but designing a future is not a straightforward intellectual exercise. While there is a need for the integration of ethical, legal and social issues, the cultural and epistemological differences between the relevant expert communities for these issues are hardly less challenging than those between physical and social scientists.

There are some fields where theoretical debates are advancing well about new conceptualisations of the interactions between environment and society and about the changing nature of more socially engaged scientific research. These include 'restored geography' and sustainability science (Demeritt, 2001, 2008; Kates *et al.*, 2001 – and further work since then by several of that paper's authors; Whatmore *et al.*, 2008; and to some extent Hulme, 2008); and ecological economics (e.g. Costanza, 1991), although this emerging field tends to revert to numerical models as the 'common language'. Another rapidly growing field that merits closer attention in the context of natural hazards and environmental risk is that of socio-ecological systems and resilience (e.g. Ostrom, 1997, Folke, 2006, Young *et al.*, 2006; Ostrom *et al.*, 2007; Liverman and Roman Cuesta, 2008). The conceptualisation of the socio-environmental system explicitly considers people in their environment, and simultaneously considers the way that the environment shapes people. Analysing these relationships requires assumptions and simplifications to be made about both society and the natural

---

[3] These guidelines are currently being updated through a consultation process.

world, so there is a strong focus on both analytic and deliberative methods for the 'validation' of models, metrics and indicators that combine or integrate components of socio-environmental systems. Often, this requires a focus on different time and spatial scales to those that apply to the physical hazards themselves. Such work includes, for example, investigating the resilience of communities vulnerable to natural hazards and studying the long-term impacts of a disaster.

Macnaghten *et al.* (2005) draw particular attention to a practical matter that we regard as vital: uncertainty assessments need to be less sequentially and more collaboratively and deliberatively constructed than they generally are at present. 'Social science' should not be the end-of-pipe bolt-on to make the risk analysis more palatable or politic; cultural, economic and societal dimensions should be part of the problem-statement stage, engaged with at the outset. The research questions that are being asked will change as a result of attending to the ways that complex social and political relationships constitute what counts as evidence, to what ends data and findings are being put, to whom findings are directed and who is excluded in such processes, and so on.

15.3.1.1 *Upstream integration: the UEKC case*

In this section we draw on the experiences of the interdisciplinary research project, Understanding Environmental Knowledge Controversies, which proposed a methodological apparatus for 'mapping' uncertain and contested science in the contexts of flood prediction, hydrological modelling and community-centred natural hazards risk assessment (Whatmore, 2009). This endeavour, of course, did not map in the normal cartographic sense, but developed an effort to expose and explore more explicitly, and sometimes experimentally, the 'entanglements of scientific knowledge claims with legal, moral, economic and social concerns.' Although this project shares many features with other action-oriented and participatory socio-environmental risk research (e.g. Petts *et al.*, 2003; Cobbold and Santema, 2008), we use it as a case study because it had an explicit focus on uncertainty in natural hazards and it documented its processes well, as well as its findings.

The three-year project, begun in 2007, examined the production and circulation of environmental knowledge, and explored the ways in which knowledge controversies could be productively utilised in the interests of public interdisciplinary science (Lane *et al.*, 2011: 16). The project focused on the science of flood risk management, locating much of its energies in the flood-prone market town of Pickering in North Yorkshire, where recent severe flood events had led to the development of a proposed flood defence plan, which had met local resistance. The project's objective was to analyse how new scientific knowledge formations about flood risk could arise through the integration of lay and academic inputs, and shape different and more effective practices for 'doing science' about potentially divisive socio-natural issues (Lane *et al.*, 2011: 17).

Core to the work of UEKC was the principle of experimentation through experience (Lane *et al.*, 2010: 24). The departure from the norm was that community engagement was not seen as an 'end-of-pipe' phase following on from the scientific scoping and decision-making process. Instead, the project supported the 'upstream' involvement of the community of interest in framing the flood risk issue. In the first instance, the project set up what it termed an Environmental Competency Group, to collectively address the community's flood risk. This group was made up of interested members of the local public, academics, flood-risk managers, a facilitator and recorders of the proceedings. They worked together with the aim of learning about the local hydrological environment and, in doing so, producing a collaborative description of the risk ecology and possible responses to it. The representation of the hazard event emerged through negotiation among the Environmental Competency Group in the recognition that 'any attempt at representing the complex social-economic-political composition of communities is itself an act of framing around a pre-conceived notion of what that composition is' (Lane *et al.*, 2011: 24). The project team emphasised five key aspects of this process, which it took to be distinctive in hazards risk science (pp. 24–26). First, it focused on the practices of knowledge production, rather than simply the results. Second, these practices manifested the risk research itself as an ongoing and collaborative process. The knowledge contributions of the lay and academic participants and the joint agency arising from the broad assemblage of people, places, techniques, and objects shaped the nature of the science itself. Third, the collaborative process of flood risk assessment focused its attention through the various events which cascaded from the collective experiences of the hazard. The result was that the flood hazard itself was framed as a multivalent event, through the various descriptions of flooding, and the different competencies of the actors involved. Fourth, rather than being narrowly defined by a more-often-than-not external academic view, the estimation of the risk was constituted by the collaborative and experiential narration of the hazard and its attendant consequences. The participants thus came to recognise their shared responsibility for the hazard and its risk assessment as an emergent property of the collaborative description with, fifth, the result that representation of the event also came to be shared as a responsibility of the community of concern, rather than simply a dissociated matter for institutional actors (i.e. regulatory bodies, advisors, etc.).

Through the iterative consultative processes, a more detailed qualitative understanding of flood-generating processes began to emerge. This multifaceted, locally informed understanding began to displace the earlier assessments made by the generalised hydrological and hydraulic models that are the workhorses of the flood risk management agencies. With the integration of the iterative, qualitative understanding generated from the collaborative consultative engagement, a new model emerged specific to the hydrological risk requirements of the local area. Termed the 'bund model' (see also Odoni and Lane, 2010), it and other iterative models were the product of the Environmental Competency Group and became scientific risk assessment objects through which the collective asserted its hazards risk response and its competence (Lane *et al.*, 2011: 26).

The UEKC involved collaboration across disciplines, bridging hydrology, human geography, social theory, meteorology, social policy and rural economy. This brought multi-dimensionality to understanding, but necessarily also demanded often contentious dialogue and negotiation between the academic disciplines, regarding how problems were to be defined, framed, analysed and what actions brought to bear to address the risks. Findings from the UEKC and similar projects are now driving a transformation in risk assessment and management methods, including views about who contributes to knowledge framing and production, and the nature and desired robustness of evidence. Most importantly, in assessing and managing the uncertainty in risk assessments, there is increasing attention to the importance of a consideration of the dynamic, knowledge-transforming implications of reflexivity (Romm, 1998) in these steadily more integrated modes of working (Lane *et al.*, 2011; Odoni and Lane, 2010; Whatmore and Landstrom, 2011).

15.3.1.2 Lessons for the framing of risk

What the UEKC project learned that became important to its risk analysis and response was that rather than taking scientific knowledge as a given, new co-produced knowledge-theoretic models emerged as collaborative products, which could be adjudicated according to community-specific needs and understandings. The UEKC project argues that 'the key consequence of co-production' (Lane *et al.*, 2011: 31) is the 'sense of knowledge' produced:

What we experienced was a shift from taking knowledge as a given, to being able to see model predictions as just one set of knowledge, against which they [i.e. interested residents] were entitled to compare and to use their own knowledge.

This enables parties to arbitrate the claims regarding risk and uncertainty made using models among their varied backgrounds of social knowledge. The value of these significant epistemic differences, as they are integrated into the predictive and descriptive scientific capacities, is manifest in 'better' models – both in the immediate sense of having better scientific representation of local flooding processes (and reduced epistemic uncertainty), and also in the broader sense of a co-developed, trustworthy and transparent model that provided a focal point for the community's engagement with the issues of flood hazard. Co-produced models thus become opportunities for innovative and targeted solutions for specific hazard risk problems, which hold the potential to be more materially effective and socially responsive than is now often the case (Odoni and Lane, 2010: 169–170).

However, this is not the way that knowledge is generally produced and transmitted. Clearly, there are implications for the role of the scientist. Becoming engaged with resident knowledge is a matter of collaborative trust. It is not simply a matter of unidirectional scientific communication. As the UEKC experiments attest, trust, which lies at the purposive heart of risk and uncertainty science, is a function of social assemblages which co-produce interested and specific strategies for decision-making and action. The purpose of effective communication (trust, engagement, safety, enhancement of well-being) is situated at the heart of the scientific process itself.

Establishing trust in models is not just about the restricted kind of validation implied by comparison of predictions with data, but a wider set of practices through which a modeller makes a model perform.
(Lane et al.*, 2011: 32)*

Deliberative, inclusive and participatory engagement is a lengthy and resource-intensive process, but it is an effective means of managing uncertainty in contested socio-environmental decision-making (e.g. O'Riordan *et al.*, 1999; Prell *et al.*, 2007; Petts *et al.*, 2010). The necessary integrations – both across contributing academic disciplines and of other contributory expertise – require new conceptualisations, shifts in the normal modes of working, and a focused attention to the differences in worldview and ontology of the different epistemic communities. This last need is often considered alien to ostensibly objective, value-neutral scientists, although those working in practical ways in natural hazards are usually well aware of the multi-dimensional real-world complexities and do indeed reflect on their methods and findings in this context.

The UEKC project is not alone in its innovations, but there is still a great deal of scope for exploring and documenting what appears to work well. Another valuable example of where this has happened is the Trustnet-In-Action (TIA) network (trustnetinaction.com). TIA began in 1997, and ended in 2006, as a European research initiative to 'bring a fresh and pluralist thinking' to environmental risk decision-making, and to address the limitations of the mainstream approaches to risk assessment and risk management. TIA documented their emerging philosophy of inclusive governance (TIA, 2007), along with frameworks and tried-and-tested patterns for inclusive, experimental deliberative processes:

[T]he TIA process is neither basic, nor applied research, but is close to research-in-action ... which is, strictly speaking, more explorative than confirmative. TIA methodology is elaborated during the process itself through the cooperative inquiry of experts and actors, with an undeniable part of contingency. Its major contribution as a pragmatic methodology is to create the durable institutional conditions for a process of governance which will place the actors in a position of influence. For that, it is necessary to conceive a new role for the experts whose science must be dedicated to an ethical objective of helping the actors to be subjects of their own life.

(TIA, 2007: 77)

In the next sections we step back from the lessons outlined from the UEKC and TIA examples to address communication and social learning as commonly understood by risk science, before returning again through the UEKC and related examples in the final section to address how co-productive efforts can help scientists meet the increasing calls for involvement in political advocacy or policy-relevant action.

15.3.2 Comprehensibility and communication

Good communication is essential if the research products of any discipline are to be of any value, and some insights from social research can be vitally important in improving the communication and comprehensibility of complex messages associated with natural hazard risks. Uncertainty in this context relates mainly to epistemic concerns – miscommunication, misreadings of the hazard situation – and its management is therefore responsive to training and technological investment.

Risk assessors routinely use concepts, mathematical techniques and probabilistic outputs – including uncertainty measures – that are often difficult for decision-makers and the public (and indeed other experts) to understand. And risk models are getting increasingly complicated. It is presumed that including more information in a risk model will not actually worsen the risk estimates it produces, but there are trade-offs associated with increasing the model complexity. As the numbers of model processes and parameters rise, the model may yield better descriptive capacities, but there is also an increase in the number of uncertain quantities needing to be estimated, and hence a rise in the overall uncertainties capable of being contained within the model's integrated system. Data paucity for model 'validation' (alert to Oreskes *et al.*'s (1994) objections to that term) is a problem dealt with elsewhere in this volume, but it is clearly imperative to ensure that users and the multidisciplinary co-developers of a risk assessment understand the models being used, and can communicate the uncertainty issues and model limitations embedded in them.

There is a need to address both *perceptual and cognitive* issues (e.g. what meanings do people give to particular words, concepts or images? How does this vary culturally? Linguistically? How does experience affect the response to information?), and the more *technical/operational* issues (how clear are the maps? Is information translated into local languages? How can complex, multi-dimensional information be presented clearly? And so on). The success of the latter often hinges on understanding of the former. Overall, there is still comparatively little research in the area of natural hazards risk and uncertainty of how comprehensible the information produced actually is. Research to date has largely focused on socio-technical risks and – with a growing focus – on climate mitigation (e.g. Moss, 2007; Nowotny *et al.*, 2001).

There have been some examples of natural science 'overkill' in the communication and public understanding of science. For instance, a recent project (Public Awareness of Flood Risk) funded by the UK Economic and Social Research Council has investigated how the Environment Agency's flood maps affect public awareness and attitudes to flood risk (Clark and Priest, 2008). It found that although the public generally like the online maps and appreciated access to this kind of information, there were issues with managing people's expectations (and frustrations) about how specific the information might be to their own situations. Like many other studies, this project also came across the problems associated with people's handling of complex and probabilistic information. Important concepts such as return frequency or percentage likelihood are not well grasped by many members of the public. More prosaically, users often found navigating the online maps to be a complicated procedure, which detracted from their ability to access flood information within the maps.

Over-elaborated hazards maps have previously been distributed widely in Guatemala at times of volcanic threat, but communities and households were found to have inadequate understanding of what the maps were meant to represent, and remained exposed to avoidable risks (W. Aspinall, pers. comm.). More fundamentally, little attempt was made, on the part of scientists, to understand that the maps and their information were themselves highly complex products of particular social contexts. They assumed representational strategies and expectations, which in different social and cultural contexts might have very different meanings to those intended. Sometimes this kind of complex information can be transmitted better, with better effect: people in Montserrat living under threat of volcanic eruptions were also found to have poor skills in map-reading, but when presented with three-dimensional visualisations of digital elevation data of their island, they demonstrated good spatial awareness and were able to engage in discussions about their future in a more informed way (Haynes *et al.*, 2007). But it needs to be understood, in any risk communication strategy, that the risk, its definition and its address all come from epistemic 'genealogies' and perspectives, which shape the knowledge and its applicability in particular ways. In the instance of the hazard maps in Guatemala, had local citizens been involved in the production of the hazard stories and their dissemination, the avoidance strategies embodied in the map might have been more effective.

Having ensured that risk communicators have the necessary understanding of the potential knowledge production and transfer challenges, widened engagement requires the targeting of audiences (e.g. Boiko *et al.*, 1996), the development of mechanisms for stakeholder or public dialogue (e.g. Rowe and Frewer, 2005) and the establishment of functional communication structures (e.g. Cornell, 2006). The social sciences will not provide magic bullets for environmental scientists engaging in hazard communication, but the tools of the trade of social science, particularly qualitative research, are useful in these three stages. Interviews, focus groups, community workshops, observant participation, ethnography, language learning, archival analysis, deep involvement over long periods of time and 'thick description' all aid in framing effective and productive engagement. Applying the good practice developed in social research to these socio-environmental hazard contexts can help in assessing and reducing uncertainty. Indeed, several recent initiatives are attempting to integrate qualitative knowledge forms that are dynamic within stakeholder and partner communities into the parameterisations, structure and outputs of formal quantitative models (e.g. Siebenhüner and Barth, 2005; TIA, 2007; Odoni and Lane, 2010; the UEKC project: Lane *et al.*, 2011; Whatmore and Landstrom, 2011). As these studies argue,

The practice of modelling ... needs to move 'upstream' into fields and farms, community halls, pubs, so that those who live with model predictions, and who have local knowledge to contribute, might become involved in generating them, and so trusting them.

(RELU, 2010: 4)

A growing concern that merits more focused research is the interplay between risk, its communication, inequity and social injustice. Mitchell and Walker (2003) reviewed the links between the UK's environmental quality and social deprivation, leading to an Environment Agency research programme, Addressing Environmental Inequalities (Stephens *et al.*, 2007; Walker *et al.*, 2007). Empirical evidence for environmental inequality is still rather limited, even in the UK, so a key recommendation of this programme was to engage in longer-term case studies, with a particular focus on cumulative or multiple impacts. In the United States, Taylor-Clark *et al.* (2010) used Hurricane Katrina as a case example through which to explore the consequences of 'communication inequalities', finding that socio-economic position and the strength of social networks had a significant impact on both hearing and understanding evacuation orders. There is also international research on communication and social justice in the context of climate adaptation and mitigation. For example, Ford *et al.* (2006) carried out a study including information-exchange methods involving photography and voice recordings to explore how aboriginal human rights can effectively be incorporated in environmental and science policy in the Canadian Arctic. This study emphasised the value of community engagement in reducing vulnerability, and has been part of a steady evolution of more publicly engaged Arctic governance which has incorporated learning central to climate change impacts and mitigation, social and cultural resilience strategies, human rights legislations for environmental protection and biodiversity preservation.

15.3.3 Behavioural change and social learning

Human behaviour is often a critical element determining the impact of socio-environmental and socio-technical risks (Gardner and Stern, 2002; OECD, 2009): the volcanic eruption would not be a problem but for the illegal farming on its slopes; the storm surge would not cause such losses but for the beachfront hotel development boom, etc. Furthermore, human behaviour can present risk assessors with a 'catch-22' situation: uncertainty, often linked to social variability, can result in maladaption and accumulated or deferred risk. Reducing this social uncertainty can be achieved, but only at the cost of a degree of 'social engineering' or dictatorial action that is currently regarded as undesirable in democratic societies.

Thus, in risk research, interest has extended into the fields of human behaviour and the processes of social change. Research that is relevant to natural hazard management comes from economics, political science, psychology, management studies, healthcare and public health, and increasingly from education for sustainable development, which brings an ethos of empowerment and a concern for ideals and principles like human rights and sustainable livelihoods. Exploring these research areas in depth is beyond the scope of this chapter; instead, we select the following as illustrative challenges: first, how knowledge affects behaviour; second, the role of research in promoting social change as a means of achieving risk reduction; and finally, the growing recognition that the interface between knowledge and real-world action requires greater critical reflexivity, both in research and in governance.

15.3.3.1 Knowledge and human behaviour

As this volume indicates, the bulk of risk research has concentrated on technological and environmental risks from a quantitative and analytic perspective. As capability has grown in the prediction of the deterministic elements of risk, and the characterisation and representation of the contingent elements, attention has turned to the human elements that previously were considered too subjective to address within this 'objective' assessment of risk. Subjectivity, or an individual person's perceptions of risk and concerns about it, is often taken to be in opposition to 'real', objective risk, with the corollary that once adequate objective knowledge is provided, people will align their perceptions to the objective assessment. Throughout this chapter we have argued that the way to deal with the diversity of human experiences, values and preferences is not to eliminate them from the analysis, nor to attempt to bring them all into alignment, but to allow their multiplicity and often fraught negotiation to shape the landscape of the knowledge formation about the risks themselves. Zinn (2010), in reviewing and promoting biographical research in the context of risk and uncertainty, argues, like we do, that people's personal identities and life experiences are an important constitutive element in environmental risk analysis. Zinn sets out the case for better integration of individual experience and institutional processes into our understanding of the management – and production – of risk and uncertainty.

The nub of the problem is that improving people's *knowledge* about environmental risks does not seamlessly improve people's *action* in response to those risks. To begin with, people are not purely rational beings. Irrationality and inconsistency are human traits, and a large body of research in the area of risk and decision-making confirms this (e.g. Otway, 1992; Kahneman and Tversky, 1996; Kahneman, 2003). Adding to this, people are exposed to many risks and opportunities simultaneously, and are always seeking some kind of life balance – 'an equilibrium of their experiences, abilities and feelings' (TIA, 2007: 25). To the extent that people do rationalise their decisions, the way that they may weight these different risks and opportunities in some kind of multiple optimisation algorithm will vary from person to person, and from context to context. In addition, an individual's scope to respond to risk and uncertainty is constrained by their social context.

In what can be considered the 'first wave' of cognitive and behavioural research in the context of environmental risks, the focus was on improving the understanding of individual interpretations (or misinterpretation) of risk in order to improve risk models. Often, this involved comparing lay people's 'mental models' or perceptions of an environmental hazard with those of experts (e.g. Pidgeon *et al.*, 1992; Atman *et al.*, 1994; Fischhoff, 1998). These mental models are efforts to map out the causalities and dependencies in a hazardous process. Essentially, the divergences between lay and expert perspectives were taken as a dimension of the risk model uncertainty, which could be corrected in a more efficient version of the risk model, or mitigated by changing the model of risk communication: by knowing better what people do not know, or do not understand, the risk manager is enabled to better direct people's behaviour towards a rational decision.

This body of research did indeed improve risk communication, but what it really high-lighted was the complexity of the ways in which individuals construct and experience their social reality, and the extent to which they rely on strategies of social learning. Rational responses to (perceived) short-term or local risks may cause a higher exposure to other risks; people are unwittingly complicit in their own risk exposure (examples are listed in the *World Disasters Report*, 2009: 19). This interplay between responding to risk and altered exposure is at the root of potential moral hazard (e.g. Laffont, 1995; Cutter and Emrich, 2006; McLeman and Smit, 2006) and societal maladaptations (Niemeyer *et al.*, 2005; Brooks *et al.*, 2009) resulting from human behaviour change. These concepts have already received attention in some risk areas, notably in flood risk and the agriculture/agrochemicals context, but are less articulated across other important natural hazards, where more attention to policy analysis and the broader political economy is required.

Slovic (2001: 1, 5) illustrates the shift to the 'second wave' of behavioural risk research, in which a greater emphasis is given to people's life-experience within the risk definition process:

Ironically, as our society and other industrialized nations have expended this great effort to make life safer and healthier, many in the public have become more, rather than less, concerned about risk. [...] Research has also shown that the public has a broad conception of risk, qualitative and complex, that incorporates considerations such as uncertainty, dread, catastrophic potential, controllability, equity, risk to future generations, and so forth, into the risk equation. In contrast, experts' perceptions of risk are not closely related to these dimensions or the characteristics that underlie them.

One goal of these new approaches is not just to understand the perceptions of risk held by people living in a hazard situation, but also to obtain a better understanding of how and why people develop their world-views, identities, preferences and behaviours (e.g. Henwood *et al.*, 2010; Zinn, 2010). The subject is not an abstract 'rational' person, but a 'concrete person who articulates and adapts the multiple dimensions of his or her identity, personality and existence', who in turn is rooted in a community in close interaction with a specific cultural and natural environment (TIA, 2007). This is a much more dynamic engagement with people as the subjects of research. It potentially confers a much richer understanding of the extent to which risk and uncertainty matter to people in their everyday lives, including the ebbs and flows of concern about risk that so profoundly affect people's practical responses to risk situations.

15.3.3.2 Social learning and social change

In the light of the recognition that individual knowledge and perceptions of risk and uncertainty are shaped by social interactions, there is growing attention to social responses to environmental risk.

The first dimension of this body of work that we want to highlight relates to society's response to the provision of information about risk. The questions of why and how a specific risk becomes a public social issue – or sometimes does *not* – are among the most perplexing challenges in risk assessment. Very often, these responses contrast sharply with the rational responses as assessed by technical risk experts. For example, people's agitated concern about the triple Measles–Mumps–Rubella vaccine in the UK was a regular feature in the news headlines for nearly a decade since the first suggestion of a link with autism was raised, despite a technically very low risk (Cassell *et al.*, 2006). In contrast, inhabitants of the volcanic island of Montserrat were found to make frequent incursions into the danger zone during red-alert periods when risk was manifestly high (Haynes *et al.*, 2007). Impacts on society and the economy can be profound, whether the social response to the risk is 'duller' and more attenuated than it should be, or is more heightened and amplified.

Kasperson *et al.* (1988) set out a conceptual framework for the 'social amplification of risk' that has proven very useful in bridging our understandings of risk perception and risk communication. Using the analogy of signal amplification in communications, they highlight the diverse 'signal processors' (such as risk scientists, risk management institutions, the media, peer groups and activists), and identify various stages in the amplification or attenuation of the risk information, including the selective filtering of (comprehensible) information; the use of 'cognitive heuristics' – conceptual short-cuts deeply embedded in the human psyche that are deployed to simplify complex problems; and of course, the multiple social interactions and conversations people engage in, to contextualise and validate their interpretations, and to collectively decide how to respond to the risk. Since then, others have looked at the phenomenon of 'herd behaviour' (including panic responses) and the role of information cascades in general socio-economic risks (e.g. Gale, 1996; Kameda and Tamura, 2007). In the socio-environmental context, Pidgeon *et al.* (2003) have reviewed the empirical and theoretical research on the social amplification of risk framework, which now addresses formal and informal information channels, the issues of individual understanding and behaviour already described above, and the behaviour-shaping forces in society – attitudes, political dynamics, social movements and so on.

In the climate context, there is a growing focus on the issues of the positive diffusion of knowledge in adaptation processes (e.g. Pelling and High, 2005; Pahl-Wostl *et al.*, 2007; Ebi and Semenza, 2008; Jabeen *et al.*, 2010; Mercer, 2010), also evidenced in the large numbers of non-governmental organisations that are particularly alert to their role as 'communities of practice', building the capability for accelerating and cross-fertilising knowledge flows into vulnerable populations. Clearly, knowledge diffusion is a requisite step in moving from individual to social learning and societal change. However, social learning based on collective memory of previous experiences is only adequate when deviations from the baseline situation are modest. When risks have high consequences (as in many natural hazards), or when new developments or requirements emerge (as in the case of climate change), then there is a need for society to change more radically or fundamentally – perhaps to a situation with no historical or geographic analogue – to avoid the worst consequences. Decision-making under these circumstances becomes a complex matter. The idea of 'conscientisation' (Freire, 2005) involves a kind of critical consciousness of the world linked to action to resolve the perceived difficulties. It has been a helpful concept in health (Kilian, 1988; Cronin and Connolly, 2007), and features increasingly in environmental change (Kuecker, 2007), partly because it is alert to the frequent internal contradictions and 'wickedness' (Rittel and Webber, 1973) of real-world problems.

15.3.3.3 Reflexivity in research and governance of risk

The limitations of risk science, the importance and difficulty of maintaining trust, and the subjective and contextual nature of the risk game point to the need for a new approach – one that focuses upon introducing more public participation into both risk assessment and risk decision-making in order to make the decision process more democratic, improve the relevance and quality of technical analysis, and increase the legitimacy and public acceptance of the resulting decisions.

(Slovic, 2001: 7)

There are many mechanisms for changing people's individual and social behaviours in the face of environmental risk. These include incentives, regulation and, of course, persuasion and education. But the idea that experts of any discipline should actually deliberately try to change the values and beliefs that drive behaviour patterns, changing social norms, is loaded with ethical and political tensions. There is a delicate balance to be maintained. Knowledge, when available, should clearly shape action, but simply providing information about risks and uncertainty does not automatically lead to an optimal change of action. The role of the scientist in driving social transformations clearly requires attention, care and – many argue – new forums.

The role of the scientist in governance is therefore also a growing research field in its own right. Patt and Dessai (2005) and Grothmann and Patt (2005), operating in the climate adaptation context, have looked at the role of experts and the effect of uncertainty in expert assessments on citizens' choices in response to perceived risks. It is essential but not necessarily sufficient to follow good practice in risk communication, and deliver training in technical or cognitive skills to those receiving the complex messages of environmental risk assessment. Marx *et al.* (2007) distinguish between analytic versus experiential learning, and argue that both can and should be developed through participatory processes, to actively engage the public in the processes of risk assessment and management. There is a broad consensus that social learning interventions for environmental risks require transparent procedures (after all, research into people's complicity in environmental disasters is potentially a delicate task), iterative cycles and oversight by individuals or agencies with no perceived strongly vested interests. Collins and Ison (2006: 11) propose a list of 'systemic variables' that need attention if changed understanding and practices are needed, demonstrating that the integration of social science techniques is proving valuable as scientific research and risk assessment processes are opened to greater public scrutiny:

The variables are: an appreciation of context; ecological constraints or conditions; institutional and organisational framings and practices; stakeholders and stakeholding; and appropriate facilitation. How a messy situation is understood and progressed with regard to these key variables is likely to determine the extent to which social learning can enable concerted action to emerge from changes in understandings and practices.

Ethical dilemmas clearly arise in this kind of 'messy situation', and researchers and risk managers are beginning to give them more explicit consideration. Pidgeon *et al.* (2008) explore the ethical issues associated with socio-cultural studies in risk research. These relate to the power asymmetry associated with expertise; the impact of the knowledge provided (or the process of information elicitation) on the participants; the repercussions for institutions involved, which may have their validity and authority questioned in the process of engagement, and so on. As they clearly state, there is no simple formula for dealing with these ethical challenges. Along with the other 'human dimensions' of multi-dimensional risks, they require places and times to be provided for deliberation and reflection.

Reflexivity has already been mentioned several times in this chapter. We draw on Romm (1998), Macnaghten *et al.* (2005), Doubleday (2007) and Scoones *et al.* (2007), who have translated the ideas of reflexivity from general social theory into the context of environmental risk and governance. Reflexivity involves a sensitivity to inputs from diverse perspectives, consciously recognising that there are alternative ways of seeing issues of concern. It involves a deliberate consideration of whether all the necessary voices are present, and are being listened to. It is a recursive and multi-directional activity of critical (and self-critical) thought and dialogue, taking place throughout the process. Since persuasion and the promotion of social change are often at the heart of responses to socio-environmental risks, there is a need for particular self-awareness on the part of the expert/researcher, at the stage of information provision, and during the negotiations that lead to the transformation. This should not be regarded as navel-gazing; it is part of the process of accounting for researchers', risk managers' and indeed all participants' involvement in the processes that shape or steer society. Reflexivity involves opening up the tacit assumptions, exposing the whole 'risk knowledge process' to public scrutiny, and as exemplified in the UEKC studies, involving those publics in the co-production of the knowledge base for societal action.

Because socio-environmental risks are characterised by a high degree of complexity, the process of change often needs to be provisional and adaptive, in the first instance. But through an experimental and iterative encounter premised on building trust competencies, the reflexive process provides the means for a structural pausing-for-thought, accommodates uncertainty and complexity into research design and output, and can result in radical transformations to how and what scientific knowledge is produced (Lane *et al.*, 2011: 33), and for whom it becomes meaningful with 'the objective of helping the actors to be subjects of their own lives' (TIA, 2007: 77).

15.4 Conclusion: understanding scientific action

In this concluding section we reflect briefly on the fact that research in natural hazards has a dual motivation: to develop greater understanding of our environment, and to inform real-world action. Activism may be a strong or unpalatable term for some. Researchers, however, are increasingly being turned to for more than pure knowledge. They are being asked to deliver real-world results with policy impacts that can demonstrably be seen to affect economic and social outcomes. Such demands are often a very far cry from labs and field research, let alone much of the formal education histories of the researchers and their teams. Some of this expectation is the result of misunderstandings on the part of those calling for such applications (politicians, policy-makers, business leaders and the media), of the power of predictive modelling to forecast future climate or related earth surface events. Indeed, one of the risks of predictive modelling is that it can lend itself to such misappropriated expectations. However, it is also the case that public and publicly accountable bodies enshrine in their own prospects – responsibilities for scientists beyond those of mere objective description.

The framing of risk research in terms of promoting societal objectives is a relatively recent phenomenon. A decade ago, the Phillips Report (2001) on bovine spongiform encephalopathy emphasised precise demarcations of the scope and relevance of scientific expertise for the advisory panel, making a sharp distinction between the science and policy domains. This contrasts markedly with the UK Chief Scientific Advisor's 2005 guidance on evidence for policy (HM Government, 2005a), which presses for wider public dialogue in cases of uncertainty over risks to human well-being or the environment, and emphasises the need for scientific input to be embedded in departmental policies and practices. To facilitate this process, most UK government departments have appointed a scientific advisor, often supported by advisory panels. The unprecedented collaborative partnership among UK research councils and government departments that the research programme Living with Environmental Change represents is predicated upon improved ongoing dialogue among research funders, researchers and government stakeholders, explicitly 'to provide for significant impact on policy and practice' (NERC, 2009).

In its report on science in the public interest, the Royal Society (2006: 10) likewise argues that the 'research community ... needs to shoulder two main responsibilities with respect to the public. The first is to attempt an accurate assessment of the potential implications for the public.... The second is to ensure the timely and appropriate communication to the public of results if such communication is in the public interest'. While the former is evidence of good ethical practice, the latter is more difficult to implement. We have spoken already of communication, but it is interesting to note that the Royal Society qualifies what it means by 'appropriate communication' to include 'indicators of uncertainty in the interpretation of results; expressions of risk that are meaningful; and comparison of the new results with public perceptions, "accepted wisdom", previous results and official advice' (Royal Society (2006: 11). We suggest that the social scientific insights that have been indicated in brief thus far, and perhaps as they manifest in the actual co-production of scientific knowledge in projects like the UEKC initiative, go a long way to addressing how natural hazards risk science can meet the Royal Society criteria of 'interpretive', 'meaningful' and integrative with other forms of knowledge and understanding.

It is thus not beyond the scope of this chapter to suggest that if natural hazards scientists were to implement truly interdisciplinary research co-production strategies around hazards risks and their uncertainty communication, like those exemplified by the UEKC project, neither the reticence about advocacy nor the loss of scientific privilege should be significant issues. Meaningful co-development of knowledge and communication of issues and their uncertainties would be built in from the start. Of course, for each specific project, the co-production strategies would differ, and some hazard situations might require a different balance of conventional scientific input than others. But adopting the principle of meaningful and relevant knowledge production in forms that better meet the challenges of accountability, transparency and impact would bring profound changes to risk management. Action, agency and – at least to some extent – political engagement would become explicit, and importantly, would be *negotiated* parts of the scientific method, whereas today these aspects might largely remain implicit. Hazards scientists should therefore, by all means, consider the extent to which they inform and shape action, and the means by which they do so.

We have, in this chapter, identified three broad challenges for uncertainty analysis in the natural hazards context. These are:

- Expanding the need for interdisciplinary working to understand the interactions between the natural and human systems that occur in emergencies and disasters.
- Involving a much wider range of knowledge inputs from the communities of interest (who bring important contributory, experiential expertise) and the academic social sciences (such as psychology, ethics, law, sociology and history, in addition to the more expected economics and political science).
- Developing theoretical and conceptual approaches that combine or integrate components of socio-environmental systems to produce knowledge-theoretic assemblage models for hazards risk analysis.

Expanding the networks and involving and developing new knowledge contributors within the scientific co-production processes demands an awareness of the many interrelated social and institutional contexts of natural hazards themselves, including their research. In turn, this requires that we focus on different time and spatial scales to those that apply to the physical hazards themselves, for the risks and uncertainties are themselves emergent properties of the socio-natural systems that face the hazards. As we have noted throughout, risk management in some sectors and domains already does this well. Notably, Europe's responses to flood risk have integrated social and political dimensions over the last one or two decades, with river basin and coastal cell management structures involving strategic transboundary and multi-administration partnerships in most countries.

Perhaps the greatest challenges, though, are those relating to the translation of knowledge to explicitly political action. The first is the building up of effective engagement with stakeholders of risk and uncertainty in natural hazards, starting with identifying who they are and what they need. The next, and arguably the most demanding, is the internal and cultural shifts that many scientists may need to face, as their societal role develops from 'detached investigator' to participant-observer or indeed activist.

Throughout this chapter we have emphasised that the co-production of knowledge – including the definition of risk and the identification and management of its uncertainty – emerges through the activity of an assemblage or interacting dynamic system of humans (who can think, make choices and act on them) and non-human factors. Scientific action (or any action for that matter), write Callon and Law (1995: 482), 'including its reflexive dimension that produces meaning, takes place in hybrid collectives'. The action is the product of the assemblage as a whole, and cannot be isolated in any one part. The elements that make up the collective expression of scientific knowledge production are an almost endless list: the hazardous events, risk scientists, related academics, lay people, policy-makers, administrators, natural processes, models, technical instruments, workshops, communication devices, maps, photographs, stories, museums, etc. Viewed in this way, the emergence of action and agency from these socio-natural assemblages goes some considerable way towards distributing responsibility and representation, in instances of activism and advocacy, from the locus of 'the scientist' to the co-productive system. We view this as an exciting development, in that increasingly literate and engaged publics, enrolled in their own enabling forms of scientific agency and self-understanding can shape, in specific but translatable ways, the social and natural meanings relevant to their lives, and their interrelated ecologies.

What we need to explore now is how this wide, powerful and interested socio-natural assemblage, different in each case and specific to particular social, cultural, political and environmental contexts, can be understood in its entirety – theoretically and methodologically. And we need to understand better how it can be asked, as a collective, to co-produce scientific understanding for hazards and their impacts. Social science has an integral role to play in the opening up of this new and potentially radical scientific method. We hope we have shown how it is already shaping the future of the risk research landscape.

Acknowledgements

The authors would like to thank Thomas Osborne, Dominik Reusser, Steve Sparks, Jonty Rougier and Lisa Hill for their important and insightful contributions.

References

Ackoff, R. L. and Emery, F. E. (1972) *On Purposeful Systems: An Interdisciplinary Analysis of Individual and Social Behaviour as a System of Purposeful Events*, London: Tavistock Publications.

Antunes, P., Santos, R. and Videira, N. (2006) Participatory decision making for sustainable development: the use of mediated modelling techniques. *Land Use Policy* 23: 44–52.

Ashley, R., Balmforth, D., Saul, A., *et al.* (2005) Flooding in the future: predicting climate change, risks and responses in urban areas. *Water Science & Technology* 52 (5): 265–273.

Aspinall, W. (2010) Opinion: a route to more tractable expert advice. *Nature* 463: 294–295.

Atman, C. J., Bostrom, A., Fischhoff, B. *et al.* (1994) Designing risk communications: completing and correcting mental models of hazardous processes, Part I. *Risk Analysis* 14: 251–262.

Bäckstrand, K. (2003) Precaution, scientisation or deliberation? Prospects for greening and democratizing science. In *Liberal Environmentalism*, ed. M. Wissenburg and Y. Levy, London and New York: Routledge.

Batty, M. and Torrens, P. M. (2005) Modelling and prediction in a complex world. *Futures* 37 (7): 745–766.

Beck, U. (1992) *Risk Society: Towards a New Modernity*, trans. M. Ritter, London: Sage Publications.

Beck, U. (1999) *World Risk Society*, Cambridge: Polity Press.

Bennett, J. (2010) *Vibrant Matter: A Political Ecology of Things*, Durham, NC and London: Duke University Press.

Best, J. (2010) The limits of financial risk management: or what we did not learn from the Asian crisis. *New Political Economy* 15 (1): 29–49.

Blaikie, N. W. H. (1991) A critique of the use of triangulation in social research. *Quality and Quantity* 23: 115–136.

Boiko, P. E., Morrill, R. L., Flynn, J., *et al.* (1996) Who holds the stakes? A case study of stakeholder identification at two nuclear weapons production sites. *Risk Analysis* **16**: 237–249.

Bonnell, V., L. Hunt and R. Biernacki (eds) (1999) *Beyond the Cultural Turn: New Directions in the Study of Society and Culture*, Berkeley, CA: University of California Press.

Brooks, N., Grist, N. and Brown, K. (2009) Development futures in the context of climate change: challenging the present and learning from the past. *Development Policy Review* **27**: 741–765.

Brown, K. and Corbera, E. (2003) A multi-criteria assessment framework for carbon mitigation projects: putting 'development' in the centre of decision-making. Tyndall Centre Working Paper 29, Tyndall Centre, Norwich. www.tyndall.ac.uk/sites/default/files/wp29.pdf (accessed 9 August 2012).

Brulle, R. J. (2010) From environmental campaigns to advancing the public dialog: environmental communication for civic engagement. *Environmental Communication: A Journal of Nature and Culture* **4**: 82–98.

Çalışkan, K. and Callon, M. (2010): Economization, part 2: a research programme for the study of markets. *Economy and Society* **39** (1): 1–32.

Callon, M. (1999) The role of lay people in the production and dissemination of scientific knowledge. *Science, Technology and Society* **4** (1): 81–94.

Callon, M. and Law, J. (1995). Agency and the hybrid collectif. *The South Atlantic Quarterly* **94** (2): 481–508.

Callon, M., Lascoumes, P. and Barthes, Y. (2009) *Acting in an Uncertain World: An Essay on Technical Democracy*, Cambridge, MA and London: MIT Press.

Cassell, J. A., Leach. M., Poltorak. M. S., *et al.* (2006) Is the cultural context of MMR rejection a key to an effective public health discourse? *Public Health* **120**: 783–794.

Checkland, P. (1999) *Systems Thinking, Systems Practice*, Chichester: John Wiley and Sons.

Clark, M. J. and Priest, S. J. (2008) Public awareness of flood risk: the role of the Environment Agency flood map. ESRC Full Research Report. www.esrcsocietytoday.ac.uk/ESRCInfoCentre/Plain_English_Summaries/environment/RES-000-22-1710.aspx (accessed 1 December 2010).

Cobbold, C. and Santema, R. (2008) *Going Dutch II: Toward a Safe and Sustainable Future of the Manhood Peninsula*. Chichester: The Manhood Peninsula Partnership. http://peninsulapartnership.org.uk/projects/going-dutch (accessed 9 August 2012).

Collins, H. M. and Evans, R. (2002) The third wave of science studies: studies of expertise and experience. *Social Studies of Science* **32**: 235–296.

Collins, H. M. and Evans, R. (2007) *Rethinking Expertise*, Chicago, IL: University of Chicago Press.

Collins, K. and Ison, R. (2006) Dare we jump off Arnstein's ladder? Social learning as a new policy paradigm. Proceedings of PATH (Participatory Approaches in Science & Technology) Conference, 4–7 June, Edinburgh. www.macaulay.ac.uk/PATHconference/PATHconference_proceeding_ps3.html (accessed 1 December 2010).

Coole, D. and Frost, S. (2010) Introducing the new materialisms. In *New Materialisms: Ontology, Agency and Politics*, ed. D. Coole and S. Frost, Durham, NC and London: Duke University Press, pp. 1–43.

Cornell, S. E. (2006) Understanding the relationship between stakeholder engagement and effectiveness and efficiency in flood risk management decision-making and delivery. In *Managing the Social Aspects of Floods*, Science report SC040033/SRZ Bristol: Environment Agency.

Costanza, R. (ed.) (1991) *Ecological Economics: The Science and Management of Sustainability*, New York and Oxford: Columbia University Press.

Costanza, R., Leemans, R., Boumans, R., *et al.* (2007) Integrated global models. In *Sustainability or Collapse: An Integrated History and Future of People on Earth*, ed. R. Costanza, L. J. Graumlich, and W. Steffen, Cambridge, MA: MIT Press, pp. 417–446.

Cox, Jr., L. A. (1999) Internal dose, uncertainty analysis, and complexity of risk models. *Environment International* **25** (6–7): 841–852.

Cronin, M. and Connolly, C. (2007) Exploring the use of experiential learning workshops and reflective practice within professional practice development for post-graduate health promotion students. *Health Education Journal* **66**: 286–303.

Crotty, J. (2009) Structural causes of the global financial crisis: a critical assessment of the new financial architecture. *Cambridge Journal of Economics* **33**: 563–580.

Cutter, S. L. and Emrich, C. T. (2006) Moral hazard, social catastrophe: the changing face of vulnerability along the hurricane coasts. *Annals of the American Academy of Political and Social Science* **604**: 102–112.

Davern, M., Jones, A., Jr., Lepkowski, J., *et al.* (2006) Unstable inferences? An examination of complex survey sample design adjustments using the Current Population Survey for health services research. *Inquiry* **43**: 283–297.

Demeritt, D. (2001) The construction of global warming and the politics of science. *Annals of the Association of American Geographers* **91**: 307–337.

Demeritt, D. (2008) From externality to inputs and interference: framing environmental research in geography. *Transactions of the Institute of British Geographers* **34**: 3–11.

Dequech, D. (2004) Uncertainty: individuals, institutions and technology. *Cambridge Journal of Economics* **28**: 365–378.

Dessai, S., Hulme, M., Lempert, R., *et al.* (2009) Climate prediction: a limit to adaptation? In *Adapting to Climate Change: Thresholds, Values, Governance*, ed. W. N. Adger, I. Lorenzoni and K. L. O'Brien, Cambridge: Cambridge University Press, pp. 64–78.

Dillon, M. (2010) *Biopolitics of Security in the 21st Century: A Political Analytics of Finitude*, London: Routledge.

Doubleday, R. (2007) Risk, public engagement and reflexivity: alternative framings of the public dimensions of nanotechnology. *Health Risk and Society* **9** (2): 211–227.

Douglas, M. (1992) *Risk and Blame: Essays in Cultural Theory*, London: Routledge.

Douglas, M. and Wildavsky, A. (1982) *Risk and Culture: An Essay on the Selection of Technical and Environmental Dangers*, Berkeley, CA: University of California Press.

Ebi, K. K. and Semenza, J. C. (2008) Community-based adaptation to the health impacts of climate change. *American Journal of Preventive Medicine* **35**: 501–507.

Eisenack, K., Scheffran, J. and Kropp, J. P. (2005) Viability analysis of management frameworks for fisheries. *Environmental Modeling and Assessment* **11** (1): 69–79. http://www.springerlink.com/index/10.1007/s10666-005-9018-2.

Ekberg, M. (2004) Are risk society risks exceptional? Proceedings of The Australian Sociological Association conference TASA 2004. www.tasa.org.au/conferencepapers04/ (accessed 1 December 2010).

Ericson, R., Barry, D. and Doyle, A. (2000) The moral hazards of neo-liberalism: lessons from the private insurance industry. *Economy and Society* **29** (4): 532–558.

Escobar, A. (2008) *Territories of Difference: Place, Movements, Life, Redes*, Durham, NC and London: Duke University Press.

Etkin, D. and Ho, E. (2007) Climate change: perceptions and discourses of risk. *Journal of Risk Research* **10**: 623–641.

European Food Safety Authority (2006) Guidance of the Scientific Committee on a request from EFSA related to uncertainties in Dietary Exposure Assessment. *The EFSA Journal* **438**: 1–54. www.efsa.europa.eu/en/efsajournal/doc/438.pdf (accessed 9 August 2012).

Evans, E. P., Ashley, R., Hall, J. W., *et al.* (2004) *Foresight: Future Flooding*, vol. 1–2, London: Office of Science and Technology.

Farrell, K. N. (2011) Snow White and the wicked problems of the West: a look at the lines between empirical description and normative prescription. *Science, Technology and Human Values* **36**: 334–361.

Fischhoff, B. (1998) Communicate unto others. *Reliability Engineering and System Safety* **59**: 63–72.

Folke, C. (2006) Resilience: the emergence of a perspective for social-ecological systems analyses. *Journal of Global Environmental Change* **16**: 253–267.

Ford, J., Smit, B. and Wandel, J. (2006) Vulnerability to climate change in the Arctic: a case study from Arctic Bay, Nunavut. *Global Environmental Change* **16**: 145–160.

Fox, C. R. and Tversky, A. (1995) Ambiguity aversion and comparative ignorance. *Quarterly Journal of Economics* **110** (3): 585–603.

Freire, P. (2005) *Education for Critical Consciousness*, New York: Continuum International Publishing Group.

Funtowicz, S. O. and Ravetz, J. R. (1993) Science for the post-normal age. *Futures* **25**: 739–755.

Gabbert, S. (2006) Improving efficiency of uncertainty analysis in complex integrated assessment models: the case of the RAINS emission module. *Environmental Monitoring and Assessment* **119**: 507–526.

Gale, D. (1996) What have we learned from social learning? *European Economic Review* **40** (3–5): 617–628.

Gardner, G. T. and Stern, P. C. (2002) *Environmental Problems and Human Behavior*, Boston, MA: Pearson Custom Publishing.

Giddens, A. (1999) Risk and responsibility. *Modern Law Review* **62** (1): 1–10.

Grothmann, T. and Patt, A. (2005) Adaptive capacity and human cognition: the process of individual adaptation to climate change. *Global Environmental Change* **15**: 199–213.

Haas, P. (1992) Introduction: epistemic communities and international policy coordination. *International Organization* **46** (1): 1–35.

Haldane, A. G. and May, R. M. (2011) Systemic risk in banking ecosystems. *Nature*. **469**: 351–355.

Harremoës, P., Gee, D., MacGarvin, M., *et al.* (2002) *The Precautionary Principle in the 20th Century*, London: Earthscan.

Hart, A., Gosling, J. P. and Craig, P. (2010) A simple, structured approach to assessing uncertainties that are not part of a quantitative assessment. Technical report, The Food and Environment Research Agency. www.personal.leeds.ac.uk/~stajpg/pub/ (accessed 9 August 2012).

Hartmann, S. (1999) Models and stories in hadron physics. In *Models as Mediators: Perspectives on Natural and Social Science*, ed. M. S. Morgan and M. Morrison, Cambridge: Cambridge University Press, pp. 326–346.

Haynes, K., Barclay, J. and Pidgeon, N. (2007) Volcanic hazard communication using maps: an evaluation of their effectiveness. *Bulletin of Volcanology* **70**: 123–138.

Heath, R. L., Palnenchar, M. J., Proutheau, S., *et al.* (2007) Nature, crisis, risk, science, and society: what is our ethical responsibility? *Environmental Communication: A Journal of Nature and Culture* **1**: 34–48.

Henderson-Sellers, A. (1993) *Future Climates of the World: A Modelling Perspective*, Amsterdam: Elsevier Scientific.

Henwood, K., Pidgeon, N., Parkhill, K., *et al.* (2010) Researching risk: narrative, biography, subjectivity. *Forum Qualitative Sozialforschung – Forum: Qualitative Social Research* **11**: 20. http://nbn-resolving.de/urn:nbn:de: 0114-fqs1001201 (accessed 9 August 2012).

HM Government (2005a) Guidelines on scientific analysis in policy making. www.berr.gov.uk/files/file9767.pdf (accessed 9 August 2012).

HM Government (2005b) Principles for public dialogue on science and technology. In *Government Response to the Royal Society and Royal Academy of Engineering Nanotechnology Report*, Annex B. www.berr.gov.uk/files/file14873.pdf (accessed 9 August 2012).

HM Government (2011) Science and Technology Committee: Third Report, Scientific advice and evidence in emergencies. www.publications.parliament.uk/pa/cm201011/cmselect/cmsctech/498/49802.htm. (accessed 4 November 2011).

HM Treasury/DTI/DfES (2004) *Science and Innovation Investment Framework 2004–2014*, London: HM Treasury. www.berr.gov.uk/files/file31810.pdf (accessed 9 August 2012).

Hordijk, L. (1991) Use of the RAINS Model in acid rain negotiations in Europe. *Environmental Science and Technology* **25**: 596–602.

House of Lords Select Committee on Science and Technology (2000) *Science and Society 3rd Report*, London: The Stationery Office.

Hulme, M. (2008) Geographical work at the boundaries of climate change. *Transactions of the Institute of British Geographers* **33**: 5–11.

Hulme, M. (2009) *Why We Disagree About Climate Change: Understanding Controversy, Inaction, and Opportunity*, Cambridge: Cambridge University Press.

ICSU (International Council for Science) (2008) *A Science Plan for Integrated Research on Disaster Risk: Addressing the Challenge of Natural and Human-Induced Environmental Hazards*, Paris: ICSU. www.icsu.org/publications/reports-and-reviews/IRDR-science-plan (accessed 9 August 2012).

Jabeen, H., Johnson, C. and Allen, A. (2010) Built-in resilience: learning from grassroots coping strategies for climate variability. *Environment and Urbanization* **22**: 415–431.

Jasanoff, S. (1990) *The Fifth Branch: Science Advisors as Policymakers*, Cambridge, MA: Harvard University Press.

Jones, R. N. (2000) Managing uncertainty in climate change predictions: issues for impact assessment. *Climatic Change* **45**: 403–419.

Kahneman, D. (2003) A perspective on judgment and choice: mapping bounded rationality. *American Psychologist* **58**: 697–720.

Kahneman, D. (2011) *Thinking, Fast and Slow*. London: Allen Lane.

Kahneman, D. and Tversky, A. (1996) On the reality of cognitive illusions. *Psychological Review* **103**: 582–591.

Kahneman, D. and Tversky, A. (eds) (2000) *Choices, Values and Frames*, New York: Cambridge University Press.

Kahneman, D., Slovic, P. and Tversky, A. (1982) *Judgment Under Uncertainty: Heuristics and Biases*, New York: Cambridge University Press.

Kameda, T. and Tamura, R. (2007) To eat or not to be eaten? Collective risk-monitoring in groups. *Journal of Experimental Social Psychology* **43**: 168–179.

Kasperson, J. X. and Kasperson, R. E. (2001) *Global Environmental Risk*, Tokyo, New York and Paris: United Nations University Press and Earthscan.

Kasperson, R. E., Renn, O., Slovic, P., et al. (1988) Social amplification of risk: a conceptual framework. Risk Analysis 8: 177–187.

Kates, R. W., Clark, W. C., Corell, R., et al. (2001) Sustainability science. Science 292: 641–642.

Kilian, A. (1988) Conscientization: an empowering, nonformal education approach for community-health workers. Community Development Journal 23: 117–123.

Klinke, A. and Renn, O. (2002) A new approach to risk evaluation and management: risk-based, precaution-based and discourse-based strategies. Risk Analysis 22 (6): 1071–1094.

Klinke, A. and Renn, O. (2006) Systemic risks as challenge for policy making in risk governance. Forum Qualitative Sozialforschung/Forum: Qualitative Social Research 7 (1): Art. 33. http://nbn-resolving.de/urn:nbn:de: 0114-fqs0601330 (accessed 9 August 2012).

Konikow, L. F. and Bredehoeft, J. D. (1992) Ground-water models cannot be validated. Advances in Water Resources 15 (1): 75–83.

Kuecker, G. D. (2007) Fighting for the forests: grassroots resistance to mining in northern Ecuador. Latin American Perspectives 34: 94–107.

Laffont, J. J. (1995) Regulation, moral hazard and insurance of environmental risks. Journal of Public Economics 58: 319–336.

Lane, S. N. (2010) Making mathematical models perform in geographical space(s). Handbook of Geographical Knowledge, ed. J. Agnew and D. Livingstone. London: Sage, pp. 228–245.

Lane, S. N., Odoni, N., Landstrom, C., et al. (2011) Doing flood risk science differently: an experiment in radical scientific method. Transactions of the Institute of British Geographers 36: 15–36.

Latour, B. (2004a) Why has critique run out of steam? From matters of fact to matters of concern. Critical Inquiry 30: 225–248.

Latour, B. (2004b) Politics of Nature: How to Bring the Sciences into Democracy (translated by C. Porter), Cambridge, MA: Harvard University Press.

Lenton, T. M., Held, H., Kriegler, E., et al. (2008) Tipping elements in the Earth's climate system. Proceedings of the National Academy of Science USA 105: 1786–1793.

Levy, J. K. and Gopalakrishnan, C. (2009) Multicriteria analysis for disaster risk reduction in Virginia, USA: a policy-focused approach to oil and gas drilling in the outer continental shelf. Journal of Natural Resources Policy Research 11: 213–228.

Liverman, D. and Roman Cuesta, R. M. (2008) Human interactions with the Earth system: people and pixels revisited. Earth Surface Processes and Landforms 33: 1458–1471.

Lo, A. W. and Mueller, M. T. (2010) Warning: physics envy may be hazardous to your wealth! arXiv. http://arxiv.org/abs/1003.2688v3 (accessed 29 November 2010).

Luhmann, N. (1993) Risk: A Sociological Theory, New York: A. de Gruyter.

Luhmann, N. (1995) Social Systems (translated by J. Bednarz), Stanford, CA: Stanford University Press.

LWEC (Living With Environmental Change) (2008) Living With Environmental Change: Brochure, version 4. www.lwec.org.uk. (accessed 29 November 2010).

MacDonell, M., Morgan, K. and Newland, L. (2002) Integrating information for better environmental decisions. Environmental Science and Pollution Research 9: 359–368.

Macnaghten, P., Kearnes, M. B. and Wynne, B. (2005) Nanotechnology, governance and public deliberation: what role for the social sciences? Science Communication 27: 268–291.

Maier, H. R., Ascough, J. C., II, Wattenbach, M., *et al.* (2008) Uncertainty in environmental decision-making: issues, challenges and future directions. In *Environmental Modelling, Software and Decision Support*, ed. A. J. Jakeman, A. A. Voinov, A. E. Rizzoli, *et al.*, Amsterdam: Elsevier Science, pp. 69–85.

Marx, S., Weber, E. U., Orove, B. S., *et al.* (2007) Communication and mental processes: experiential and analytic processing of uncertain climate information. *Global Environmental Change* **17**: 1–3.

McDowell, L. (2011) Making a drama out of a crisis: representing financial failure, or a tragedy in five acts. *Transactions of the Institute of British Geographers* **36**: 193–205.

McLeman, R. and Smit, B. (2006) Vulnerability to climate change hazards and risks: crop and flood insurance. *Canadian Geographer – Geographe Canadien* **50**: 217–226.

Mearns, L. O., Hulme, M., Carter, R., *et al.* (2001) Climate scenario development. In *Climate Change 2001: The Scientific Basis*, ed. J. T. Houghton, Y. Ding, D. Griggs, *et al.* Cambridge: Cambridge University Press, pp. 739–768.

Meijer, I. S. M., Hekkert, M. P., Faber, J., *et al.* (2006) Perceived uncertainties regarding socio-technological transformations: towards a framework. *International Journal of Foresight and Innovation Policy* **2**: 214–240.

Mercer, J. (2010) Disaster risk reduction or climate change adaptation: are we reinventing the wheel? *Journal of International Development* **22**: 247–264.

Millstone, E. (2009) Science, risk and governance: radical rhetorics and the realities of reform in food safety governance. *Research Policy* **38**: 624–636.

Mitchell, G. and Walker, G. (2003) *Environmental Quality and Social Deprivation*, Bristol: Environment Agency.

Mokrech, M., Hanson, S., Nicholls, R. J., *et al.* (2009) The Tyndall coastal simulator. *Journal of Coastal Conservation: Planning and Management* **15** (3): 325–335.

Morgan, M. G. and Henrion, M. (1990) *Uncertainty: A Guide to Dealing with Uncertainty in Quantitative Risk and Policy Analysis*, New York: Cambridge University Press.

Moss, R. (2007) Improving information for managing an uncertain future climate. *Global Environmental Change* **17**: 4–7.

Moss, R. H. and Schneider, S. H. (2000) Uncertainties in the IPCC TAR: guidance to authors for more consistent assessment and reporting. In *Third Assessment Report: Cross-cutting Issues Guidance Papers*, ed. R. Pauchuri, T. Tanagichi and K. Tanaka, Geneva: World Meteorological Organisation.

National Research Council (1996) *Understanding Risk: Informing Decisions in a Democratic Society*, Washington, DC: National Academy Press.

NERC (2009) Living with Environmental Change (LWEC) Partners' Board Terms of Reference. www.nerc.ac.uk/research/programmes/lwec/management.asp (accessed 9 August 2012).

Niemeyer, S., Petts, J. and Hobson, K. (2005) Rapid climate change and society: assessing responses and thresholds. *Risk Analysis* **25**: 1443–1456.

Nowotny, H., Scott, P. and Gibbons, M. (2001) *Re-Thinking Science: Knowledge and the Public in an Age of Uncertainty*, Cambridge: Polity Press.

Odoni, N. and Lane, S. N. (2010) Knowledge-theoretic models in hydrology. *Progress in Physical Geography* **34**: 151–171.

OECD (2009) Draft policy handbook on natural hazard awareness and disaster risk reduction education. www.oecd.org/dataoecd/24/51/42221773.pdf (accessed 1 December 2010).

O'Malley, P. (2004) *Risk, Uncertainty, and Government*, London: Cavendish Press/ Glasshouse.

Oreskes, N. (2003) The role of quantitative models in science. In *Models in Ecosystem Science*, ed. C. D. Canham, J. J. Cole, and W. K. Lauenroth, Princeton, NJ: Princeton University Press, pp. 13–31.

Oreskes, N. and Belitz, K. (2001) Philosophical issues in model assessment. In *Model Validation: Perspectives in Hydrological Science*, ed. M. G. Anderson and P. D. Bates, London: John Wiley and Sons, pp. 23–41.

Oreskes, N., Shrader-Frechette, K. and Belitz, K. (1994) Verification, validation, and confirmation of numerical models in the earth sciences. *Science* 263: 641–646.

O'Riordan, T., Burgess, J. and Szerszynski, B. (1999) *Deliberative and Inclusionary Processes: A Report from Two Seminars*, Norwich: CSERGE, University of East Anglia.

Ostrom, E., Janssen, M. and Anderies, J. (2007) Robustness of social-ecological systems to spatial and temporal variability. *Society and Natural Resources* 20: 307–322.

Ostrom, V. (1997) *The Meaning of Democracy and the Vulnerability of Democracies*, Ann Arbor, MI: University of Michigan Press.

Otway, H. (1992) Public wisdom, expert fallibility: toward a contextual theory of risk. In *Social Theories of Risk*, ed. S. Krimsky and D. Golding, Westport, CT: Praeger, pp. 215–228.

Owens, S. (2005) Making a difference? Some reflections on environmental research and policy. *Transactions of the Institute of British Geographers* 30: 287–292.

Pahl-Wostl, C., Craps, M., Dewulf, A., *et al.* (2007) Social learning and natural resources management. *Ecology and Society* 12: Art. 5.

Patt, A. and Dessai, S. (2005) Communicating uncertainty: lessons learned and suggestions for climate change assessment. *Compte Rendus Geosciences* 337: 427–441.

Pelling, M. and High, C. (2005) Understanding adaptation: what can social capital offer assessments of adaptive capacity? *Global Environmental Change: Human and Policy Dimensions* 15: 308–319.

Peschard, I. (2007) The value(s) of a story: theories, models and cognitive values. *Principia* 11 (2): 151–169.

Petersen, A. C., Cath, A., Hage, M., *et al.* (2011) Post-normal science in practice at the Netherlands Assessment Agency. *Science, Technology and Human Values* 36 (3): 362–388.

Petts, J. and Brooks, C. (2006) Expert conceptualisations of the role of lay knowledge in environmental decision-making: challenges for deliberative democracy. *Environment and Planning A* 38: 1045–1059.

Petts, J., Homan, J. and Pollard, S. (2003) *Participatory Risk Assessment: Involving Lay Audiences in Environmental Decisions on Risk*, Bristol: Environment Agency.

Petts, J., Pollard, S. J. T., Rocks, S. A., *et al.* (2010) Engaging others in environmental risk assessment: why, when and how. University of Birmingham and Cranfield University, UK.

Phillips Report (2001) *The BSE Inquiry: The Report (The Inquiry into BSE and CJD in the United Kingdom)*, London: The Stationery Office.

Phillips, L. and Stock, A. (2003) *Use of Multi-Criteria Analysis in Air Quality Policy*, London: Defra/Environment Agency. www.airquality.co.uk/reports/cat09/0711231556_MCDA_Final.pdf (accessed 9 August 2012).

Pidgeon, N. and Frischhoff, B. (2011) The role of social and decision sciences in communication uncertain climate risks. *Nature: Climate Change* 1: 35–41.

Pidgeon, N. F., Hood, C., Jones, D., *et al.* (1992) Risk perception. In *Risk Analysis, Perception and Management*, London: The Royal Society, pp. 89–134.

Pidgeon, N., Kasperson, R. E. and Slovic, P. (2003) *The Social Amplification of Risk*, Cambridge: Cambridge University Press.

Pidgeon, N., Simmons, P., Sarre, S., *et al.* (2008) The ethics of socio-cultural risk research. *Health, Risk and Society* **10**: 321–329.

Pilkey, O. H. and Pilkey-Jarvis, L. (2007) *Useless Arithmetic: Why Environmental Scientists Cannot Predict the Future*. Durham, NC: Duke University Press.

Popper, K. (2002) *The Poverty of Historicism*, Oxford: Routledge.

Prell, C., Hubacek, K., Reed, M., *et al.* (2007) If you have a hammer everything looks like a nail: 'traditional' versus participatory model building. *Interdisciplinary Science Reviews* **32**: 1–20.

Rademaker, O. (1982) On modelling ill-known social phenomena. In *Groping in the Dark: The First Decade of Global Modeling*, ed. In D. Meadows, J. Richardson and G. Bruckmann, Chichester: John Wiley and Sons, pp. 206–208.

Randall, D. A. and Wielicki, B. A. (1997) Measurements, models, and hypotheses in the atmospheric sciences. *Bulletin of the American Meteorological Society* **78**: 399–406.

Regan, H. M., Colyvan, M. and Burgman, M. (2002) A taxonomy and treatment of uncertainty for ecology and conservation biology. *Ecological Applications* **12**: 618–628.

Reichert, P. and Borsuk, M. E. (2005) Does high forecast uncertainty preclude effective decision support? *Environmental Modelling and Software* **20** (8): 10.

RELU (Rural Economy and Land Use Programme) (2010) Models, decision-making and flood risk: doing simulation modelling differently. In *RELU Policy and Practice Notes*. Note 22, October 2010.

Renn, O., Webler, T., Rakel, H., *et al.* (1993) Public participation in decision making: a three step procedure. *Policy Sciences* **26**: 189–214.

Ricci, P. F., Rice, D., Ziagos, J., *et al.* (2003) Precaution, uncertainty and causation in environmental decisions. *Environment International* **29**: 1–19.

Rittel, H. and Webber, M. (1973) *Dilemmas in a General Theory of Planning*, Amsterdam: Elsevier Scientific Publishing Company.

Roberts, K. A., Dixon-Woods, M., Fitzpatrick, R., *et al.* (2002) Factors affecting uptake of childhood immunisation: a Bayesian synthesis of qualitative and quantitative evidence. *Lancet* **360**: 1596–1599.

Romm, N. R. A. (1998) Interdisciplinarity practice as reflexivity. *Systemic Practice and Action Research* **11**: 63–77.

Rose, N. (2006) *The Politics of Life Itself: Biomedicine, Power, and Subjectivity in the Twenty-First Century*, Princeton, NJ: Princeton University Press.

Rowe, G. and Frewer, L. J. (2005) A typology of public engagement mechanisms. *Science, Technology and Human Values* **30**: 251–290.

Rowe, W. D. (1994) Understanding uncertainty. *Risk Analysis* **14**: 743–750.

Royal Society/FSA (2006) Social science insights for risk assessment. Findings of a workshop held by the Royal Society and the Food Standards Agency on 30 September 2005. http://royalsociety.org/Content.aspx?id=5231 (accessed 10 August 2012).

Scoones, I., Leach, M., Smith, A., *et al.* (2007) *Dynamic Systems and the Challenge of Sustainability*, Brighton: STEPS Centre.

Shrader-Frechette, K. (1991) *Risk and Rationality*, Berkeley, CA: University of California Press.

Shrader-Frechette, K. (2010) Analysing public perception in risk analysis: how the wolves in environmental injustice hide in the sheep's clothing of science. *Environmental Justice, Participation, and the Law* **3** (4): 119–123.

Siebenhüner, B. and Barth, V. (2005) The role of computer modelling in participatory integrated assessments. *Environmental Impact Assessment Review* **25**: 367–389.

Slovic, P. (2001) Trust, emotion, sex, politics, and science: surveying the risk-assessment battlefield. *Risk Analysis* **19** (4): 689–701.

Smil, V. (2003) *Energy at the Crossroads: Global Perspectives and Uncertainties*, Cambridge, MA: MIT Press.

Solo, T. M., Godinot, M. and Velasco, O. (2003) Community participation in disaster management: reflections on recent experiences in Honduras and Nicaragua. In *Pensando en Voz Alta IV: Innovadores Estudios de Caso sobre Instrumentos Participativos*, ed. R. Senderowitsch, Washington, DC: World Bank, pp. 26–35. http://siteresources.worldbank.org/INTLACREGTOPHAZMAN/Resources/ArticlefromTOLIV.pdf (accessed 9 August 2012).

Stephens, C., Willis, R., Walker, G. P., et al. (2007) *Addressing Environmental Inequalities: Cumulative Environmental Impacts*, Bristol: Environment Agency.

Sterman, J. D. (2011) Communicating climate change risks in a skeptical world. *Climatic Change* **108**: 811–826.

Stern, P. C. and Fineberg, H. V. (1996) *Understanding Risk: Informing Decisions in a Democratic Society*, Washington, DC: National Academies Press.

Stirling, A. and Mayer, S. (2001) A novel approach to the appraisal of technological risk: a multicriteria mapping study of a genetically modified crop. *Environment and Planning C: Government and Policy* **19**: 529–555.

Taylor-Clark, K. A., Viswanath, K. and Blendon, R. J. (2010) Communication inequalities during public health disasters: Katrina's wake. *Health Communication* **25**: 221–229.

Taylor-Gooby, P. and Zinn, J. O. (eds) (2006) *Risk in Social Sciences*, Oxford: Oxford University Press.

TIA (Trustnet-in-Action) (2007) European Commission final report. www.trustnetinaction.com/IMG/pdf/TIA-Final_Report.pdf (accessed 4 November 2011).

Titus, J. and Narayanan, V. (1996) The risk of sea level rise: a Delphic Monte Carlo analysis in which twenty researchers specify subjective probability distributions for model coefficients within their respective areas of expertise. *Climatic Change* **33**: 151–212.

Turnpenny, A., Jones, M. and Lorenzoni, I. (2011) Where now for post-normal science? A critical review of its development, definitions, and uses. *Science, Technology, and Human Values* **36** (3): 287–306.

Tversky, A. and Kahneman, D. (1991) Loss aversion in riskless choice: a reference dependent model. *Quarterly Journal of Economics* **107** (4): 1039–1061.

van Asselt, M. B. A. and Rotmans, J. (2002) Uncertainty in integrated assessment modelling: from positivism to pluralism. *Climate Change* **54**: 75–105.

van der Sluijs, J. P. (1996) Integrated assessment models and the management of uncertainties, IIASA working paper no. WP 96–119, Laxenburg, Austria.

Voils, C., Hasselblad, V., Cradell, J., et al. (2009) A Bayesian method for the synthesis of evidence from qualitative and quantitative reports: the example of antiretroviral medication adherence. *Journal of Health Services Research and Policy* **14**: 226–233.

Walker, G., Damery, S., Petts, J., et al. (2007) *Addressing Environmental Inequalities: Flood Risk, Waste Management and Water Quality in Wales*, Bristol: Environment Agency.

Weber, M. (1946) *Essays in Sociology*, Translated and edited by H. H. Gerth and C. W. Mills, Oxford: Oxford University Press.

Whatmore, S. J. (2002) *Hybrid Geographies: Natures, Cultures, Spaces*, London: Routledge.

Whatmore, S. J. (2009) Mapping knowledge controversies: science, democracy and the redistribution of expertise. *Progress in Human Geography* **33**: 587–598.

Whatmore, S. J. and Landstrom, C. (2011) Flood apprentices: an exercise in making things public. *Economy and Society* **40** (4): 582–610.

Whatmore, S., Landström, C. and Bradley, S. (2008) Democratising science. In *Science & Public Affairs*, London: British Association for the Advancement of Science, pp. 17–18. www.britishscienceassociation.org/NR/rdonlyres/20040D8F-AFEB-4AF4-A8C3-795596798A12/0/FinalversionSPA_June08v4.pdf (accessed 9 August 2012).

Wilby, R. L. and Dessai, S. (2010) Robust adaptation to climate change. *Weather* **65**: 180–185.

World Disasters Report (2009) Geneva: International Federation of Red Cross and Red Crescent Societies.

Yearley, S. (1999) Computer models and the public's understanding of science: a case-study analysis. *Social Studies of Science* **29**: 845–866.

Young, O. R., Berkhout, F., Gallopin, G., *et al.* (2006) The globalization of socio-ecological systems: an agenda for scientific research. *Global Environmental Change* **16**: 304–316.

Zinn, J. O. (2010) Biography, risk and uncertainty: is there common ground for biographical research and risk research? *Forum Qualitative Sozialforschung – Forum: Qualitative Social Research* **11**: 15. http://nbn-resolving.de/urn:nbn:de: 0114-fqs1001153 (accessed 9 August 2012).

Zintzaras, E., Kanellou, A., Trichopoulou, A., *et al.* (1997) The validity of household budget survey (HBS) data: estimation of individual food availability in an epidemiological context. *Journal of Human Nutrition and Dietetics* **10**: 53–62.

16

Natural hazards and risk: the human perspective

H. S. CROSWELLER AND J. WILMSHURST

More and more people worldwide are living in areas at risk from the impact of natural hazards (Dilley *et al.*, 2005). The most publicised events are rapid-onset, high-impact hydro-meteorological and geophysical events such as hurricanes, tornadoes, earthquakes, volcanoes, flooding, wildfires, tsunamis and landslides. These are often catastrophic and attract global attention (e.g. the Asian tsunami in December 2004). However, hazards also take the form of slow-onset and chronic events, such as drought, heat waves, coastal erosion and other such threats brought about by climate variability and changing landscapes. Ideally, people would not live in areas at high risk of natural hazards. However, a variety of factors, including socio-economic, political, cultural and personal, drive populations to continue to live or even migrate to these areas. It is human factors that are of interest here, with a specific focus on psychological aspects. We seek in this chapter to address the question of why people place their lives at risk by continuing to live in areas of high hazard. In doing so, we will review some relevant bodies of both theoretical and applied literature, as well as utilising illustrative examples of hazard-related behaviours in order to build a better understanding of such behaviours. We go on to offer suggestions as to the direction of future research and possible applications for the reviewed material, which may be of use to practitioners as well as researchers working in the field of disaster risk reduction (DRR). We start by looking at some areas within psychology that may help us to make sense of some of the choices made by those living at risk.

16.1 The psychology of decision-making

While there are numerous factors which may force people to live in areas of high hazard (e.g. on a flood plain or close to a volcano), many people have made a conscious decision to live in such locations. Comprehending how humans make decisions may help understand why people choose and continue to live in risky areas. There is a large literature on human decision-making, primarily based on the economics and psychology literature. This has been built mostly on laboratory-based experimental work and seeks to explain the mechanics of the higher cognitive processing using quantitative statistical methodologies.

Risk and Uncertainty Assessment for Natural Hazards, ed. Jonathan Rougier, Steve Sparks and Lisa Hill. Published by Cambridge University Press. © Cambridge University Press 2013.

Despite the plethora of laboratory studies contributing to the theoretical knowledge of decision-making, humans typically make decisions in a complex environment in which they are influenced by many additional factors that cannot easily be replicated in a laboratory. Many decision-making theories (e.g. the utility theories) suggest that making a decision is a rational, information-processing task. However, there has more recently been a greater focus on providing a more comprehensive explanation of how people actually cope with difficult and complex decisions in the real world, not least because very often a course of action is the result of a process that has taken place sub- or unconsciously and therefore cannot be brought to the surface and analysed in the same way as more visible, rational processes. For example, the decision by a large number of community members not to evacuate from the flanks of Galeras volcano in Colombia, when scientists reported that activity had increased, can appear to be a rather irrational response to a dangerous situation. Clear warnings were given and all residents had full access to shelters a short distance from their homes, so it could be seen as a very simple choice between moving to safety or putting themselves at unnecessary risk. In reality, however, it was found that for many of them the decision was made based on a range of variables far broader than those that may have seemed immediately relevant to an observer, i.e. a simple case of safety over danger in relation to a possible eruption. Some had decided to believe that there was no danger to them despite the increased activity, so they could continue with their lives undisrupted. These people felt that disruption was more distressing to them than the uncertainty of the scientific information. Others felt that they did not fully trust the source of the information, and there were also those who felt that the risks to their health posed by the evacuation shelters were worse than those posed by the volcano. In a second example, many people in 'tornado alley' in the United States every year during tornado season choose not to take shelter when a tornado warning is put in place. Shelters are plentiful and accessible and the risks are advertised clearly. In this context, the reasons given during field interviews ranged from not believing that a tornado was actually on the ground unless or until it was witnessed first-hand, to actively wanting to see or film one as a type of adrenaline sport. Often, though, people would give a reason when asked but then add that in actual fact they did not really know why they responded the way that they did, even joking that they know their behaviour was foolish and dangerous. It is these reactions, the reasons for which even the actors themselves often have little conscious insight, that pose the greatest challenge to researchers in this field and to which psychology may have the most to offer.

The real world, as opposed to the laboratory, often creates situations (like those in the examples above) in which a high emotional content is likely and where the consequences of decisions may be far reaching for the individual. Thus, there has been a move towards the study of naturalistic decision-making (see Cook *et al.*, 2008). As an example, Mann *et al.* (1998) developed a decision-making scale based on real-world evidence, which identified four decision-making styles: vigilance, buck-passing, procrastination and hyper-vigilance. The first of these is viewed as an adaptive style in that it relates to a tendency to weigh up all of the available options, while the others could all be seen as maladaptive since they are essentially ways in which making a definitive decision can be avoided. These maladaptive strategies provide evidence of defensive avoidance (denial), which is a coping strategy under stress. The perceived timescales on which many natural hazards work also plays a role in deciding when, or whether, to take precautions (Wachinger and Renn, 2010). The model of Mann *et al.* (1998) also includes a measure of decision-making confidence. People, as a general rule, tend to be overconfident in the decisions they make regardless of the success of the outcome, thus confidence influences how people make choices.

Literature with a natural hazards context further explains how some of the characteristics in human decision-making are manifest. Dibben's (2008) study of communities residing near Mount Etna volcano in Italy showed that some fairly affluent members of society chose to move closer to the volcano, as the quality of life was thought to be better than living in a city. Kelman and Mather (2008) listed numerous benefits to living close to a volcano, such as a reliable freshwater supply, mining and quarrying, potentially fertile soils and tourism, to name a few. It could also be that the volcanic risk was thought to be low – for example, by virtue of the fact that other people already lived in its proximity (the so-called bandwagon effect) or that volcanic activity at Mount Etna typically has limited impact and affects very few people. In this case, it may be that the volcanic risk was not even considered as a factor in their decision to move or, if it was, it was not thought to be important enough when compared with other, more day-to-day, considerations.

Unlike in a risk assessment, which concentrates on a single hazard, people must balance a wide range of factors with their associated benefits and costs. They typically do not do this in a quantitative manner, but instead often employ a heuristic, defined as a form of mental shortcut (see Tversky and Kahneman, 1974). One such heuristic is the 'affect' heuristic, whereby people make judgements based on their general feeling about a situation (Finucane *et al.*, 2000), which can be influenced, for example, by a person's previous experience with the hazard. Loewenstein *et al.* (2001) suggested that these feelings can, at times, override what logically might be considered to be the wisest course of action. Beliefs about risk are likely to be more difficult to change once they have been developed as heuristics (Frewer, 1999). Also of relevance is the 'availability' heuristic, which refers to the ease with which an example of the hazard can be recalled. Although such a cognitive shortcut can be beneficial, it can lead to overestimation of frequently reported risks and underestimation of less frequently reported risks (Xie *et al.*, 2003).

Theories of attitude are relevant to decision-making since, if people's attitudes can be modified, then their decisions and subsequent behaviours may change. Thus, changing attitudes and perceptions through various forms of persuasive communication will lead to changing levels of engagement in protective behaviours. In reality, this relationship is repeatedly found to be far from straightforward. The existence of gaps between attitudes and behaviour has been widely identified and discussed in the psychological literature (e.g. Ajzen, 2005; Crano and Prislin, 2006; Fishbein and Ajzen, 2010). While this issue is starting to be addressed in the natural hazards literature (e.g. Boura, 1998; Paton, 2003; Webb and Sheeran, 2006; Lavigne *et al.*, 2008), it is still far from being operational in the field.

As discussed earlier, we know that people make decisions using complex mental processes and that often these decisions are made in contexts in which many factors are at play.

By understanding the interplay between internal and external variables a more complete picture of the reasons behind seemingly contradictory and sometimes dangerous decisions can be built.

16.2 Place attachment

Place attachment has been identified as an important theme within the environmental psychology literature. The term is used interchangeably with community attachment with the distinction between them being unclear; both may relate either to attachment to a physical location (the house, street, neighbourhood or town/village) or to people (family, neighbours, wider community) or both.

Research by Billig (2006) looked at place and home attachment of Jewish settlers in the Gaza Strip during hostilities that posed a risk to the settlers' lives. Despite the danger, many settlers chose to stay, related to a strong feeling of attachment to place, an ideological view of the land and human relationships with the earth, strong religious beliefs and perceived levels of low risk. Some or all of these factors may exist for people exposed to other types of environmental risk, including natural hazards. Altman and Low (1992) found that a bond to a place can increase a person's sense of self-esteem and happiness, and can therefore be resistant to any incentives offered to them to move. At Galeras volcano in Colombia the government has had little success in encouraging residents to move, despite offering financial incentives. Inhabitants of the region surrounding Mount Merapi, Indonesia, returned to their homes soon after government attempts to resettle them in Sumatra as the unfamiliar place seemed more risky; they preferred instead the cultural familiarity of 'home' and utilised pre-existing coping mechanisms for dealing with the volcanic threat (Laksono, 1988).

Davis *et al.* (2005), in their study on volcanic risk perception at Vesuvius and Etna, Italy, included a survey element related to residents' 'sense of community', which is intertwined with people's attachments to place. Residents in the area surrounding Etna demonstrated much stronger community bonds than those around Vesuvius, and also rated their ability to protect themselves from a future eruption more highly. Riad and Norris (1998) also found that stronger community ties (referred to as 'social embeddedness') led to lower evacuation rates in response to hurricane warnings. They discussed the idea of social influence, which is divided into two components, namely social comparison processes (the actions of neighbours) and disaster-related support (discussing with others about the warning and possible actions). Being open to social influence can affect an individual's likelihood to heed evacuation warnings.

Self-concept is essentially an individual's understanding of who they are and provides a useful way to examine cultural differences in how people's relationships with place and communities may affect how they evaluate and respond to threats in the natural environment. Further, research conducted across different cultures and different hazard types shows that the concept provides a more complete picture of responses to natural hazard risk (Wilmshurst, 2010). For example, in a review on research into self and culture in the

United States and Japan, Markus and Kitayama (1991) raised interesting questions about cognition, emotion and motivation. They demonstrated that in a more collective culture, such as Japan, individuals seek to identify themselves as part of a collective, while in the United States the tendency is much more strongly towards independence and individuality.

In addition to the concepts of the 'individual' and 'collective' selves described above, Sedikides and Brewer (2001) identified a third construal covering a 'relational' self-concept. This categorisation moved away from the restrictive dichotomy of individual versus collective used by Markus and Kitayama by acknowledging a tendency in some cultures to identify oneself primarily through a network of individual relationships and the associated roles. This 'relational' self is in contrast to seeing oneself as one component in a large collective, as is the case in the 'collective' self as observed in cultures such as China and Japan. This third 'relational' construal is likely to be found in cultures where extended family networks are given a high level of importance, such as in many parts of Central and South America. This is particularly relevant for natural hazard risk research in those communities who live in isolated conditions and rely heavily on interdependence to meet the challenges that face them. This was certainly borne out in a study conducted in an isolated rural community in Belize facing threats not only from hurricanes but also from a number of development issues including depleting natural resources, poor transport links and poor access to education (Wilmshurst, 2010). In this community, reliance on family and neighbourhood relationships emerged as a key aspect in their resilience to the many threats to their welfare.

Self-concept literature is only one way to understand such issues, but recognition of the importance of such differences helps us to understand responses in the context of communities, acknowledging that people do not make choices in isolation but alongside and in relation to others. For example, in a more individualistic culture it may be that responses reflect the motivation to protect a different set of interests than in one where an individual's place in their wider community is more important than personal safety. Whether this is attributed to 'community attachment', 'place attachment', 'social embeddedness' or some kind of a collective self-concept, this response clearly shows that something other than' protection of individual life motivates the choice to stay in a high-risk location.

16.3 Perception of risk

Risk perception is the term used to describe the more subjective judgement of risk. There is a large body of literature on general risk perception (almost 1000 publications in the 1980s alone, one of the most influential being that by Slovic, 1987) and a growing body of literature in relation to natural hazards in particular (see Wachinger and Renn, 2010 for a summary). The notion of 'risk tolerability' for example, which suggests that there is a level of risk that people are willing to tolerate or 'accept', originated from work by Sowby (1965) and Starr (1969) in relation to the lack of public acceptance of new technologies such as nuclear power. This traditional approach to risk suggests that if we understand people's relationship to risk in general (i.e. whether they are risk 'tolerant' or risk 'averse') then this will help to predict how they will respond to risk from natural hazards. However, the way in which people react to natural phenomena could also be seen as a function of how they relate to the natural world rather than just a function of how they respond to risk in general (Torry, 1979; Hewitt, 1983; Susman *et al.*, 1983). It is therefore important to find out in what ways people, communities or cultures relate to the natural world around them and make their own assessments of the dangers and threats that it poses.

If an assumption is made that measures of risk perception indicate the true extent to which members of a given community appreciate the likelihood of a hazard event occurring, then it is a small step to assume that education would be the most useful intervention. Here it is assumed that the education concerns the level of risk calculated objectively by experts. If, however, earlier exposure to information about the risks has led to an uncomfortable realisation that current circumstances are not providing sufficient protection or mitigation, and clear options for action are not available, then this leaves the person with the choice of either changing their behaviour or their beliefs. This psychological phenomenon relating to holding contradictory ideas is known as cognitive dissonance.

The theory of cognitive dissonance proposes that people have a motivational drive to reduce dissonance by either changing their attitudes, beliefs and behaviours, or by justifying or rationalising their existing ones (Festinger, 1957). According to Festinger, the existence of dissonance will motivate the person to try to reduce the dissonance and achieve consonance. When dissonance is present, in addition to trying to reduce it, the person will actively avoid situations and information which would likely increase the dissonance. In other words, a person will employ some kind of psychological defence in order to 'rationalise' their decisions. Since the psychological defences resulting from cognitive dissonance will have an impact on people's attitudes and behaviours, this influences how they respond to the various risk management methods. Those with particularly strongly held or extreme attitudes are likely to be particularly resistant to change (Terpstra *et al.*, 2009). This could be one explanation for the behaviour of community members on the volcano in Colombia, who reported that they did not feel that there was any danger and that evacuating caused them greater distress. Cognitive dissonance can lead to a tendency to surround oneself with others holding similar viewpoints and to ignore any information which may seem to conflict with such views. This could help explain why much of the work conducted in an attempt to invoke behavioural changes, particularly using fear messages, has been largely unsuccessful (as an example, see Moser and Dilling, 2004, and their work on climate change).

Cognitive dissonance has been found to be present in populations at risk from a number of natural hazards; for example, volcanoes (Chester *et al.*, 2002; Gregg *et al.*, 2004a; Dibben, 2008), drought (Caldwell and Boyd, 2009) and flooding (Armas and Avram, 2009). In choosing to live with a repeated threat, people may experience cognitive dissonance and will therefore change their attitudes and/or beliefs rather than their behaviour. One such strategy could be denial of the risk, or conversely, may include greater engagement in awareness and preparedness. The coping strategy adopted by any individual will be motivated by a range of factors, both individual (e.g. personality, prior experience) and social (e.g. culture, social identity), and it is these complex interactions which are the most difficult to understand. Denial is commonly seen to occur in response to elevated feelings of fear, especially when it is felt that there is no available course of action to avoid the threat (e.g. Mulilis and Duval, 1995). There have also been cases where the general public do not want to acknowledge the hazard, or for it to be discussed, because it could have a negative impact on their livelihoods (e.g. on the volcanic island of Santorini in Greece where the tourism industry is a key employment sector potentially at risk from a future volcanic eruption (Dominey-Howes and Minos-Minopoulos, 2004)). The risk may also be denied due to strong religious beliefs; it may be believed that 'God' will keep people safe during a hazard event, or that 'He' would prevent it from occurring in the first place. Alternatively, a person's faith could result in the risk being accepted as part of 'God's will'.

Slovic *et al.* (2000), in their assessment of how the theory of bounded rationality (Simon, 1959) can be applied to the risk perceptions of natural hazards, found that residents in a flood-prone area (using the work of Kates, 1962) had two distinct responses to the inherent unpredictability of natural phenomena. The most common reaction was to view the floods as repetitive or cyclical phenomena, thereby removing the randomness and taking comfort from the deterministic behaviour of the hazard. The other response, in complete contrast, denied the determinability of natural phenomena and instead placed it as under the influence of a higher power (e.g. God, scientists or the government). Either way, the residents did not need to trouble themselves with the problem of dealing with uncertainty (Slovic *et al.*, 2000: 7). Similar patterns were also found by research carried out by Renn (2008) on perceptions of flood risk in Germany; many residents believed that, following a '100-year flood', it would take another 100 years before the next flood would occur and therefore there was no need to prepare.

The findings of Slovic *et al.* (2000) are likely to be influenced by how long people have been resident in an area. Living in a particular location for a long time increases the likelihood of having had direct experience of the hazard, and this should strongly influence an individual's perception of the risk (Glassman and Albarracin, 2006). This has the potential to invoke a number of 'cognitive biases', such as a normalisation bias (Mileti and O'Brien, 1992), which can occur when a person's experience of a hazard is relatively minor but is taken as typical. This is particularly problematic with respect to volcanic eruptions where the magnitude and even the style of eruption can change dramatically from one eruption to the next. Therefore, this kind of bias is likely to reduce a person's perceived need for information and preparation, believing that they will be able to cope with a future event as they did in the past (Paton and Johnston, 2001). This would have a bearing on the effectiveness of future outreach efforts, as the information may not be seen as necessary. Crosweller (2009), in her research on two volcanic Caribbean islands, found that prior experience was one of the main influences on residents' perceptions and beliefs about volcanic hazards (see also Miceli *et al.*, 2008; Siegrist and Gutscher, 2006 for flood risk examples where experience is a strong predictor for risk perception). While it might be expected that previous experience of a hazard would increase awareness and knowledge of the potential risks, this research found the opposite to be the case with residents on the island, with experience being more likely to create beliefs which could be deemed to be 'risky'. This has also been noted in numerous other studies relating not only to volcanic hazards but other natural hazards, as well as health issues (Gladwin and Peacock, 1997; Breakwell, 2000; Gregg *et al.*, 2004b; Grothmann and Reusswig, 2006; Wagner, 2007; Barberi *et al.*, 2008). A further consideration in the human perception of risk is trust, as this can influence both people's beliefs and actions.

16.4 Trust

Risk perceptions are found to be influenced by trust (Siegrist and Cvetkovich, 2000; Siegrist *et al.*, 2005), and trust has been found to be particularly important under conditions of uncertainty (Siegrist and Cvetkovich, 2000) and where there is a reliance on others for information (Earle, 2004). Trust, however, is a multifaceted, complex concept with no clear definition. There has been much discussion within the literature as to the various 'dimensions' of trust, with the following being suggested by Poortinga and Pidgeon (2003): competence, credibility, reliability, care, fairness, openness and value similarity. Given this, the measurement of trust is difficult but nevertheless important. Therefore, studies such as Eiser *et al.* (2007), while carried out in a different risk context, can be very useful. This study found, in relation to contaminated land, that general levels of trust in the local authorities were low, but was higher in areas where the authorities had been more open and transparent in their communication of the risks. The main predictors of trust in relation to the authorities were found to be a perceived willingness to communicate openly and a perception that they had the respondent's interests as their main consideration. Due to the value of the above study in its own context, similar trust measures were explored in relation to natural hazard risks (Wilmshurst, 2010). The perception that authorities had the respondent's interests at heart was also found to be the most important predictor of trust. In this study, agents (e.g. government, scientists, media) and dimensions of trust were analysed together, with the findings showing that dimensions or agents alone are not as valuable in explaining trust as is an examination of the interactions between them. In other words, it was not the case that having one's interests at heart or knowing the most about the risk, for example, came out as more important than each other or any of the other dimensions of trust. Instead, the importance of a particular dimension of trust (e.g. knowledge about the risk) would vary depending on the agent (e.g. scientists).

Messages are believed to be judged on the credibility of the source as well as the message content (National Research Council, 1989). Paton *et al.* (2010) showed how people's relationships with information sources influence their interpretation of the value of the information. Research on the communication of volcanic risk in the Caribbean (Crosweller, 2009) found that communicators fell into three distinct groups as far as trust was concerned: official sources (this included the government, disaster officials, Red Cross, media and emergency services officers); personal sources (such as family and friends); and scientific sources. Personal sources were rated more highly in cases where residents of the island had previous experience of volcanic hazards; this could have consequences for communication via official sources, particularly if their messages conflict with those of personal sources. Variation in the attribution of trust to different groups was also found by Haynes *et al.* (2008) in terms of volcanic hazards, and Eiser *et al.* (2009) in relation to contaminated land.

There are further complications with regard to trust where evacuation or other significant action is required. Due to the inherent uncertainty in hazard predictions, it is almost impossible to guarantee the occurrence of the hazard as predicted. Instances where a warning is given but the event does not happen result in a 'false alarm'. Not only is there likely to be a high economic cost resulting from a false alarm (commonly associated with evacuation from a high-hazard area), but also a loss of trust between authorities and at-risk communities. It has been demonstrated that it is much easier to lose trust than it is to earn it (the 'asymmetry principle'; Slovic, 1993) and that trust is a key attribute in risk communication, particularly when direct experience is lacking (Renn and Levine, 1991; Paton, 2008). With this in mind, there is a great desire on the part of both scientists and risk managers to minimise the chance of false alarms. Although natural science research is constantly striving to provide more accurate predictions of when and where a particular hazard will strike, there will always be an element of aleatory uncertainty that cannot be accounted for (see Chapter 2). In order to limit the chances that a false alarm is seen as incompetence, which leads to a loss of trust, efforts need to be made to increase public awareness and understanding of such uncertainty, so that there is an appreciation that any prediction comes with some uncertainty attached. Although Frewer and Salter (2007) claimed that 'lay people do distinguish between different kinds of uncertainty', and are 'more accepting of uncertainty resulting from inadequacies in scientific process' this is not evident in all natural hazards situations (see, e.g. Fiske, 1984). Communication of uncertainty, however, is far from straightforward, particularly considering the 'lay' person is generally poor at understanding probabilities in terms of the chances of an event actually occurring (Gigerenzer, 2002).

In summary, a person's experience and their exposure to a hazard will evolve and therefore the way in which this influences their future decision-making and behaviour in relation to natural hazards is likely to change. People's perceptions, attitudes and responses also change during a crisis itself, especially when faced with the reality of the threat. Overlying each of these issues is the role of trust. This can dramatically influence the way in which information is received, risks are perceived and behaviours shown both before and during a crisis. This dynamism is a major contributor in the complexity of 'predicting' human responses in the face of risk.

16.5 Preparedness for hazardous events

In understanding human responses to risk there is also much of value to be learned from the field of health psychology in that it has spent several decades seeking to understand how individuals respond to health risks. While such risks are generally more of an individual concern, if applied with care they can have much to offer. For example, the protection motivation theory (PMT) proposed by Rogers (1983) attempts to provide clarity about our responses to fear and the effects of persuasive communications, with an emphasis on how cognitive processes mediate behavioural change. PMT makes the distinction between two major conceptual processes, namely, threat appraisal and coping appraisal. If, during the threat appraisal process, the hazard is deemed to pose little danger, no further action will be taken. However, if the threat is deemed large enough, then the coping appraisal stage will be initiated. Dependent on the level reached by the coping appraisal, the individual will either intend a non-protective response if the coping appraisal is low (e.g. risk denial, wishful thinking, fatalism) or a protective response, 'protective motivation', if the coping appraisal is high (Grothmann and Reusswig, 2006). However, protection motivation does not necessarily lead to actual behaviour taking place (Hurnen and McClure, 1997). There may be barriers, such as lack of knowledge, time or physical resources not anticipated at the time of intention forming, which prevent an action from being taken. In support of this theory, Rimal and Real (2003) found that by increasing both a person's awareness of a risk (threat appraisal) as well as what action they could take to minimise the risk (coping appraisal) people were more likely to report intentions to adopt more risk-reducing behaviours. Although originating in health, this theory has been used to investigate preparedness for a number of natural hazards: earthquakes (Mulilis and Lippa, 1990), floods (Grothmann and Reusswig, 2006), hurricanes (Franck, 2008) and wildfires (Martin *et al.*, 2008).

The likelihood of someone taking some kind of preparatory or mitigative action against a risk can depend on their perceived locus of control (Rotter, 1954). Those individuals with an external locus of control would have a high reliance on others, particularly officials, to take responsibility for risk reduction. Those with an internal locus of control would be aware that individuals can take risk-reducing actions. Shifting from the former to the latter should encourage self-preparedness (Hurnen and McClure, 1997). Such a shift may be achieved through the provision of information on the risk and possible protective measures that individuals can take. However, there is a difficult balance to be struck between providing enough reassurance so as not to trigger a state of panic in the population, but to stimulate enough interest, or arousal, so as not to become complacent. This is described as the reassurance–arousal paradox (Otway and Wynne, 1989) and is particularly pertinent for hazards with long repose periods such as volcanoes and earthquakes. Terpstra *et al.* (2009) found that trust in flood protection measures put in place in the Netherlands reduced people's perceptions of flood likelihood and magnitude, and hence also reduced their intentions to prepare. Moreover, information intended to be reassuring (such as the presence of a comprehensive monitoring network) may result in risk compensation (Adams, 1995) where an increase in the perceived level of safety can result in a decline in the perceived need to adopt protective measures. This accentuates the need to be able to communicate that hazard predictions are not infallible. For example, investigations conducted by Valery (1995) after the Kobe earthquake in Japan found that many citizens knew about the risks and how to prepare, but believed that the science was so advanced that they would receive a warning before the earthquake struck, and so there was no real need to prepare. This is similar to the process of risk transference, which typically refers to the practice whereby the burden of the risk is reduced through taking out insurance.

A related phenomenon, which is often found in populations, is optimistic bias (Weinstein, 1980). This cognitive bias is illustrated by the belief that your personal level of preparedness or safety (or some other measure) is higher than others in your community, but often with no evidence that this is in fact the case. In a recent study by Harvatt et al. (2011), many respondents believed themselves to be less at risk from flooding and sea-level rise than others in the region. While some people will indeed be situated in areas of lower risk, this cannot be the case for everyone. This belief is thought to be a barrier not only to accepting the risk but also to taking actions to reduce the risk. Spittal et al. (2005), in their study of earthquake risks in Wellington, New Zealand, found that, while many of their participants displayed optimistic bias in relation to preparedness for and personal injury from an earthquake, they judged that their homes were more likely to be damaged than others, thus highlighting that optimism does not necessarily manifest itself across all aspects. The existence of optimistic bias can be difficult to tackle since it can have knock-on effects for risk communication efforts. People may believe that such communications are directed more at others than to them, as the others are more in need of the information. Further, awareness of the existence of information (e.g. a leaflet explaining what to do in the event of a crisis) can sometimes be confused with actually knowing, or indeed practising, that information and hence leads to feelings of optimism concerning one's level of knowledge (e.g. Ballantyne et al., 2000).

16.6 Other avenues of research

There has been recent work in a number of natural hazards using participatory methods in an attempt to involve residents in the risk management and planning process from the start (see Cronin et al., 2004 for application in a volcanic situation; Fukuzono et al., 2006 and Stanghellini and Collentine, 2008 for flood risk areas). In order for these participatory methods to have a chance of being successful, they need to be truly participatory (i.e. take into account the views and concerns of all stakeholders equally) and therefore avoid being superficial. This ensures that the process is not simply seeking endorsement of decisions already made by those in positions of official decision-making power, as this can negatively impact the relationship between the 'experts' and 'lay' people (Wachinger and Renn, 2010). There has been a great deal of work carried out using participatory work to empower communities in relation to problem solving in issues that impact upon them directly. This approach has been very successful in helping to bridge the gaps between theory and practice and between 'experts' and those who they are trying to help (see, for example, Fals Borda, 2001).

Also, the role of mythology and religion has recently emerged as an important area for more in-depth exploration (e.g. Chester, 2005), especially among indigenous communities with cultural practices that are more closely interwoven with the natural surroundings. Gregg et al. (2008) found that the religious beliefs of some residents near Kilauea volcano on the Hawaiian island of Kauai meant they had a strong negative reaction to the mitigation practice of bombing the lava flows as they see the volcano as the home of the goddess Pele.

There are clearly also many context-specific factors that will have an impact upon people's beliefs and behaviours, and these will interact with the internal processes of response and decision-making discussed here. For example, the priority given to natural hazard risks will depend to a certain degree on the other risks posed by the environment in which a person lives. High crime rates, prevalence of disease and poor access to sustainable livelihood opportunities can all have the effect of diluting the amount of attention and resources given to a threat that may be very low probability, even if potentially very high impact.

Beyond the individual there are issues relating to political and financial constraints in relation to allocation of resources and response in a crisis, vulnerability of communities due to economic and social status or levels of education, poor collaboration and communication between risk management agencies and scientific groups, or within groups themselves, to name but a few. These topics have not been discussed in this chapter because of its focus on the individual, but these are all potential reasons why, despite the best scientific knowledge, disasters may still occur.

16.7 Mitigating natural disasters

Given that large numbers of humans live in areas at threat from natural hazards and, for a variety of reasons, this situation will continue, how can the risk to these individuals be more effectively mitigated?

Professionals working in disaster management have for some time been calling for more attention to be paid to the issues of mitigation and preparedness, and for a need to move away from historically reactive approaches to disasters. In cases where people are living in what has been assessed to be a high-risk area, educational outreach programmes are often used as tools to 'better inform' people of the risk. However, while there is evidence that successful responses correlate strongly with the degree to which policies of hazard reduction are already in place before the onset of the hazard event, it has also been shown that it can be difficult to shift risk perceptions without direct experience. Therefore, simply providing information is unlikely to change someone's assessment of the risk (e.g. Parry *et al.*, 2004 in relation to food poisoning; Paton *et al.*, 2001 in relation to volcanic hazards).

Two major international publications in recent years (the Hyogo Framework for Action, 2007 – United Nations International Strategy for Disaster Risk Reduction Strategy, 2007 – and the International Council for Science plan, 2008) have recognised the requirement for a more integrated approach to tackling the losses from natural hazards. There is a need to add to the process of assessing the risk in a scientific framework and then educating people about it by including something that is more inclusive, and involving those people faced with the risk. The more integrated approach suggested by the United Nations Development Programme (2007) clearly requires multidisciplinary, international collaboration. Most of this work has so far been initiated by the physical sciences with a focus on increased efficiency in predicting events that may lead to disaster. Disaster management professionals have also been working with non-governmental organisations (NGOs) and policy-makers to improve the communication of risk and preparatory measures in order to help communities to protect themselves more effectively. In the academic literature, social scientists have been calling for greater attention to be paid to social and human behavioural factors in DRR since the 1970s (Kates, 1971; White, 1973; Burton *et al.*, 1978).

Despite a better understanding of natural hazard systems within the physical sciences, there is international recognition of the need for the integration of social and physical sciences. This is evident by the Hyogo Framework for Action (United Nations International Strategy for Disaster Risk Reduction Strategy, 2007), which comprises a 'global blueprint' to reduce vulnerabilities to natural hazards over a ten-year time span (of which 168 governments worldwide have become signatories). It is also borne out by the recent publication of the International Council for Science report, 'A science plan for integrated research on disaster risk' (International Council for Science, 2008). This report addresses the question: 'Why, despite advances in the natural and social science of hazards and disasters, do losses continue to increase?' (p. 7). It calls for integrated research across the physical and social sciences, as well as across academic/policy-maker boundaries, with one of the principal stated objectives being to 'understand decision-making in complex and changing risk contexts' (p. 6). While the wording here focuses on decision-making, there is a recognition that many factors may contribute to how people arrive at such decisions, and this is a key area for development. This could be achieved by a more systematic review of the factors that influence the way in which people make decisions and choose behaviours in the context of the threats posed by natural hazards.

In the majority of cases, hazard and risk assessments are carried out by scientists and risk managers because between them they hold the majority of the expertise on the hazards and the people and items at risk, and the resources available for allocation to risk management strategies and their implementation. For this reason, historically there has been a tendency to focus research efforts on the physical science in order to understand better and predict the behaviour of natural phenomena. Research into the human responses to such phenomena has, until recently, received much less attention. As a result, risk management strategies incorporate advice and information provided by physical scientists on a range of factors, such as the likelihood of a hazard event occurring and the physical consequences of such an event. This includes the impacts on communities themselves, as well as on land used for farming or elements of ecosystems upon which humans rely. Scientific information is used alongside information gathered by the risk managers, usually with the first priority being given to preventing loss of life, and the second to economic losses. Risk management planning is naturally constrained by the resources available, the wider political agenda and the systems in which risk managers must operate.

These systems are commonly designed following extensive consultation with experts across a range of disciplines. This will usually include those with expertise specific to the hazard, such as meteorologists, seismologists or volcanologists. These experts can provide information on the predictability of a hazard event and its likely consequences. In addition, those with expertise in gauging the impacts of such events may be called upon. For example, engineers and urban planners, who are able to provide information as to the likely impact on buildings and infrastructures; geographers and biologists, who can advise as to the impact on agriculture and ecosystems; and medical advisers concerning the potential health consequences for those affected by an event. The time and attention given to such processes will vary enormously depending on political will and available resources.

'Risk management' is itself a broad term, and while the general meaning of the term is clear, it encompasses a range of people and organisations. These include national and local government agencies, NGOs such as the Red Cross, the civil authorities and emergency responders and local community leaders. The individuals and organisations involved will vary according to the cultural, economic, political and social context. Further, there is an increasing focus on the integration of DRR into the wider field of development, and this acknowledges that economic factors are extremely important in effective disaster management (Collins, 2009). However, events such as Hurricane Katrina in the United States illustrate that disasters are not exclusively a problem for the 'global South',[1] but can also occur in nations with greater financial resources. Factors such as political will and the allocation of resources are clearly also of great importance.

Despite increasing attention and resources being allocated to DRR around the world in rich and poor nations alike, risk management planning for natural hazards continues to be carried out on the basis of a series of often poorly evidenced assumptions about how humans respond both to risks and to hazard events once they occur. Whenever a hazard event turns into a disaster, the same questions are asked and many of them centre on the behaviour of those at risk. When people do not behave in the way expected by scientists and risk managers, it is often assumed that these people did not know or understand the risk, or did not know how to respond to the event when it occurred. There is often an assumption that responses other than those considered as rational by scientists and risk managers are due to a lack of knowledge, understanding or ability for rational decision-making. While in many cases these are certainly contributing factors to a hazard event becoming a disaster, anecdotal evidence repeatedly shows that, even when a comprehensive educational process has been included in mitigation and preparedness strategies, and adequate warning has been given, many people choose not to respond in a manner that ensures their personal safety.

16.8 Concluding comments

The number of people living at risk from natural hazards is on the increase as a consequence of population growth and urbanisation. The frequency of some hazards may also be increasing, perhaps due to climate change. While our understanding of the physical hazards is improving, this is only one aspect of disasters and emergency management. The primary driver behind trying to predict hazards accurately is to reduce the loss of life, yet human

---

[1] The 'global South' refers to those countries with a Human Development Index (HDI) score below 0.8, most of which are located in the Southern Hemisphere.

behaviour is also uncertain in the face of risk and is liable to variation over time and as external factors change.

Human decision-making is much more complex than simply a 'rational' information-processing task carried out by individuals. Perceptions may differ depending on the type of risk, the risk context, the personality of the individual and the social context (Wachinger and Renn, 2010: 8). Therefore, the definition of what is a 'common sense' response to a given scenario will vary greatly across individuals and groups and comprise what are often unexpected, even counter-intuitive, behaviours. It is often these behaviours that risk managers try to change through providing more information about the risk, but this often does not have the desired effect. Taking the time to understand the roots of these beliefs and behaviours can go a long way in helping to find ways of tackling them, where this is necessary.

The concept of place attachment (to either a physical location or community) may explain some of the resistance by local populations to being moved from a hazardous area, either in advance of an event or emergency evacuation during the event itself. However, cultural differences may also account for the manner in which people respond to environmental threats. This is a key factor in the need to consider individuals as part of a larger community such as society (this social dimension is discussed further in Chapter 15). At the same time, it is important to acknowledge that the link between perception and behaviour is not straight-forward, and that an apparent perception of low risk does not necessarily equate to a lack of awareness of the hazard.

Cognitive dissonance and denial are common psychological defences found to occur in sections of at-risk populations and can go some way towards explaining the inaction of some to natural hazard events. The theory of optimistic bias has also been seen to be evident in many natural hazards cases where an individual perceives that their knowledge or prepared-ness is better than it really is and often better than those around them. This can be one reason why efforts to educate the population about the risks they face seem to have little effect. Considering the lack of literature on these phenomena in relation to natural hazard risks, there are no suggestions as to how to combat them. As noted earlier, though, simply providing more information about the risk will not necessarily have any effect and may, in some circumstances, exacerbate the issue by pushing people further into denial. Any communications should be designed only once an understanding of the beliefs of the population concerned has been gained and therefore the messages can be designed to complement these beliefs and not contradict them.

Through numerous natural hazard studies, the influence of prior experience on risk perception has been found to be significant. This is particularly liable to change, not just through further interaction with the hazard but also a lack of experience of it. Even the occurrence of other hazards, which may divert attention away from the original hazard of concern, will have an effect. It may be necessary in some cases to draw renewed attention to a threat and while this may achieve the goal of alerting people to the potential dangers it may well at the same time serve to renew the stress that had previously been alleviated through these defence mechanisms. Such examples highlight the need for great sensitivity in seeking a balance between maintaining a healthy level of awareness in order to ensure an appropriate and timely response when an alert is given, while at the same time allowing a certain level of defence against the chronic stress that may be caused by living in a long-term state of heightened arousal.

Similarly, the concept of trust has been shown to be important and can have dramatic influences on people's responses to hazards. It is therefore imperative to determine what communication channels are utilised and those actors which are considered most trustworthy. The difficulty of communicating uncertainty in particular is highlighted and this is a key area for further research in order to reduce the chance of false alarms eroding trust, but also the need to communicate that science is not infallible and therefore people need to take some responsibility for preparation themselves. Whether this will occur will also depend on the way in which people make attributions of responsibility for their safety and that of their community. The PMT explains how a person not only needs to feel sufficiently threatened by something in order to act, but also to feel able to do something to reduce the threat. Again, this has implications for the way in which communication about potential risks is conducted.

Clearly, the human dimension of DRR is a rich topic for new research and it probably needs greater attention than it has so far been given, both in the academic arena and by those involved in a more operational understanding of how hazards events can so rapidly become disasters and how this can in the future be more effectively averted. Humans are not necessarily 'rational', but such irrationality must not immediately be assumed to be a negative thing; in many cases it is simply a response to a different set of priorities and an alternative way of reaching decisions about how best to manage their safety and that of those around them. In the context of hazard management, this can be seen in cases where mandatory evacuation and or resettlement, while implemented for the immediate protection of life, are in conflict with the deeply embedded need to remain close to and protect community ties, land and property. Any approach predicated on the assumption that people behave in unexpected ways due to a simple lack of education, an inability to assess accurately the risks or to make rational decisions is at best naive. At worst, it could contribute to an even greater division between those whose job it is to study and manage the potential impacts of natural hazard events and those who live alongside them in their daily lives. The work to be done here by social scientists should not be *for* these people but *with* them because, while the need for expertise in tried and tested methodologies, transferable skills and lessons learned across different hazard contexts is not in doubt, no-one is more of an expert in the needs and wishes of their communities than the people living in them.

We stated at the start of the chapter that we aimed to offer an overview of the relevant theoretical and applied literature and to offer some suggestions not only for future research but also for practitioners working in the field of DRR. We posed the question 'Why do people place their lives at risk by continuing to live in areas of high hazard?' It is clear from the breadth and volume of literature on this subject that there is no clear answer to this question and that the contributing factors are many and diverse. Some are contextual and can be observed and understood, if not changed, more easily, such as poverty, migration and poor infrastructure. Others, which are the focus of this chapter, are much less easy to define as they are driven by invisible factors that live largely inside the minds of individuals and collectives.

Yet this dilemma is not confined to the subject matter dealt with here. Many conferences are being held, research projects carried out and books written asking the very same question in relation to the many other ways in which we, as a species, make strange and seemingly irrational choices in relation to our personal and collective safety. Even when the risks are outlined to us clearly and alternatives are offered, many of us choose, at the individual level, to smoke, to eat the wrong foods or to drive without wearing seatbelts. At the collective level, we continue to pollute the environment, destroy habitats and watch on as whole communities live in poverty and squalor. For this reason, it is important that in seeking to answer the most challenging questions about human behaviour, we remember not only to look into the specific context, but also outwards to all those who are asking similar questions in other contexts. It is for this reason that there is much for those in the DRR community to learn, in particular from those working in the fields of health psychology, environmental psychology and social psychology, who are seeking to understand the very same aspects of what appear to be very deeply embedded, but hopefully not inaccessible, aspects of the human condition.

Acknowledgements

We thank the editors for numerous useful discussions and comments on this manuscript. It was also greatly improved with reviews by Jan Noyes and Jenni Barclay. The authors were supported by a European Research Council Advanced Grant.

References

Adams, J. (1995) *Risk*, London: UCL Press.

Ajzen, I. (2005) *Attitudes, Personality and Behaviour*, 2nd edn, Maidenhead: Open University Press.

Altman, I. and Low, S. M. (1992) *Place Attachment*, New York: Springer-Verlag.

Armas, I. and Avram, E. (2009) Perception of flood risk in Danube Delta, Romania. *Natural Hazards* **50** (2): 269–287.

Ballantyne, M., Paton, D., Johnston, D. J., *et al.* (2000) *Information on Volcanic and Earthquake Hazards: The Impact on Awareness and Preparation*, Lower Hutt, New Zealand: Institute of Geological and Nuclear Sciences.

Barberi, F., Davis, M. S., Isaia, R., *et al.* (2008) Volcanic risk perception in the Vesuvius population. *Journal of Volcanology and Geothermal Research* **172** (3–4): 244–258.

Billig, M. (2006) Is my home my castle? Place attachment, risk perception, and religious faith. *Environment and Behaviour* **38**: 248–265.

Boura, J. (1998) Community fireguard: creating partnerships with the community to minimise the impact of bushfire. *Australian Journal of Emergency Management* **13** (3): 59–64.

Breakwell, G. M. (2000) Risk communication: factors affecting impact. *British Medical Bulletin* **56** (1): 110–120.

Burton, I., Kates, R. W. and White, G. F. (1978) *The Environment as Hazard*, New York: Oxford University Press.

Caldwell, K. and Boyd, C. P. (2009) Coping and resilience in farming families affected by drought. *Rural Remote Health* **9** (2): 1088.

Chester, D.K. (2005) Theology and disaster studies: the need for dialogue. *Journal of Volcanology and Geothermal Research* **146** (4): 319–328.

Chester, D.K., Dibben, C.J.L. and Duncan, A.M. (2002) Volcanic hazard assessment in western Europe. *Journal of Volcanology and Geothermal Research* **115** (3–4): 411–435.

Collins, A.E. (2009) *Disaster and Development*, London: Routledge.

Cook, M.J., Noyes, J.M. and Masakowski, Y. (2007) *Decision Making in Complex Systems*. Aldershot: Ashgate.

Crano, W.D. and Prislin, R. (2006) Attitudes and persuasion. *Annual Review of Psychology* **57**: 345–374.

Cronin, S.J., Gaylord, D.R., Charley, D., *et al.* (2004) Participatory methods of incorporating scientific with traditional knowledge for volcanic hazard management on Ambae Island, Vanuatu. *Bulletin of Volcanology* **66** (7): 652–668.

Crosweller, H.S. (2009) An analysis of factors influencing volcanic risk communication on two islands in the Lesser Antilles. Unpublished PhD thesis, University of East Anglia.

Davis, M.S., Ricci, T. and Mitchell, L.M. (2005) Perceptions of risk for volcanic hazards at Vesuvio and Etna, Italy. *The Australasian Journal of Disaster and Trauma Studies* **1**: n.p.

Dibben, C.J.L. (2008) Leaving the city for the suburbs: the dominance of 'ordinary' decision making over volcanic risk perception in the production of volcanic risk on Mt Etna, Sicily. *Journal of Volcanology and Geothermal Research* **172** (3–4): 288–299.

Dilley, M., Chen, R.S., Deichmann, U., *et al.* (2005) *Natural Disaster Hotspots: A Global Risk Analysis*, Washington, DC: World Bank. http://www.preventionweb.net/files/1100_Hotspots.pdf

Dominey-Howes, D. and Minos-Minopoulos, D. (2004) Perceptions of hazard and risk on Santorini. *Journal of Volcanology and Geothermal Research* **137** (4): 285–310.

Earle, T.C. (2004) Thinking aloud about trust: a protocol analysis of trust in risk management. *Risk Analysis* **24**: 169–183.

Eiser, J.R., Stafford, T., Henneberry, J., *et al.* (2007) Risk perception and trust in the context of urban brownfields. *Environmental Hazards* **7** (2): 150–156.

Eiser, J.R., Stafford, T., Henneberry, J., *et al.* (2009) 'Trust me, I'm a scientist (not a developer)': perceived expertise and motives as predictors of trust in assessment of risk from contaminated land. *Risk Analysis* **29** (2): 288–297.

Fals Borda, O. (2001) Participatory (action) research in social theory: origins and challenges. In *The Handbook of Action Research*, ed. P. Reason and H. Bradbury. London: Sage, pp. 27–37.

Festinger, L. (1957) *A Theory of Cognitive Dissonance*, London: Tavistock Publications.

Finucane, M.L., Alhakami, A., Slovic, P., *et al.* (2000) The affect heuristic in judgments of risks and benefits. In *The Perception of Risk*, ed. P. Slovic, London: Earthscan, pp. 413–429.

Fishbein, M.F. and Ajzen, I. (2010) *Predicting and Changing Behaviour: The Reasoned Action Approach*, New York: Psychology Press.

Fiske, R.S. (1984) Volcanologists, journalists, and the concerned local public: a tale of two crises in the Eastern Caribbean. In *Explosive Volcanism: Inception, Evolution, and Hazards*, ed. Geophysics Study Committee, Washington, DC: National Academy Press, pp. 170–176.

Franck, T. (2008) A behavioural model of hurricane risk and coastal adaptation. Proceedings of the International Conference of the System Dynamics Society, 20 July, Athens, Greece.

Frewer, L.J. (1999) Public risk perceptions and risk communication. In *Risk Communication and Public Health*, ed. P. Bennett and K. Calman, Oxford: Oxford University Press, pp. 20–32.

Frewer, L. and Salter, B. (2007) Societal trust in risk analysis: implications for the interface of risk assessment and risk management. In *Trust in Cooperative Risk Management: Uncertainty and Scepticism in the Public Mind*, ed. M. Siegrist, T. C. Earle and H. Gutscher. London: Earthscan, pp. 143–159.

Fukuzono, T., Sato, T., Takeuchi, Y., *et al.* (2006) Participatory flood risk communication support system (Pafrics). In *A Better Integrated Management of Disaster Risks: Toward Resilient Society to Emerging Disaster Risks in Mega-cities*, ed. S. Ikeda, T. Fukuzono, and T. Sato, Tokyo: Terrapub and NIED, pp. 199–211.

Gigerenzer, G. (2002) *Reckoning with Risk*, London: Allen Lane.

Gladwin, H. and Peacock, W. G. (1997) Warning and evacuation: a night for hard houses. In *Hurricane Andrew: Ethnicity, Gender, and the Sociology of Disasters*, ed. W. G. Peacock, B. H. Morrow, H. Gladwin, London: Routledge, pp. 52–74.

Glassman, L. R. and Albarracin, D. (2006) Forming attitudes that predict future behaviour: a meta-analysis of the attitude–behaviour relation. *Psychological Bulletin* **132**: 778–822.

Gregg, C. E., Houghton, B. F., Paton, D., *et al.* (2004a) Community preparedness for lava flows from Mauna Loa and Hualalai volcanoes, Kona, Hawai'i. *Bulletin of Volcanology* **66**: 531–540.

Gregg, C. E., Houghton, B. F., Johnston, D. M., *et al.* (2004b) The perception of volcanic risk in Kona communities from Mauna Loa and Hualalai volcanoes, Hawai'i. *Journal of Volcanology and Geothermal Research* **130**: 179–196.

Gregg, C. E., Houghton, B. F., Johnston, D. M., *et al.* (2008) Hawaiian cultural influences on support for lava flow hazard mitigation measures during the January 1960 eruption of Kilauea volcano, Kapoho, Hawaii. *Journal of Volcanology and Geothermal Research* **172** (3–4): 300–307.

Grothmann, T. and Reusswig, F. (2006) People at risk of flooding: why some residents take precautionary action while others do not. *Natural Hazards* **38** (1): 101–120.

Harvatt, J., Petts, J. and Chilvers, J. (2011) Understanding householder responses to natural hazards: flooding and sea-level rise comparisons. *Journal of Risk Research* **14** (1): 63–83.

Haynes, K. A., Barclay, J., Pidgeon, N. (2008) The issue of trust and its influence on risk communication during a volcanic crisis. *Bulletin of Volcanology* **70** (5): 605–621.

Hewitt, K. (1983) The idea of calamity in a technocratic age. In *Interpretation of Calamities*, ed. K. Hewitt, Boston, MA: Allen & Unwin Inc., pp. 263–286.

Hurnen, F. and McClure, J. (1997) The effect of increased earthquake knowledge on perceived preventability of earthquake damage. *The Australasian Journal of Disaster and Trauma Studies* **3**: n.p.

International Council for Science (ICSU) (2008) *A Science Plan for Integrated Research on Disaster Risk: Addressing the Challenge of Natural and Human-induced Environmental Hazards*. Paris: ICSU.

Kates, R. W. (1962) Hazard and choice perception in flood plain management. University of Chicago, Department of Geography, Research Paper No. 78.

Kates, R. W. (1971) Natural hazard in human ecological perspective: hypotheses and models. *Economic Geography* **47** (3): 438–451.

Kelman, I. and Mather, T. A. (2008) Living with volcanoes: the sustainable livelihoods approach for volcano-related opportunities. *Journal of Volcanology and Geothermal Research* **172** (3–4): 189–198.

Laksono, P. M. (1988) Perception of volcanic hazards: villagers versus government officials in Central Java. In *The Real and Imagined Role of Culture in Development: Case Studies from Indonesia*, ed. M. R. Dove, Honolulu: University of Hawaii Press, pp. 183–200.

Lavigne, F., De Coster, B., Juvin, N., *et al.* (2008) People's behaviour in the face of volcanic hazards: perspectives from Javanese communities, Indonesia. *Journal of Volcanology and Geothermal Research* **172** (3–4): 273–287.

Loewenstein, G. F., Weber, E. U., Hsee, C. K., *et al.* (2001) Risk as feelings. *Psychological Bulletin* **127** (2): 267–286.

Mann, L., Burnett, O., Radford, M., *et al.* (1998) The Melbourne decision making questionnaire: an instrument for measuring patterns for coping with decision conflict. *Journal of Behavioural Decision Making* **10**: 1–19.

Markus, H. R. and Kitayama, S. (1991) Culture and the self: implications for cognition, emotion and motivation. *Psychological Review* **98** (2): 224–253.

Martin, I. M., Bender, H. W. and Raish, C. (2008) Making the decision to mitigate risk. In *Wildfire Risk: Human Perceptions and Management Implications*, ed. W. E. Martin, C. Raish and B. Kent, Washington, DC: Resources for the Future, pp. 117–141.

Miceli, R., Sotgiu, I. and Settanni, M. (2008) Disaster preparedness and perception of flood risk: a study in an alpine valley in Italy. *Journal of Environmental Psychology* **28**: 164–173.

Mileti, D. S. and O'Brien, P. W. (1992) Warnings during disaster: normalising communicated risk. *Social Problems* **39** (1): 40–57.

Moser, S. C. and Dilling, L. (2004) Making climate hot: communicating the urgency and challenge of global climate change. *Environment Magazine* **46** (10): 32–46.

Mulilis, J.-P. and Duval, T. S. (1995) Negative threat appeals and earthquake preparedness: a person-relative-to-event (PrE) model of coping with threat. *Journal of Applied Social Psychology* **25** (15): 1319–1339.

Mulilis, J.-P. and Lippa, R. (1990) Behavioural change in earthquake preparedness due to negative threat appeals: a test of protection motivation theory. *Journal of Applied Social Psychology* **20**: 619–638.

National Research Council (NRC) (1989) *Improving Risk Communication*, Washington, DC: National Academy Press.

Otway, H. and Wynne, B. (1989) Risk communication: paradigm and paradox. *Risk Analysis* **9** (2): 141–145.

Parry, S. M., Miles, S., Tridente, A., *et al.* (2004) Differences in perception of risk between people who have and have not experienced salmonella food poisoning. *Risk Analysis* **24** (1): 289–299.

Paton, D. (2003) Disaster preparedness: a social-cognitive perspective. *Disaster Prevention and Management* **12** (3): 210–216.

Paton, D. (2008) Risk communication and natural hazard mitigation: how trust influences its effectiveness. *International Journal of Global Environmental Issues* **8** (1/2): 1–16.

Paton, D. and Johnston, D. M. (2001) Disasters and communities: vulnerability, resilience and preparedness. *Disaster Prevention and Management* **10** (4): 270–277.

Paton, D., Johnston, D. M., Bebbington, M. S., *et al.* (2001) Direct and vicarious experience of volcanic hazards: implications for risk perception and adjustment adoption. *Australian Journal of Emergency Management* **15** (4): 58–63.

Paton, D., Sagala, S., Okada, N., *et al.* (2010) Making sense of natural hazard mitigation: personal, social and cultural influences. *Environmental Hazards* **9** (2): 183–196.

Poortinga, W. and Pidgeon, N. F. (2003) Exploring the dimensionality of trust in risk regulation. *Risk Analysis* **23** (5): 961–972.

Renn, O. (2008) *Risk Governance: Coping with Uncertainty in a Complex World*, London: Earthscan.

Renn, O. and Levine, D. (1991) Credibility and trust in risk communication. In *Communicating Risks to the Public: International Perspectives*, ed. R. E. Kasperson and P. J. M. Stallen, Dordrecht: Kluwer Academic Publishers, pp. 175–218.

Riad, J. M. and Norris, F. H. (1998) Hurricane threat and evacuation intentions: an analysis of risk perception, preparedness, social influences and resources. University of Delaware Disaster Research Center Preliminary Paper 271. http://dspace.udel.edu:8080/dspace/handle/19716/107 (accessed 22 December 2011).

Rimal, R. N. and Real, K. (2003) Perceived risk and efficacy beliefs as motivators of change: use of the Risk Perception Attitude (RPA) framework to understand health behaviours. *Human Communication Research* **29** (3): 370–399.

Rogers, R. W. (1983) Cognitive and psychological processes in fear appeals and attitude change: a revised theory of protection motivation. In *Social Psychophysiology*, ed. J. Cacioppo and R. Petty, New York: Guilford Press.

Rotter, J. B. (1954) *Social Learning and Clinical Psychology*, New York: Prentice-Hall.

Sedikides, C. and Brewer, M. B. (2001) *Individual Self, Relational Self, Collective Self*, Philadelphia, PA: Psychology Press.

Siegrist, M. and Cvetkovich, G. (2000) Perception of hazards: the role of social trust and knowledge. *Risk Analysis* **20** (5): 713–719.

Siegrist, M. and Gutscher, H. (2006) Flooding risks: a comparison of lay people's perceptions and expert's assessments in Switzerland. *Risk Analysis* **26** (4): 971–979.

Siegrist, M., Gutscher, H. and Earle, T. (2005) Perception of risk: the influence of general trust, and general confidence. *Journal of Risk Research* **8** (2): 145–156.

Simon, H. A. (1959) Theories of decision-making in economics and behavioural science. *The American Economic Review* **49** (3): 253–283.

Slovic, P. (1987) Perception of risk. *Science* **236** (4799): 280–285.

Slovic, P. (1993) Perceived risk, trust, and democracy. *Risk Analysis* **13** (6): 675–682.

Slovic, P., Kunreuther, H. and White, G. (2000) Decision processes, rationality and adjustment to natural hazards. In *The Perception of Risk*, ed. P. Slovic, London: Earthscan, pp. 1–31.

Sowby, F. D. (1965) Radiation and other risks. *Health Physics* **11** (9): 879–887.

Spittal, M. J., McClure, J., Siegert, R. J., *et al.* (2005) Optimistic bias in relation to preparedness for earthquakes. *Australian Journal of Disaster and Trauma Studies* **1**: n.p.

Stanghellini, L. P. S. and Collentine, D. (2008) Stakeholder discourse and water management: implementation of the participatory model CATCH in a Northern Italian alpine sub-catchment. *Hydrology and Earth System Sciences* **12**: 317–331.

Starr, C. (1969) Social benefit versus technological risk. *Science* **165**: 1232–1238.

Susman, P., O'Keefe, P. and Wisner, B. (1983) Global disasters, a radical interpretation. In *Interpretation of Calamities*, ed. K. Hewitt, Boston, MA: Allen & Unwin Inc., pp. 263–286.

Terpstra, T., Lindell, M. K. and Gutteling, J. M. (2009) Does communicating (flood) risk affect (flood) risk perceptions? Results of a quasi-experimental study. *Risk Analysis* **29** (8): 1141–1155.

Torry, W. I. (1979) Hazards, hazes and holes: a critique of the environment as hazard and general reflections on disaster research. *Canadian Geographer* **23** (4): 368–383.

Tversky, A. and Kahneman, D. (1974) Judgment under uncertainty: heuristics and biases. *Science* **185**: 1124–1131.

United Nations Development Programme (2007) Mainstreaming Disaster Risk Reduction. *CPR Newsletter* **4** (3). http://www.undp.org/cpr/newsletters/2007_4/article3.htm. (accessed 16 December 2011).

United Nations International Strategy for Disaster Risk Reduction Secretariat (ISDR) (2007) Hyogo framework for action 2005–2015: building the resilience of nations and communities to disasters. http://preventionweb.net/go/1037 (accessed 16 December 2011).

Valery, N. (1995) Fear of trembling: a survey of earthquake engineering. *The Economist* **22**: 3–12.

Wachinger, G. and Renn, O. (2010) Risk perception and natural hazards. CapHaz-Net WP3 Report. http://caphaz-net.org/outcomes-results/CapHaz-Net_WP3_Risk-Perception.pdf (accessed 5 December 2011).

Wagner, K. (2007) Mental models of flash floods and landslides. *Risk Analysis* **27** (3): 671–682.

Webb, T. L. and Sheeran, P. (2006) Does changing behavioural intentions engender behaviour change? A meta-analysis of the experimental evidence. *Psychological Bulletin* **132** (2): 249–268.

Weinstein, N. D. (1980) Unrealistic optimism about future life events. *Journal of Personality and Social Psychology* **39** (5): 806–820.

White, G. F. (1973) Natural hazards research. In *Directions in Geography*, ed. R. J. Chorley, London: Meuthen & Co., pp. 193–216.

Wilmshurst, J. (2010) Living with Extreme weather events: an exploratory study of psychological factors in at-risk communities in the UK and Belize. Unpublished PhD thesis, University of Sheffield.

Xie, X., Wang, M. and Xu, L. (2003) What risks are Chinese people concerned about? *Risk Analysis* **23** (4): 685–695.

Index